[illegible]
5 July '94

Biological Relationships between Africa and South America

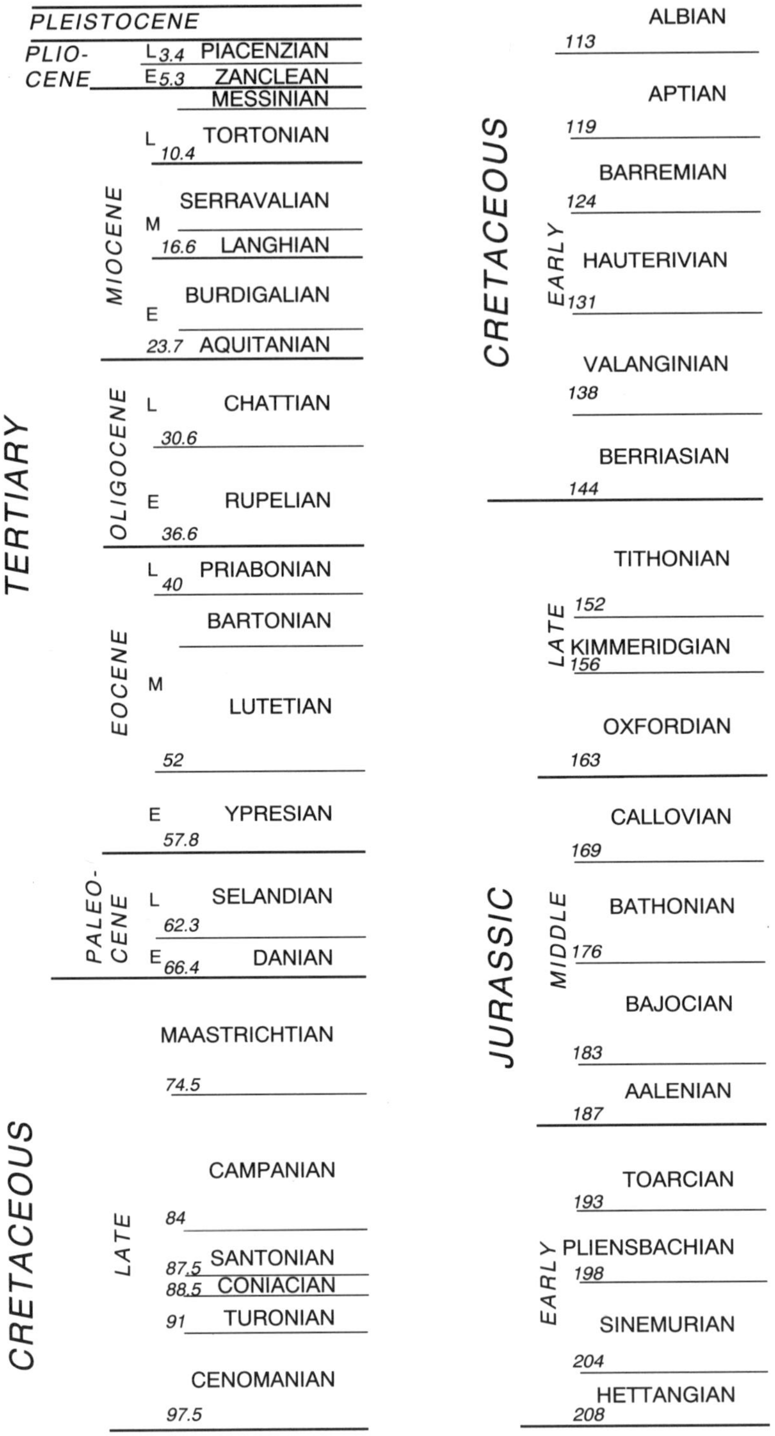

The geologic time scale from the Jurassic to the present.

Biological Relationships between Africa and South America

Peter Goldblatt, editor

Yale University Press

New Haven and London

Designed by Sonia Scanlon.
Set in Times Roman type by Keystone Typesetting, Inc., Orwigsburg, Pennsylvania.
Printed in the United States of America by Thomson-Shore, Inc., Dexter, Michigan.

Library of Congress Cataloging-in-Publication Data
Biological relationships between Africa and South America / Peter Goldblatt, editor.
p. cm.
"Proceedings of the 37th Annual Systematics Symposium, held at Missouri Botanical Gardens, 4–6 October 1990"—P.
Includes bibliographical references (p.) and indexes.
ISBN 0-300-05375-4
1. Biogeography—Africa—Congresses. 2. Biogeography—South America—Congresses. 3. Gondwana (Geology)—Congresses. I. Goldblatt, Peter.
QH194.B56 1993
574.96—dc20 93-18210
CIP

A catalogue record for this book is available from the British Library.

The paper in this book meets the guidelines for permanence and durability of the Committee on Production Guidelines for Book Longevity of the Council on Library Resources.

10 9 8 7 6 5 4 3 2 1

Contents

Foreword

The chapters in this volume, encompassing geology, climatology, and biology, were assembled to address the broad subject of the biological relationships between Africa and South America from their breakup in the Cretaceous until the present. It is a sign of the maturity of these biogeographical sciences that they have moved beyond the merely descriptive, and now present their results instead in an overtly theoretical framework, testable by new experimental and analytic methods. These comparative studies take on a disproportionate urgency as we struggle to learn about the organisms and ecosystems of tropical regions before they disappear at the hand of an expanding human population.

Since the early voyages of exploration by Europeans in the fifteenth and sixteenth centuries, biologists and natural historians have been fascinated by the remarkable differences between the biotas of Africa and South America. The theory of evolution ostensibly provided the raison d'être for these differences, but until quite recently this was true only in broad and superficial terms. The advent and explosive development of plate tectonic theory provided new impetus and direction to comparative studies, and indeed the geological evidence now provides the essential theoretical framework within which biological studies can proceed. As necessary background, then, this book includes an excellent detailed account of the complex geological history of the separation of Africa from South America since the Cretaceous, as well as reconstructions of the paleoclimates that existed when the separation took place and subsequently.

The development over the past 30 years of cladistic methods of phylogeny reconstruction has allowed systematic biologists to transform their descriptive and more or less intuitive accounts of relationships among organisms into explicit phylogenetic hypotheses. The chapters in this volume include a broad range of analyses of various groups of plants, insects, fish, amphibians, reptiles, birds, and mammals. As the phylogenies of more groups of organisms with African–South American distributions are analyzed in the context of the developing geological and climatological models, it is becoming possible to combine those phylogenies into models of the evolution of the whole biota, or major parts of it, in the region that was West Gondwana.

The results of these broad comparative analyses demonstrate the degree to which timing of major radiations, ability of a given group to disperse, and the influence of biota from other regions have shaped the evolution of organisms with African–South American disjunctions. This excellent volume provides the most complete and lucid account of the historical biogeography of West Gondwana, yet like any truly useful scientific compendium, it raises as many questions as it answers. As demonstrated by many of the authors, a comprehensive resolution of the biogeographic puzzles posed here must await the integration of these data and analyses with those involving Australasia, Antarctica, South Asia, and ultimately the northern continents. The accumulation of additional phylogenetic analyses of groups in this region, and the continued refinement of the models of geological separation, biotic dispersal, and in situ radiations promise an exciting future to the comparative study of the biota of Africa and South America.

PETER H. RAVEN

Preface

This volume represents the proceedings of the 37th Annual Systematics Symposium, held at the Missouri Botanical Garden, 4–6 October 1990. The symposium was part of a series sponsored by the Missouri Botanical Garden since 1954. The subject of this symposium reflects the Garden's strong interest in tropical biology and its specialization in the Neotropics and secondarily in Africa. The remarkable biodiversity of the Neotropics and the comparative poverty of the tropical African flora and to some extent its forest fauna has long been a subject of interest. The aim of this symposium was to focus on this question and to compare the biota of the two continents in the light of the current information concerning plate tectonics and climatic change as it relates to Africa and South America since the early Cretaceous, when the single continent of West Gondwana began its separation into two. Although directed primarily to students of biogeography, paleontology, and comparative biology, this volume will be of interest to all biologists interested in the biota of Africa and South America.

I thank the many people who took part in the symposium as well as those who assisted in the preparation of this volume. I especially thank the speakers for their participation in the symposium and for their gracious cooperation with the assembly of chapters for publication. The symposium was supported by the United States National Science Foundation (grant BSR 89-18138) and the Missouri Botanical Garden. I thank Dale E. Johnson of the Missouri Botanical Garden and Jean Thomson Black of Yale University Press for their advice and assistance in bringing the preparation of the volume to completion.

Part I

Introduction and Plate Tectonics

Chapter 1

Biological Relationships between Africa and South America: An Overview

Peter Goldblatt

I. Introduction

The striking differences in animal and plant life at both tropical and temperate latitudes across the earth make biogeography a source of endless fascination and wonder, and the underlying reasons for these differences provide a constant stimulus to scientists. With the acceptance of the theory of continental drift and its transformation into the science of plate tectonics, the anomalous distributions of many organisms no longer need support the theory. Instead, increasingly accurate dating of continental shifts and accompanying latitudinal position and growing knowledge of global changes in climate and sea level are providing a basis for the explanation and understanding of the differences in the world's biota. These developments now provide a framework for reevaluating the continuing debate about the biological relationships between Africa and South America, two continents that for a period of geological time composed a single isolated landmass, West Gondwana.

West Gondwana—the remaining large fragment of the supercontinent of Gondwanaland after the separation in the Jurassic of India, Australia, and Antarctica—began to fragment in the early Cretaceous. In the Valanginian, about 135 million years ago (Ma), rifting at its southern end marked the birth of the proto–Atlantic Ocean, and the interior of the continent became the site of cataclysmic geological changes as the process of rifting apart progressed. This was a critical time in the evolution of some of today's dominant land and freshwater organisms. Through the Aptian, 119–113 Ma, Africa and South America were connected north of the Niger

 ISBN 0-300-05375-4

Delta, but the South Atlantic was an increasing water barrier to the south. Complete separation of the continents was achieved between 106 and 84 Ma (more exact dating is difficult—see Chapter 2). Late Albian marine sediments on the São Paulo plateau show the region closed to the North Atlantic, but by Turonian-Conician time, 88–91 Ma, a permanent connection was established between the North and South Atlantic. Pitman and co-workers (Chapter 2) suggest that by between 95 and 100 Ma (Cenomanian) Africa and South America were effectively isolated from each other except to a few easily dispersed terrestrial organisms. Thus by the Turonian, most of Africa and South America were probably separated by some 200 km of ocean, the distance that now separates Africa and Madagascar. Although this distance clearly constitutes an effective barrier to the dispersal of most land plants, terrestrial animals, and freshwater organisms, what is now the northeast corner of Brazil remained relatively close to the western bulge of Africa for some time after complete separation, allowing at least limited interchange (Raven and Axelrod 1974). The shared biota of Africa and South America began to diverge after the Albian as they were influenced by increasing isolation, different episodes of tectonic activity, local climatic events, and biotic migrations from other continents.

The angiosperms, the predominant land plants today, had differentiated by the beginning of the breakup of West Gondwana but were still a minor element of the land flora, constituting less than 1 percent of pollen assemblages (Crane 1987). The primary diversification of the flowering plants, a group first recognized unambiguously in the fossil record in the Hauterivian-Barremian of the early Cretaceous, appears to have taken place across northern West Gondwana—then astride the equator—in the Aptian, at the same time that separation of Africa and South America was nearing completion. By the time separation was complete, the angiosperms were fairly well diversified and on their way to becoming dominant. In the late Mesozoic, when angiosperm diversification was well underway, mammals and birds were still minor groups of vertebrates in a world where dinosaurs still ruled. They did not begin their major radiations until after the end of the Cretaceous, when the dinosaurs became extinct.

Part of the key to understanding today's vast biotic differences between Africa and South America is to determine how the environments of the two continents changed as Africa moved north to contact Eurasia, while South America remained more or less in place latitudinally and in effect constituted a giant island until finally linked with North America via the Isthmus of Panama. The landbridges have probably had at least as much significance to the biota of the two continents as have different patterns of climatic change. These landbridges are briefly described below and are examined in greater depth in the following chapters.

Significant links between South and North America began to develop via island arcs across the eastward-moving Caribbean Plate (Chapter 2). The proto-Antillean arc may have provided a path for the dispersal of some dinosaur groups from Laurasia into South America in the late Cretaceous (Rage 1981) and must have allowed for exchange of other organisms, including marsupial mammals (Carroll 1988) and some angiosperms (Gentry 1982), between the two continents. The connection became increasingly tenuous in the early Cenozoic as the islands of the Caribbean Plate moved further northeastward with respect to South America. There is weak evidence for the existence of an island-arc filter bridge across the Rio Grande Rise–Walvis Ridge from the Albian until the early Paleocene. The effects, if any, of this bridge are obscure. Geological evidence for secondary island arcs across the central Atlantic is lacking, but the possibility of their existence along the Ascension Fracture Zone is raised by Maisey (Chapter 14) as a pathway for interchange of some organisms after the final separation of West Africa and Brazil.

The end of the Cretaceous was marked by the cataclysmic collision of a large asteroid with the earth. At this time Africa and South America were as far apart as the latter was from North America. Dinosaurs became extinct worldwide, and both terrestrial and marine faunas changed drastically. Plants were also dramatically affected (Tschudy 1984; Wolfe 1987), at least locally in North America where data are available, although fewer extinctions of major lineages occurred. For animals and plants, the start of the Tertiary represented a major chance for radiation and diversification. Island chains across the Caribbean Plate linked North and South America and provided a significant path for dispersal in the Upper Oligocene (Deseadan) at a time of lowered sea levels (see Chapter 12). Africa remained isolated from landmasses to the south and east, except for Madagascar, which probably maintained its position relative to Africa throughout the Cenozoic. To the north, links between Africa and Laurasia across the narrow western Tethys continued to be made and broken. The Adrian (Apulian) microplate may have rafted African organisms to southern Europe or may have constituted a landbridge allowing biotic interchange in the mid to late Cretaceous.

The counterclockwise rotation of Africa in the Paleogene brought the northern part of the continent into the intertropical zone of reduced precipitation and, in combination with rafting of fragments of the northern part of the continent against Laurasia, created temporary landbridges in the Miocene and Pliocene (see Chapter 13).

Local and global climate changes were at least as significant as the establishment of landbridges. Africa was increasingly affected by seasonal monsoon precipitation, whereas South America remained wetter due to the influence of trade winds carrying moisture across the continent (see Chapters 15 and 16). Global cooling became significant after the separation of South America from Antarctica in the mid-Oligocene (see Chapter 2) and the establishment of circum-Antarctic oceanic circulation. Cold ocean currents and upwelling along the west coasts of both continents had aridifying landward effects and led to the development of the Namib Desert in Africa and the Atacama Desert in South America. The post-Miocene effects of cold-current desiccation had more severe effects in Africa, reaching across the equator into West Africa at times of climatic extreme. In South America the rise of the Andes along the west coast, beginning in the Miocene, limited aridity to the west of the continent, and the principal effect of polar glacial extension elsewhere on the continent was probably a decrease in temperature (see Chapter 16). The well-known paucity of the tropical African flora may be a direct consequence of the absence of a comparable barrier to the spread of aridity and consequent higher extinction rates. However, in southern South America the rise of the southern Andes created a rainshadow to moisture from westerlies. The resulting environmental stresses constituted a driving force in evolution (see Chapter 12). In both continents Mediterranean climates became established toward the end of the Miocene. High-pressure systems induced by cold offshore currents prevented summer precipitation, but their northerly shift in winter allowed the westerlies to bring winter rain to latitudes between 30° south and 35° south.

As an introduction to this volume, some of the major changes in the flora and fauna of the two continents are reviewed below. Coverage is necessarily brief, and some groups of organisms are not treated at all, due to limitations of space and unavailability of authorities for some taxa.

II. Angiosperms

Evidence from many sites around the world make it clear that the flowering plants were a minor element of the earth's floras until the Aptian of the mid-Cretaceous (see Crane 1987 for a review). Their place of origin remains obscure, but a growing body of data suggests that their major diversification took place in West Gondwana across the equatorial belt, through what is now Brazil, to Israel (Doyle 1978). The earliest angiosperm pollen types, granular and semitectate-reticulate monosulcate grains (Ward et al. 1989), appear in southern Laurasia and northern West Gondwana almost simultaneously, but typically dicotyledonous triaperturate grains appear first in tropical West Gondwana and diversify there before they appear in North America and Eurasia in the Albian (Doyle et al. 1982; Crane 1987). The pattern of dispersal and changes in frequency of angiospermous fossils thus accords best with spread from a northern West Gondwana center. Through the early Cretaceous the northern and tropical part of West Gondwana comprised a single floristic province, what are now southern South America and southern Africa another, and temperate Laurasia two more (see Chapters 3 and 4).

Tectonic activity in the early Cretaceous continued as the proto-Atlantic Ocean extended north with the northern and southern halves uniting effectively in the Cenomanian. The time of separation of West Gondwana appears to be important for considerations of angiosperm biogeography because it was in the Aptian that major angiosperm diversification occurred, and angiosperms rose from being a minor (less than 1%) floristic component to comprising about 50 percent of the flora at the beginning of the Albian (Crane 1987). Members of a few extant plant families can be recognized confidently by the Albian, and regional floristic divergence is evident by the middle of the Upper Cretaceous (see Chapter 4). Records are poor or nonexistent for southern West Gondwana until the latest Cretaceous. In northern South America, however, Gunneraceae (*Tricolpites*) and Proteaceae (*Proteacidites*) can be distinguished in the Turonian. By Maastrichtian time, northeast and northwest South America constituted separate floristic regions sharing palms (Arecaceae) and Ulmaceae. The characteristic Laurasian *Aquilapollinites* (probably referable to Hamamelidales or Fagales) appeared in the Venezuelan basin at this time (see Chapter 4). The appearance of *Aquilapollinites* may be the result of the landbridge that developed as the Caribbean Plate inserted itself between North and South America in the Campanian (see Chapter 2).

To the south, the only Cretaceous angiosperm record is *Nothofagus* (Fagales), the southern beech, in the Campanian of Patagonia. This genus, of an order of apparently northern origin, was already present in Australia in the Santonian, where its presence so early remains puzzling. Paleogene records in southern South America suggest a mixed flora in the Eocene—of *Casuarina* (now extinct there), Gunneraceae, *Nothofagus,* palms, Podocarpaceae, Proteaceae, and Sapindaceae-Cupanieae. In sharp contrast to floras to the north, the strong similarity with contemporary Australian paleofloras seems to indicate the establishment of an austral migration route via high-latitude but nonglaciated Antarctica at this time. The broad similarities in the modern temperate floras of Australia and South America thus date from at least the Paleogene and indicate that cool temperate South America had become a distinct floristic region very early.

Africa lay some 15° farther south in the Cretaceous than at present, but we know little of what must have been its temperate flora. The presence of Lactoridaceae in southern Africa in the Turonian-Campanian (Zavada and Benson 1987) gives a tantalizing hint of the angiosperm flora of the subcontinent. Lactoridaceae, a primitive angiosperm family, is now known from a single species in the Juan Fernández Islands off Chile. Southern African plant communities in the late Cretaceous have enough in common with those of southern South America and Australia-Antarctica to be treated as a single floristic province, but *Nothofagus,* so characteristic of the Australia and Antarctica, is absent at African sites investigated (see Chapter 3). Shared gymnosperms represent pre-Cretaceous floristic links, and Proteaceae and Restionaceae, recorded at some southern sites, may have arrived there independently from the tropics lying to the north rather than dispersing from Australia or Antarctica. A Paleocene site in western southern Africa has yielded Ericaceae and Thymelaeaceae as well as Restionaceae and other plants that suggest a seasonally dry climate. However, coastal records (Chapter 3) from the early Miocene of South Africa indicate moist forest with *Ascarina* (Chloranthaceae), Casuarinaceae, *Podocarpus,* Winteraceae, and many palms. Among these only *Podocarpus* occurs there today. Sarcolaenaceae, now restricted to Madagascar, is also represented in this southern flora. The degree to which this community can be taken as evidence of a shared worldwide cool temperate flora (see Chapter 3) is questionable. It seems at least as likely that their similarities were largely acquired independently by direct migration from nearby tropical areas.

Although space does not allow for a detailed analysis of more than a few plant groups, in this volume Annonaceae, one of the oldest angiosperm families, is contrasted with Asteraceae, one of the most recent and specialized. The first undisputed record of Annonaceae is from the Maastrichtian of Colombia, but the family had probably diversified by the mid-Cretaceous. Present patterns of distribution suggest that the family had its primary center in West Gondwana (Raven and Axelrod 1974), although it is now well represented in the Asian tropics (see Chapter 5). Asteraceae in contrast appears to be of Tertiary age, and although represented by abundant pollen records in the Oligocene, it is confirmed in the Eocene (see Chapter 4). Distribution of extant members of the basal tribe Barnadesieae suggest a Pacific origin for the family (see Chapter 6), but there seems no doubt that Cretaceous tectonic activity was too early to have influenced the geography of Asteraceae (Raven and Axelrod 1974). Most of the other large tribes have wide ranges and are relatively easily dispersed, a factor contributing to the success of the family (Chapter 6).

For angiosperms as a whole, it appears likely that some 27 orders, including at least 80 large fairly primitive to moderately specialized families, dispersed directly between Africa and South America before the continents separated (Raven and Axelrod 1974). In many cases subfamilies or tribes and even a few genera of these families were probably in existence as the continents broke apart. Among the shared families, the poor representation in Africa of some key elements—including Monimiaceae, Lauraceae, Proteaceae-Macadamieae, Arecaceae, and Theaceceae—is likely due to later impoverishment. Most of these families are better represented in Madagascar, which implies that they were also at one time in continental Africa.

III. Animals

A. Insects

Even though insects are winged and presumably more easily dispersed than many other animals, their distribution was affected by tectonic movements. Because of their greater geological age, vicariance events are at lower taxonomic levels for insects than for other animals. For the wasp family Vespidae, Carpenter (Chapter 7) demonstrates how rigorous analysis can be used to show divergence correlated with the separation of Gondwana as a whole, as well as separation of West Gondwana. McKey and Davidson's (Chapter 19) consideration of ant-plant symbioses in Africa and the Neotropics delves into insect relationships and the evolution of these associations. In every case they examined, the symbioses had evolved independently on the two continents, suggesting that these interrelationships evolved after the separation of the two continents.

B. Fishes

Long used as biogeographic evidence for continental drift, the distribution of some freshwater fish can now be correlated in detail with known drift events. Lungfish (Lepidosirenidae) are a classic example of African–South American vicariance, and evidence reviewed by Lundberg (Chapter 8) is compatible with this model, which can also be applied to doradoid catfish among the Siluriformes. The order Siluriformes may date from the early Cretaceous and may have rapidly become cosmopolitan.

Among Characiformes, independent radiation from stocks in West Gondwana in the Cretaceous seems consistent with cladistic phylogenies. The order is confined to Africa and South America except for stocks in Central America and Texas (post-Isthmian dispersants) and fossils in Eocene-Oligocene of western Europe. The latter, allied to African characiforms, presumably reached Europe via direct late-Cretaceous land connections. The record suggests a drift-vicariance pattern of evolution in the group (see Chapter

8), with several subgroups shared between the two continents. Lopsided extinction is likely to have occurred, with drier Africa experiencing many more losses. Cichlids (Cichlidae), now spectacularly well represented in the deep freshwater lakes of tropical Africa, and also present in South America, India, and Madagascar, may not have been directly affected by the separation of Africa from South America. Their relatively late appearance in the New World fossil record makes marine dispersal from African stock as likely as vicariance due to continental drift (Chapter 8).

C. Amphibians and Reptiles

The historical biogeography of two of the three extant groups of Amphibia—Gymnophiona and Anura—was influenced by the breakup of the supercontinent of Gondwana (see Chapter 9). The third group, Caudata, has an extensive record in Laurasia but no fossils from Gondwanan continents; the presence of this order in North Africa and Central and Andean South America today is presumably the result of migration across island arcs or landbridges, primarily established in the Miocene and later.

Among the 17 living families of Neobatrachia, to which most living Anura belong, 6 are endemic or largely so to South America. Of these, Dendrobatidae appears to be the sister group of the African Arthroleptidae. Three more families are endemic to Africa and surrounding islands. The lopsided diversity of anurans in Africa and South America is striking. Sub-Saharan Africa, with an area nearly comparable to South America, has fewer than half the number of species—606 versus 1334 (see Chapter 9)—which is best explained by the disparity of xeric and humid environments on the two continents. Especially significant is the small extent of African montane forest, a habitat especially suited to amphibians.

A contrasting distribution is evident among living Reptilia. This illustrates their preference for drier habitats, which are more widespread in Africa. Among living Reptilia, pelomedusid turtles appear to exhibit a classic pattern of cladogenesis associated with the separation of Africa and South America, although this is by no means certain (see Chapters 10 and 14). The distribution of most other turtle groups, including the tortoises, is probably attributable to migration into Africa or South America after their separation (Chapter 10). The arrival of tortoises in South America from North America during the Miocene is debatable, and dispersal from Africa may have occurred across ocean barriers. Lizard taxa shared between Africa and South America include Scincidae and Gekkonidae, both of which include many more African representatives. Among Iguania, the fragmentation of West Gondwana seems to have had important consequences, although there are competing hypotheses about relationships (see Chapter 10). South American iguanas may be the sister group of the Madagascan oplurids; mainland African oplurids are now extinct (Estes 1983). In general, Africa seems to bear a sister-group relationship to South America and Madagascar, although the patterns are often complex. Intercontinental relationships of the gekkonines are more complex. Most New World representatives may have their closest affinities with African taxa that diverged after continental separation.

The phylogeny of snakes is poorly

known. A Gondwanan origin for the transatlantic distribution of Typhlopidae and Booidae seems likely, however, although the groups are pantropical (Chapter 10). Opposed hypotheses about the biogeography of the advanced snakes—including, for example, the arrival of Colubridae in South America over water from Africa versus their southward migration from North America—render conclusions about the role of the separation of West Gondwana in the evolution of snakes premature.

Dinosaur biogeography seems barely relevant to West Gondwanan history. However, the Campanian to early Tertiary landbridge linking North and South America, discussed above, was important in the major interchange of late Cretaceous fauna, which included the dinosaur families Hadrosauridae, Avisauridae, Ceratopsidae, and Titanosauridae (Rage 1981; Chapter 12 in this volume).

D. Birds

Although the inclusion of a chapter dealing with so mobile a group as the birds might seem incongruous in a book that examines some of the effects of the separation of West Gondwana, the avifaunas of the two continents are very different (see Chapter 11). And some of the differences, including the rhea-ostrich and Neotropical cuckoo–Old World cuckoo divergences, have been attributed to the mid-Cretaceous separation of Africa and South America (Sibley and Ahlquist 1990). The fossil record of ratites in both Northern and Southern hemispheres in the early Tertiary makes the dating of the rhea-ostrich divergence uncertain; ostriches may have evolved from Laurasian stock and migrated into Africa, where they are recorded first in the Miocene (Chapter 11). The fragmentary fossil record renders the biogeography of New and Old World cuckoos equally open to question.

As provocative as questions of biogeography are comparisons of the bird communities on the two continents. South America has a richer avifauna (a reflection of post-Miocene African extinctions?), but African savanna birds are more diverse (Chapter 11). Africa is notably poorer in forest species, whereas the Mediterranean ecosystems of both continents have bird faunas of about equal size.

E. Mammals

Not only were the mammals a minor vertebrate group in the Cretaceous, but they are so poorly known from Africa at the time of the fragmentation of West Gondwana that virtually nothing is known of the evolutionary consequences of this event in Africa. Both didelphid marsupials and primitive placentals were present in South America in the Cretaceous and underwent major radiation there in the late Cretaceous and Paleogene. The remarkably good fossil record in South America provides a detailed insight into the autochthonous evolution of the mammals of the continent and has led Marshall and Sempere (Chapter 12) to conclude that the rainshadow of the developing southern Andes formed a stress environment responsible for accelerated speciation and extinction rates and, further, that crustal movements induced or monitored land mammal macroevolution during the later Cenozoic. This hypothesis is supported by Vrba (Chapter 13), using data from African fossil mammals and other organisms.

Marsupials, but not placentals, probably

dispersed from South America via Antarctica to Australia where they are first recorded in the Oligocene (Carroll 1988). The occurrence in late-Cretaceous North America of marsupials similar to those in South America indicates that there were connections between the two continents. It seems reasonable to assume the presence of shared marsupial stock in Africa in the mid-Cretaceous, although records are lacking.

Africa was never as isolated as South America although exchange of faunas between Africa and adjacent Eurasia was less frequent than between Eurasia and North America. Artiodactyl and perissodactyl ungulates and Carnivora arrived in Africa in waves from the north. Proboscideans, hyraxes, and aardvarks evolved in Africa; only the proboscideans seem to have become significant beyond that continent. At least four episodes of direct migration from western Asia into northeast Africa occurred after the beginning of the Miocene, involving ungulates in particular (see Chapter 13). Spreading aridity in Africa and the extension of grassland habitats coincided with the introduction of these stocks and allowed for their subsequent radiation. Early primate groups were widespread in Africa, but they were absent from Australasia and from South America. Ancestors of anthropoid Catarrhini, which appear in Africa in the Oligocene, gave rise to the South American Playrrhini, first recorded there in the late Oligocene. How they reached that continent is puzzling. Rafting from Africa seems unlikely, but dispersal via an island filter bridge across the Caribbean, together with caviomorph rodents and land tortoises (Chapter 12), leaves open the question of how they reached the Caribbean, for there are no North American records of the group. The early history of the anthropoid primates has not been established, but their African radiation, beginning in the Miocene, is the subject of close study.

Establishment of the Panama land connection linking North and South America circa 2.5 Ma set the stage for the Great American Faunal Interchange. Vrba's habitat theory (Chapter 13) proposes a novel explanation for the lopsided exchange in which North American migrants were more numerous and successful than their South American counterparts. The interchange coincided with major climatic deterioration, spreading aridity, and polar glaciation. Newly created open habitats in South America were rapidly colonized by northern immigrants already adapted to those conditions, whereas fewer native South American mammals were adapted to these habitats.

IV. Ecosystems and Communities

Comparative studies of ecosystems on a continental scale are few, and only lowland forest floras (see Chapter 17) and savannas (see Chapter 18) are included here. For additional ecological comparisons, readers are referred to Brown 1973 (ant faunas), Jago 1973 (grasshoppers), Marlier 1973 (limnology of the Congo and Amazon rivers), and Roberts 1973 (fishes). Other chapters in the present volume provide valuable ecological comparisons—for example, of amphibians (Chapter 9). A novel view of a remarkably well preserved and varied fossil assemblage is reviewed by Maisey (Chapter 14). The Santana Formation of northeastern Brazil includes remains of fishes and reptiles as well as arthropods (especially insects), molluscs, and some land plants. It thus gives us an unusually complete picture of a central West Gondwanan Cretaceous ecosystem until the Albian. The assemblage should give us pause to consider how poor our knowledge is of the paleobiota of West Gondwana. The strangeness of most of the organisms indicates much extinction, and records of taxa previously only known from other parts of the world make it clear that hypotheses about vicariance events are constantly subject to revision.

I suspect that the forests of the warm tropics and elements of southern temperate forests have the most direct link to the Cretaceous separation of Africa and South America. Africa's tropical forests are today restricted in extent, although more of its landmass lies within the tropics than any other continent and its flora is less diverse than that of the New World or tropical Asia. African forests have also been considered comparatively poor in lianas and epiphytes. Continental African forests have an unexpectedly low diversity of such characteristic moist tropical groups as orchids, palms, Lauraceae, Myristicaceae, and Myrtaceae.

Is Africa the "odd man out," as Richards (1973) suggested? In some respects, perhaps, but as Gentry (Chapter 17) shows, lowland tropical African forest communities resemble those of South America remarkably closely. As expected, genera shared between the two continents are few, as are shared endemic families. Such data led Thorne (1973) to conclude that continental drift was not a significant factor in explaining the distribution of seed plants on the two continents. This ignores the question of divergence. Although we do not yet have phylogenies for tropical plant families to compare related genera (or sections and tribes) on both continents, patterns of familial and generic distribution led Raven and Axelrod (1974) to conclude that direct migration in the Cretaceous and subsequent diversification best explains the modern geography of much of the African and South American flora. The marked differences between their floras are frequently due to widespread extinction, especially in Africa. The evidence for these extinctions has been steadily accumulating and includes, for example, Miocene records in southern Africa of Casuarinaceae, Sarcolaenaceae, and Winteraceae, reviewed by Coetzee (Chapter 3).

A comparison of tropical lowland community composition provides a useful alternative way to compare the floras of the two continents (see Chapter 17). One striking result is that familial composition in many study sites across South America and Africa, including Madagascar, is virtually identical, and shared genera range from 30 percent to nearly 50 percent. Fabaceae is the most speciose family on both continents, contrasting with the predominance of Dipterocarpaceae in comparable tropical Asian situations. Wet forest sites in Madagascar and Africa are equally diverse and comparisons with Neotropical lowland sites provide few noteworthy differences in structure, diversity, and higher-level taxonomic composition. An aspect not included in Gentry's survey are ant-plant and plant-ant symbioses, the diversity of which is much greater in the Neotropics (see Chapter 19). As McKey and Davidson conclude, this is probably related to greater overall diversity of the Neotropics, the larger variety of habitats, and lower extinctions rates than in Africa.

Comparison of lowland forest, the oldest

type of plant community, is most relevant to considerations of continental drift. Portions of both continents experienced cycles of ecological stress—including desiccation or cooling or both, plus uplift—that were imposed on a forested landscape. Their woodlands, steppes, and Mediterranean vegetation belts developed independently and cannot be expected to have more than limited common floral elements. Indeed their few similarities in composition are more striking than their differences.

The effects of climatic change and tectonic movement affected Africa and South America in profoundly different ways. The two halves of what was a single continent until the mid-Cretaceous are now populated by such different biota that it is difficult to discern the effects of the separation some 105 Ma. Cenozoic tectonic and climatic change combined with migration from other parts of the world seem considerably more significant than the effects of divergence following the separation of the continents, more so for most animal groups than for plants. The debate over the biological relationships between Africa and South America will continue. As new evidence from the fossil record comes to light and as our understanding of plant and animal phylogeny improves, so will our appreciation of intercontinental biological relationships.

Literature Cited

Brown, W. L. 1973. A comparison of the Hylean and Congo-West African rain forest ant faunas. Pp. 161–186 *in* B. J. Meggers, E. S. Ayensu, and W. D. Duckworth (eds.), *Tropical Forest Ecosystems in Africa and South America: A Comparative Review.* Washington, D.C.: Smithsonian Institution Press.

Carroll, R. L. 1988. *Vertebrate Paleontology and Evolution.* New York: W. H. Freeman.

Crane, P. R. 1987. Vegetational consequences of the angiosperm diversification. Pp. 107–144 *in* E. M. Friis, W. G. Chaloner, and P. R. Crane (eds.), *The Origins of Angiosperms and Their Biological Consequences.* Cambridge: Cambridge University Press.

Doyle, J. A. 1978. Origin of angiosperms. Ann. Rev. Ecol. Syst. 9:365–392.

Doyle, J. A., S. Jardiné, and A. Doerenkamp. 1982. *Afropollis,* a new genus of early angiosperm pollen, with notes on the Cretaceous palynostratigraphy and paleoenvironments of northern Gondwana. Bull. Cent. Rech. Explor.-Prod. Elf-Aquitaine 6:39–1117.

Estes, R. 1983. The fossil record and early distribution of lizards. Pp. 365–398 *in* A. Rodin and K. Miyata (eds.), *Studies in Evolutionary Biology: Essays in Honor of Ernest Howard Williams.* Cambridge, Mass.: Harvard University Press.

Gentry, A. H. 1982. Neotropical floristic diversity: Phytogeographical connections between Central and South America, Pleistocene climatic fluctuations, or an accident of the Andean orogeny? Ann. Missouri Bot. Gard. 69:557–593.

Jago, N. D. 1973. The genesis and nature of tropical forest and savanna grasshopper faunas, with special reference to Africa. Pp. 187–196 *in* B. J. Meggers, E. S. Ayensu, and W. D. Duckworth (eds.), *Tropical Forest Ecosystems in Africa and South America: A Comparative Review.* Washington, D.C.: Smithsonian Institution Press.

Marlier, G. 1973. Limnology of the Congo and Amazon rivers. Pp. 223–238 *in* B. J. Meggers, E. S. Ayensu, and W. D. Duckworth (eds.), *Tropical Forest Ecosystems in Africa and South America: A Comparative Review.* Washington, D.C.: Smithsonian Institution Press.

Meggers, B. J., E. S. Ayensu, and W. D. Duckworth (eds.). 1973. *Tropical Forest Ecosystems in Africa and South America: A Comparative Review.* Washington, D.C.: Smithsonian Institution Press.

Rage, J. C. 1981. Les continents peri-atlantiques au Crétacé Supérieur: migrations des faunes continentales et problèmes paléogéographiques. Cretaceous Res. 2:65–84.

Raven, P. H., and D. I. Axelrod. 1974. Angiosperm biogeography and past continental movements. Ann. Missouri Bot. Gard. 61:539–673.

Richards, P. W. 1973. Africa, the "odd man out." Pp. 21–26 *in* B. J. Meggers, E. S. Ayensu, and W. D. Duckworth (eds.), *Tropical Forest Ecosystems in Africa and South America: A Comparative Review.* Washington, D.C.: Smithsonian Institution Press.

Roberts, T. R. 1973. Ecology of fishes in the Amazon and Congo basins. Pp. 239–254 *in* B. J. Meggers, E. S. Ayensu, and W. D. Duckworth (eds.), *Tropical Forest Ecosystems in Africa and South America: A Comparative Review.* Washington, D.C.: Smithsonian Institution Press.

Sibley, C. G., and J. E. Ahlquist. 1990. *Phylogeny and Classification of Birds: A Study in Molecular Evolution.* New Haven, Conn.: Yale University Press.

Thorne, R. F. 1973. Floristic relationships between tropical Africa and tropical America. Pp. 27–48 *in* B. J. Meggers, E. S. Ayensu, and W. D. Duckworth (eds.), *Tropical Forest Ecosystems in Africa and South America: A Comparative Review.* Washington, D.C.: Smithsonian Institution Press.

Tschudy, R. H. 1984. Palynological evidence for change in continental floras at the Cretaceous Tertiary boundary. Pp. 315–337 *in* W. A. Berggren and J. A. van Couvering (eds.), *Catastrophes and Earth History.* Princeton, N.J.: Princeton University Press.

Ward, J. V., J. A. Doyle, and C. L. Hotton. 1989. Probable granular magnoliid angiosperm pollen from the early Cretaceous. Pollen & Spores 31:113–132.

Wolfe, J. A. 1987. An overview of the origins of the modern vegetation and flora of the northern Rocky Mountains. Ann. Missouri Bot. Gard. 74:785–803.

Zavada, M. S., and J. M. Benson. 1987. First fossil evidence for the primitive angiosperm family Lactoridaceae. Amer. J. Bot. 74:1590–1594.

Chapter 2

Fragmentation of Gondwana: The Separation of Africa from South America

W. C. Pitman III, S. Cande, J. LaBrecque, and J. Pindell

I. Introduction

We wish in this chapter to provide a geographic framework for the dispersal of the Pangea continents as these events relate to the timing of possible migration routes between Africa and South America. Because migratory routes may have been indirect—via Antarctica and North America—our scenario describes the separation of South America from North America and from Antarctica as well as the separation of South America from Africa. From the initiation of rifting and breakup in the Triassic to the present is our focus. The picture would not be complete, however, without considering briefly the pre-Triassic (Paleozoic) plate tectonic and paleogeographic relations among South America, Africa, and North America. The data from which these pre-Triassic relations have been derived are mostly geologic, paleomagnetic, and paleontologic (see Rowley and Pindell 1989 for an extensive summary).

It is now clear that Africa and South America were sutured together as a single continent throughout the Paleozoic, finally separating in the Mesozoic. It also now appears that before they separated, Africa and South America collided with North America. This collision, beginning in the Mississippian, folded up the Appalachian along the northeast segment of the suture; South America and Africa then continued to rotate clockwise into North America, causing diachronous suturing that becomes younger southwestward to final suturing along the Marathon Mountains by the Middle Permian. Thus by the Middle Permian, suturing was complete to form one supercontinent.

 ISBN 0-300-05375-4

II. Plate Tectonics

Before the development of the plate tectonic hypothesis, estimates of the timing of the dispersion of the Pangea continents were based mostly on paleontologic data. The advent of plate tectonics (Hess 1962; Dietz 1961) and in particular the recognition of the significance of magnetic lineation—mid-ocean ridges and fracture zones (Vine and Mattews 1963; Wilson 1963; Morgan 1968)— have allowed marine scientists to reconstruct the timing and path of relative motion of sections of Pangea as the supercontinent rifted and dispersed. (For a thorough discussion of the kinematics of plate tectonics, see Cox and Hart 1986.) Magnetic lineations form a pattern that is generally symmetric with respect to the spreading ridge axis (figure 2.1). Lineations of the same age but from opposite sides of the ridge axis may be fitted together like continental margins. The lineations are formed as the magmatic material in the mantle rises into the narrow rifting zone. As this material cools, it becomes magnetized in the direction of the earth's magnetic field at that time and place. This is called *thermal remanent magnetization*. Once the rocks have cooled, the magnetization is extremely stable, preserving for millions of years the record of the direction and intensity of the magnetic field at the time and place of cooling (see McElhinney 1973 for a thorough discussion of paleomagnetism).

Because the earth's magnetic field reverses polarity abruptly but nonperiodically, the lineations appear as parallel stripes of alternating normally and reversely magnetized oceanic lithosphere. The pattern is easily mapped with towed magnetometers. The lineations are young at the ridge axis and increase in age away from the ridge. By correlating the parts of the lineation pattern with magnetic reversal histories determined from dated subaerial lava flows and exposed dated marine strata and from numerous age determinations from deep-sea bore holes, the age of the magnetic lineations has been determined and a magnetic polarity time scale developed (figure 2.2; see also Kent and Gradstein 1986 and Harland et al. 1989). Thus, the lineations, which represent old ridge axes, may be used to determine paleogeographic arrangements.

The offsets in the active ridge axes, called transform faults, are parallel to the directions of relative motion. The fracture zones, which are topographic extensions of the transform faults and are seismically inactive, may be regarded as "fossil" transform faults. They show the direction of relative motion in the past (figure 2.1). The fracture zone and transform fault traces thus show precisely the path along which relative motion has taken place. The configuration of a pair of continents such as South America and Africa for a point in time can be constructed by fitting together magnetic lineations of that age from opposite sides of the ridge axis, using the fracture zone and transform fault traces as constraints to the path (Pitman and Talwani 1972; McKenzie and Sclater 1971; Dewey et al. 1973). Implicit in this treatment is the assumption that the separating plates were internally rigid—that no internal shear or other significant deformation occurred within the separating plate. Only if this condition is satisfied do the reconstructions have paleogeographic significance.

The relative motion of the lithospheric plates for any instant can be described as the

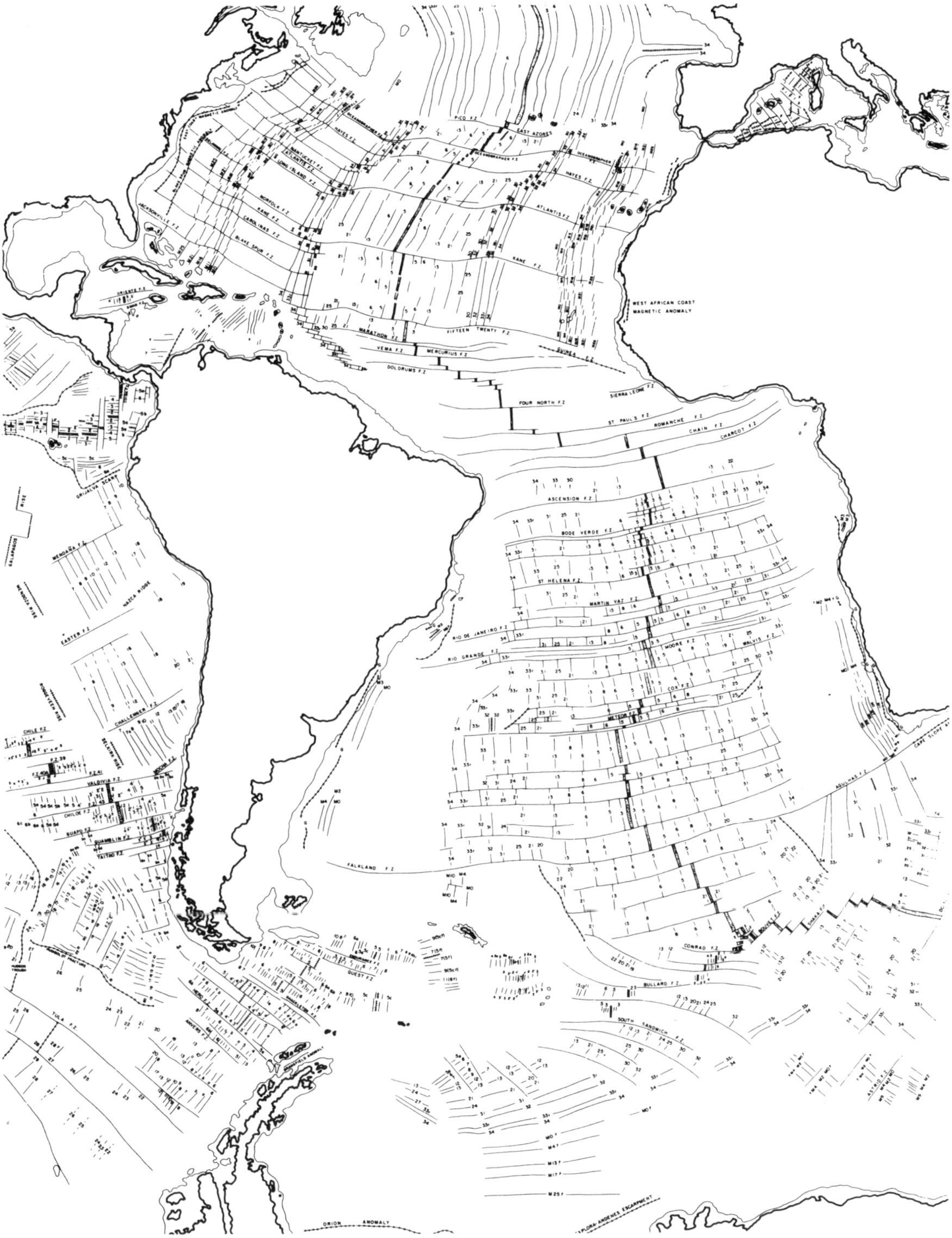

Figure 2.1. The magnetic lineations and fracture zone traces of the North and South Atlantic, Scotia Sea, and Caribbean Sea. The active ridge axis is shown by the broad speckled line down the approximate middle of the Atlantic and extending eastward toward the Indian Ocean and westward toward the Scotia Sea. The ridge axis is offset by transform faults and fracture zones. The lineations are identified by anomaly numbers. The equivalent ages and geologic stages and systems are given in figure 2.2. Modified from Cande et al. (1989).

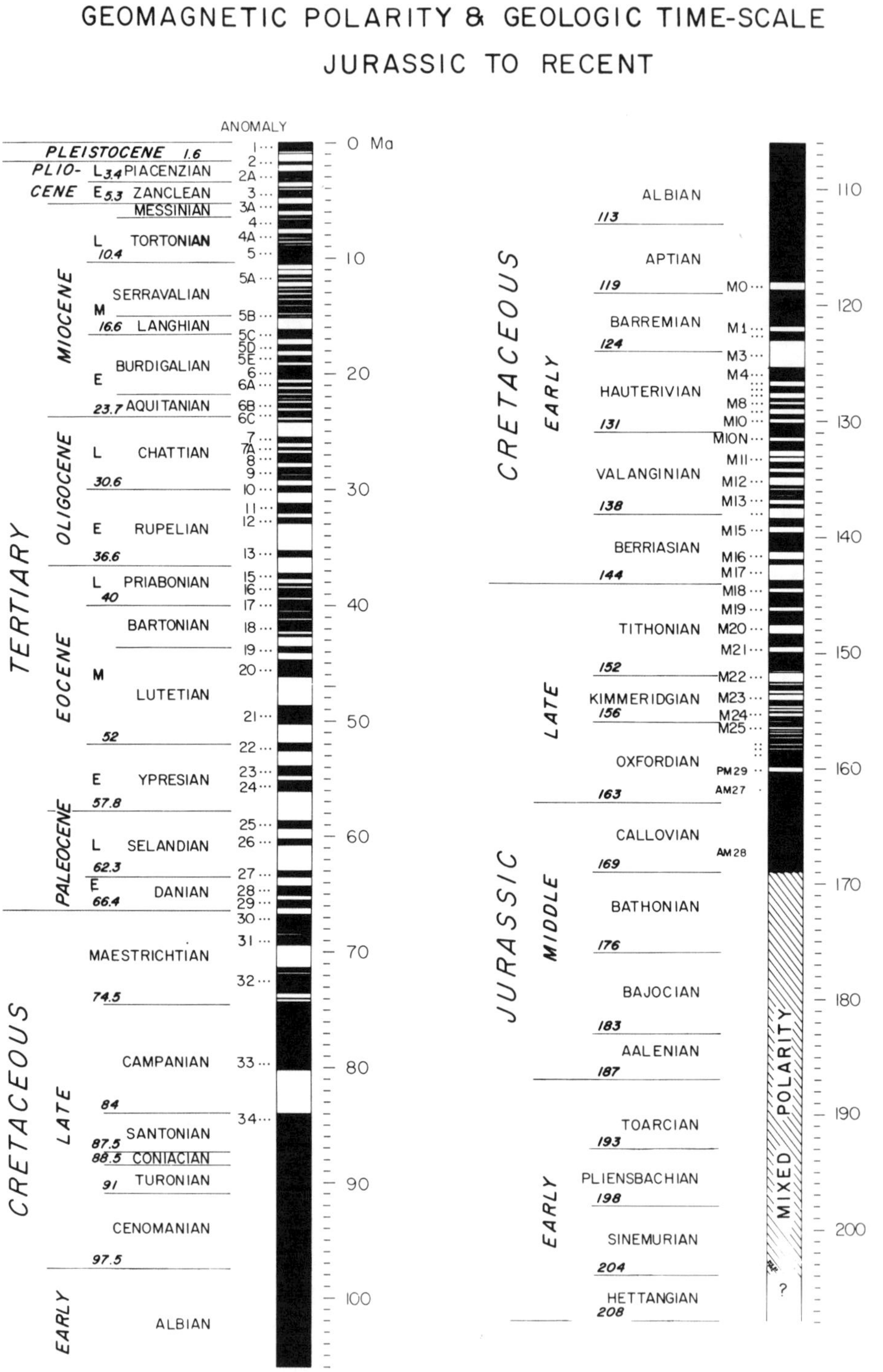

Figure 2.2. The geomagnetic polarity and geologic time scale from the Jurassic to the present. Shaded areas represent periods of normal magnetic polarity; open spaces represent periods of reversed magnetic polarity. The anomaly numbers are listed to the left of the columns; the age in Ma to the right. Adapted from Kent and Gradstein (1985) for anomalies 1 to M29 and from Handschumacher et al. (1988) for anomalies M30 to M37.

rotation of one plate with respect to the other about a pole. The transform faults that offset the active ridge axes are parallel to the path of instantaneous relative motion; hence these transforms lie along small circles that are concentric to the pole of instantaneous relative motion (Morgan 1968). Although in theory the poles of instantaneous motion could move dramatically with time, the smooth curvilinear shape of the fracture zones (figure 2.1) shows that the poles are in fact rather stable for long periods of time (LePichon 1968). The pole that at present describes the separation, at the northern mid-Atlantic ridge, of North America from Eurasia lies in northern Siberia. Analysis of magnetic lineations and fracture zone shapes has enabled scientists to derive sequences of poles and angular rates of motion that describe the history of drift and separation of all the Gondwana continents, thus facilitating fairly precise paleogeographic reconstructions in most cases (see, for example, Pitman and Talwani 1973; McKenzie and Sclater 1971; Cande et al. 1988; Klitgord and Schouten 1986; and Ladd 1976).

The initial stages of rifting at the initiation of continental separation are often lengthy and complex (Dewey et al. 1973). During the rifting stage the continental lithosphere is stretched and attenuated, with accompanying volcanism and rift valley formation (see Bally 1981 for a review of these processes). Even during this stage the relative motion of the bounding continental masses may be approximated by rotation of one plate with respect to the other about a pole. It is not until oceanic lithosphere is formed, with the indicative magnetic lineations, that true continental separation can be assumed to have occurred.

During the rifting phase, as with all subsequent phases, the rate of extension or separation is slowest at that part of the boundary nearest the pole of relative motion and increases along the boundary away from the pole. The transition from rifting and attenuation of the continental lithosphere to formation of the oceanic lithosphere and the symptomatic magnetic lineations occurs first along that part of the boundary farthest from the pole; and the transition tends to move diachronously toward the pole, giving the appearance of a migrating rifting process (LaBrecque and Rabinowitz 1977).

The magnetic lineations and fracture zones in the South Atlantic describe a simple opening between South America and Africa, with the qualification that during the Early Cretaceous phase of rifting there was probably extension within Africa along the Benue Trough (Dewey and Burke 1974; Pindell and Dewey 1982). Thus, for a brief time Africa was not one but rather two rigid plates.

To the north, the geographic history of possible connections between North and South America is far more complex. The magnetic lineations and fracture zone pattern of the North and South Atlantic have been interpreted to give a clear picture of the timing and path of separation of South America with respect to North America since the earliest Jurassic (Ladd 1976; Pindell et al. 1988). During this interval, however, the interaction of various blocks between North America and South America, the episodic presence of island arcs, and, finally, beginning in Upper Cretaceous (Campanian), the insertion of the Caribbean Plate from the Pacific have imposed the possibility of the episodic presence of corridor and filter connections between North and South America. It is only up through the Early Jurassic and since the Pliocene that we can be certain of established corridor connection between the two continents.

The Scotia Sea, between South America and Antarctica, is another Caribbean-like plate. The Scotia Plate contains a pattern of clearly and easily identifiable magnetic anomalies, all of Upper Tertiary age. And thus although the plate at present and for at least the past 30 million years has been thrust eastward with respect to South America, the slow rate of this motion (less than one centimeter per year) and the youthfulness of the Scotia Sea suggest that South America separated from the Palmer Peninsula in the Middle to Late Oligocene.

III. The Fragmentation of Pangea and Gondwana

The fragmentation of Pangea began in the Upper Triassic–Lower Jurassic with rifting between Laurasia (North America and Eurasia) and Gondwana (South America, Africa, Madagascar, India, Antarctica, and Australia). This initial phase of separation was characterized by rifting, volcanism, and the attenuation of continental lithosphere (Dewey et al. 1973). The zone of rifting and eventual separation extended from the Tethys westward between Africa and Spain, south between Africa and North America, and westward north of Yucatán separating North America from South America.

It is not clear whether the next phase can be distinguished from that just described or whether they are continuous. Magnetic anomalies as old as M25 (156 Ma) have been found off of northwest Australia (Larson 1975), and anomalies as old as M22 (152 Ma) have been found west of Madagascar (Norton and Sclater 1979) and north of Madagascar (Rabinowitz et al. 1983; Coffin 1985).

Starting in the east, the rift passed north of Australia; at this time a continental fragment, later to be incorporated into Asia (suggested in Larson 1975 to have been Tarim), was rifted from northwest Australia (Larson 1975). The rift system extended westward, passing north of India and then south along the east coast of Africa, separating Madagascar, India, and Australia from Africa (Rabinowitz et al. 1983; Coffin 1985; Norton and Sclater 1979). Norton and Sclater (1979) propose that the rift continued along the southeast coast of Africa, separating Antarctica from Africa and South America and forming part of the Weddell Sea, with Antarctica moving away as a single plate. If so, the pole of relative motion was located in the vicinity of the Palmer Peninsula (at approximately 64° S and 64° W.). This would be analogous to the observation that the present-day active North Atlantic rift system terminates at its pole of relative motion in northern Siberia. Alternatively, LaBrecque (1986) proposed that East and West Antarctica moved as two separate plates at this time, with a seaway appearing along the boundary between the plates. The magnetic lineation in the Weddell Sea (Bergh 1977; LaBrecque 1986) and off southwest Africa (LaBrecque and Hayes 1979; Bergh and Barrett 1980; LaBrecque 1986) constrains relative motion that most surely took place between East and West Antarctica before anomaly 34 time (84 Ma) and subsequent to the early Jurassic initiation of rifting (LaBrecque 1986). It is perhaps during that time that some relative motion took place along the proposed suture between East and West Antarctica.

The next phase of separation began just before anomaly M11 time, 135 Ma, in the Early Cretaceous (Valanginian). Again beginning in the east, the line of separation passed between Australia and India (Larson 1977), between India and Antarctica, and up through the South Atlantic between Africa and South America (Pindell et al. 1988) to the equatorial region. This was the beginning of the separation of Africa from South America (Larson and Ladd 1973; Rabinowitz and LaBrecque 1979).

The early phase of rifting between Africa and South America appears to have lasted from about 135 Ma (middle Valanginian) to about 106 Ma (middle Albian) (Rabinowitz and LaBrecque 1979). Rabinowitz and LaBrecque assumed that both Africa and South America acted as rigid plates during this interval. They estimated that the pole of relative motion for this interval was located in the vicinity of the Niger Delta and that Africa rotated counterclockwise about the pole with respect to South America. This scenario allows extension along the margin to the south of the pole but requires compression to the north along the northern coast of Brazil. In an alternative scenario, based on the observation of Dewey and Burke (1974), the Benue Trough in west equatorial Africa underwent rifting at this same time (figure 2.3a). This model requires that for the interval Africa acted as two plates; that part of Africa north of the Benue rift zone remained rigidly attached to South America, whereas the part of Africa south of the Benue rift zone rotated away from South America as shown in figure 2.3b (Pindell and Dewey 1982; Pindell et al. 1988) opening up a narrow restrictive South Atlantic that was terminated to the north by the still-joined continent masses of South America and northern Africa and restricted to the south by the near contact of the continental Falkland Plateau and the Agulhas Plateau (figure 2.3c). During this phase of early opening the synrift environment may have been much like that of the present-day African rift system.

Because there were no magnetic polarity reversals between anomaly 34 time (84 Ma) and MO (115 Ma), no magnetic lineations from oceanic lithosphere were formed dur-

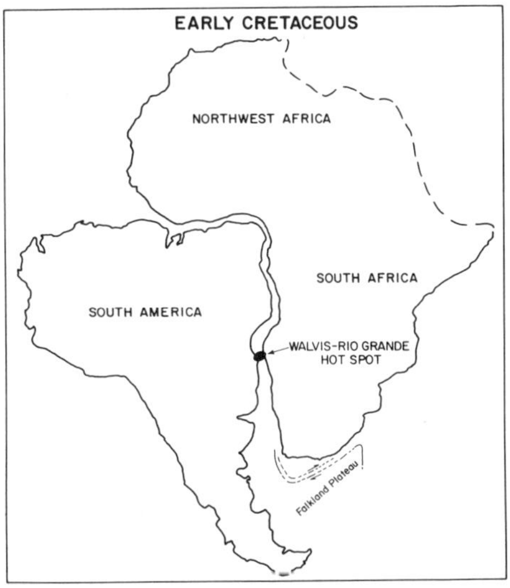

Figure 2.3a. The relative configuration of Africa and South America in the earliest Cretaceous (that is, the prerift configuration). Note that the fit of Africa to South America in the region of the north Brazilian coast is tighter than that shown in Bullard et al. (1965). This is achieved by treating Africa as two plates separated by the Benue Trough and its northeast equivalent rift zone. The fit of Africa south of the Niger Delta to South America is similar to that of Bullard et al. The split of Africa reflects the hypothesis that during the early rifting stage Africa acted as two separate plates. Adapted from Pindell et al. (1988).

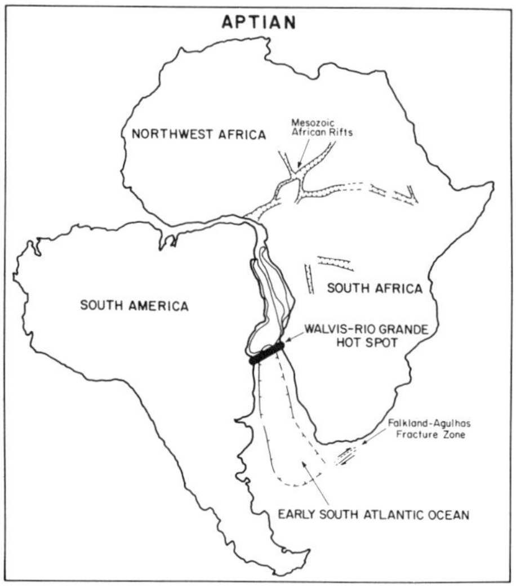

Figure 2.3b. The relative position of Africa with respect to South America at about Aptian, before separation in the equatorial Atlantic. The South Atlantic south of the Niger Delta has opened up by rifting that entered the Benue Trough. During this phase of separation, Africa north of the Niger Delta has remained fixed with respect to South America, in the same relative position as shown in figure 2.3a. Adapted from Pindell et al. (1988).

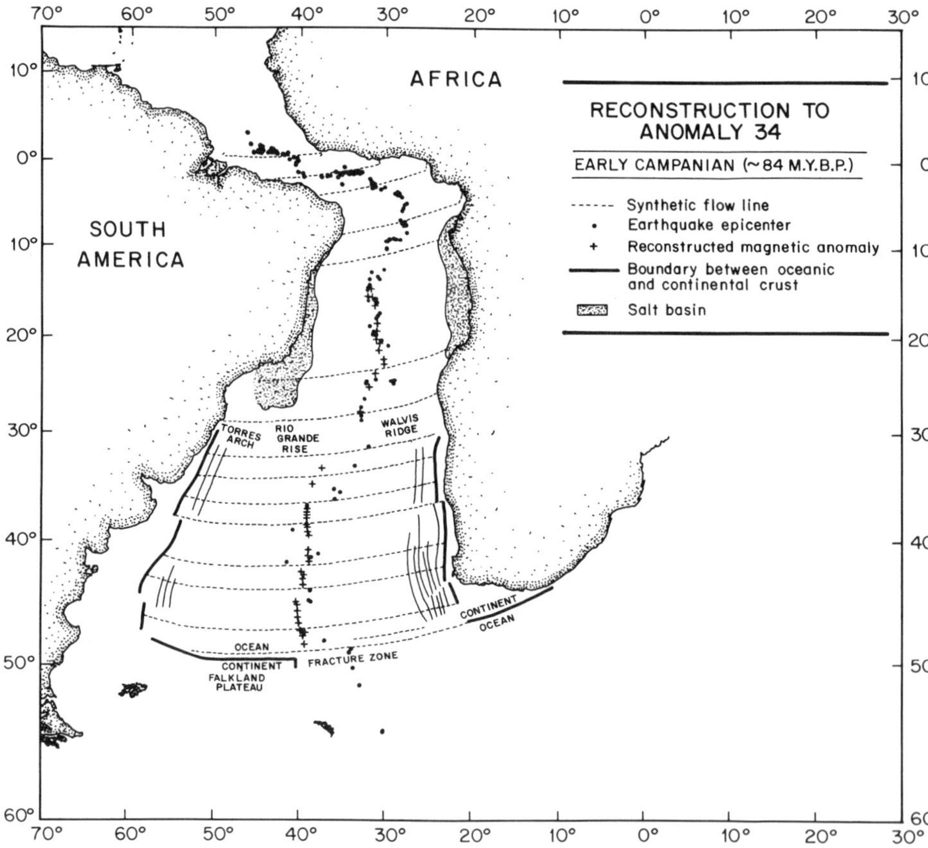

Figure 2.3c. The position of the continents at anomaly 34 time, 84 Ma (Upper Santonian). Africa is now completely separated from South America. Modified from Rabinowitz and LaBrecque (1979).

ing that interval (Helsley and Steiner 1969; Irving and Couillard 1973), and we are forced to rely only on fracture zone trends to determine the direction of relative motion, with little or no way to determine the timing of the motion. It is believed that at about 106 Ma (during quiet zone time) the pole of relative motion for Africa–South America shifted northward and Africa acted as one rigid plate rotating away from South America about a pole located in the central Atlantic (Rabinowitz and LaBrecque 1979; Cande et al. 1988). The next stage of which we are certain is at anomaly 34 time (84 Ma; figure 2.3c). At that time South America and Africa had completely separated. But the point in time of final separation must be between 84 Ma (figure 2.3c) and 106 Ma (figure 2.3b). Because transform margins such as that of northern Brazil often have associated seaward ridges (Bonatti and Chermak 1981) and because seaward platform fragments at these margins, once subaerial, usually subside (often to great depths), the time of final separation cannot be determined on the basis of geography. Sediments from Deep-Sea Drilling Project site 356 on the São Paulo Plateau show that at Late Albian time the region was closed to the North Atlantic but intermittently open to the far south (Premoli-Silva and Boersma 1977). It was during this period, Late Aptian to Albian, that there was major salt deposition around the margin of the South Atlantic, north of the Walvis Ridge. By Turonian-Coniacian time—about 90 Ma—the presence of Tethyan species indicates a permanent north-south connection. Because it takes some time after separation to fully develop a competent marine connection, we estimate that by 100 to 95 Ma (Late Albian to Middle Cenomanian) Africa and South America were effectively separated for migration except for a few specialized species. To the south, for a long time, as Africa and South America separated, the continental promontory of the Falkland Plateau remained adjacent to the African promontory of the Agulhas Plateau and probably formed an effective barrier to marine exchange between the South Atlantic Ocean and the waters of the Antarctic Region. Deep-sea drilling results from legs 36 and 71 on the Falkland Plateau suggest that shallow-marine conditions were prevalent beginning in at least the mid-Jurassic and that the area has since been continuously marine (Barker and Dalziel 1977; Ludwig and Krasheninkov et al. 1983). It appears that the Falkland Plateau–Agulhas Plateau complex was not an effective terrestrial migratory path after rifting but continued to be a partial barrier to water exchange between the South Atlantic and the very southern oceans. By the late Aptian free water exchange to the south was established (Supko and Perch-Nielson 1977).

IV. The Separation of North America from South America

As stated above, rifting and separation between Gondwana and Laurasia began in the Upper Triassic–Lower Jurassic (Pitman and Talwani 1972). Although this resulted in a long period of almost continuous separation (Early Jurassic to Early Campanian) of South America with respect to North America (figure 2.4), the only traces of this separation in terms of magnetic lineations and fracture zones are a few unidentifiable magnetic lineations in the Gulf of Mexico (Pindell and Barrett 1990) and the Jurassic crust

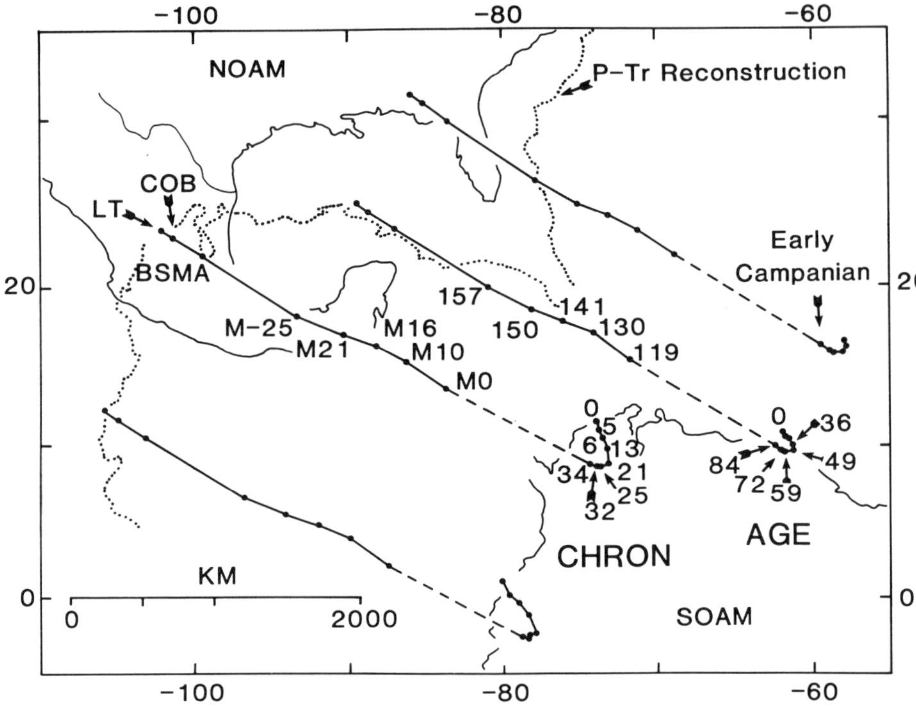

Figure 2.4. The dotted lines (COB) are the continent ocean boundary of North America in the Gulf of Mexico and off of the East Coast. This line also shows the Permo-Triassic (P-Tr) position of Africa with respect to North America and of South America with respect to North America. The four lines extending from upper left to lower right show the calculated path of drift of South America with respect to North America from the Late Triassic (LT) to the present (0). The dashed segments are paths of uncertainty. The numbers along the second line down give the age (AGE) in millions of years before present (Ma). The alpha-numeric designations along the third line down are anomaly numbers (from figure 2.2).

between Trinidad and the Demerara Rise off the Guyana margin, which also has no identifiable magnetic lineations. This is because the Caribbean Plate has been inserted between the two continents—beginning in about Campanian and continuing to present—effectively subducting the oceanic lithosphere formed during the separation (Wilson 1986b; Pindell 1985). However, by using the magnetic lineation and fracture zone traces in the North and South Atlantic to determine the relative motion of both North America and, separately, South America with respect to Africa, South America–North America relative motion can be resolved (Ladd 1976; Pindell 1985; Pindell et al. 1988). These data and the geology of the Caribbean and gulf and their surroundings have been used by Pindell (1985) and Pindell and Barrett (1990) to derive the scenario that follows.

At the time of the Lower Jurassic rifting and separation of Gondwana-Laurasia, the rift zone passed from the Atlantic between

Africa and North America westward between the North American and South American plates, along a line north of Yucatán (forming the Gulf of Mexico), and into the Pacific along a set of transforms in western Mexico. At this time elements of western Mexico were moved southeastward, maintaining contact with South America (figure 2.5a). Continental fragments in Florida slid southeastward to form the basement of the south Florida shelf and the western Bahama platform. By Valanginian time (figure 2.5b), the basic system of continental blocks had been delineated and the Gulf of Mexico had been formed. Separation was taking place between Yucatán and South America. In effect the Yucatán Block was now part of the North American plate. The ridge-transform system separated North America from South America, effectively severing all routes for terrestrial or shallow-marine migration except at an Aleutian-type island-arc system that may have existed in the area labeled "Proto–Greater Antilles" in figure 2.5b.

The separation continued through the Cenomanian (figure 2.5c). North America and South America were still separated by an actively spreading ridge and transform system. See also figure 2.4, which shows southeastward movement of South America with respect to North America continuing but terminating by anomaly 34 time (84 Ma; Santonian-Campanian). During this interval—before 84 Ma and subsequent to the earliest Valanginian—while this ridge transform system was active, probably the only possible migratory paths would have been through an Aleutian-type island-arc system that may have been formed to the south and west (figures 2.5b and 2.5c). This would have been at best an episodic filter bridge.

By Campanian time (80 Ma; figure 2.5d) the Caribbean Plate had begun to insert itself from the Pacific (Wilson 1966b; Pindell 1985; Pindell and Barrett 1990), with the Aves Ridge and Greater Antilles, which are constructual edifices made of island arcs and subduction zone complexes, pushing at its leading edge (Pindell 1985; Pindell and Barrett 1990). The Caribbean Plate has continued to move relatively eastward through the gap until the present (figures 2.5d–h). The Aves Ridge and Greater Antilles probably formed an arcuate connection between northwestern South America and Central America, which may have provided a corridor for migration between North and South America until about the Late Eocene (just after 49 Ma; figure 2.5f). During the Campanian to Middle Eocene interval, this corridor consisted entirely of island arc, accretionary wedge, reefal, and carbonate terranes all undergoing orogenic deformation: thrusting to the east at the Aves Ridge and Lesser Antilles, and transpressional shear to the north in the Greater Antilles. Also during this interval (Cenomanian to Campanian) sea level was high, perhaps at a maximum (Pitman 1978; Haq et al. 1987; Kominz 1982). Because the potential corridor was at a convergent margin, it was constantly subjected to tectonic stresses and motion as well as to sea level fluctuations and may therefore have frequently been impaired—passing from a pure corridor to a filter corridor to an impasse. Furthermore, species migrating from the north or the south may have reached some of the islands only to be cut off from behind or ahead and become isolated. The possibility nonetheless exists that at times during the interval there were corridor connections between the continents.

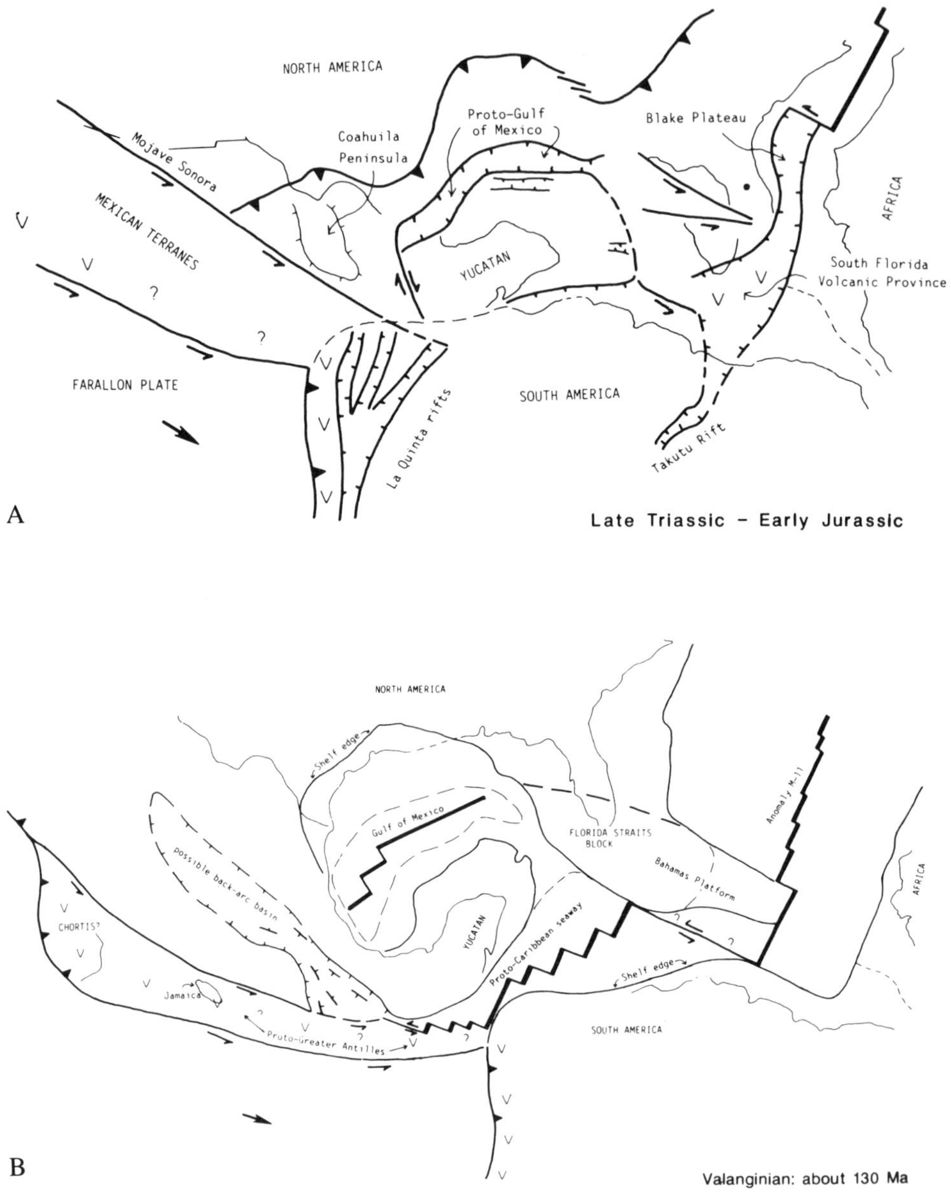

Figure 2.5. Parts A through H of figure 2.5 show the paleogeography of the Gulf of Mexico–Caribbean region at a sequence of discrete points in geologic time. Separation of North America from Africa begins in the Late Triassic–Early Jurassic (A), but true separation of South America from North America may not have taken place until Valanginian (B). Although episodic filter connections may have been in place via island arcs, connections were probably quite tenuous (B and C). Beginning in Campanian (D), the Caribbean Plate was inserted from the Pacific, probably providing more stable and broader (if episodic) connections between South America and North America via the Aves Ridge and Greater Antilles (D and E). By mid-Eocene the Greater Antilles

Figure 2.5 ***Continued.***

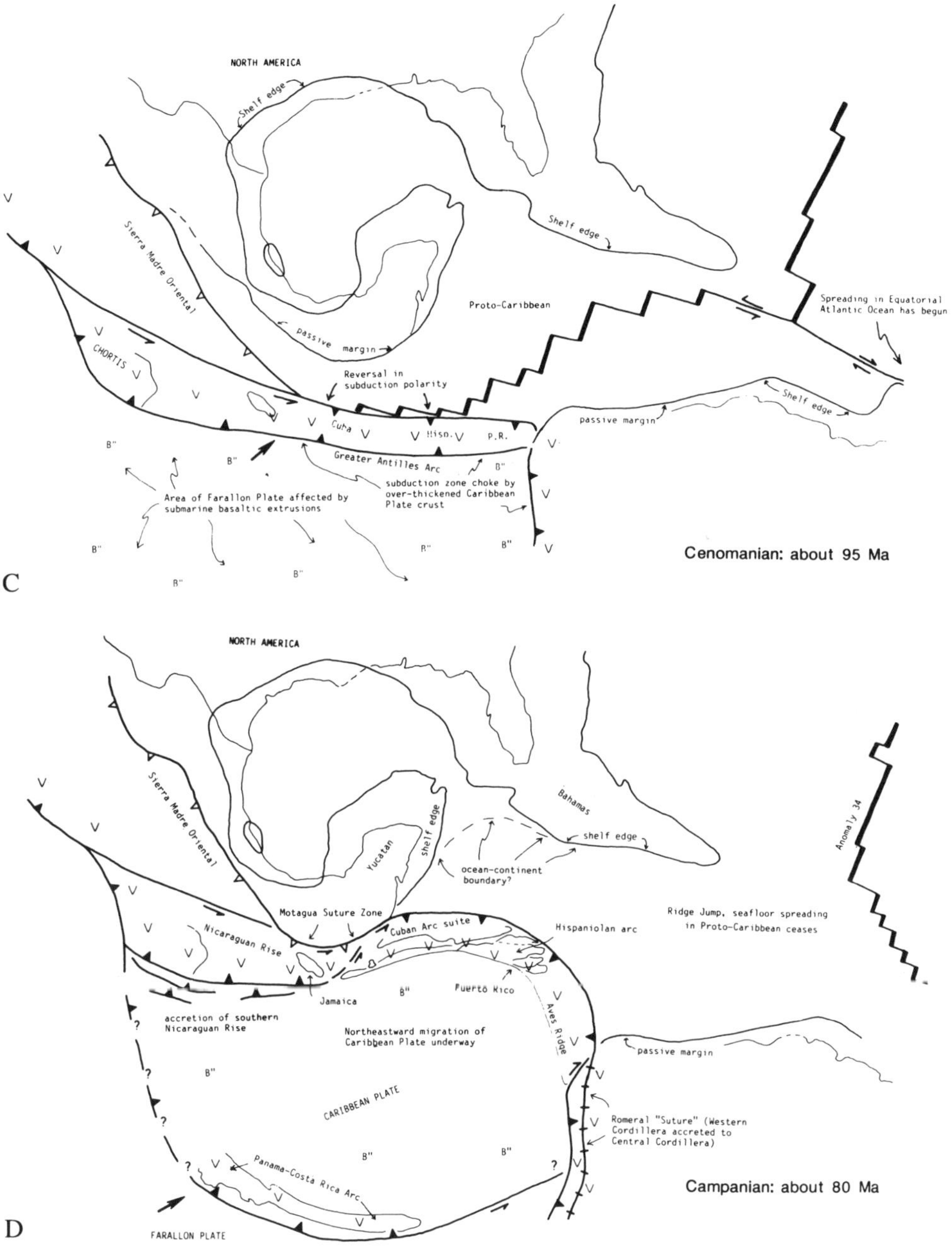

themselves began to be fragmented by shear (F), and subsequent connections were probably episodic to nonexistent. By mid-Miocene (G) the island-arc system that was to become the Isthmus of Panama corridor was established as it exists today (H). Adapted from plate 1 in Pindell and Barrett (1990).

Figure 2.5 ***Continued.***

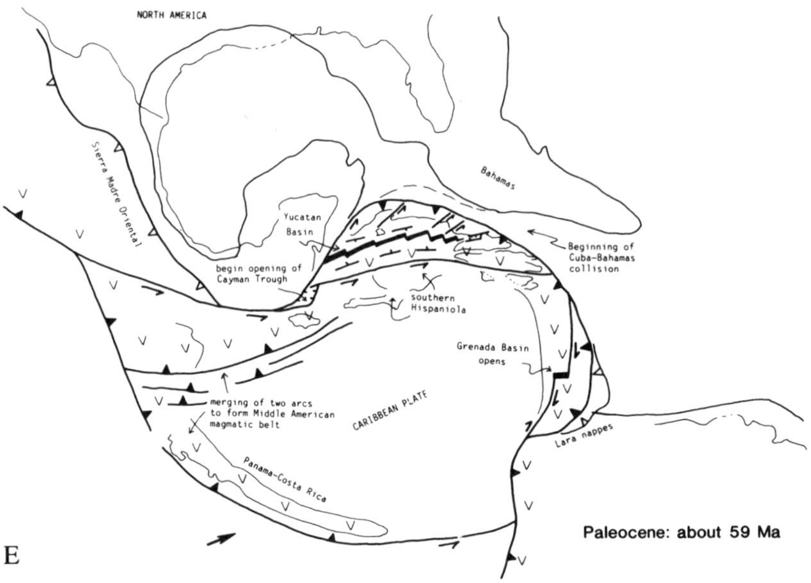

E

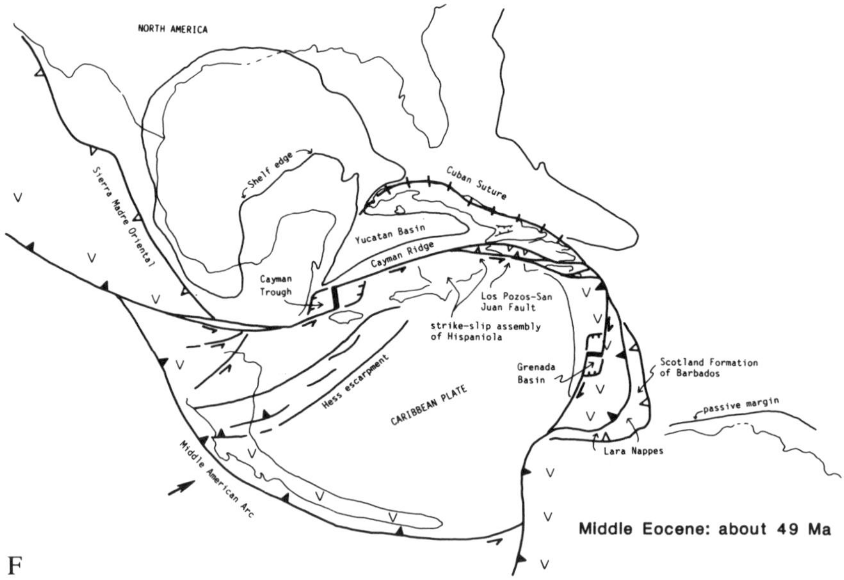

F

Figure 2.5 ***Continued.***

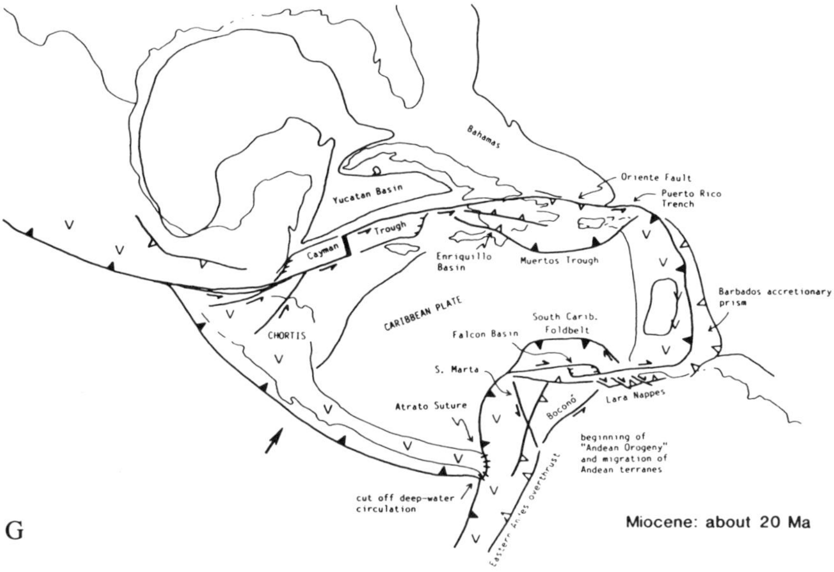

G

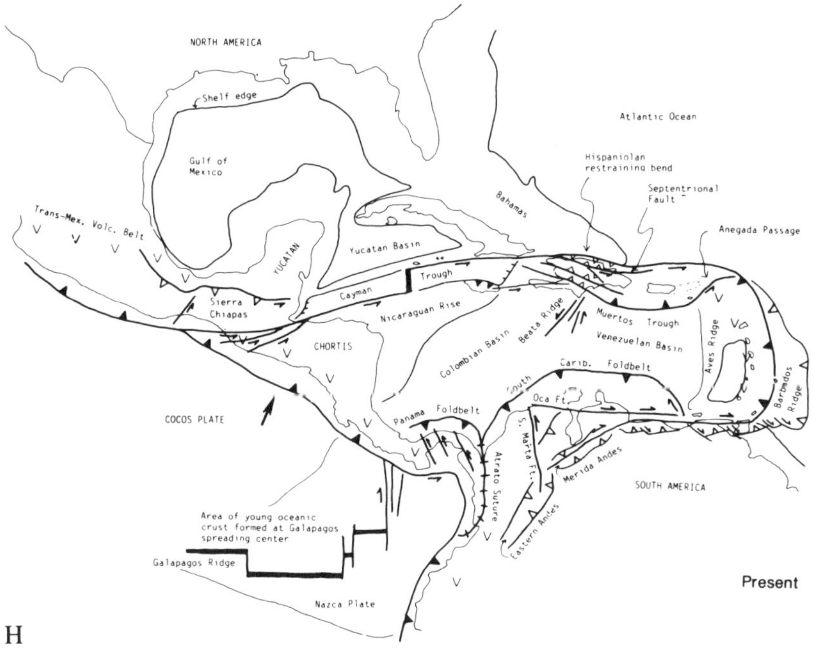

H

After the late Middle Eocene, the relatively eastward movement of the Caribbean continued and significant shear began to take place within the Greater Antilles, separating the elements—Cuba, Hispaniola, and Puerto Rico—along the northern Caribbean transform (Pindell and Barrett 1990; Pindell et al. 1988). The elements of the bridge thus became progressively elongated and separated (figures 2.5f and 2.5g), and we can postulate that between Middle to Late Eocene and Middle Miocene migratory paths between North and South America were less probable, or at best amounted to episodic filter bridges.

Beginning in the mid-Miocene (figure 2.5g), the island-arc–volcanic-arc system that was to become the Isthmus of Panama may have made contact with the northwest corner of South America, thus establishing an Aleutian-type connection between South America and North America (Pindell and Barrett 1990). By the mid-Pliocene the Isthmus of Panama had been uplifted and the Panama corridor as we know it today had been established.

V. Scotia Sea: Separation of South America from Antarctica

It is clear from the magnetic lineation data (figure 2.1) that by the early Cretaceous Africa had become separated from both East and West Antarctica. The timing of separation of South America from West Antarctica is more difficult to determine because the geology of West Antarctica is largely buried in glacial ice and because it is uncertain how much subduction has taken place at the northern and southern edges of the Scotia Sea. Wilson (1966b) first suggested that the Scotia Sea had been injected from the west between Antarctica and South America, analogous to the insertion of the Caribbean Plate. But at the same time, in contrast with the Caribbean, clearly defined young anomaly lineations attest to the youthfulness of the Scotia Sea, so that insertion probably plays a minor role in the history of separation between South America and Antarctica.

A systematic set of Tertiary lineations has been identified in the Scotia Sea (figure 2.1) in the area between 25° W and 75° W and 52° S and 62° S (Barker 1972a, 1972b; Barker and Burrell 1977; LaBrecque and Rabinowitz 1977; Barker and Hill 1981; LaBrecque 1986; see also figure 2.1). These lineations reveal three basic spreading patterns. To the east, immediately west of the Scotia Arc, is an active spreading center that strikes north-south across the Scotia Sea and ranges in age from the present (0 Ma) to anomaly 4 (7 Ma). Immediately to the west is a zone of poorly identified anomalies ranging in age from perhaps 10 to 15 Ma or 25 to 30 Ma. Further west is a much more easily identifiable anomaly pattern with an offset ridge system striking in a northeast direction. Anomalies range from about anomaly 5 or 4a at the axis (9 to 8 Ma) to anomaly 10 (30 Ma) at the northern margin of the Scotia Sea. At present there is a transform system along the southern tip of South America, and the Scotia Sea has moved eastward with respect to South America at about 0.8 centimeters per year (Pelayo and Wiens 1989) since anomaly 10 (30 Ma), which would give 240 kilometers of eastward movement of the Scotia Sea relative to South America and Antarctica since 30 Ma. This is a relatively small lateral motion. Considering that the oldest anomalies in the Scotia Sea date from the mid-Oligocene (anomaly 10), we conclude that at that time elements of the Palmer Peninsula and West Antarctica were rifted from South America, causing a final separation of Antarctica from South America. Two cautionary notes must be kept in mind. First, some subduction may have occurred along the north and northwest side of the Palmer Peninsula and more tentatively along the south coast of South America; thus, anomalies or pieces of ocean older than those now visible in the Scotia Plate may have been subducted. Second, West Antarctica is thought to be composed of a number of microplates (Dalziel 1982). It is therefore not clear whether or when this region formed a continuous migratory route between Antarctica and South America. We tentatively conclude that 30 Ma (mid-Oligocene) probably represents the most recent time of full separation between Antarctica and South America.

VI. Conclusions

In summary, the separation of Gondwana from Laurasia began in the Late Triassic–Early Jurassic. The line of separation extended from Tethys southward between Africa and North America (opening up what was to become the central North Atlantic) and westward between Yucatán and North America and reaching the Pacific via a series of transform faults and ridges through western Mexico. Separation of Africa and North America was probably complete early in this stage. Connection between South America and North America may have been maintained during this stage via a set of southeastward-moving blocks that comprised western Mexico and Yucatán (see figure 2.5a).

By Valanginian time North America was probably completely separated from South America. The South Atlantic was opening up along a line south from the Niger Delta. Connection between South America and Africa remained intact north of the Niger Delta. Northern Africa began to separate from northern South America by the mid-Albian, and Africa, as one plate, moved away from South America. We estimate that separation was almost complete by the uppermost Albian.

By the Campanian, South America and North America had ceased drifting apart and, except for a few hundred kilometers of convergence since the Eocene, the relative position (figure 5.4) has since been rather stable. In the Campanian, the insertion of the Caribbean Plate, with the Aves Ridge and the Greater Antilles, between South America and North America possibly reestablished sporadic corridors and filter connections between the two continents (figure 2.5d). The continued eastward insertion of the Caribbean Plate (figures 2.5d and 2.5e) continued to provide a mechanism for connection between the continents until the mid-Eocene, when the eastward movement of the Caribbean had progressed so far as to begin to separate by shear the large Greater Antilles rocks (figure 2.5f).

By mid-Miocene (figure 2.5g), the Panama island arc may have provided a filter causeway, and by the Pliocene the present Isthmus of Panama was uplifted and in place. To the south, it appears that rifting and spreading initiated in the mid-Oligocene caused the final separation of South America and West Antarctica and at the same time opened up the Drake Passage and allowed for the establishment of the Circum Antarctic current. This current thermally isolated Antarctica sufficiently to allow extensive permanent glacial buildup (Kennett et al. 1974).

Literature Cited

Bally, A. W. 1981. Atlantic-type margins. Pp. 1.1–1.48 *in Geology of Passive Continental Margins: History, Structure and Sedimentologic Record*. Tulsa, Okla.: American Association of Petroleum Geologists.

Barker, P. F. 1972a. Magnetic lineations in the Scotia Sea. Pp. 17–26 *in* R. J. Adie (ed.), *Antarctic Geology and Geophysics*. Cambridge: Cambridge University Press.

———. 1972b. A spreading centre in the east Scotia Sea. Earth Planet. Sci. Letts. 15:123–132.

Barker, P. F., and J. Burrell. 1977. The opening of the Drake Passage. Marine Geol. 35:15–34.

Barker, P. F., and I. W. D. Dalziel. 1977. Evolution of the southwestern Atlantic Ocean Basin. Pp. 995–104 *in* P. F. Barker, I. W. D. Dalziel, et al. (eds.), *Initial Reports of the Deep Sea Drilling Project 36,* Washington, D.C.: U.S. Government Printing Office.

Barker, P. F., and I. A. Hill. 1981. Back-arc extension in the Scotia Sea. Philos. Trans. Soc. London, Ser. A, 300:249–262.

Bergh, H. W., 1977. Mesozoic seafloor off Dronning Maud Land, Antarctica. Nature 269:686–687.

Bergh, H. W., and D. W. Barrett. 1980. Agulhas Basin magnetic bight. Nature 287:591–595.

Bonatti, E., and A. Chermak. 1981. Formerly emerging crustal blocks in the equatorial Atlantic. Tectonophysics 72:165–180.

Bullard, E. C., J. E. Everett, and A. G. Smith. 1965. The fit of the continents around the Atlantic. In *A Symposium on Continental Drift,* Royal Soc. London Phil. Trans. Ser. A, 258:41–51.

Cande, S. C., J. L. LaBrecque, and W. F. Haxby. 1988. Plate kinematics of the South Atlantic chron C34 to present. J. Geophys. Res. 93:13,479–13,492.

Cande, S. C., J. L. LaBrecque, R. L. Larson, W. C. Pitman, III, X. Golovchenko, and W. F. Haxby. 1989. Magnetic lineations of the world's ocean basins (map). Tulsa, Okla.: American Association of Petroleum Geologists.

Coffin, M. F. 1985. "Evolution of the conjugate East African–Madagascar margins and the western Somali Basin." Ph.D. diss., Columbia University.

Cox, A., and R. B. Hart. 1986. *Plate Tectonics: How It Works*. Palo Alto, Calif.: Blackwell Scientific.

Dalziel, I. W. D. 1982. West Antarctica: Problem child of Gondwanaland. Tectonics 1:3–19.

Dewey, J. F., and K. Burke. 1974. Hot spots and continental break-up: Implications for collisional orogeny. Geol. 2:57–60.

Dewey, J. F., W. C. Pitman III, W. B. F. Ryan, and J. Bonin. 1973. Plate tectonics and the evolution of the Alpine system. Geol. Soc. Amer. Bull. 84:3137–3180.

Dietz, R. S. 1961. Continent and ocean basin evolution by spreading of the sea-floor. Nature 190:854–857.

Handschumacher, D. W., W. W. Sager, T. W. C. Hilde, and D. R. Bracey. 1988. PreCretaceous evolution of the Pacific plate and extension of the geomagnetic polarity reversal timescale. Tectonophysics 155:365–380.

Haq, B. U., J. Hardenbol, and P. R. Vail. 1987. Chronology of fluctuating sea levels since the Triassic. Science 235:1156–1167.

Harland, W. B., R. L. Armstrong, A. V. Cox, L. E. Craig, A. G. Smith, and D. G. Smith. 1989. *A Geologic Time Scale*. Cambridge: Cambridge University Press.

Helsley, C. E., and M. B. Steiner. 1969. Evidence for long intervals of normal polarity during the Cretaceous period. Earth Planet. Sci. Letts. 5:325–332.

Hess, H. H. 1962. History of the ocean basins: Petrologic studies. Pp. 599–620 *in* J. Engel and L. Engel (eds.), *A Volume to Honor A. F. Buddington*. Boulder, Colo.: Geological Society of America.

Irving, E., and G. W. Couillard. 1973. The Cretaceous normal polarity interval. Nature Phys. Sci. 244:10–11.

Kennett, J. P., R. E. Houtz, P. B. Andrews, A. R. Edwards, V. A. Gostin, M. Hajos, M. A. Hampton, D. G. Jenkins, S. V. Margolis, A. T. Ovenshine, and K. Perch-Nielsen. 1974. Development of the Circum-Antarctic Current. Science 186:144–147.

Kent, D. V., and F. M. Gradstein. 1985. A Cretaceous and Jurassic geochronology. Geol. Soc. Amer. Bull. 96:1419–1427.

Klitgord, K., and H. Schouten. 1986. Plate kinematics of the central Atlantic. Pp. 351–377 *in* B. E. Tucholke and P. R. Vogt, (eds.), *The Geology of North America.* Vol. M, *The Western Atlantic Region.* A Decade of North American Geology Series. Boulder, Colo.: Geological Society of America.

Kominz, M. 1982. Oceanic ridge volumes and sea level change: An error analysis. Pp. 109–129 *in* J. Schlee (ed.), *Interregional Unconformities and Hydrocarbon Accumulation.* AAPG Memoir. Tulsa, Okla.: American Association of Petroleum Geologists.

LaBrecque, J. L. 1986. *South Atlantic Ocean and Adjacent Antarctic Continental Margin.* Atlas 13, *Ocean Margin Drilling Program.* Regional Atlas Series, sheet 20. Woods Hole, Mass.: Marine Science International.

LaBrecque, J. L., and P. Barker. 1981. The age of the Weddell Basin. Nature 290:489–492.

LaBrecque, J. L., and D. E. Hayes. 1979. Seafloor spreading history of the Agulhas Basin. Earth Planet. Sci. Letts. 45:411–428.

LaBrecque, J. L., and P. D. Rabinowitz. 1977. Magnetic anomalies: Argentine continental margin, Argentine Basin, North Scotia Ridge, Falkland Plateau. Special Map Ser., no. 826. Tulsa, Okla.: American Association of Petroleum Geologists.

Ladd, J. W. 1976. Relative motion of South America with respect to North America and Caribbean tectonics. Geol. Soc. Amer. Bull. 87:969–976.

Ladd, J. W., G. O. Dickson, and W. C. Pitman III. 1973. The age of the South Atlantic. Pp. 555–573 *in* A. E. M. Nairn and F. G. Stehli (eds.), *The Ocean Basins and Margins: The South Atlantic.* New York: Plenum.

Larson, R. L. 1975. Late Jurassic sea-floor spreading in the eastern Indian Ocean. Geol. 3:69–72.

———. 1977. Early Cretaceous breakup of Gondwanaland off western Australia. Geol. 5:57–60.

Larson, R. L., and J. W. Ladd. 1973. Evidence for the opening of the South Atlantic in the Early Cretaceous. Nature 246:210–212.

Larson, R. L., and W. C. Pitman III. 1972. World-wide correlation of Mesozoic magnetic anomalies and its implications. Geol. Soc. Amer. Bull. 83:3645–3662.

Le Pichon, X. 1968. Sea-floor spreading and continental drift. J. Geophys. Res. 73:3661–3697.

Ludwig, W. J., and V. A. Krasheninkov et al. 1983. *Initial Reports of the Deep Sea Drilling Project 71.* Washington, D.C.: U.S. Government Printing Office.

McElhinny, M. W. 1973. *Paleomagnetism and Plate Tectonics.* Cambridge: Cambridge University Press.

McElhinny, M. W., and B. J. Embleton. 1976. The paleoposition of Madagascar: Remanence and magnetic properties of Late Paleozoic sediments, Earth Planet. Sci. Letts. 31:101–112.

McKenzie, D., and J. G. Sclater. 1971. The evolution of the Indian Ocean since the Late Cretaceous. Geophys. J. R. Astr. Soc. 25:437–528.

Morgan, W. J. 1968. Rises, trenches, great faults, and crustal blocks. J. Geophys. Res. 73:1959–1982.

Norton, I. O., and J. G. Sclater. 1979. A Model for the evolution of the Indian Ocean and the breakup of Gondwanaland. J. Geophys. Res. 84:6803–6830.

Pelayo, A. M., and D. A. Wiens. 1989. Seismotectonics and relative plate motions in the Scotia Sea region. J. Geophys. Res. 94:1293–1320.

Pindell, J. L. 1985. "Plate tectonic evolution of the Gulf of Mexico and Caribbean region." Ph.D. diss., University of Durham.

Pindell, J., and J. F. Dewey. 1982. Permo-Triassic reconstruction of western Pangea and the evolution of the Gulf of Mexico/Caribbean region. Tectonics 1:179–211.

Pindell, J. L., and S. F. Barrett. 1990. Geologic evolution of the Caribbean: A plate tectonic perspective. *In* J. E. Case and G. Dengo (eds.), *The Geology of North America.* Vol. H, *The Caribbean Region.* A Decade of North American Geology Series. Boulder, Colo.: Geological Society of America.

Pindell, J. L., S. C. Cande, W. C. Pitman III, D. B. Rowley, J. F. Dewey, J. LaBrecque, and W. Haxby. 1988. A plate-kinematic framework for models of Caribbean evolution. Tectonophysics 155:121–138.

Pitman, W. C., III. 1978. Relationship between eustacy and stratigraphic sequences of passive margins. Geol. Soc. Amer. Bull. 89:1389–1403.

Pitman, W. C., III, and J. R. Heirtzler. 1966. Magnetic anomalies over the Pacific-Antarctic ridge. Science 154:1164–1171.

Pitman, W. C., III, and M. Talwani. 1972. Seafloor spreading in the North Atlantic. Geol. Soc. Amer. Bull. 83:619–646.

Premoli Silva, I., and A. Boersma. 1977. Cretaceous planktonic foraminiferas: DSDP Leg 39 (South Atlantic). Pp. 615–641 *in* P. R. Supko, K. Perch-Nielsen et al. (eds.), *Initial Reports of the Deep Sea Drilling Project 39.* Washington, D.C.: U.S. Government Printing Office.

Rabinowitz, P. D., and J. L. LaBrecque. 1977. The isostatic gravity anomaly: Key to the evolution of the ocean-continent boundary at passive continental margins. Earth Planet. Sci. Letts. 35:145–150.

———. 1979. The Mesozoic South Atlantic Ocean and evolution of its continental margins. J. Geophys. Res. 84:5973–6002.

Rabinowitz, D. P., M. F. Coffin, and D. Falvey. 1983. The Madagascar-Africa separation. Science 220:67–69.

Rowley, D. B., and J. L. Pindell. 1989. End Paleozoic–Early Mesozoic western Pangean reconstruction and its implications for the distribution of pre-Cambrian and Paleozoic rocks around Meso-America. Precambrian Res. 42:411–444.

Shackleton, N. J., and J. P. Kennett. 1975. Paleotemperature history of the Cenozoic and the initiation of Antarctic glaciation: Oxygen and carbon isotope analyses in DSDP Sites 277, 279, and 281. Pp. 743–756 in P. R. Supko and K. Perch-Nielson et al. (eds.), *Initial Reports of the Deep Sea Drilling Project 29.* Washington, D.C.: U.S. Government Printing Office.

Supko, P. R., and K. Perch-Nielson. 1977. General synthesis of Central and South Atlantic drilling results: Leg 39 Deep Sea Drilling Project. Pp. 1099–1132 *in* P. R. Supko and K. Perch-Nielson et al. (eds.), *Initial Reports of the Deep Sea Drilling Project 39.* Washington, D.C.: U.S. Government Printing Office.

Vine, F. J. 1966. Spreading of the ocean floor: New evidence. Science 154:1405–1415.

Vine, F. J., and D. H. Matthews. 1963. Magnetic anomalies over oceanic ridges. Nature 199:947–949.

Wilson, J. T. 1963. Hypothesis of Earth's behavior. Nature 198:925–929.

———. 1966a. Did the Atlantic close and then re-open? Nature 211:676–681.

———. 1966b. Are the structures of the Caribbean and Scotia Arc regions analogous to ice rafting? Earth Planet. Sci. Letts. 1:335–338.

Part II

Plants

Chapter 3

African Flora since the Terminal Jurassic

J. A. Coetzee

I. Introduction

The origin and evolution of the African flora are closely correlated with the geological and climatic history of the continent, and especially with the contacts Africa has had with other landmasses. Biogeographic patterns and paleontological data, subsequently supported by palynological studies, long ago convinced biologists of the existence of these former contacts. During recent decades, intensive studies of ocean-floor spreading and plate tectonics have provided important reconstructions of former fits and separations of land masses and a time scale for these events (see Chapter 2).

I attempt here to elucidate the evolution of the African flora and vegetation since the time of the breakup of Gondwanaland using the available paleogeographic information. An earlier synthesis of this nature was compiled by Axelrod and Raven (1978). The main focus of this chapter will be on the angiosperms; the fossil remains of these plants, in particular their fossil pollen, are providing a steadily growing picture of the vicissitudes of their long history.

II. The Jurassic

Worldwide climates were warm to hot and wet, with surface sea water temperatures of

Grateful thanks are due to the Missouri Botanical Garden for the invitation to take part in the Symposium and to E. M. van Zinderen Bakker for valuable discussions. I am also indebted to the Geological Survey of South Africa for providing cores from the Southwest Cape and to the Council for Scientific and Industrial Research for financial support for the research. I also wish to thank Sandra Venter for typing the manuscript.

 ISBN 0-300-05375-4

14°C at 75° S latitude, in the Late Jurassic (Frakes 1979:168). The floras that characterized this period consisted mainly of conifers, cycads, ferns, seedferns, ginkgos, lycopods, horsetails, and bryophytes. *Classopollis,* among other gymnosperms, had a wide distribution in North and South America, North Africa, Europe, Asia, and Australia (Roche 1974). This taxon, together with *Eucommiidites* (Bennettitales), predominated in the tropical climate of the time (Kuyl et al. 1955). These authors consider these zonosulcates and their distribution to be important in connection with angiosperm origins.

At the breakup of Gondwanaland in the Late Jurassic, the Stormberg lavas were deposited over southern Africa and the uplift of the subcontinent varied relatively greatly between about 1800 and 2350 meters prior to rifting. The escarpment that originated during the rifting climatically isolated the elevated central plains from the coastal lowlands. At this stage the long African cycle of erosion that would last until the early Miocene set in. A hot and humid climate existed in northern and central southern Africa, while warm, semi-arid conditions prevailed in the south (Maud and Partridge 1987). Arid conditions along the Namibian coast are suggested by the occurrence of aeolian sandstones, although these may be of Lower Cretaceous age (Reyment and Dingle 1987). Floras in the meantime were changing, and in the Upper Jurassic, sporomorph spectra of Europe and Africa–South America were already different from each other (Hughes 1969; Jardiné et al. 1974a). This change marked the end of a time of diverse gymnosperm and pteridophyte flora.

III. The Cretaceous

This period is of particular interest because it includes the first appearance of the angiosperms and their early evolution and rapid dispersal. The Cretaceous palynomorphs of this group are relatively well known and have been exhaustively reviewed by Herngreen and Chlonova (1981).

The warm and humid conditions of the Jurassic changed to much warmer and drier climates in the Cretaceous. During a climatic optimum within the Albian-Santonian interval, temperatures were 10°–15°C warmer than today at latitudes between 45° N and 70° S (Frakes 1979), and sea levels were rising (Vail et al. 1977). Beyond these latitudes the climates were warm to cool-temperate. A global arid zone existed between 45° N and 55° S, except in western Europe (Frakes 1979:178). Progressive cooling set in after the climatic optimum, with marked fluctuations during the Maastrichtian (Frakes 1979).

Microfloral provinces resulting from this climatic situation have been described and variously delimited by different authors (Brenner 1976; Herngreen and Chlonova 1981; Herngreen and Duenas Jimenez 1990; Chapter 4, this volume). In this chapter I will mainly use the terminology of Herngreen and Chlonova, although I sometimes refer to *Northern Gondwana* and *Southern Gondwana* as designed by Brenner (1976).

A. Early Cretaceous (Pre-Albian)

In the Neocomian, northwest Africa and South America were still joined, and in the Aptian, the Agulhas and Falkland plateaus were still united (Chapter 2). The northern part of the embryonic South Atlantic was cut off by the subaerial Walvis–Rio Grande barrier, and conditions comparable to the Miocene Messinian Salinity Crisis of the Mediterranean existed. Large-scale salt accumulation took place in this basin, and the general opinion is that the climate around this hypersaline sea must have been similar to that of the present Red Sea, which is surrounded by hyperarid desert (Reyment and Dingle 1987). These conditions are reflected in the microfossil assemblages.

The southern tip of Africa was situated near 50° S latitude (Smith et al. 1981), and northern South America, West Africa, the Sahara, Egypt, and Israel lay in the paleoequatorial zone, where scientists have found important palynological information on the early angiosperms. In view of the limited information beyond the paleoequatorial regions, Herngreen and Chlonova (1981) provisionally distinguish the west African–South American (WASA) province (called Northern Gondwana in Brenner 1976) and the Gondwana province (called Southern Gondwana by Brenner). The northern boundary of the WASA province extends through northern South America, Suriname, Tunisia, Egypt, and Israel, while the northern limit of the Gondwana province in Africa and South America lies south of the Angola and Espírito Santo basins. The Gondwana province also includes Australia, New Zealand, Antarctica, India, and Madagascar.

The microfloras that characterize the WASA province are mainly from Brazil and western Africa (Senegal to the Gabon and Congo basins; see Chapter 4). A number of elements of these floras are also present in the Sahara, along the Tethyn belt to Israel, in Italy, and in the southern Alps. An admixture of southern Laurasian taxa in these regions is evidence of easy interchange in relatively humid cool climates across the Tethys (Jardiné et al. 1974b; Herngreen and Chlonova 1981; Hochuli 1981; Schrank 1987; Uwins and Batten 1988; Doyle et al. 1977, 1982; Kora and El Beialy 1989). The presence in Italy of WASA elements like *Dicheiropollis etruscus* can be ascribed to the drift of the Apulian Plate from Africa in the Middle Jurassic and its return to the southern border of the Tethys, carrying with it Wealden floras that were introduced into the Sahara (Jardiné et al. 1974a). In Europe and North America, the Wealden microfloras are characterized by bisaccates (Pinaceae, possibly Podocarpaceae), Taxodiaceous-Cupressaceous and Araucariaceous pollen, and a diversity of fern spores, especially of Schizaeaceae. *Classopollis* and inaperturates are rare. The WASA microfloras, by contrast, are characterized by the predominance of *Classopollis, Araucariacites, Exesipollenites* (Bennettitales), *Eucommiidites, Dicheiropollis etruscus,* and *Ephedripites.* Bisaccates are rare or absent, as are ferns, which are of low density (Doyle et al. 1977; Herngreen and Chlonova 1981; Thusu et al. 1988). This type of assemblage, with a prominence of *Ephedripites,* suggests a hot and dry climate, in contrast to the subtropical and humid Laurasian climates (Doyle et al. 1982).

Of special significance in pre-Albian pollen assemblages is the first appearance of angiosperm pollen in a number of localities in the world and especially in Africa–South America and Israel (Brenner 1976; Doyle et al. 1977, 1982; Brenner and Bickoff 1990;

Burger 1990). The oldest record now available is from the Late Valanginian–Late Hauterivian of Israel (Brenner and Bickoff 1990). This taxon reflects a gnetalean-magnolialean common ancestry and predates the previous oldest records of angiosperms such as *Clavatipollenites hughesii* from the Barremian of England (Kemp 1968) *C.* cf. *hughesii,* probably of pre-Albian age, from the Potomac Group of Maryland (Doyle and Hickey 1976), and cf. *C. hughesii* from probably the Barremian of Gabon (Doyle et al. 1977; Burger 1990).

Clavatipollenites also occurs in the Barremian of equatorial Africa with other monosulcates like *Stellatopollis, Liliacidites, Retimonocolpites,* and *Peromonolites;* tricolpates; the peculiar microfossil *Afropollis jardinus;* and other *Afropollis* species (Jardiné and Magloire 1963; Brenner, 1968; Doyle et al. 1977, 1982). These pollen types, together with tricolporates, even at species level appear in similar sequence in equatorial Africa–South America and in southern Laurasia, where they are not so abundant and where trans-Tethyn interchange of this group is indicated. According to Brenner (1976:29), tricolpates reached Australasia in Albian to Cenomanian times (see also Burger 1990) and arrived in the present high latitudes in the Late Cretaceous.

Brenner and Bickoff (1990), on the one hand, support Stebbins (1974) in concluding that the major evolutionary advancement of tricolpates occurred during Late Aptian semi-arid times in northern Gondwana and Israel. Doyle et al. (1982), on the other hand, show that the early angiosperms in fact occupied widely different paleoenvironments created by geological events, and occurred as abundantly in the wet areas of the equatorial belt as in the arid Gabon Congo region. Widespread macrofossil evidence has also recently indicated that the oldest known angiosperms were coastal plants (Burger 1990).

In the Southern Gondwana microfossil province of southern Africa, microfossil evidence is scarce. In pre-Albian times, southern South America, southern Australia, and New Zealand were directly in contact with Antarctica, and Africa (linked to the Falkland Plateau at the Agulhas Bank), Madagascar, and India were closely aligned with that continental mass. Floral exchange with southern Africa was possible except via the Falkland-Agulhas plateau complex (Chapter 2). By comparing the microfossil evidence from the Algoa Basin (Scott 1976) and the Deep-Sea Drilling Project (DSDP) site 361 in the Cape Basin (McLachlan and Pieterse 1978) with that from Antarctica and other Gondwana segments (Truswell 1982; Dettmann 1989), including Madagascar (Herngreen et al. 1982), we can surmise that forests rich in Podocarpaceae and araucarians covered South Africa in pre-Albian times. The pollen assemblages from the above-mentioned southern regions also contain sporomorphs of bennettitaleans, ginkgos, taeniopterids, ferns, and lycopods, as well as *Clavatipollenites.* It is significant that *Clavatipollenites,* an angiosperm of probable northern origin, was already present in many southern lands at this time. It has been recorded from the Barremian-Aptian of Antarctica, Australia, and the Falkland Plateau; the Late Aptian of New Zealand (Dettmann, 1989); and the Early Albian of Patagonia (Chapter 4). Antarctica seems to have been a source of origin for some early angiosperms and a dispersal route for many others (Dettmann 1989). Burger (1990) points to the alternative possibility of a migration route for most of the earliest angiosperms from a paleotropic East Asian region to Australasia via Indonesia.

B. Middle Cretaceous (Albian to Cenomanian)

Important oceanographic changes took place during this period (see Herngreen and Chlonova 1981). The link between the Falkland Plateau and the Agulhas Bank was finally severed by Late Aptian times (Chapter 2). Salt accumulation in the South Atlantic ceased in the Early Albian, and the middle Late Albian saw a regression that continued into the Cenomanian (Reyment and Dingle 1987).

The Cenomanian marks the time of major transgressions, which in southern Africa affected the southwest coast and the Angola region. In the north the sea crossed the Sahara from the Tethys toward the Ahaggar-Tibesti massifs spreading to Egypt. At the same time the South Atlantic advanced northward. In Cenomanian-Turonian times these seas were totally linked and lined by lakes and swamps (Reyment and Dingle 1987). Microfossil evidence indicates that the climate was changing from dry and warm to humid and warm. Continental links with South America still existed via the Rio Grande–Walvis Ridge and between Brazil and Nigeria (Chapter 2).

Two microfloral provinces for this period are distinguished by Herngreen and Chlonova (1981): the ASA (African–South American) province and the Gondwana province (the Northern and Southern Gondwana provinces of Brenner 1976; figure 3.1). The ASA province has now been extended to the Far East (Herngreen and Duenas Jimenez 1990).

The axis of the ASA province is paleoequatorial, running through northwestern Brazil to western Africa. The Northern Gondwana (ASA) microfloral province (the *Galeacornea* Province of Srivastava 1978) is characterized by a diversity of elater-bearing genera, with the highest representation in the axial zone. Here such taxa as *Galeacornea* and *Elaterocolpites* are identical in the sediments of Peru and West Africa (Brenner 1976; Chapter 4, this volume). Also occurring in the axial zone are higher representations of the *Ephedripites* complex comprising *Ephedripites, Equisetosporites,* and *Steevsipollenites* (Herngreen and Chlonova 1981) than in the earlier WASA province. Angiosperm pollen taxa are also remarkably abundant here and are highly diversified.

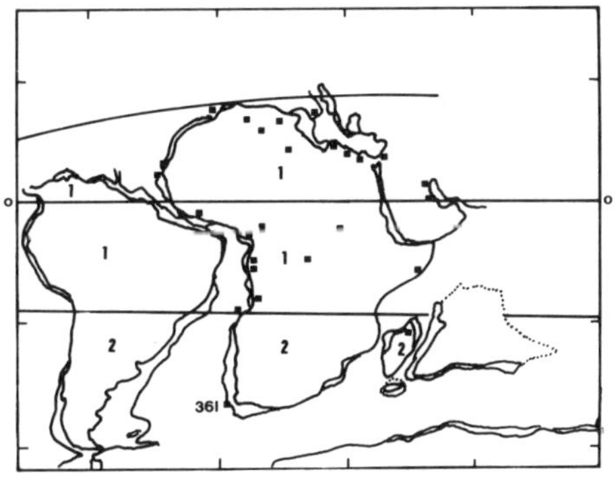

Figure 3.1. Mid-Cretaceous (Albian-Cenomanian) microfloral provinces of Africa and South America (after Herngreen and Duenas Jimenez 1990). 1, Northern Gondwana (ASA) province; 2, Southern Gondwana province. Localities of borehole sites in Africa and Madagascar are indicated by black squares. (See figure 4.2 for fossil sites in South America.)

Among these are the *Afropollis* complex, tricolpates, reticulate monocolpates like *Retimonocolpites* and *Liliacidites*. Also present are *Cretacaeiporites,* a polyporate genus, and *Hexaporotricolpites,* which dominates the Brazil and Congo basins. Bisaccates and trisaccates are absent, and the only gymnosperms present are the nonsaccate *Classopollis, Araucariacites,* and *Callialasporites*. These taxa become rare to absent in the post-Cenomanian of the Ivory Coast, Mali-Niger, northeast Nigeria, the Sahara, Sudan, and the Angola Basin (Schrank 1990). Cycadophyte pollen is also more abundant than in the WASA province, while pteridophyte spores are rare or absent. This type of assemblage, with an abundance of Ephedroids and Cycadophytes and relative scarcity of fern spores, in the axial section of ASA points to an arid to semi-arid tropical climate.

Typical ASA elements were also present near the northern and southern boundaries of the province. In many localities in northeast Africa (Nubian Sandstones of Egypt, Libya, central and northern Sudan, and eastern Somalia), elaterate-*Afropollis-Cretacaeiporites* communities were recorded, although the microfloras of these regions as a whole differ qualitatively from the contemporaneous floras of the axial regions (Doyle et al. 1982; Schrank 1990). Abundant pteridophyte spores indicate wet conditions northward beyond the axial regions. Transitions of ASA floras to the Boreal microfossil province have been indicated in Israel, Saudi Arabia, and Iraq (Herngreen and Chlonova 1981).

Farther south in the Angola Basin (figure 3.1), pollen evidence indicates the dominance of ephedroids, cycadophytes, angiosperms, and *Classopollis*. Pteridophytes and bisaccates are very rare, but *Hexaporotricolpites, Elateroplicites,* and *Cretacaeiporites* are present. These assemblages imply tropical hot, semi-arid conditions and affinities with the axial regions of ASA (Morgan 1978). According to Morgan (1978), the climate of Angola must have been affected by the cyclic restrictions of the basin until the final separation of Africa and South America. The aridity of these regions is corroborated by the isohyets of low values developed by Parrish and co-workers (1982) for the Middle Cretaceous of this area. The only microfossil evidence available for the Middle Cretaceous of the Southern Gondwana province in southern Africa is from the DSDP site 361 in the Cape Basin (McLachlan and Pieterse 1978). Very little also is known for this period from South America, the Falkland Plateau, Antarctica (except the Antarctic Peninsula), or India (Dettmann 1989). The characteristics of the ASA microfloral province, as in the case of WASA, are therefore based mainly on the palynological evidence from southeastern Australia, New Zealand, and the Antarctic Peninsula. As in the Early Cretaceous, bisaccates of the Podocarpaceae predominate, *Microcachryidites* is common, and the nonsaccates *Classopollis, Araucariacites,* and *Callialasporites* are present (*Callialasporites* was not recorded at site 361), together with some diverse ferns. The dominance of *Classopollis* pollen at site 361 is attributed to the possible existence of a local community or restricted provenance. According to McLachlan and Pieterse (1978), however, this assemblage may be indicative of a harsh climate here. In the Cenomanian of these southern lands, angiosperms (mainly diverse tricolpates) suddenly occur in great numbers.

C. Late Cretaceous

Microfossil records covering the various stages of the period are relatively scarce, especially in the southern part of Africa. During the trans-Saharan transgressions (Cenomanian-Turonian and Maastrichtian), when the warm Tethys joined the South Atlantic, the inland sea again influenced the environment as fluviatile and lacustrine regions surrounded the seaway. The regressive stages (Early Turonian, Coniacian, Santonian) were characterized by huge brackish lakes with their associated habitats. These changes are reflected in the microfossil record. At the end of the Turonian, by which time angiosperms had further diversified, many archaic gymnosperms became extinct worldwide. For the Coniacian, Santonian, and Campanian (together with the Senonian), Herngreen and Chlonova (1981) distinguish the Palmae province (the *Constantinisporis* province of Srivastava 1978) of the equatorial regions and the *Nothofagidites* province of the Southern Gondwana region (Chapter 4).

Characteristic of the Palmae province is the greater diversity of angiosperm pollen and a significant strong development of Palmae. Apparently the first divergence of microfloras between Africa and South America dates from the uppermost Cenomanian and lowermost Turonian, when certain microfossils show restricted distribution, for example, the ephedroid-like types such as *Steevesipollenites nativensis* and *S. amphoriformis,* which appear to be confined to northern South America. Later in the Upper Senonian, endemisms are apparent, like the thick walled *Crassitricolporites brasiliensis, Crassitricolpites,* and *Crassitriapertites* in North Brazil and *Andreisporis, Constantinisporis,* and *Victorisporis* in Gabon, West Africa (Herngreen and Chlonova 1981). Some parallel trends in the microfloras of western Africa and northern South American basins are also evident, for example, the definite decrease of *Ephedripites* pollen types during the Senonian and the appearance of a rich diversity of angiosperm pollen in the Upper Senonian. Among the angiosperm pollen are distinctive species like *Proteacidites dehaani, Cupanieidites* spp. (Sapindaceae) and *Triorites* spp., as well as *Auriculidites reticulatus, Retidiporites magdalensis,* and *Buttinia andreevi.*

Among the diverse Palmae that came into existence is the mangrove palm, *Spinizonocolpites echinatus,* the fossil pollen of which is identical to the living mangrove palm, *Nypa fruticans* (Herngreen and Chlonova 1981), which is now restricted to Ceylon, northeast India, Indonesia, New Guinea, and northeast Australia, in ever-wet conditions. It had a wide distribution in the past, with a previous oldest record in the Senonian of Borneo and tropical western Africa (Muller 1981); recently the microfossil record has been extended to the Early Cretaceous (Aptian) of Westphalia (Germany) (Jahnichen 1990). This taxon became extinct in tropical Africa in the Eocene (Germraad et al. 1968). *Spinizonocolpites echinatus,* together with the related *Proxapertites,* attained a pantropical distribution in the Maastrichtian (van Hoeken-Klinkenberg 1966; Germraad et al. 1968; Jan du Chéne 1980; Muller 1981; Salard-Cheboldaeff 1981; Schrank 1984).

In northeast Africa the Palmae province,

including the above mangrove taxa, is transitional to the Boreal *Normapolles* and *Aquilapollenites* provinces (Herngreen and Chlonova 1981; Muller 1981; Schrank 1987). Salard-Cheboldaeff (1981) recorded microfossils of many palms, including *Nypa,* in the Maastrichtian of Cameroon and provided the first microfossil evidence for Lowland tropical rainforest there. She also recorded vegetation types in which Ulmaceae, Restionaceae, and Proteaceae occurred.

The present distribution and fossil record of palms led Moore (1973) to believe that this family originated in West Gondwanaland (Africa–South America and part of Antarctica) and that massive extinctions occurred in Africa in the Tertiary or Pleistocene. That possibility is addressed later in this chapter.

The Late Cretaceous *Nothofagidites* province of the Southern Gondwana region is defined by the presence of the characteristic *Nothofagidites* pollen, the oldest record of which comes from the Campanian of Antarctica and Australia (Dettmann 1989). Also included in this province are South America (Fuego and Patagonia) and the rest of Australasia. *Nothofagidites* has not yet been recorded from southern Africa and it seems clear that that part of the continent was not part of this province. Africa was well separated from the other continents by Late Santonian times (84 Ma), and by Campanian times there was probably no contact by which this southern taxon could have reached Africa. However, some of the long-ranging austral gymnosperm taxa associated with this province have been recorded at the few African sites investigated (figure 3.2).

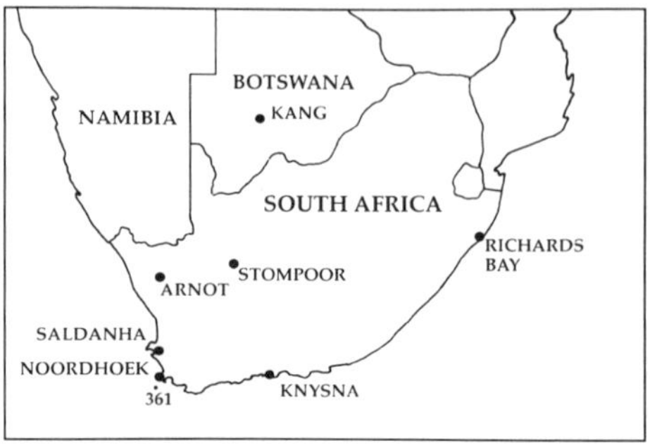

Figure 3.2. Late Cretaceous and Tertiary pollen study sites in southern Africa.

In the northwest sector of South Africa the pollen spectra of the Stompoor site (29°51′ S, 21°50′ E) near Prieska give a good insight into the Mid-Senonian paleoenvironment of this region (Scholtz 1984a). Many tricolpates were present, as well as occasional pollen grains of tropical African plants like *Constantinisporis, Andreisporis, Hexaporotricolpites,* and *Fenestriorites.* The Stompoor microflora was distinctly different from the Palmae province of West Africa and was dominated by *Araucariacites.* Examination of Lower to Mid-Senonian sediments from the KN70 and KN49 sites in Botswana shows that *Ephedripites* was the most abundant microfossil there (Scholtz 1984b). These assemblages point to a hot, dry climate with probable marked seasonality. At DSDP site 361, *Classopollis,* which had been dominant in these regions, declined sharply in the Late Cretaceous, while the austral bisaccates and trisaccates with some *Araucariacites* are persistent (McLachlan and Pieterse 1978). The assemblages also contain diverse fern taxa and angiosperms. *Zlivisporites blanensis* of Gabon is present, as are *Constantinisporis, Andreisporis, Hexaporotricolpites, Fenestriorites* and *Cupaniedites.* These pollen assemblages indicate that a mixed warm-temperature humid vegetation existed in the southwestern Cape. As in the case of the northwestern sector of South Africa, similarities between the west African floras and those at site 361 are few. Scholtz (1984a) points to the uniqueness of the vegetation in southern Africa and suggests that by the Late Cretaceous a vegetation gradient already existed between the more xeric northwestern sector of southern Africa and the temperate-humid southwestern regions, and between the northwestern sector and the subtropical Richards Bay region. This zonation is corroborated by the paleoclimatic maps of Parrish and co-workers (1982), which show that an area of consistent low rainfall existed in the northwestern sector of southern Africa from the Mid-Cretaceous to the Maastrichtian. This includes the area of the Angola Basin which, as previously mentioned, was characterized by a tropical semi-arid vegetation from Albian to Senonian times (Morgan 1978). The maps also support the microfossil evidence for a much higher precipitation in the southwestern Cape, the Natal Coast, and tropical Africa during this long period.

Migration of plants and animals between South America and Africa was still possible via the Rio-Grande Rise–Walvis Ridge until about the Coniacian (Reyment and Dingle 1987). Farther north a landbridge existed between the Brazilian Bulge and Nigeria until the end of the Cretaceous (Rand and Mabesoone 1982). Taxa with Late Cretaceous origins in the southern Gondwana regions could have reached Africa from Australasia and Antarctica via South America (see Dettmann 1989).

The numerous links between the angiosperm floras of East Africa and Madagascar can no longer all be explained by direct land connections in the Mid-Cretaceous and long-distance dispersals. Knowledge of the separation of this island from Africa in the Late Jurassic (160 Ma) and its southward movement by 120 Ma (Rabinowitz et al. 1983; Chapter 2, this volume) has considerably changed the theory about its biological contacts. Only one migration route between Africa and Madagascar remained open until Aptian-Albian times: from the south when links still existed with South America, Antarctica, India, Australasia, Ninetyeast Ridge, and the Mascarene Plateau. These old connections are demonstrated by present phytogeographic patterns of archaic angiosperm and austral gymnosperm taxa and their fossil record (Coetzee and Muller 1984).

Many tropical families common to Africa–South America could spread along the dry Tethys borders as far as Southeast Asia (Axelrod 1975; Axelrod and Raven 1978; see also Burger 1990 for ancient routes). Transatlantic contact between the Mediterranean and California sclerophyll vegetation was possible (Axelrod 1975: 77–79). Relatives of this vegetation occur on the Canary and Azores islands.

IV. The Cretaceous-Tertiary Boundary

By the end of the Cretaceous angiosperm domination was well established, but at the Cretaceous-Tertiary boundary, complex paleogeographic or other events caused major changes in the biotas of the earth. Massive phytoplankton and faunal extinctions were recorded, and many floras disappeared from Africa (as described by Muller [1981] for central Africa, by Salard-Cheboldaeff

[1981] for Cameroon, and by Meon and Donze [1988] and Meon [1990] for northwestern Tunisia, where a change to European species is evident). Meon and Donze (1988) dismiss hypotheses of catastrophic events as a cause for the extinctions because their pollen evidence indicates that the extinctions occurred over a long time.

At this boundary a major regression is recorded in eastern and southern Africa and in the southwestern Cape (Reyment and Dingle 1987). Marine microfossils provide evidence for a significant drop in ocean temperature (Stapleton 1975). Sharp fluctuations in ocean temperature have been recorded for the Maastrichtian, with a short-lived decrease at the Cretaceous-Tertiary boundary (Douglas and Savin 1973), when the Cretaceous aridity was also ending.

According to Axelrod and Raven (1978), the relations between the tropical floras of South America and Africa were by this time mainly at the family level. By the early Tertiary, most of the floras had developed a relatively modern aspect, and 139 present families have fossil records in the Late Cretaceous and Early Tertiary (Muller 1981).

The past 65 million years, since the beginning of the Cenozoic era, are significant not only for the evolution of humans and other mammals and the progressive modernization and diversification of the floras but for the major climatic trends of the Tertiary and Quaternary periods.

V. The Tertiary Period

Younger geological formations yield more information for the reconstruction of paleoenvironments and former vegetation patterns. On the continents, erosion cycles and tectonic and mountain-building processes provide details on the emerging relief, and sedimentary sequences are of great value for understanding climatic evolution. New knowledge of the world's oceans includes a wealth of data on sea levels, ocean currents, and especially paleotemperatures, which were haltingly dropping during the Tertiary (Chapter 15). From the Eocene onward, the major variations in the earth's temperature can be correlated with geological events in the Antarctic region, which led to the isolation of the continent in the Paleogene, and with the progressive development of the ice sheet. This climatic situation changed dramatically in the Pliocene, when the long sequence of northern glaciations steered the earth's temperature regime.

A. The Paleogene Subperiod

In the ice-free globe of the Paleocene and early Eocene, ocean temperatures in the equatorial regions were 4°C higher than they had been during the Maastrichtian cooling, even in the higher latitudes. The temperature gradient from the poles to the equator was about 4–6°C, compared with 12°C at present (Frakes 1979). During the Paleogene the Antarctic Ocean changed from a temperate ocean with a temperature of 20°C to an Antarctic sea with a temperature of only 7°C. Strengthened atmospheric and oceanic circulation accompanied this decrease in temperature. In the Oligocene, glaciers began to develop on the high Antarctic mountains. The present oceanic circulation pattern dates to Eocene-Oligocene times, when the powerful Circum-Antarctic Current originated and isolated the Antarctic water from that of other oceans (Chapter 2).

These changes in the energy budget of the globe had a dramatic influence on the paleoecology of the entire earth and especially on the southern continents. The main consequence of the drops in temperature was the spreading of aridity and the subsequent impoverishment of the floras. The ongoing separation of the Gondwana landmasses also affected the biota, causing greater isolation and differentiation. By Eocene times the pollen spectra of Africa and South America showed distinct differences, and by Oligocene times the migration route between South America and Antarctica was severed (Chapter 2). The consequence of the earlier migrations along the austral route (between South America and Australia, excluding Africa) is that there is greater similarity between present floras of South America and Australasia than between those of Africa and either of these landmasses (Axelrod and Raven 1978).

Paleogene paleobotanical records for Africa are scarce, and not much can be added to the suggested zonation of the vegetation provided by Axelrod and Raven for that period. In their Cretaceous-Paleocene map, northern Africa, including the Sahara up to 10–20° N paleolatitude, is shown as covered with lowland rainforest (figure 6A in Axelrod and Raven 1978). Pollen evidence from Cameroon confirms this zonation (Salard-Cheboldaeff 1981). Northern Africa received high relative precipitation, according to the isohyet maps of Parrish and associates (1982). However, microfossil assemblages from the Paleocene-Eocene boundary point to a significant change in the lowland forest. They indicate the co-existence of dry and humid forest. Among the fossil pollen taxa recorded are Poaceae, Apocynaceae (*Rauvolffia*), Mimosaceae (*Parkia*), and such pollen types as Bombacaceae, Euphorbiaceae, Guttiferae, and Sapotaceae (Salard-Cheboldaeff 1981). The present Congo Basin was savanna woodland stretching as far as 35° S paleolatitude. East and west of this woodland may have been subtropical rainforest. The northern extremity of the continent was covered with scleraphyll woodland, but the southern coastal region probably supported temperate rainforest (Axelrod and Raven 1978). Pollen assemblages from the DSDP site 361 indicate that in the Late Cretaceous mixed warm-temperate forest existed there (McLachlan and Pieterse 1978). Recent pollen evidence from the sediments of the Arnot Kimberlite pipe in Namaqualand (figure 3.2) indicate the occurrence during the Paleocene of an open-canopied dry subtropical forest with an understory of Ericaceae, Epacridaceae, Thymelaeaceae, and Restionaceae. This group of families may have direct bearing on the evolution of the present Capensis floral region (Scholtz 1985).

The presence of *Araucariacites australis* in the Arnot paleofloras is evidence of their strong austral links. The angiosperms in these paleofloras seem to have a greater relationship with the Paleocene of southeast Australia than with tropical Africa. Taxa in common with Australia are Casuarinaceae, Ericaceae, Epacridaceae, Gunneraceae, Proteaceae and Restionaceae. The oldest record for most of these taxa is from the Maastrichtian. It is significant that the oldest records for Proteaceae are from the Campanian-Maastrichtian of Antarctica–southern Australasia (Dettmann 1989) and the Maastrichtian of West Africa–South America (*Propylipollis* [Proteacidites] *de-*

haani; Muller 1981). The oldest record for Restionaceae is from the Maastrichtian of Senegal (Jardiné and Magloire 1963). These two taxa are prominent in the present Capensis region and seem to be relative latecomers in the southernmost region of Africa. Gunneraceae were first recorded in the Turonian of Peru, which may indicate a northern origin (Dettmann 1989; Jarzen and Dettmann 1989). Casuarinaceae seem to be of younger origin, with the oldest record from the Paleocene of New Zealand. The existence of migratory routes in the Late Cretaceous could explain the presence of the above families in the Paleocene of South Africa, except for the Casuarinaceae, which appeared in the fossil record after the complete isolation of Africa. Raven and Axelrod (1974: 616) suggest that this family may have dispersed to Australia from Africa in the Mid-Cretaceous. The above microfossil assemblages from the Arnot pipe in Namaqualand may support the Late Cretaceous Stompoor microfossil evidence that the southwestern sector has experienced a long history of dry climates.

B. The Neogene Subperiod

In the later Early and Mid-Miocene, marked reversals of temperature began (Shackleton and Kennett 1975). The face of Africa was also changing through geological events. (For a general outline of the changes of vegetation of Oligocene-Miocene times, see Axelrod and Raven 1978: figure 6B.) It is suggested that savanna woodland gradually replaced the lowland rainforest in the central and northern Sahara and in the center of southern Africa. The lowland rainforest belt shifted southward and occupied the whole latitudinal belt between the present southern Sahara and the southern border of Zaire. Microfossil evidence of Miocene age from Cameroon indicates the further development of this forest (Salard-Cheboldaeff 1981). The assemblages contain Balanophoraceae, Lythraceae, and Myristicaceae and point to a warm and increasingly humid climate. Lowland tropical rainforests were probably also present in eastern Africa (Hamilton 1968). According to Andrews and van Couvering (1975) in Axelrod and Raven (1978), this belt of tropical rainforest was interrupted during the Miocene uplift of the continental divide and was locally replaced by woodland and grassland. Many Ethiopian sites have produced fossil evidence of probable early Miocene age suggesting savanna woodland vegetation. Grassy woodlands were already widespread 23–17 Ma (Retallack et al. 1990).

In southern Africa the African cycle of erosion continued until the end of the Early Miocene (Maud and Partridge 1987). At this time subtropical laurel forest is considered to have covered the Mediterranean coastal area, Mozambique, and the southeast coast of South Africa as far as the southwestern Cape (Axelrod and Raven 1978: figure 6C). More recent microfossil evidence from Noordhoek on the Cape Peninsula (figure 3.2), however, indicates that subtropical rainforest existed there in the Early Miocene (Coetzee 1978a, 1978b, 1980, 1983, 1986; Coetzee and Muller 1984). This rich and diverse forest had a strong Southern Gondwana gymnosperm component, with *Podocarpus, Microcachrys,* and *Widdringtonia.* Other elements of considerable phytogeographic significance were recorded: Winteraceae, Casuarinaceae, *Ascarina,*

Cupaniopsis, Sparganiaceaepollenites, Sarcolaenaceae, and many palm taxa. These elements are now extinct on the African continent but may shed light on the present disjunct distribution pattern of their derivatives in the Southern Hemisphere. They support the hypothesis of a common flora that existed in the Cretaceous in South America, Antarctica, Africa, and Madagascar (Coetzee and Muller 1984). A microfossil record of Lactoridaceae from the Turonian-Campanian off the southwest African coast is further evidence of this common flora. At present, this family is confined to the Juan Fernández Islands off the coast of Chile (Zavada and Taylor 1986; Zavada and Benson 1987). The present humid forest of Madagascar, containing derivatives of the extinct suite of taxa recorded at Noordhoek, is a modified relict of this ancient flora, as is the present Knysna forest of the southern Cape Coast (figure 3.2; Coetzee and Muller 1984). The Miocene lignites of the Knysna forest contain Casuarinaceae, Sarcolaenaceae, and extinct Palmae (Coetzee, unpublished data). As mentioned above, migration to Madagascar could only have taken place from the south. The origin and migration of these archaic taxa are discussed in Raven and Axelrod 1974, Axelrod 1975, Truswell 1982, Coetzee and Muller 1984, Coetzee and Praglowski 1988, Dettmann 1989, and van der Ham 1990.

As in the case of the Paleocene flora of Arnot (Scholtz 1985), the rich Miocene microflora of the southwestern Cape are more closely linked to those of Australia, New Zealand, and Ninetyeast Ridge than to those of the tropics (Coetzee and Muller 1984). Some of these links are not easy to explain because of advanced plate movement before the strong and diverse introduction of angiosperms into Australia during Campanian-Maastrichtian times (Dettmann 1989; Burger 1990). It is significant in this regard that many of the angiosperm taxa have their oldest records in the Tertiary.

During a global warm interval at the Early to Middle Miocene boundary the tropical lowland rainforest of the Congo Basin and the savanna woodland in the Sahara retreated. In the Middle Miocene the closure of the Tethys resulted in a drier climate that must have had a degrading influence on the rainforest of the Sahara.

High sea levels were recorded for the Miocene climatic optimum of 17–15 Ma (Vail et al. 1977). In southwestern Africa a transgression that occurred at 16 Ma is considered to have brought on a monsoonal climate for this region (Hendey 1981). Fossil faunal assemblages along the Namibian coast are indicative of savanna and riverine woodland and confirm the widespread existence of this vegetation by the Middle Miocene. In the Saldanha region, pollen data support the evidence for the opening of the forest vegetation at this time (Coetzee and Rogers 1982).

The establishment of the Antarctic ice-sheet by 14 Ma (Shackleton and Kennett 1975) was part of a fundamental change in the paleoenvironment of the entire Southern Hemisphere. One of the consequences of this development was the production of cold water, which affected the whole earth. During the cold episodes accompanying the establishment of the ice sheet, the northwestern Sahara experienced very dry conditions when strong trade winds deposited dust in the central Atlantic (Sarnthein et al. 1982: 589; Diester-Haas and Chamley 1982). Upwelling in the Canary Current probably contributed to the arid conditions. In Kenya drier conditions are indicated by fossil records of grassy woodlands and woody grasslands supporting diverse antelope, giraffe, rhinoceros, elephants, and apes 14 Ma (Retallack et al. 1990). In the southwestern Cape, subtropical rainforests still predominated (Coetzee 1986).

There is evidence that in the earliest Late Miocene the temperature of the Antarctic Ocean dropped sharply by 7°C, initiating upwelling in the Canary Current. The climate of the western region of the present desert became arid as trade winds strengthened (Sarnthein et al. 1982). A similar process was taking place in the Benguela Current, where the spasmodic introduction of cold water had started in the Middle to Late Oligocene (Siesser 1978). In the early Late Miocene the upwelling was intensified. The wooded grassland of the Namib region was replaced by desert. The process was aggravated by the strengthening of the subtropical anticyclone (van Zinderen Bakker and Mercer 1986). Ward and co-workers (1983) are, however, of the opinion that the main Namib Sand Sea developed during the Pliocene.

The Late Miocene cold period was followed by a warm episode from 9 to about 6 Ma (Shackleton and Kennett 1975; Stein and Sarnthein 1984: figure 2). In Ethiopia, lignite of 8 Ma age contains a wealth of microfossils that indicate that dense lowland forest grew at paleoaltitudes of 900–1000 meters. This forest has strong present Guineo-Congolian affinities, and the pollen assemblage indicates that true savanna and open environments were not yet firmly established as of 8 Ma on the northern plateau (Yemane et al. 1987).

In the terminal Miocene at 5.2 Ma a drastic cooling of global temperature caused a drop in sea level of about 70 meters. This event isolated the Mediterranean, causing the Messinian Salinity Crisis of 5.7–4.6 Ma (van Zinderen Bakker and Mercer 1986: 225). During the cold Messinian cycles, Europe and northern and eastern Africa were probably arid, although Tiedeman and coworkers (1989) point to a humid climate. These incongruous findings can be explained by accepting that the warm cycles of the Messinian event were more humid. During these more favorable periods, the savanna described by Hsü (1974) and Hsü and co-workers (1977) could have spread in the enormous region surrounding the Mediterranean.

At the southern end of the continent the subtropical rainforest was eliminated and by the Early Pliocene was entirely replaced by sclerophyllous vegetation, or fynbos (Coetzee and Muller 1984). This complete floral change may shed light on the impoverishment of the African flora during the Tertiary in comparison with Madagascar, Southeast Asia, and South America.

It has been assumed that the present Mediterranean climate in the southwestern Cape, with dry summers and humid winters, originated during the Early Pliocene (Hendey 1981), causing the further spread and diversification of fynbos. The origin and evolution of fynbos may be linked to the drier Paleocene flora of Arnot (Scholtz 1985); for other hypotheses in this connection see Levyns (1964), Goldblatt (1978), Taylor (1978), and Axelrod and Raven (1978). This unique vegetation, because of its species richness and endemicity, has been designated a floral kingdom.

Highly impoverished relicts of the Early Tertiary rainforests have survived to the present in localized patches in the southwestern Cape mountains, notably at Knysna. These forests are part of the Afromontane Archipelago floristic region, which extends northward along the eastern escarpment as far as the Ethiopian Highlands and from the mountains of Sierra Leone and Cameroon to the mountains of Somalia (White 1981). The biotic relationships of this complex montane region have been studied by several scientists, including A. Aubréville (1949), O. Hedberg (1951), E. M. van Zinderen Bakker, Jr. (1971), and F. White (1978, 1981). More palynological evidence is needed, however, to elucidate the history of this element of the vegetation.

Other evidence for Early Pliocene changes to semi-aridity in the southwestern Cape is reflected in the fossil faunas of the Saldanha embayment (Tankard and Rogers 1978; Hendey 1981). After the end of the Messinian Salinity Crisis, the Mediterranean Basin was filled again and the Indian Ocean flooded the Red Sea, 4–3.5 Ma (Hsü et al. 1977). The ensuing humid climate encouraged the spread of woody vegetation in the northern and northeastern semi-arid Sahara. The sequence of progressively arid periods that followed were not directly related to glacial events in the Northern Hemisphere (Tiedemann et al. 1989).

At the beginning of the Upper Pliocene (3.5–1.8 Ma) there was a cold episode in the Southern Hemisphere, while in the north climatic cycles of the Milankovitch type increased in amplitude, especially from 3.2 Ma onward. Colder phases, accompanied by stronger trade winds and upwelling, caused severe aridity in the northern and central Sahara (Tiedemann et al. 1989). Large ice sheets developed in the Northern Hemisphere. In deep sea cores off the northwestern coast of Africa, Stein and Sarnthein (1984: figure 2) recorded colder episodes at 2.5, 2.0 and 1.2 Ma, when the Sahara Desert was finally established. Fossil pollen studies in the Ethiopian mountains yielded important evidence for a cooling close to the Gaus-Matuyama boundary (2.4 Ma; Bonnefille 1983). Pollen data also indicate distinct vegetation changes in East Africa at 2.5, 1.9, and near 1.6 Ma—grass pollen increased and arboreal pollen disappeared (Vincens 1987; Bonnefille 1981).

VI. The Quaternary

During the Quaternary the amplitude of the Milankovitch cycles increased and evolved into a glacial-interglacial sequence, the climates of which dramatically changed the environments on the African continent. The scope of this chapter allows only a concise description of the vegetation patterns during two extreme situations of the last glaciation—the Last Glacial Maximum (LGM) and the Holocene Climatic Optimum (HCO).

A. The Last Glacial Maximum

The sea surface temperature, and consequently the evaporation, of the seas and oceans surrounding the African continent was during LGM times much lower than at present (Flohn and Nicholson 1980). A drop in temperature of 4–9.5°C has been registered, and glaciations occurred on the high mountains (the Atlas, Hoggar, and East African mountains). The increased aridity was aggravated by much stronger atmospheric circulation (Newell et al. 1981; Talbot 1984). Repeated glacial conditions generated extinction, selection for drought resistance, isolation, and speciation (Axelrod and Raven 1978).

Environmental conditions in the wide Sahara region changed considerably. The forests and macchia of north Africa were, except for small refuges, replaced by grassland and steppe (Agwu and Beug 1984). The 5400-kilometer southern boundary of the Sahara, the Sahel, shifted more than 300 kilometers southward.

The changes in the vegetation of East Africa are best known from 16 pollen cores of the high mountains, the results of which have been reviewed by Hamilton (1972, 1982). In pollen diagrams from Sacred Lake on Mount Kenya, which can be considered as representative, the pollen spectra of LGM age show that the mountain vegetation had shifted about 1000–1100 meters downward. This represents a drop in temperature of 5.1–8.8°C (Coetzee 1967) or 6°C (Hamilton 1982). Livingstone (1967) did not accept these pollen data as evidence for lowering of temperature and has instead argued in favor of pedological differences, but critical revision of Livingstone's pollen sequences by Hamilton (1972) and a reevaluation of the results from Mount Kenya (van Zinderen Bakker and Coetzee 1972) have shown that the decrease of temperature must be accepted.

The climatic change caused the near total disappearance of forest vegetation from the hills and lower mountain slopes in East Africa. The vegetation that covered the plateau surrounding the mountains is not well known, but the aridity demonstrated by the low sea levels suggests that the plateau must have been occupied by dry grassland and semi-desert.

In Central Africa the ice-age aridity had a devastating effect on the vast rainforest biome. Palynological studies of sediments from the submerged Zaire Delta and the Gulf of Guinea show that the rainforest was mostly replaced by dry grassland and savanna (Caratini and Giresse 1979; Frédoux and Tastet 1988; Lanfranchi and Schwartz 1990). Some refuges of the forest survived along the Guinea coast and in eastern Zaire (figure 3.4) (Hamilton 1976; updated by Maley 1987: figure 1).

Extensive data on environmental changes in southern Africa are available (Vogel 1984; Tyson 1986; Deacon and Lancaster 1988), but the Quaternary environmental changes there are not yet well understood. Two different schools of thought prevail: shifts of the westerly belt toward the equator, accompanied by increased cyclonic winter rainfall (van Zinderen Bakker 1967, 1982; Nicholson and Flohn 1980; Cockcroft et al. 1987), or changes in the tropical circulation system (Harrison 1988a, 1988b). The field data have recently been summarized by Partridge and co-workers (1990).

Many authors accept that Namibia (except the northern Namib Desert) and the Kalahari were arid. Others, however, presume that during the LGM the entire Kalahari was wetter than today (Partridge et al. 1990). Another controversy exists about the LGM climate of the southern and eastern part of South Africa: some authorities indicate aridity (ibid.), while others describe the climate of many sites as subhumid (Vogel 1984; Deacon and Lancaster 1988). Many more accurate dates of the LGM and a more pre-

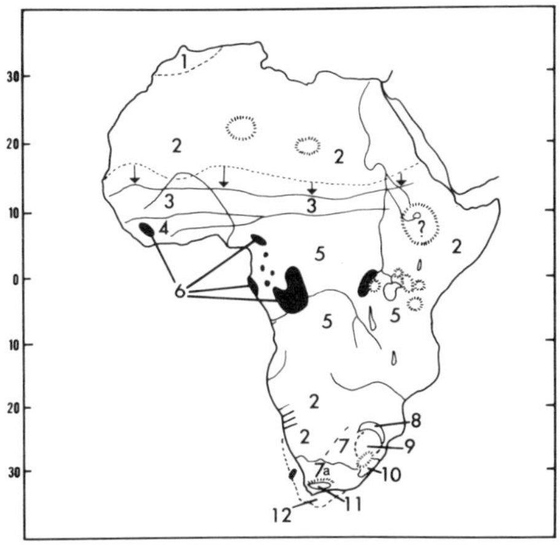

Figure 3.3. Schematic representation of the Last Glacial Maximum vegetation of Africa: *1*, refugia of forest and macchia; *2*, desert and semi-desert; *3*, Sahel (arrows indicate displacement); *4*, semi-deciduous forest; *5*, arid grassland and savanna; *6*, refugia of rainforest and savanna; *7*, cold grassland and lakes; *7a*, grassy karoo and lakes; *8*, montane forest; *9*, grassland and fynbos elements; *10*, coastal forest; *11*, fynbos; *12*, exposed shelf with cold grassland. (After Maley 1987)

cise interpretation of the field data will be necessary before this latter problem can be solved. A possible reconstruction of the LGM vegetation pattern of southern Africa is presented in figure 3.3.

B. The Holocene Climatic Optimum

Climatic changes leading to the HCO set in during Late Glacial times. When the temperature rose after 14,000 B.P., the rain-bringing Atlantic monsoon was re-established. The vegetation of the greater part of Africa obtained its present pattern, in which the biomes are arranged in a concentric fashion around the Gulf of Guinea, ranging from tropical rainforest in the center to hyper-arid deserts in the periphery (figure 3.4).

The area of the montane forest expanded considerably and reached a more or less continuous distribution between 14,000 and 12,000 B.P. (Moreau 1966; Coetzee and van Zinderen Bakker 1970: figure 3).

At the southern boundary of the Sahara, the savannas of the Sahel began to shift northward around 12,500 B.P. and became gradually richer in woody species (Hooghiemstra 1988). Well-dated evidence from 54 limnological sites, supported by data from 16 pollen sequences from the southern Sahara and the Sahel, testify to two (and sometimes three) wet phases with high lake levels between 9000 and 3000 B.P. (Lézine and Casanova 1989; Dupont and Hoo-

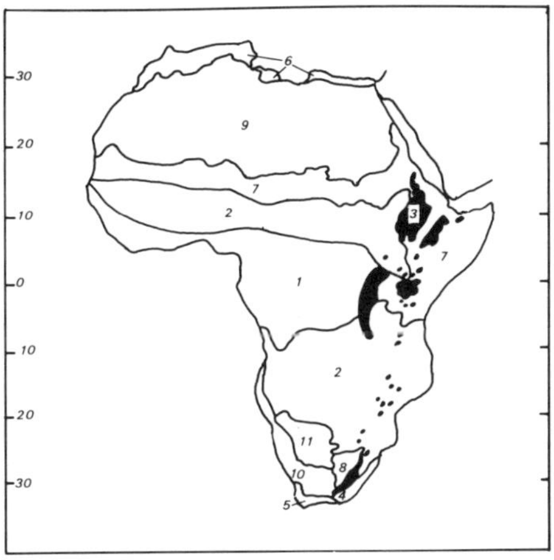

Figure 3.4. Present distribution of vegetation in Africa: *1*, lowland rainforest; *2*, savanna woodland; *3*, montane vegetation; *4*, coastal forest; *5*, fynbos; *6*, sclerophyll vegetation; *7*, thorn scrub and succulent woodland; *8*, grassland; *9*, Sahara; *10*, Namib-Karoo; *11*, Kalahari. (After Axelrod and Raven 1978: figure 6D)

ghiemstra 1989; Lézine 1989a, 1989b). During the first Holocene humid phase (the HCO), which lasted from about 9500–9000 to about 7500 B.P., vegetation of the southern margin of the desert moved 400 to 500 kilometers northward (Schulz 1987).

During the most favorable period of the HCO, between approximately 8500 and 7000 B.P., rainfall reached its highest level and the temperature rose roughly to modern conditions. The tropical rainforest again occupied its present area. The tree line ascended to about its present position (Roche et al. 1988).

On the East African plateau around Pilkington Bay, woodland of *Celtis, Alchornea,* Moraceae, and Oleaceae dominated the scene from 9500 to 6000 B.P. (Kendall 1969). Comparable change took place around the northern end of Lake Tanganyika, where miombo reached maximum distribution during the HCO (Vincens 1989).

In southern Africa, the interior and southern Namibia received more summer rainfall during the Early Holocene (Butzer 1984; Tyson 1986). Optimum climatic conditions prevailed from about 8000 to 6000 B.P. Different types of savannas with tropical affinities covered the Transvaal, southern and middle Botswana, and southeastern Namibia (Scott 1982; Beaumont et al. 1984). These savannas were probably more open in the eastern and southern Orange Free State (Scott 1986), where a thornveld covered the vast plains (Horowitz et al. 1978; Scott 1984). Several large lakes occurred here (Deacon and Lancaster 1988: 107). South of the Orange River the Karoo vegetation was mixed with much grass (Bousman et al. 1988). Forest spread along the southern coast between 7000 and 2600 B.P. (Martin 1968; Scholtz 1986). Around the Boomplaas Cave, warm-dry conditions were indicated by xeric woodland (Scholtz 1986).

VII. General Concluding Remarks

The present diverse phytochoria of Africa are manifestations of a long history of paleogeographic changes since the time of the separation of the Gondwana continents. Much more paleobotanical evidence, especially from southern Africa, is required to understand this vegetation pattern. Unlike South America and Australasia, which could still exchange biota via Antarctica until Eocene times, Africa was isolated by 84 Ma. A considerable part of the flora, therefore, developed autogenously. This may be one of the reasons for the greater similarities between the floras of temperate Australia and South America than between the floras of either of these landmasses and Africa. The early separation of Africa may also account for the limited number of present austral taxa.

Most of the earliest angiosperms were concentrated in pre-Albian times in the axial part of the paleoequatorial regions of northeastern South America and western Africa. Angiosperm taxa were identical on both sides of the South Atlantic and occur in the same stratigraphic sequence. From here they radiated in similar sequence and with progressive diversification northward across the Tethys and southward toward Antarctica and Australasia by different migratory routes. As the distance between South America and Africa increased, the similarities in floras dwindled, and in the Late Cretaceous were only at family level.

In southern Africa, pollen evidence indicates that since Senonian times, arid conditions prevailed in the northwestern sector, while subtropical rainforest existed in the southwestern extremity of the continent. This subtropical forest vegetation which survived into the Miocene in the southwestern Cape contained archaic elements such as *Ascarina, Casuarina,* Winteraceae, *Microcachrys, Cupaniopsis,* Sarcolaenaceae, and Sparganiaceae, which became extinct in Africa by the Early Pliocene. These taxa are of phytogeographic significance as they shed light on present disjunct distributions of their derivatives in other Gondwana segments.

During its northward drift into new paleolatitudes, Africa underwent profound geological events—as a result of which new habitats were created and speciation could take place. New taxa were introduced when the African Plate united with Eurasia in the Middle Miocene (17–18 Ma).

During the Late Cenozoic the African biota were affected by the stepwise cooling of the earth, which caused a range of progressive aridifications. This climatic evolution resulted in repeated extinctions as well as speciation, especially of xerophytes in the huge deserts and semi-deserts.

The origin of the present-day climatic system, beginning in the Pliocene, spurred the development of the present vegetation pattern. The so-called Atlantic monsoon favors the Congo Basin and to a certain extent central Africa, while the tropical jet prevents the northward migration of summer rain into the Sahara region.

In the southwestern Cape the replacement of the subtropical rainforest by fynbos by the early Pliocene was an important event in the floral history of this region. The spread of fynbos in the southwestern Cape may be related to the existence of fynbos elements in the Paleocene of Namaqualand. This Capensis floral kingdom forms the antipode of the Mediterranean macchia. Both enjoy cyclonic winter rain, which in the southwestern Cape dates to the early Pliocene.

By the end of the Tertiary, deserts and arid, semi-arid, and savanna vegetation were established on the continent, but the Quaternary glacial-interglacial climates caused major shifts in the vegetation pattern. During LGM times the temperature decrease was 4°–9.5°C and aridity was widespread. During the Holocene the climate became warmer and more humid, and the present vegetation pattern was established.

Literature Cited

Agwu, C. O. C., and H. J. Beug. 1984. Palynologische Untersuchungen an marinen Sedimenten vor der West-afrikanishcen Kuste. Palaeoecol. Africa 16:37–52.

Andrews, P., and J. A. H. van Couvering. 1975. Paleoenvironments in the East African Miocene. Pp. 62–103 *in* F. S. Szalay (ed.), *Approaches to Primate Paleobiology,* Vol. 5. Basel: Karger.

Aubréville A. 1949. *Climats, forêts et desertification de l'Afrique tropicale.* Paris: Société d'Editions.

Axelrod, D. I. 1975. Plate tectonics and problems of angiosperm history. Mém. Mus. Natn. Hist. Nat. 88:72–85.

Axelrod, D. I., and P. H. Raven. 1978. Late Cretaceous and Tertiary vegetation history of Africa. Pp. 77–130 *in* M. J. A. Werger (ed.), *Biogeography and Ecology of Southern Africa.* The Hague: W. Junk.

Beaumont, P. B., E. M. van Zinderen Bakker, and J. C. Vogel. 1984. Environmental changes since 32 000 B.P. at Kathu Pan, northern Cape. Pp. 329–338 *in* J. C. Vogel (ed.), *Late Cainozoic Palaeoclimates of the Southern Hemisphere.* Rotterdam: Balkema.

Bonnefille, R. 1981. Palynlogical evidence for continental climatic change in East Africa at 2.5 and 1.7 MY. Terra Cognita (Spec. Iss.):102–103.

———. 1983. Evidence for a cooler and drier climate in the Ethiopian uplands towards 2.5 MY ago. Nature 302:487–491.

Bousman, C. B., T. C. Partridge, L. Scott, S. E. Metcalfe, J. C. Vogel, H. Seaman, and J. S. Brink. 1988. Palaeoenvironmental implications of late Pleistocene and Holocene valley fills in Blydefontein Basin, Noupoort, C.P. South Africa. Palaeoecol. Africa 19:43–67.

Brenner, G. J. 1968. Middle Cretaceous spores and pollen from northwestern Peru. Pollen & Spores 10:341–383.

———. 1976. Middle Cretaceous floral provinces and early migrations of angiosperms. Pp. 23–47 *in* C. Beck (ed.), *Origin and Early Evolution of Angiosperms*. New York: Columbia University Press.

Brenner, G. J., and I. Bickoff. 1990. Palynology of the Palaeoequatorial, Lower Cretaceous Kurnub Group of Israel. Unpublished manuscript.

Burger, D. 1990. Early Cretaceous angiosperms from Queensland, Australia. Rev. Palaeobot. Palynol. 65:153–163.

Butzer, K. W. 1984. Late Quaternary environments in South Africa. Pp. 235–264 *in* J. C. Vogel (ed.), *Late Cainozoic Palaeoclimates of the Southern Hemisphere*. Rotterdam: Balkema.

Caratini, C., and P. Giresse. 1979. Palynological contribution to the study of continental and marine environments in the Congo at the end of the Quaternary. C. R. Acad. Sci. D 288(4):379–382.

Cockcroft, M. J., M. J. Wilkenson, and P. D. Tyson. 1987. The application of the present-day climatic model to the Late Quaternary in southern Africa. Climatic Change 10:161–181.

Coetzee, J. A. 1967. Pollen analytical studies in East and southern Africa. Palaeoecol. Africa 3:1–146.

———. 1978a. Climatic and biological changes in southwestern Africa during the Late Cainozoic. Palaeoecol. Africa 10:13–19.

———. 1978b. Late Cainozoic palaeoenvironments of Southern Africa. Pp. 114–127 *in* E. M. van Zinderen Bakker (ed.), *Antarctic Glacial History and World Palaeoenvironments*. Rotterdam: Balkema.

———. 1980. Tertiary environmental changes along the southwestern African coast. Palaeontol. Afr. 23:197–203.

———. 1983. Intimations on the history of the Tertiary floras and vegetation of Southern Africa. Bothalia 14:345–354.

———. 1986. Microfloral elements in the Neogene of the southwestern Cape Province and their phytogeographic significance. Palaeoecol. Africa 17:21–30.

Coetzee, J. A., and J. Muller. 1984. The phytogeographic significance of some extinct Gondwana pollen types from the Tertiary of the southwestern Cape (South Africa). Ann. Missouri Bot. Gard. 71:1088–1099.

Coetzee, J. A., and J. Praglowski. 1988. Winteraceae pollen from the Miocene of the southwestern Cape (South Africa). Grana 27:27–37.

Coetzee, J. A., and J. Rogers. 1982. Palynological and lithological evidence for the Miocene Palaeoenvironment in the Saldanha region (South Africa). Palaeogeogr. Palaeoclimatol. Palaeoecol. 39:71–85.

Coetzee, J. A., and E. M. van Zinderen Bakker. 1970. Palaeoecological problems of the Quaternary of Africa. S. Afr. J. Sci. 66: 78–84.

Deacon, J., and N. Lancaster. 1988. *Late Quaternary Palaeoenvironments of Southern Africa*. Oxford: Clarendon.

Dettmann, M. E. 1989. Antarctica: Cretaceous cradle of austral temperate rainforests? Pp. 89–105 *in* J. A. Crame (ed.), *Origins and Evolution of the Antarctic Biota*. Geol. Soc. Special Publication no. 47.

Diester-Haass, L., and H. Chamley. 1982. Oligocene and post-Oligocene history of sedimentation and climate off northwest Africa (DSDP site 369). Pp. 529–544 *in* U. von Rad, K. Hinz, M. Sarnthein, and E. Seibold (eds.),

Geology of the Northwest African Continental Margin. Berlin: Springer.

Douglas, R. G., and S. M. Savin. 1973. Oxygen and carbon isotope analyses of Cretaceous and Tertiary foraminifera from the central north Pacific. Initial Rep. DSDP 17:591–605.

Doyle, J. A., and L. J. Hickey. 1976. Pollen and leaves from the Mid-Cretaceous Potomac Group and their bearing on early angiosperm evolution. Pp. 139–206 *in* C. B. Beck (ed.), *Early Evolution of Angiosperms*. New York: Columbia University Press.

Doyle, J. A., P. Biens, A. Doerenkamp, and S. Jardiné. 1977. Angiosperm pollen from the pre-Albian lower Cretaceous of equatorial Africa. Bull. Cent. Rech. Explor.-Prod. Elf-Aquitaine 1:451–473.

Doyle, J. A., S. Jardiné, and A. Doerenkamp. 1982. *Afropollis,* a new genus of early angiosperm pollen, with notes on the Cretaceous palinostratigraphy and palaeoenvironments of northern Gondwana. Bull. Cent. Rech. Explor.-Prod. Elf-Aquitaine 6:39–117.

Dupont, L. M., and H. Hooghiemstra. 1989. The Saharan-Sahelian boundary during the Brunhes Chron. Acta Bot. Neerl. 38(4): 405–415.

Flohn, H., and S. Nicholson. 1980. Climatic fluctuations in the arid belt of the 'Old World' since the last glacial maximum: Possible causes and future implications. Palaeoecol. Africa 12:3–21.

Frakes, L. A. 1979. *Climates throughout Geologic Time*. Amsterdam: Elsevier.

Frédoux, A., and J. P. Tastet. 1988. Stratigraphie et paléoclimatologie de la marge septentrionale du Golfe de Guinée depuis 200 000 ans. Pp. 175–183 *in* C. Caratini (ed.), *Palynologie, Ecologie, Paléoécologie*. Actes X Symp. Ass. Palynol. Langue Franç. Inst. Pondichéry.

Germraad, J. H., C. A. Hopping, and J. Muller. 1968. Palynology of Tertiary sediments from tropical areas. Rev. Palaeobot. Palynol. 6:189–348.

Goldblatt, P. 1978. An analysis of the flora of Southern Africa: Its characteristics, relationships and origins. Ann. Missouri Bot. Gard. 65:369–436.

Hamilton, A. C. 1968. Some plant fossils from Bukwa. Uganda J. 32:157–164.

———. 1972. The interpretation of pollen diagrams from Highland Uganda. Palaeoecol. Africa 7:45–149.

———. 1976. The significance of patterns of distribution shown by forest plants and animals in tropical Africa for the reconstruction of upper Pleistocene palaeoenvironments: A review. Palaeoecol. Africa 9:63–97.

———. 1982. *Environmental History of East Africa: A Study of the Quaternary.* London: Academic.

Harrison, M. S. J. 1988a. The components of analogue concepts of southern African Quaternary climatic variations: A critique. Palaeoecol. Africa 19:283–292.

———. 1988b. The status of general circulation model simulations of southern African Quaternary climates. Palaeoecol. Africa 19:293–303.

Hedberg, O. 1951. Vegetation belts of the East African mountains. Svensk Bot. Tidskr. 45:141–202.

Hendey, Q. B. 1981. Palaeoecology of the Late Tertiary fossil occurrences in 'E' Quarry, Langebaanweg, South Africa, and a reinterpretation of their geological context. Ann. S. Afr. Mus. 84:1–104.

Herngreen, G. F. W., and A. F. Chlonova. 1981. Cretaceous microfloral provinces. Pollen & Spores 23:441–555.

Herngreen, G. F. W., and H. Duenas Jimenez. 1990. Dating of the Cretaceous Une Formation, Colombia, and the relationship with the Albian-Cenomanian African–South American microfloral province. Rev. Palaeobot. Palynol. 66:345–359.

Herngreen, G. F. W., A. Randrianasolo, and J. W. Verbeek. 1982. Micropalaeontology of Albian to Danian strata in Madagascar. Micropalaeontology 28:97–109.

Hochuli, P. A. 1981. North Gondwanan floral elements in Lower to Middle Cretaceous sediments of the southern Alps (southern Switzerland, northern Italy). Rev. Palaeobot. Palynol. 35:337–358.

Hooghiemstra, H. 1988. Changes of major wind belts and vegetation zones in NW Africa 20 000–5 000 yr B.P. as deduced from a marine pollen record near Cap Blanc. Rev. Palaeobot. Palynol. 55:101–140.

Horowitz, A., C. G. Sampson, L. Scott, and J. C. Vogel. 1978. Analysis of the Voigspost site, O.F.S., South Africa. S. Afr. Archaeol. Bull. 33:152–159.

Hsü, K. J. 1974. The Miocene desiccation of the Mediterranean and its climatical and zoogeographical implications. Naturwissenschaften 61:137–142.

Hsü, K. J., L. Montadert, D. Bernoulli, M. B. Cita, A. Erikson, R. E. Garrison, R. B. Kidd, F. Mélierés, C. Müller, and R. Wright. 1977. History of the Mediterranean salinity crises. Nature 267:399–403.

Hughes, N. F. 1969. Jurassic and Cretaceous pollen and spores. Pp. 311–329 *in* R. H. Tschudy and R. Scott (eds.), *Aspects of Palynology.* New York: John Wiley & Sons.

Jahnichen, H. 1990. The oldest record of *Nypa burtini*. Internat. Org. Palaeobot. Newsletter 42: 12.

Jan du Chéne, R. E. 1980. Palynological study of the Tar Sand, Upper Maestrichtian of S.W. Nigeria. Proceedings Fourth International Palynological Conference, Lucknow 2: 794–799.

Jardiné, S., and L. Magloire. 1963. Palynologie et stratigraphie du Crétacé des Bassins du Sénégal et du Côte d'Ivoire. Mdm. B. R. G. M. 32:187–245.

Jardiné, S., P. Biens, and A. Doerenkamp. 1974a. *Dicheiropollis etruscus,* un pollen caractéristique du Crétacé, inférieur afro-sudamericain. Conséquences pour l'évaluation des unités climatiques et implications dans la dérive des continents. Sci. Géol. Bull. 27: 87–100.

Jardiné, S., S. G. Kiesek, and Y. Reyre. 1974b. L'Individualisation progressive du continent africain vue à travers les données palynologiques de l'Ere secondaire. Sci. Géol. Bull. 27:69–85.

Jarzen, D. M., and M. E. Dettmann. 1989. Taxonomic revision of *Tricolpites reticulatus* Cookson ex Couper, 1953, with notes on the biogeography of *Gunnera.* Pollen & Spores 31:97–112.

Kemp, E. M. 1968. Probable angiosperm pollen from British Barremian to Albian strata. Palaeontology 11:421–434.

Kendall, R. L. 1969. An ecological history of the Lake Victoria basin. Ecol. Monogr. 39:121–176.

Kora, M., and S. El Beialy. 1989. Early Cretaceous palynomorphs from Gabal Musaba Salama area, southwestern Sinai, Egypt. Rev. Palaeobot. Palynol. 58:129–138.

Kuyl, O. S., J. Muller, and H. T. Waterbolk. 1955. The application of palynology to oil geology with reference to Western Venezuela. Geologie Mijnb., n.s. 17 (3): 49–76.

Lanfranchi, R., and D. Schwartz (eds.). 1990. Paysages quaternaires de l'Afrique centrale atlantique. Paris: ORSTOM.

Levyns, M. R. 1964. Migrations and origins of the Cape Flora. Trans. R. Soc. S. Afr. 37:85–107.

Lézine, A. M. 1989a. Late Quaternary vegetation and climate of the Sahel. Quat. Res. 32:317–334.

———. 1989b. Le Sahel: 20 000 ans d'histoire de la végétation. Bull. Soc. Géol. France 5:35–42.

Lézine, J. M., and J. Casanova. 1989. Pollen and hydrological evidence for the interpretation of past climates in tropical West Africa during the Holocene. Quat. Sci. Rev. 8:45–55.

Livingstone, D. A. 1967. Postglacial vegetation of the Ruwenzori mountains in equatorial Africa. Ecol. Monogr. 37:25–52.

McLachlan, I. R., and E. Pieterse. 1978. Preliminary palynological results: Site 361, leg 40, Deep Sea Drilling Project. Initial Rep. DSDP 40:857–881.

Maley, J. 1987. Fragmentation de la forêt dense humide Africaine et extension des biotopes mantagnards au Quaternaire récent. Palaeoecol. Africa 18:307–334.

Martin, A. R. H. 1968. Pollen analysis of Groenvlei lake sediments, Knysna (South Africa). Rev. Palaeobot. Palynol. 7:107–144.

Maud, R. R., and T. C. Partridge. 1987. Regional geomorphic evidence for climatic change in Southern Africa since the Mesozoic. Palaeoecol. Africa 18:337–348.

Méon, H. 1990. Palynologic studies of the Cretaceous-Tertiary boundary interval at El Kef outcrop, northwestern Tunisia: Palaeogeographic implications. Rev. Palaeobot. Palynol. 65:85–94.

Méon, H., and P. Donze. 1988. Etude palynologique du passage Crétacé-Tertiare dans la région du Kef (Tunisie NW): L'Environnement végétal terrestre et son évolution. Inst. Fr. Pondichéry, Trav. Sec. Tech. 25:237–250.

Moore, H. E. 1973. Palms in the tropical forest ecosystems of Africa and South America. Pp. 63–88 *in* B. J. Meggers, E. S. Ayensu, and W. D. Duckworth (eds.), *Tropical Forest Ecosystems in Africa and South America: A Comparative Review.* Washington, D.C.: Smithsonian Institution.

Moreau, R. E. 1966. *The Bird Faunas of Africa and Its Islands.* New York: Academic.

Morgan, R. 1978. Albian to Senonian palynology of site 364, Angola Basin. Init. Rep. Deep Sea Drill. Proj. 40:915–951.

Muller, J. 1981. Fossil pollen records of extant angiosperms. Bot. Rev. 47:1–146.

Newell, R. E., S. Gould-Stewart, and J. C. Chung. 1981. A possible interpretation of palaeoclimatic reconstructions for 18 000 B.P. for the region 60° N to 60° S, 60° W to 100° E. Palaeoecol. Africa 13:1–19.

Nicholson, S. E., and H. Flohn. 1980. African environmental and climatic changes and the general atmospheric circulation in the late Pleistocene and Holocene. Climatic Change 2:313–348.

Parrish, J. T., A. M. Ziegler, and C. R. Scotese. 1982. Rainfall patterns and the distribution of coals and evaporites in the Mesozoic and Cenozoic. Palaeogeogr. Palaeoclimatol. Palaeoecol. 41:31–66.

Partridge, T. C., D. M. Avery, G. A. Botha, J. S. Brink, J. Deacon, et al. 1990. Late Pleistocene and Holocene climatic change in Southern Africa. S. Afr. J. Sci. 86:302–306.

Rabinowitz, P. D., M. F. Coffin, and D. Falvey. 1983. The separation of Madagascar and Africa. Science 220:67–69.

Rand, H. H., and J. M. Mabesoone. 1982. North Eastern Brazil and the final separation of South America and Africa. Palaeogeogr. Palaeoclimatol. Palaeoecol. 39:71–85.

Raven, P. H., and D. I. Axelrod. 1974. Angiosperm biogeography and past continental movements. Ann. Missouri Bot. Gard. 61:539–673.

Retallack, G. J., D. P. Dugas, and E. A. Bestland. 1990. Fossil soils and grasses of a Middle Miocene East African grassland. Science 247:1325–1328.

Reyment, R., and R. V. Dingle. 1987. Palaeogeography of Africa during the Cretaceous period. Palaeogeogr. Palaeoclimatol. Palaeoecol. 59:93–116.

Roche, E. 1974. Paléobotanique, paléoclimatologie et dérive des continents. Sci. Géol. Bull. 27:10–24.

Roche, E., G. Bikwemu, and C. Ntaganda. 1988. Evolution du paléoenvironnement quaternaire au Rwanda et au Burundi. Pp. 105–123 *in* C. Caratini (ed.), *Palynologie, Ecologie, Paléoécologie.* Actes X Symp. Ass. Palynol. langue Franç. Inst. Pondichéry.

Salard-Cheboldaeff, M. 1981. Palynologie Maestrichtienne et Tertiare du Cameroun: Resultats botaniques. Rev. Palaeobot. Palynol. 32:401–439.

Sarnthein, M., J. Thiede, U. Pflaumann, H. Erlenkeuser, D. Futterer, B. Koopman, H. Lange, and E. Seibold. 1982. Atmospheric and oceanic circulation patterns off northwest Africa during the past 25 million years. Pp. 545–604 *in* U. von Rad, K. Hinz, M. Sarnthein, and E. Seibold (eds.), *Geology of the Northwest African Continental Margin.* Berlin: Springer.

Scholtz, A. 1984a. Interim report on the Stompoor project. Palynol. Lab. Dept. Archaeol. University of Stellenbosch, South Africa. 1–14.

———. 1984b. Preliminary report on the palynological analysis of the KN 70 and KN 49 pipes, Kang district, southeastern Botswana.

Palynol. Lab. Dept. Archaeology University Stellenbosch, South Africa. 1–9.

———. 1985. The palynology of the upper lacustrine sediments of the Arnot pipe, Banke, Namaqualand. Ann. S. Afr. Mus. 95:1–109.

———. 1986. Palynological and palaeobotanical studies in the southern Cape. Master's thesis, University of Stellenbosch, Stellenbosch, South Africa.

Schrank, E. 1984. Organic-geochemical and palynological studies of a Dakhla shale profile (Late Cretaceous) in southeast Egypt. Berliner Geowiss. Abh. 50:189–207.

———. 1987. Palaeozoic and Mesozoic palynomorphs from north Africa (Egypt and Sudan) with special reference to Late Cretaceous pollen and Dinoflagellates. Berliner Geowiss. Abh. 75:249–310.

———. 1990. Palynology of the clastic Cretaceous sediments between Dongola and Wadi Muqaddam, northern Sudan. Berliner Geowiss. Abh. (A) 120.1:149–168.

Schulz, E. 1987. Die holozäne Vegetation der zentralen Sahara (N-Mali, N-Niger, SW-Libyen). Palaeoecol. Africa 18:143–161.

Scott, L. 1976. Palynology of Lower Cretaceous deposits from the Algoa Basin (Republic of South Africa). Pollen & Spores 18:563–609.

———. 1982. A Late Quaternary pollen record from the Transvaal Bushveld, South Africa. Quat. Res. 17:339–370.

———. 1984. Palynological evidence for Quaternary palaeoenvironments in southern Africa. Pp. 65–80 *in* R. G. Klein (ed.), *Southern African Palaeoenvironments and Prehistory.* Rotterdam: Balkema.

———. 1986. Pollen analysis and palaeoenvironmental interpretation of Late Quaternary sediment exposures in the eastern Orange Free State, South Africa. Palaeoecol. Africa 17:113–122.

Shackleton, N. J., and J. P. Kennett. 1975. Palaeotemperature history of the Cenozoic and the initiation of Antarctic glaciation: Oxygen and carbon isotope analyses in DSDP sites 277, 279, 281. Initial Rep. DSDP 29:743–755.

Siesser, W. G. 1978. Aridification of the Namib Desert: Evidence from oceanic cores. Pp. 105–113 *in* E. M. van Zinderen Bakker (ed.), *Antarctic Glacial History and World Palaeoenvironments.* Rotterdam: Balkema.

Smith, A. S., A. M. Hurley, and J. C. Briden. 1981. Phanerozoic palaeocontinental world maps. Cambridge: Cambridge University Press.

Srivastava, G. K. 1978. *Cretaceous Spore-Pollen Floras: A Global Evaluation.* Biol. Mem. 3:1–130.

Stapleton, R. P. 1975. Planktonic foraminifera and calcareous nannofossils at the Cretaceous-Tertiary contact in Zululand. Palaeontol. Afr. 18:53–69.

Stebbins, G. L. 1974. *Flowering Plants: Evolution above the Species Level.* Cambridge, Mass.: Harvard University Press.

Stein, R., and M. Sarnthein. 1984. Late Neogene events of atmospheric and oceanic circulation offshore northwest Africa: High resolution record from deep-sea sediments. Palaeoecol. Africa 16:9–36.

Talbot, M. R. 1984. Late Pleistocene rainfall and dune building in the Sahel. Palaeoecol. Africa 16:203–214.

Tankard, A. J., and J. Rogers. 1978. Late Cenozoic palaeoenvironments on the west coast of Southern Africa. J. Biogeogr. 5:319–337.

Taylor, H. C. 1978. Capensis. Pp. 171–229 *in* M. J. A. Werger (ed.), *Biogeography and Ecology of Southern Africa,* Vol. 1. The Hague: Junk.

Thusu, B., J. G. L. A. van der Eem, A. El-Mehdawi, and F. Bu-Argoub. 1988. Jurassic-Early Cretaceous palynostratigraphy. Pp. 171–213 *in* A. El-Arnauti, B. Owens, and B. Thusu (eds.), *Subsurface Palynostratigraphy of Northeast Libya.* Benghazi, Libya (SPLAJ): Garyounis University Publications.

Tiedemann, R., M. Sarnthein, and R. Stein. 1989. Climatic changes in the western Sahara: Aeolo-Marine sediment record of the last 8 million years (sites 657–661). Proc. Ocean Drilling Prog., Sci. Results 108:241–277.

Truswell, E. M. 1982. Antarctica: The vegetation of the past and its climatic implications. Aust. Met. Mag. 30:169–173.

Tyson, P. D. 1986. *Climatic Change and Vari-*

ability in Southern Africa. Oxford: Oxford University Press.

Uwins, P. J. R., and D. J. Batten. 1988. Early to Mid-Cretaceous palynology of northeast Libya. Pp. 215–257 *in* A. El-Arnauti, B. Owens, and B. Thusu (eds.), *Subsurface Palynostratigraphy of Northeast Libya.* Benghazi, Libya (SPLAJ): Garyounis University Publications.

Vail, P. R., R. M. Mitchum, and S. Thompson. 1977. Seismic stratigraphy and global changes in sea level. Part 4, Global cycles of relative changes of sea level. Mem. Am. Assoc. Petrol. Geol. 26:83–97.

Van der Ham, R. W. J. M. 1990. Nephelieae pollen (Sapindaceae): Form, function and evolution. Ph.D. diss., Rijksherbarium/Hortus Botanicus, Leiden.

Van Hoeken-Klinkenberg, P. M. J. 1966. Maastrichtian, Palaeocene and Eocene pollen and spores from Nigeria. Leidse Geol. Med. 28:37–48.

Van Zinderen Bakker, E. M. 1967. Upper Pleistocene stratigraphy and Holocene ecology on the basis of vegetation changes in sub-Saharan Africa. Pp. 125–147 *in* W. W. Bishop and J. D. Clark (eds.), *Background to Evolution in Africa.* Chicago: University of Chicago Press.

———. 1982. African palaeoclimates 18 000 years B.P. Palaeoecol. Africa 15:77–99.

Van Zinderen Bakker, E. M., and J. A. Coetzee. 1972. A re-appraisal of Late Quaternary climatic evidence from Tropical Africa. Palaeoecol. Africa 7:151–181.

Van Zinderen Bakker, E. M., and J. H. Mercer. 1986. Major Late Cainozoic events and palaeoenvironmental changes in Africa viewed in a world wide context. Palaeogeogr. Palaeoclimatol. Palaeoecol. 56:217–235.

Van Zinderen Bakker, E. M., Jr. 1971. Ecological investigations on the ravine forests of the eastern Orange Free State (South Africa). Master's thesis, Department of Botany, University of Orange Free State, Bloemfontein, South Africa.

Vincens, A. 1987. Environnements botanique et climatique des hominides de l'Est Turkana, Kenya, entre 2.0 et 1.4 millions d'années: Apport de la palynologie. Palaeoecol. Africa 18:257–269.

———. 1989. Paléoenvironnements du Bassin Nord-Tanganyika (Zaire, Burundi, Tanzania) au cours des 13 derniers mille ans: Apport de la palynologie. Rev. Palaeobot. Palynol. 61:69–88.

Vogel, J. C. (ed.). 1984. *Late Cenozoic Palaeoclimates of the Southern Hemisphere.* Rotterdam: Balkema.

Ward, J. D., M. K. Seely, and N. Lancaster. 1983. On the antiquity of the Namib. S. Afr. J. Sci. 79:175–183.

White, F. 1978. The Afromontane region. Pp. 463–513 *in* M. J. A. Werger (ed.), *Biogeography and Ecology of Southern Africa.* The Hague: Junk.

———. 1981. The history of the Afromontane archipelago and the scientific need for its conservation. Afr. J. Ecol. 19:33–54.

Yemane, K. 1984. Sédimentologie et palynologie du bassin néogene de Chilga (Ethiopia): Implications paléoclimatiques et tectoniques. Thesis. lab. Géol. Quat. Luminy, Marseilles (abstract).

Yemane, K., M. Taieb, and H. Faure. 1987. Limnogeologic studies on an intertrappean continental deposit from northern Ethiopian Plateau (37°03′ E, 12°25′ N). J. Afr. Earth Sci. 6(1):91–101.

Zavada, M. S., and J. M. Benson. 1987. First fossil evidence for the primitive angiosperm family Lactoridaceae. Amer. J. Bot. 74:1590–1594.

Zavada, M. S., and T. N. Taylor. 1986. Pollen morphology of Lactoridaceae. Pl. Syst. Evol. 154:31–39.

Chapter 4

South American Paleofloras

Edgardo Juan Romero

Introduction

In this chapter, I review the plant fossil record in South America from the Early Cretaceous through the Late Miocene and discuss its relation to the flora of Africa. I do not discuss here the Pliocene and Quaternary, when glacial climates produced changes in the patterns of vegetation, because they are not relevant to the understanding of historical relationships with Africa.

South America and Africa shared similar floras in the late Paleozoic, 340–248 million years ago (Ma), and had in common lycophytes, sphenophytes, ferns, and certain conifers. The presence on both continents of genera like *Glossopteris* and *Gangamopteris* indicates the existence of the continent of Gondwanaland and supports the model of paleoclimates and paleogeography postulated by Wegener (1924:47). During most of the Mesozoic (248–120 Ma), tree ferns, pteridosperms, cycads, and conifers indicate a similar relationship between South America and Africa (Delevoryas 1973).

The relations between late Mesozoic (120–60 Ma) plants from the two continents were not studied until recently, when geophysical evidence of continental drift showed the need for correlations of sedimentary basins on either side of the opening South Atlantic. The paleobotany of the late Mesozoic is also interesting because this period coincides with the origin and early evolution of angiosperms. Pioneer studies of the Africa–South America relationship were made by Muller (1966), Jardiné and co-workers (1974), Regali and co-workers (1974a, 1974b), Srivastava (1978), and many others. They compare associations of

 ISBN 0-300-05375-4

pollen grains and spores in South America and Africa. The evidence was reviewed and discussed by Herngreen and Chlonova (1981), in a paper that set the paradigm of plant relationships during the Cretaceous. Vakhrameev (1988) supported these ideas by discussing Mesozoic phytogeography on the basis of leaf imprints and pollen grains.

No one has closely compared tertiary fossil plants of both continents, although general discussions on the subject or partial reviews for each continent are available (Axelrod and Raven 1978; Coetzee 1980, 1981, Chapter 3 this volume; Coetzee and Muller 1985; Raven and Axelrod 1974; Romero 1978, 1986; Wolfe 1985).

The relations among living plants of the two continents have been the subject of much systematic and biogeographic research (Brenan 1978; Goldblatt 1978; Thorne 1973; and several papers in Meggers et al. 1973). Herngreen and Chlonova (1981) established several microfloral provinces in three areas of the world—north temperate, tropical belt, and south temperate (figure 4.1). I shall follow a slightly different division (figure 4.1), indicated by data from South America that make it possible to highlight in considerable detail both the advent of the angiosperms and some phytogeographical changes that occurred during the Tertiary. I continue to use the term *province*, al-

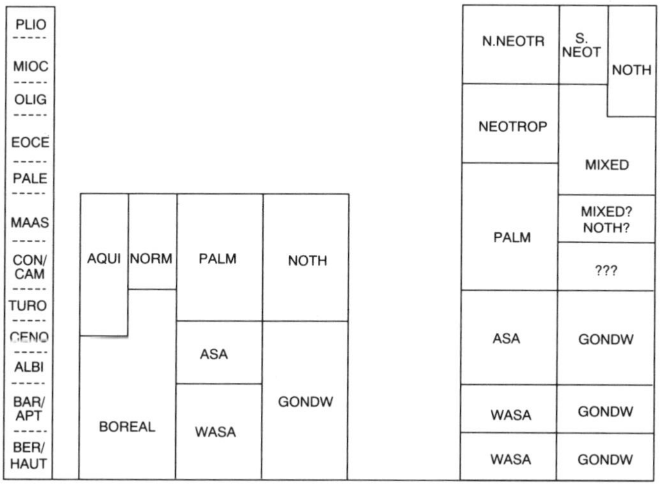

Figure 4.1. Microfloral provinces during the Cretaceous and Tertiary. *Far left*, ages and epochs. *Left*, provinces of the world according to Herngreen and Chlonova (1981); the three columns represent north temperate, equatorial, and south temperate provinces. *Right*, provinces of South America, as described in this chapter; the two columns represent equatorial and south temperate provinces. Aqui = Aquilapollenites, Norm = Normapoles, WASA = West African–South American, ASA = African–South American, Gondw = Gondwana, Noth = Nothofagidites, Neo = Neotropical.

though it may best be reserved for areas with specific or occasionally generic endemism (Takhtajan 1986:3).

I will review in detail information accumulated mainly from studies of pollen grains, using megafossil records (leaf imprints, casts of seeds, petrified trunks or seeds) to supplement the palynological data, in this first attempt at a broad synthesis. (Determinations of some megafossils have been questioned recently and should be the subject of careful analysis.) The use of pollen grains studies may facilitate comparisons of South American flora with African flora because the most important Tertiary evidence from Africa is palynological.

The study of pollen grains has an additional advantage in that it can show almost continuous sequences, giving positive evidence of the presence of taxa for long periods and allowing evaluation of the absence of certain taxa during particular periods. I shall not, however, discuss absences here, for two reasons. First, most of the tropical plants to be considered are pollinated by insects, or sometimes by other animals, and produce small amounts of poorly dispersed pollen. Little pollen is available for fossilization, therefore, and the record may become discontinuous. Second, the palynological record, especially for the southern part of South America, is often not of continuous sequences but of isolated samples. These samples give an idea of the flora at that moment, as the megafloras do, but they do not show changes through time.

Fossil pollen grains are recorded in the literature as extinct taxa of unknown affinities or are sometimes assigned to living families, genera, or species. To facilitate comparison of the record in different locations and discussion of changes in the flora, I have considered only those records that can be assigned to a living family. Pteridophytes have been disregarded, because their assignment is limited to few taxa and is doubtful in some cases.

Fossil sites in South America are mainly coastal, and can be grouped into three areas—Northwestern, Eastern, and Southern—roughly separated by the Guiana and Brazilian shields (figure 4.2). The Northwestern and Southern areas are depositional basins that remained active during most of the Cretaceous and Early Tertiary (Riccardi 1987; Zambrano 1981; Chapter 2, this volume). They have marine, continental, and coastal sediments, are rich in petroleum, and became inactive with the rise of the Andes. The Eastern area is a group of depositional basins developed after the separation from Africa and open to the Atlantic, with coastal and lacustrine sediments in inland and platform areas (Petri 1987). The three areas were intensively searched for oil, and much of the paleontological information was obtained through that search.

I. Earliest Cretaceous

The flora of South America in the earliest Cretaceous, shortly before the appearance of the angiosperms, was not much different from that of the earlier Mesozoic. It was composed of pteridophytes (including many tree ferns), pteridosperms, ginkgos, bennettitaleans, cycads, and conifers. The ferns produced an array of trilete and monolete spores, and the gymnosperms produced monosulcate, bisaccate, and other kinds of pollen grains.

The flora of the northern part of South

Figure 4.2. Fossil study sites. *1.* Doubinger and Marocco 1976; *2.* Elsik 1964, 1966; *3.* Brenner 1968, Elsik and Tanikaimoni 1970, Lammons 1970; *4.* Berry 1927, 1937; *5.* Berry 1929a; *6.* Berry 1929b, 1932; *7 and 8.* Doubinger 1973, 1976; *9.* Sole de Porta 1971, 1972a, 1972b; *10.* Muller et al. 1987; *11.* Berry 1924, van der Hammen 1954, 1956, 1957, van der Hammen and Garcia Mutis 1965; Schuler and Doubinger 1970; *12.* Gonzalez Guzman 1967; *13.* Germeraad et al. 1967, Tschudy 1970; *14.* Doubinger 1973; *15.* Berry 1920; *16 and 17.* Germeraad et al. 1968; *18.* Berry 1936; *19.* van der Hammen and Wymstra 1964, Leidelmeyer 1966; *20.* Daemon 1975; *21.* van der Hammen and Wymstra 1964; *22.*Herngreen 1975b; *23.* Regali et al. 1974a, 1974b; *24.* Lima 1982; *25.* Lima 1976a, 1976b, 1978a, 1978b; *26.* Regali et al. 1974a, 1974b; *27.* Lima and Coelho 1987; *28.* Mauri 1936; *29.* Regali et al. 1974a, 1974b; *30 and 31.* Muller 1966; *32.* Regali et al. 1974a, 1974b; *33.* Herngreen 1972, 1975b, Muller 1966, Regali et al. 1974a, 1974b; *34.* Lima and Salard-Cheboldaeff 1981; *35.* Muller 1966; *36 and 37.* Regali et al. 1974a, 1974b; *38.* Lima and Silva Amador 1985; *39.* Lima et al. 1985; *40.* Lima et al. 1986; *41.* Quatrocchio and Volkheimer 1986; *42.* Anzotegui and Garralla 1986; *43.* Palma Heldt 1983; *44.* Doubinger 1972; *45.* Doubinger and Chotin 1975, Takahashi 1977a, 1977b; *46.* Palma Heldt 1980; *47.* Stough 1968, Troncoso and Barrera 1977, 1980; *48.* Volkheimer and Salas 1976; *49.* Volkheimer et al. 1976; *50.* Volkheimer et al. 1976; *51.* Gurstein 1990; *52.* Volkheimer and Sepulveda 1976; *53.* Volkheimer and Salas 1975; *54.* Archangelsky and Romero 1974a, 1974b, Romero 1973; *55.* Baez et al. 1991; *56.* Frenguelli 1943, Romero and Hickey 1976; *57.* Archangelsky 1973, Archangelsky and Zamaloa 1986, Petriella 1972; *58.* Archangelsky 1973, Archangelsky and Zamaloa 1986; *59.* Archangelsky 1973, Archangelsky and Zamaloa 1986, Barreda 1989; *60.* Archangelsky and Gamerro 1965a, 1965b, 1965c, 1965d, 1967, Romero and Archangelsky 1986; *61.* Romero 1977, Romero and Castro 1986, Romero and Zamaloa 1985, *62.* Freile 1972; *63.* Fasola 1969; *64.* Menendez and Caccavari de Filice 1975; *65.* Vergel and Durango de Cabrera 1988, Zamaloa and Romero 1990.

Table 4.1. Earliest Cretaceous (144–125 Ma): Berriasian, Valanginian, and Hauterivian Ages

Microfloral Province	Characteristic Flora
WASA	Predominance of Classopollis. *Araucariacites, Ephedripites.* Scarce bisaccates and spores. *Dicheiropollis etruscus.*
Gondwana	Fern spores. Bisaccates and trisaccates. *Microcachrydites. Araucariacites.*
Boreal	Fern spores. Bisaccates. No trisaccates.

Sources: Doubinger and Marocco 1976; Lima and Coelho 1987; Mueller 1966; Regali et al. 1974a, 1974b; Volkheimer et al. 1976.

America had a particular composition that Heerngreen and Chlonova (1981) called the West African–South American (WASA) Microfloral Province. Evidence from pollen grains and leaf remains shows that it is more or less equivalent to Vakhrameev's (1988) Equatorial Region. The WASA province was characterized by a predominance of rimulate pollen, especially the monoaperturate *Classopollis,* produced by members of the now-extinct Cheirolepidaceae. It also had *Araucariacites, Ephedripites,* and *Dicheiropollis etruscus,* a probably gymnosperm, exclusive to and characteristic of the province.

The southern part of the continent was a part of the Gondwana Microfloral Province (Herngreen and Chlonova 1981), or the Patagonian Province in the Notal Region (Vakhrameev 1988). It had bisaccate pollen grains and trilete spores as dominants, and trisaccate pollen (*Microcachridites, Dacrycarpites*), which is the characteristic feature of the province. *Araucariacites* and *Monosulcates* were also present.

In the Northern Hemisphere, the Boreal Microfloral Province was dominated by trilete spores, as well as bisaccate (*Alisporites*), inaperturate, and monosulcate pollen grains, and lacked trisaccate grains.

II. Late Early Cretaceous

The oldest South American angiosperms have been reported from the WASA Microfloral Province. They are represented by *Reticulatisporites jardinus, Tucanopollis crisopolensis, Afropollis* sp., and *Stellatopollis bituberensis*—small reticulate pollen grains found in the Eobarremian (Regali 1987a, 1987b; Regali and Viana 1989). The oldest stratigraphic zones with angiosperm pollen grains are in Brazil and Africa (Muller et al. 1987:21), although "*Inaperturopollenites*" *crisopolensis* (an improper name) is found in northwest South America since the late Barremian (Muller et al. 1987:table II). The Barremian-Aptian limit is defined in Brazil (Regali and Viana 1989) by the first occurrence of *Clavatipollenites hughesii,* another small reticulate, colpate pollen grain.

In the Gondwana Province, *Clavatipol-*

Table 4.2. Late Early Cretaceous (125–113 Ma): Barremian and Aptian Ages

Microfloral Province	Characteristic Flora
WASA	*Classopollis. Araucariacites.* Scarce bisaccates and spores. Psilate tricolpates. Reticulatasporites.
Gondwana	Trilete spores. Bisaccates and trisaccates. *Araucariacites.* Clavatipollenites. Stephanocolpates.
Boreal	Trilete spores. Bisaccates, no trisaccates. *Clavatipollenites,* Monocotilenoid monosulcates.

Sources: Archangelsky and Gamerro 1965a, 1965b, 1965c, 1965d, 1967; Doubinger and Marocco 1976; van der Hammen and Wymstra 1964; Lammons 1970; Lima 1976a, 1976b, 1978a, 1978b, 1982; Muller 1966; Regali et al. 1974a, 1974b; Romero and Archangelsky 1986; Volkheimer et al. 1976; Volkheimer and Sepulveda 1976.

lenites hughesii, the earliest angiosperm recorded there, appears in the late Barremian or early Albian of Patagonia (Archangelsky and Gamerro 1967). It was the only angiosperm in an otherwise rich flora of more than one hundred species of pteridophytes and gymnosperms represented by pollen, spores, leaf imprints, cuticles, and mummified fructifications (Archangelsky 1973). Leaves of one species of angiosperm were found in the assemblage (Romero and Archangelsky 1986). *Clavatipollenites hughesii* was also reported in the Barremian-Albian of England and North America (Couper 1958; Brenner 1963), thus adding to the similarities of the Gondwana and Boreal microfloral provinces.

The angiosperms *Stephanocolpites mastandreai* and *Huitrinipollenites transitorius* are also found in the Gondwana Province (Volkheimer and Salas 1975). Both had reticulate, polycolpate pollen grains. The second was considered a synonym of *Asteropollis,* a genus described in the Boreal Province. *Clavatipollenites* and *Asteropollis* have been thought to be related to *Hedyosmum* and *Ascarina* (Chloranthaceae) because of their similar morphology (Muller 1981); a determination to the family level, however, does not indicate true evolutionary relations.

III. Early Late Cretaceous

By the Aptian and Albian, marine transgressions and regressions significantly altered the composition of continental basins and inland seas in northern South America (Srivastava 1978). The flora changed substantially, and their pollen grains constitute an assemblage named the African–South American (ASA) Microfloral Province by Herngreen and Chlonova (1981). I add the Turonian Age in the duration of this province because data from South America suggest that this age was transitional—similar in

Table 4.3. Early Late Cretaceous (113–88 Ma): Albian, Cenomanian, and Turonian Ages

Microfloral Province	Characteristic Flora
ASA	Elater-bearing: *Elatereocolpites, Galeacornea.* Poliplicates. Diversified angiosperms. No bisaccates. *Hexaporotricolpites. Triorites africaensis.* Tetradites. First pollen record of living families: Gunneraceae, Proteaceae.
Gondwana	Pteridophytic spores, trisaccate pollen grains, early angiosperms. No triporates or *Nothofagus.*
Boreal	Trilete spores. Bisaccate, non trisaccate. Tricolporoidates. Normapolles (since Early Cenomanian).

Sources: Archangelsky et al. 1991; Brenner 1968; Daemon 1975; van der Hammen and Wymstra 1964; Herngreen 1975a, 1975b; Lima 1976a; Muller 1966; Regali et al. 1974a, 1974b; Volkheimer and Salas 1975, 1976.

events to the preceding ages. The dominant pollens were bizarre simple-aperturate, psilate, elater-bearing grains like *Elaterocolpites, Elaterosporites,* and *Galeacornea.* Also frequent were a diverse array of polyplicates like *Gnetaceopollenites,* and *Steevispollenites.* The bisaccate gymnosperms, scarce before the Aptian, were absent in the early Late Cretaceous. The angiosperms became more diversified.

In the Aptian, *Reticulatisporites jardinus* is more frequent, *Retimonocolpites* and *Liliacidites,* thought to be monocotyledons, are present, as are colporoidate pollen, the polyporate *Cretaceidites* sp., and the tricolpodiorate *Hexaporotricolpites* (Herngreen and Chlonova 1981).

During the Cenomanian, new morphological types include *Triorites africaensis,* a widespread triporate, and *Tetradites* sp., a tetrad. It has been suggested that the first might belong to Proteaceae and the second to Fabaceae. Muller (1981) concluded that *T. africaensis* "may belong to an ancestral complex of Proteaceous pollen types," but this remains to be confirmed. Muller (1982) also noted that *Tetradites* pollen grains "suggest affinities with *Mimosa,*" but this also awaits confirmation; it would considerably extend the range of Mimosoideae.

The first record of a member of a living family was described in Turonian sediments from Peru by Brenner (1968) as *Tricolpites reticularis,* in the Gunneraceae. Jarzen and Dettmann (1989) made a detailed review of this species, using new materials and electronic microscope techniques, and confirmed its assignment to Gunneraceae. Additional triporate types, including *Echitriporites triangulariformis* and *Proteacidites sigalli,* show the early differentiation of the family Proteaceae.

Southern South America was a part of the Gondwana Province, but its flora is known only from a late Albian to Cenomanian assemblage from southern Patagonia (Arch-

Table 4.4. Middle Late Cretaceous (88–73 Ma): Coniacian, Santonian, and Campanian Ages

Microfloral Province	Characteristic Flora
Palm	Angiosperms dominant (tricolporate, tricolpate). Poliplicates scarce. Amarylidaceae, Araucariaceae, Arecaceae (5/6), Sapindaceae.
Nothofagidites	Unknown.
Aquilapollenites	Angiosperms dominant. Projectites. Deciduous.
Normapolles	Angiosperms dominant. Tricolpoidates. Evergreen.

Sources: Elsik 1964, 1966; Elsik and Thanikaimoni 1970; Herngreen 1975b; Lima et al. 1986; Muller 1966; Regali et al. 1974a, 1974b.

angelsky et al. 1991). This flora has austral affinities and is characterized by early angiosperms. The upper part of the site may be early Turonian. It had common genera with the Northern Temperate Provinces (*Clavatipollenites* sp., pteridophyte spores), as well as southern endemics (*Phymopollenites panosus,* trisaccate Podocarpaceae). Neither triporate pollen grains nor *Nothofagidites* have been found.

The evidence on relations of early Late Cretaceous flora to living plants is sparse at this time, but it seems likely that within Rosiidae a few families, including Gunneraceae and Proteaceae, had differentiated. The family Gunneraceae had a wide tropical distribution through the Late Cretaceous and lower Tertiary and is now better represented in tropical Africa than in tropical America. It is also present in temperate areas of the southern continents. Proteaceae has a fossil record that starts in the tropics, although the family is predominantly temperate today. Notably, the fairly primitive *Helicia* is restricted to the Asian tropics (Willis 1973).

IV. Middle Late Cretaceous

Pollen grains of palms are found in northern South American sediments of this age, and they are abundant and to some extent diversified. Heerngreen and Chlonova (1981) therefore call this the Palm Microfloral Province, and Vakhrameev (1988) calls it the Equatorial (Palmaceous) Region. Principal types of palm pollen are *Spinizonocolpites echinatus, S. baculatus, Proxapertites* sp., *Longapertites* sp., *Mauritidites lehmanii,* and *Psilamonocolpites medium.* Elater-bearing pollen types persist as a minor component. Other types are frequent, for example, *Striatricolporites* or *Araucariacites.* The family Sapindaceae is represented by *Cupaneidites* sp. (Herngreen 1975a) and *C. reticularis* (Regali et al. 1974a, 1974b). The record of Sapindaceae is almost continuous since the Cretaceous, although during the Tertiary in South America it is restricted to southern lands. The family is now pantropical.

Southern South America lacks pollen records from the middle Late Cretaceous. Herngreen and Chlonova (1981) postulate that it may have belonged to the Nothofagidites Province; this remains to be demonstrated.

During this time there were in the Northern Hemisphere two microfloral provinces: Aquilapollenites and Normapolles. They take their names from their dominant taxa. which are extinct, bizarre angiosperms. Srivastava (1978, 1981) shows that they have a latitudinal distribution, boreal the first, meridional the second. Krutzsch and Pacltova (1988) suggest that the Aquilapollenites Province represented a deciduous forest, whereas the Normapolles was evergreen.

V. Latest Cretaceous

In South America during the Maastrichtian, several plant families are present for the first time, showing advanced evolution in at least two subclasses of dicotyledons. Some regionalism is also detected within South America, and immigration from other continents is found for the first time.

Within the area of the Palm Microfloral Province of northern South America, two regions can be differentiated during the Maastrichtian. The Northwest Region has Aquifoliaceae, *Aquilapollenites,* Humiriaceae, Moraceae, and Myrtaceae. The East Region has Onagraceae, Proteaceae, and Sapindaceae. They share Arecaceae and Ulmaceae. The record of Aquifoliaceae is based on discoveries of *Ilexpollenites* (Muller et al. 1987), a distinctive pollen type that became widespread in the late Cretaceous (Martin 1977). Humiriaceae is represented by *Saccoglottis cipacoquensis* (Berry 1924), from the Guaduas Formation in Colombia, a fruit cast with good morphological diagnostic features. Arecaceae are the best represented family, especially in the Northwest Region, where pollen grains

Table 4.5. Latest Cretaceous (73–65 Ma): Maastrichtian Age

Microfloral Province	Characteristic Flora
Palm	Aquifoliaceae, *Aquilapollenites,* Arecaceae (2/3), Humiriaceae, Moraceae, Myrtaceae, Onagraceae, Proteaceae, Sapindaceae, Ulmaceae.
Nothofagidites	Trilete spores. Nothofagaceae, Olacaceae, Podocarpaceae, Proteaceae.
Aquilapollenites	Migration among provinces.
Normapolles	Migration among provinces.

Sources: Archangelsky and Romero 1974a, 1974b; Berry 1924, 1927; Doubinger 1972; Germeraad et al. 1968; van der Hammen 1954, 1956, 1957; van der Hammen and Garcia Mutis 1965; Herngreen 1975b; Mauri 1936; Muller et al. 1987; Regali et al. 1974a; Romero 1973; Sole de Porta 1971, 1972a, 1972b; Tschudy 1970.

and fruit casts have been found. The Eastern Region record is also a fruit cast. Moraceae is represented by a fruit cast, *Genomoites zipaquinensis*. This record antedates the first palynological record, which is from the Eocene. Myrtaceae is represented by *Myrteacidites* sp. (Herngreen 1975b; Muller 1981). Ulmaceae (*Ulmodeipites krempi*) first appears in South America in the Maastrichtian, though it is an old family, recorded in tropical latitudes of Borneo in the Turonian. Onagraceae is represented by *Jussitriporites* sp. (Regalli et al. 1974a, 1974b); this genus remained in the East Region until the Oligocene. Sapindaceae is present with the genus *Cupaneidites,* more abundant in the Maastrichtian than in the previous age.

Among the above families, several are pantropical to subtropical or warm temperate, including Aquifoliaceae, Sapindaceae, Ulmaceae, Arecaceae, and Moraceae. Others, although worldwide in distribution, are centered in South America: Humiriaceae, Myrtaceae, and Onagraceae. These three families are recorded for the first time in the Maastrichtian, whereas the pantropicals had been present since the Coniacian. This is logical because pantropical families probably had their main radiation while the continents were still united. Families centered in South America could have evolved more recently, when some barriers to exchange with Africa were present.

The evolution of the subclasses of angiosperms had advanced by the end of the Maastrichtian; the fossil evidence shows the differentiation of three close families in the superorders Malvanae and Urticanae in the Dilleniidae, and also eight related families in the superorders Myrtanae, Fabeae, and Proteanae in the Rosidae (table 4.6).

Table 4.6. Families differentiated in South America at the end of the Cretaceous (65 Ma)

Hamamelidae
  Urticales: Ulmaceae, Moraceae
Dilleniidae
  Malvales: Sterculiaceae
Rosidae
  Fabales: Leguminosae
  Proteales: Proteaceae
  Haloragales: Gunneraceae
  Myrtales: Myrtaceae, Onagraceae
  Celastrales: Aquifoliaceae
  Linales: Humiriaceae
  Sapindales: Sapindaceae

During most of the Cretaceous, the four worldwide provinces (Palms, Nothofagidites, Normapolles, and Aquilapollenites) remained separated, but toward the late Maastrichtian, the withdrawal of epicontinental seas opened ways for exchanges among them. The exchanges were particularly intense between the Normapolles and Aquilapollenites provinces, but almost no exchange occurred with the Nothofagidites Province. In South America, the only record of migration is reported by Muller and coworkers (1987), who list *Aquilapollenites* species in the Maastrichtian of the Venezuelan Basin. In southernmost South America the Maastrichtian is poorly recorded. The genus *Nothofagidites* was reported in southernmost Patagonia close to the Strait of Magellan, and in northwest Patagonia (Archangelsky and Romero 1974a), but nothing is known about other elements of the flora.

Table 4.7. Paleocene (65–55 Ma)

Microfloral Province	Characteristic Flora
Palm	Annonaceae, *Aquilapollenites,* Araceae, Arecaceae (4/5), Bombacaceae (2/3), Ctenolophonaceae Melastomataceae, Olacaceae, Onagraceae, Polygalaceae, Proteaceae, Restioniaceae, Ulmaceae.
Mixed	Anacardiaceae, Aquifoliaceae, Araucariaceae, Arecaceae (6/9), Caryophyllaceae, Casuarinaceae, Cochlospermaceae, Combretaceae, Cunoniaceae, Ephedraceae, Gunneraceae, Haloragaceae, Loranthaceae, Malvaceae, Myricaceae, Myrtaceae, Nothofagaceae (1/6), Olacaceae, Onagraceae, Pandanaceae, Podocarpaceae (5/5), Polygalaceae, Proteaceae (3/5), Restionaceae, Rubiaceae, Sapindaceae, Symplocaceae, Tiliaceae, Ulmaceae.

Sources: Archangelsky 1973; Archangelsky and Zamaloa 1986; Doubinger 1972, 1973; Doubinger and Chotin 1975; Freile 1972; Frenguelli 1943; Germeraad et al. 1968, van der Hammen and Garcia Mutis 1965; Leidelmeyer 1966; Muller et al. 1987; Palma Heldt 1980, 1983; Petriella 1972; Quatrocchio and Volkheimer 1986; Regali et al. 1974a; Sole de Porta 1971, 1972a, 1972b; Takahashi 1977a, 1977b; Tschudy 1970.

VI. Paleocene

After the Maastrichtian, the flora of the world changed considerably. Many modern families had differentiated, and the old groups (Normapolles, *Aquilapollenites,* elatereous pollen) had become relictual or extinct. The major continents were relatively isolated, and local conditions influenced the development of their floras. In South America there was a great diversification of taxa, and 36 modern families have been identified.

The Palm Province was still present in northern South America, in both the Brazilian and Northwest basins. The flora included a good representation of Arecaceae, with four genera and five species, and Bombacaceae, with two genera and three species. *Ctenolophon,* now restricted to tropical Africa and Asia, was present. The migration of *Aquilapollenites* species was still in progress, and some have been reported from the Brazilian Basin. Melastomataceae and a probable Poaceae (*Monoporites annulatus;* Regali 1974a) are recorded for the first time.

In the southern part of South America, Paleocene fossils, preserved as pollen grains, leaf and fruit imprints, and petrified trunks, are abundant and fairly well known. They are exposed in the Golfo San Jorge and Southern basins, both open to the Atlantic Ocean; the Concepcion Basin, open to the Pacific Ocean; and at least one inland basin in northern Argentina. In this southern area, Romero (1978, 1986) defined the Paleoflora

Mixta, a phytogeographic province with an apparent mixture of species related to taxa now living in the Neotropics or the subantarctic forests. Arecaceae are the dominant family (6 genera, 9 species), but Nothofagaceae (1/6) and Podocarpaceae (5/5) are important. Proteaceae, Anacardiaceae, Gunneraceae, and Sapindaceae are also present. Casuarinaceae had *Casuarinidites coenozoicus* and *Haloragacidites harrissi* (pollen grains) and *Casuarina patagonica* (imprints of cones). Cochlospermaceae are known from leaf and fruit imprints. A paleoenvironmental reconstruction (Petriella and Archangelsky 1975) in the region of the Golfo San Jorge, based on paleoxylological and palynological studies, suggests the presence of mangrove communities, swamps, humid and mountain forests, and sclerophyllous communities.

The presence of Araucariaceae, Gunneraceae, Podocarpaceae, and Cunoniaceae (affinities to *Weinmannia;* Petriella and Archangelsky 1975) should be noted as the possible antecedent to the disjunction existing today between the Araucarian District of the Subantarctic Region in Chile and the Araucarian Paranense District of the Neotropical Region in southern Brazil.

The presence of Casuarinaceae, Myrtaceae, and Proteaceae, together with some megafossils not considered in table 4.7—for example, Akaniaceae and Proteaceae (Durango de Cabrera and Romero 1986; Gandolfo et al. 1988; Romero and Hickey 1976; Romero et al. 1986)—suggests a relation with the floras of the Australian Region other than that established through the Nothofagidites Province.

In southernmost Argentina, near the border with Chile, a problematic association was recorded (Freile 1972). It had the genus *Nothofagidites;* a few genera and species of Podocarpaceae, including trisaccate pollen grains; several species of trilete spores; and *Anacolosidites* (Olacaceae), a genus of Paleotropical affinities.

A number of the families recorded in the Paleocene of South America have a pantropical to subtropical distribution: Anacardiaceae, Annonaceae, Araceae, Arecaceae, Combretaceae, Myricaceae, Olacaceae, Rhizophoraceae, Rutaceae, Sapindaceae, Zamiaceae. Others have a similar distribution but are stronger in South America: Bombacaceae, Loranthaceae, Melastomataceae. The presence of these categories of families during the Latest Cretaceous was discussed above. Starting in the Paleocene, however, new groups of families are present in South America: tropical or subtropical families that are weakly represented in Africa (Tiliaceae, Elaeocarpaceae, Symplocaceae); Old World families (Pandanaceae, Casuarinaceae); extratropical families (Restionaceae, Cunoniaceae); and even a temperate family (Caprifoliaceae). The presence of these new groups can be explained by the growing distance between Africa and South America and their increasing proximity to other continents. Therefore, many of the new species that are differentiated in this period belong to families with a more equilibrated worldwide distribution. It is also important that not only tropical but also extratropical and temperate families were shared with the mentioned continents. This means that by the Paleocene, temperate modern families were starting to diversify and expand their distributional area. This is also consistent with the worldwide climatic cooling trend reported since the Paleocene (Wolfe 1975).

Most of the above families can be presumed to have originated in the Brazilian-African region of Gondwanaland (Raven and Axelrod 1974), but Myricaceae, Symplocaceae, Caprifoliaceae, and Rhizophoraceae may have originated in Laurasia. The diversification of the Dilleniidae (table 4.6) can be traced by the appearance of Cochlospermaceae, Tiliaceae, Bombacaceae, and Malvaceae in the superorder Malvanae.

VII. Eocene

Forty-eight modern families, distributed in two main regions or provinces, are recorded from Eocene deposits. These families are rather different from those of the Paleocene. Arecaceae are still important in the northern province (4 genera, 7 species), but other, more modern, families are present for the first time and become even more abundant than the Arecaceae: Leguminosae (6/6), Malpighiaceae (4/4), Annonaceae (3/6), Bombacaceae (3/3), and Bignoniaceae

Table 4.8. Eocene (55–38 Ma)

Microfloral Province	Characteristic Flora
Neotropical	Anacardiaceae, Annonaceae (3/6), Arecaceae (4/7), Asteraceae, Bignoniaceae (2/2), Bombacaceae (3/3), Caprifoliaceae, Chenopodiaceae, Combretaceae, Ericaceae, Euphorbiaceae, Ephedraceae, Humiriaceae, Guttiferae, Leguminosae (6/6), Malpighiaceae (4/4), Malvaceae, Meliaceae, Melastomataceae, Moraceae, Myrtaceae, Olacaceae, Onagraceae (2/2), Poaceae, Podocarpaceae (1/2), Polygalaceae, Proteaceae, Rhizophoraceae, Rubiaceae, Solanaceae, Sterculiaceae, Ulmaceae.
Mixed	Anacardiaceae (2/3), Apiaceae, Aquifoliacae, Araliaceae, Araucariaceae, Arecaceae (1/1), Asteraceae, Brassicaceae, Casuarinaceae, Ephedraceae, Ericaceae, Gunneraceae, Leguminosae, Loranthaceae, Malvaceae (1/3), Myricaceae, Myrtaceae (1/2), Nothofagaceae (1/11), Onagraceae, Podocarpaceae (5/10), Proteaceae (2/6), Rubiaceae, Sapindaceae (3/3), Solanaceae, Tiliaceae, Winteraceae.

Sources: Archangelsky and Romero 1974a, 1974b; Baez et al. 1991; Berry 1920, 1927, 1929a, 1929b, 1932, 1936, 1937; Doubinger 1972, 1976; Gonzalez Guzman 1967; Gurstein 1990; van der Hammen 1957; Herngreen 1972, 1975a, 1975b; Leidelmeyer 1966; Lima 1982; Lima and Salard-Cheboldaeff 1981; Menendez and Caccavari de Felice 1975; Muller 1966; Muller et al. 1987; Regali et al. 1974a, 1974b; Romero 1973, 1977; Romero and Hickey 1976; Romero and Castro 1986; Romera and Zamaloa 1985; Stough 1968; Troncoso and Barrera 1977, 1980; Vergel and Durango de Cabrera 1988; Zamaloa and Romero 1990.

(2/2). Also, the first records in this province of Poaceae and Asteraceae are from the Eocene. The presence of such families indicates a modernization of the flora, which has changed in composition and can justifiably be considered a province distinct from the Palm Province of the Paleocene. I have named it the Neotropical Province.

In the southern part of the continent, the Mixed flora remains, with its distinguishing apparent mixture of cold temperate and tropical elements. It is rather different, however, from the Paleocene flora that occupied the same area. Arecaceae, though extending to Tierra del Fuego, is represented by a single species. Sapindaceae (3/3) and Anacardiaceae (2/3) are now more abundant tropical families. Nothofagaceae and Podocarpaceae are frequent (1/11 and 5/10), as is Proteaceae (2/6). The Fabaceae are also present, but as a minor element.

The group of families that bear a suggested relationship to the Araucarian disjunction in southern Brazil and Chile (see Paleocene section above) is still present in the Mixed paleoflora, with the addition of Winteraceae.

Fabaceae are dominant in the Neotropical Province and secondary in the Mixed Province. The opposite distribution is shown by Proteaceae and Podocarpaceae, which are very poor in the Neotropics and restricted to the Brazilian basin, but are dominant in the Mixed Province. Onagraceae has two species belonging to two genera in the Neotropical Province, and only one species in the Mixed Province. The Mixed Province species, *Crassiorites australis,* was found with viscin threads attached to the apertures, a characteristic of modern Onagraceae. In the Asteraceae, pollen grains from the Brazilian basins have tentatively been determined to be *Echitricolpites* sp. (Lima and Silva Amador 1985); pollen grains from the Mixed Province have tentatively been identified as "Pollen tipo Compuesta" (Palma Heldt 1983). Confirmation should be relatively easy in both cases, because the illustrations are clear enough to allow assignation to the family.

Most of the families listed in table 4.8 belong mainly to the group of pantropical to subtropical plants as outlined in the Maastrichtian and Paleocene. Newcomers to this group are Araliaceae, Guttiferae, Meliaceae, Sterculiaceae, and Euphorbiaceae. The second group, also outlined in the Maastrichtian and Paleocene, similar to the first but more centered in South America, includes most of the families present during the Paleocene plus the new Bignoniaceae and Malphighiaceae. Pantropicals that are rare in Africa now include the Winteraceae. There are also other groups of families, such as the cosmopolites (Brassicaceae, Solanaceae), and many subtropical and temperate families, like Asteraceae, Caprifoliaceae, Chenopodiaceae, and Ericaceae (poor in tropical Africa). Several families with species adapted to dry environments are present in this epoch, especially in the Brazilian basins—for example, Ephedraceae, Chenopodiaceae, and to some extent Casuarinaceae and Proteaceae.

VIII. Oligocene

The two microfloral provinces of the Eocene are also present during the Oligocene. In addition, pollen and megafloral fossils allow the definition of a new province, Nothofagidites, occupying small southernmost

Table 4.9. Oligocene (38–25 Ma)

Microfloral Province	Characteristic Flora
Neotropical	Aquifoliaceae, Arecaceae (4/4), Bombacaceae, Chenopodiaceae, Convolvulaceae, Ephedraceae (2/2), Euphorbiaceae (2/2), Guttiferae, Lecythidaceae, Leguminosae (4/4), Malpighiaceae (4/4), Malvaceae, Meliaceae, Myrtaceae, Myricaceae (2/2), Nymphaceae, Olacaceae, Orchidiaceae, Poaceae, Podocarpaceae (2/2), Polygalaceae, Proteaceae (2/2), Rhizophoraceae, Sapotaceae, Symplocaceae, Thymelaceae, Tiliaceae, Ulmaceae.
Mixed	Anacardiaceae (2/3), Apiaceae, Araucariaceae, Arecaceae, Chenopodiaceae, Ephedraceae, Euphorbiaceae, Gunneraceae, Haloragaceae, Leguminosae, Malvaceae (2/2), Myrtaceae (1/2), Nothofagaceae (1/9), Poaceae, Podocarpaceae (5/9), Proteaceae, Restoniaceae, Rubiaceae (2/2), Solanaceae, Sparganiaceae, Symplocaceae, Tiliaceae, Violaceae, Winteraceae.
Nothofagidites	Araucariaceae, Myrtaceae, Nothofagaceae, Podocarpaceae, Proteaceae.

Sources: Barreda 1989; Doubinger 1976; Fasola 1969; Germeraad et al. 1968; Gurstein 1990; van der Hammen and Wymstra 1964; Lima and Silva Amador 1985; Lima et al. 1985; Menendez and Cacavari de Felice, 1975; Muller et al. 1987; Regali et al. 1974a, 1974b; Schuler and Doubinger 1970.

areas, and with species found only in the Subantarctic Region.

The Neotropical Province has about the same families as before, with Arecaceae, Leguminosae, and Malphigiaceae dominant (4 genera with 4 species each), and several other families with two genera each: Ephedraceae, Euphorbiaceae, Myricaceae, Podocarpaceae, and Proteaceae. Orchidaceae, Thymelaeaceae, and Lecythidaceae are recorded for the first time.

The Mixed Province also has a composition similar to that of the Eocene; Podocarpaceae, Anacardiaceae, Rubiaceae, and Malvaceae have more than two genera described. The Mixed Province in the Oligocene contained many taxa belonging to the wooded savannas of the Chaco Domain, including Anacardiaceae, Malvaceae, Leguminosae, Chenopodiaceae, Poaceae, Arecaceae, and the arid elements Chenopodiaceae and Ephedraceae. I suggest that the differentiation of the type of vegetation that would later give rise to the wooded savannas of the present Chaco Domain and to other arid associations of the Chaco and Andino-Patagonian domains (Cabrera and Willink 1973) began at this time. The Neotropical

Province, however, may have been the forerunner of the more humid Caribbean, Amazonian, and Guayanan domains.

The Nothofagidites Province, poorly documented in the Paleocene and Eocene, is now more evident, on the basis of small leaves preserved as imprint fossils in Tierra del Fuego, southern Argentina, and northwest Patagonia. They belong to Nothofagaceae, Araucariaceae, Podocarpaceae, Myrtaceae, and Proteaceae. The number of families recorded in the continent for the Oligocene is 39. Most of these are families that today have a pantropical to subtropical distribution, making this the first group of families outlined in Maastrichtian and Paleocene times. Significantly, during the Oligocene no new family was added to this group, which suggests that the isolation of South America was enough to prevent genetic exchange needed to share a pantropical family. A second group, outlined in the Maastrichtian and Paleocene times, is that with families with a pantropical distribution but mainly centered in South America, comprising Bombacaceae, Malphigiaceae, Myrtaceae, and one family—Lecythidaceae—that was added during the Oligocene. A third group, plants weakly represented in South Africa, is present only in the Mixed Province. It includes Winteraceae, which was already present during the Eocene; Tiliaceae, which was in the Brazilian Basin during the Eocene; and Symplocaceae, which is a new family for South America. Thymelaeaceae, a family better represented in Africa, did not reach South America until the Oligocene. Available evidence suggests that the Thymelaeaceae differentiated in the African continent after South America started drifting away, and not until the Oligocene did some genera reach the latter continent. A fourth group of subtropical and temperate families are represented by Chenopodiaceae, Ephedraceae, Proteaceae, Restionaceae, and Haloragaceae. A fifth group of cosmopolitan families have the addition of Poaceae and Violaceae. The growing number of cosmopolitan families is consistent with the evolution of modern taxa, easily dispersed under Late Tertiary environmental conditions.

IX. Miocene

During the Miocene three provinces can still be distinguished, but there is a change in the flora, with more modern families becoming abundant and many being recorded for the first time. For this reason I prefer to call the former Mixed Province the South Neotropical Province, and the previous Neotropical the North Neotropical.

In the North Neotropical Province, 17 families are recorded, and there are no clear dominants. Aquifoliaceae is the only family with two genera and two species. In the South Neotropical Province, Asteraceae is the most differentiated family (five genera and five species), followed by Anacardiaceae, Malvaceae, and Leguminosae. Chenopodiaceae has only one genus with three species. Amaranthaceae is recorded for the first time. The families that today are found in the wooded savannas of the Chaco Domain are present, and one of them (Anacardiaceae, 4/4) is more diversified in the Miocene than in the Oligocene. The Nothofagidites Province is known only by imprints of leaves and shows a reduction in its area compared to the Oligocene.

In none of the three provinces are there

Table 4.10. Miocene (25–5 Ma)

Microfloral Province	Characteristic Flora
North Neotropical	Acanthaceae, Aquifoliaceae (2/2), Arecaceae, Asteraceae, Bombacaceae, Chenopodiaceae, Euphorbiaceae, Guttiferae, Lecythidaceae, Leguminosae, Malphigiaceae, Melastomataceae, Myrtaceae, Poaceae, Rhizophoraceae, Solanaceae.
South Neotropical	Amaranthaceae, Anacardiaceae (4/4), Apiaceae (2/2), Aquifoliaceae, Asteraceae (5/5), Chenopodiaceae (1/3), Ephedraceae, Euphorbiaceae, Goodeniaceae, Haloragaceae, Leguminosae (3/5), Lythraceae, Malvaceae (4/4), Myrtaceae, Onagraceae (1/2), Poaceae, Podocarpaceae (3/4), Polygalaceae, Proteaceae, Sapindaceae, Solanaceae, Ulmaceae, Verbenaceae.
Nothofagidites	Araucariaceae, Myrtaceae, Nothofagaceae, Podocarpaceae, Proteaceae.

Sources: Anzotegui and Garralla 1986; Barreda 1989; Gurstein 1990; van der Hammen and Wymstra 1964; Regali et al. 1974a, 1974b.

additions to the group of pantropical families, neither those more centered in South America nor those less developed in Africa. Instead, new cosmopolitan families (Verbenaceae, Goodeniaceae, Apiaceae) and xeric families are present. The xeric group is relatively strong, with the presence of Chenopodiaceae, Amaranthaceae, and, to some extent, Proteaceae and Asteraceae; it is mainly concentrated in the South Neotropical Province.

Although outside the scope of this paper, records of Pleistocene age in Rondônia, Brazil (van der Hammen 1972), show increasing desertification during at least some of the Quaternary. It may have been during this time that the arid zones of the Chaco Domain expanded to northeast Brazil, reaching what is now the Caatinga.

Summary

Since the Cretaceous and through the Tertiary periods, two main phytogeographical areas ("Microfloral Provinces") occupied South America. An equatorial area (with the successive provinces of WASA, ASA, Palm, Neotropical, and North Neotropical) was affiliated with Africa. A southern area (with the successive provinces of Gondwana, Nothofagidites, Mixed, and South Neotropical) was probably related to Antarctica, Australia, and sometimes to Africa. The angiosperms were first found in the late Early Cretaceous in the Brazilian basins. They spread rather rapidly—in about 4 million years—to other basins of the WASA Province and to the Gondwana Province.

In the early Late Cretaceous, bizarre extinct pollen grains were present in the equatorial ASA Province, which occupied northern South America and Central Africa. In the middle Late Cretaceous, bizarre, extinct angiosperm pollen grains were also found in temperate northern Aquilapollenites and Normapoles provinces. Such rare pollen grains have not been recorded in the southern Nothofagidites Province.

The oldest South American fossils that can be assigned to a living family are 90 million years old. By the close of the Cretaceous (65 Ma), 11 modern families, mainly Rosidae, had differentiated. During the Maastrichtian and Paleocene, new and modern families rose to dominance. In the equatorial Palm Province palms were abundant, and in the Mixed Province there was an apparent mixture of tropical and temperate taxa. The southernmost flora is poorly known and was apparently dominated by *Nothofagidites*.

The ancestors of some taxa that now have a disjunct distribution in the Araucarian Districts in southern Brazil and central Chile are found from the Paleocene in many localities of the Mixed Province. Other taxa found in the Mixed Province suggest close relations with the flora of the present phytogeographical Australian Region (not to be confused with the present Subantarctic Region, with *Nothofagus* that occupies parts of Australia).

During the Eocene and Oligocene, further modernization of the flora gave the Neotropical and Mixed provinces their neotropical character. It is suggested that the Neotropical Province gave origin to the rain forests and mesic vegetation of present South America, and the Mixed Province to the savannas and drier vegetation.

Families that are today considered pantropicals have their oldest fossil records from the Coniacian to the Eocene (83–38 Ma). Those considered tropical but centered mainly in South America enter the record between the early Maastrichtian and the late Oligocene (73–25 Ma). This is considered consistent with the increasing separation of the continents. Subtropical, extratropical, and even temperate families, only weakly developed in Africa, are recorded in South America since the Paleocene (65 Ma), and cosmopolitans since the Eocene (55 Ma). This is consistent with the more equidistant position of South America relative to other continents, the climatic deterioration during the Tertiary, and the diversification of modern families.

Literature Cited

Anzotegui, L. M., and S. S. Garralla. 1986. Estudio palinológico de la Formación Paraná (Mioceno Superior) (Pozo Josefina, Provincia de Santa Fe, Argentina). 1, Descripciones sistemáticas. Facena 6:101–178.

Archangelsky, S. 1967. Estudio de la Fm. Baquero, Cretácico Inferior de Santa Cruz, Argentina. Rev. Mus. La Plata, n.s., 5:63–171.

———. 1973. Palinología del Paleoceno de Chubut. I, Descripciones sistemáticas. Ameghiniana 10:339–399.

Archangelsky, S., and J. C. Gamerro. 1965a. Estudio palinológico de la Formación Baqueró (Cretácico) Provincia de Santa Cruz, I. Ameghiniana 4:159–167.

———. 1965b. Estudio palinológico de la Formación Baqueró (Cretácico) Provincia de Santa Cruz, II. Ameghiniana 4:201–207.

———. 1965c. Estudio palinológico de la Formación Baqueró (Cretácico) Provincia de Santa Cruz, III. Ameghiniana 4:229–236.

———. 1965d. Estudio palinológico de la Formación Baqueró (Cretácico) Provincia de Santa Cruz, IV. Ameghiniana 4:363–370.

———. 1967. Spore and pollen types of the Lower Cretaceous in Patagonia (Argentina). Rev. Paleobot. Palynol. 1:211–217.

Archangelsky, S., and E. J. Romero. 1974a. Los registros más antiguos del polen de *Nothofagus* (Fagaceae) de Patagonia (Argentina y Chile). Bol. Soc. Bot. México 33:13–30.

———. 1974b. Polen de Gimnospermas (Coniferas) del Cretácico Superior y Paleoceno de Patagonia. Ameghiniana 11:217–236.

Archangelsky, S., and M. C. Zamaloa. 1986. Nuevas descripciones palinológicas de las Formaciones Salamanca y Bororó, Paleoceno del Chubut (República Argentina). Ameghiniana 23:35–46.

Archangelsky, S., G. A. Jalfin, E. S. Bellosi, and C. Perrot. 1991. Palynology and alluvial facies from the Middle Cretaceous of Patagonia, subsurface of San Jorge Basin, Argentina. Cretaceous Events. In press.

Axelrod, D. I., and P. H. Raven. 1978. Late Cretaceous and Tertiary vegetation history of Africa. Pp. 77–130 *in* M. J. A. Werger (ed.), *Biogeography and Ecology of Southern Africa.* The Hague: W. Junk.

Baez, A. M., M. C. Zamaloa, and E. J. Romero. 1991. Nuevos hallazgos de microfloras y anuros paleógenos en el Noroeste de Patagonia: Implicancias paleoambientales y paleobiogeográficas. Ameghiniana 27:83–94.

Barreda, V. D. 1989. "Palinología estratigráfica de las sedimentitas terciarias del Patagoniano en los alrededores de la ciudad de Comodoro Rivadavia, Provincias de Chubut y Santa Cruz." Ph.D. diss., University of Buenos Aires F.C.E.N.

Berry, E. W. 1920. Tertiary fossil plants from Venezuela. Proc. U.S. Natl. Mus. 59:553–579.

———. 1924. Fossil fruits from the eastern Andes of Colombia. Bull. Torrey Bot. Club 51:61–67.

———. 1927. Petrified fruits and seeds from Oligocene of Peru. Pan-Am. Geol. 51:241–244.

———. 1929a. Eocene plants from Restin Formation of Peru. Pan-Am. Geol. 51:241–244.

———. 1929b. Fossil fruits in the Ancon sandstone of Ecuador. J. Paleontol. 3:298–301.

———. 1932. A new palm from the Upper Eocene of Ecuador. J. Wash. Acad. Sci. 22:327–329.

———. 1936. Tertiary fossil plants from Venezuela, II. Proc. U.S. Natl. Mus. 83:335–360.

———. 1937. The Parinas sandstone of Peru. Johns Hopkins Univ. Stud. Geol. 12:99–106.

Brenan, J. P. M. 1978. Some aspects of the phytogeography of tropical Africa. Ann. Missouri Bot. Gard. 65:437–478.

Brenner, G. J. 1963. The spores and pollen of the Potomac Group of Maryland. Bull. Md. Dept. Geol. Mines 27:1–25.

———. 1968. Middle Cretaceous spores and pollen from northeastern Peru. Pollen & Spores 10:341–383.

Cabrera, A. L., and A. Willink. 1973. Biogeografía de América Latina. Serie Biología 13. Washington, D.C.: Organización de Estados Americanos.

Coetzee, J. A. 1980. Tertiary environmental changes along the southwestern African coast. Palaeontol. Afr. 23:197–203.

———. 1981. A palynological record of very primitive angiosperms in Tertiary deposits of the southwestern Cape Province, South Africa. S. Afr. J. Sci. 77:341–343.

Coetzee, J. A., and J. Muller. 1985. The phytogeographic significance of some extinct Gondwana pollen types from the Tertiary of Southwestern Cape (South Africa). Ann. Missouri Bot. Gard. 71:1088–1099.

Couper, R. A. 1958. British Mesozoic microspores and pollen grains: A systematic and stratigraphic study. Paleontographica 103(abt. B): 75–179.

Daemon, R. F. 1975. Contribuição a datação da Formação Alter do Chão Bacia do Amazonas. Rev. brasil. Geoc. 5:77–84.

Delevoryas, T. 1973. Postdrifting Mesozoic floral evolution. Pp. 9–19 *in* B. J. Meggers, E. S. Ayensu, and W. D. Duckworth (eds.), *Tropical Forest Ecosystems in Africa and South America: A Comparative Review.* Washington, D.C.: Smithsonian Institution Press.

Doubinger, J. 1972. Evolution de la flore (pollens et spores) au Chili central (Arauco), du Crétace supérieur au Miocène. C. R. Soc. Biogeogr. 427:18–25.

———. 1973. Pollen et spores du bassin houillier paleocène du Cerrejon (province de Guajira, Colombie). C. R. 96 Congr. Nat. Soc. Savantes 5:253–262.

———. 1976. Quelques associations sporopolliniques du Tertiaire de Colombie. Actes 97 Congr. Nat. Soc. Savantes 4:9–18.

Doubinger, J., and P. Chotin. 1975. Etude palynologique de lignites tertiaires du basis D'Arauco-Concepcion (Chili). Rev. Esp. Micropaleontol. 7:549–565.

Doubinger, J., and R. Marocco. 1976. Découverte d'une microflore Wealdienne (Neocomien) dans la Region de Cuzco (sud du Pérou). Première datation des Grès Huancané. Sci. Géol. Bull. 29:79–89.

Durango de Cabrera, J., and E. J. Romero. 1986. *Roupala patagonica* n. sp. de Laguna del Hunco (Paleoceno) Provincia Chubut, Argentina. Actas V Congr. Argent. Paleontol. Bioestrat. 3:121–124.

Elsik, W. C. 1964. A new sporomorph genus from eastern Peru. Pollen & Spores 6:601–604.

———. 1966. New sporomorph genera from the Upper Cretaceous of Peru. Pollen & Spores 8:553–564.

Elsik, W. C., and G. Thanikaimoni. 1970. *Bomarea lyncina* (Amaryllidaceae) and *Auriculiidites* Elsik. Pollen & Spores 12:177–180.

Fasola, A. 1969. Estudio palinológico de la Fm Loreto (Terciario medio), provincia de Magallanes, Chile. Ameghiniana 6:3–19.

Freile, C. 1972. Estudio palinológico de la Formación Cerro Dorotea (Maastrichtano-Paleoceno) de la Provincia de Santa Cruz, I. Rev. Mus. La Plata, n.s., 6:39–63.

Frenguelli, J. 1943. Restos de *Casuarina* en el Mioceno en El Mirador, Patagonia Central. Rev. Mus. La Plata, 8:349–354.

Gandolfo, M. A., M. Dibbern, and E. J. Romero. 1988. *Akania patagonica* n.sp. and additional material on *A. americana* Romero & Hickey (Akaniaceae) from Paleocene sediments of Patagonia. Bull. Torrey Bot. Club 115:88–95.

Germeraad, J. H., C. A. Hopping, and J. Muller. 1968. Palynology of Tertiary sediments from tropical areas. Rev. Paleobot. Palynol., 6:89–348.

Goldblatt, P. 1978. An analysis of the flora of southern Africa: Its characteristics, relationships, and origins. Ann. Missouri Bot. Gard. 65:369–436.

Gonzalez Guzman, A. E. 1967. *A Palynological Study on the Upper Los Cuervos and Mirador Formation (Lower and Middle Eocene; Tibu area, Colombia).* Leiden: E. J. Brill.

Gurstein, G. R. 1990. Palinología estratigráfica del Terciario de la Cuenca del Colorado, República Argentina. 1, Especies terrestres de la perforación Nadir no. 1. Rev. Esp. Micropaleontol. 21:31–61.

Hammen, Th. van der. 1954. El desarrollo de la flora Colombiana en los períodos geológicos. 1, Maestrichtiano hasta Terciario más inferior (Una investigación palinológica de la formación de Guaduas y equivalentes). Bol. Geol. 2:49–106.

———. 1956. Description of some genera and species of fossil pollen and spores. Bol. Geol. 4:111–117.

———. 1957. Estratigrafía palinológica de la sabana de Bogotá. Bol. Geol. 5:187–203.

———. 1972. Changes in vegetation and climate in the Amazon Basin and surrounding areas during the Pleistocene. Geologie Mijnb. 51:641–643.

Hammen, Th. van der, and C. Garcia Mutis. 1965. The Paleocene pollen flora of Colombia. Leidse Geol. Med. 35:105–116.

Hammen, Th. van der, and T. A. Wymstra. 1964. A palynological study on the Tertiary and Upper Cretaceous of British Guiana. Leidse Geol. Med. 30:183–241.

Herngreen, G. F. W. 1972. Some new pollen grains from the Upper Senonian of Brazil. Pollen & Spores 14:97–112.

———. 1975a. Palynology of Middle and Upper Cretaceous strata in Brasil. Mededel. R. Geolog. Dienst, n.s., 26:39–91.

———. 1975b. An Upper Senonian pollen assemblage of borehole 3-PIA-10-AL state of Alagoas, Brazil. Pollen & Spores 17:93–140.

Herngreen, G. F. W., and A. F. Chlonova. 1981.

Cretaceous microfloral provinces. Pollen & Spores 23:441–555.

Jardiné, S., P. Bienes, and A. Doerenkamp. 1974. *Dicheiropollis etruscus,* un pollen caracteristique du Crétacé inférieur Afro-Sudamericain, conséquences pour l'evaluation des unités climatiques et implications dans la dérive des continents. Sci. Géol. Bull. 27:87–100.

Jarzen, D. M., and M. E. Dettmann. 1989. Taxonomic revision of *Tricolpites reticulatus* Cookson ex Couper, 1953, with notes on the biogeography of *Gunnera* L. Pollen & Spores 31:97–112.

Krutzsch, W., and B. Pacltova. 1988. Paleophytogeography of the Normapolles and the Triprojectacites (Aquilapollenites) groups in the Upper Cretaceous and Lower Tertiary. J. Palynol. 23–24:217–223.

Lammons, J. M. 1970. *Pentapsis,* a new palynomorph genus from the Cretaceous (Aptian) of Peru. Micropaleontology 16:175–178.

Leidelmeyer, P. 1966. The Paleocene and Lower Eocene pollen flora of Guyana. Leidse Geol. Med. 38:49–70.

Lima, M. R. 1976a. "*Crotonipollis,*" a new pollen genus from Santana Formation, Cretaceous of northeastern Brazil. Bol. Asoc. Latinoamer. Paleobot. Palinol. 3:14–20.

———. 1976b. O gênero *Classopolis* e as bacias mesozoicas do nordeste do Brasil. Ameghiniana 13:226–234.

———. 1978a. Estudio palinologico preliminar de um folhelho betumiinoso da Formação Missão Velha. Chapada do Araripe. Bol. I. G., Univ. São Paulo 9:140–143.

———. 1978b. Palinologia da Formação Santana (Cretaceo do nordeste do Brasil). Ph.D. diss., Instit. Geociênc., University of São Paulo.

———. 1981. Palinologia do Mesozoico Brasileiro: Una síntese. Pp. 445–460 *in* Comité Sudamericano del Jurásico y Cretácico (eds.), *Cuencas Sedimentarias del Jurásico y Cretácico de América del Sur,* vol. 2 Buenos Aires.

———. 1982. Palinologia da Formação Codó, Maranhao. Bol. I. G., Univ. São Paulo 13:116–128.

Lima, M. R., and M. P. C. Coelho. 1987. Estudo palinologico da sondagem estratigrafica de Lagoa do Forno, Bacia do Rio do Peixe, Cretaceo do Nordeste do Brasil. Bol. I. G., Univ. São Paulo 18:67–83.

Lima, M. R., and M. Salard-Cheboldaeff. 1981. Palinologie des Bassins de Gandarela et Fonseca (Eocene de l'état de Minas Gerais, Bresil). Bol. I. G. Univ. São Paulo 12:33–54.

Lima, M. R., and E. da Silva Amador. 1985. Analise palinologica de sedimentos da Formação Resende Terciario do estado do Rio de Janeiro, Brasil. Depart. Nac. Produção Min., Coletanea de Trabalhos Paleontol., ser. Geol. 27, seção Paleontol. Estrat. 1:371–378.

Lima, M. R., M. Salard-Cheboldaeff, and K. Suguio. 1985. Etude palynologique de la formation Tremembe, Tertiaire du Basin de Taubat (Etat de São Paulo, Brasil), d'apres les Echantillons du Sondage N° 42 du CNP. Depart. Nac. Produção Min., Coletanea de Trabalhos Paleontol., ser. Geol. 27, seção Paleontol. Estrat. 1:379–393.

Lima, M. R., S. Mezzalina, R. Dino, and A. R. Saad. 1986. Descoberta de microflora em Sedimentos do grupo Bauru, Cretaceo do Estado de São Paulo. Rev. Inst. Geoc. 7:5–9.

Martin, H. A. 1977. The history of *Ilex* (Aquifoliaceae) with special reference to Australia: Evidence from pollen. Austr. J. Bot. 25:655–673.

Mauri, C. J. 1936. O Cretaceo de Sergipe. Div. Geol. Mineral., Minist. Agric., Brasil, Monogr. 11:44–47.

Meggers, J., E. S. Ayensu, and W. D. Duckworth (eds.). 1973. *Tropical Forest Ecosystems in Africa and South America: A Comparative Review.* Washington, D.C.: Smithsonian Institution Press.

Menendez, C. A., and M. A. Caccavari de Filice. 1975. Las especies de *Nothofagidites* (polen fosil de *Nothofagus*) de sedimentos Terciarios y Cretácicos de Estancia La Sara, norte de Tierra del Fuego, Argentina. Ameghiniana 12:165–183.

Muller, J. 1966. Palynological investigations of Cretaceous sediments in northeastern Brazil. Pp. 123–136 *in* J. E. van Hinte (ed.), *Proceedings of the Second West African Micropaleon-*

tol. Colloquium, Ibadan, Nigeria. Leiden: E. J. Brill.

———. 1981. Fossil pollen records of extant angiosperms. Bot. Rev. 47:1–142.

Muller, J., E. Di Giacomo, and A. W. van Erve. 1987. *A Palynological Zonation for the Cretaceous Tertiary and Quaternary of Northern South America.* Contribution Series 19. College Station, Tex.: American Association of Stratigraphic Palynologists.

Palma Heldt, S. 1980. Nuevos antecedentes en el estudio palinológico de los mantos carboníferos del Terciario de Arauco-Concepción, Chile. Bol. I. G. Univ. São Paulo 11:161–168.

———. 1983. Estudio palinológico del Terciario sedimentario de Lonquimay, Provincia de Malleco, Chile. Rev. Geol. Chile 18:55–75.

Petri, S. 1987. Cretaceous paleogeographic maps of Brazil. Palaeogeogr. Palaeoclimatol. Palaeoecol. 59:117–168.

Petriella, B. T. P. 1972. Estudio de maderas petrificadas del Terciario inferior del área Central de Chubut (Co Bororó). Rev. Mus. La Plata, n.s., 6:159–254.

Petriella, B. T. P., and S. Archangelsky. 1975. Vegetación y ambientes en el Paleoceno de Chubut. Actas I Congr. Argent. Paleontol. Bioestrat. 2:257–270.

Quatrocchio, M. E., and W. Volkheimer. 1986. Microflora de los Estratos Limítrofes entre Cretácico y Terciario en las localidades de Tilián y Corralito, Cuenca del Grupo Salta. Descripciones Sistemáticas. Actas IV Congr. Argent. Paleontol. Bioestrat. 3:109–120.

Raven, P., and D. Axelrod. 1974. Angiosperm biogeography and past continental movements. Ann. Missouri Bot. Gard. 61:539–673.

Regali, M. 1987a. O genero *Afropolis* Doyle, Jardiné et Doerenkamp, 1982 e su distribuição no Cretaceo do Brasil. Anais X Congr. Brasil. Paleontol. 2: 605–635.

———. 1987b. Palinomorfos do Barremiano/Albiano brasileiros. Parte I. Anais X Congr. Brasil. Paleontol. 2: 637–645.

Regali, M., and S. Gonzaga. 1985. Palinocronoestratigrafia da bacia Potiguar-Rio Grande do Norte. Brasil. Ser. Geol. 27. Seção Paleontol. Estrat. 2:443–460.

Regali, M., and C. F. Viana. 1989. *Sedimentos do Neojurasico-Eocretaceo do Brasil.* Rio de Janeiro: Petrobras.

Regali, M., N. Uesugui, and A. Silva Santos. 1974a. Palinologia dos sedimentos meso-cenozoicos do Brasil, I. Bol. Tec. Petrobras 17:177–191.

———. 1974b. Palinologia dos sedimentos meso-cenozoicos do Brasil, II. Bol. Tec. Petrobras 17:263–301.

Riccardi, A. C. 1987. Cretaceous paleogeography of southern South America. Palaeogeogr. Palaeoclimatol. Palaeoecol. 59:169–195.

Romero, E. J. 1973. Polen fosil de "*Nothofagus*" ("*Nothofagidites*") del Cretácico y Paleoceno de Patagonia. Rev. Mus. La Plata 7:291–303.

———. 1977. *Polen de Gimnospermas y Fagáceas de la Formación Río Turbio (Eoceno) Santa Cruz, Argentina.* Buenos Aires: Fecic Editorial.

———. 1978. Paleoecología y paleofitogeografía de las Tafofloras del Cenofítico de Argentina y áreas vecinas. Ameghiniana 15:209–227.

———. 1986. Paleogene phytogeography and climatology of South America. Ann. Missouri Bot. Gard. 73:449–461.

Romero, E. J., and S. Archangelsky. 1986. Early Cretaceous angiosperm leaves from southern South America. Science 234:1580–1582.

Romero, E. J., and M. T. Castro. 1986. Material fúngico y granos de polen de angiospermas de la Formación Río Turbio (Eoceno), Provincia de Santa Cruz, República Argentina. Ameghiniana 23:101–118.

Romero, E. J., and J. Hickey. 1976. A fossil leaf of Akaniaceae from Paleocene beds in Argentina. Bull. Torrey Bot. Club 103:126–131.

Romero, E. J., and M. C. Zamaloa. 1985. Polen de angiospermas de la Formación Río Turbio (Eoceno), Provincia de Santa Cruz, República Argentina. Ameghiniana 22:43–51.

Romero, E. J., M. Dibbern, and M. A. Gandolfo. 1986. Revisión de *Lomatia bivascularis* (Berry) Frenguelli (Proteaceae) del Yacimiento de la Laguna del Hunco (Paleoceno) Pcia.

de Chubut. Actas IV Congr. Argent. Paleontol. Bioestrat. 3:125–130.

Schuler, M., and J. Doubinger. 1970. Observation palynologiques dans le bassin d'Amaga (Colombie). Pollen & Spores 12:427–451.

Sole de Porta, N. 1971. Algunos géneros nuevos de polen procedentes de la Formación Guaduas (Maastrichtiense-Paleoceno) de Colombia. Stvdia Geol. 2:133–143.

———. 1972a. Contribución a la palinología del nivel de Lutitas y Arenas (Campaniense) en Guataqui (Valle del Magdalena). Colombia. Rev. Esp. de Micropal. 4: 225–250.

———. 1972b. Palinología de la Formación Cimarrona (Maastrichtiense) en el Valle Medio del Magdalena, Colombia. Stvdia Geol. 4: 103–142.

Srivastava, S. K. 1978. Cretaceous spore-pollen floras: A global evaluation. Biol. Mem. 3:1–130.

———. 1981. Evolution of Upper Cretaceous phytogeoprovinces and their pollen flora. Rev. Paleobot. Palynol. 35: 155–173.

Stough, J. B. 1968. Palynomorphs from South America. Univ. Kansas, Paleontol. Contrib., Paper 32: 1–12.

Takahashi, K. 1977a. Palynology of the Lower Tertiary Concepcion Formation, Central Chile. Trans. Proc. Paleontol. Soc. Japan, n.s., 106:71–88.

———. 1977b. Upper Cretaceous palynoflora from Quiriquina Island, Chile. Bull. Fac. Liberal Arts, Nagasaki Univ. Nat. Sci. 17:29–53.

Takhtajan, A. 1986. *Floristic Regions of the World.* Berkeley, Calif.: University of California Press.

Thorne, R. F. 1973. Floristic relationships between tropical Africa and tropical America. Pp. 27–48 *in* B. J. Meggers, E. S. Ayensu, and W. D. Duckworth (eds.), *Tropical Forest Ecosystems in Africa and South America: A Comparative Review.* Washington, D.C.: Smithsonian Institution Press.

Troncoso, A., and E. Barrera. 1977. Estudio palinológico de tres Testigos del Pozo Río Blanco No. 1 (Osorno, Chile). Actas II Congr. Chileno, Arica 3:1–14.

———. 1980. Polen del Eoceno de Osorno (Chile). Bol. Mus. Nac. Hist. Nat. Chile 37:179–203.

Tschudy, R. H. 1970. Two new pollen genera (Late Cretaceous and Paleocene) with possible affinity to the Illiciaceae. Contributions to Paleontology, U.S. Geological Survey. Professional Paper 643-B:1–13.

Vakhrameev, V. A. 1988. Jurassic and Cretaceous floras and climates of the Earth (in Russian). Trans. Acad. Sci. USSR 430:1–214.

Vergel, M. M., and J. Durango de Cabrera. 1988. Palinología de la Formación Cullen (Terciario) de las inmediaciones de Canadón Beta, Tierra del Fuego, República Argentina. Actas V. Congr. Geol. Chileno 2:227–245.

Volkheimer, W., and A. Salas. 1975. Die ältesten Angiospermen-Palynoflora Argentiniens von der Typuslokalität der unterkretazischen Huitrin-Folge des Neuquen-Beckens: Mikrofloristische Assoziation und biostratigraphische Bedeutung. N. Jb. Geol. Paläont. Mh. 7: 424–436.

———. 1976. Estudio palinológico de la Formación Huitrin Cretácico de la cuenca Neuquina, en su localidad tipo. Actas VI Congr. Geol. Argent. 1:433–456.

Volkheimer, W., and E. Sepulveda. 1976. Biostratigraphische Bedeutung und mikrofloristische Assoziation von *Cyclusphaera psilata* n. sp., einer Leitform aus der Unterkreide des Neuquen-Beckens (Argentinien). N. Jb. Geol. Paläont. Mh. 2: 97–108.

Volkheimer, W., M. Quattrochio, A. Salas, and E. Sepulveda. 1976. Caracterización palinológica de formaciones del Jurásico superior y Cretácico inferior de la Cuenca Neuquina (República Argentina). Actas VI Congr. Geol. Argent. 1:593–608.

Wegener, A. 1924. *Die Klimate der geologischen Vorzeit.* Berlin: Borntraeger.

Willis, J. C. 1973. *A Dictionary of the Flowering Plants and Ferns.* New York: Cambridge University Press.

Wolfe, J. A. 1975. Some aspects of plant geography of the northern hemisphere during the Late Cretaceous and Tertiary. Ann. Missouri Bot. Gard. 62:264–279.

———. 1985. Distribution of major vegetational

types during the Tertiary. Pp. 357–375 *in The Carbon Cycle and Atmospheric CO_2: Natural Variations Archean to Present,* Geophysical Monogr. 32. Washington, D.C.: American Geophysical Union.

Zamaloa, M. C., and E. J. Romero. 1990. Some spores and pollen from the Cullen Formation (Upper Eocene to Middle Oligocene), Tierra del Fuego, Argentina. Palynology 14:123–133.

Zambrano, J. J. 1981. Distribución y evolución de las cuencas sedimentarias en el continente Sudamericano durante el Jurásico y el Cretácico. Pp. 9–44 *in* Comite Sudamericano del Jurásico y Cretácico (eds.), *Cuencas Sedimentarias del Jurásico y Cretácico de América del Sur,* vol. 1. Buenos Aires.

Chapter 5

Annonaceae: A Primitive Dicot Family with an Ancient Center in Africa–South America

George E. Schatz and

Annick Le Thomas

I. Introduction

"The classification of the Annonaceae into tribes is a difficult problem and I do not think there can ever be a satisfactory division in spite of some admirable attempts" (Sinclair 1955).

Undoubtedly all who work in the Annonaceae have at one time or another expressed sentiments similar to J. Sinclair's and despaired of ever achieving a satisfactory infrafamilial classification. Although the family is well circumscribed, the reticulate nature of character expression has confounded the elucidation of relations among the genera and has resulted in largely artificial groupings based on often ambiguous criteria, particularly aestivation and inflorescence position (Sinclair 1955; Fries 1959; Hutchinson 1964). Evolutionary trends in character advancement—for example, reduction, loss, or fusion of parts—do not necessarily constitute valid synapomorphies with which to reconstruct phylogeny. Character convergences are unfortunately rife within the family.

During the past two decades, however, scientists have begun to reevaluate the confusing patterns of macromorphological character distribution in light of new palynological (Walker 1971a; Le Thomas 1980–81) and karyological data (Morawetz 1984a, 1984b, 1986a, 1986b, 1988; Morawetz and

We wish to thank Professor Dr. Ph. Morat for his gracious hospitality at the Laboratoire de Phanérogamie of the Museum National d'Histoire Naturelle, Dr. P. Goldblatt for useful discussions on cytology, and Dr. J. Doyle for helpful suggestions on the chapter. Support for GES in Paris was provided by E.P.H.E., and in St. Louis by the W. Alton Jones Foundation.

 ISBN 0-300-05375-4

Waha 1985; Morawetz and Le Thomas 1988; Okada and Ueda 1984; Sauer and Ehrendorfer 1984), which have provided new means of establishing the primitive Annonaceous archetype and so have allowed a clearer assignment of polarites to character states. Various informal realignments have been proposed as a result (Walker 1971a; Hesse et al. 1985; Morawetz 1988). Many of the new data are derived from African and Neotropical taxa and bear directly on the question of relations between the two regions.

Although the Annonaceae are generally acknowledged to be among the most primitive of extant flowering plants, their age of origin remains unclear. The oldest pollen record positively assigned to the family (*Foveomorphomonocolpites*) dates from the Maastrichtian of Colombia (Sole de Porta 1971) and resembles modern-day *Malmea,* although the modern genus exhibits some advanced features. Because the separation of South America from Africa was complete by the uppermost Albian (Chapter 2), an affirmative answer to whether that separation constituted a vicariance event within already established lines of the family would require a significantly earlier origin for the Annonaceae. Given the probable poor preservation of early angiosperm pollen and the difficulties associated with its identification (Walker and Walker 1984), as well as the primitive nature of pollen of numerous extant Annonaceae, one could certainly postulate an early Cretaceous origin for the family, and thereby potentially attribute some modern phytogeographic patterns to the initial separation of West Gondwana. Although most present links between Africa and South America remain obscure, an analysis of possible relations reveals probable ancient sister-groups, as well as isolated clades, convergences, and parallel trends, all of which contribute to our understanding of infrafamilial classification. In addition, the distribution of primitive character states indicates a possible African–South American origin for the Annonaceae. Overall patterns of species diversity echo the relative impoverishment of the African tropical flora in general and point to more dynamic, recent taxogenesis as having occurred in South America.

II. The Delimitation of "Primitive Angiosperms" and the Position of Annonaceae

A. Magnoliaceae as the sister family? *Mischogyne* and *Toussaintia* in Africa

In a recent cladistic analysis aimed at resolving relations at the base of the angiosperm taxonomic tree (Donoghue and Doyle 1989), Annonaceae are grouped with six other families of the Magnoliales of Cronquist (1981)—Canellaceae, Degeneriaceae, Eupomatiacae, Himantandraceae, Magnoliaceae, and Myristicaceae, but excluding Austrobaileyaceae, Lactoridaceae, and Winteraceae—at the very base of the tree. These families differ from the rest of the angiosperms primarily by their retention of monosulcate pollen that lacks an endexine. Whereas Walker (1976a, 1976b) grouped these seven families into his Magnoliales, placing Annonaceae in a suborder with Canellaceae and Myristicaceae, the recent analysis presents Annonaceae and Magnoliaceae as possible sister groups—as did Dunal (1817), the first monographer of the family. Annonaceae and Magnoliaceae are united by the shared presence of benzylisoquinoline alkaloids and large pollen size (but see Ward et al. 1989 for a discussion of large pollen as a possible primitive state). The most recent review of sieve element plastid ultrastructure in the Magnoliiflorae (Behnke 1988) presents a confusing, nearly unresolvable distribution of plastid types among the seven families that lack endexine. Although some species of Magnoliaceae contain S-type plastids, other species, along with Degeneriaceae and all Annonaceae thus far surveyed, have P-type plastids with a single large protein crystal. Annonaceae exhibit more advanced wood anatomy than the other families that lack an endexine, with short vessels and nearly transverse end walls containing simple rather than scalariform perforations (van der Wijk and Canright 1956).

Annonaceae differ from Magnoliaceae principally in their lack of stipules, their possession of usually two distinct whorls of three petals each versus the spiral insertion of an indeterminate number of tepals, and their usually convex as opposed to an elongate receptacle. Several poorly known African Annonaceae are intriguing in their closer resemblance to Magnoliaceae, which are wholly absent from Africa today. One West African species of *Toussaintia, T. hallei* Le Thomas, exhibits up to 10 petals borne in two or three irregular whorls and may thus represent a transitional state between spiral and whorled petal insertion (Le Thomas 1969). Both *Toussaintia* and *Mischogyne,* a probably monotypic genus from Angola (*M. elliotianum* [Engl. & Diels] R. E. Fries is perhaps better retained in *Uvariastrum*), exhibit elongate receptacles, in effect androgynophores, with stamens inserted perpendicularly. The stamens of *M. michelioides* Exell are extremely aberrant within Annonaceae, consisting almost entirely of long linear anthers that lack the characteristic apical connective prolongation and closely resembling *Michelia* (Magnoliaceae) (Exell 1932). The pollen of *T. hallei* Le Thomas, borne in tetrads, is very unusual and is difficult to interpret in light of its highly reduced exine. The exine consists solely of a small number of foliations, sometimes just one, in direct contact with a thick intine (Le Thomas, 1980). The pollen of *Mischogyne michelioides* has not yet been examined with transmission election microscopy (TEM). Clearly, additional work is warranted on both of these enigmatic African genera.

B. The Original Angiosperm Pollen Type?

Among those basal angiosperm families that lack an endexine, only Annonaceae, Magnoliaceae, and Degeneriaceae include at least some members that exhibit what has been proposed as the original pollen type of the angiosperms—that is, boat-shaped, large- to medium-sized, noninterstitiate (homogeneous to weakly granular infratectal layer), and with a psilate to microperforate or foveolate tectum (Walker 1976a, 1976b; Walker and Walker 1984). Until recently, the earliest accepted Early Cretaceous angiosperm pollen forms possessed clearly advanced features, including a columellar interstitium, reticulate tectum, and, most often, endexine in the apertural region. However, an additional pollen type, *Lethomasites* (Ward et al. 1989), more closely resembling the hypothesized original angiosperm type (and the Magnolialean families above), has been described from the Early Aptian (Potomac Group) of Delaware and Virginia. This discovery lends support to Walker and Walker's (1984) assertion for the primacy of such pollen. Although Ward et al. (1989) hesitated to assign *Lethomasites* to any extant Magnolialean family, pollen types that share numerous features with *Lethomasites* and that can confidently be assumed to be the most primitive pollen types within the family are particularly well represented by a number of African and Neotropical genera of Annonaceae. Walker (1971a) united several of these genera, along with three Asian genera, into his "Fusaea tribe," and Hesse and co-workers (1985) have suggested possible inclusion of two additional African and Madagascan genera, but clearly it is dangerous to base a phylogeny on a single suite of plesiomorphic pollen characters. It is nevertheless useful to note the especially primitive pollen types in Africa and the Neotropics, among which at least one case of close intercontinental relationship seems probable.

C. Primitive Pollen Types in Africa and the Neotropics

The simplest pollen types within Annonaceae are particularly well represented in Africa, and perhaps the most primitive among them is the West African *Piptostigma* (Le Thomas 1980–81). Its solitary, monosulcate grains are strongly boat-shaped and exhibit an exine structure completely lacking inner basal layer organization, with only a homogeneous tectum penetrated by rare perforations overlaying a poorly differentiated granular interstitium. Chromosome data are lacking for *Piptostigma,* and the presumably advanced cauliflorous inflorescence positions, as well as the reduction in size of the outer petal whorl (a relatively rare state in the family, in contrast to the more frequent reduction of the inner whorl), render the affinities of *Piptostigma* uncertain. Other primitive pollen types in Africa include *Polyalthia stuhlmannii* (Engl.) Verdc. and *Lettowianthus stellatus* Diels, both from East Africa, and *Cleistopholis,* extending from West Africa to Uganda. These pollen types show the beginnings of inner layer organization in the form of lamellar foliations. The majority of *Polyalthia* species, apart from a sizable radiation on Madagascar (Schatz and Le Thomas 1990), occur in Asia, and all show more advanced pollen—either inaperturate or monosulcate, with well-developed columellae. Neither *Cleistopholis,* for which the rare base number $x = 7$ has been recorded (Mangenot and Mangenot 1962), nor *Lettowianthus,* one of several isolated endemic East African genera, for which chromosome data are lacking, possess characters that could unequivocally signify a link with other genera.

A more likely connection exists between the African genera *Meiocarpidium* and *Polyceratocarpus* and the mostly Neotropical genus *Anaxagorea,* three species of which are disjunct in Asia (Maas and Westra 1984–85). *Anaxagorea* is unique within Annonaceae in possessing a more or less dry, explosively dehiscent fruit containing two seeds. Although both *Meiocarpidium* and *Polyceratocarpus* possess indehiscent fruits with numerous seeds in two rows, their pollen exine structure is virtually identical to that of *Anaxagorea:* a massive continuous tectum overlaying a granular infratectal layer consisting of variable-diameter granules agglomerated into irregular shapes and fused in the innermost region to form a thick, nonfoliated basal layer. Furthermore, *Anaxagorea* and *Meiocarpidium* share the peculiarity of pollen grains that empty their cytoplasmic contents on contact with water—a result of the extrusion of the outer layer of a stratified intine (Hesse et al. 1985). *Meiocarpidium* and *Polyceratocarpus* differ primarily by the presence of peltate scales in the former, a condition approached by the stellate hairs of some *Anaxagorea* species. Both diploids and a hexaploid on the hypothesized original base number $x = 8$ are recorded for *Anaxagorea;* chromosome data for *Meiocarpidium* and *Polyceratocarpus* are still lacking.

III. Taxa Shared between Africa and South America

Of approximately 66 genera and 1244 species of Annonaceae present in Africa (including Madagascar) and the New World, only 2 genera (*Annona* and *Xylopia*) and a single species (*Annona glabra* L.) are shared in common (tables 5.1 and 5.2).

A. *Xylopia* and Related (and Perhaps Not-So-Related) Genera

Xylopia, the only pantropical genus of Annonaceae, is the largest genus in Africa (including Madagascar) and is comparably well represented in the Neotropics. The genus is rather homogeneous throughout its range and is characterized by a combination of both plesiomorphic and derived characters. *Xylopia* is among the relatively few genera of the family to have retained dehiscent fruits, and the numerous seeds have either a sarcotesta or an aril—all of which presumably are primitive features. Although pollen is shed in loose tetrads to polyads from locellate anthers, and the individual grains are

Table 5.1. African and Madagascan genera of Annonaceae and their approximate number of species.

Ambavia Le Thomas, 2
Annickia Van Setten & Maas, 10 (formerly *Enantia* Oliver)
Annona L. s. l., 8 (incl. *Anonidium* Engl. & Diels, 4)
Artabotrys R. Br., 31*
Asteranthe Engl. & Diels, 2
Cleistochlamys Oliver, 1
Cleistopholis Pierre ex Engl., 4
Dennettia Baker f., 1
Dielsiothamnus R. E. Fries, 1
Friesodielsia Steenis, 7*
Greenwayodendron Verdc., 2
Hexalobus A. DC., 4
Isolona Engl., 19
Letestudoxa Pellegrin, 2*
Lettowianthus Diels, 1
Meiocarpidium Engl. & Diels, 1
Mischogyne Exell, 1
Mkilua Verdc., 1
Monanthotaxis Baillon, 56* (incl. *Exellia* Boutique, 1; *Gilbertiella* Boutique, 1)
Monocyclanthus Keay, 1
Monodora Dunal, 15
Neostenanthera Exell, 5 (incl. *Boutiquea* Le Thomas, 1)
Ophrypetalum Diels, 1
Pachypodanthium Engl. & Diels, 3
Piptostigma Oliver, 14
Polyalthia Blume, 21
Polyceratocarpus Engl. & Diels, 7
Pseudartabotrys Pellegrin, 1*
Sphaerocoryne (Boerl.) Ridley, 5*
Toussaintia Boutique, 3*
Uvaria L., 69* (incl. *Afroguatteria* Boutique, 2; *Balonga* Le Thomas, 1)
Uvariastrum Engl., 9
Uvariodendron, R. E. Fries, 15
Uvariopsis Engl., 13
Xylopia L., 61

*Genus in which liana species occur.

Table 5.2. New World genera of Annonaceae and their approximate number of species.

Anaxagorea Martius, 21
Ancana F. Muell./*Guamia* Merr.?, 1
Annona L. s. l., 173* (incl. *Raimondia* Saff., 4; *Rollinia* A. St. Hil./*Rolliniopsis* Saff., 46)
Asimina Adans., 8
Bocagea A. St. Hil., 2
Bocageopsis R. E. Fries, 4
Cardiopetalum Schldl., 3 (incl. *Froesiodendron* R. E. Fries, 2)
Cremastosperma R. E. Fries, 19
Cymbopetalum Benth., 23
Deeringothamnus Small, 2
Desmopsis Saff., 18
Diclinanona Diels, 3
Duckeanthus R. E. Fries, 1
Duguetia A. St. Hil., 82
Ephedranthus S. Moore, 5
Fusaea (Baill.) Saff., 2
Guatteria Ruiz & Pavon, 279* (incl. *Guatteriella* R. E. Fries, 2; *Guatteriopsis* R. E. Fries, 6)
Heteropetalum Benth., 2
Hornschuchia Nees, 6
Malmea R. E. Fries, 18
Onychopetalum R. E. Fries, 4
Oxandra A. Rich., 30
Porcelia Ruiz & Pavon, 7
Pseudephedranthus Aristeg., 1
Pseudoxandra R. E. Fries, 10
Ruizodendron R. E. Fries, 1
Sapranthus Seemann, 7
Stenanona Standley, 10 (incl. *Reedrollinsia* J. W. Walker, 1)
Tetrameranthus R. E. Fries, 6
Tridimeris Baillon, 3
Trigynaea Schldl., 5
Unonopsis R. E. Fries, 40
Xylopia L., 51

*Genus in which liana species occur.

inaperturate with proximal reduction of exine, all presumably derived features, it retains primitive exine architecture. The pollen has a thick, smooth, continuous tectum and poorly differentiated granular interstitium and lacks lamellar foliations in the basal layer, which is formed by the fusion of basal granular elements. The overall homogeneity of *Xylopia* has discouraged attempts at infrageneric classification thus far, with the most recent enumeration of species strictly on a geographical basis (Fries 1959). It is noteworthy, however, that ovaries with both uniseriate and biseriate ovules, and hence monocarps with both a single row and two rows of seeds, exist throughout the range of the genus.

Two additional African genera, *Neostenanthera* and *Boutiquea,* are undoubtedly closely related to *Xylopia.* They share similar pollen, also borne in loose tetrads and with the exine almost entirely solid with only very small cavities (thus reflecting the further fusion of original granular elements), as well as overall floral morphological similarity, including the transversely septate thecae. They differ from *Xylopia* by virtue of reduction to a single ovule. In *Neostenanthera* this ultimately results in an indehiscent berry-like fruit, whereas in *Boutiquea* further specialization is manifested by sessile, contiguous monocarps giving the appearance of a syncarp (Le Thomas 1965; see also the discussion of syncarpy below). *Xylopia* has been found to be uniformly diploid—$2n = 16$ (Mangenot and Mangenot 1962; Morawetz 1984a, 1986a; Morawetz and Le Thomas 1988)—as has *Neostenanthera* (Mangenot and Mangenot 1962), lending additional support for a close relationship.

In the Neotropics, a possible relation of *Xylopia* to *Fusaea* and closely related *Duckeanthus* has been suggested based upon similar pollen borne in loose tetrads (Walker 1971a; Hesse et al. 1985) and, in the case of *Fusaea*, a chromosomal complement of $2n = 16$ (Morawetz 1986a). The anthers are non-locellate, however, and pollen ultrastructural data are still lacking for both genera. Although *Fusaea* exhibits syncarpy, the fruit is uniquely subtended by a collar, the expanded formerly male portion of the receptacle. In addition, *Fusaea* is the only genus of Annonaceae that has outer staminodes. The fruit of *Duckeanthus* is insufficiently known. In spite of superficial pollen similarities, the affinities of *Fusaea* and *Duckeanthus* remain unclear.

B. *Annona* sensu lato

The sole species shared by Africa and the Neotropics, *Annona glabra* L., is a wide-ranging coastal species known in the Neotropics from both Pacific and Atlantic coasts, from Florida to Brazil and Ecuador, and from Senegal to the Congo in West Africa. As currently circumscribed, the genus *Annona* is wholly Neotropical, with the exception of *A. glabra* and three African species. One of these, *A. senegalensis* Pers., occurs in Madagascar, although it may only have reached there since human colonization. Subgeneric classification places *A. glabra* as the sole member of section *Phelloxylon*, whereas the other three African species are assigned to subsection *Papillosae* of section *Helogenia*. Among those sections of *Annona* with six petals (*Annona, Macrantha, Ulocarpus, Campicola, Psammogenia, Phelloxylon*, and *Helogenia*), section *Phelloxylon* is defined by the valvate nature of the inner petal whorl, purportedly equal in size to the outer whorl (Fries 1959), and by carpels that exhibit greater fusion at anthesis (Safford 1914). In contrast, section *Helogenia*, containing both Neotropical and the three endemic African species, has an inner whorl of petals sufficiently reduced in size to render the assessment of valvate versus imbricate seemingly impossible. In fact, the size of the inner petals of *A. glabra* is variable and, moreover, usually distinctly smaller than the outer whorl, and the degree of fusion of the carpels is no more pronounced at anthesis than it is in other *Annona* species. Karyological investigations of 11 Brazilian species of *Annona* showed *A. glabra* to be tetraploid ($2n = 28$) with four distinct nucleolar organizing regions (NORS), suggesting that *A. glabra* may be a comparatively young allopolyploid (Morawetz 1984b). Given the probable ease of dispersal, one could postulate a more recent arrival of *A. glabra* in Africa and subsequent radiation of derivative inland taxa. Clearly, karyological examination of the three endemic savanna species of *Annona*, as well as African populations of *A. glabra*, would be desirable to test such a hypothesis.

Pollen of *Annona* exhibits several advanced features, including a foliated basal layer, well-developed columellae, and a reticulate tectum. It is most often borne in tetrads. Pollen of the African *Anonidium*, which like *Annona* has syncarpous fruits, conforms in all respects to that of *Annona*, and in particular, its enormous size is most similar to the primitive sections of *Annona*. The type of the genus, *Anonidium mannii* (Oliv.) Engl. & Diels, was in fact originally described as an *Annona;* the basis for its segregation as a distinct genus by Engler and Diels (1900) was its many-flowered, elongate inflorescence and the presence of both unisexual (staminate) and hermaphroditic flowers. Staminate flowers also occur in several species of *Annona* section *Annona,* however, and the *Anonidium* inflorescence, albeit elongate and pendulous, is the standard rhipidium encountered throughout the family. Similar broad amplitude in inflorescence length exists in the Neotropical genus *Duguetia,* in which the section *Geanthemum* includes species with flagelliflorous inflorescences, a further stage in specialization over the trunciflorous inflorescences of *Anonidium* (Fries 1959). Despite the absence of cytological data for any of the four currently recognized species of *Anonidium,* one can convincingly argue for their submersion into *Annona.*

Exceptions to tetrads in *Annona* are the solitary grains produced by the presumably advanced West Indian–Mexican species in sections *Annonula* and *Annonella.* The presence of both very loose tetrads and solitary grains in sections *Saxigena* and *Annonella* clearly indicates the secondarily derived nature of solitary pollen in the genus—a reversal that calls into question the distinctiveness of the related Neotropical genera *Rollinia* (with syncarpous fruits indistinguishable from *Annona*) and *Rolliniopsis* (Safford 1916). (*Rolliniopsis* exhibits the further reversal to apocarpous fruits; in a forthcoming monograph of *Rollinia, Rolliniopsis* will be submerged into *Rollinia* s.l., the secondarily derived apocarpous fruits apparently having evolved several times independently [Maas, personal communication, 1990].) Although grains are solitary, the pollen exine structure of *Rollinia* s.l. is identical to *Annona* monads. Furthermore, *Rollinia* exhibits a polyploid series similarly based on the rare base number $x = 7$, with the diploid karyotype identical to *Annona,* though with heterochromatin differences at the polyploid levels (Morawetz 1986b). *Rollinia* s.l. has laterally expanded, propeller-like outer petals that are partially fused at their base. When the complete range of variation in corolla morphology of species currently recognized as belonging to *Annona* is considered, it is not unreasonable, especially in light of the additional concordance in pollen ultrastructure and karyology, to accommodate *Rollinia* s.l. within *Annona.* Indeed, Fries (1943) suggested that *Rollinia* was derived from *Annona* section *Oligantha,* which exhibits similar fusion of inner and outer petal whorls, an identical opening through which pollinators gain access to the androecium/gynoecium, and the beginnings of lateral outer petal expansion into wings that are barely, if at all, surpassed by those in *Rollinia cardiantha* Diels and *R. sphaerantha* R. E. Fries.

In a similar spirit of avoidance of paraphyly, the Neotropical genus *Raimondia* should probably also be subsumed within an *Annona* s.l. Erected by Safford (1913) on the basis of unisexual flowers borne on short shoots and stamens that lack the expanded truncate anther connective of *Annona, Raimondia* comprises several mid-elevation Andean syncarpous species. In spite of Walker's (1971a) tentative assertion that *Raimondia* is "palynologically somewhat distinct within the [Annona] tribe, especially with regard to its prominent tectal perforations," pollen tetrads of *Raimondia* are by no means outside the range of an inclusive *Annona* s.l. Fries (1943) speculated that *Raimondia* could be derived from *Annona* section *Atta,* which is characterized by linear, triquetrous outer petals similar to those of *Raimondia* and includes several mid-elevation Andean species.

IV. Functional (and True) Syncarpy and Its Bearing on Relations among African and Neotropical Annonaceae

The overwhelming majority of species of Annonaceae have apocarpous fruit, with the individual carpels often termed monocarps. Monocarps are usually indehiscent and berry-like; the rare dehiscence of carpels is often correlated with a showy aril. Dehiscence is presumably a primitive trait, and indeed, as discussed above, does correlate in several cases with primitive pollen types (*Anaxagorea, Xylopia;* although in sharp contrast is the highly specialized pollen of *Cymbopetalum* [Walker 1971b]). Unequivocal trends in carpel specialization can be traced within genera or groups of genera, and include the reduction from biseriate to uniseriate ovules and from many ovules to a single ovule. A further specialization, nearly always associated with reduction to a single ovule, is the fusion of adjacent lateral carpel walls to form a syncarpous (or aggregate) fruit (figure 5.1).

The advantages and functional significance of syncarpy, including facilitation of ovule fertilization and seed dispersal, have been discussed by Carr and Carr (1961) and Endress (1982). Some workers (Le Thomas 1981; Deroin 1985) have adopted a terminology to distinguish between fruits resulting from the fusion of open carpels, forming a unilocular compound ovary with parietal placentation, as in *Isolona* and *Monodora* (Guédès and Le Thomas 1981; Deroin 1985)—which they refer to as *true syncarps*—and fruits resulting from the fusion of closed carpels along their lateral walls (which are therefore plurilocular, e.g., *Annona*)—which are designated *pseudosyncarps*. Insofar as there is a continuum (even within a genus, e.g., *Duguetia*) in the degree of fusion of adjacent carpel walls, from only slight fusion at their base (or, in several species of *Duguetia,* no fusion at all) to complete fusion, application of precise terminology is problematic. The existence of such a continuum provides justification for the merging of genera for which segregation has been based primarily on the presence or absence of carpel fusion, if indeed other characters are in close agreement. Analogous situations exist in Magnoliaceae and Winteraceae, where recent workers have expanded generic limits to accommodate the full range of variation from apocarpous to syncarpous taxa: *Magnolia* s.l. (Magnoliaceae), including *Alcimandra, Aromadendron,* and *Talauma* (Nooteboom 1985); and *Zygogynum* s.l. (Winteraceae), including *Bubbia* and *Exospermum* (Vink 1985). In some cases, monocarps of an apocarpous fruit, that is, one in which the individual carpels are free all the way to their base at the point of attachment to the receptacle, may nonetheless be so tightly contiguous as to give the appearance of syncarpy (e.g., *Boutiquea*). Indeed, in the context of dispersal, such a fruit is different from one with distinctly stalked individual carpels, and is instead functionally equivalent to one in which there is complete carpel fusion; it could, therefore, be termed a *functional syncarp.* Notwithstanding the fundamental difference

Figure 5.1. Syncarpous fruits of Annonaceae. A, *Annona muricata* L., market in Veracruz, Mexico; B, *Anonidium mannii* (Oliver) Engl. & Diels (*Halle & Le Thomas 109*), Belinga, Gabon; C, *Duguetia panamensis* Standley (*Schatz & Grayum 1113*), Gandoca, Costa Rica; D, *Pachypodanthium staudtii* (Engl. & Diels) Engl. & Diels (*Gentry 33310*), M'Passa Field Station, Gabon; E, *Tetrapetalum volubile* Miq. (*Schatz 1418*), Danum Valley Field Center, Sabah, Malaysia; F, *Isolona* sp. (*Miller et al. 3414*), Marojejy, Madagascar.

between syncarps derived from a unilocular gynoecium with parietal placentation and those resulting from a plurilocular gynoecium, a careful analysis of all those Annonaceae with functional syncarps is necessary to determine whether taxa sharing such fruits are actually closely related or merely examples of convergence under selection for probable dispersal by large vertebrates.

With that caveat in mind, an analysis of the distribution of functional syncarpy in Annonaceae reveals that it is most common in the Neotropics (over 250 species); considerably less so in Africa and Madagascar (about 50 species), where, however, the only unilocular syncarpy has evolved; and curiously, nearly absent from the Asian tropics (4 species). In rejecting the genus *Ararocarpus,* which was apparently based on an aberrant specimen of *Meiogyne virgata* with a greater than normal number of carpels, Sinclair (1958) proposed a mechanism whereby sessile, or nearly so, carpels might "fuse" upon expansion if space were limiting. Space can become limiting in conjunction with the contraction of the receptacle, as illustrated by the Schisandraceae, in which the principal difference between the genera *Schisandra* and *Kadsura* (treated by Dunal [1817] as a member of the Annonaceae by virtue of its aggregate fruit) is the elongate, cylindrical receptacle (torus) of *Schisandra* versus the short, obovate receptacle of *Kadsura* (Smith 1947). Carpels remain distinct in the fruit of *Schisandra,* whereas the contracted torus of *Kadsura* dictates that carpels be contiguous in fruit, resulting in a functional syncarp. In Asia, the functional syncarps of *Fissistigma lanuginosum* (Hook.f. & Thoms.) Merr., *Goniothalamus ridleyi* King from peninsular Malaysia, and *Uvaria sphenocarpa* Hook.f. & Thoms. from Sri Lanka are all relative oddities within their respective apocarpous genera. The fruit of *Tetrapetalum volubile* Miq., a climber perhaps related to *Uvaria,* was recently first collected from eastern Sabah on Borneo (*Schatz 1418*). It exhibits a syncarp resembling some Neotropical species of *Duguetia.* Thus, functional syncarps have evolved at least four times independently in Asia.

Similarly, in Africa and the Neotropics, functional syncarps occur in several independent lines. As discussed above, taxa placed in *Annona* s.l., including those with secondarily derived apocarpous fruits (*Rolliniopsis*), show additional corroborating characters of pollen exine structure and karyology to support the hypothesis of a common syncarpous ancestor for the group. *Uvaria scabrida* Oliv., with a wide distribution throughout West Africa from Liberia to Angola, represents a second independent origin of functional syncarpy within the genus. In a situation analogous to that of the anomalous Asian syncarps and *Uvaria scabrida* in Africa, similar floral morphology and virtually identical pollen characteristics argue for the submersion of *Boutiquea* into *Neostenanthera.* The syncarpous genus *Letestudoxa* (and possibly *Pseudartabotrys,* for which fruit is not yet known) has a pollen exine structure somewhat different from other syncarpous taxa in Africa and the Neotropics: inaperturate monads with a discontinuous granular tectum but a foliated basal layer. Its relations remain unclear, although its pollen is similar to that of Neotropical syncarpous *Duguetia,* differing principally in its development of the foliated basal layer. In addition, *Letestudoxa* lacks the stellate hairs and peltate scales of *Duguetia* and differs in its lianaceous habit. *Pachypodanthium* also shows a possible relation in Africa to *Duguetia. Pachypodanthium* possesses an extremely reduced exine consisting solely of isolated spinules embedded directly in the intine and is therefore difficult to relate to other inaperturate types, including *Duguetia,* which also exhibits considerable exine reduction. In addition to pollen exine structure differences, the two genera can be distinguished on the basis of wood anatomy, *Pachypodanthium* having few (2–5) vessels/mm^2 and *Duguetia* having numerous (20–40) vessels/mm^2 (van der Wijk and Canright 1956). The presence of stellate hairs (but no scales) in *Pachypodanthium,* as well as similarity in overall habit and aspect, lend support for a possible sister relationship with *Duguetia.* As stated in Section II.A. above, pollen characteristics, the presence of outer staminodes, and the unique basal collar seem to indicate an independent origin of syncarpy for Neotropical *Fusaea.*

The only cases in Annonaceae of syncarpy resulting from a unilocular compound ovary with parietal placentation (Deroin 1985; see Leins and Erbar 1980, 1982, for a dissenting opinion) occur in *Isolona* and *Monodora* in Africa, with *Isolona* also reaching Madagascar (Le Thomas 1988). Among the families most closely related to Annonaceae, only Canellaceae possess a similarly advanced gynoecium. Coincidentally, similar advanced syncarpy also occurs in Madagascar in Winteraceae, represented by *Takhtajania,* which has a unilocular bicarpellate ovary (Leroy 1977). Both *Isolona* and *Monodora* exhibit further specialization in floral structure (*Isolona* with the petals fused into a single whorl; *Monodora* with brightly colored, intricately marked petals), but neither possesses particularly advanced pollen. Le Thomas (1981) has suggested that *Isolona* shows a critical transitional stage between granular and columellar exine, and between monad and tetrad pollen, the tetrad stage of *Isolona* being prolonged with a concomitant reduction in the proximal exine. With only minor differences in ovary development and a shared chromosomal complement of $2n = 16$, it seems reasonable, in spite of differences in pollen and floral morphology, to assume a single origin of unilocular syncarpy for the two genera.

In summary, then, in addition to the four independent occurrences in Asia, functional syncarps have originated at least three (and possibly four) times exclusively in Africa (*Uvaria scabrida; Neostenanthera* (*Boutiquea*); *Isolona/Monodora;* and possibly *Letestudoxa*); once independently in the Neotropics (*Fusaea/Duckeanthus*); and twice more with the resulting taxa distributed in both the Neotropics and Africa (*Annona* s.l.; *Duguetia/Pachypodanthium,* also possibly including *Letestudoxa*).

V. African or Neotropical Endemic Taxa with Unclear Affinities or Alliances to Asian Taxa

A. Genera with a Base Number of $x = 7$

In addition to *Annona* s.l., four groups of African or Neotropical Annonaceae share the base number $x = 7$ and might, considering the infrequency of this number, be interrelated. Ehrendorfer and co-workers (1968) and Walker (1971a) initially suggested $x = 7$ as the base number for the family, but subsequent cytological work (summarized in Morawetz 1986a) has shown this number to be rare, occurring in only 6 of the more than

60 genera for which counts are available. The base numbers $x = 8$ and $x = 9$ are nearly equally represented among the remaining genera counted. Because several genera or clades with primitive features and pantropical distribution (*Xylopia, Anaxagorea/Meiocarpidium*) are based on $x = 8$, Morawetz (1986a) suggests that $x = 8$ is the original base number for the family.

Among the genera based on $x = 7$, the only Asian representative, *Mezettia,* is rather advanced in floral features—including reduction to a single carpel and only 9–12 stamens—and exhibits a karyotype peculiar by virtue of the small size of the satellite portion of the satellite chromosomes (Okada and Ueda 1984). The lack of any clear affinities for *Cleistopholis* has already been mentioned above. In Madagascar, *Ambavia* presents a number of unique features consistent with its possible long period of isolation, including uniquely shaped monosulcate pollen with extremely asymmetrical, poorly differentiated granular exine—the proximal half much reduced with a perforate tectum and a thick intine and the distal half containing the sulcus with thick smooth exine and thin intine beneath (Le Thomas 1980)—and carpels with open ventral slits (Morawetz and Le Thomas 1988).

In the Neotropics, *Tetrameranthus* exhibits both diploid ($2n = 14$) and tetraploid ($2n = 28$) species (Morawetz 1986b), and is well delimited by its exclusively spiral leaf arrangement and four-merous flowers (Westra 1985). Spiral phyllotaxy in the family is not uncommon on orthotropic shoots, and thus *Tetrameranthus* may illustrate the paedomorphic absence of normally distichous plagiotropic branches. Pollen morphology provides further proof of the isolated nature of the genus. The monads have a single rounded aperture (sulcus), a thick stratified intine, a thin lamellated inner exine layer, and thick outer homogenous layer with a smooth tectum, with exine globules occasionally present on the inner surface of the lamellae (Hesse and Waha 1984). Although *Tetrameranthus* shares some karyological features with the African genus *Uvariopsis,* including its continuous condensing behavior (Morawetz 1986b), both the chromosome number ($2n = 16$) and the pollen (tetrads with a foliated basal layer and incipient columellar interstitium) are quite different, and militate against a close relationship. The Neotropical genus *Guatteria* s.l., the largest genus in the family, is equally isolated from the rest of the family. All species of *Guatteria* thus far examined are $2n = 28$, with a karyotype indicative of a high degree of diploidization and rearrangement; it is therefore probably the sole example of a paleopolyploid in the Annonaceae (Morawetz and Waha 1985). Pollen morphology is likewise unusual, with a zonate condition that includes complex folding in the apertural zone and an extremely reduced exine.

B. Parallel Trends in Exine Development

Another group of genera, wholly Neotropical with the exception of the African genus *Annickia* (formerly *Enantia*), were placed together by Walker (1971a) in his Malmea tribe. They are characterized by solitary, monosulcate, boat-shaped pollen with a more or less well developed columellar exine. Subsequent ultrastructural studies (Waha 1985; Morawetz and Waha 1986) have shown that several genera exhibit a granular interstitium (*Malmea, Unonopsis*), another shows transitions to columellae (*Bocageopsis*), and still others have well-developed columellae (*Cremastosperma, Ephedranthus, Oxandra*). Such a series parallels similar exine specialization in African and Madagascan *Polyalthia* (Le Thomas and Lugardon 1976; Le Thomas 1980–81; Schatz and Le Thomas 1990) and in the African monosulcate genera (*Piptostigma, Cleistopholis, Cleistochlamys, Annickia*), and appears to be a fundamental path in the early evolution of angiosperm pollen (Walker and Skvarla 1975; Walker 1976b). Similar stages in the evolution of columellae from a granular interstitium have been documented in the primitive Madagascan genus *Mauloutchia* (Myristicaceae) (Walker and Walker 1981). Inasmuch as a columellar interstitium may have evolved from primitively granular types numerous times independently within the family, caution must be exercised in proposing relationships on the basis of advanced pollen features alone. Although Rogstad and Le Thomas (1989) considered several of the Neotropical genera with monosulcate, columellar pollen (*Oxandra, Pseudoxandra*) as possible outgroups for the Asian *Polyalthia hypoleuca* complex, they concluded that it is more reasonable to propose the Madagascan species of *Polyalthia* with such pollen (groups B and C of Schatz and Le Thomas 1990) as the most probable sister group.

C. The Mesoamerican-Asian Connection

Another direction in pollen evolution in the Annonaceae, the loss of the aperture to form inaperturate (and possibly related disulcate) monads varying in shape and intine distribution, is exhibited by *Polyalthia* in both Madagascar and Asia (Waha and Hesse 1988; Schatz and Le Thomas 1990) and by a group of endemic Australian genera (Waha and Morawetz 1988). In the Neotropics, a group of essentially endemic Mesoamerican genera (*Desmopsis, Sapranthus, Stenanona, Tridimeris*) exhibit identical disulcate and inaperturate forms (Schatz 1987; Waha and Hesse 1988) and furthermore agree in chromosome number ($2n = 18$ [Morawetz 1988; Schatz, unpublished data] with the exception of *Desmopsis* $2n = 16$ [Walker 1972]). This suggests a relictual Mesoamerican-Asian connection, informally designated by Morawetz (1988) as the *Sapranthus* tribe.

VI. Diversity and Habit

Reflecting similar patterns in overall plant diversity in the Neotropics as compared to Africa (Chapter 17), the Annonaceae comprise fewer than half as many species in Africa (including Madagascar, which accounts for over one quarter of the species in the region) as in the Neotropics, despite slightly greater generic diversity in Africa (tables 5.1 and 5.2). In groups where a definite close relation can be cited, only *Xylopia* shows comparable diversity in both regions; in other genera, the representatives in Africa are conspicuously depauperate (*Annona* s.l., *Pachypodanthium*). Karyological data seem to indicate more dynamic (recent) modes of speciation in Neotropical Annonaceae, as witnessed by polyploid series in *Anaxagorea; Annona* s.l.; *Cymbopetalum; Tetrameranthus;* and *Duguetia,* including polyploid races of *D. furfuracea* (St. Hil.) Benth. & Hook. (Morawetz 1984a, 1986a). In contrast, polyploidy is thus far virtually absent in Africa and Asia. Morawetz (1986a) has noted a correlation between polyploid species and races in extreme habitats (savannas) in South America and their diploid relatives occupying presumably more stable humid forests. In light of widespread past aridification of the African continent (Chapter 3), the savanna taxa of African Annonaceae should be surveyed for possible polyploid series.

Whereas in the Neotropics the bulk of the diversity is accounted for by endemic or nearly endemic elements (the isolated genus *Guatteria; Annona* s.l.), in Africa a significant proportion of the overall diversity is contributed by genera shared with Asia (*Artabotrys; Polyalthia, Uvaria,* and the group of poorly delimited genera that includes *Friesodielsia, Monanthotaxis,* and *Sphaerocoryne;* see tables 5.1 and 5.2). With the exception of *Polyalthia,* this Asian connection results in perhaps the most striking difference between the Neotropical and African (as well as Asian) Annonaceae—namely, the nearly complete absence of liana species in the Neotropics. Less than 1 percent of the Neotropical species are lianas (seven species of *Annona* and one *Guatteria*), whereas over 40 percent of the African-Madagascan species are lianas (tables 5.1 and 5.2), as are 32 percent of the peninsular Malaysian species (Sinclair 1955). Lianaceous genera in Africa and Asia exhibit more advanced pollen features, including a columellar interstitium (*Artabotrys*), inaperturate grains (*Uvaria, Letestudoxa, Pseudartabotrys*), or extremely reduced exine (*Friesodielsia, Monanthotaxis*) (Le Thomas 1980–81). The near absence of lianas in the Neotropics can perhaps best be attributed to chance historical factors, with the separation of South America from Africa preceding the appearance of the liana habit in ancestors of the above genera.

VII. The Breakup of West Gondwana: A Vicariance Event for Early Annonaceae?

The overwhelming conclusion from the above analysis of possible relations between extant African and Neotropical Annonaceae is one of strong independent diversification on each continent after separation. Nevertheless, several intercontinental relations appear to predate the breakup of West Gondwana. *Meiocarpidium/Polyceratocarpus* in Africa and *Anaxagorea* in the Neotropics and Asia appear to be vicariant members of an ancient lineage, which today retains pollen among the most primitive in the family. The occurrence of *Anaxagorea* in Asia but not in Africa raises the question of possible extinction in Africa. Although *Xylopia* also retains several primitive features and may well have been among those taxa present at the time of separation, it is not necessary to invoke vicariance to explain its current pantropical distribution. High vagility of arillate disseminules and similar patterns of variation throughout its range argue for more recent dispersion.

The other two lineages that appear to predate separation both exhibit the advanced feature of syncarpic fruits: *Annona* s.l. (represented initially only by *Anonidium* in Africa) and *Duguetia/Pachypodanthium.* Such simple syncarpy involving only the fusion of lateral carpel walls was therefore an early development in the family, if not in the angiosperms in general, as evidenced by the fossil fruiting axis *Lesqueria* (Crane and Dilcher 1984) from the upper Albian-middle Cenomanian, whose tight cluster of numerous dehiscent follicles could easily be envisaged as a proto-syncarp.

VIII. Conclusions: Is There a Phylogeny in the Future for Annonaceae?

Annonaceae, among the most primitive of the angiosperms, possess by far the most diverse range of pollen types, floral morphology, and habit, and thus provide special insight into the early evolution of flowering plants. Recent pollen ultrastructural and karyological studies have significantly improved our understanding of infrafamilial relationships and will undoubtedly continue to do so as more data accumulate. However, consistent with the probable age of the family, certain isolated taxa, several of which occur in South America and Africa, exhibit combinations of symplesiomorphic (for example, primitive pollen types) and autapomorphic (for example, base number of x = 7) character states that render the discovery of their affinities especially problematic. In such cases, new molecular techniques hold particular promise for providing additional shared derived character states upon which to base an internal phylogeny for the family. In the absence of molecular data, it is likely that the Annonaceae will continue stubbornly, as is also befitting the privileges accorded to age, to refuse to divulge their last ancient secrets.

Literature Cited

Behnke, H.-D. 1988. Sieve-elements plastids, phloem protein, and evolution of flowering plants. III, Magnoliidae. Taxon 37:699–732.

Carr, S. G. M., and D. J. Carr. 1961. The functional significance of syncarpy. Phytomorphology 11:249–256.

Crane, P. R., and D. L. Dilcher. 1984. Lesqueria: An early angiosperm fruiting axis from the mid-Cretaceous. Ann. Missouri Bot. Gard. 71:384–402.

Cronquist, A. 1981. *An Integrated System of Classification of Flowering Plants*. New York: Columbia University Press.

Deroin, T. 1985. Contribution à la morphologie comparée du gynécée des Annonaceae-Monodoroideae. Bull. Mus. Natn. Hist. Nat., Paris, 4th sér., sect. B, Adansonia 7:167–176.

Donoghue, M. J., and J. A. Doyle. 1989. Phylogenetic analysis of angiosperms and the relationships of Hamamelidae. Pp. 17–45 *in* P. R. Crane, and S. Blackmore (eds.), *Evolution, Systematics, and Fossil History of the Hamamelidae*. Vol. 1. Oxford: Clarendon.

Dunal, M. F. 1817. *Monographie de la famille des Anonacées*. Paris: Treuttel and Wurtz.

Ehrendorfer, F., F. Krendl, E. Habeler, and W. Sauer. 1968. Chromosome numbers and evolution in primitive Angiosperms. Taxon 17:337–353.

Endress, P. K. 1982. Syncarpy and alternative modes of escaping disadvantages of apocarpy in primitive Angiosperms. Taxon 31:48–52.

Engler, A., and L. Diels. 1900. I. Übersicht über die bekannten Gattungen der Annonaceen und Beschreibung einiger neuen Gattungen dieser Familie aus dem tropischen Afrika. Notizbl. Bot. Gart. Berlin-Dahlem 3(23):45–59.

Exell, A. W. 1932. Mr. John Gossweiler's plants from Angola and Portuguese Congo. J. Bot. 70 (Suppl.):213–214.

Fries, R. E. 1943. Einige Gesichtspunkte zur systematischen Gruppierung der amerikanischen Annonaceen-Gattungen. Ark. Bot. 30A(8):1–31.

———. 1959. Annonaceae. Pp. 1–171 *in* A. Engler and K. Prantl, *Die Natürlichen Pflanzenfamilien,* 2d ed., vol. 17 a II. Berlin: Duncker and Humblot.

Guédès, M. and A. Le Thomas. 1981. Le gynécée syncarpe de Monodora (Annonacées-Monodoroidées). Compt. Rend. Hebdomadaires Séanc. Acad. Sci., sér. 3, 292:1025–1028.

Hesse, M., W. Morawetz, and F. Ehrendorfer. 1985. Pollen ultrastructure and systematic affinities of *Anaxagorea* (Annonaceae). Pl. Syst. Evol. 148:253–285.

Hesse, M., and M. Waha. 1984. Sporoderm characters of *Tetrameranthus duckei* (Annonaceae) and their systematic implications. Pl. Syst. Evol. 147:323–326.

Hutchinson, J. 1964. *The Genera of Flowering Plants: Dicotyledones.* Oxford: Clarendon Press.

Leins, P., and C. Erbar. 1980. Zur Entwicklung der Blüten von *Monodora crispata* (Annonaceae). Beitr. Biol. Pflanzen 55:11–22.

———. 1982. Das monokarpellate Gynoeceum von *Monodora crispata* (Annonaceae). Beitr. Biol. Pflanzen 57:1–13.

Leroy, J.-F. 1977. A compound ovary with open carpels in Winteraceae (Magnoliales): Evolutionary implications. Science 196:977–978.

Le Thomas, A. 1965. Un nouveau genre africain d'Annonacées, *Boutiquea* Le Thomas. Adansonia, n.s., 5:531–535.

———. 1969. Annonaceae. Pp. 1–371 *in* A. Aubreville (ed.), *Flore du Gabon,* vol. 16. Paris: Museum National d'Histoire Naturelle.

———. 1980–81. Ultrastructural characters of the pollen grains of African Annonaceae and their significance for the phylogeny of primitive Angiosperms. Part 1, Pollen & Spores 22:267–342; Part 2, Pollen & Spores 23:5–36.

———. 1988. Les structures reproductives des Magnoliales africaines et malgaches: significations phylogéniques. Monogr. Syst. Bot. Missouri Bot. Gard. 25:161–174.

Le Thomas, A., and B. Lugardon. 1976. De la structure grenue à la structure columellaire dans le pollen des Annonacées. Adansonia, n.s., 15:543–572.

Le Thomas, A., W. Morawetz, and M. Waha. 1986. Pollen of paleo- and neotropical Annonaceae: Definition of the aperture by morphological and functional characters. Pp. 375–388 *in* S. Blackmore and I. K. Ferguson (eds.), *Pollen and Spores: Form and Function.* London: Academic.

Maas, P. J. M., and L. Y. Th. Westra. 1984–85. Studies in Annonaceae. II, A monograph of the genus *Anaxagorea.* A. St. Hil. Bot. Jahrb. Syst. Part 1, 105:73–134; Part 2, 105:145–204.

Mangenot, S., and G. Mangenot. 1962. Enquête sur les nombres chromosomiques dans une collection d'espèces tropicales. Rev. Cytol. Biol. Veg. 25:411–447.

Morawetz, W. 1984a. Karyological races and ecology of the Brazilian *Duguetia furfuracea* as compared with *Xylopia aromatica.* Flora 175:195–209.

———. 1984b. Karyologie, Ökologie und Evolution der Gattung *Annona* (Annonaceae) in Pernambuco, Brasilien. Flora 175:435–447.

———. 1986a. Remarks on karyological differentiation patterns in tropical woody plants. Pl. Syst. Evol. 152:49–100.

———. 1986b. Systematics and karyoevolution in Magnoliidae: *Tetrameranthus* as compared with other Annonaceae genera of the same chromosome number. Pl. Syst. Evol. 154:147–173.

———. 1988. Karyosystematics and evolution of Australian Annonaceae as compared with Eupomatiaceae, Himantandraceae, and Austrobaileyaceae. Pl. Syst. Evol. 159:49–79.

Morawetz, W., and A. Le Thomas. 1988. Karyology and systematics of the genus *Ambavia* and other Annonaceae from Madagascar. Pl. Syst. Evol. 158:155–160.

Morawetz, W., and M. Waha. 1985. A new pollen type, C-banded and fluorochrome counterstained chromosomes, and evolution in *Guatteria* and related genera. Pl. Syst. Evol. 150:119–141.

———. 1986. Pollen-Ultrastruktur und Systematik bei *Cremastosperma* und *Oxandra* (Annonaceae). Sitzungsber. Osterr. Akad. Wiss., Math.-Naturw. Kl., Abt. I, 195:309–314.

Nooteboom, H. P. 1985. Notes on Magnoliaceae with a revision of *Pachylarnax* and *Elmerillia* and the Malesian species of *Manglieta* and *Michelia.* Blumea 31:65–121.

Okada, H., and K. Ueda. 1984. Cytotaxonomical studies on Asian Annonaceae. Pl. Syst. Evol. 144:165–177.

Rogstad, S. H., and A. Le Thomas. 1989. Pollen characters of the *Polyalthia hypoleuca* complex (Annonaceae): Their significance in establishing monophyly and candidate outgroups. Bull. Mus. Natl. Hist. Nat., Paris, 4th sér., sect. B, Adansonia 11:257–278.

Safford, W. E. 1913. *Raimondia,* a new genus of Annonaceae from Colombia. Contr. U.S. Natl. Herb. 16(5): 217–219.

———. 1914. Classification of the genus *Annona* with descriptions of new and imperfectly known species. Contr. U.S. Natl. Herb. 18:i–ix, 1–68.

———. 1916. *Rolliniopsis,* a new genus of Annonaceae from Brazil. J. Wash. Acad. Sci. 6:197–204.

Sauer, W., and F. Ehrendorfer. 1984. Notes on the karyosystematics of Annonaceae. Pl. Syst. Evol. 146:47–55.

Schatz, G. E. 1987. "Systematic and ecological studies of Central American Annonaceae." Ph.D. diss., University of Wisconsin.

Schatz, G. E., and A. Le Thomas. 1990. The genus *Polyalthia* Blume (Annonaceae) in Madagascar. Bull. Mus. Natl. Hist. Nat., Paris, 4th sér., sect. B, Adansonia 12:113–130.

Sinclair, J. 1955. A revision of the Malayan Annonaceae. Gard. Bull. Straits Settlem., ser. 3, 14:149–516.

———. 1958. *Ararocarpus:* A monstrosity. Gard. Bull. Straits Settlem., ser. 3, 17:93–95.

Smith, A. C. 1947. The families Illiciaceae and Schisandraceae. Sargentia 7:1–224.

Sole de Porta, N. 1971. Algunos géneros nuevos de polen procedentes de la Formación Guadas (Maastrichtiense-Paleoceno) de Colombia. Stvdia Geol. (Salamanca) 2:141–142.

Vink, W. 1985. The Winteraceae of the Old World. V. *Exospermum* links *Bubbia* to *Zygogynum.* Blumea 31:39–55.

Waha, M. 1985. Ultrastruktur und systematische Bedeutung des Pollens bei *Bocageopsis, Ephedranthus, Malmea,* und *Unonopsis* (Annonaceae). Pl. Syst. Evol. 150:165–177.

Waha, M., and M. Hesse. 1988. Aperture types within *Sapranthus* and *Polyalthia* (Annonaceae). Pl. Syst. Evol. 161:135–146.

Waha, M., and W. Morawetz. 1988. Pollen evolution and systematics in Annonaceae with special reference to the disulcate Australian endemic genera. Pl. Syst. Evol. 161:1–12.

Walker, J. W. 1971a. Pollen morphology, phytogeography, and phylogeny of the Annonaceae. Contr. Gray Herb. 202:1–132.

———. 1971b. Unique type of Angiosperm pollen from the family Annonaceae. Science 172:565–567.

———. 1972. Chromosome number, phylogeny, phytogeography of the Annonaceae and their bearing on the (original) basic chromosome number of Angiosperms. Taxon 21:57–65.

———. 1976a. Comparative pollen morphology and phylogeny of the Ranalean complex. Pp. 241–299 *in* C. B. Beck (ed.), *Origin and Early Evolution of Angiosperms.* New York: Columbia University Press.

———. 1976b. Evolutionary significance of the exine in the pollen of primitive Angiosperms. Pp. 251–308 *in* I. K. Ferguson and J. Muller (eds.), *The Evolutionary Significance of the Exine.* London: Academic.

Walker, J. W., and J. J. Skvarla. 1975. Primitively columellaless pollen: A new concept in the evolutionary morphology of Angiosperms. Science 187:445–447.

Walker, J. W., and A. G. Walker. 1981. Comparative pollen morphology of the Madagascan genera of Myristicaceae (*Mauloutchia, Brochoneura,* and *Haematodendron*). Grana 20:1–17.

———. 1984. Ultrastructure of Lower Cretaceous Angiosperm pollen and the origin and early evolution of flowering plants. Ann. Missouri Bot. Gard. 71:464–521.

Ward, J. V., J. A. Doyle, and C. L. Hotton. 1989. Probable granular Magnoliid Angiosperm pollen from the early Cretaceous. Pollen & Spores 31:113–132.

Westra, L. Y. Th. 1985. Studies in Annonaceae. IV, A taxonomic revision of *Tetrameranthus* R. E. Fries. Proc. Kon. Ned. Akad. Wetensch., ser. C, 88:449–482.

Wijk, R. W. van der, and J. E. Canright. 1956. The anatomy and relationships of the Annonaceae. Trop. Woods 104:1–24.

Chapter 6

Intercontinental Relationships of African and South American Asteraceae: A Cladistic Biogeographic Analysis

Kåre Bremer

I. Introduction

The transatlantic distribution of numerous plant groups is frequently explained by the separation of Africa and South America from their former union in Gondwanaland. Turner (1977) invoked continental drift as an explanation for the distribution of the flowering plant family Asteraceae. Yet, other authors have argued that although the distribution of older, so-called primitive flowering plant families may be the result of continental drift (e.g., Schuster 1976), the distribution of more advanced and presumably younger families such as the Asteraceae did not result from the breakup of Gondwanaland (Raven and Axelrod 1974). The results of the study discussed in this chapter, which examined not only the presumed age of the family but also a cladistic biogeographic analysis of distribution patterns, lead me to the latter conclusion.

The study concerns other continents in addition to Africa and South America. The mere occurrence of the same plant group in Africa and South America is no evidence for a close biogeographic connection between the two areas. In this case, biogeographic analyses are analogous to phylogenetic analyses. Two organisms, or areas, are always more or less related, but in order to show that they are each other's closest relatives, they must be compared with other organisms, or areas, respectively. There is no point in discussing the relation of two areas, like Africa

I thank A. Anderberg, B. Bremer, J. M. Carpenter, C. J. Humphries, C. Jeffrey, P. O. Karis, and R. D. M. Page for useful suggestions. This study was supported by a travel grant from the Missouri Botanical Garden and a Swedish Natural Science Research Council grant NFR 2401-108.

 ISBN 0-300-05375-4

and South America, without considering other areas, because all areas, or continents, are related (Nelson and Platnick 1981).

In this vein, I have analyzed Asteraceae distributions not only between Africa and South America but also among these continents and the northern hemisphere, tropical Southeast Asia, and Australia. I have done this in two steps. First, I have surveyed intercontinental sister group relations. Second, I have undertaken a number of cladistic biogeographic analyses of Asteraceae groups with intercontinental distribution patterns. Before I turn to the subjects of intercontinental sister group relationships and area cladograms, however, I shall elaborate on the spatial origin of the Asteraceae, as implied from recent cladistic analyses of the basal phylogeny of the family.

II. Origin of the Asteraceae

Recent work on the phylogeny of the Asteraceae has shown that there is a basal dichotomy between the subfamily Barnadesioideae and the rest of the family (Jansen and Palmer 1987; Bremer 1987). The Barnadesioideae comprise the one tribe Barnadesieae with nine genera and 89 species in South America, mainly distributed along the Andes. The remaining, major part of the Asteraceae is currently divided into two large subfamilies, the Cichorioideae and the Asteroideae. The subfamily Cichorioideae is a provisional, paraphyletic taxon; in addition to a number of monophyletic tribes it comprises a paraphyletic complex of largely plesiomorphic genera of the tribe Mutisieae.

In order to explore the early evolution of the Asteraceae, Karis and co-workers (1992) analyzed morphological data from the Mutisieae, as well as representatives of other tribes. The Barnadesioideae were used as an outgroup. The Karis group's analysis suggests that the early branches of the Asteraceae phylogeny comprise a number of generally plesiomorphic Mutisieae-Gochnatiinae genera representing successive sister groups to the rest (figure 6.1).

Among the "archaic," relict genera of the basal complex are *Stenopadus* (figure 6.2) and its immediate relatives (not included in the cladogram) from the Guyana Highlands in Venezuela, *Hesperomannia* from the Ha-

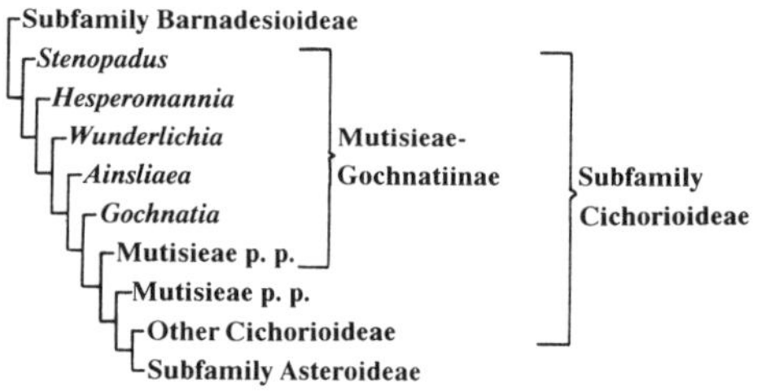

Figure 6.1. Basal part of Asteraceae cladogram from Karis et al. (1992).

Figure 6.2. *Stenopadus connellii* (N. E. Br.) Blake, an archaic, relict species of the Asteraceae from Mt. Roraima, Guyana Highlands, Venezuela. Photograph by J. Steyermark.

waiian Islands, *Wunderlichia* from Brazil, *Ainsliaea* from Eastern Asia, and *Gochnatia* from throughout South America and part of Southeast Asia, where a few species occur. Most of these genera are shrubs or trees with entire alternate leaves and large discoid capitula. Many Barnadesioideae are similar, although that subfamily is in itself highly variable and apomorphic in several features.

A. A Center of Origin?

The distributions of the first branches of the Asteraceae phylogeny are plotted in figure 6.3. Some of the nodes that are resolved in figure 6.1 are united into a tetratomy in figure 6.3, since these nodes are vaguely supported and the exact positions of the genera involved are unclear (Karis et al. 1992). Nevertheless, if the cladogram by Karis and co-workers is correct in its main features, the pattern shown in figure 6.3 suggests the spatial origin of the Asteraceae.

The origin of the Asteraceae is sometimes pinpointed to northern South America or the northern Andes (Raven and Axelrod 1974; Turner 1977). The distribution of the smaller sister group in the basal dichotomy of the family, the Barnadesioideae, along the Andes has been considered to support this hypothesis (Jansen et al. 1991). However, the Barnadesioideae are actually distributed throughout most of South America, excluding the Amazon Basin. They are not restricted to northern South America or the northern Andes. In agreement with the general criticism of the center of origin concept, I find that there is no support for such a narrow center of origin for the Asteraceae (Croizat et al. 1974; Humphries and Parenti 1986).

Under a strict vicariance model, nothing can be said about the spatial origin of the Asteraceae; the sister group to the South American Barnadesioideae is worldwide in

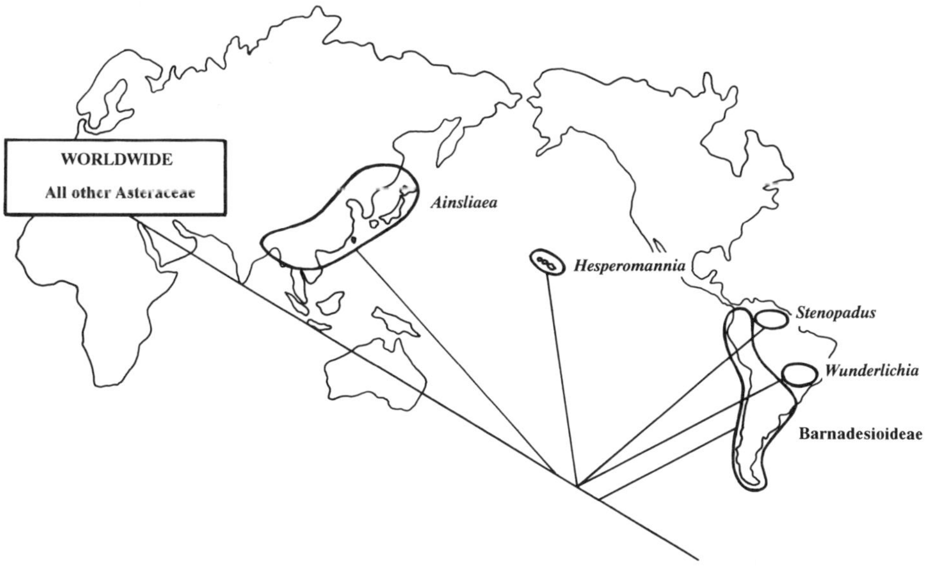

Figure 6.3. The basal branches in the Asteraceae phylogeny, as suggested in Karis et al. (1992).

distribution, and under this model there is no need to assume a more restricted distribution for the ancestral species of the family. However, considering that species are normally not cosmopolitan, it may be justified to assume that the ancestral species were more restricted in distribution and that the Asteraceae must have attained their worldwide presence by dispersal from a more restricted area. The basal phylogeny of the family can be resolved into a series of successive sister groups with a Pacific distribution (figure 6.3). Hence, it is possible that the ancestors of the Asteraceae and the first members of the family were part of a Pacific distribution pattern involving South America but excluding Africa.

Plate tectonics of the Pacific area are complicated, involving a number of possible rafts traversing the Pacific and ancient, slowly migrating island areas like the Hawaiian Islands (Nur and Ben-Avraham 1981; Humphries and Parenti 1986). The changing geography of the Pacific may have influenced the early distribution of the Asteraceae, but the initial separation of the major continents of Gondwanaland was a much earlier event, presumably too early to have affected the distribution of this family. Raven and Axelrod (1974) assumed no more than a mid-Oligocene age for the family and stated that the migrations of Asteraceae must be seen "in the light of present geography." They consulted the fossil record in estimating the age of the family. Reliable Asteraceae fossils are restricted to pollen, which becomes abundant in the Oligocene (Muller 1981), although there are scattered earlier records (Turner 1977; Chapter 4, this volume). Contrary to Raven and Axelrod, Turner (1977) argued that the family originated in the mid-Cretaceous—early enough for continental drift to promote its distribution around the world, a distribution attained "as a result of plate tectonics." Turner correctly observed that the microfossil record does not tell us that the family originated or differentiated during the Oligocene; the Asteraceae may very well be much older. This does not necessarily mean, however, that the Asteraceae originated before the breakup of Gondwanaland, as Turner assumed. The analysis of distribution patterns presented below does not support that scenario.

III. Intercontinental Sister Group Relationships

Croizat (1958, 1964) developed historical biogeography by introducing the concepts of tracks and generalized tracks. He connected disjunctions with lines, "tracks," combining supposedly related areas. Repeatedly occurring tracks were then summarized into generalized tracks, indicating area relationships. A similar approach has been adopted here in order to explore intercontinental links not simply by counting shared taxa (e.g., Good 1974) but by identifying intercontinental sister group relationships within the Asteraceae. By surveying distributions and relationships as far as they are known within the Asteraceae family—if necessary down to the generic or the species level—African Asteraceae can be linked with sister groups outside Africa (appendix: table 6.1) and South American Asteraceae with sister groups outside South America (appendix: table 6.2). There may be errors and omissions, though presumably not to the extent that they should affect the general conclusions drawn below.

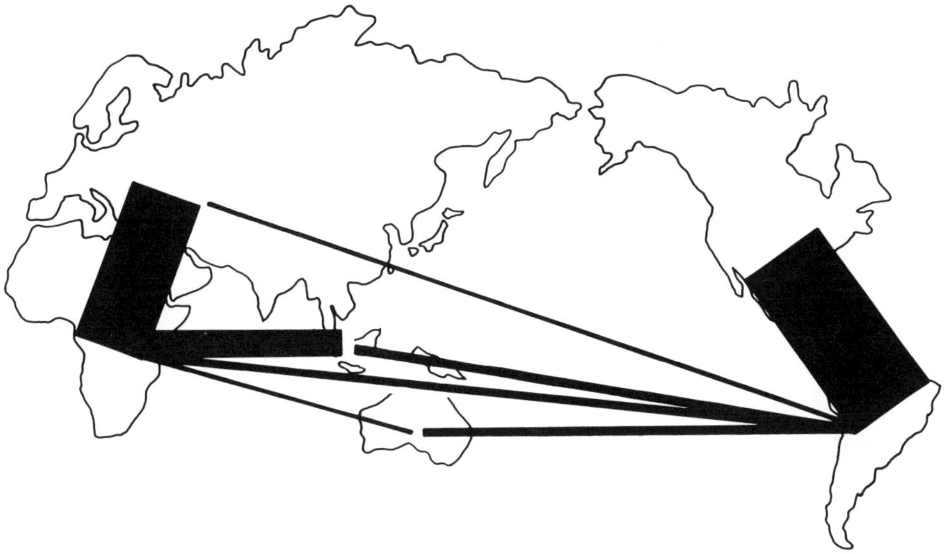

Figure 6.4. Intercontinental sister-group relationships, summarized from tables 6.1 and 6.2 in the appendix. The thickness of the bars corresponds to the number of relationships included. The thick bar between Africa and Eurasia includes sister-group relationships to taxa distributed in Eurasia and those in Eurasia plus North America, and the thick bar between South America and North America includes sister-group relationships to taxa in North America and those in North America plus Eurasia. A few sister-group relationships involving India (*Baccharoides, Blepharispermum*) and northern tropical Australia (*Coleocoma*) were omitted because they could not unequivocally be assigned to any of the six major areas. Ambiguous area relationships were also omitted.

The proportions of identified intercontinental sister-group relationships in the Asteraceae are shown in figure 6.4. In this figure, the areas have been simplified to six major continental regions: (1) Africa excluding the North African–Mediterranean region, (2) Eurasia including the North African–Mediterranean region, (3) tropical Southeast Asia, (4) Australia, (5) South America, and (6) North America. Taxa distributed throughout tropical Asia—that is, in Southeast Asia as well as in South India or Ceylon—were grouped with taxa restricted to Southeast Asia; the additional distribution in India or Ceylon is irrelevant here. The aim is to estimate the amount of intercontinental biogeographic relationships within the Asteraceae.

In this comparison, numerous sister group relationships to the northern hemisphere and a large number between Africa and Southeast Asia become apparent. The others seem to represent mere exceptions. All tribes except the exclusively South American Barnadesieae and Liabeae contribute to the southern-northern hemisphere sister-group relationships. Thirty-three relationships between Africa and the northern hemisphere and 40 between South America and the northern hemisphere are evident. In Africa, the tribes Cardueae, Lactuceae, Inuleae, Gnaphalieae, and Anthemideae contribute most of those relationships. There are connections of southern and tropical Africa to the North African–Mediterranean region, to Arabia and Southwest Asia, and to Eurasia or the northern hemisphere in general. There are no clear differences among these patterns; together they suggest a strong biogeographic connection between Africa and Eurasia in the Asteraceae. In South America, the majority of the northern hemisphere relationships come from the Astereae, Senecioneae, Helenieae, Tageteae, and Heliantheae. Most involve connections between South America and the Mexican–southwestern United States area, but there is a continuous variation to more general South American–North American or South American–northern hemisphere relationships. The repeated patterns suggest a strong biogeographic connection between South America and North America in the Asteraceae.

The only other strong intercontinental biogeographic link in the Asteraceae—12 sister-group relationships—is between Africa and Southeast Asia. These involve links in the paleotropical Asteraceae genera, particularly from the Vernonieae and Astereae, but also from the Plucheeae, Inuleae, Senecioneae, and Heliantheae.

There are seven African–South American sister group relationships. Of these, four must be regarded as dubious, whereas others may represent isolated cases of weedy dispersal. There is, therefore, no strong biogeographic connection between Africa and South America in the Asteraceae. The same conclusion applies to the other intercontinental tracks shown in figure 6.4. They are supported by 2–4 sister-group relationships, representing isolated cases: Africa–Australia, 2; South America–Eurasia, 2; South America–Southeast Asia, 4; and South America–Australia, 4 sister-group relationships. The South American–Australian relationships include *Leptinella* of the Anthemideae-Matricariinae, which are part of the cladistic biogeographic analyses presented below. The other cases are commented upon in the appendix.

When considered together, the South American–Southeast Asian and the South American–Australian links all seem to be part of a more general Pacific distribution pattern within the Asteraceae, exemplified by the Blennospermatinae (Senecioneae), the *Cotula* group (Anthemideae-Matricariinae), *Centipeda, Lagenifera/Myriactis,* and *Podocoma/Ixiochlamys* (Astereae). This pattern would furnish an interesting case for further biogeographic analysis.

A. African–South American Sister-Group Relationships

The putative African–South American sister group relationships include (1) *Gerbera/Chaptalia* of the Mutisieae, (2, 3) *Pseudognaphalium* and *Achyrocline* of the Gnaphalieae, (4) the subtribe Senecioninae of the Senecioneae, (5, 6) *Aspilia* and *Acmella* of the Heliantheae, and (7) *Stomatanthes* of the Eupatorieae.

Hansen (1990) presented a cladistic analysis of the *Gerbera* group (Mutisieae). Several equally parsimonious solutions were suggested, but in all solutions the South American genus *Chaptalia* came out as the sister group to part of the Old World genus *Gerbera,* that is, the three sections *Piloselloides, Lasiopus,* and *Pseudoseris*. Sections *Piloselloides* and *Lasiopus* are almost entirely African, except for one widespread species, *G. piloselloides,* which occurs in Africa as well as in Asia. Section *Pseudoseris* is restricted to Madagascar. Although Hansen's cladogram is based on few characters, from it we see that the South American *Chaptalia* seems to be nested within the paraphyletic, mainly African genus *Gerbera.* Hence, *Gerbera/Chaptalia* may provide a real sister-group relationship between Africa and South America.

Pseudognaphalium and *Achyrocline* are two genera closely related to *Helichrysum* of the tribe Gnaphalieae (Anderberg 1990). Both have species in Africa as well as in South America. This suggests that *Pseudognaphalium* and *Achyrocline* contain African–South American sister-group relationships among their species. However, *Pseudognaphalium* and *Achyrocline* were stated to be vaguely delimited from the widespread genus *Helichrysum,* which is also distributed in Eurasia. The three genera may be conceived as one monophyletic group within which relationships are poorly understood (Anderberg 1990). Because this complex is widely distributed not only in Africa and South America but also in Eurasia, African–South American sister-group relationships cannot yet be positively demonstrated for *Pseudognaphalium* and *Achyrocline*.

The subtribe Senecioninae is considered here because it is overwhelmingly concentrated in Africa and South America. The largest genus, *Senecio,* is almost worldwide in distribution even if the genus is taken in a restricted sense. Most *Senecio* species actually occur in temperate South America and in tropical and southern Africa. There are also many other Senecioninae genera with poorly understood relationships, restricted either to Africa or to South America. It is therefore possible that the Senecioninae, as well as the genus *Senecio* itself (C. Jeffrey, personal communication, 1990), comprise a number of African–South American sister group relationships, although this remains to be conclusively demonstrated.

The genus *Aspilia,* distributed in Africa and South America, is a case similar to that of *Pseudognaphalium* and *Achyrocline*. *Aspilia* also seems to be a poorly delimited genus; it is part of a larger complex that also includes *Blainvillea, Melanthera,* and *Wedelia* (Wild 1965). The complex has a pantropical distribution, and because cladistic relationships within the complex are unknown, an African–South American sister-group relationship cannot yet be positively demonstrated among the species of *Aspilia.*

Acmella is a pantropical genus with most species in South America. A few species are either pantropical or restricted to Africa or tropical Asia. Jansen (1985) presented a manually compiled cladogram of *Acmella.* Numerous equally parsimonious alternatives exist, but it is possible that Jansen's implication that the Old World species are nested within the bulk of South American species is correct. In Jansen's solution, the African *A. caulirhiza* and the American *A. oppositifolia* are sister species. Many *Acmella* species are weedy, easily dispersed, and widely distributed, suggesting that the intercontinental sister-group relationships may have individual dispersal explanations.

The same interpretation may apply to *Stomatanthes,* the last item in the table of African–South American sister-group relationships. *Stomatanthes* is a mainly South American genus with a few species in Africa (King and Robinson 1987).

It must be concluded that putative transatlantic sister-group relationships within the Asteraceae in most cases require confirmation from more detailed cladistic analyses. Those that hold up seem to represent isolated cases possibly with separate dispersal explanations, rather than results of more general biogeographic events.

IV. Cladistic Biogeography

Cladistic biogeography was first practiced by Rosen (1978, 1979) in his now-classic studies on vicariance biogeography of Central American fishes. Rosen substituted taxa with their distribution areas in order to construct area cladograms. He compared area cladograms from different groups and argued that congruence between area cladograms indicated that they reflected the historical relationships of the areas. The problems with missing areas, widespread taxa, redundant distributions (the same area represented by more than one branch in the cladogram), and incongruence between area cladograms have subsequently been considered in detail, particularly by Nelson and Platnick (1981). Nelson and Platnick developed component analysis to construct area cladograms from taxon cladograms. The journals *Cladistics* and *Systematic Zoology* contain many examples of cladistic biogeographic analyses. The recent reviews by Page (1989a, 1990a, 1990b) are particularly recommended; see also Chapter 7.

To explore intercontinental area relationships within the Asteraceae, I have attempted a cladistic biogeographic analysis. Few Asteraceae cladograms involving taxa from several continents are available. For this study, six groups were analyzed: (1) Mutisieae *sensu stricto,* (2) Plucheeae, (3) Gnaphalieae-Loricariinae, (4) Gnaphalieae-Gnaphaliinae, (5) Gnaphalieae-Cassiniinae, and (6) Anthemideae-Matricariinae (figures 6.5–6.10; Karis et al. 1992; Anderberg 1990, 1991b; Bremer and Humphries 1993; see comments in the appendix).

Jansen's (1985) *Acmella* study and Keeley and Turner's (1990) *Vernonia* analysis were omitted because the data yield excessive numbers of equally parsimonious solutions. Jansen's *Spilanthes* analysis comprises six species only; the *Spilanthes* cladogram is thus difficult to compare with those of large tribes and subtribes with hundreds of species. However, the *Spilanthes* area cladogram is congruent with all area cladograms presented in figure 6.12.

Hypotheses of the phylogeny within the six groups are shown in figures 6.5–6.10. They are partly well supported, partly less well supported. Presumably they will eventually be revised, and the revisions will affect the corresponding area cladograms. The following analysis is offered as a pilot study, bearing in mind the limitations of the underlying cladograms.

A. Component Analysis

A quick examination of figures 6.5–6.10 reveals that the cladograms contain several widespread taxa and redundant distributions. A set of rules is available for handling the problems that arise when such cladograms are reduced to area cladograms. These rules are generally referred to as Assumptions 0, 1, and 2 (Nelson and Platnick 1981; Humphries and Parenti 1986; Zandee and Roos 1987; Page 1990a, 1990b). Page has shown that Assumption 2 is the most

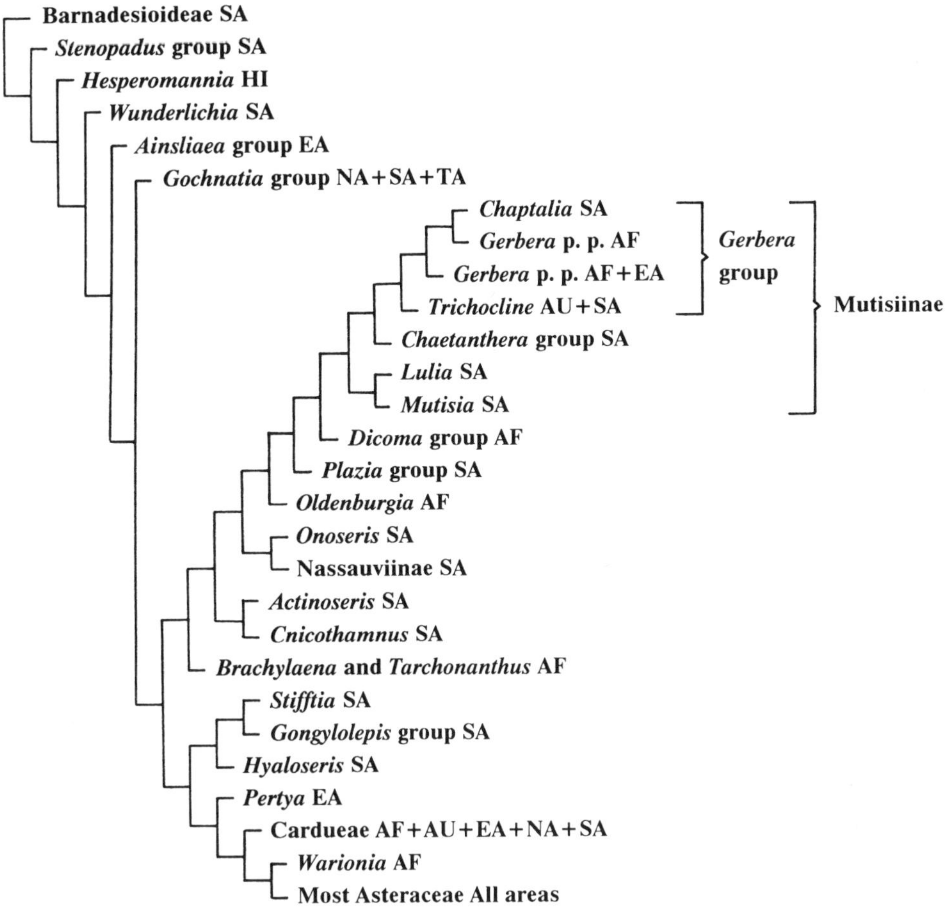

Figure 6.5. Mutisieae cladogram, from Karis et al. (1992). AF = Africa, AU = Australia, EA = Eurasia, HI = Hawaiian Islands, NA = North America, SA = South America, TA = tropical Southeast Asia.

inclusive approach; Assumptions 0 and 1 are subsets of Assumption 2. Assumption 2 involves the fewest ad hoc presumptions and allows for dispersal. In the following analyses, Assumption 2 was consistently used to construct the area cladograms.

The area cladograms and their strict consensus trees were generated with Page's (1989b) program COMPONENT. Because cladograms with too many taxa or areas result in more area cladograms than can be saved in available memory, areas had to be simplified to a limited number. Africa, for example, was used for both southern African and tropical African taxa. This simplification seems justified here because there is no attempt to analyze detailed area interrelationships. In some cladograms it was necessary to use smaller additional areas, such as India, Madagascar, Mauritius, North

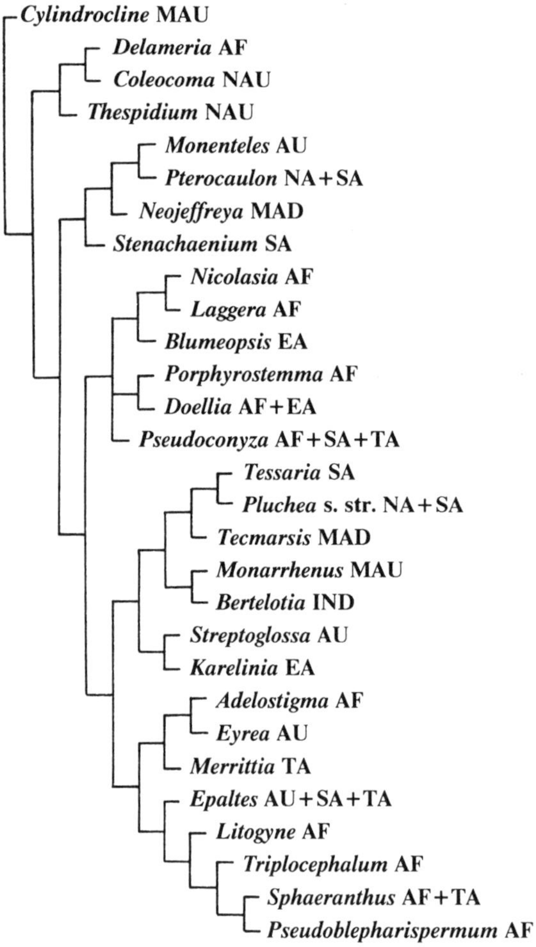

Figure 6.6. Plucheeae consensus tree from Anderberg (1991b). AF = Africa, AU = Australia, EA = Eurasia, IND = India, MAD = Madagascar, MAU = Mauritius, NA = North America, NAU = North Australia, SA = South America, TA = tropical Southeast Asia.

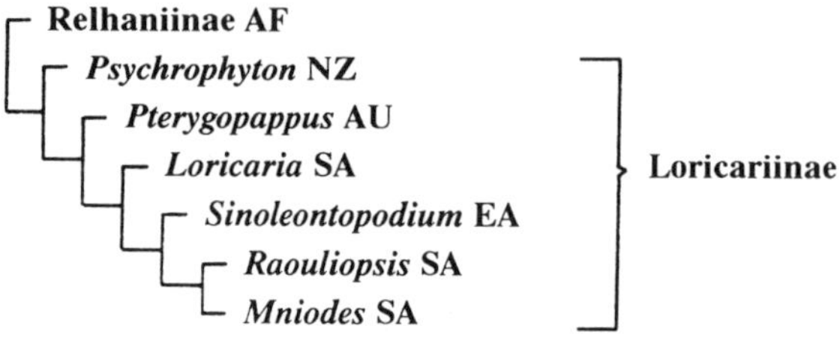

Figure 6.7. Loricariinae + Relhaniinae cladogram from Anderberg (1990). AF = Africa, AU = Australia, EA = Eurasia, NZ = New Zealand, SA = South America.

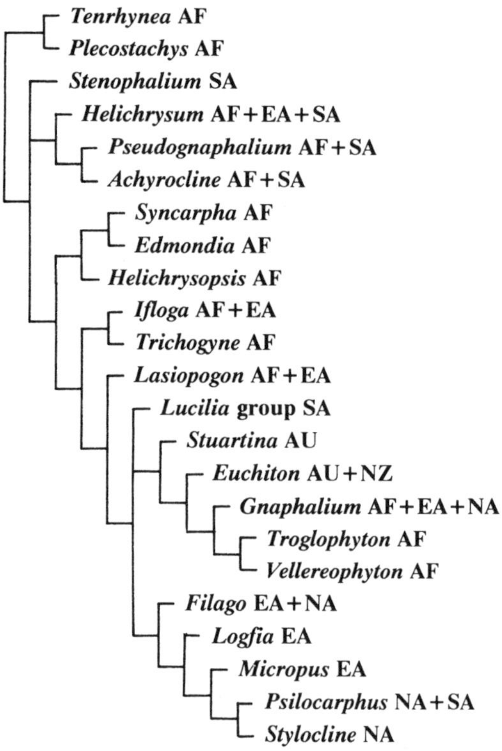

Figure 6.8. Gnaphaliinae consensus tree from Anderberg (1990). AF = Africa, AU = Australia, EA = Eurasia, NA = North America, NZ = New Zealand, SA = South America.

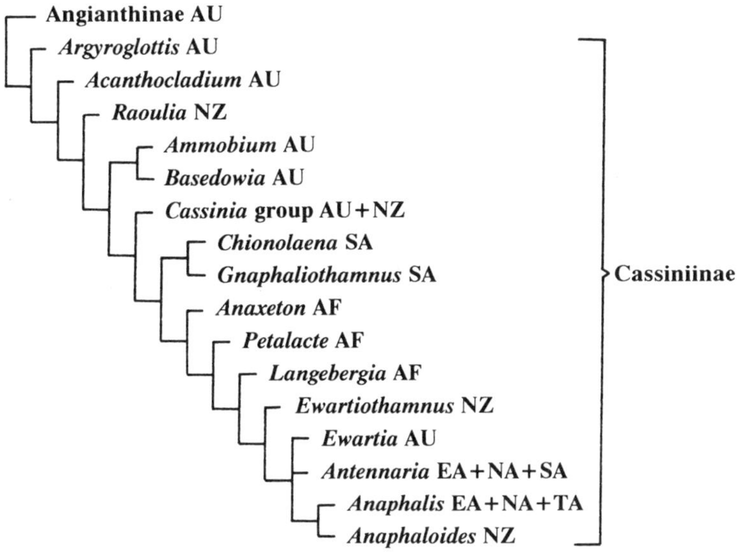

Figure 6.9. Cassiniinae + Angianthinae cladogram from Anderberg (1990). AF = Africa, AU = Australia, EA = Eurasia, NA = North America, NZ = New Zealand, SA = South America, TA = tropical Southeast Asia.

Australia, and New Zealand. These areas are present in only a few cladograms and consequently were usually omitted when area cladograms from the different groups were compared.

Excessively widespread taxa also lead to large numbers of possible area cladograms. For example, it is impossible to analyze the entire cladogram in figure 6.5 under Assumption 2 because of the widespread taxa Cardueae (five areas) and "Most Asteraceae" (all seven areas). Hence, the analysis of that cladogram was restricted to the major monophyletic part of the Mutisieae—that is, from *Brachylaena* and *Tarchonanthus* to *Chaptalia.*

COMPONENT accepts dichotomous taxon cladograms only. If there are polytomies in the cladograms, all inherent dichotomous resolutions need to be analyzed. This puts another practical limit on the analysis of some groups. Thus the entire Gnaphaliinae sensu Anderberg (1990) could not be analyzed. A limited number of genera, involving several polytomies at the base of Anderberg's consensus tree, had to be excluded. A few genera of the *Filago* group were also excluded; all had redundant distributions and their removal did not affect the construction of the area claodgrams. The major part of the Gnaphaliinae is included in figure 6.8, which is a strict consensus tree containing two trichotomies. Since there are three dichotomous resolutions in each trichotomy, there are 3 × 3 = 9 dichotomous resolutions, all of which were analyzed. Similarly, all three resolutions inherent in the trichotomies in figures 6.6, 6.9, and 6.10 were analyzed.

The Plucheeae consensus tree in figure

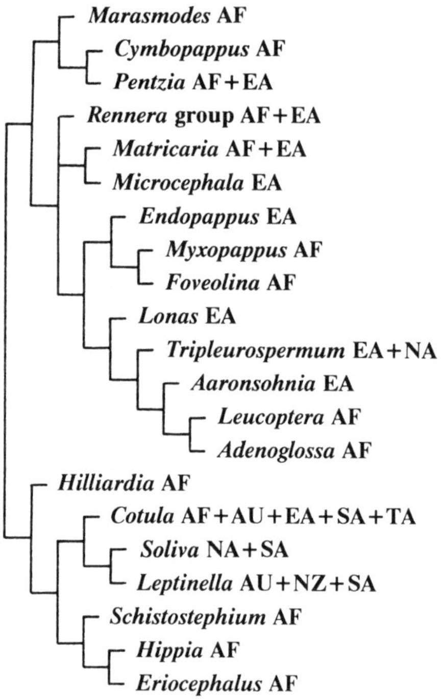

Figure 6.10. Matricariinae cladogram from Bremer and Humphries (1993). AF = Africa, AU = Australia, EA = Eurasia, NA = North America, NZ = New Zealand, SA = South America, TA = tropical Southeast Asia.

6.6 is a combinable component consensus tree (Bremer 1990) of six equally parsimonious cladograms (A. Anderberg, personal communication, 1990). The strict consensus tree presented by Anderberg (1991b) is more collapsed, and the combinable component consensus is preferred here because it is more informative yet does not conflict with any of the individual cladograms (Bremer 1990). Using the combinable component consensus tree, which includes just one trichotomy, resulted in only three resolutions to be analyzed. Analysis of all the resolutions in the individual cladograms, which contain polytomies as well, or of all the resolutions inherent in the much-collapsed strict consensus tree would not have been possible.

The analyses of the Mutisieae, Loricariinae, Cassiniinae, and Matricariinae were based on individual taxon cladograms (figures 6.5, 6.7, 6.9, and 6.10). As explained above, the Gnaphaliinae analysis was based on the dichotomous resolutions of the strict consensus tree, whereas the Plucheeae analysis was made with the resolutions of a combinable component consensus tree. The latter approach would have been preferred in the Gnaphaliinae as well, but a combinable component consensus tree could not be constructed because the individual taxon cladograms were not available.

Mutisieae (figure 6.5). The major part of the Mutisieae, from *Brachylaena* and *Tarchonanthus* to *Chaptalia,* reduces into 5 possible area cladograms. All are shown in figure 6.11, together with their strict consensus tree, which is totally unresolved. Thus, there is no pattern of area relationships derivable solely from the Mutisieae cladogram in figure 6.5.

Plucheeae (figure 6.6). The Plucheeae consensus tree yields 43 possible area cladograms. All have Africa and North Australia as sister areas. It is the only area relationship retained in the strict consensus tree, which is otherwise unresolved.

Loricariinae (figure 6.7). The Loricariinae cladogram is uncomplicated; it gives a single area cladogram (figure 6.12A).

Gnaphaliinae (figure 6.8). Figure 6.8, which is part of Anderberg's (1990) strict consensus tree, produces 253 area cladograms. The strict consensus tree of all the equally possible area cladograms is totally unresolved.

Cassiniinae (figure 6.9). In this cladogram, the *Cassinia* group consists of six genera (Anderberg 1990). The six genera all occur in Australia, and some of them are also found in New Zealand. The *Cassinia* group was entered as a single taxon in the analysis in order to avoid polytomies within the group; polytomies in taxon cladograms are not accepted by COMPONENT. The Cassiniinae cladogram generates 339 area cladograms. The strict consensus tree is totally unresolved.

Matricariinae (figure 6.10). In the Matricariinae cladogram (Bremer and Humphries 1993), the *Rennera* group consists of five genera, two from Africa and three from the Mediterranean region. As with the Cassiniinae, the generic group was entered as a single taxon in order to avoid polytomies. The Matricariinae cladogram yields 108 area cladograms, with a totally unresolved strict consensus tree.

Widespread taxa and redundant distributions have allowed for several equally possible area cladograms in all groups except the Loricariinae. The strict consensus trees are unresolved, and no resolved pattern of area relationships has been recovered from any of the six individual analyses other than that of the Loricariinae. If there is a pattern common to two or more of the six groups, however, it may be revealed by a shared-tree comparison of area cladograms from the groups. The area cladograms shared by two or more groups are fewer than those of the individual groups, and the consensus tree of the shared area cladograms may be at least partly resolved. Area cladograms shared between groups may thus suggest a pattern reflecting their common biogeographic history.

B. Common Patterns

To search for common patterns, the six sets of area cladograms were compared using COMPONENT's Shared Tree option. For example, when the 5 area cladograms from the Mutisieae were compared with the 43 area cladograms from the Plucheeae, 2 shared area cladograms were found. All combina-

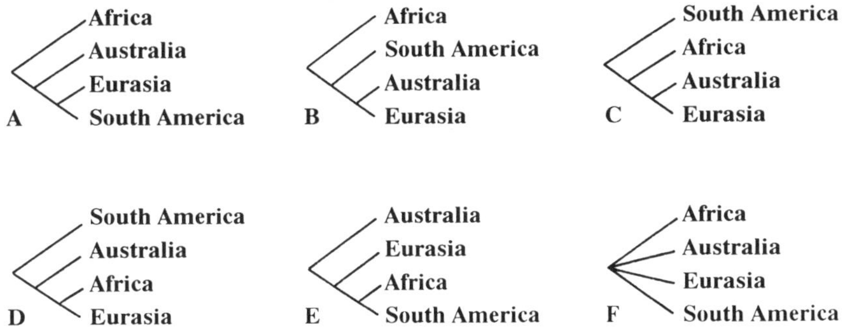

Figure 6.11. *A–E,* the five resolved area cladograms derivable from the Mutisieae cladogram in figure 6.5; *F,* their unresolved strict consensus tree.

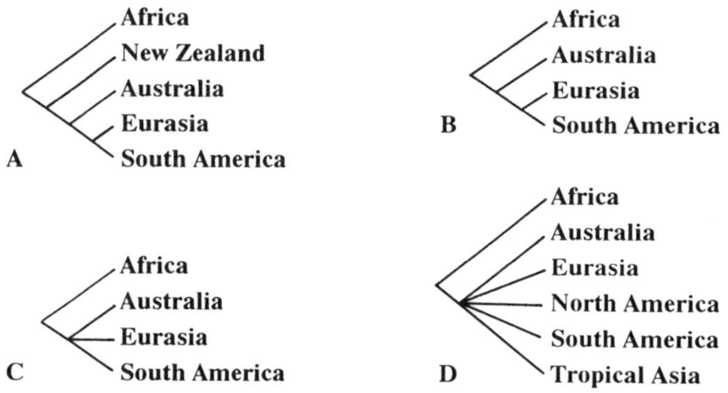

Figure 6.12. *A,* single area cladogram from Loricariinae cladogram in figure 6.7; *B,* single area cladogram shared between Mutisieae, Plucheeae, and Loricariinae; *C,* strict consensus tree of the 2 area cladograms shared between Mutisieae, Plucheeae, and Gnaphaliinae; *D,* strict consensus tree of the 43 area cladograms shared between Plucheeae and Gnaphaliinae.

tions of the six sets of area cladograms were examined for shared area cladograms. When two or more sets of area cladograms are compared using the Shared Tree option, unique areas are omitted. The shared area cladograms and the consensus trees derived from them (figure 6.12) thus only contain those areas present in all compared sets of area cladograms.

The Mutisieae, Plucheeae, and Loricariinae share one area cladogram (figure 6.12B). This is also the one area cladogram that was found when the Loricariinae area cladogram was compared with those of either the Mutisieae or the Plucheeae. In other words, the area cladogram derived from the Loricariinae (figure 6.12A, excluding New Zealand) is also among the area cladograms of the Mutisieae (figure 6.11A) and Plucheeae. Furthermore, with New Zealand excluded, the arrangement in figure 6.12B is also among the area cladograms of the Gnaphaliinae. With New Zealand included, however, the Loricariinae cladogram in figure 6.12A is incompatible with all Gnaphaliinae area cladograms.

The strict consensus tree of the two area cladograms shared by the Mutisieae, Plucheeae, and Gnaphaliinae is shown in figure 6.12C. The same consensus tree is gained from the area cladograms shared between the Mutisieae and Plucheeae. If the Mutisieae and Gnaphaliinae are compared, however, the shared area cladograms involve so many possibilities that the strict consensus tree is totally unresolved. The Plucheeae and Gnaphaliinae share six areas; the strict consensus tree of their shared area cladograms is shown in figure 6.12D.

Note that the four area cladograms in figure 6.12 are all compatible with one another. They suggest that all areas except Africa are more closely related to each other than any is to Africa. This pattern is incompatible with all area cladograms of the Cassiniinae and Matricariinae.

The Mutisieae, Gnaphaliinae, and Cassiniinae share one area cladogram, that of figure 6.11C. If the Mutisieae and Gnaphaliinae are examined, the strict consensus tree of the shared area cladograms is totally unresolved. The same applies to the comparison of Mutisieae and Cassiniinae, and the comparison of Gnaphaliinae and Cassiniinae. When all three are examined together, however, the pattern in figure 6.11C emerges. This arrangement is present also among the Matricariinae area cladograms, although those cladograms are mutually incompatible with the area cladograms of the Gnaphaliinae and Cassiniinae; in those pairwise comparisons there are more areas involved than the four areas of figure 6.11, and there are no shared area cladograms.

The Mutisieae and Matricariinae share a number of area cladograms but have a totally unresolved strict consensus tree. All other comparisons—that is, Cassiniinae compared with either Plucheeae, Loricariinae, or Matricariinae, or Matricariinae compared with either Plucheeae, Loricariinae, or Gnaphaliinae—are unsuccessful; there are no shared area cladograms in these comparisons.

Comparison of the area cladograms indicates that there are several incompatible patterns. The common patterns sustained by most of the area cladograms are those of figure 6.12, and they are supported by the Plucheeae and Loricariinae, as well as by some of the area cladograms from the Mutisieae (figure 6.11A) and Gnaphaliinae. A different pattern, that of figure 6.11C, is supported by other area cladograms from the Mutisieae and by area cladograms from the Gnaphaliinae, Cassiniinae, and Matricariinae.

V. Conclusions

It would be premature to suggest that figure 6.12 reflects real biogeographic events. Instead, it should be seen simply as one possible set of area relationships among genera of the Asteraceae. Figure 6.12 seems incompatible with the pattern of intercontinental sister group relationships shown in figure 6.4. However, figure 6.4 is a summary of more recent intercontinental relationships, generally at the species level. This is evident from the details given in the appendix, from which figure 6.4 was derived.

Identification of general biogeographic patterns within the Asteraceae requires many more cladograms than are now available. In addition to this lack of cladistic information, there is also a methodological problem of delimiting the size of the taxon cladograms to be analyzed. In this study, the six groups were circumscribed as widely as possible for COMPONENT analysis. This approach induces many redundant distributions, and there is a danger of including fundamentally different patterns in different parts of the same cladogram. Another approach would be to delimit cladograms as narrowly as possible without reducing the number of areas included. For example, the Mutisieae analysis could be confined to the *Gerbera* group (see figure 6.5); but by that method much information inherent in the remaining part of the Mutisieae cladogram is lost. Another methodological problem concerns the delimitation of areas. These issues deserve much more attention in cladistic biogeography.

In spite of all the uncertainties involved, I venture a few conclusions from the present study. First, South America was somehow involved in the origin of the Asteraceae and was part of the ancestral area of the family. Africa was probably not part of the ancestral area, and with few exceptions, Africa lacks old relict Asteraceae genera. Instead, the original distribution of the Asteraceae seems to be Pacific. Second, there are very few sister-group relationships between African and South American Asteraceae, with those found representing individual exceptions rather than parts of more general patterns. Most intercontinental sister-group relationships in the Asteraceae are between South and North America and between Africa and Eurasia. Third, there are several incompatible intercontinental biogeographic patterns in the family, but among the area cladograms shared by different groups, none was found with Africa and South America as sister areas. Fourth, the above conclusions suggest no causal relationship between continental separation of Africa and South America and the distribution of the Asteraceae.

Appendix

Table 6.1. African Asteraceae (excluding North Africa), with sister groups outside Africa (or in North Africa).

African Asteraceae	Sister Groups
Mutisieae	
Gerbera p. p.	*Chaptalia*
South and Tropical Africa	South and Central America
Gerbera p. p.	*Gerbera* p. p.
South and Tropical Africa	Central Asia
Cardueae	
Carduus p. p.	*Carduus* p. p.
East Africa	North Africa, Eurasia
Centaurea p. p.	*Centaurea* p. p.
East Africa	North Africa, Eurasia
Cirsium p. p.	*Cirsium* p. p.
East Africa	North Africa, Eurasia
Ochrocephala	*Centaurea* group
Tropical Africa	North Africa, Eurasia
Volutaria p. p.	*Volutaria* p. p.
East Africa	North Africa, Eurasia
Lactuceae	
Crepis p. p.	*Crepis* p. p.
Tropical and South Africa	North hemisphere
Dianthoseris	Crepidinae p. p.
East Africa	?–North hemisphere or Africa
Lactuca p. p.	*Lactuca* p. p.
Tropical and South Africa	North hemisphere
Launaea p. p.	*Launaea* p. p.
Tropical and South Africa	Eurasia
Picris p. p.	*Picris* p. p.
Tropical Africa	Eurasia

Tolpis p. p.	*Tolpis* p. p.
Tropical Africa	Eurasia
Vernonieae	
Baccharoides p. p.	*Baccharoides* p. p.
Tropical Africa	India
Distephanus p. p.	*Distephanus* p. p.
Tropical Africa, Madagascar	Southeast Asia
Ethulia p. p.	*Ethulia* p. p.
Tropical Africa	Tropical Asia
Vernonia (Centrapalus) p. p.	*Vernonia (Centrapalus)* p. p.
Tropical Africa, Madagascar	Tropical Asia
Vernonia (Gymnanthemum) p. p.	*Vernonia (Gymnanthemum)* p. p.
Tropical Africa, Madagascar	Tropical Asia
Arctoteae	
Arctotis	*Cymbonotus*
South Africa	Australia
Platycarpha	*Gundelia*
Namibia	SW Asia
Plucheeae	
Delamerea	*Coleocoma*
East Africa	North Australia
Laggera and *Nicolasia*	*Blumeopsis*
South and Tropical Africa	Central Asia
Porphyrostemma	*Doellia*
Tropical Africa	SW Asia, Northeast Africa
Adelostigma	*Eyrea*
Tropical Africa	Australia
Inuleae	
Anisopappus p. p.	*Anisopappus* p. p.
Tropical and South Africa	Central Asia
Blumea p. p.	*Blumea* p. p.
Tropical Africa	Tropical Asia
Geigeria p. p.	*Geigeria* p. p.
South Africa	North Africa
Inula p. p.	*Inula* p. p.
Tropical Africa	Eurasia
Pulicaria p. p.	*Pulicaria* p. p.
Tropical Africa	Eurasia
Gnaphalieae	
Relhaniinae	Loricariinae
South Africa	?–South America, East Asia, New Zealand, Tasmania
Leysera p. p.	*Leysera* p. p.
South Africa	North Africa

(*continued*)

Table 6.1. *Continued.*

African Asteraceae	Sister Groups
Helichrysum p. p.	*Helichrysum* p. p.
South and Tropical Africa	Eurasia
Pseudognaphalium p. p.	*Pseudognaphalium* p. p.
South and Tropical Africa	?–South America
Achyrocline p. p.	*Achyrocline* p. p.
Tropical Africa	?–South America
Ifloga p. p.	*Ifloga* p. p.
South Africa	Mediterranean
Lasiopogon p. p.	*Lasiopogon* p. p.
South Africa	Mediterranean
Gnaphalium p. p.	*Gnaphalium* p. p.
Tropical and South Africa	North hemisphere
Astereae	
Dichrocephala p. p.	*Dichrocephala* p. p.
Tropical and South Africa	Tropical Asia
Grangea p. p.	*Grangea* p. p.
Tropical Africa	Tropical Asia
Microglossa p. p.	*Microglossa* p. p.
Tropical and South Africa	Tropical Asia
Nolletia p. p.	*Nolletia* p. p.
South Africa	North Africa
Psiadia p. p.	*Psiadia* p. p.
Tropical and South Africa	Tropical Asia
Anthemideae	
Inulanthera	*Gonospermum*
South and Tropical Africa	Canary Islands
Artemisia p. p.	*Artemisia* p. p.
South Africa	North hemisphere
Pentzia p. p.	*Pentzia* p. p.
South Africa	North Africa
Rennera and *Oncosiphon*	Matricariinae p. p.
South Africa	Mediterranean
Myxopappus and *Foveolina*	*Endopappus*
South Africa	North Africa
Leucoptera and *Adenoglossa*	*Aaronsohnia*
South Africa	Mediterranean
Cotula p. p.	*Cotula* p. p.
South Africa	?–South hemisphere, rarely in North Africa
Cotula p. p.	*Cotula* p. p.
South Africa	North Africa
Senecioneae	
Senecionineae p. p.?	Senecioninae p. p.?
Mainly Tropical and South Africa	?–South America
Emilia p. p.	*Emilia* p. p.
Tropical Africa	Tropical Asia

Gynura group p. p.	*Gynura* group p. p.
Tropical Africa	Tropical Asia, rarely in SW Asia and North Australia
Othonna group p. p.	*Othonna* group p. p.
South and Tropical Africa	SW Asia
Synotis group p. p.	*Synotis* group p. p.
Tropical and South Africa	?–Eurasia, Tropical Asia
Calenduleae	
Calenduleae p. p.	*Calendula*
South Africa, rarely in North Africa	Mediterranean
Helenieae	
Hypericophyllum	Helenieae, p. p.?
Tropical Africa	?
Welwitschiella	Helenieae p. p.?
Tropical Africa	?
Heliantheae	
Acmella p. p.	*Acmella* p. p.
Tropical and South Africa	South and North America
Aspilia group p. p.	*Aspilia* group p. p.
Tropical Africa	?–South America
Athroisma p. p.	*Athroisma* p. p.
Tropical Africa	Tropical Asia
Blepharispermum p. p.	*Blepharispermum* p. p.
Tropical Africa	India, Ceylon
Guizotia	Heliantheae p. p.?
Tropical Africa	?
Hoffmanniella	Heliantheae p. p.?
Tropical Africa	?
Sclerocarpus p. p.	*Sclerocarpus* p. p.
Tropical Africa	Mexico, Central America
Eupatorieae	
Stomatanthes p. p.	*Stomatanthes* p. p.
Tropical and South Africa	South America

Note: "?" after the names of sister groups indicates uncertainty with respect to the sister-group relationship; "?" before and replacing area statements indicates uncertainty with respect to the sister-group distribution.

Table 6.2. South American Asteraceae with sister groups outside South America.

South American Asteraceae	Sister Groups
Mutisieae	
Gochnatia p. p.	*Gochnatia* p. p.
South and Central America	Southeast Asia

(*continued*)

Table 6.2. *Continued.*

South American Asteraceae	Sister Groups
Trichocline p. p.	*Trichocline* p. p.
South America	Australia
Chaptalia	*Gerbera* p. p.
South and Central America	South and Tropical Africa
Adenocaulon p. p.	*Adenocaulon* p. p.
Argentina, Chile	East Asia or North and Central America
Cardueae	
Centaurea p. p.	*Centaurea* p. p.
South America, mainly Chile	North hemisphere
Centaurodendron	*Centaurea* group
Juan Fernandez Islands	?–North hemisphere, South America
Lactuceae	
Agoseris p. p.	*Agoseris* p. p.
Argentina, Chile	North America
Hypochaeris p. p.	*Hypochaeris* p. p.
South America	Eurasia
Microseris p. p.	*Microseris* subg. *Microseris* p. p.
Chile	North America
Picrosia	Microseridinae p. p.
South America	North America
Vernonieae	
Vernonia p. p.	*Vernonia* p. p.
South and Central America	North America
Plucheeae	
Pterocaulon p. p.	*Pterocaulon* p. p.
South America	?–North America, Tropical Asia, Australia
Epaltes p. p.	*Epaltes* p. p.
South America	?–Tropical Asia, Australia
Gnaphalieae	
Raouliopsis and *Mniodes*	*Sinoleontopodium*
South America, Andes	China
Pseudognaphalium p. p.	*Pseudognaphalium* p. p.
South America	?–South and Tropical Africa
Achyrocline p. p.	*Achyrocline* p. p.
South America	?–Tropical Africa
Psilocarphus p. p.	*Psilocarphus* p. p.
South America	North America
Antennaria p. p.	*Antennaria* p. p.
South America	North hemisphere
Astereae	
Baccharis p. p.	*Baccharis* p. p.
South America	North America
Centipeda p. p.	*Centipeda* p. p.
South America	?–Tropical Asia, Australia
Egletes p. p.	*Egletes* p. p.
South America	North America

Erigeron p. p.	*Erigeron* p. p.
South America	North hemisphere
Grindelia p. p.	*Grindelia* p. p.
South America	North America
Gutierrezia p. p.	*Gutierrezia* p. p.
South America	North America
Haplopappus	*Hazardia*
South America	North America
Laennicia p. p.	*Laennicia* p. p.
South America	North America
Lagenifera p. p.	*Lagenifera* p. p.
South and Central America	?–Tropical Asia, Australia, New Zealand
Myriactis p. p.	*Myriactis* p. p.
South America	Tropical Asia
Podocoma	*Ixiochlamys*
South America	Australia
Solidago p. p.	*Solidago* p. p.
South America	North America
Anthemideae	
Artemisia p. p.	*Artemisia* p. p.
South America	North hemisphere
Cotula p. p.	*Cotula* p. p.
South America	?–South hemisphere
Leptinella p. p.	*Leptinella* p. p.
South America	Australia
Senecioneae	
Abrotanella p. p.	*Abrotanella* p. p.
South America	Australia, New Zealand
Blennosperma p. p.	*Blennosperma* p. p.
Chile	North America, California
Senecionineae p. p.?	Senecioninae p. p.?
Mainly South America	?–Tropical and South Africa
Acrisione	*Arnoglossum* and *Rugelia*
Chile	North America
Erechtites p. p.	*Erechtites* p. p.
South America	North America
Gynoxys group	Tussilagininae p. p.
South Ameria, Andes	North and Central America
Helenieae	
Bahia p. p.	*Bahia* p. p.
South America	North America
Gaillardia p. p.	*Gaillardia* p. p.
South America	North America
Helenium p. p.	*Helenium* p. p.
South America	North America
Hymenoxys p. p.	*Hymenoxys* p. p.
South America	North America
Schkuhria p. p.	*Schkuhria* p. p.

(*continued*)

Table 6.2. *Continued.*

South American Asteraceae	Sister Groups
South America	North America
Villanova p. p.	*Schkuhria* p. p.
Soúth America	North America
Coreopsideae	
Cosmos p. p.	*Cosmos* p. p.
South America	North America
Tageteae	
Pectis p. p.	*Pectis* p. p.
South America	North America
Porophyllum p. p.	*Porophyllum* p. p.
South America	North America
Tagetes p. p.	*Tagetes* p. p.
South America	North America
Heliantheae	
Heliantheae p. p.	Heliantheae p. p.
South America	North America
Acmella p. p.	*Acmella* p. p.
South America	North America
Acmella p. p.	*Acmella* p. p.
South and North America	Tropical and South Africa
Acmella p. p.	*Acmella* p. p.
South America	Tropical Asia
Aspilia group p. p.	*Aspilia* group p. p.
South America	?–Tropical Africa
Flaveria p. p.	*Flaveria* p. p.
South America	North America
Jaumea p. p.	*Jaumea* p. p.
South America	North America
Madia p. p.	*Madia* p. p.
Chile	North America
Spilanthes p. p.	*Spilanthes* p. p.
South America, West Indies	Southeast Asia
Stachycephalum p. p.	*Stachycephalum* p. p.
Argentina	Mexico
Eupatorieae	
Eupatorieae p. p.	Eupatorieae p. p.
South America	North America
Eupatorium p. p.	*Eupatorium* p. p.
South America	North hemisphere
Stomatanthes p. p.	*Stomatanthes* p. p.
South America	Tropical and South Africa

Note: "?" after the names of sister groups indicates uncertainty with respect to the sister-group relationship; "?" before area statements indicates uncertainty with respect to the sister-group distribution.

Comments on Tables 6.1 and 6.2

Mutisieae. The majority of the Mutisieae occur in South America. There are also several genera in Africa, a smaller group of genera in eastern Asia, and scattered occurrences of genera in North America, central Asia, and Australia. With the former subtribe Barnadesiinae now raised to a separate subfamily (Bremer and Jansen 1992), three subtribes remain in the Mutisieae—the paraphyletic and artificial Gochnatiinae and the putatively monophyletic Mutisiinae and Nassauviinae (Cabrera 1961, 1977). The last two are mainly South American, though the Mutisiinae also comprise the partly Old World *Gerbera* group.

Generic interrelationships within the Mutisieae are not well understood, although some conclusions may be drawn from the cladogram by Karis and co-workers (1992; figure 6.5). The *Gerbera*/*Chaptalia* case in tables 6.1 and 6.2 is one example of an African–South American relationship (Hansen 1990).

The tables include a number of intercontinental relationships at the species level within the genera *Gerbera, Gochnatia, Trichocline,* and *Adenocaulon*. *Gerbera* has been treated by Hansen (1985a, 1985b, 1988, 1990); it comprises a number of genera with African-Asian links. *Gochnatia* is a rather large, probably unnatural genus distributed throughout South America and the West Indies (Cabrera 1971). There are also a few species in Southeast Asia. However, the classification, and hence species interrelationships within *Gochnatia,* is uncertain. *Trichocline* is a South American genus with one species in Australia, the only representative of the tribe on that continent (Zardini 1975). *Adenocaulon* has been classified in several tribes, but is currently accepted in the Mutisieae (Bittmann 1990). It has one species in Chile, one in Central America, one in North America, and two in eastern Asia.

The peculiar southern African genus *Oldenburgia* deserves special comment. Bond (1987), in her revision of the genus, could not identify any close relative to *Oldenburgia* among the African Mutisieae. She suggested instead that *Oldenburgia* is related to the South American Mutisieae genera *Wunderlichia, Chimantaea* (a derivative of *Stenopadus*), and *Cnicothamnus*. All of these genera are shrubby or arborescent with large leaves. However, the analysis of Karis and co-workers (1992) suggests that habit and floral similarities between *Oldenburgia* and the three South American genera are symplesiomorphies and that the sister group of *Oldenburgia* consists of a large part of the tribe Mutisieae rather than any of these genera.

Cardueae. The tribe Cardueae occurs predominantly in the northern hemisphere, with a few representatives in Africa, South America, and Australia. The African representatives are all species of large Eurasian or northern hemisphere genera. The only genus endemic to Africa is *Ochrocephala* (Dittrich 1983), which is part of the large, mainly Eurasian *Centaurea* generic group. In South America there are a few species of *Centaurea,* and the genus *Centaurodendron* is endemic to the Juan Fernandez Islands. The sister group of *Centaurodendron* is probably part of the predominantly Eurasian *Centaurea* group.

Lactuceae. The genera of the Lactuceae show a general pattern similar to that found in the Cardueae. The Lactuceae tribe is predominantly northern hemisphere, with some widely distributed genera (*Sonchus*) and few southern hemisphere genera. As in the Cardueae, there are a number of Eurasian genera with species in Africa. The monotypic *Dianthoseris* from the East African mountains is part of the mainly northern hemisphere subtribe Crepidinae (Stebbins 1953), which is the *Crepis* subgroup in Jeffrey's (1966) informal classification. Jeffrey suggested that *Dianthoseris* may be related to the Asian *Dubyaea* and *Soroseris*.

Most South American Lactuceae genera display the usual pattern of South American–North American sister-group relationships, exemplified by *Agoseris, Microseris* (Chambers 1955), and *Picrosia* (Stebbins 1953). The genus *Hypochaeris* provides one of the two South American–Eurasian links in tables 6.1 and 6.2. The genus has a peculiar disjunct distribution, with several species in the Mediterranean region and several in temperate South America. The relatives of *Hypochaeris* are all Eurasian in distribution, and the occurrence of many unmistakable *Hypochaeris* species in South America remains to be explained. Since there are 30–40 clearly distinct species involved, the South American presence is hardly a case of recent introduction.

Dendroseris from the Juan Fernandez Islands and *Thamnoseris* from San Ambrosio Island are two peculiar island genera of the Lactuceae. They are probably sister genera, but their shared sister group is unknown. Jeffrey (1966) placed them within his *Sonchus* group, which is almost worldwide in distribution.

Vernonieae. The Vernonieae is a pantropical tribe. The large genus *Vernonia* has numerous species in Africa and South America and is usually considered as a prime example of an African–South American connection in the Asteraceae. *Vernonia* is not restricted to Africa and South America, however, but is also widely distributed in tropical Asia and in North America. Jones (1977, 1979, 1981) and Jeffrey (1988) indicate that South American species of *Vernonia* are more closely related to North American than to Old World *Vernonia* species, and that African species of *Vernonia* are more closely related to tropical Asian than to New World *Vernonia* species. Jeffrey suggested a reclassification of all Old World species into a number of different subgenera or genera. Several of those taxa, listed in table 6.1, are distributed in Africa and tropical Asia, suggesting a series of paleotropical sister-group relationships within this tribe.

Keeley and Turner (1990) presented a preliminary cladistic analysis of *Vernonia.* They suggest that the New World *Eremosis* (Mexico and Central America) and *Critoniopsis* (South America) are related to the Old World *Strobocalyx* (Africa, Madagascar, and Southeast Asia) and *Urceolatae* (Africa). However, their data yield more than 1000 equally parsimonious cladograms (Bremer, unpublished analysis) and the strict consensus tree is uninformative with respect to the sister groups of these taxa and to intercontinental sister group relationships in general. Further analysis of the Vernonieae, using chloroplast DNA restriction site data, is now being undertaken (Keeley, unpublished data) and may throw light on African–South American sister group relationships within the tribe.

Arctoteae. This is a small southern African tribe with one genus in Australia (*Cymbonotus*) and one genus in southwestern Asia (*Gundelia*). The Australian genus belongs to the otherwise exclusively southern African subtribe Arctotidinae (Norlindh 1977). The Asian *Gundelia* is usually classified in a subtribe of its own. *Platycarpha,* a Namibian member of the Arctoteae (Karis, unpublished data), is a possible sister group to *Gundelia.*

Plucheeae. Anderberg (1989, 1990, 1991a, 1991b) divided the traditional Inuleae *sensu lato* into three tribes: Plucheeae, Gnaphalieae, and Inuleae *sensu stricto.* The Plucheeae are pantropical but also widely distributed in temperate areas. Anderberg's (1991b) analysis of generic interrelationships within the Plucheeae are summarized in figure 6.6. Intercontinental sister-group relationships suggested by Anderberg's analysis are listed in tables 6.1 and 6.2. There are also some intercontinental relationships at the species level within the genera *Epaltes, Pterocaulon,* and *Sphaeranthus.* The last genus occurs in Africa and tropical Asia. *Epaltes* and *Pterocaulon* are more widespread. Because mutual relationships among the species of *Epaltes* and *Pterocaulon* from the different continents are unknown, the corresponding area relationships are obscure.

Inuleae. The Inuleae *sensu stricto* is an Old World tribe with species distributed mainly in Eurasia and Africa (Anderberg 1991a).

Gnaphalieae. This is the largest tribe now separated from the former Inuleae *sensu lato.* It is worldwide, with many intercontinental relationships. Anderberg (1990) divided the Gnaphalieae into five subtribes—Loricariinae, Relhaniinae, Gnaphaliinae, Cassiniinae, and Angianthinae—and presented cladograms and consensus trees of generic interrelationships for each. The Loricariinae comprise a few genera from South America, New Zealand, Tasmania, and China (figure 6.7). According to Anderberg, their sister group is the Relhaniinae from South Africa. The Gnaphaliinae is the largest and most widely distributed subtribe. Generic interrelationships of a major part of the Gnaphaliinae are shown in figure 6.8. The Cassiniinae are mainly Australian, with few representatives in South America, southern Africa, Eurasia, or tropical Asia (figure 6.9). The sister group of the Cassiniinae is the entirely Australian Angianthinae.

There are several genera of the Gnaphalieae that show intercontinental sister groups at the species level. Most of these links are between South America and North America or between Africa and Eurasia. The genera *Pseudognaphalium* and *Achyrocline* are dubiously distinct from the more widespread *Helichrysum* (Anderberg 1990), and the putative connection between Africa and South America must be regarded as spurious. The Loricariinae provide one of the rare South American–Asian connections. According to Anderberg, the sister group of the two South American genera, *Raouliopsis* and *Mniodes,* is the Chinese genus *Sinoleontopodium.*

Astereae. The Astereae are worldwide, with many genera in all continents. Generic interrelationships within the tribe are not well understood, although there are a number of well-established generic groups, most of which have a restricted distribution. The only subtribal classification is an old, purely artificial subdivision (Hoffmann 1890). Several New World genera are distributed in both South and North America—for example, *Egletes* (Shinners 1949), *Grindelia* (Steyermark 1934), *Gutierrezia* (Lane 1985; Solbrig 1966), and *Laennicia* (Zardini 1981). The genus *Haplopappus* in its restricted sense is confined to South America, with a possible sister group in the North American *Hazardia* (Brown and Clark 1982). *Lagenifera* has a Pacific distribution (Cabrera 1966). *Myriactis* is a Southeast Asian genus related to *Lagenifera;* one South American species of *Lagenifera* has been transferred to *Myriactis* (Velez 1981). *Centipeda* is another genus of the Astereae with a Pacific distribution. The South American *Podocoma* may have its closest relative in the Australian *Ixiochlamys* (Dunlop 1980). Several African genera also have species in tropical Asia, for example, *Dichrocephala* and *Grangea* (Fayed 1979).

Anthemideae. The Anthemideae are predominantly Eurasian and African, and the tribe has been monographed by Bremer and Humphries (1993). They divided the Anthemideae into 12 subtribes, most of which are restricted to Eurasia or the northern hemisphere. The subtribes Ursiniinae and Thaminophyllinae are African. Their precise sister groups within the Anthemideae are unknown, and subtribal interrelationships within the Anthemideae are uncertain.

Several genera display the common African–Eurasian sister-group relationships, for example, *Inulanthera* and *Gonospermum* (Gonosperminae; Källersjö 1985) and *Artemisia* (Artemisiinae). All the ensuing taxa under Anthemideae in tables 6.1 and 6.2 belong to subtribe Matricariinae. This subtribe includes also the southern hemisphere *Cotula* group. The main part of Bremer and Humphries's (1993) cladogram of the Matricariinae is shown in figure 6.10.

Senecioneae. The worldwide Senecioneae can be divided into the subtribes Blennospermatinae, Senecioninae, and Tussilagininae (Bremer, unpublished data). The Blennospermatinae comprise four genera—*Blennosperma, Crocidium, Abrotanella,* and *Ischnea*—which together occur in an interesting Pacific distribution (Nordenstam 1977). Most of the Senecioneae belong to the Senecioninae and the Tussilagininae, also known as the senecioid and cacalioid complexes (Nordenstam 1977, 1978; Jeffrey and Chen 1984).

The Senecioninae are markedly concentrated in tropical and southern Africa and South America. The general Senecioninae distribution is in the southern hemisphere with many fewer representatives in Eurasia, North and Central America, and Australasia. Although worldwide in distribution, an overwhelming majority of *Senecio* species occur in temperate South America and in tropical and southern Africa. This pattern suggests that the Senecioninae may provide a transatlantic link between Africa and South America. Unfortunately, relationships among genera of the Senecioninae from Africa and South America are poorly understood. With a better knowledge of generic interrelationships there should be further biogeographic information from this subtribe.

The Tussilagininae are mainly from the northern hemisphere. The majority occur in Asia and North and Central America, except the *Brachyglottis* group from Australasia, and the *Gynoxys* group and a few other genera from South America.

A number of generic groups with intercontinental sister-group relationships are listed in tables 6.1 and 6.2. The paleotropical *Gynura* group consists of *Gynura* itself, *Kleinia,* and *Solanecio* (Jeffrey 1986). The Old World *Synotis* group comprises *Austrosynotis, Delairea,* and *Mikaniopsis* in Africa; *Synotis* in Eurasia; and *Cissampelopsis* in tropical Asia (Jeffrey 1986; Nordenstam and Rechinger 1989). There are also some Madagascan synotoid genera (C. Jeffrey, personal communication, 1990). The mainly South African *Othonna* group has sometimes been associated with the Andean genus *Werneria* (e.g., Hoffmann 1890). However, *Werneria* has its sister group among the South American species of *Senecio* (V. A. Funk, personal communication, 1990). The *Gynoxys* group consists of *Gynoxys, Cabreriella,* and *Scrobicaria,* all with a mainly Andean distribution (Bremer, unpublished data); they comprise most of the Tussilagininae genera from South America. Another South American Tussilagininae genus, *Acrisione,* is probably related to the North American *Arnoglossum* and *Rugelia* (Abdel-Aal et al. 1988).

Calenduleae. This small tribe is mainly South African, with one genus, *Calendula,* restricted to the Mediterranean region.

Helenieae. The Helenieae, as circumscribed by Bremer (1987) are concentrated in southwestern North America. There are a number of genera with species also in South America. Two genera of the Helenieae, *Hypericophyllum* and *Welwitschiella,* occur in tropical Africa. Their sister groups are unknown.

Coreopsideae. The Coreopsideae (Turner and Powell 1977; Robinson 1981, as subtribe) are widely distributed but with a concentration of species in the New World. Unfortunately, generic interrelationships within the tribe are almost unknown. A better knowledge of the phylogeny of this tribe could eventually provide more biogeographic information on intercontinental relationships.

Tageteae. The small tribe Tageteae is restricted to the New World and concentrated in Mexico (Strother 1977). It includes several South American–North American connections.

Heliantheae. The Heliantheae *sensu stricto,* as circumscribed by Bremer (1987), is a large and widely distributed tribe, but most of the taxa are found in the New World. A large number of genera provide sister-group relationships between South and North America.

The pantropical genera *Spilanthes* and *Acmella* have been monographed by Jansen (1981, 1985), who presented cladograms of species interrelationships. Within *Spilanthes,* the tropical Asian and South American species are clearly more closely related to each other than either is to the single African species. The *Acmella* phylogeny is very uncertain, yet it may provide African–South American connection (Jansen 1985). *Aspilia,* a genus with species in South America and in Africa, is part of a larger group with a pantropical distribution (Wild 1965). A more detailed analysis of interrelationships within that group is needed to clarify transatlantic connections. In addition to the pantropical *Aspilia* and relatives, there are two other pantropical Heliantheae genera, *Eclipta* and *Enydra,* but the exact species interrelationships, which could provide biogeographic information for these genera, are unknown.

Athroisma and *Blepharispermum* are paleotropical genera formerly classified in the Inuleae but recently transferred to the Heliantheae (Eriksson 1990). *Guizotia* and *Hoffmaniella* are the only exclusively African Heliantheae genera. Their sister-group relationships are unknown. *Sclerocarpus,* with species in tropical Africa and Mexico, provides the only African–North American sister-group relationship in table 6.1 (it was omitted from figure 6.4 for clarity).

Eupatorieae. The Eupatorieae are almost entirely restricted to the New World. They provide many sister-group relationships between South and North America, but these are not listed in detail in table 6.2. Two genera, *Adenostemma* and *Mikania,* are pantropical (King and Robinson 1987), but species interrelationships providing biogeographic information from these genera are not available. The genus *Stomatanthes* occurs in South America and Africa (King and Robinson 1987), providing one of the few identified sister-group relationships between the two continents.

Literature Cited

Abdel-Aal, M., F. Bohlmann, T. Sarg, M. El-Domiaty, and B. Nordenstam. 1988. Oplopane derivatives from *Acrisione denticulata.* Phytochemistry 27:2599–2602.

Anderberg, A. 1989. Phylogeny and reclassification of the tribe Inuleae (Asteraceae). Canad. J. Bot. 67:2277–2296.

———. 1990. Taxonomy and phylogeny of the tribe Gnaphalieae (Asteraceae). Opera Bot. 104:1–195.

———. 1991a. Taxonomy and phylogeny of the tribe Inuleae (Asteraceae). Pl. Syst. Evol. 176:75–123.

———. 1991b. Taxonomy and phylogeny of the tribe Plucheeae (Asteraceae). Pl. Syst. Evol. 176:145–177.

Bittmann, M. 1990. Die Gattung *Adenocaulon* (Compositae). II, Ökologie, Verbreitung und Systematik. Candollea 45:493–518.

Bond, P. 1987. A revision of *Oldenburgia* (Asteraceae-Mutisieae). S. Afr. J. Bot. 53: 493–500.

Bremer, K. 1987. Tribal interrelationships of the Asteraceae. Cladistics 3:210–253.

———. 1990. Combinable component consensus. Cladistics 6:369–372.

Bremer, K., and C. J. Humphries. 1993. Generic monograph of the Asteraceae-Anthemideae. Bull. Brit. Mus. Nat. Hist. (Bot.). In press.

Bremer, K., and R. K. Jansen. 1992. A new subfamily of the Asteraceae. Ann. Missouri Bot. Gard. 72:414–415.

Brown, K., and W. D. Clark. 1982. Taxonomy of *Haplopappus* sect. *Gymnocoma* (Compositae). Syst. Bot. 7:199–213.

Cabrera, A. L. 1961. Compuestas Argentinas. Clave para la determinacion de los géneras. Rev. Mus. Argent. Cienc. Nat. Bot. 2:291–362.

———. 1966. The genus *Lagenifera* (Compositae). Blumea 14:285–308.

———. 1971. Revisión del género *Gochnatia* (Compositae). Rev. Mus. La Plata, Secc. Bot. 12:1–160.

———. 1977. Mutisieae. Systematic review. Pp. 1039–1066 *in* V. H. Heywood, J. B. Harborne, and B. L. Turner (eds.), *The Biology and Chemistry of the Compositae.* London: Academic.

Chambers, K. L. 1955. A biosystematic study of the annual species of *Microseris.* Contr. Dudley Herb. 4:207–312.

Croizat, L. 1958. *Panbiogeography.* Published by the author, Caracas, Venezuela.

———. 1964. *Space, Time, Form: The Biological Synthesis.* Published by the author, Caracas, Venezuela.

Croizat, L., G. Nelson, and D. E. Rosen. 1974. Centers of origin and related concepts. Syst. Zool. 31:291–304.

Dittrich, M. 1983. *Ochrocephala,* eine neue afrikanische Compositengattung der Subtribus Centaureinae. Bot. Jahrb. Syst. 103:467–480.

Dunlop, C. R. 1980. A revision of *Ixiochlamys* (Asteraceae: Astereae). J. Adelaide Bot. Gard. 2:241–252.

Eriksson, T. 1990. Re-instatement of the genus *Leucoblepharis* Arnott (Asteraceae-Heliantheae). Bot. Jahrb. Syst. 112:167–191.

Fayed, A. 1979. Revision der Grangeinae. Mitt. Bot. Staatssamml. München 15:425–576.

Good, R. 1974. *The Geography of the Flowering Plants, 4th ed.* London: Longman.

Hansen, H. V. 1985a. A taxonomic revision of the genus *Gerbera* (Compositae, Mutisieae) sections *Gerbera, Parva, Piloselloides* (in Africa), and *Lasiopus.* Opera Bot. 78:1–36.

———. 1985b. Notes on *Gerbera* sect. *Pseudoseris* (Compositae-Mutisieae). Nord. J. Bot. 5:451–453.

———. 1988. A taxonomic revision of the genera *Gerbera* sect. *Isanthus, Leibnitzia* (in Asia), and *Uechtritzia.* Nord. J. Bot. 8:61–76.

———. 1990. Phylogenetic studies in the *Gerbera* complex (Compositae, tribe Mutisieae, subtribe Mutisiinae). Nord. J. Bot. 9:469–485.

Hoffmann, O. 1890. Compositae. Pp. 87–391 *in* A. Engler and K. Prantl (eds.), *Die natürlichen Pflanzenfamilien,* vol. 4, pt. 5. Berlin.

Humphries, C. J., and L. R. Parenti. 1986. *Cladistic Biogeography.* Oxford Monographs on Biogeography 2. Oxford: Clarendon.

Jansen, R. K. 1981. Systematics of *Spilanthes* (Compositae-Heliantheae). Syst. Bot. 6:231–257.

———. 1985. The systematics of *Acmella* (Asteraceae-Heliantheae). Syst. Bot. Monogr. 8:1–115.

Jansen, R. K., and J. D. Palmer. 1987. A chloroplast DNA inversion marks an ancient evolutionary split in the sunflower family (Asteraceae). Proc. Natl. Acad. Sci. U.S.A. 84:5818–5822.

Jansen, R. K., H. J. Michaels, and J. D. Palmer. 1991. Phylogeny and character evolution in the Asteraceae based on chloroplast DNA restriction site mapping. Syst. Bot. 16:98–115.

Jeffrey, C. 1966. Notes on Compositae. I, The Cichorieae in east tropical Africa. Kew Bull. 18:427–486.

———. 1986. The Senecioneae in East Tropical Africa. Kew Bull. 41:873–943.

———. 1988. The Vernonieae in East Tropical Africa. Notes on Compositae: V. Kew Bull. 43:195–277.

Jeffrey, C., and Y.-L. Chen. 1984. Taxonomic studies on the tribe Senecioneae (Compositae) of eastern Asia. Kew Bull. 39:205–454.

Jones, S. B. 1977. Vernonieae: Systematic review. Pp. 503–521 *in* V. H. Heywood, J. B. Harborne, and B. L. Turner (eds.), *The Biology and Chemistry of the Compositae.* London: Academic.

———. 1979. Synopsis and pollen morphology of *Vernonia* (Compositae: Vernonieae) in the New World. Rhodora 81:425–447.

———. 1981. Synopsis and pollen morphology of *Vernonia* (Compositae: Vernonieae) in the Old World. Rhodora 83:59–75.

Källersjo, M. 1985. Fruit structure and generic delimitation of *Athanasia* L. and related South African genera (Asteraceae-Anthemideae). Nord. J. Bot. 5:527–542.

Karis, P. O., M. Källersjö, and K. Bremer. 1992. Phylogenetic analysis of the Cichorioideae (Asteraceae) with emphasis on the Mutisieae. Ann. Missouri Bot. Gard. 79:416–427.

Keeley, S. C., and B. L. Turner. 1990. A preliminary cladistic analysis of the genus *Vernonia* (Vernonieae: Asteraceae). Pl. Syst. Evol. 4 (Suppl.): 45–66.

King, R. M., and H. Robinson. 1987. The genera

of the Eupatorieae (Asteraceae). Monogr. Syst. Bot. Missouri Bot. Gard. 22:1–581.

Lane, M. A. 1985. Taxonomy of *Gutierrezia* (Compositae: Astereae) in North America. Syst. Bot. 10:7–28.

Muller, J. 1981. Fossil pollen records of extant angiosperms. Bot. Rev. (Lancaster) 47:1–142.

Nelson, G., and N. I. Platnick. 1981. *Systematics and Biogeography: Cladistics and Vicariance*. New York: Columbia University Press.

Nordenstam, B. 1977. Senecioneae and Liabeae: Systematic review. Pp. 799–830 *in* V. H. Heywood, J. B. Harborne, and B. L. Turner (eds.), *The Biology and Chemistry of the Compositae*. London: Academic.

———. 1978. Taxonomic studies in the tribe Senecioneae (Compositae). Opera Bot. 44:1–83.

Nordenstam, B., and K. H. Rechinger. 1989. Compositae Tribus Senecioneae. Pp. 41–126 *in* K. H. Rechinger (ed.), *Flora Iranica 164*. Graz: Akademische.

Norlindh, T. 1977. Arctoteae: Systematic review. Pp. 943–959 *in* V. H. Heywood, J. B. Harborne, and B. L. Turner (eds.), *The Biology and Chemistry of the Compositae*. London: Academic.

Nur, A., and Z. Ben-Avraham. 1981. Lost Pacifica continent: A mobilistic speculation. Pp. 341–358 *in* G. Nelson and D. E. Rosen (eds.), *Vicariance Biogeography: A Critique*. New York: Columbia University Press.

Page, R. D. M. 1989a. Comments on component-compatibility in historical biogeography. Cladistics 5:167–182.

———. 1989b. *COMPONENT, Release 1.5*. Published by the author, Auckland, New Zealand.

———. 1990a. Component analysis: A valiant failure? Cladistics 6:119–136.

———. 1990b. Temporal congruence and cladistic analysis of biogeography and cospeciation. Syst. Zool. 39:205–226.

Raven, P. H., and D. I. Axelrod. 1974. Angiosperm biogeography and past continental movements. Ann. Missouri Bot. Gard. 61:539–673.

Robinson, H. 1981. A revision of the tribal and subtribal limits of the Heliantheae (Asteraceae). Smithsonian Contr. Bot. 51:1–102.

Rosen, D. E. 1978. Vicariant patterns and historical explanation in biogeography. Syst. Zool. 27:159–188.

———. 1979. Fishes from the uplands and intermontane basins of Guatemala: Revisionary studies and comparative geography. Bull. Amer. Mus. Nat. Hist. 162:267–376.

Schuster, R. M. 1976. Plate tectonics and its bearing on the geographical origin and dispersal of angiosperms. Pp. 48–138 *in* C. B. Beck (ed.), Origin and Early Evolution of Angiosperms. New York: Columbia University Press.

Shinners, L. H. 1949. Revision of the genus *Egletes* Cassini North of South America. Lloydia 12:239–250.

Solbrig, O. T. 1966. The South American species of *Gutierrezia*. Contr. Gray Herb. 197:3–42.

Stebbins, L. G. 1953. A new classification of the tribe Cichorieae, family Compositae. Madrono 12:65–81.

Steyermark, J. A. 1934. A monograph of the North American species of the genus *Grindelia*. Ann. Missouri Bot. Gard. 21:433–608.

Strother, J. L. 1977. Tageteae: Systematic review. Pp. 769–783 *in* V. H. Heywood, J. B. Harborne, and B. L. Turner (eds.), *The Biology and Chemistry of the Compositae*. London: Academic.

Turner, B. L. 1977. Fossil history and geography. Pp. 21–39 *in* V. H. Heywood, J. B. Harborne, and B. L. Turner (eds.), *The Biology and Chemistry of the Compositae*. London: Academic.

Turner, B. L., and A. M. Powell. 1977. Helenieae: Systematic review. Pp. 699–737 *in* V. H. Heywood, J. B. Harborne, and B. L. Turner (eds.), *The Biology and Chemistry of the Compositae*. London: Academic.

Velez, M. C. 1981. Karpologische Untersuchungen an amerikanischen Astereae (Compositae). Mitt. Bot. Staatssamml. München 17:1–170.

Wild, H. 1965. The African species of the genus *Melanthera* Rohr. Kirkia 5:1–17.

Zandee, M., and M. C. Roos. 1987. Component-compatibility in historical biogeography. Cladistics 3:305–332.

Zardini, E. M. 1975. Revisión del género *Trichocline* (Compositae). Darwiniana 19:618–733.

———. 1981. Contribuciones para una monografia del género *Conyza* Less. II, Rehabilitación del género *Laennicia* Cass. Darwiniana 23:159–169.

Part III

Animals

Chapter 7

Biogeographic Patterns in the Vespidae (Hymenoptera): Two Views of Africa and South America

James M. Carpenter

I. Introduction

Austral distributions are now widely thought to be a result of continental drift, but the composition of the succeeding continental fragments is disputed. In view of the interest in transantarctic distributions in particular, it is remarkable how few corroborated cladograms of austral organisms are available to evaluate these disputes (Patterson 1981). More are required for the critical reconstruction of southern biotic history, and analysis of the wasp family Vespidae can make a contribution. These wasps are present in Africa and South America, as well as Australia, North America, and Eurasia, and so their relationships and distribution are pertinent to broader questions of biotic interrelationships.

In this chapter I present a preliminary biogeographic analysis of two subfamilies of the Vespidae that show an African–South American connection, the Masarinae and Polistinae. The results illustrate a common problem: the pattern for each subfamily is quite different. The relationship between Africa and South America corresponds to an austral connection for Masarinae but is part of an Old World/New World dichotomy for Polistinae. I conclude by suggesting a possible reconciliation of these two patterns.

II. Material and Methods

A. Taxa

The Masarinae and Polistinae are not very closely related. The cladogram (figure 7.1)

I thank Rod Page for assistance with the biogeographic analyses, and Kåre Bremer, Jim Liebherr, and Norman Platnick for critical comments on the chapter. This work was supported by NSF grants BSR-8508055 and BSR-8817608.

 ISBN 0-300-05375-4

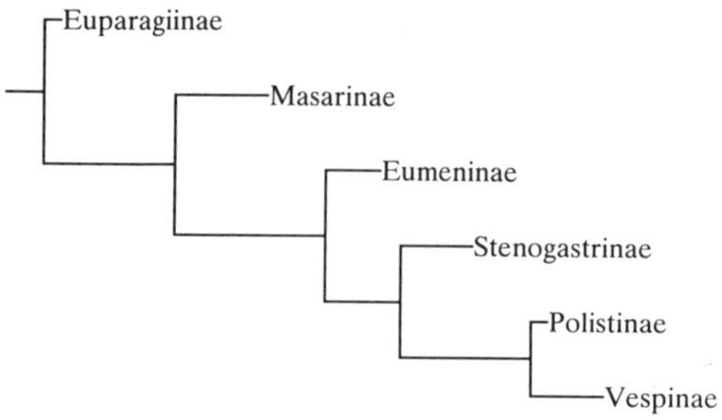

Figure 7.1. Cladogram of extant vespid subfamilies according to Carpenter (1981).

is based on the first cladistic analysis of the Vespidae (Carpenter 1981), which resulted in a reclassification: the three families and nine subfamilies recognized at the time were reduced to one family and six subfamilies. I did not provide a character matrix for numerical analysis in 1981. Carpenter and Rasnitsyn (1990: table 1) scored the informative characters but did not encompass all of the variation, for the most part omitting autapomorphies. A matrix that includes more of the character variation is presented in table 7.1. Figure 7.1 has a length of 65, a consistency index (Kluge and Farris 1969) of 0.87, and a retention index (Farris 1989) of 0.73 for the matrix.

In subsequent publications, I reclassified the genera of Vespinae (Carpenter 1987a) and Stenogastrinae (Carpenter 1988b), made a preliminary analysis of eumenine genera (Carpenter and Cumming 1985), and treated the Cretaceous Vespidae in a paper (Carpenter and Rasnitsyn 1990) that included the description of a new subfamily, the extinct Priorvespinae (figure 7.2). The distributions of four of the seven subfamilies are not pertinent to the question of the biogeographic interrelationships of Africa and South America: the fossils of Priorvespinae were found in Central Asia, the Stenogastrinae are endemic to the Oriental Region, the Vespinae are Holarctic and Oriental, and extant Euparagiinae are Nearctic (fossil Euparagiinae have been found in Central Asia and Botswana; see Brothers 1992). Of the three subfamilies with relevant distributions, the Eumeninae fails the next essential criterion for biogeographic analysis—the phylogenetic relationships of its constituent taxa are largely unknown (Carpenter and Cumming 1985 dealt in detail only with the Nearctic genera). But the Masarinae and Polistinae have been subjected to phylogenetic analysis, and the relationships of their genera are sufficiently known to allow investigation of historical biogeography, as detailed below.

1. Masarinae The Masarinae is the only group of solitary wasps that provision nests with nectar and pollen, as bees do—hence the appellation *pollen wasps* (Houston 1984). Almost 300 species are described, and they are notable biogeographically for having a disjunct austral distribution, the most pronounced disjunct distribution in the Vespidae (table 7.2). Endemism is predominant, with only three genera found in more

Table 7.1. Characters for vespid subfamilies and tribes.

CHARACTERS

(1) Forewing longitudinal plaiting: absent, 0; present 1.
(2) Forewing cu-a position: distad of fork of M+CuA, 0; basad, 1.
(3) Forewing cu-a curve: present, 0; straight, 1.
(4) Forewing subdiscal cell: not produced dorsoapically, 0; produced, 1.
(5) Forewing submarginal cells: two, 0; one, 1.
(6) Forewing recurrent veins: received in second and third submarginal cells, 0; received in second submarginal cell, 1.
(7) Forewing marginal cell: truncate, 0; narrowed, 1; pointed, 2.
(8) Forewing preaxillary excision: present, 0; absent, 1.
(9) Hindwing cell number: three, 0; two, 1.
(10) Hindwing jugal lobe: long, 0; short, 1; absent, 2. Nonadditive.
(11) Hindwing cu-a: transverse, 0; angled with A, 1; aligned with A, 2. Nonadditive.
(12) Hindwing CuA: diverging distad of cu-a, 0; diverging at cu-a, 1; diverging basad of cu-a, 2. Nonadditive.
(13) Hindwing A: free abscissa present, 0; absent, 1.
(15) Ligula: as short as prementum, broad, 0; longer, attenuate, 1.
(16) Acroglossal buttons: absent, 0; present, 1.
(17) Posterior lingual plate: narrow, 0; broad, 1; absent, 2. Nonadditive.
(18) Occipital carinae: two, 0; one, complete to mandible, 1; one, incomplete, 2. Nonadditive.
(19) Hypostomal apodemes: absent, 0; present, 1.
(20) Labrum: narrow, concealed, 0; broad, exposed, 1.
(21) Clypeus: dorsally straight, ventrally emarginate, 0; dorsally bisinuate, 1; ventrally pointed, 2; ventrally pointed with lateral angles produced, 3. Nonadditive.
(22) Male antennae: thirteen articles, 0; twelve, 1.
(23) Pronotal lobe: close to tegula, 0; separated by several times its length, 1.
(24) Secondary spiracular entrance: absent, 0; present, 1.
(25) Mesoscutal lamella: present adjoining tegula, 0; reduced, 1.
(26) Parategula: absent, 0; present, 1.
(27) Scutellum: rounded, 0; pointed, 1.
(30) Trochantellus: present, 0; absent, 1.
(32) Hindcoxa: smooth, 0; carinate, 1.
(33) Claws: simple, 0; toothed, 1; bifid, 2.
(34) Metasomal tergum and sternum I: unfused, 0; fused, 1.
(35) Metasomal retraction: absent, 0; present, 1.

(continued)

Table 7.1. ***Continued.***

CHARACTERS

(37)	Parameral spines: present, 0; short, 1.
(38)	Parameral spine curvature: absent, 0; strongly recurved, 1.
(39)	Volsella: cuspis truncate, digitus broad, 0; cuspis rounded, digitus acute, 1.
(40)	Volsellar apodeme: absent, 0; present, 1
(41)	Aedeagus: broad, blunt, 0; narrow, attenuate, 1.
(42)	Van der Vecht's gland: absent, 0; present, 1.
(43)	Ovariole number: three per ovary, 0; six per ovary, 1.
(44)	Larval mandibles: tridentate, 0; bidentate, 1.
(45)	Larval labrum: wide, 0; narrow, 1; dorsally bisinuate and ventrally bilobed, 2. Nonadditive.
(46)	Larval clypeus: medially ventral to mandibular bases, 0; dorsal, 1.
(47)	Larval pronotum: bare, 0; bristled, 1.
(48)	Larval spiracular atrium: without processes, 0; with processes, 1.
(49)	Larval provisions: insects, 0; pollen, 1.
(50)	Sociality: solitary, 0; progressive provisioning, 1; permanently sterile workers, 2; long-term monogyny, 3.

TAXA

Euparagiinae	01010001000000000000000000100000010000000000000
Gayellini	00100000010100100101000100101110000000001000100
Masarini	00101100012210100110100101101000000000000000100
Eumeninae	10100110011200101000000010012110000000002000000
Stenogastrinae	00100120111201022002010000001110010100010100 1
Polistinae	10100120011200111003001000000111001010011000 2
Vespinae	10100120021200111000000000010111001011010001 3

Source: Modified from Carpenter 1981.

Note: The numbers in parentheses correspond to the original character numbers; only groundplan states are scored. The scorings for the occipital carinae (18), hindcoxal carina (32), and claws (33) have been corrected following Carpenter and Cumming (1985) and Carpenter (1989). Terminology for wing venation follows Carpenter and Rasnitsyn (1990). Multistate characters are treated as additive except as noted.

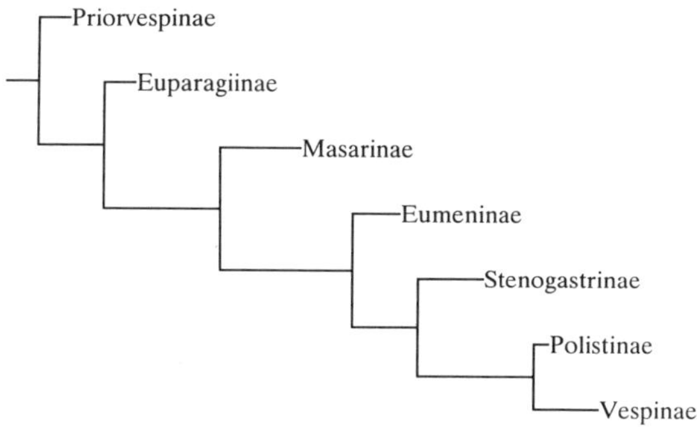

Figure 7.2. Cladogram of vespid subfamilies according to Carpenter and Rasnitsyn (1990).

Table 7.2. Distributions for the genera of Masarinae.

Taxon	Label	Distribution
Gayella	NT	Andean Chile, Andean and Patagonian Argentina, Bolivia
Paramasaris	NT	Central America, Amazonia, southeastern Brazil, northeastern Argentina
Paragia	OZ	Australia
Metaparagia	OZ	Australia
Ceramius	AFR, PA	Cape Province of South Africa, Namibia, Mediterranean to trans-Caucasus
Ceramiopsis	NT	Northern Argentina, Bolivia, Paraguay, southwestern Brazil
Trimeria	NT	Northern Argentina to southeastern Brazil, coastal Chile and Peru
Masaris	PA	Mediterranean, Central Asia
Pseudomasaris	NA	Southwestern and western United States to northern Mexico
Jugurtia	AFR, PA	Southern Africa, northern Nigeria, Burkina Faso, Gambia, Senegal, Mediterranean, trans-Caucasus, Iran
Celonites	AFR, PA	Cape Province of South Africa, Senegal, Kenya(?), Ethiopia, Djibouti, Mediterranean, Central Asia, Mongolia, Iran, Pakistan
Quartinia	AFR, PA	Southern Africa, Mediterranean, Central Asia, Mongolia, northwestern India

Note: Labels for areas of endemism defined: *NT,* Neotropical Region; *OZ,* Australia; *AFR,* Afrotropical Region; *PA,* Palearctic (Mediterranean) Region; *NA,* Nearctic Region.

than one of the areas of endemism described below.

Richards (1962) provided a dendrogram (figure 7.3) and traditionally based classification for the genera of this subfamily. The Masarinae were treated by Richards as a subfamily of his family Masaridae, which also included the Gayellinae and Euparagiinae. He arranged the 17 genera into two tribes, Masarini and Paragiini, and several genus groups. My subfamily analysis (Carpenter 1981) showed this system to be unnatural, for Richards's family Masaridae was paraphyletic. This is shown by comparison of the different positions of Euparagiinae in figures 7.1 and 7.3. Richards's subfamilies Gayellinae and Masarinae were shown to be sister groups and were reduced in rank to tribes of a revised Masarinae, which was placed in an expanded family Vespidae. Richards's tribe Paragiini in the Masarinae was also shown to be paraphyletic. The major character upon which it was based is *lack* of the emargination of the clypeus, presence of which characterizes Richards's tribe Masarini. Thus, Paragiini exhibits the plesiomorphic state and is a classic taxonomic wastebasket. These tribes were also synonymized. I have recently extended cladistic analysis to the genera in the Masarinae. I gave an analysis of the genera and species of Gayellini in Carpenter (1989), and present here results for the other tribe, Masarini, drawn from a manuscript in preparation.

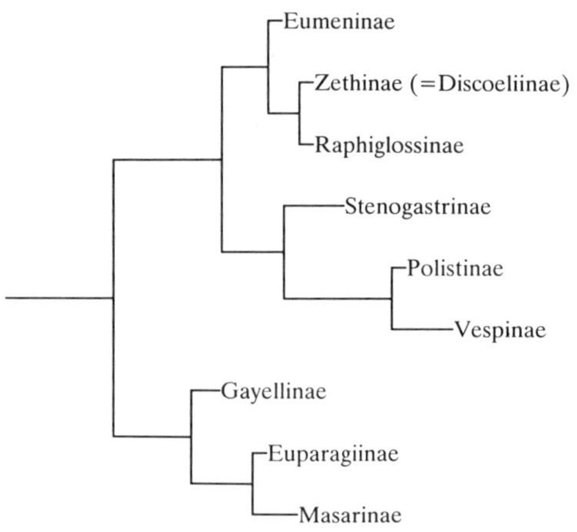

Figure 7.3. Dendrogram of vespid subfamilies according to Richards (1962: figure 34). Masarinae, Euparagiinae, and Gayellinae were included in the family Masaridae; Raphiglossinae, Zethinae (as Discoeliinae), and Eumeninae formed the family Eumenidae; Vespinae, Polistinae, and Stenogastrinae formed the family Vespidae.

The cladogram of figure 7.4 resulted from exact analysis of about 50 characters with the Hennig86 computer program by J. S. Farris (1988). Synonymies, most of which are necessary in order to eliminate paraphyly, include seven genera (van der Vecht and Carpenter 1990).

The only general treatment of the biogeography of the pollen wasps is that of Bequaert (1929), who concluded that their disjunct distribution did not support austral terrestrial connections but rather corresponded to ecological preference for arid regions. I disputed this interpretation in Carpenter (1981) and explicitly referred to their distribution as Gondwanian. I will return to this below.

2. *Polistinae* The paper wasps (Polistinae) form the largest subfamily of social wasps, with about 800 described species distributed primarily in tropical regions, especially South America. Some Polistinae are found in temperate climes; most of these are species of the cosmopolitan genus *Polistes*.

The paper wasps were arrayed by Rich-

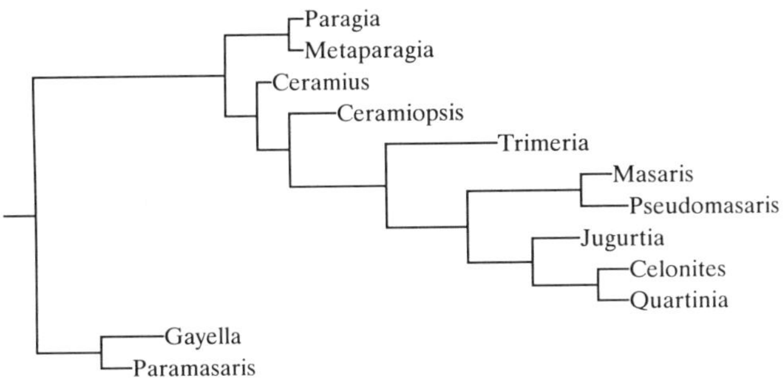

Figure 7.4. Cladogram of masarine genera according to Carpenter (1989; In prep.).

ards (1962, 1978) into three tribes: Polistini for *Polistes* and *Sulcopolistes,* Ropalidiini for *Ropalidia,* and Polybiini (properly Epiponini) for all of the remaining 26 genera. I pointed out that this tribal arrangement was not natural (Carpenter 1981). The major character upon which it is based is the shape of the propodeal orifice, the external opening for the suspensory ligament of the metasoma. An elongate, narrow opening is supposed to characterize Polistini and Ropalidiini, and a rounded one the Polybiini. Not only does the state vary in all the groups, however, but the state of a rounded orifice is plesiomorphic (Carpenter 1981). I concluded that the Polybiini is paraphyletic and I synonymized the tribes. Since that time I have studied the generic relationships in more detail.

In Carpenter (1991), I presented a matrix of 31 informative characters for 29 genera and an approximate analysis with the Hennig86 microcomputer program. The analysis produced ambiguous results: 304 cladograms of consistency index 0.48. Successive approximations character weighting (Farris 1969; see also Carpenter 1988a) reduced the number of solutions to 30. All of the cladograms differed only in alternative resolutions of the genera within a single lineage, and there is a precedent for considering that lineage a tribe, for which the name Epiponini is available (Carpenter 1991). All that is necessary for present purposes is to have some resolution of the basal relationships of Polistinae, because the disagreement among the cladograms is on placement of taxa that all have the same, essentially Neotropical, distribution (tables 7.3 and 7.4). If the epiponine lineage is treated as a single taxon, and the character matrix is reduced to just the pertinent characters (table 7.5), a single cladogram results from exact analysis (figure 7.5). I therefore employ this cladogram in what follows. The length is 21, the consistency index is 0.76, and the retention index is 0.61. A revised tribal classification that reflects this cladogram is proposed in table 7.3.

Van der Vecht (1965) suggested that all the social wasp subfamilies originated in the Indo-Malayan tropics, because Polistinae, Vespinae, and Stenogastrinae are found together only there. As mentioned above, Vespinae is basically Laurasian but also occurs in the Oriental tropics, whereas

Table 7.3. Revised tribal classification for Polistinae.

Tribe	Composition
Polistini	*Polistes (=Sulcopolistes)* [a]
Mischocyttarini (new tribe)	*Mischocyttarus*
Ropalidiini	*Ropalidia, Parapolybia, Polybioides, Belonogaster*
Epiponini (=Polybiini)	*Apoica, Agelaia (=Stelopolybia),* [b] *Angiopolybia, Pseudopolybia, Parachartergus, Leipomeles, Marimbonda, Chartergellus, Nectarinella, Protopolybia (=Pseudochartergus),* [c] *Polybia, Protonectarina, Charterginus, Chartergus, Brachygastra, Synoecoides, Epipona, Synoeca, Asteloeca, Clypearia, Occipitalia, Metapolybia*

[a]Van der Vecht and Carpenter (1990) and Carpenter (1991).
[b]Carpenter and Day (1988).
[c]Carpenter and Wenzel (1990).

Table 7.4. Distributions for the tribes of Polistinae.

Taxon	Label	Distribution
Polistini	NT, OZ, AFR, EUR, ASIA, NA, OR	Cosmopolitan except New Zealand
Mischocyttarini	NT, NA	Argentina to southeastern and western United States, British Columbia
Ropalidiini		
Ropalidia	OZ, AFR, OR	Australia, sub-Saharan Africa, Arabian Peninsula, Oriental Tropics
Parapolybia	ASIA, OR	Oriental Tropics, China, Korea, Japan, Iran
Polybioides	AFR, OR	Equatorial Africa, Oriental Tropics
Belonogaster	AFR, OR	Sub-Saharan Africa, Arabian Peninsula, India
Epiponini	NT, NA	Argentina to southwestern United States (*Polybia* and *Brachygastra* occur in Texas and Arizona)

Note: The genera of Ropalidiini are listed separately because their distributions differ for the areas of endemism defined here. Labels for the areas of endemism defined: *NT,* Neotropical Region; *OZ,* Australia; *AFR,* Afrotropical Region; *EUR,* Eurasia; *ASIA,* Eastern Asia; *NA,* Nearctic Region; *OR,* Oriental Region.

Table 7.5. Informative characters for the basal lineages of Polistinae.

CHARACTERS

(1) Antennal articles: twelve in female/thirteen in male, 0; eleven in female/twelve in male; 1.

(3) Palpal formula: six-segmented maxillary palpi and four-segmented labial palpi, 0; five and three, 1.

(12) Pronotal fovea: present, 0; absent, 1.

(13) Pronotal dorsal carina: present, 0; absent, 1.

(16) Dorsal groove: present, 0; absent, 1.

(17) Scrobal sulcus: present, 0; absent, 1.

(18) Epicnemium: carina present, 0; absent, 1.

(19) Scutal lamella: developed adjoining tegula, 0; reduced, 1.

(24) Metasomal petiole: absent, 0; Segment I less than half the width of Segment II, 1; linear, 2. Epiponini is coded as "?" because of variation within the tribe, much of which is nonapplicable.

(25) Thyridium: linear, transverse, and basal, 0; not transverse or basal, 1.

(27) Volsella: cuspis and lamina not fused, 0; fused, 1.

(28) Larval mandible: tridentate, 0; bidentate, 1.

(29) Meconium extraction: not extracted before adult emergence, 0; extracted through back of cell, 1; cells repaired by adult secretion, 2.

(31) Swarm-founding: absent, 0; present, 1.

TAXA

Polistes	00000000000000
Ropalidia	00101101111120
Parapolybia	00000011111120
Polybioides	11010010211111
Belonogaster	11111110211110
Mischocyttarus	00000010111100
Epiponini	00000010?11101

Source: Carpenter 1991.

Note: The numbers in parentheses correspond to the original character numbers. Multistate characters are treated as additive.

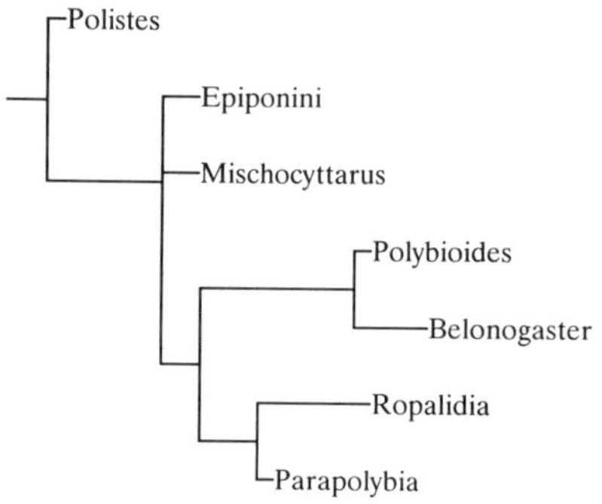

Figure 7.5. Cladogram for the Polistinae based on the data in table 7.5.

Stenogastrinae is confined to the Oriental tropics, occurring from India to New Guinea. Within Polistinae, the genus *Polistes* is cosmopolitan, 4 other genera are confined to the Old World, and 22 others are found only in the New World; they all occur primarily in tropical areas (table 7.4). Van der Vecht proposed that the paper wasps dispersed from the Indo-Malayan region via Beringea to South America. Richards (1971) later elaborated on this, suggesting that the dispersal occurred after the Polistinae had diversified at the tribal level. He was then forced to construct elaborate dispersal scenarios in order to explain why Polistini are cosmopolitan, why *Ropalidia* is found only in Africa, tropical Asia and Australia, and why his tribe Polybiini is found in South America, Africa, and Asia but not in Australia. I pointed out that the supposed Indo-Malayan center of origin was based on the fallacious assumption that a center of diversity is necessarily a center of origin, and that when these tribes are viewed as parts of one evolutionary lineage, the broad pattern of paper wasps is Gondwanian (Carpenter 1981).

B. Biogeographic Methods

Historical biogeography remains to some extent a science in search of a paradigm (Carpenter 1987b), but considerable progress has been made in recent years. In contemporary analytic biogeography, distributional data are treated as character states and are mapped onto cladograms. The resulting patterns are then analyzed under a variety of assumptions about the connections between the areas, in order to produce a general hypothesis, or area cladogram. The assumptions comprise different procedures for dealing with such problems as widespread taxa and missing or redundant areas, and may be viewed as different appreciations of the relative roles of dispersal or vicariance in producing present-day distributions. Plainly, assuming either dispersal or vicariance to the exclusion of the other a priori may be misleading, but that is precisely what is done by some recently advocated procedures. The analytical procedure I employ is component analysis under Assumption 2 of Nelson and Platnick (1981). As discussed by Page (1990), component analysis with Assumption 2 is the most realistic method for dealing with widespread taxa because it allows for dispersal. Component analysis has been criticized by Wiley (1987, 1988a, 1988b), who advocated the analysis of distributional data directly by Wagner parsimony, by analogy with cladistic character analysis. As Platnick (1988) and Page (1990) demonstrated, Wiley completely misunderstood component analysis, and Wiley's own approach necessarily assumes a priori that distributions arose by vicariance. Fit of distributions to the cladogram is all that can be obtained from one given data set, that is, an area cladogram. Dispersal or vicariance can legitimately be added as causal explanations only after comparison with patterns in other groups and with geologic data. I will contrast these various approaches, thereby illustrating the justification for Assumption 2, before proceeding to the results.

The assumptions that have been formalized are Assumptions 1 and 2 of Nelson and Platnick (1981), Assumption 0 of Zandee and Roos (1987), and what Wiley termed "Brooks Parsimony Analysis" (Wiley 1987, 1988a, 1988b; see also Brooks 1990). These assumptions are rules for converting taxon cladograms to area cladograms. If every area harbored only an endemic taxon, the construction of area cladograms would be a trivial process, for one could simply replace the taxon names with the names of the areas. Three phenomena complicate both the construction and the comparison of area cladograms. The first is missing areas, those found on one cladogram but not another. The second is widespread taxa, those occurring in more than one area, each of which harbors an endemic taxon. And the third is redundant distributions, the occurrence of an area at two or more places on a single cladogram. Opinion is divided as to how to handle these problems. For example, Zandee and Roos's (1987) Assumption 0 treats a missing area as primitive absence by the taxa whose cladogram lacks the area. In the case of widespread taxa, Kluge (1988) advocated that they be discarded from an analysis, while Mickevich (1981) suggested that overlapping distributions be combined until they formed non-overlapping areas. By contrast, the widespread taxa are treated as synapomorphies of the relevant areas by Assumption 0 and by Wiley's method. Regarding redundant distributions, Zandee and Roos (1987) did not clearly specify any procedure for handling this problem, while Assumption 1 treats each occurrence as potentially informative (Nelson and Platnick 1981; Page 1988, 1990).

Page summarized the differences among the formalized procedures and showed that Brooks Parsimony Analysis and Assumption 0 are nearly identical (Page 1990: table 1). They differ only in the treatment of missing areas, where the area cladograms allowed by Assumption 0 are a subset of those allowed under Brooks Parsimony Analysis and Assumptions 1 and 2, all of which treat missing areas as uninformative. Widespread taxa become synapomorphies for the areas inhabited by such taxa under Brooks Parsimony Analysis and Assumption 0, whereas Assumption 1 allows the area cladogram to be paraphyletic for the areas inhabited by the widespread taxon. Assumption 2 treats each occurrence of the taxon separately, and the areas can therefore "float" on the solution cladogram. Thus, a taxon can be in several widely separated areas on the final area cladogram, something no other method readily allows. If the areas have no close historical connection, the possibility, of course, is that the widespread taxon dispersed. This is certainly the most realistic way of handling widespread taxa. For such taxa, Assumption 2 permits the most solutions, while those under Assumption 1 are a subset of those under Assumption 2, and those under Assumption 0 and Brooks Parsimony Analysis are in turn a subset of those under Assumption 1. Finally, in the case of redundant distributions, Page (1990: table 1) attributed to Assumption 0 and Brooks Parsimony Analysis a procedure for combining the taxa showing the redundant distributions; Brooks (1990:16) has since advocated treating each redundant occurrence as a separate area. Assumption 1 treats each occurrence the same, as a relict, while Assumption 2 again permits more possibilities. Each occurrence of redundant endemic taxa is treated as a *different* starting point for analysis. Redundancy caused by widespread taxa is ignored, hence allowing for dispersal (Page 1988: 269–270).

The essential differences among these methods can be captured by viewing them as the differences between types of character state trees (Page 1988). Under Assumption 0 and Brooks Parsimony Analysis, area cladograms are constructed as if the underlying taxon cladograms were ordered as linear variables. Under Assumption 1, the taxon cladograms are additive variables, but under Assumption 2 they are nonadditive. Under Assumption 2, areas inhabited by one taxon are not necessarily closely connected on the solution area cladogram. This avoids possible confounding of proximity and relationship, unlike with other methods. The greater number of possibilities allowed under Assumption 2 increases the likelihood that one of the solutions is historically correct (Platnick and Nelson 1978: figure 8; Page 1988, 1989a, 1990).

1. Area Delimitation Historical biogeography of the Masarinae and the Polistinae was investigated using the COMPONENT program (Page 1989b). The cladograms (figures 7.4 and 7.5) and the distributions of each taxon were used to construct area cladograms. The distributions were defined as areas of endemism and were delimited on the basis of endemism of the *species* included in the taxa in figures 7.4 and 7.5. Overlapping distributions of species were combined until nonoverlapping areas of endemism resulted. As a consequence, distributions differing in details are still comparable. Five areas are so defined for Masarinae (table 7.2), and seven for the Polistinae (table 7.4). The names for the areas of endemism are arbitrary, for the areas were defined solely by taxon distribution, but the names are meant to evoke the classical zoogeographic regions. In the Masarinae, Neotropical is primarily Chilean and Patagonian, Australia is that entire continent, and Afrotropical is really primarily South African. The Palearctic is essentially Mediterranean, and the Nearctic comprises the southwestern United States and northern Mexico. The areas of endemism for Polistinae are not identical to those of the Masarinae, in spite of the names, and three different areas—Eurasia, eastern Asia, and the Oriental region—are defined. Australia in the case of Polistinae is primarily wet tropics, as are the Neotropical and Afrotropical regions. The Nearctic region is transcontinental North America. Some polistine genera are found in the Oriental Tropics, where no Masarinae occur. Europe, the Mediterranean region, and eastern Asia overlap for species of the cosmopolitan *Polistes* but not for the other genera, with only the otherwise Oriental *Parapolybia* also occurring in "Asia." Eastern Asia (Iran to China and Japan) is therefore delimited as a separate area, for it overlaps in disparate ways with Eurasia (Europe and the Mediterranean) and the Oriental Tropics, and so does not readily combine with either.

Partitioning of some of these areas is obviously desirable and will certainly affect the

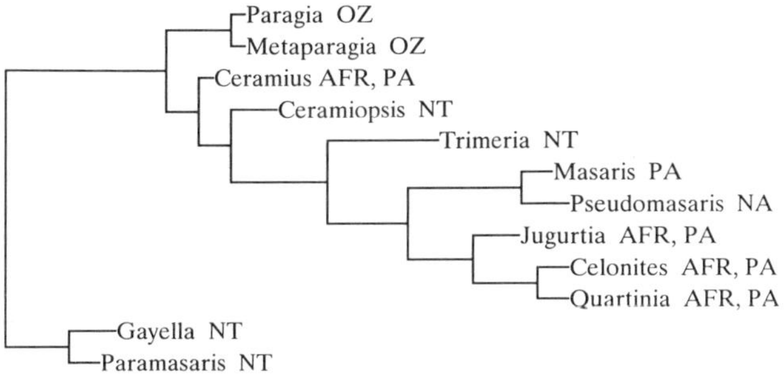

Figure 7.6. Taxon-area cladogram for Masarinae.

interpretations, but it must await elucidation of species-level phylogenetic relationships. The areas are mapped onto the cladograms in figures 7.6 and 7.7. Note the number of redundant and widespread distributions. For the Masarinae, Africa and the Neotropics appear in several places, and several genera have species in both the Palearctic and South Africa. For the Polistinae, all taxa are widespread and all distributions show redundancy.

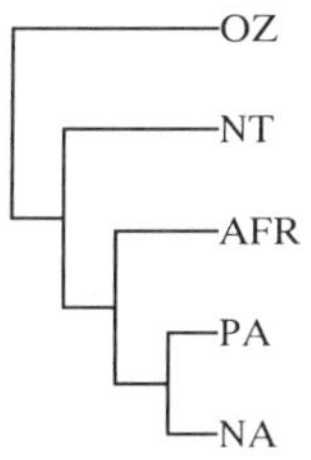

Figure 7.7. Area cladogram for Masarinae constructed using Assumption 2 and Nelson consensus (Page 1989a, modified from Nelson 1979).

III. Results

Component analysis under Assumption 2 results in three area cladograms for the Masarinae. Figure 7.8 is the "general cladogram" for these three area cladograms. The general cladogram is a type of consensus tree that maximizes resolution; the strict consensus tree for the three area cladograms (Nelson 1979; compare to Page 1988, 1989a) resolves only PA + NA, and the Adams consensus tree (Adams 1972) resolves only AFR + (PA + NA). On figure 7.8, the Neotropics are more closely connected to Africa than either area is to Australia, but Africa is still more closely con-

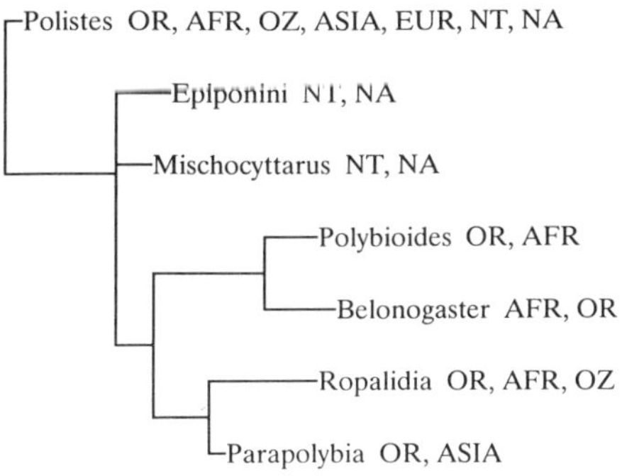

Figure 7.8. Taxon-area cladogram for Polistinae.

nected to the Palearctic and the Nearctic. For comparative purposes I generated area cladograms under Assumptions 0 and 1. The single tree under Assumption 0 differs only in treating Africa and the Palearctic as more closely connected than either is to North America. Under Assumption 1, the general cladogram calculated from three area cladograms does not resolve the connections between Africa, the Palearctic, and North America.

The results of component analysis for the Polistinae are shown in figure 7.9. Version 1.5 of COMPONENT accepts only fully bifurcating input cladograms, but R. D. M. Page has given me access to a prerelease upgrade of the program that does not have this limitation. Assumption 2 was implemented in this program, although with redundancy treated as under Assumption 1. The "general cladogram" for the nine polistine area cladograms generated under Assumption 2 (figure 7.9; this is identical to the strict and Adams consensus trees) is very different from the pattern shown by masarines. Basically, there is an Old World–New World split, with the

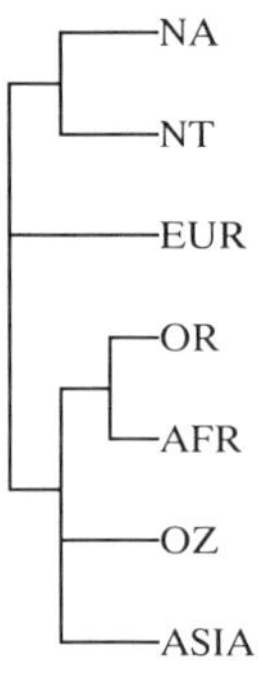

Figure 7.9. Area cladogram for Polistinae constructed using Assumption 2 and Nelson consensus.

position of Europe uncertain. In contrast to any of the masarine area cladograms, note particularly the remote connection between Africa and the Neotropics; Africa is most closely connected to the Oriental Tropics. On the masarine area cladograms, Australia is connected to the base, whereas North America is connected to the Palearctic (Mediterranean) region. Again for comparison, I also generated polistine area cladograms under Assumptions 0 and 1, using COMPONENT version 1.5 and each possible resolution of the trifurcation of the polistine cladogram (figure 7.5). The single tree under Assumption 0 resolves Australia as being more closely connected to Africa + the Oriental Tropics than is eastern Asia, and North and South America as being more closely connected than Europe to the rest of the Old World. Under Assumption 1, the general cladogram calculated from 1455 area cladograms does not resolve the connections among Africa, the Oriental Tropics, eastern Asia, and Australia, while Europe is resolved as more closely connected to the rest of the Old World than are North and South America.

The difference between the masarine and polistine area cladograms (figures 7.8 and 7.9) can be quantified using the items-of-error method developed by Nelson and Platnick (1981; see Page 1990). This approach determines how many extra branches and terminal taxa would be necessary to make the area cladograms completely congruent if they were mapped upon one another. The items of error could thus be viewed as a count of the unsampled or extinct taxa and cladogenetic events necessary to reconcile different hypotheses. For the comparison of the masarine and polistine area cladograms, I treated the five masarine areas as equivalent to their polistine counterparts (so that PA and EUR correspond). The items-of-error count for the comparison is 46. Well, is that a little or a lot? To answer this question, I used COMPONENT to generate all the possible fully bifurcating trees for the seven areas. The 10,395 resulting trees were then assessed for items of error. The mean items-of-error count was 34.4, and over 9000 of the trees had fewer items of error than the actual count. The masarine and polistine area cladograms are about as different as possible.

IV. Conclusion

The question then remaining is how the disparity between these two patterns can be explained. As far as vespids are concerned, both Africa and South America are evidently composite regions, and therein lies a possible key. This seeming incongruence may still reflect a single pattern. Recall that the Masarinae are phylogenetically older than the Polistinae, as illustrated by their relatively more basal position on the subfamily cladograms (figures 7.1 and 7.2). The family Vespidae arose before the breakup of Gondwanaland: the oldest fossils assignable to an extant subfamily, the Euparagiinae, are Lower Cretaceous, as are those of the extinct Priorvespinae (Carpenter and Rasnitsyn 1990). The biogeographic pattern of the next subfamily to diversify, the Masarinae, corresponds basally to vicariance dating from the earliest stages of the breakup of Gondwanaland, that is, while South America, Africa, and Australia were connected. Unlike the Masarinae, the paper wasps do not occur in southern South America; their Gondwanian connection is via tropical Africa–tropical South America, and is therefore consistent with a South American–African vicariant event during a later stage of the breakup of Gondwanaland, after Australia had rifted away. The social wasps thus may have arisen after the breakup of Gondwanaland was well advanced (see Wenzel 1990 on a possible Cretaceous social wasp nest). North America may have been reached subsequently by dispersal from the Mediterranean of pollen wasps, or perhaps the connection between these regions, also found in other groups of organisms, indicates a more fundamental pattern. North America and Australia may have been reached after breakup via dispersal in the paper wasps, or perhaps the suggestion of an Afrotropical-Oriental connection by Schuh and Stonedahl (1986) is supported by this pattern of distribution. But as for the connection between present-day Africa and South America, the two area cladograms apparently do not really conflict. They simply reflect different aspects of the connection. And as for the details of other, subsequent area connections where the two subfamilies disagree, since the other present continents are also geologically composite areas, further work may show that even these details do not truly conflict. Considering their different relative ages and the fact that paper wasps are primarily forest creatures while pollen wasps are found mostly in semi-arid habitats, the disparity between their area cladograms is perhaps not surprising. Determining the generality of either pattern will require similar investigations by other specialists, which I close by urging.

Literature Cited

Adams, E. N. 1972. Consensus techniques and the comparison of taxonomic trees. Syst. Zool. 21:390–397.

Bequaert, J. 1929. A new *Pseudomasaris* from California, with some considerations on the masarid wasps (Hymenoptera). Psyche 36:61–88.

Brooks, D. R. 1990. Parsimony analysis in historical biogeography and coevolution: Methodological and theoretical update. Syst. Zool. 39:14–30.

Brothers, D. J. 1992. The first Mesozoic Vespidae from the Southern Hemisphere, Botswana (Hymenoptera). J. Hym. Res. 1:119–124.

Carpenter, J. M. 1981 (1982). The phylogenetic relationships and natural classification of the Vespoidea (Hymenoptera). Syst. Ent. 7:11–38.

———. 1987a. Phylogenetic relationships and classification of the Vespinae (Hymenoptera: Vespidae). Syst. Ent. 12:413–431.

———. 1987b. Review of *Cladistic Biogeography* by C. J. Humphries and L. R. Parenti. Cladistics 3:384–386.

———. 1988a. Choosing among multiple equally parsimonious cladograms. Cladistics 4:291–296.

———. 1988b. The phylogenetic system of the Stenogastrinae (Hymenoptera: Vespidae). J. New York Ent. Soc. 96:140–175.

———. 1989 (1988). The phylogenetic system of the Gayellini (Hymenoptera: Vespidae; Masarinae). Psyche 95:211–241.

———. 1991. Phylogenetic relationships and the origin of social behavior in the Vespidae. Pp. 7–32 *in* K. G. Ross and R. W. Matthews (eds.), *The Social Biology of Wasps*. Ithaca: Cornell University Press.

Carpenter, J. M., and J. M. Cumming. 1985. A character analysis of the North American potter wasps (Hymenoptera: Vespidae; Eumeninae). J. Nat. Hist. 19:877–916.

Carpenter, J. M., and M. C. Day. 1988. Nomenclatural notes on Polistinae (Hymenoptera: Vespidae). Proc. Ent. Soc. Wash. 90:323–328.

Carpenter, J. M., and A. P. Rasnitsyn. 1990. Mesozoic Vespidae. Psyche 97:1–20.

Carpenter, J. M., and J. W. Wenzel. 1990 (1989). Synonymy of the genera *Protopolybia* and *Pseudochartergus* (Hymenoptera: Vespidae; Polistinae). Psyche 96:177–186.

Farris, J. S. 1969. A successive approximations approach to character weighting. Syst. Zool. 18:374–385.

———. 1988. Hennig86, version 1.5. Port Jefferson Station, New York.

———. 1989. The retention index and the rescaled consistency index. Cladistics 5:417–419.

Houston, T. F. 1984. Bionomics of a pollen-collecting wasp, *Paragia tricolor* (Hymenoptera: Vespidae: Masarinae), in Western Australia. Rec. West. Aust. Mus. 11:141–151.

Kluge, A. G. 1988. Parsimony in vicariance biogeography: A quantitative method and a Greater Antillean example. Syst. Zool. 37:315–328.

Kluge, A. G., and J. S. Farris. 1969. Quantitative phyletics and the evolution of anurans. Syst. Zool. 18:1–32.

Mickevich, M. F. 1981. Quantitative phylogenetic biogeography. Pp. 209–222 *in* V. A. Funk and D. R. Brooks (eds.), *Advances in Cladistics: Proceedings of the First Meeting of the Willi Hennig Society.* New York: New York Botanical Garden.

Nelson, G. 1979. Cladistic analysis and synthesis: Principles and definitions, with a historical note on Adanson's *Familles des Plantes* (1763–1764). Syst. Zool. 28:1–21.

Nelson, G., and N. I. Platnick. 1981. *Systematics and Biogeography: Cladistics and Vicariance*. New York: Columbia University Press.

Page, R. D. M. 1988. Quantitative cladistic biogeography: Constructing and comparing area cladograms. Syst. Zool. 37:254–270.

———. 1989a. Comments on component-compatibility in historical biogeography. Cladistics 5:167–182.

———. 1989b. COMPONENT, release 1.5. Auckland, New Zealand.

———. 1990. Component analysis: A valiant failure? Cladistics 6:119–136.

Patterson, C. 1981. Methods of paleogeography. Pp. 446–489 *in* G. Nelson and D. E. Rosen (eds.), *Vicariance Biogeography: A Critique*. New York: Columbia University Press.

Platnick, N. I. 1988. "Systematics, evolution and biogeography: A Dutch treat." Review of *Systematics and Evolution: A Matter of Diversity,* edited by P. Hovenkamp. Cladistics 4:308–313.

Platnick, N. I., and G. Nelson. 1978. A method of analysis for historical biogeography. Syst. Zool. 27:1–16.

Richards, O. W. 1962. *A Revisional Study of the Masarid Wasps (Hymenoptera, Vespoidea).* London: British Museum (Natural History).

———. 1971. The biology of the social wasps (Hymenoptera, Vespidae). Biol. Rev. 46:483–528.

———. 1978. *The Social Wasps of the Americas Excluding the Vespinae*. London: British Museum (Natural History).

Schuh, R. T., and G. M. Stonedahl. 1986. Historical biogeography in the Indo-Pacific: A cladistic approach. Cladistics 2:337–355.

Vecht, J. van der. 1965. The geographical distribution of the social wasps (Hymenoptera, Vespidae). Proc. XII Int. Congr. Ent., London 1964:440–441.

Vecht, J. van der, and J. M. Carpenter. 1990. A catalog of the genera of the Vespidae (Hymenoptera). Zool. Verhand. Leiden 260:1–62.

Wenzel, J. W. 1990. A social wasp's nest from the Cretaceous Period, Utah, USA, and its biogeographical significance. Psyche 97:21–29.

Wiley, E. O. 1987. Methods in vicariance biogeography. Pp. 283–306 *in* P. Hovenkamp (ed.), *Systematics and Evolution: A Matter of Diversity.* Utrecht: Institute of Systematic Botany.

———. 1988a. Parsimony analysis and vicariance biogeography. Syst. Zool. 37:271–290.

———. 1988b. Vicariance biogeography. Annu. Rev. Ecol. Syst. 19:513–542.

Zandee, M., and M. C. Roos. 1987. Component-compatibility in historical biogeography. Cladistics 3:305–332.

Chapter 8

African–South American Freshwater Fish Clades and Continental Drift: Problems with a Paradigm

John G. Lundberg

I. Introduction

Not long ago, discussions of biotic relationships between Africa and South America were dominated by the issue of biological evidence for or against continental drift (e.g., Mayr et al. 1952; Darlington 1965). Modern freshwater fishes were leading subjects in those discussions. Based on modern fish distributions, Regan (1922) had insisted on a union of southern continents years before plate tectonics revolutionized earth science. Indeed, the apparent close phylogenetic relations within certain Recent groups of African–South American freshwater fishes has been noted in the scientific literature for over a century: Günther (1880), Eigenmann (1909, 1910), Myers (1938, 1967), Gosline (1944, 1975), Schaeffer (1952), Darlington (1957), Géry (1969),

I wish to thank Drs. Peter Goldblatt and Peter Raven for inviting me to participate in the 37th Annual Systematics Symposium at the Missouri Botanical Garden. Donald J. Stewart provided me with useful and challenging criticism on all aspects of African–South American fish biogeography, and he shared the stage with me at the Symposium to offer his knowledge and insight on the ecologies of African and South American fishes. I am especially indebted to Lucinda McDade for her detailed review of this chapter. For helpful comments on the manuscript or on the phylogenetics of various taxa, I thank Paulo Buckup, Xiaoping Chen, Brian Dyer, Cristina Cox Fernandes, John Friel, Doug Futuyma, John Maisey, Brent Mishler, Lynn Parenti, Melanie Stiassny, and Mark Westneat. To be sure, not all of these people agree with all of my conclusions and points of view.

I am especially grateful to Maury, Charlotte, and Emanuel Barnes of Camden, Arkansas, for permission to study the material of "titanoglanis," the large Eocene catfish that was collected on their property.

 ISBN 0-300-05375-4

Roberts (1973), Patterson (1975), Briggs (1979, 1987), Cracraft (1974), Chardon (1967), Novacek and Marshall (1976), Greenwood (1983). Recurring mention is made in this literature of seven extant fish groups shared by Africa and South America: lungfishes, osteoglossomorphs, characins, catfishes, cichlids, nandids, and cyprinodonts. More recently, the polypterids, galaxiids and synbranchids have been added to this list (Gayet and Meunier 1991; Rosen 1974; Rosen and Greenwood 1976). There are also a number of extinct Cretaceous taxa shared by West Africa and Brazil (Patterson 1975; Chapter 14). In the parlance of contemporary cladistic vicariance biogeography, the foregoing fish taxa have been offered as evidence for an "area cladogram" in which Africa and South America are most closely related.

Various hypotheses have been advanced to account for African–South American fish relationships. Eigenmann (1909) supported von Ihering's (1900) concept of "Archhelenis," an intermediate land mass or bridge connecting stable African and South American continents and allowing riverine dispersal of strictly freshwater groups. Darlington (1957), also working within a stablist framework, proposed long distance dispersals of Southeast Asian ancestral stocks through the northern continents into Africa and South America, followed by extinctions in the North.

In contrast to the foregoing, biogeographers today accept the fact of continental drift, which is supported by extensive geological and geographical evidence, in addition to biological information. Africa and South America were broadly united until early Cretaceous, about 135 Ma (Valanginian), when their separation began in the south (Chapter 2). Disunion of the continents was slow, with their final parting occurring sometime during the middle of the Cretaceous (Albian to Santonian), between 106 and 84 Ma (Chapter 2), and they have since been increasingly isolated by a widening South Atlantic Ocean. The most common current hypotheses for African–South American freshwater fish relations invoke initial nonmarine dispersals from various southern (Gondwanan) centers of origin, followed by continental drift-based vicariance (e.g., Regan 1922; Myers 1967; Géry 1969; Gosline 1975; Novacek and Marshall 1976; Briggs 1979; but see Chapter 14). Drift/vicariance hypotheses differ mainly in the suggested sites (e.g., Southeast Asia, Africa, South America) and timings (early to late Mesozoic) of taxonomic origins and the directions and timings of previcariance dispersals. None of the proposed general drift-vicariance distributional histories of fishes has achieved wide acceptance among ichthyologists, although most would agree that drift played some role.

With continental mobilism and its vicariant consequences in place, the principal question for biogeographers is no longer that of biotic evidence for or against drift. Rather we should seek to determine how many of the African–South American sister-group pairs can be simultaneously explained by the Cretaceous drift-vicariance event. Are alternative explanations possible or even necessary for some taxa? A more fundamental question still confronts ichthyologists (and, in fact, those who study most other animals and plants): What are the direct phylogenetic relations between the groups found in the inland waters of Africa and of South America? Given our mobilist worldview, biogeographers will surely continue to develop drift-vicariance models to explain the taxonomic relationships between the modern African and South American faunas. However, alternative models of distributional history must also be considered in cases where the drift-vicariance model leads to conflicts among data or far-fetched scenarios. In all such endeavors it is essential that the taxa under consideration are phylogenetically relevant, that is, that they belong to monophyletic groups. It is also necessary to apply paleontological and earth history data within phylogenetic frameworks to provide the best estimates of ages of taxonomic origins.

In this chapter, I review the latest opinions on the phylogenies and ages of origin of the living freshwater fishes shared by Africa and South America. Within these historical contexts, each group will be evaluated in relation to African–South American drift-vicariance and alternative models of distributional history.

II. Methods

A. Freshwater Fishes

This chapter focuses on related African and South American fishes whose distributions are generally assumed to involve freshwater dispersal. Species and higher taxa of fishes are often grouped according to their observed present and presumed historical habitat and physiological salinity tolerances (Myers 1938; Darlington 1957). Obligatory, or primary, freshwater fishes spend their entire lives in freshwater habitats and are physiologically incapable of coping with sea water. Patterson (1975) has emphasized the importance of recognizing the phylogenetic level at which subgroups within a clade show strictly freshwater distributions. Monophyletic groups of primary freshwater fishes, such as characins and lepidosirenid lungfishes, have a special place in historical biogeography, because they require fresh water for dispersal and thus are closely linked to the evolution of land forms and watersheds. Secondary freshwater fishes, such as cyprinodontiforms and cichlids, are usually limited to fresh water, especially for reproduction, and have their greatest diversity and abundance there, but occasional individuals or member species can be found regularly in coastal saltwater habitats. Secondary groups are often suspected of being marine dispersants. Peripheral fishes are marine groups that include species with individuals that may move sporadically into fresh water and, rarely, species that are permanent fresh water residents. The ichthyofaunas of both Africa and South America include many peripheral groups with persistent freshwater populations or species (e.g., herrings, anchovies, needlefishes, puffers, pipefishes, gobioids, soles). Such taxa are generally treated as marine dispersants, although they should not automatically be discounted in intercontinental comparisons and drift-vicariance hypotheses unless their closest intragroup relatives are known to be resident marine taxa.

B. Phylogenetic Systematics

Precise hypotheses of phylogenies are prerequisites for framing and testing hypotheses in biogeography (Nelson and Platnick 1981; Humphries and Parenti 1986). This ideal is not nearly achieved for the large African and South American ichthyofaunas, and the present task requires that we begin by assessing to what degree it is realized for each group. Emphasizing studies that are explicitly phylogenetic cladistic in perspective, I review the phylogenetics of the ten freshwater fish groups that have been suggested as having direct African–South American relationships: Dipnoi, Polypteridae, Osteoglossomorpha, Characiformes, Siluriformes, Cyprinodontiformes, Nandidae, Cichlidae, Galaxiidae, and Synbranchidae. In particular, a search was made in each case for subgroups for which a hypothesis of a direct ancestral freshwater African–South American distribution is most parsimonious. Such subgroups may be either strictly monophyletic, involving any sister-group pair that is ancestrally African–South American, or paraphyletic, involving any ancestrally African–South American clade that also includes one or more subgroups now distributed elsewhere because of subsequent drift-vicariant processes or dispersal.

C. Biogeography

Cladistic vicariance biogeography aims to identify historical relationships among *areas* of biotic endemism that have resulted from correlated evolution of the earth and its biota (Nelson and Platnick 1981; Humphries and Parenti 1986). The core method of vicariance biogeography seeks common patterns of geographic distribution by comparing phylogenetic hypotheses and distribution data for diverse taxa. Historical relation between areas is indicated by repeated patterns of geographic distribution of sister-group pairs. The evolutionary process underlying common area and biotic relations is generally assumed to be vicariance—that is, fragmentation of ancestral biotas caused by formation of barriers to dispersal. However, coincident or correlated dispersals of diverse taxa following the removal of barriers can also produce repeated or general patterns of biotic relationship (Savage 1982). Independent dispersals of individual taxa across barriers are generally disfavored as a priori explanations of biotic relations among areas.

Use of the vicariance approach is an essential step in historical biogeography. For the particular case of Africa and South America, however, we are well beyond the initial stage of discovery of biological evidence for a pre–South Atlantic connection between Africa and South America. Geologists and geographers have provided biogeographers with a continental drift-vicariance model for these areas that is exceptionally well documented by independent data. Restricting an analysis of African–South American biotic relations to cladistic vicariance biogeography would be repetitious and simplistic. The task is now to investigate the phylogenetic biogeographies of the putative African–South American clades in order to identify those best explained by the drift-vicariance model and those for which alternative models, such as dispersal, might be equally or more likely. Thus for each African–South American clade, I have specified outgroups and their distributions as far as possible and used fossil evidence, when available, to estimate the minimum ages of taxic origin and possibly wider paleodistributions. For groups without a significant fossil record, I consider the implications of the drift-vicariance model for ages of origin and for paleodistributions of related taxa. When relevant, I consider the bearing of modern ecology and physiology of the Afro-South American groups on the drift-vicariance model and its alternatives.

D. Paleontology

The analysis and interpretation of paleontological data in phylogeny reconstruction and biogeography have been critically reviewed in recent years. Among others, Patterson (1981, 1983) and Humphries and Parenti (1986) have rightfully argued against the interpretation of oldest known fossils as *direct* ancestors, and as indicators of *absolute* ages and centers of taxonomic origins. These justifiable caveats must not detract from the value of paleontological data in biogeography and phylogenetic systematics. Fossils obviously contribute knowledge of past geographical distributions and of extinction events. Paleodistributional information is particularly significant when the provenance of a fossil lies beyond the range of its modern relatives. Indications of extinction in exotic localities are important for full understanding of the histories of biotas that may be highly complex owing to combinations of vicariance, dispersal, and extinction (Grande 1985). The Cenozoic fossil records of fishes now extirpated from the Saharan region and Lake Edward Basin in Africa (Greenwood 1974) and the Magdalena River region in South America (Lundberg and Chernoff 1992) provide clear examples of these points.

Paleontological data can play a special role in dating the biotic relations between Africa and South America. Cladistic analysis yields hypotheses on the relative ages of nested monophyletic groups within larger clades. Accurately dated fossils placed in a phylogenetic framework yield estimates of minimum absolute ages of origins of taxa and character transformations (ages of "origin" and "differentiation" of Hennig 1966:161; also Grande 1985; Patterson 1975; Ashlock 1974). Inference of minimum absolute ages can be critically important for constraining hypotheses on correlation or causation among evolutionary, biogeographic, and earth history events. If, for example, fossil evidence shows that African–South American sister taxa originated before continental separation, then the formation of the South Atlantic marine barrier could not have had a causative role in their divergence. The method for using paleontological data in estimating ages is outlined by a hypothetical example in figure 8.1. A phylogenetic tree is given in which taxa are recognized and supported by derived character states (synapomorphies). The placement of a fossil group on a phylogenetic tree is also based on the parsimonious interpretation of all available derived characters. It is perhaps tediously obvious that the oldest fossil of a group provides a direct estimate of its own minimum ages of origin and differentiation (development of apomorphies) (F at **t**). Further, because sister groups originate synchronously by cladogenesis of their common ancestor, the fossil *F* provides an estimate of the minimum age of origin of its sister groups (A at **t**). The fossil reveals nothing about either the time of differentiation or the time of diversification (cladogenesis) within its sister group; these events may have occurred at any later time. It also follows that the common ancestor, X, of this sister group pair, F + A, existed and had differentiated prior to the minimum age set by F. From the minimum age of X + F + A, the minimum age of origin of their sister group, B, is at least greater than **t**. This argument proceeds down the phylogenetic tree as far as relationships are resolved.

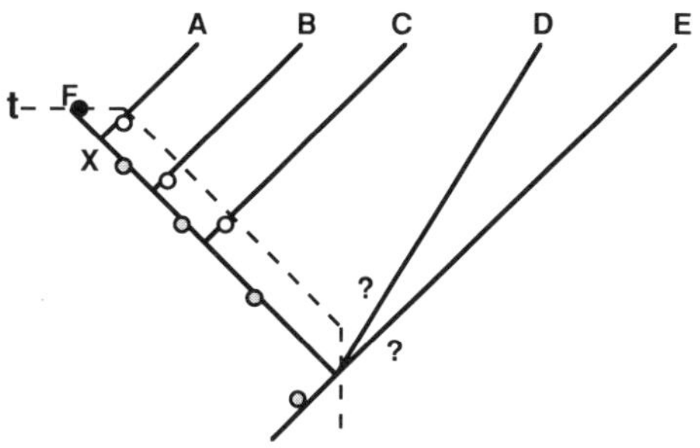

Figure 8.1. Paleontological estimation of minimum ages of taxonomic origin and character transformation. This hypothetical phylogeny includes a cladistically placed fossil, F, dated at age **t**. All circles mark taxonomic lineages inferred to be as old as or older than the fossil. Closed circle marks the fossil itself. Shaded circles mark levels of character differentiation (apomorphy) inferred to be at least as old as the fossil. Open circles mark sublineages with indeterminate levels of character transformation and diversification into subgroups. Question marks show basal unresolved groups that cannot be individually dated by F.

E. Vicariance-based Estimation of Taxonomic Ages

Vicariance biogeographic models, which are independent of paleontological data, also provide estimates of dates of taxonomic origins. Extant vicariant sister-group pairs must be of the same age as the earth history event that fragmented their ancestral range. Like age estimates based on dated fossils, vicariance-based age estimates apply indirectly to outgroups of the sister-group pair if these are fully resolved. Reliability of vicariance-based age estimates depends on the accuracy of the date of barrier formation, just as paleontology-based estimates depend on accuracy of ages of fossils. Both approaches assume a framework of correct relationships. There is, however, an important difference between the two approaches. The paleontological method gives a minimum age estimate that is expected to be superseded by discovery of older fossils. The vicariance-based method estimates absolute age of taxonomic origin because it is tied to a process-related event. Given an estimate of age derived by either method, if older fossils cladistically related to one or the other sister group were discovered, they would take priority in estimating the minimum age of the sister groups and would at the same time falsify the younger vicariance event as the process causing their origin. In such a case, the vicariance method would have underestimated the absolute age of origin of the groups. Alternatively, if dispersal after barrier formation, rather than vicariance, were the process underlying the allopatric distribution of the sister groups, the vicariance method would overestimate their age.

F. Models of Distributional History

I propose the following framework for thinking about biogeographical processes. It is well documented that a once broadly united African–South American landmass was divided into separate African and South American continents. Pitman estimates (Chapter 2) that the final break occurred between 106 and 84 Ma. Given a clade of freshwater fishes, *C*, containing taxa *C*1 and *C*2, with present-day distributions on Africa and South America, respectively, there are four biogeographic explanations, or models, for their distribution:

1. Simple drift-vicariance. The ancestral distribution of *C* encompassed the African–

South American landmass. The origin of $C1$ in Africa and $C2$ in South America was a result of divergence in isolation after separation of Africa and South America at least 84 Ma.

2. Predrift intracontinental speciation. The ancestral distribution of C encompassed the African–South American landmass. The origin of $C1$ and $C2$ predated the final separation of Africa and South America and resulted from an undetermined speciation mechanism. The predrift distributions of $C1$ and $C2$ may have been vicariantly allopatric on Africa–South America, in subregions that later were parts of the separated continents. Or, if $C1$ and $C2$ were sympatric on the African–South American landmass, there must have been postdrift extinction of $C1$ in South America or $C2$ in Africa, or both.

3. Postdrift dispersal. (a) The ancestral distribution of C was restricted to Africa. The origin of $C1$ in Africa and $C2$ in South America resulted from divergent speciation following a dispersal event from Africa to South America after continental separation. (b) The reverse of (a) with respect to ancestral distribution and direction of dispersal.

4. Indirect dispersal pathways. The distribution of C did not include either Africa or South America; $C1$ and $C2$ originated in Africa and South America, respectively, after independent dispersals from the ancestral distribution. $C1$ and $C2$ are likely most closely related to different phylogenetic subunits of C.

With only the modern distributions and hypothesized relation of $C1$ and $C2$ in hand, we are led by the principle of parsimony to prefer the drift-vicariance model of distributional history. This model involves the fewest assumptions about evolutionary and biogeographic processes. However, additional information from systematics, paleontology, or ecology might alter the balance of parsimony in favor of other hypotheses. To be consistent with the drift-vicariance model, the paleontological record should indicate or not contradict (1) a minimum age of origin greater than 84 Ma for the stem group C, (2) minimum ages of origin of $C1$ and $C2$ no older than 84 Ma, and (3) paleodistributions of these groups in Africa and South America. The discovery of fossils of $C1$ or $C2$ significantly older than 84 Ma would refute the drift-vicariance and postdrift dispersal models in favor of the predrift intracontinental speciation model. The indirect dispersal pathway model would be favored by evidence from analysis of relationships (with or without fossils) indicating that the ancestral distribution of C was not African–South American. The postdrift dispersal model might challenge the drift-vicariance model in at least three circumstances, which are not mutually exclusive. First, dispersal may be more or less likely if modern or paleontological evidence indicates that the fishes in question were capable of entering saltwater habitats. Second, the drift-vicariance model requires predrift ages of origin of C and all successive outgroups. If there is no fossil record of these outgroups predating 84 Ma, the drift-vicariance model might imply unexpectedly ancient ages of origin for many higher clades. Third, if many successive outgroups of C have endemic distributions in only modern South America or Africa, the drift-vicariance model would imply a corresponding number of extinctions within those outgroups on the continent from which they are now absent.

III. Phylogeny, Paleontology, and Biogeography

A. Lepidosirenidae

The lungfishes (Dipnoi) are a monophyletic group that appeared in the Devonian, became diverse and wide-ranging in the remaining Paleozoic and Mesozoic, and are represented by a few extant species in South America, Africa, and Australia. *Lepidosiren paradoxa* is the only extant species in South America, where it ranges throughout the central Amazon and Paraguay basins. The African *Protopterus* includes four modern and two nominal fossil species (Marshall 1987a; Greenwood 1974), with an overall range through the Nilo-Sudan, Upper and Lower Guinean, Zairean, East Coast, and Zambezi ichthyofaunal provinces (Roberts 1975).

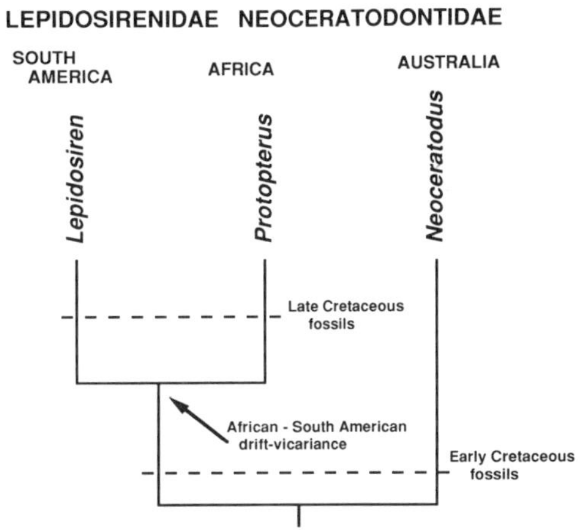

Figure 8.2. Hypothesis of relationships among extant lungfishes (Dipnoi), based on Miles (1977) and Marshall (1987b).

The hypothesis of relations among the extant lungfishes shown in figure 8.2 is well supported (Miles 1977; Marshall 1987b). Placement of the South American *Lepidosiren* and African *Protopterus* in a single family, Lepidosirenidae, instead of separate families emphasizes their sister-group relationship. The Australian lungfish, *Neoceratodus* (Neoceratodontidae), is the extant sister group of the Lepidosirenidae and is known from fossils as old as early Cretaceous (Marshall 1987a). Certain extinct ceratodontid dipnoans, some as old as Triassic, are thought to be most closely related to *Neoceratodus* (Patterson 1975). If this is correct, such early lungfishes establish a much more ancient estimate for the age of the lepidosirenid stem group. Also, Martin (1984) has suggested a close relationship of two late Cretaceous, African "Ceratodus" species with *Protopterus*. Because many of the Mesozoic ceratodontids are found in marine and estuarine deposits, a marine habitat is possible for the immediate common ancestor of *Neoceratodus* and Lepidosirenidae. The foregoing has led some authors to express the opinion that early lepidosirenids underwent intercontinental marine dispersal (e.g., Schultze 1991). However, even if lepidosirenids and neoceratodontids are related to marine ceratodontids, *Protopterus* and *Lepidosiren* are most parsimoniously treated as being a strictly freshwater sister-group pair.

The available fossil records of *Lepidosiren* and *Protopterus* are limited to their current continental ranges in South America and Africa, respectively, and as far as is known, these fossil lungfishes had essentially modern morphologies. *Lepidosiren* fossils are recorded in South America from various Tertiary sites and one late Cretaceous (or Paleocene) site (Fernandez et al.

1973; Sigé 1968). Silva Santos (1985) described *Lepidosiren megalos* as a species distinct from the modern *L. paradoxa,* based on a large skull of upper Miocene age from Brazil. *Lepidosiren* remains from the Miocene La Venta fauna located in the Magdalena region of Colombia indicate a wider distribution in the past. Similarly, fossils identified as *Protopterus* are of Tertiary and late Cretaceous age (Marshall 1987a; de Broin et al. 1974). Pleistocene fossils record *Protopterus* from Lake Malawi, from where it has since been extirpated (Greenwood 1974).

The intercontinental biogeography of the Lepidosirenidae appears to be straightforward. The phylogeny and African–South American distribution of the lepidosirenid clade fit a simple drift-vicariance model. The known fossil record is consistent with this hypothesis in suggesting at least an early Cretaceous origin of the lepidosirenid stem group, before the opening of the South Atlantic. The drift-vicariance model also correlates well with lepidosirenid biology. These fishes exhibit specialized aestivation habits and osmoregulatory physiologies for life in warm, shallow, often anoxic and temporary fresh waters. Marine dispersal is out of the question for these animals. The simple drift-vicariance model would be falsified by discovery of *Lepidosiren* or *Protopterus* fossils that predate the African-South American split. As yet, however, there are no fossil *Lepidosiren* or *Protopterus* older than late Cretaceous. Because the Dipnoi are such an archaic group with a cosmopolitan distribution, it might be suggested that the current lepidosirenid Afro-South American pattern is an accidental relict of an older, more complex intercontinental distribution, as is that of *Neoceratodus* on Australia. There is, however, no evidence that *Lepidosiren* or *Protopterus* ever occurred on continents other than their present ones. In sum, it seems that the Lepidosirenidae merit special status as the most convincing example of a clade that has undergone African–South American vicariance (table 8.1).

B. Polypteridae

Today polypterids, the reedfishes or bichirs, are restricted to African inland waters. Polypterids are uncertainly related to other living fishes but the prevailing opinion is that they are a relatively basal clade among osteichthyans or actinopterygians (Nelson 1984). Therefore, although their fossil record is scant, reedfishes are thought to be an ancient group that should have been present on the united African–South American landmass and possibly on other Gondwanan lands. This expectation is fulfilled by the recent discovery of late Cretaceous and Paleocene remains of polypterids in Bolivia. Gayet and Meunier (1991) assign to the Polypteridae a new genus, *Dajetella,* based on scales, fin spines, and vertebrae. The authors cite histological and morphological features of these elements as being diagnostic.

The oldest *Dajetella* fossils are of Maastrichtian age and, therefore, postdate the final opening of the South Atlantic. The oldest African polypterid fossils are probably a little older than the South American ones but may still be younger than the time of complete continental separation. Nevertheless, from their likely basal phylogenetic position among bony fishes, the polypterid stem group should be inferred to predate the drift event. This lineage, therefore, conforms to the drift-vicariance model. The main difference

Table 8.1. Possible clades of South American and African freshwater fishes and their suggested biogeographic histories.

Higher Taxon	South American Taxa	African Taxa	Biogeographic Model
LEPIDOSIRENIFORMES			
LEPIDOSIRENIDAE	(*Lepidosiren*	*Protopterus*)	1
POLYPTERIFORMES			
POLYPTERIDAE	(†*Dajetella*	*Polypterus*, *Calamoichthys*)	1
OSTEOGLOSSIFORMES			
ARAPAIMIDAE	(*Arapaima* [†*Laeliichthys*	{*Heterotis*, †*Paradercetis*}])	2
CHARACIFORMES			
?Unnamed	(CTENOLUCIIDAE	HEPSETIDAE)	1?
?SERRASALMINAE	(SERRASALMINAE [part?]	†*Sindacharax*)	1?
SILURIFORMES			
DORADOIDS	([DORADIDAE, AUCHENIPTERIDAE, AGENEIOSIDAE]	MOCHOKIDAE)	1
?Unnamed	(LORICARIOIDEI	AMPHILIIDAE [part?])	1?
CYPRINODONTIFORMES			
APLOCHEILOIDEI	(RIVULIDAE	APLOCHEILIDAE)	1 or 3b
POECILIIDAE	(POECILIINAE, *Fluviphylax?*	APLOCHEILICHTHYINAE)	1 or 3b
PERCIFORMES			
CICHLIDAE	([Neotropical cichlids]	[most African cichlids])	1 or 3a
NANDIDAE	(*Polycentrus*, *Monocirrhus*	*Polycentropsis*, ?*Afronandus*)	1 or 3
SYNBRANCHIFORMES			
?SYNBRANCHIDAE	(*Ophisternon*	*Ophisternon* [part]?)	3a or 4
SALMONIFORMES			
?GALAXIIDAE	(some galaxiids?	*Galaxias zebratus*)	3 or 4

Note: Daggers indicate fossil taxa. Parentheses delineate putative monophyletic groups. Question marks indicate highly speculative phylogenetic relationships based on preliminary analyses. Biogeographic models, as outlined in Section II. F, are: *1*, African–South American drift-vicariance; *2*, Predrift intracontinental speciation; *3a*, Postdrift dispersal from Africa to South America; *3b*, Postdrift dispersal from South America to Africa; *4*, Indirect dispersal pathways between Africa and South America.

between polypterids and lepidosirenids is that the reedfishes became extinct in South America sometime after early Paleocene.

C. Osteoglossomorpha

Osteoglossomorphs are a clade of distinctive teleosts that includes the osteoglossids, hiodontids, notopterids, and mormyriforms. As dated by the extinct Lycopteridae (figure 8.3A), osteoglossomorphs are at least late Jurassic in age (Greenwood 1970). Osteoglossomorphs have received considerable attention from phylogeneticists and biogeographers (Nelson 1969; Greenwood 1970, 1973; Patterson 1975, 1981; Taverne 1979; Lauder and Liem 1983). They are always included in discussions of Gondwanan biogeography, because several subgroups are scattered among southern tropical freshwater regions: *Heterotis, Pantodon,* and Mormyriformes in Africa; Notopteridae in Africa and Asia; *Scleropages* in Austral-Asia; and *Arapaima* and *Osteoglossum* in the Neotropics. However, the occurrence of both modern and extinct osteoglossomorphs at widely scattered places in the Northern Hemisphere, including some marine habitats, complicates reconstruction of osteoglossomorph history. My review of osteoglossomorph phylogeny and biogeography centers on the two monophyletic subgroups in the order Osteoglossiformes that may have direct African–South American relationships: Arapaimidae, and Osteoglossidae plus Pantodontidae.

The Arapaimidae (= Heterotinae of some authors) exhibit a straightforward African–South American pattern of distribution (figure 8.3B). There are two living species: *Heterotis niloticus,* with a wide range in the Nilo-Sudan fish province, and *Arapaima gigas,* presently distributed throughout the Amazon lowlands and Guianas. *Heterotis* is unknown in the fossil record. *Arapaima* is known from fossils of Miocene age from the Río Acre region of Brazil (Taverne 1979) and the La Venta fauna of Colombia (Lundberg and Chernoff 1992). Two fossil taxa are assigned to the Arapaimidae by Taverne (1979) and Silva Santos (1985): *Laeliichthys,* from the late early Cretaceous (Aptian, ca. 110 Ma) of Brazil, and *Paradercetis,* from late Cretaceous (Cenomanian, about 95 Ma) rocks in west Africa. Because both fossils occur in deposits that are possibly brackish, the constraint of obligatory freshwater distribution for the stem group of the living arapaimids might not be met, but this does not alter the African–South American distribution of family.

As shown in figure 8.3B, *Laeliichthys* is explicitly hypothesized by both Taverne and Silva Santos to be the sister group of *Paradercetis* + *Heterotis*. This sets a minimum Aptian age for the latter pair, for the *Arapaima* lineage, and for the whole family. *Paradercetis* indicates a minimum Cenomanian age of origin for its sister group, *Heterotis.* This phylogenetic and temporal framework implies that the cladistic separation of *Arapaima* and *Heterotis* predated the complete parting of South America and Africa. From this, I infer that the origin of these extant lineages did not result from final African–South American drift-vicariance parting but instead was the result of speciation on the still-united continent (table 8.1). The same predrift argument may hold for the sister pair *Laeliichthys* + (*Paradercetis* + *Heterotis*) that are at least ten million years

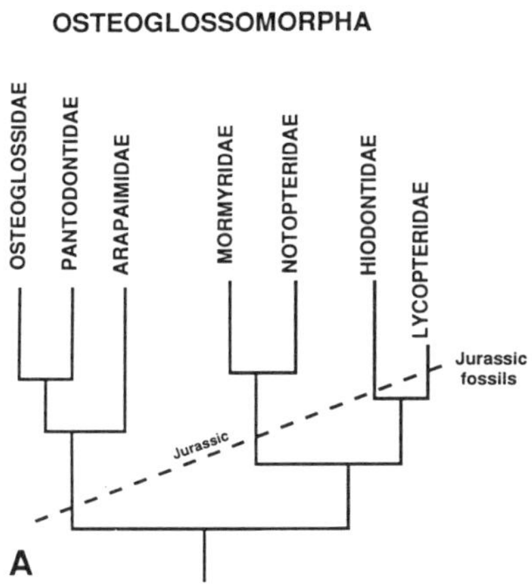

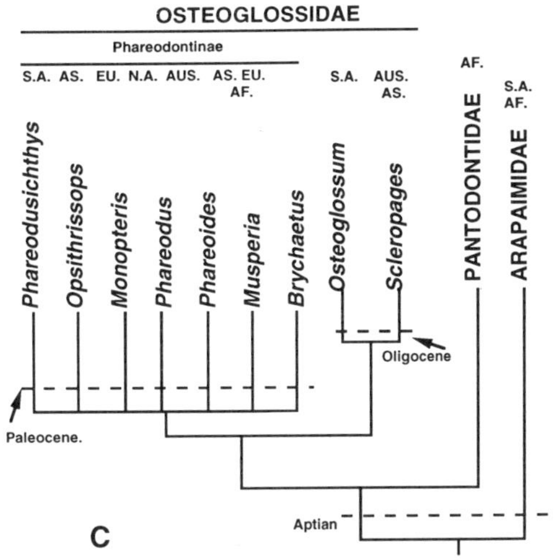

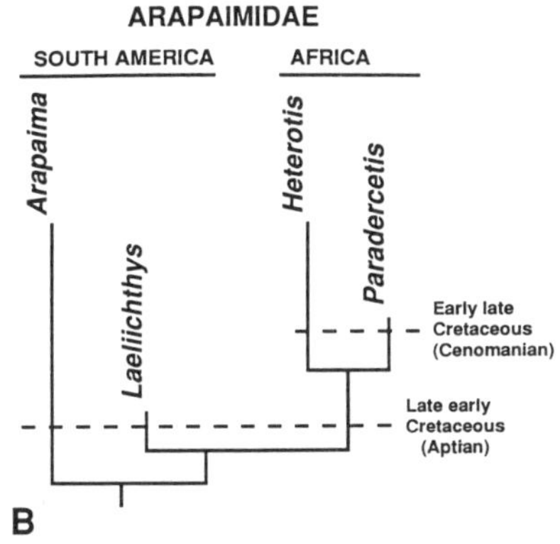

Figure 8.3. Osteoglossomorph relationships, from Lauder and Liem (1983) and Taverne (1979). *A*, Osteoglossomorpha; *B*, Arapaimidae; *C*, Osteoglossidae, Pantodontidae, and Arapaimidae.

older than the youngest estimate for final separation.

Osteoglossidae contains two living genera: *Osteoglossum*, with one species widespread in the lowland Amazon and another in the upper Rio Negro, and *Scleropages*, with one species in lowland rivers of southeast Asia and two in northern Australia. As diagrammed in figure 8.3C, the Recent sister group of the Osteoglossidae is the unique African butterflyfish *Pantodon* (Pantodontidae), which suggests a possible direct African–South American relationship. However, fossil Osteoglossidae considerably extend the geographic range of this clade beyond Africa and South America. As shown in figure 8.3C, Taverne (1979) groups six extinct genera in an osteoglossid subfamily Phareodontinae. Gayet (1991) described another phareodontine, *Phareodusichthys*, from the Paleocene of South America. Of the phareodontines, *Brychaetus*, *Monopteris*, and *Opsithrissops* are marine. Taverne also tentatively places *Foreichthys*, from the Eocene marine Monte Bolca fauna of Italy, as the sister group to either Osteoglossidae alone or Osteoglossidae + Pantodontidae. An additional African osteoglossomorph fossil, *Singida jacksonoides* (Greenwood and Patterson 1967), from the Miocene of Tanzania, does not appear to be closely related to either *Heterotis* or *Pantodon* (Greenwood 1974; Taverne 1979) and has uncertain relations to extant groups.

Within the Osteoglossidae, it has been proposed that *Scleropages* reached Southeast Asia and Australia via marine dispersal pathways (e.g., Briggs 1987). From *Laeliichthys,* the oldest fossil osteoglossiform, I infer that at least the stem group of the Osteoglossidae plus Pantodontidae existed in the late early Cretaceous. This stem group was, therefore, old enough to have had its distribution disrupted by continental drift. A drift-vicariance model has been applied to the Osteoglossidae by Cracraft (1974), who suggested that the *Osteoglossum* and *Scleropages* are related through an immediate ancestor that ranged from South America across East Antarctica to Australia. The final separation of Australia, Antarctica, and South America was possibly as late as Eocene (Woodburn and Zinsmeister 1984). The fossil record of the osteoglossines is consistent with this scenario. The stem group of *Osteoglossum* plus *Scleropages* is minimally dated at Paleocene by the oldest phareodontines, *Brychaetus* and *Phareodusichthys*. In addition, fossils of *Scleropages* provide a minimum date of Oligocene for origin of the living osteoglossine genera, after the final separation of Australia, Antarctica, and South America. A drift-vicariance model for South American and Australian taxa has been suggested by Keast (1972) for marsupials and galaxiid fishes (see below).

The extinct osteoglossid subfamily Phareodontinae severely complicates the foregoing suggestions. In light of the widely scattered localities and unresolved relationships of the phareodontines (figure 8.3C) the distributions and habitats of the basal groups of the osteoglossid-pantodon clade and the Osteoglossidae remain obscure. The wide geographic separation and the meager fossil record of the extant species show that a great deal of information is lacking. Further, I agree with Patterson (1975) that the likely marine habitat of *Brychaetus* and other extinct osteoglossid relatives suggests that marine dispersal may have played a role in their early history. Within a dispersalist mode, we lack unambiguous phylogenetic evidence for a direct exchange of osteoglossids or osteoglossids + pantodontids between Africa and South America.

Gayet and Meunier (1983) and Gayet (1991) proposed that minute fossil scale fragments from the late Cretaceous of Bolivia are most similar to those of the North American Eocene phareodontine genus *Phareodus*. Gayet (in Muizon et al. 1983) also listed undescribed fossils from the Bolivian late Cretaceous as Phareodontinae and possible Hiodontidae. Gayet (1987) speculated that these fossils indicated coastal or freshwater ichthyofaunal connections between North and South America during the Cretaceous, connections perhaps facilitated by the hypothesized "Pacifica" continent. Without detailed comparative anatomical descriptions of these fossils and rigorous phylogenetic analysis of whatever characters they may preserve, we cannot place much credence on this speculation.

In summary, neither of the two potentially African–South American groups of osteoglossomorph fishes is readily explained by the simple drift-vicariance model (table 8.1). The Arapaimidae clearly have an African–South American relation, but they owe their divergence to an unknown predrift speciation event. The osteoglossid-pantodontid clade is too poorly resolved to fit any precise biogeographic model.

D. Otophysi

Characins, catfishes, electric (knife) fishes, carps, and loaches compose a well-known higher teleostean lineage named Otophysi (Ostariophysi in older literature). This clade accounts for about 28 percent of modern fish species (Nelson 1984), and most of these live exclusively in fresh water. I begin with a general treatment of the Otophysi because of their overwhelming importance in the African and South American faunas and in the history of biogeographic study of continental fishes. Individual treatments of characiforms and siluriforms follow.

The characins (Characiformes) are the largest group of freshwater fishes in South America and are a very important faunal component in Africa. Fossils also record the group in the early Tertiary of Europe. Catfishes (Siluriformes) make up major portions of the Asian, African, and South American faunas. In addition, there are prominent siluriform families in the fresh waters of Europe and North America and in tropical to subtropical coastal marine waters. The knife fishes (Gymnotiformes) are an endemic Neotropical clade. A fair representation of barbs and carps and a few loaches (Cypriniformes) are found in Africa, but cypriniforms are most diverse in Asia and are important in European and North American faunas. Cypriniforms are entirely lacking from the Neotropics.

Otophysan monophyly is supported by several synapomorphies, of which the best known are odd auditory specializations of the inner ear, anterior vertebrae (the Weberian complex), and gas bladder. Greenwood and co-workers (1966) and Rosen and Greenwood (1970) found related modifications of the gas bladder and other structures in a small, morphologically heterogeneous group, the Gonorynchiformes. The Gonorynchiformes includes two modern Indo-Pacific species (*Chanos chanos, Gonorynchus gonorynchus*), placed in monotypic families. Several fossils, the oldest of which are early Cretaceous, are assigned to the Chanidae and the Gonorynchidae (Patterson 1975). There are also 24 extant gonorynchiform species in the African freshwater fauna—one placed in the monotypic family Phractolaemidae, the others in Kneriidae. Rosen and Greenwood (1970) united the gonorynchiforms with characins, catfishes, cypriniforms, and knife fishes into Ostariophysi and coined the new name Otophysi for the more traditional assemblage (i.e., without gonorynchiforms). Despite the unfortunate name change for such a widely known group, the hypothesis of sister taxon relationship between the Gonorynchiformes and Otophysi has held up under scrutiny (figure 8.4).

Many biogeographers have offered hypotheses on the distributional history of the Otophysi, and all of these necessarily consider possible direct African–South American relationships of characins and catfishes. Two recent general treatments are Novacek and Marshall (1976) and Briggs (1979). These and earlier papers lack a fully resolved phylogenetic framework for otophysan groups and also invoke unproven assumptions about phylogenetic processes. Before 1981, the prevailing phylogenetic

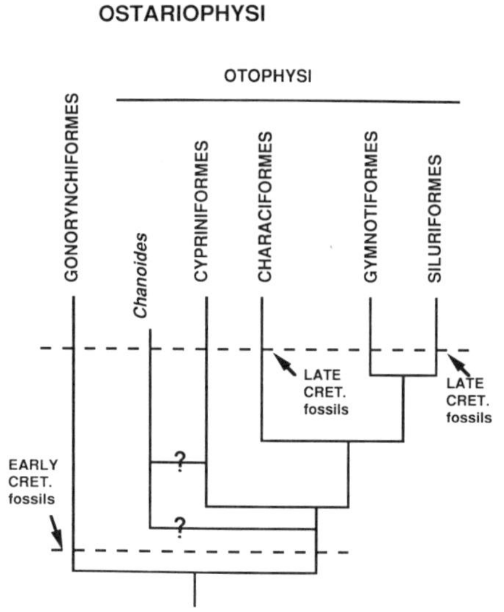

Figure 8.4. Hypothesis of relationships among Ostariophysi, based on Fink and Fink (1981) and Patterson (1975).

concept for otophysans was that characiforms were relatively primitive (i.e., gradistically more like otophysan ancestors) and were more closely related to cypriniforms than to siluriforms. Gymnotiforms were generally viewed as derivatives of characin-like ancestors within South America. The differences among the biogeographic hypotheses for the otophysans largely involve the place and time of origin of subgroups and the directions of their dispersal before the vicariance events that separated continental areas and faunas. Novacek and Marshall (1976) assumed that relatively "primitive" groups survive close to their center of origin. They therefore surmised that the Otophysi arose in South America, where there are an abundance of characins plus relatively primitive diplomystid catfishes. Briggs (1979) assumed the opposite—that relatively "primitive" groups will disperse from centers of origin and be replaced at the centers by more advanced lines. Thus, Briggs proposed that otophysans originated in tropical Asia, where the supposedly more advanced cyprinoids and some advanced catfishes presently occur. In either case, the expectations are that the characins and catfishes spread over Africa and South America before the opening of the South Atlantic, that gymnotiforms originated after the split, and that cypriniforms may have originated before the time of continental separation but were somewhere else. Because of the lack of finely resolved phylogenetic hypotheses, there has been no speculation about which characin and catfish subgroups or stem groups spread over the African–South American landmass before continental fragmentation.

A cladistic study of the Otophysi by Fink and Fink (1981) significantly departed from traditional concepts of interrelations (figure 8.4). A clade uniting Gymnotiformes and Siluriformes was proposed to be the sister lineage of Characiforms, and these three were proposed to be the sister taxon of the Cypriniformes. The hypothesis of relations developed by Fink and Fink brought a new level of precision to thinking about higher-level otophysan biogeography. It is an explicit statement on the *relative* timing of origins of the major otophysan groups. However, the lack of detail on intraordinal relationships dissuaded Fink and Fink from suggesting a general hypothesis on the distributional history of the Otophysi.

Guesses about the absolute age of origin of the Otophysi and its subgroups extend as far back as the Triassic (Myers 1967), but there is no convincing evidence to support such an ancient age (Patterson 1975). The oldest unequivocal Otophysan fossils are late Cretaceous catfishes and possibly characiforms, whose systematic position within their orders is unknown (see below). These Cretaceous otophysans provide the oldest minimum estimate for ages of origin of the Siluriformes, Gymnotiformes, Characiformes, and Cypriniformes, as well as Otophysi as a whole, but they suggest nothing about the levels of diversification within any of the four orders. Patterson (1984) analyzed relationships of the Eocene fossil *Chanoides* and found that it is probably either the sister group of the Cypriniformes or the sister group of the Otophysi as a whole. In either position, *Chanoides* does not alter the age estimate provided by the Cretaceous catfishes. Moving outward to the Gonorynchiformes, early Cretaceous fossils (Patterson 1975) permit the inference of the minimum age of origin of the Otophysi, including *Chanoides*. Thus far then, the fossil record combined with Fink and Fink's phylogenetic reconstructions allow the inference that the otophysan stem group had originated by the time of separation of Africa and South America. Although paleontological evidence is lacking, it has been generally expected that the characiform and siluriform stem groups originated and dispersed across Africa and South America before late Cretaceous.

I now turn to a variety of Cretaceous fishes that are described and discussed in a series papers by the French paleontologist M. Gayet. In Gayet's view, these taxa significantly alter the traditional picture of Ostariophysan relationships and distributional history. *Lusitanichthys characiformis,* from Portugal in marine rocks of early late Cretaceous (Cenomanian) age, is treated by Gayet (1981, 1982a, 1985, 1986) as a characiform with a primitive type of Weberian complex. *Salminops ibericus,* also from the marine Cenomanian of Portugal, is interpreted by Gayet (1985, 1986) as a characiform. *Ramallichthys orientalis,* from Israel in marine Cenomanian strata, is hypothesized by Gayet (1982b, 1986) to be a basal or stem-group cypriniform with a primitive Weberian complex. *Molinichthys ibericus,* based on a fragmentary tooth-bearing bone interpreted as a pharnygeal jaw (fifth ceratobranchial), is from the late Cretaceous (Maastrichtian) of Bolivia. Gayet (1982c, 1986, 1991) suggested that *Molinichthys* is a primitive cypriniform; if true, this extends the range of the Cypriniformes to South America.

On the basis on these fossils, Gayet (1982a) proposed an indirect dispersal model (see Section II.F.4) that hypothesizes a marine origin of the basal groups of characiforms and cypriniforms and separate dispersals into the fresh waters of various continents. Gayet's view does not require a widespread ancestral biota containing characiforms (and cypriniforms!) in the Cretaceous fresh waters of the united African–South American landmass. Patterson (1975) had similarly framed a hypothesis of ancient ostariophysan biogeography based on unspecified marine dispersal pathways. Gayet (1982a), however, stands alone in suggesting dispersal of cypriniforms out of South America—followed by their extinction there—to North America and then Asia.

Gayet's work on the history of otophysans has been criticized on several fronts by Patterson (1984) and Fink and co-workers (1984), who have questioned some of Gayet's interpretations of anatomical structure, phylogenetic analysis, and derivative hypotheses on biogeography. Gayet (1986) has responded to some of these criticisms. Many of the complex issues raised in these exchanges are clouded by absence of detailed anatomical description and illustration, no doubt resulting in part from the contingencies of preservation. Gayet's critics, however, have not examined all the specimens upon which her descriptions were based. Extensive comparison of these fossil specimens to modern material is needed, and these comparisons must address conflicts among putative synapomorphies suggested by researchers with different points of view. Until the basic character data of these contentious fossils are clearly understood and their phylogenetic placement as firmly set as possible, it is difficult to accept Gayet's indirect dispersal hypotheses of early otophysan biogeographic history. In order to demonstrate a phylogenetic basis for any such hypothesis, one must show that there are unique sister-group relations between separate endemic continental otophysan subgroups and different Cretaceous marine otophysans.

E. Characiformes

The characins are central to considerations of ichthyofaunal relations between Africa and South America. Characins are confined to the fresh waters of these continents, except for one New World species that ranges north to Texas, and some Tertiary fossils in western Europe. Despite a few scattered brackish water records of individual characins, this vast group is confined to freshwater habitats, and long-distance marine dispersal seems out of the question. The oldest putative characin fossils are isolated teeth from late Cretaceous to early Paleocene deposits in Bolivia. Gayet (1982d, 1991) referred these teeth to the modern characid family groups and even some genera, *Hoplias, Triportheus* and *Rhoadsia.* If correctly identified, these fossils would be highly significant for dating characiforms diversification. Fink and co-workers (1985) questioned Gayet's generic placement of these fossil teeth but we agree that they appear to belong to characiforms. Without a more resolved placement, Gayet's Bolivian fossils can only indicate a late Cretaceous minimum age of origin of the Characiformes, the same conclusion inferred indirectly from the oldest fossil siluriforms. Independently and without drawing on fossil evidence, Bussing (1985) implied a late Cretaceous to Paleocene age of diversification of characiforms, based on distributions and degrees of divergence of Central American genera.

The family group classification of characins is in transition as its subgroups are subjected to phylogenetic analysis. The former practice of recognizing a single family Characidae (Eigenmann and Allen 1942; Weitzman 1962) has given way to very different schemes with 14 (Géry 1977) to 16 (Greenwood et al. 1966) families. I anticipate changes in the higher level phylogenetic classification of characins as well.

The African characiform fauna encompasses about 150 living species, currently placed in four families: Hepsetidae, Distichodontidae, Citharinidae, and Characidae. Characins are present in all African ichthyological provinces except the Maghreb, Abyssinian, and Cape, and the group is most diverse in the Zairean, Guianan and Nilo-Sudan provinces (Roberts 1975). Roberts

(1969) considered the monotypic Hepsetidae to be one of the most primitive lineages of characins, and noted its similarity to the South American Ctenoluciidae. Fink and Fink (1981) and Buckup (1991) have interpreted the similarities between hepsetids and ctenoluciids as synapomorphies that strengthen the hypothesis of hepsetid-ctenoluciid relationship. Vari (1979) presented synapomorphies supporting monophyly of the citharinids plus distichodontids (including "ichthyoborids"). The nominal family Characidae has an African–South American distribution. Although recent authors are of the opinion that the Characidae is not monophyletic (Weitzman and Fink 1983; Buckup 1991), African Characidae share synapomorphies (Roberts 1969; Vari 1979; Brewster 1986) and are grouped, following Vari, into the subfamily Alestiinae (including *Hydrocinus*). Fossil characins of the African alestiine clade are found in western Europe in rocks of Eocene to Oligocene age (Capetta et al. 1972; Gaudant 1980). Thus, the African characin fauna presently appears to contain three monophyletic clades: distichodontid-citharinids, alestiines, and hepsetids.

The South American characin fauna is richly diverse, with perhaps 900–1000 living species assigned to twelve nominal Neotropical families (Greenwood et al. 1966). Discovery and description of new characin species is ongoing. Characiforms are represented in all of Géry's (1969) ichthyogeographic provinces of South America and are most diverse in the Amazonian, Orinocoan, and Paraguay-Parana. Several of the South American families have representatives in isthmian Central America, and the nominal characid genus *Astyanax* just reaches southern Texas. Central American characins are generally considered to trace back to South American ancestors (Myers 1966; Bussing 1985). Although all American characiform families except the nominal Characidae are Neotropical endemics, they apparently are not a monophyletic group. The possibility of a close South American ctenoluciid and African hepsetid relation was noted above. Vari (1983) detailed cladistic evidence supporting the monophyly and interrelations of the South American families Prochilodontidae, Curimatidae, Anostomidae, and Chilodontidae. In the latest literature, the relationships among the other New World characin families are unresolved, as are those of the roughly two dozen nominal family-group taxa (Géry 1977) placed in or close to the American Characidae.

The foregoing is not an auspicious phylogenetic framework for a detailed hypothesis on the distributional history of characiforms between Africa and South America. The current classification of the nominal family Characidae implies just one African–South American characiform relationship: between Alestiinae and whatever group of New World characids may be suggested as its sister taxon. This, however, implies no greater level of African–South American relationship than we would infer from the recognition of the Characiformes as a natural taxon, and it is possibly misleading, given the uncertain monophyly of the Characidae and the uncertainties of suprafamilial phylogeny. Buckup (1991) has made a promising, though preliminary, cladistic study of the higher level phylogeny of characiforms based on synapomorphies. Buckup's working hypothesis must be considered tentative, because it includes only about 30 represen-

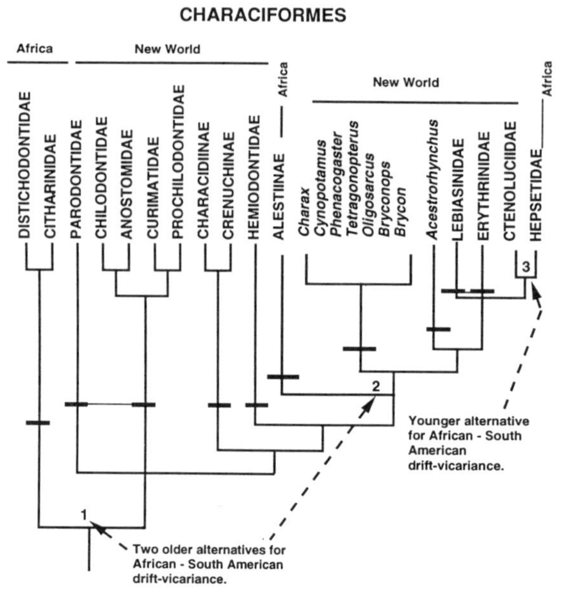

Figure 8.5. Hypothesis of relationships among Characiformes, based on Buckup (1991). The younger alternative applies the drift-vicariance hypothesis to event "*3*," and leads to the inference that lineages labelled with tick marks (eight to ten, depending on resolution of polytomies) had an ancestral African–South American distribution and subsequent extinction in Africa or South America. Either older alternative (applying the drift-vicariance hypothesis to event "*1*" or "2") would imply marine dispersal of one or two descendant lineages from South America to Africa.

tative taxa, and its character base is not yet published. Nevertheless, this hypothesis suggests three levels of Afro-South American characiform relationship (figure 8.5): (1) between the African citharinid-distichodontid clade and all other characiforms (both South American and African), (2) between the African alestiines and a subgroup of American and African characiforms, and (3) between the African Hepsetidae and the South American Ctenoluciidae (already hinted at by Roberts [1969]).

The implications of Buckup's hypothesis for characiform intercontinental biogeography are striking. If we take a strict vicariance view and disallow dispersals of characiforms across a widening South Atlantic marine barrier, the cladistic event labelled 3 in figure 8.5 is explained by subdivision of the hepsetid-ctenoluciid ancestral line after the separation of Africa and South America. This hypothesis concomitantly provides a predrift, late early Cretaceous minimum date of origin for many currently endemic Neotropical subgroups of characiforms and for the basal dichotomy in the order (figure 8.5). It also suggests that these several groups originally had African–South American distributions. This raises the question—likely unanswerable—of why more characiform subgroups (six to eight, according to Buckup's hypothesis) now confined to the New World do not have close relatives in the African fauna (see also *Sindacharax* below). A high and remarkably lopsided rate of characiform extinction in Africa is implied.

Two alternative vicariance scenarios relate the phylogenetically older cladistic events (*1* or *2* in figure 8.5) to African-South American drift-vicariance. As a consequence of the first (event *1*) at least two post-drift, marine dispersal events from South America to Africa are required to explain the distributions of the hepsetids and alestiines. The second scenario requires marine dispersal of hepsetids only. Given that marine dispersal is unlikely for characiforms both older alternatives are problematic.

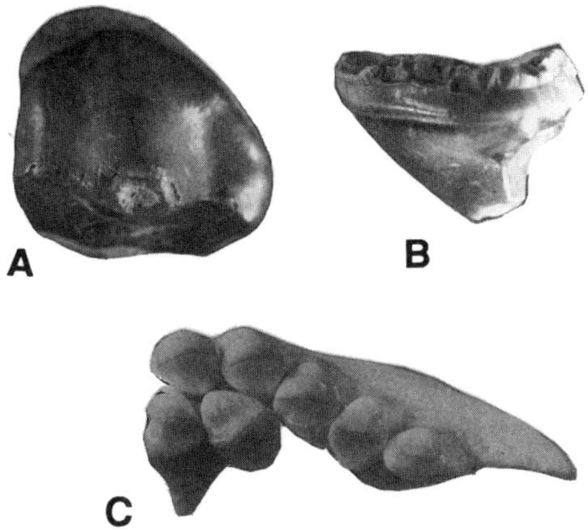

Figure 8.6. Isolated tooth of *Sindacharax leppersoni* (10.5 mm in anteroposterior dimension, 13.3 mm in transverse dimension): *A*, occlusal view, anterior side facing upward, and *B*, lateral view, anterior side facing to right. *C*, Right premaxilla in occlusal view of *Colossoma macropomum* (36.3 mm in length).

The fossil record of characiforms has thus far contributed little to concepts of their intercontinental biogeography. However, Greenwood and Howes (1975) described unusual characiform teeth of Miocene to Lower Pleistocene age from Africa that suggest either remarkable morphological convergence or another important African–South American biotic relationship. These fossils, *Sindacharax lepersonnei* (from Zaire) and *S. deserti* (from Egypt), show little resemblance to the teeth of any extant African characiform but instead are most similar to the teeth of such South American Serrasalminae as *Colossoma* and *Piaractus* (figure 8.6). The Neotropical Serrasalminae are a monophyletic taxon that include specialized herbivorous forms like *Colossoma* as well as the equally specialized, carnivorous piranhas (*Serrasalmus, Pygocentrus*). A detailed phylogenetic hypothesis of this subfamily has been developed by Machado-Allison (1983; figure 8.7). Fossil evidence from the New World shows that some of the highly apomorphic genera of serrasalmines, like *Colossoma* and the piranhas, were differentiated by at least 13 Ma (Lundberg et al. 1986; Lundberg In press). *Sindacharax* shares with the large herbivorous serrasalmines synapomorphic similarities of dentition form, including broadly rounded, stublike teeth each bearing a single cusp along the middle of a transverse crest. If *Sindacharax* truly belongs to the serrasalmine clade, the historical implications are most interesting. Specifically, because serrasalmines are exclusively freshwater fishes, I would favor the African–South American drift-vicariance model to explain

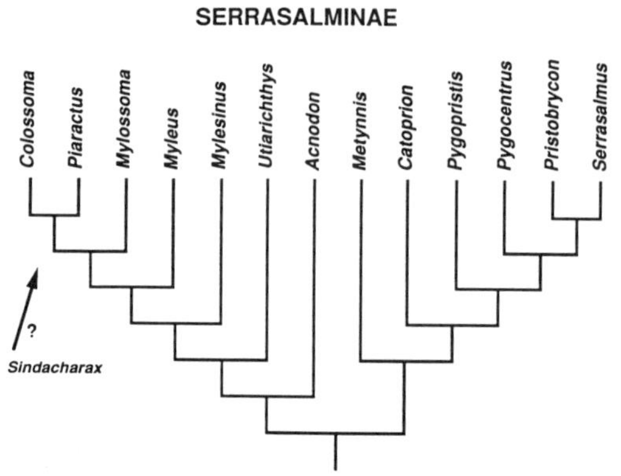

Figure 8.7. Hypothesis of relationships among Serrasalminae, based on Machado-Allison (1983). The African fossil *Sindacharax* is most similar to *Colossoma* and *Piaractus*.

their distribution. As with the hepsetid-ctenoluciid pair, this would imply a minimum age of origin for the highly apomorphic serrasalmine clade about the middle of the Cretaceous (greater than 84 Ma). This would in turn suggest an equivalent or even earlier origin for some additional characin subgroups and would also provide an example of extirpation in Africa of one trans–South Atlantic clade. We can only hope that additional collecting will yield more complete remains of *Sindacharax* that will clarify its phylogenetic position.

In summary, tentative hypotheses of characiform intrarelations suggest multiple African–South American clades, not a single sister-group pair. Following our expectation that characiforms in the past, as now, were freshwater dispersants, the African–South American drift-vicariance model may explain some trans–South Atlantic sister-group pairs. However, the drift-vicariance model implies that several Neotropical characiform subgroups were represented before drift in Africa, where they are now extinct.

F. Siluriformes

About 35 families of catfishes are provisionally recognized, together comprising over 2000 species. Siluriforms are most diverse in species and morphology in tropical South America, Africa, and Asia. However, catfishes live or have lived in the inland or coastal waters of all continents, including Antarctica, where fossils of Eocene age are found on Seymour Island (Grande and Eastman 1986). Most families of catfishes are primary freshwater fishes, but four clades occur in coastal and shallow-water marine habitats: Ariidae (worldwide tropical and temperate regions, with a few species living permanently in fresh water), Plotosidae (Indo-Pacific, with some permanent freshwater species), Aspredininae (northeastern South America, the sister group of the freshwater Bunocephalinae), and an undescribed extinct species informally called "titanoglanis" (Eocene Gulf Coast of North America; see below). The marine habitat is considered to be derived among catfishes because of the more general, strictly freshwater habitat of the majority of subgroups, including the relatively primitive and cladistically basal families Diplomystidae and Hypsidoridae, as well as other otophysan outgroups.

Monophyly of the Siluriformes is strongly supported by many synapomorphies (Fink and Fink 1981 and references therein). Although there is general agreement on the broad outline of the higher-level phylogeny of catfishes (figure 8.8), many questions remain. The southern South American Diplomystidae are the plesiomorphic sister group of the remaining catfishes or Siluroidei (Grande 1987). Within the Siluroidei, the middle Eocene *Hypsidoris* (Hypsidoridae) from North America is the primitive sister taxon of all other catfishes or Siluroidea (Grande 1987). Two of the largest nominal families, the South American Pimelodidae and the African-Asian Bagridae, have been vaguely portrayed as the "ancestral" groups for the Old World and Neotropical catfish radiations. In a cladistic sense, it is possible that the pimelodids and bagrids are paraphyletic, although monophyletic subgroups have been diagnosed within each family (e.g., Lundberg et al. 1991; Bailey and Stewart 1984).

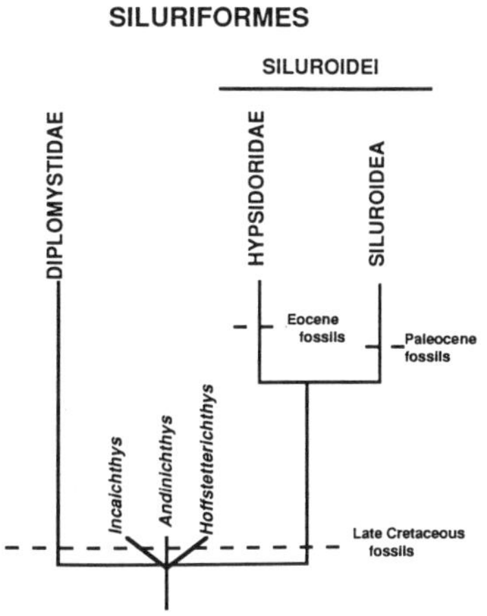

Figure 8.8. Hypothesis of relationships among Siluriformes, based on Grande (1987), with late Cretaceous fossils (Gayet 1990) included in a basal cladistic polytomy.

The latest attempt to provide a comprehensive higher-level classification of catfishes is that of Chardon (1968). Because Chardon's approach was explicitly gradistic, many of his proposed taxa are neither monophyletic nor based on character evidence that can be construed as synapomorphic. It is not surprising, however, to find that a few of Chardon's groups are equivalent or nearly so to clades that are supported by phylogenetic criteria—for example, Diplomystoidei, doradoids, Clarioidea, and Loricarioidei (in part). Regan (1911) recognized similar entities in his classic paper on catfish classification.

Although catfishes are mentioned in virtually all treatments of African–South American ichthyofaunal relations, lack of a detailed cladistic framework for catfish phylogeny has impeded the development of specific historical hypotheses. No monophyletic taxa of family or lower rank of strictly freshwater catfishes occur in both Africa and South America. In Africa, six families of freshwater catfishes plus the Ariidae are represented by over 400 species. Three of these six families, Bagridae, Clariidae, and Schilbidae, also occur in Asia. The others are endemic—Malapteruridae, Amphiliidae, and Mochokidae. Fourteen endemic families of freshwater siluriforms plus several ariids live in the Neotropics, with a species total that likely exceeds 1200. Of the strictly freshwater families, only the Pimelodidae has any representation outside of the Neotropics, with some species of *Rhamdia* in Mexico. A few other families have representatives in isthmian Central America; as with the characins, these are generally thought to be descended from ancestors that dispersed out of South America (Myers 1966; Bussing 1985).

The earliest known fossil siluriforms are in freshwater and marine deposits of late Cretaceous (Campanian and Maastrichtian) age. Most are fragmentary fin spines and shoulder girdle and skull elements from South America: Argentina (Cione and Lafitte 1980; Cione et al. 1985; Cione 1987), Bolivia (Wenz 1969; Gayet in Muizon et al. 1983), and Brazil (Gayet and Brito 1989). Although workers have assigned some of the fossils to modern families, these specimens have a relatively plesiomorphic structure that disallows a precise statement on their relationship (figure 8.8). In general, these remains resemble structures found in various modern species of Ariidae, Ictaluridae, Bagridae, Pimelodidae, and others. There is no basis in shared derived character states for Gayet's identifications (in Muizon et al. 1983) of Bolivian fossil catfishes as North American *Rhineastes* and Ictaluridae. Cione (1987) tentatively referred some Cretaceous pectoral spine fragments to the Diplomystidae, noting that in several features the fossils are nearly identical to the modern *Olivaichthys viedmensis*. Although the notion that diplomystids were differentiated by the late Cretaceous does not strain credulity, the fossil spines and those of *Olivaichthys* have a relatively primitive character and may not be distinguishable from spines expected of basal stem groups or found in some other extant catfishes. Cione (1987) also tentatively identified some spines as pertaining to the Ariidae, but again it is not certain that the character evidence involves synapomorphies. Gayet (1988, 1990, 1991) described partial skulls and anterior vertebrae of catfishes from the late Cretaceous or Paleocene of Bolivia and names these *Andinichthys, Hoffstetterichthys,* and *Incaichthys*. Because of the apparently plesiomorphic nature of the fossil skulls and vertebrae, Gayet (1990) wisely does not attempt to place the new taxa within an extant family or infer their relationships. Late Cretaceous (Maastrichtian) isolated otoliths from marine deposits in South Dakota were assigned to the Siluriformes by Frizzell (1965). Although these otoliths bear shape and size similarities to the enlarged otoliths of marine ariid catfishes, these structures have not been subjected to phylogenetic character analysis; it would therefore be premature to identify them as ariids.

The late Cretaceous records of siluriforms signal at least a Campanian age of origin for the order and, by inference, for the other orders of the Otophysi. This minimum age postdates the final separation of Africa and South America, but of course these groups may have originated earlier, for the otophysan stem group is at least early Cretaceous in age. The late Cretaceous catfish fossils, because they cannot yet be phylogenetically placed, reveal nothing about the level of diversification within the Siluriformes by the end of the Cretaceous (figure

8.8). The oldest fossil catfishes identified as belonging to an extant family are late Paleocene Ictaluridae (Lundberg 1975; Grande and Lundberg 1988). All we know of the higher interrelations of the Ictaluridae is that they belong to the Siluroidea. Their Paleocene fossils suggest only that ictalurids were coeval with the diplomystids, the hypsidorids, and an unspecified group of Siluroidea (figure 8.8). Thus, according to the phylogenetic, distributional, and temporal information in figure 8.8, the Siluroidea, Siluroidei and Siluriformes all resolve, uninformatively, to an ancestral cosmopolitan range in the late Cretaceous. If there are any unique siluriform relations between the African and South American catfish faunas, they must be at lower taxonomic levels. Three such possibilities are considered: Ariidae, doradoids, and Amphiliidae-Loricarioidei.

The Ariidae are the only family represented in the fresh waters of both South America and Africa. Unfortunately, our poor understanding of the alpha taxonomy and intrarelations of ariids does not permit hypotheses to be drawn on their possible trans–South Atlantic affinities. Resident freshwater ariid species occur in Central America and Mexico, Australia and New Guinea, and Madagascar. Except for *Arius gigas* in the Niger, however, the ariid species that move into the main rivers of Africa and South America are euryhaline coastal marine fishes. If African–South American sister-group pairs of ariids are ever identified, any explanation of their distribution would therefore have to consider the possibility of marine dispersal.

A putative catfish clade that combines the endemic African family Mochokidae and the South American endemics Doradidae, Auchenipteridae, and Ageneiosidae is informally referred to herein as the doradoids, following Günther (1864) and Chardon (1968). Ferraris (1988) rearranged the South American doradoid genera into the families Doradidae, Auchenipteridae, and Centromochlidae (including the ageneiosids); Curran (1989), however, maintains the traditionally recognized subgroups. In either arrangement of the Neotropical members, the suggestion of a higher-level relation between these South American and African catfishes is long standing (Günther 1864; Regan 1911; Chardon 1968; Royero 1988; Walsh 1990) and is based on synapomorphic characteristics (see caption to figure 8.9). The doradoids thus provide the only published example of a possible direct African–South American relationship among catfishes. This relationship can be parsimoniously explained by the simple drift-vicariance model (figure 8.9). The doradoid case is possibly complicated, however, by additional evidence for relation among the doradoids, the Ariidae, and the undescribed fossil "titanoglanis"—all three collectively referred to herein as "arioids" (figure 8.9). The Eocene fossil catfish "titanoglanis" comes from marine coastal deposits of southeastern Arkansas. This taxon shares the highly apomorphic arrangement of nuchal plates with the doradoids, but in each clade the shapes of the individual plates are unique (figures 8.10 and 8.11). The Ariidae are provisionally placed as the sister taxon to "titanoglanis" plus the doradoids (figure 8.9). Under this hypothesis, the two closest outgroups of the freshwater doradoids are marine fishes, but this does not imply that the ancestral doradoids per se were marine. Because all modern and Tertiary fossil doradoids are fresh-

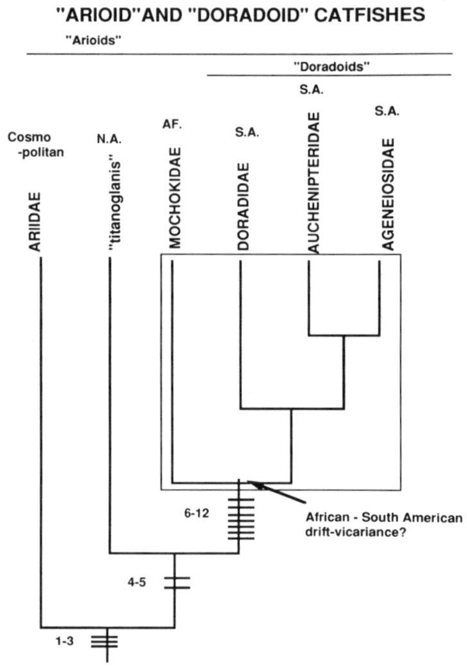

Figure 8.9. Hypothesis of relationships and synapomorphy pattern among the major groups of "arioid" and "doradoid" catfishes. Numbered sequence of tick marks corresponds to the following list of synapomorphic characters that diagnose three nested levels of relationship. 1. Basioccipital with short lateral processes articulating with ossified transcapular ligament. 2. Epioccipital with a ligamentous sheet or osseous process reaching posteriorly to fourth parapophysis. 3. Anterior limb of fourth transverse process modified to form an "elastic spring apparatus" controlled by anterior supracarinales muscle (uncertain in "titanoglanis"). 4. Posterior edge of supraoccipital truncated, not drawn out to form a process. 5. Middle nuchal plate with anterolateral processes contacting posttemporal-epioccipital region of skull (apparently reversed in some species within Mochokidae and Ageneiosidae). 6. Distal end of elastic spring (anterior limb of fourth transverse process) expanded to form a disc. 7. Depressor dorsalis muscle of second dorsal-fin lepidotrich, or spine, inserts on base of first lepidotrich, or spinelet (uncertain in "titanoglanis"). 8. First dorsal-fin lepidotrich (spinelet) with greatly elongated and recurved limbs (uncertain in "titanoglanis"). 9. First and second dorsal-fin basals tightly bound or sutured to fourth, fifth, and sometimes sixth neural spines. 10. Epioccipital with a large superficial dermal component in the skull roof (present also in some ariids). 11. Tripus with a recurved transformator process that enters the peritoneal tunic of swim bladder (uncertain in "titanoglanis"). 12. Compound centrum of Weberian complex strongly sutured to exoccipitals. Characters compiled from Regan (1911), Chardon (1968), Royero (1987), Ferraris (1988), Curran (1989), Walsh (1990), and personal observation.

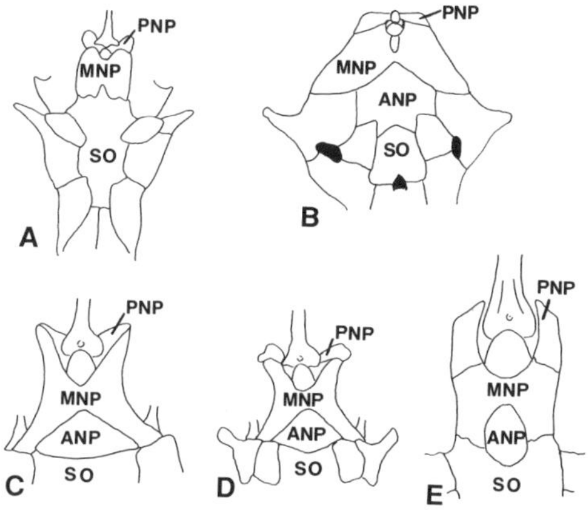

Figure 8.10. Dorsal views of occipital and nuchal region skeletal patterns in the major groups of arioid and doradoid catfishes; anterior toward bottom of page. *A*, Ariidae, based on *Arius proops; B*, "titanoglanis"; *C*, Auchenipteridae, based on *Parauchenipterus galeatus; D*, Doradidae, based on *Leptodoras linneli; E*, Mochokidae, based on *Synodontis membranaceus. C–E* based on Royero (1987). See figure 8.9, characters 4 and 5, for explanation of synapomorphy pattern. *ANP*, anterior nuchal plate; *MNP*, middle nuchal plate; *PNP*, posterior nuchal plate; *SO*, supraoccipital.

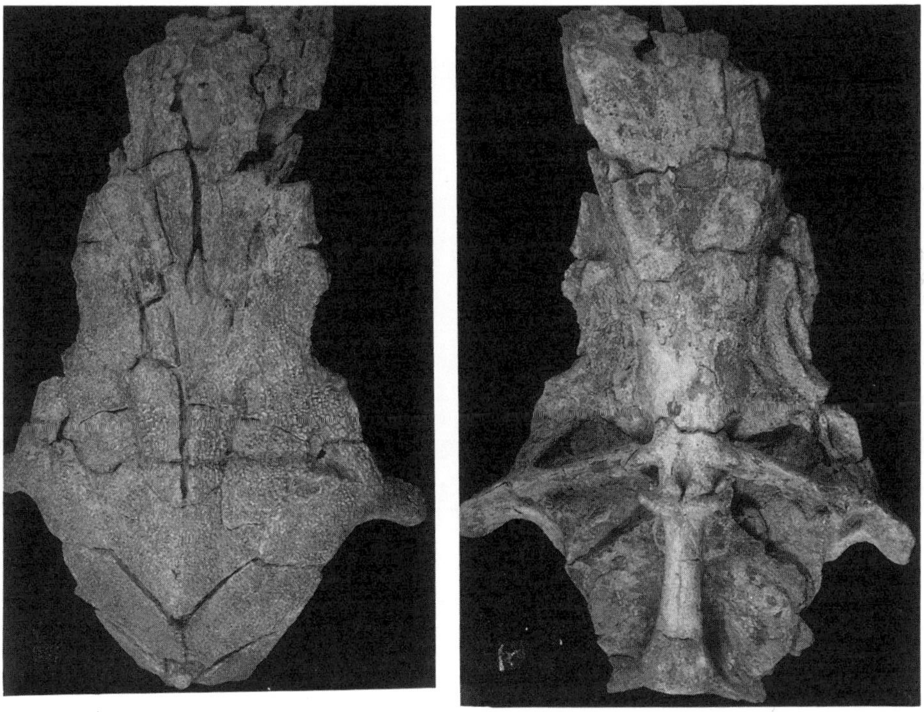

Figure 8.11. Skull of the undescribed fossil catfish "titanoglanis," from Eocene coastal deposits of Camden, Arkansas; anterior toward top of page. Left, dorsal view; right, ventral view. Maximum length of specimen, 59 cm.

water fishes, the ancestral distribution of the doradoids parsimoniously remains as freshwater African–South American, and this freshwater distribution appears to be a relatively apomorphic feature among arioid catfishes.

The African endemic family Amphiliidae is uncertainly related to other catfish families. Similarities between some rheophilic amphiliids and Asian sisorid catfishes have been noted by Skelton (1988). Regan (1911) vaguely regarded amphiliids as bagrid relatives. Bailey and Stewart (1984) transferred the genera *Leptoglanis* and *Zaireichthys* from the Bagridae to Amphiliidae. Skelton (1988) presented data in support of this alignment for Amphiliidae and *Leptoglanis* and further suggested their close relationship to *Australoglanis,* a nominal bagrid. In striking contrast to the foregoing suggestions of Old World relationships, Pinna (1989) illustrated derived features shared by some amphiliids and Neotropical loricarioid catfishes. This last group includes more than 1000 species assigned among the Nematogenyidae, Trichomycteridae, Callichthyidae, Scoloplacidae, Loricariidae and Astroblepidae (Chardon 1968; Baskin 1972; Howes 1983; Schaefer and Lauder 1983; Schaefer et al. 1989). Pinna's preliminary results may eventually lead to a well-supported hypothesis of an additional trans–South Atlantic relationship of catfishes, but at present it is highly tentative.

In summary, two of the three potentially African–South American groups of catfishes—Ariidae and Amphiliidae-Loricaroidei—are too poorly understood phylogenetically to be analyzed rigorously. However, from what we know of modern ariid physiology, it seems that a dispersal model must be considered for any possible African–South American subgroup. The doradoid group exhibits a straightforward African–South American pattern and can be hypothesized to fit the simple drift-vicariance model, without implications of unexpected patters of wider distributions and extinctions of related catfish groups.

G. Cyprinodontiformes

Cyprinodontiforms, or killifishes, and their relatives number about 800 extant freshwater and brackish water species. The group has an overall wide distribution in North America, Europe, Southeast Asia, Africa, and South America. These fishes are notably absent from the temperate Orient, Australia, and Antarctica. Parenti (1981) referred to this distribution as "reduced Pangaean." The traditionally recognized families (see, e.g., Greenwood et al. 1966) included four viviparous New World groups (Poeciliidae, Anablepidae, Goodeidae, Jenynsiidae) and the nearly cosmopolitan Cyprinodontidae (killifishes). The cyprinodontids frequently have been listed as a group with possible direct freshwater relations between Africa and South America, although the coastal brackish water habitats of many species have led most authors to mention the possibility of their marine dispersal.

Parenti's (1981, 1984a, 1984b) studies of the killifishes provide some of the most resolved phylogenetic hypotheses and biogeographic analyses of African–South American piscine relationships. Those aspects of her phylogenetic conclusions that principally involve African and South American members are shown in figure 8.12. Among the cyprinodontiforms, Parenti (1981) sug-

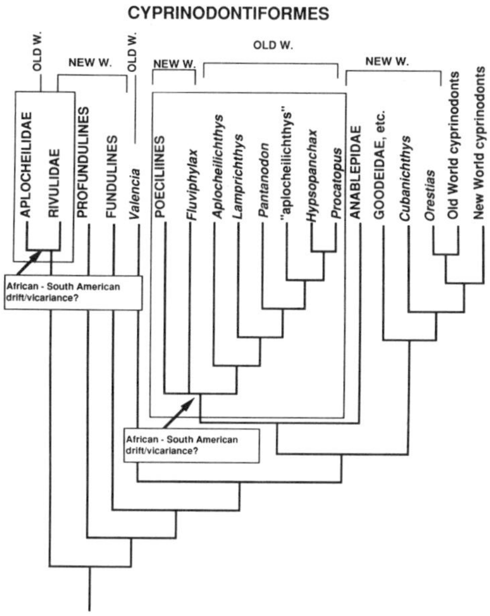

Figure 8.12. Hypothesis of relationships among Cyprinodontiformes, based on Parenti (1981). Groups enclosed by rectangles are those affected by African–South American drift-vicariance.

gests two direct African–South American relationships. The first involves the Aplocheiloidei, within which there is a basal dichotomy into Old World and New World family-level clades—Aplocheilidae and Rivulidae (figure 8.12). The second involves a family group named Poeciliidae by Parenti, including the New World Poeciliinae (corresponding to the more traditional family of that name), as well as the South American *Fluviphylax* and the African–Madagascan Aplocheilichthyinae (figure 8.12). Using Parenti's phylogenetic framework as the best estimate of the phylogeny of African–South American cyprinodontiforms, I consider alternative explanations of their historical biogeography.

Under the African–South American drift-vicariance scenario (figure 8.12), an age of 84–106 Ma is estimated for the relationships between (1) Aplocheilidae and Rivulidae and (2) Poeciliinae (plus *Fluviphylax?*) and Aplocheilichthyinae. If the origins of these groups are correctly related to African–South American separation, I infer that most of the basic clades of cyprinodontiforms existed by late early Cretaceous. This age estimate lacks the direct support of paleontological evidence. The oldest fossil cyprinodont is *Prolebias,* from the Oligocene of Europe (Parenti 1981). A member of the cladistically distal subfamily Cyprinodontinae, *Prolebias* sets a minimum age of origin in the early Tertiary for all of the basal clades of cyprinodontiforms shown in figure 8.12. It is generally thought that these Tertiary fossils underestimate the age of origin of the Cyprinodontiformes. Myers (1938), Kosswig (1943), and Sethi (1978) suggested a late Triassic origin for cyprinodontiforms in the Tethys Sea, followed by dispersal to their present range. Parenti (1981) hypothesized, based on the "reduced Pangean" distribution of the killifishes, that they must be of late Triassic age to have achieved their present distribution via vicariance processes.

There are reasons to question the foregoing vicariance-based estimates of times of origin of cyprinodontiform fishes that are considerably older than middle Cretaceous. The first is paleontological, and I find it compelling, even though it rests on negative evidence. If the Cyprinodontiformes had originated by late Triassic, their nested higher clades Atherinomorpha and Acanthomorpha must also have. The oldest known fossil records for these taxa are, respectively, early Tertiary (Carroll 1988) and basal late Cretaceous (Patterson 1990). Acanthomorpha is an immense clade encompassing 21 of the 35 orders of teleostean fishes. The lack of paleontological support for very old acanthomorph, atherinomorph, and cyprinodontiform origins and diversification could be due to massive sampling error in the fossil record. There is no way to know unless older fossils are discovered. This, however, seems unlikely, given the rich supply of late Triassic to middle Cretaceous fossil fishes representing many higher actinopterygian clades from a variety of habitats around the globe. In this material there is no signal from the Acanthomorpha.

Second, if the timing of cyprinodontiform phylogenetic events is based on the available timetable of continental drift, inconsistencies arise. For example, within the Old World poeciliid subfamily Aplocheilichthyinae, the genus *Pantanodon* has one species in east Africa and its sister species on Madagascar (Parenti 1981:451, figure 81). Also among the Old World Aplocheilidae, the genus *Aplocheilus* (as treated in Parenti 1981:583, figure 92) is entirely Asiatic, ranging from India to the Indo-Malay Archipelago. Parenti places *Aplocheilus* as the closest relative of *Pachypanchax,* with species in east Africa, Madagascar, and the Seychelles. The remaining aplocheilids are African endemics. A strict drift-related vicariance model for *Pantanodon* and the aplocheilids, based on the separation of India-Madagascar from east Africa, yields age estimates of 140–165 Ma for these cladogenetic events. These drift-based, middle Mesozoic estimates of relatively terminal clades directly conflict with the African–South American vicariance models that give mid-Cretaceous minimum estimates for the relatively older, more basal dichotomies within the Poeciliidae and Aplocheilidae.

Because continental drift-vicariance models do not completely explain the intercontinental sequence of phylogenetic events in cyprinodontiform history, alternative models should be considered. The coastal marine distribution of several cyprinodontiforms suggests the possibility that salt water was not a barrier to these fishes in the past as it is now. If African–South American drift-vicariance models are maintained for the Aplocheiloidei (Aplocheilidae and Rivulidae) and Poeciliidae (Poeciliinae, *Fluviphylax,* and Aplocheilichthyinae), dispersal explanations must be found for the younger *Pantanodon* and Old World aplocheilid intercontinental distributions. The hierarchical pattern of geographic distributions within each of these clades suggests dispersal away from an ancestral African range to Madagascar and India. Dispersal scenarios for the African–South American Aplocheiloidei and Poeciliidae can also be polarized within the cyprinodontiform phylogenetic framework. Because the New World distribution is more general, I infer that the direction of movement for both clades was from the New World to Africa. The dispersal scenarios cannot alone provide a time estimate for the implied intercontinental movements of these fishes.

In summary, there are two African–South American clades of cyprinodontiform fishes—Aplocheiloidei and Poeciliidae. Both groups are relatively basal within the available framework of cyprinodontiform relations. The distribution of these groups may follow the African–South American drift model, and they would thereby be minimally dated by that vicariance event. However, younger sister-group pairs of cyprinodontiforms with distributions on older Gondwanan fragments (India and Madagascar) cannot be explained by the current continental drift timetable. The Mesozoic dates of cyprinodontiform origin and diversification suggested by the early Gondwanan drift timetable imply a higher-level radiation of advanced teleosts that is not indicated in the fossil record. Dispersal models for cyprinodontiform fishes do not contradict knowledge of their ecology and physiology. As polarized by the hierarchic pattern of cyprinodontiform distributions, dispersals are inferred to have been from the New World to Africa, and from Africa to Madagascar and India.

H. Cichlidae

The Cichlidae are an advanced family of spiny-rayed, perciform fishes placed in the suborder Labroidei, along with wrasses and parrot fishes (Labridae), sergeant majors (Pomacentridae), and surf-perches (Embiotocidae). The range of cichlids includes South and Central America, Mexico north to the Rio Grande, some of the Greater Antilles, Africa, Syria, Iran, Madagascar, southernmost India, and Sri Lanka. Because of a continuing high wave of new discoveries in Africa and South America, estimates of species numbers of cichlids now exceed 1000. Much of the species and morphological diversity of cichlids is found in their species flocks in the east and central African lakes. The numbers of endemic species that have evolved in these lakes are astounding (Greenwood 1984): Malawi has more than more than 200 cichlid species (Lewis and coworkers [1986] speculate that there may be 500 to 1000 Malawi cichlids), Victoria has at least 200, and Lake Tanganyika has about 140.

Cichlids generally are confined to freshwater habitats, but some, notably members of the Indian–Sri Lankan genus *Etroplus,* can live in brackish water. All of the labroid fishes, identified as the closest relatives of cichlids, are marine (Liem and Greenwood 1981). Cichlids are therefore considered to be secondary freshwater fishes, and biogeographers generally raise the possibility of marine or coastal dispersal.

The phylogeny and higher level classification of the Cichlidae is unsettled but is being actively investigated by contemporary workers using an explicitly cladistic approach. Stiassny (1981) reviewed and added to the evidence for monophyly of the family. Liem and Greenwood (1981), Kaufman and Liem (1982), and Stiassny and Jensen (1987) have examined the relationships of cichlids with other Labroidei (labrids, pomacentrids, embiotocids). Stiassny (1991) combines and extends several recent phylogenetic analyses of the Cichlidae (Cichoki 1976; Greenwood 1986; Kullander 1983; Oliver 1985; Stiassny 1982, 1987). The result of Stiassny's synthesis is presented in figure 8.13. I repeat her caveat that the data set underlying this cladogram contains considerable homoplasy, and some alternative tree topologies do not differ greatly in terms of parsimony.

In South America, there are cichlid fossils (*Macracara*) of late Oligocene or early Miocene age, and in Africa, the oldest cichlid remains are of Oligocene age (Van Couvering 1982). In neither case are phylogene-

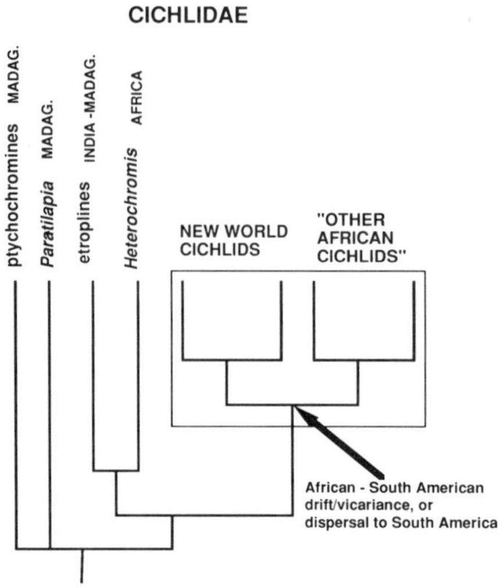

Figure 8.13. Hypothesis of relationships among Cichlidae, based on Stiassny (1991). Groups enclosed by rectangles are those affected by African–South American drift-vicariance.

tic relationships resolved, but, as far as known, these "earliest" cichlids had essentially modern morphologies (Stiassny 1991). These fossils probably underestimate the age of origin of the Cichlidae, but this is not out of line with the earliest known records for many other monophyletic groups of perciforms (Patterson 1990).

I use Stiassny's phylogenetic framework as the best available estimate of cichlid phylogeny in which to consider alternative drift-vicariance and dispersal explanations of their historical biogeography. I focus first on a drift-vicariance model. The African–South American sister-group pair including the monophyletic Neotropical cichlids and the majority of the African ones can be related to formation of the South Atlantic marine barrier in the late Cretaceous. In the drift-vicariance model, no matter how the basal trichotomy is resolved, the cladogenetic event that divided the Indo-Madagascar "etropline" clade from the African *Heterochromis* can be hypothesized to have resulted from the separation of India-Madagascar from Africa. This separation is thought to have occurred between 140 and 165 Ma (middle Mesozoic). Although no dispersal across the South Atlantic or Indian Ocean is implied by a strict drift-vicariance explanation, I believe that there is a problem with the ages it implies for this and related taxa.

The drift/vicariance model requires an early Cretaceous to late Jurassic minimum date of taxonomic origin and diversification of the cichlids. Stiassny (1991) proposes that cichlids originated in the early Cretaceous and underwent considerable differentiation before continental breakup. Such age estimates for cichlids, and necessarily their higher clades, seem unexpectedly ancient given their distal position in the overall phylogeny of teleosts and are not corroborated by paleontologic evidence. The oldest known fossil records for the nested set of higher taxa including the cichlids are as follows: Cichlidae, late Oligocene (A. L. Cione, personal communication, 1990; Van Couvering 1982); Labroidei, early Tertiary (Carroll 1988); Percomorpha, Paleocene (Patterson 1990); Acanthomorpha, early late Cretaceous (Patterson 1990). As noted in the discussion of cyprinodontiforms, the Acanthomorpha is an extremely large clade and would probably have been sampled much earlier in the paleontological record if its diversification had produced the Cichlidae by at least 84 Ma. The Mesozoic record of fishes is diverse in actinopterygians from a variety of habitats and globally widespread localities but contains no acanthomorphs until near the end of the Cretaceous; one explanation for this is that they did not yet exist.

Doubt about the accuracy of age estimates for cichlids and acanthomorphs derived from drift-vicariance models, coupled with knowledge of the marine habits of at least some living cichlids, lead me to concur with Greenwood (1983), Briggs (1987), and others in considering the alternative of intercontinental marine dispersal for these fishes. The arrangement of Stiassny's cladistic framework implies that the New World distribution is relatively derived from an ancestral Old World (likely African) one (figure 8.13). A minimum date for cichlid dispersal across the South Atlantic is set by late Oligocene or early Miocene fossil cichlids in the New World. The directions of possible earlier or basal cichlid dispersals among India, Madagascar, and Africa cannot be specified. Because all potential cichlid outgroups among the Labroidei are widespread marine fishes, the ancestral habitat and distribution of the cichlid stem group are uncertain.

To sum up the Cichlidae, it appears to me that drift-vicariance explanations imply an age of origin that is far too old in light of available fossil evidence. Models of transatlantic cichlid dispersal can be polarized by distributions within the family and do not violate what we know about cichlid biology.

I. Nandidae

Among the Nandidae, a perciform family, the fishes of interest here are the "leaf fishes." These include two South American monotypic genera, *Monocirrhus* and *Polycentrus,* two west African monotypic genera, *Polycentropsis* and *Afronandus,* and the south Asian genus *Nandus,* with two species. Generally, nandids are small fishes of forest streams, but *Nandus* enters brackish water habitats. The south Asian genera *Badis* and *Pristolepis* are sometimes included as separate subfamilies in the Nandidae. With or without these small groups, the higher-level relationships of the leaf fishes are unresolved.

The relations among the nandid genera are also unresolved. The African nandid genera plus *Nandus* are grouped by Nelson (1984) separately from the South American genera, but this arrangement is not based on cladistic analysis. Liem (1970) identified plesiomorph to apomorph morphological trends across the series *Nandus–Afronandus–Polycentrus–Polycentropsis–Monocirrhus,* but he disclaimed this sequence as being phylogenetic. Cracraft (1974) cites Liem (personal communication) as believing that *Polycentropsis* is the sister group of *Polycentrus* and *Monocirrhus.* Assuming monophyly of the family, any hypothesis of leaf fish intrarelationships will contain at least one direct relationship between Africa and South America.

There are no known fossil nandids but the paleontological record for the higher taxa to which the group belongs is like that for the Cichlidae. A predrift early Cretaceous origin for the family seems unlikely, given the lack of fossils of perciforms and acanthopterygians older than Paleocene and late Cretaceous, respectively. Thus for the Nandidae, as for the cichlids, a drift-vicariance model involving the middle Mesozoic separation of India from Africa and the later Cretaceous separation of Africa and South America imply unacceptably ancient ages of origin for the family and its basic subgroups. Alternatively, postdrift dispersals of nandids through marine habitats seem more likely. No polarized dispersal model for the leaf fishes can be proposed without a resolved phylogenetic hypothesis, and no minimally dated dispersal model can be specified without a fossil record.

J. Galaxiidae

These are smeltlike fishes of the temperate Southern Hemisphere with a circum-Antarctic distribution. Although several species occur only in inland waters, galaxiids are not primary freshwater fishes. The greatest taxonomic diversity of galaxiids is in Australia, Tasmania, and New Zealand. A single species of *Galaxias* occurs in South Africa, and in southern South America there are three species of *Galaxias* and the monotypic *Brachygalaxias*. The group is also represented on oceanic islands—New Caledonia, Lord Howe, Chatham, and the Falklands. One species, *Galaxias maculatus,* has an ample distribution around the southern continents, including Australia, New Zealand, and South America. Some species have a juvenile marine phase; *G. maculatus* breeds in estuaries, and its larvae range far into the southern oceans (McDowell 1978, 1981).

Begle (1991) located synapomorphies supporting the monophyly of galaxiids and some other southern fishes, but the phylogeny of galaxiid genera and species is poorly resolved. Rosen (1978) proposed a partially resolved phylogeny, based on a single putative synapomorphy, that sets the South African *G. zebratus* apart as the plesiomorphic sister taxon to all other galaxiids. If Rosen was correct, the galaxiid relationship between Africa and South America cannot be monophyletic. It is also unlikely that the South American galaxiid assemblage forms a monophyletic subgroup, because it includes the widespread *G. maculatus* as well as the endemic *Brachygalaxias*. The fossil record does not offer any indication of the existence of the Galaxiinae before the Pliocene (Carroll 1988), but this surely underestimates their age of origin.

Rosen (1974, 1978) and Croizat and coworkers (1974) included the galaxiids in a group of clades sharing a general "panaustral Gondwanian" biotic pattern of distribution among the southern continents. The chironomid midges that those authors include in this fauna have closer phylogenetic relationships between South America and Australia than either has with South Africa, and these relations are congruent with the temporal sequence of southern continental separations. Application of the African-South American drift-vicariance model to the Galaxiidae in the context of Rosen's partial phylogeny dates their basal dichotomy between 84 and 106 Ma, and the South American–Australian and similar relations would be younger than that.

McDowell (1978, 1981) and Campos (1989) argue against a drift-vicariance model for the galaxiids in favor of a distributional history dominated by marine dispersal. McDowell emphasizes the congruent south-temperate distribution pattern of galaxiids, southern lampreys, retropinnids, aplochitonids, and prototroctids, all of which contain species that migrate between marine and fresh water. McDowell and Campos also argue that the widespread distribution and lack of geographic differentiation of *G.*

maculatus do not suggest ancient separation by drift-related vicariant events. The biology of galaxiids clearly indicates that marine habitats have not presented barriers for them. For any biogeographic model, a more completely resolved picture of galaxiid phylogeny is badly needed in order to understand the relations among the strictly freshwater species on different southern landmasses, but at present, drift-related vicariance seems unlikely.

K. Synbranchidae

The synbranchid swamp eels are not true eels but instead are considered an apomorphic acanthopterygian group. Synbranchids have a wide distribution in the tropics in Asia, Australia, Africa, and the Neotropics, including South and Central America and Trinidad. Rosen (1976) referred to their distribution pattern as Gondwanian, except for the Central American and West Indian elements. Rosen and Greenwood (1976) produced a partially resolved phylogeny of synbranchids that implies one or perhaps two direct New World–Old World relationships in the genera *Ophisternon* and *Synbranchus* (figure 8.14). It is unclear if there are any direct sister-group pairs uniting African and South American members within this group. In any case, the fact that synbranchids are commonly found in estuaries suggests that synbranchid history will not be convincingly explained by African–South American drift-vicariance. Based on available phylogenetic information among these fishes, any intercontinental dispersal of synbranchids would appear to have been from the Old World to the New, possibly through Africa.

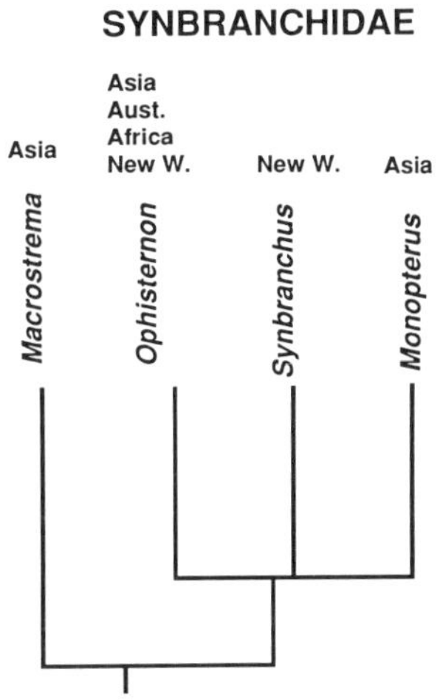

Figure 8.14. Hypothesis of relationships among Synbranchidae, based on Rosen and Greenwood (1976).

IV. Conclusions

Two developments in historical natural science have provided the impetus for critical reviews of the biotic relationships between Africa and South America: (1) a timetable for the separation of the two continents, based on geological and geographical data (Chapter 2) and (2) an increasing number of estimates of phylogenies, based on cladistic methods, for the taxa presumed to have been affected by the continental drift event. In the literature of ichthyology and biogeography, ten extant higher taxa have been proposed as containing subgroups with direct phylogenetic and biogeographic relationships between Africa and South America: Dipnoi, Polypteridae, Osteoglossiformes, Characiformes, Siluriformes, Cyprinodontiformes, Cichlidae, Nandidae, Galaxiidae, and Synbranchidae. I examined each of these higher taxa in as much detail as available systematic analyses permit and found support ranging from strong cladistic evidence to vague speculation for 13 groups that might have directly ancestral African–South American relationships (table 8.1). Not included are the Osteoglossidae + Pantodontidae and Ariidae, which were dismissed because of lack of phylogenetic and biogeographic resolution. Except for the well-corroborated lepidosirenid lungfishes, *Lepidosiren* and *Protopterus,* all of the African–South American groups in table 8.1 need further testing, and most need enhanced cladistic resolution.

Among the 13 putative African–South American clades, only 3, the lepidosirenid lungfishes, the polypterid reedfishes, and the doradoid catfishes, can be fit to the simple African–South American drift-vicariance model without raising unexpected biogeographic or biochronologic problems. The fossil record of the Lepidosirenidae is consistent with the hypothesis that the common ancestor of *Lepidosiren* and *Protopterus* inhabited the fresh waters of the African–South American landmass before the opening of the South Atlantic Ocean and the modern sister genera differentiated in isolation after continental separation. The known fossil record and presumed remote phylogenetic relations of polypterids suggest that their stem group was in place on the African–South American landmass at the time of fragmentation. Following their vicariance, the South American polypterid lineage suffered extinction sometime after the early Paleocene. For the catfishes, a number of late Cretaceous fossils suggest an age of siluriform diversification approaching the time of African–South American drift. Therefore, although the known fossil record of catfishes does not directly indicate that an African–South American doradoid stem group existed before drift, the discovery of evidence for such a group would not be unexpected. Perhaps the loricarioid-amphiliid catfish and characiform cases will also fit the drift-vicariance model, but the suggested clades need further support. In any case, it appears that the long-recognized, common African–South American pattern shared by several freshwater groups is the result of different kinds of biogeographic processes operating at different times in earth history, and probably not single-event vicariance alone.

The Arapaimidae do not fit the African–South American drift-vicariance model. Based on the phylogenetic positions and ages of related fossil taxa, the modern arapaimid genera *Heterotis* and *Arapaima* are estimated to have ages of origin that predate ultimate separation of the two continents. Some drift-related event preceding the final separation may have affected this clade.

Contrarily, it can be argued that the African–South American clades of Cichlidae, Nandidae, Synbranchidae, and perhaps Aplocheiloidei and Poeciliidae are too young to have been affected by drift-vicariance. This position is based on negative evidence—the fossil record does not indicate the origin of higher outgroups in the richly diverse clades Acanthomorpha, Percomorpha, or Atherinomorpha until after African–South American drift. If the African–South American cichlid, nandid, synbranchid, and cyprinodontiform relationships are not explained by drift, postdrift dispersal across a marine habitat is an alternative. Physiological and ecological considerations of modern members of these taxa, as well as the galaxiids, do not rule out, and might even encourage, marine dispersal explanations of their present intercontinental distributions. Under a dispersal model, the Old World distributions of within-family outgroups for the Neotropical cichlids and perhaps synbranchids suggest that the direction of dispersals was toward the Neotropics.

Even if the cyprinodontiforms diversified as far back as middle Mesozoic, simple Gondwanian drift-vicariance cannot provide a general explanation of their distribution, assuming that the available phylogenetic framework is correct. In this taxon, there are cases of phylogenetically young clades exhibiting endemic distributions in Gondwanian fragments (India and Madagascar) that are older than the African–South American fragments occupied by phylogenetically older and more inclusive groups. Such seemingly self-contradictory vicariant patterns strongly suggest that marine dispersal has played some role in cyprinodontiform history. Again, physiological and ecological data indicate that marine habitats are not a barrier to cyprinodontiforms and that marine dispersal hypotheses therefore should be considered. Within the available cladistic framework, New World cyprinodontiform subgroups are phylogenetically more general, suggesting that the direction of dispersals was toward the Old World, opposite of that inferred for cichlids.

The characiforms are perhaps the most enigmatic of the African–South American groups. Because the order is monophyletic, there must be at least one monophyletic subgroup pair straddling the South Atlantic. For characins, as we understand their biology, marine habitats are formidable barriers; transoceanic dispersal hypotheses, even across a much narrower early South Atlantic, should be disfavored. However, within the developing yet tentative phylogenetic picture of Characiformes, the drift-vicariance model carries some unexpected and problematical implications. First, it implies a mid-Cretaceous or earlier age for a high degree of diversification of subgroups, for which there is no fossil evidence until middle Tertiary. Second, the vicariance model suggests that at least six to eight now-endemic South American subgroups and two now-endemic African clades were present on the united continent at the time of separation. Because there is no reason to assume that these stem groups were restricted to the South American or the African portion of the united landmass, the model leads to an oddly lopsided postdrift extinction scenario for characiform clades in Africa. The Mio-Pleistocene African characiform *Sindacharax,* if it really belongs to the Neotropical Serrasalminae, may record one such extinction event. At present, trans–South Atlantic characiform distribution cannot be simply explained by either the drift-vicariance or the dispersal model.

I must mention two other taxa, Mormyriformes and Gymnotiformes, for which the drift vicariance model requires extinction such as that exhibited by the Polypteridae. These groups should have been present on the united African–South American continent, fragmented with the drift event, and extirpated on one or the other continent. The endemic African mormyriform electric fishes are osteoglossomorphs and thus might be expected to be old enough to have been Gondwanian. There is no fossil record of these fishes outside of Africa, and their phylogenetic relationships to African-Asian Notopteridae and North American

Hiodontidae suggest complicated wide-ranging distributions of stem groups. The endemic South American gymnotiform electric fishes, as the putative sister group to catfishes, should be ancient enough to have been on united Africa–South America, but there is no indication of them outside of the Neotropics. Other than to call attention to the absence of these clades from one continent or another, all one can do is list such possible explanations as differential extinction, restricted predrift ranges on the united continent, or mismatches between the timing of drift and taxonomic origins.

Many uncertainties continue to cloud hypotheses of African–South American biotic relationships. It is clear enough that the fish faunas of Africa and South America have multiregional relationships. South America shares monophyletic groups with Central America (e.g., several subgroups of characins, catfishes, electric fishes, and cyprinodontiforms) and Australasia (Osteoglossidae). Africa shares several clades with Asia and in some cases Europe (e.g., subgroups of cypriniforms, catfishes, anabantoids, and mastacembelids). In addition, Africa and South America each have a diverse sampling of marine groups, either as sporadic visitors or, more interestingly, as secondary freshwater residents. This biotic complexity would seem at once to preclude explanation by a single biogeographic event. Furthermore, it appears that even those taxa that have a straightforward African–South American distribution cannot all be successfully related to vicariance. Some African–South American fish diversification was prompted by drift-related isolation, but the rest is independent of drift and related earlier or later events. Despite the appealing simplicity and generality of having a single paradigmatic vicariance explanation for all trans–South Atlantic relations, such an approach is erroneous. Whereas uncertainties remain about many particulars, an important general conclusion for historical biogeography emerges from this review (see also Grande 1985). Large biotas have complex, piecemeal histories, and any method of analysis that a priori ignores such complexity therefore can yield only incomplete and likely incorrect results.

Literature Cited

Ashlock, P. 1974. The uses of cladistics. Ann. Rev. Ecol. Syst. 5:81–100.

Bailey, R. M., and D. J. Stewart. 1984. Bagrid catfishes from Lake Tanganyika, with a key and descriptions of new taxa. Misc. Publ. Mus. Zool., Univ. Michigan 168:1–41.

Baskin, J. N. 1972. Structure and relationships of the Trichomycteridae. Ph.D. diss., City University of New York.

Begle, D. P. 1991. Relationships of the osmeroid fishes and the use of reductive characters in phylogenetic analysis. Syst. Zool. 40(1):33–53.

Brewster, B. 1986. A review of the genus *Hydrocynus* Cuvier 1819 (Teleostei: Characiformes). Bull. Brit. Mus. (Nat. Hist.) Zool. 50:163–206.

Briggs, J. C. 1979. Ostariophysan zoogeography: An alternative hypothesis. Copeia 1979:111–118.

———. 1987. *Biogeography and Plate Tectonics*. Amsterdam: Elsevier Science.

Broin, F. de, E. Buffetaut, J. C. Koniguer, J. C. Rage, D. Russell, P. Taquet, C. Vergnaud-Grazzini, and S. Wenz. 1974. La faune de vertébrés continentaux du gisement d'In Becten (Sénonien du Niger). C. R. Acad. Sci., Paris, D, 279:469–472.

Buckup, P. A. 1991. The Characiinae: A phylogenetic study of the South American darters and their relationships with other characiform fishes. Ph.D. diss., University of Michigan.

Bussing, W. A. 1985. Patterns of distribution of the Central American ichthyofauna. Pp. 453–473 *in* F. G. Stehli and S. D. Webb, (eds.), *The Great American Biotic Interchange*. New York: Plenum.

Campos, H. 1989. Protein variations in *Galaxias maculatus* (Salmoniformes: Galaxiidae) from Chile, Australia and New Zealand. Abstracts of the 1989 Annual Meeting of the American Society of Ichthyologists and Herpetologists, San Francisco State University, p. 72.

Cappetta, H., D. E. Russell, and J. Braillon. 1972. Sur la découverte de Characidae dans l'Eocene inférieur Français (Pisces, Cypriniformes). Bull. Mus. Natn. Hist. Nat., Paris, 3d sér., 51:1–51.

Carroll, R. L. 1988. *Vertebrate Paleontology and Evolution*. New York: W. H. Freeman.

Chardon, M. 1967. Réflexions sur la dispersion des Ostariophysi à la lumière de recherches morphologiques nouvelles. Ann. Soc. Roy. Zool. Belgique, 97:175–186.

———. 1968. Anatomie comparée de l'appareil de Weber et des structures connexes chez lez Siluriformes. Ann. Mus. Roy. Afr. Centr., Sci. Zool. 169:1–227.

Cichoki, F. 1976. Cladistic history of the cichlid fishes and reproductive categories of the American genera. Ph.D. diss., University of Michigan.

Cione, A. L. 1987. The late Cretaceous fauna of Los Alamitos, Patagonia, Argentina. Part II, The fishes. Rev. Mus. Argent. Cienc. Nat. 3:111–120.

Cione, A. L., and G. Lafitte. 1980. El primer siluriforme (Osteichthyes, Ostariophysis) del Cretácico de Patagonia. Consideraciones sobre el área de diferenciación de los Siluriformes. Aspectos biogeográficos. Act. II Congr. Argent. Paleontol. Bioestratig. y I Congr. Latinoamer. Paleontol. 2:35–46.

Cione, A. L., S. M. Pereira, R. Alonso, and J. Arias. 1985. Los bagres (Osteichthyes, Siluriformes) de la Formación Yacoraite (Cretácico tardío) del Noroeste argentino: Consideraciones biogeográficas y bioestratigraficas. Ameghiniana 21:294–304.

Cracraft, J. 1974. Continental drift and vertebrate distribution. Ann. Rev. Ecol. Syst. 5:215–262.

Croizat, L., G. Nelson, and D. E. Rosen. 1974. Centers of origin and related concepts. Syst. Zool. 23:265–287.

Curran, D. J. 1989. Phylogenetic relationships among the catfish genera of the family Auchenipteridae (Teleostei: Siluriformes). Copeia 1989:408–419.

Darlington, P. J. 1957. *Zoogeography: The Geographical Distribution of Animals*. New York: John Wiley & Sons.

———. 1965. *Biogeography of the Southern End of the World*. Cambridge: Harvard University Press.

Eigenmann, C. H. 1909. The fresh-water fishes of Patagonia and an examination of the Archiplata-Archhelenis theory. Pp. 227–374 *in Reports of the Princeton University Expeditions to Patagonia, 1896–1899,* Part 3. Princeton, N.J.: Princeton University.

———. 1910. The origin of the fish-fauna of the fresh waters of South America. Proceed. 7th Intern. Zool. Congress, Boston 1907. Advance print, pp. 1–2.

Eigenmann, C. H., and W. R. Allen. 1942. *Fishes of Western South America*. Lexington, Ky.: University Press of Kentucky.

Fernandez, J., R. Pascual, and P. Bondesio, 1973. Restos de *Lepidosiren paradoxa* (Osteichthyes, Dipnoi) de la Formación Lumbrera (Eogeno, Eoceno?) de Jujuy: Consideraciones estratigráficas, paleoecologicas y paleozoogeograficas. Ameghiniana 10:152–172.

Ferraris, C. J. 1988. The Auchenipteridae: Putative monophyly and systematics, with a classification of the neotropical doradoid catfishes (Ostariophysi: Siluriformes). Ph.D. diss., City University of New York.

Fink, S. V., and W. L. Fink. 1981. Interrelationships of the ostariophysan fishes (Teleostei). J. Linn. Soc. (Zool.) 72:297–353.

Fink, S. V., P. H. Greenwood, and W. L. Fink. 1984. A critique of recent work on fossil ostariophysan fishes. Copeia 1984:1033–1041.

Frizzell, D. L. 1965. Otoliths of new fish (*Vorhisia vulpes*, n. gen., n. sp., Siluroidei?) from Upper Cretaceous of South Dakota. Copeia 1965:178–181.

Gaudant, J. 1980. *Eurocharax tourainei* nov. gen., nov. sp. (Poisson téléostéen, Ostariophysi): Nouveau Characidae fossile des "Calcaires à Bythinies" du Var. Geobios 13:683–703.

Gayet, M. 1980. Hypothèse sur l'origine des Ostariophysaires. C. R. Acad. Sci., Paris, D, 290:1197–1199.

———. 1981. Contribution á l'étude anatomique et systématique de l'ichthyofaune cénomanienne du Portugal. Deuxieme partie, Les Ostariophysaires. Comun. Serv. Geol. Portugal 67: 173–190.

———. 1982a. Considération sur la phylogénie et la paléobiogéographie des Ostariophysaires. Geobios, Mém. spéc. 6:39–52.

———. 1982b. Cypriniforme ou gonorhynchiforme? *Ramallichthys,* nouveau genre du Cénomanien inférieur de Ramallah (Monts de Judée). C. R. Acad. Sci., Paris, D, 295:405–407.

———. 1982c. Cypriniformes crétacés en Amérique du Sud. C. R. Acad. Sci., Paris, D, 295:661–664.

———. 1982d. Découverte dans le Crétacé supérieur de Bolivie des plus anciens Characiformes connus. C. R. Acad. Sci., Paris, D, 294:1037–1040.

———. 1985. Contribution à l'étude anatomique et systématique de l'ichthyofaune cénomanienne du Portugal. Troisième partie: Complément à l'étude des ostariophysaires. Comun. Serv. Geol. Portugal 71:91–118.

———. 1986. About ostariophysan fishes: A reply to S. V. Fink, P. H. Greenwood and W. L. Fink's criticisms. Bull. Mus. Natn. Hist. Nat., Paris, 4th sér., sect. C, 3:393–409.

———. 1987. Consideraciones preliminares sobre la paleobiogeografía de los Osteoglossomorpha. IV Congr. Latinoamer. Paleontol. Bolivia 1:379–398.

———. 1988. Le plus ancien crane de siluriforme: *Andinichthys bolivianensis* nov. gen., nov. sp. (Andinichthyidae nov. fam.) du Maastrichtien de Tiupampa (Bolivie) C. R. Acad. Sci., Paris, D, 307:833–836.

———. 1990. Nouveaux Siluriformes du Maastrichtien de Tiupampa (Bolivie) C. R. Acad. Sci., Paris, 2d sér., 310:867–872.

———. 1991. "Holostean" and teleostean fishes of Bolivia. Pp. 453–494 *in* R. Suarez-Soruco (ed.), *Fosiles y Facies de Bolivia,* Vol. 1: *Vertebrados.* Rev. Téc. Yaci. Petrol. Fisc. Bol. 12(3–4):357–718.

Gayet, M., and P. Brito. 1989. Ichthyofaune nouvelle du Crétacé supérieur du groupe Bauru (états de São Paulo et Minas Gerais, Brésil). Geobios 22(6):841–847.

Gayet, M., and F. J. Meunier. 1983. Ecailles actuelles et fossiles d'Ostéoglossiformes (Pisces, Teleostei). C. R. Acad. Sci. Paris, D, 297:867–870.

———. 1991. Polypteridae (Pisces, Cladistia, Polypteriformes) from the late Cretaceous and early Paleocene of Bolivia. Pp. 495–498 *in* R. Suarez-Soruco (ed.), *Fosiles y Facies de Bolivia,* Vol. 1: *Vertebrados.* Rev. Téc. Yaci. Petrol. Fisc. Bol. 12(3–4):357–718.

Géry, J. 1969. The fresh-water fishes of South America. Pp. 828–848 *in* E. J. Fittkau et al. (eds.), *Biogeography and Ecology in South America.* Monographiae Biologicae 19. The Hague: W. Junk.

———. 1977. *Characoids of the World.* Neptune City, N.J.: TFH.

Gosline, W. A. 1944. The problem of the derivation of the South American and African freshwater fish fauna. Ann. Acad. Brasil. Ci. 16:211–223.

———. 1975. A reexamination of the similarities between the freshwater fishes of Africa and South America. Pp. 146–155 *in Biogéographie et liaisons intercontinentales au cours du Mésozoïque.* Mém. Mus. Natn. Hist. Nat. Paris, n.s. (A) Zool. 88.

Grande, L. 1985. The use of paleontology in systematics and biogeography, and a time control refinement for historical biogeography. Paleobiology 11:234–243.

———. 1987. Redescription of *Hypsidoris far-*

sonensis (Teleostei: Siluriformes) with a reassessment of its phylogentic position. J. Vert. Paleontol. 7:24–54.

Grande, L., and J. T. Eastman. 1986. A review of the Antarctic ichthyofaunas in light of new fossil discoveries. Paleontology 29:113–137.

Grande, L., and J. G. Lundberg. 1988. Revision and redescription of the genus *Astephus* (Siluriformes: Ictaluridac) with a discussion of its phylogenetic relationships. J. Vert. Paleontol. 8:139–171.

Greenwood, P. H. 1970. On the genus *Lycoptera* and its relationship with the family Hiodontidae (Pisces, Osteoglossomorpha). Bull. Brit. Mus. (Nat. Hist.) Zool. 16:257–285.

———. 1973. The interrelationships of the Osteoglossomorpha. J. Linn. Soc. (Zool.) 53 (Suppl. 1):307–332.

———. 1974. Review of the Cenozoic freshwater fish faunas of Africa. Ann. Geol. Surv. Egypt 4:211–232.

———. 1983. The zoogeography of African freshwater fishes: Bioaccountancy or biogeography? Pp. 179–199 *in* R. W. Sims, J. H. Price and P. E. S. Whalley (eds.), *Evolution, Time and Space: The Emergence of the Biosphere*. London: Academic.

———. 1984. African cichlid fishes and evolutionary theory. Pp. 141–154 *in* A. A. Echelle and I. Kornfield (eds.), *Evolution of Fish Species Flocks*. Orono: University of Maine at Orono Press.

———. 1986. The pars jugularis and the intrarelationships of cichlid fishes (Labroidei, Teleostei). J. Nat. Hist. 20:949–974.

Greenwood, P. H., and G. J. Howes. 1975. Neogene fossil fishes from the Lake Albert–Lake Edward rift (Zaire). Bull. Brit. Mus. (Nat. Hist.) Geology 26:69–127.

Greenwood, P. H., and C. Patterson. 1967. A fossil osteoglossoid fish from Tanzania (East Africa). J. Linn. Soc. (Zool.) 47:211–223.

Greenwood, P. H., D. E. Rosen, S. H. Weitzman, and G. S. Myers. 1966. Phyletic studies of teleostean fishes, with a provisional classification of living forms. Bull. Amer. Mus. Nat. Hist. 131:339–455.

Günther, A. G. L. C. 1864. *Catalogue of Fishes of the British Museum,* Vol. 5. London.

———. 1880. *An Introduction to the Study of Fishes*. Edinburgh: A. and C. Black.

Hennig, W. 1966. *Phylogenetic Systematics*. Urbana, Ill.: University of Illinois Press.

Howes, G. J. 1983. Problems in catfish anatomy and phylogeny exemplified by the Neotropical Hypophthalmidae (Teleostei: Siluroidei). Bull. Brit. Mus. (Nat. Hist.) Zool. 45:1–39.

Humphries, C. J., and L. R. Parenti. 1986. *Cladistic Biogeography.* Oxford: Clarendon.

von Ihering, H. 1900. The history of the neotropical region. Science, n.s., 12:857–864.

Jordan, T. E., and M. P. Gareweg. 1989. Tectonic evolution of the late Cenozoic Central Andes (20°–33° S). Pp. 193–207 *in* Z. Ben-Avraham (ed.), *The Evolution of the Pacific Ocean Margins*. Oxford Monographs in Geology and Geophysics 6. London: Oxford.

Kaufman, L., and K. F. Liem. 1982. Fishes of the suborder Labroidei (Pisces: Perciformes): Phylogeny, ecology, and evolutionary significance. Breviora. 472:1–19.

Keast, A. 1972. Continental drift and the evolution of the biota on southern continents. Pp. 23–87 *in* A. Keast, F. C. Erk, and B. Glass (eds.), *Evolution, Mammals, and Southern Continents*. Albany: SUNY Press.

Kosswig, C. 1943. Über Tethysrelikte in der türkischen Fauna. C. R. Ann. Arch. Soc. Turque. Sci. Phys. Nat. 10:30–47.

Kullander, S. O. 1983. Review of the South American Cichlidae. Ph.D. diss., University of Stockholm, Sweden.

Lauder, G. V., and K. F. Liem. 1983. The evolution and interrelationships of the actinopterygian fishes. Bull. Mus. Comp. Zool. 150:95–197.

Lewis, D., P. Reinthal, and J. Trendall. 1986. *A Guide to the Fishes of Lake Malawi National Park*. Gland, Switzerland: World Wildlife Fund.

Liem, K. F. 1970. Comparative functional anatomy of the Nandidae (Pisces: Teleostei). Fieldiana: Zool. 56:1–166.

Liem, K. F., and P. H. Greenwood. 1981. A

functional approach to the phylogeny of the pharyngognath teleosts. Amer. Zool. 21:81–101.

Lundberg, J. G. 1975. The fossil catfishes of North America. Papers in Paleontology 11. Ann Arbor: University of Michigan, Museum of Paleontology.

———. In press. Fishes of the Miocene La Venta Fauna: additional taxa and their paleobiotic implications. *In* R. Kay et al. (eds.), *A History of Neotropical Fauna: Vertebrate Paleontology of the Miocene of Tropical South America.* New York: Springer Verlag.

Lundberg, J. G., and B. Chernoff. 1992. A Miocene fossil of the Amazonian fish *Arapaima* (Teleostei, Arapaimidae) from the Magdalena River region of Colombia: Biogeographic and evolutionary implications. Biotropica 24(1): 2–14.

Lundberg, J. G., A. Machado-Allison, and R. F. Kay. 1986. Miocene characid fishes from Colombia: Evolutionary stasis and extirpation. Science 234:208–209.

Lundberg, J. G., A. H. Bornbusch, and F. Mago-Leccia. 1991. *Gladioglanis conquistador* N. Sp. from Ecuador with diagnoses of the subfamilies Rhamdiinae Bleeker and Pseudopimelodinae N. Subf. (Siluriformes: Pimelodidae). Copeia 1991:190–201.

McDowell, R. M. 1978. Generalized tracks and dispersal in biogeography. Syst. Zool. 27:88–104.

———. 1981. The relationships of Australian freshwater fishes. Pp. 1251–1273 *in* A. Keast (ed.), *Ecological Biogeography of Australia.* Monographiae Biologicae 41. The Hague: W. Junk.

Machado-Allison, A. 1983. Estudios sobre la subfamilía Serrasalminae (Teleostei-Characidae). Parte 1, Discusion sobre la condicion monofilética de la subfamilía. Acta Biol. Venez. 11:145–196.

Marshall, C. R. 1987a. A list of fossil and extant dipnoans. J. Morph. 1986 (Suppl. 1):15–23.

———. 1987b. Lungfish: Phylogeny and parsimony. J. Morph. 1986 (Suppl. 1):151–162.

Martin, M. 1984. Deux Lepidosirendiae (Dipnoi) Crétacés du Sahara, *Protopterus humei* (Priem) et *Protopterus protopteroides* (Tabaste). Paläont. Zeits. 58:265–277.

Mayr, E., et al. 1952. The problem of land connections across the South Atlantic, with special reference to the Mesozoic. Bull. Amer. Mus. Nat. Hist. 99:79–258.

Miles, R. S. 1977. Dipnoan (lungfish) skulls and the relationships of the group: A study based on new species from the Devonian of Australia. J. Linn. Soc. (Zool.) 61:1–328.

Muizon, C. de, M. Gayet, A. Lavenu, L. Marshall, B. Sige, and C. Villarroel. 1983. Late Cretaceous vertebrates, including mammals, from Tiupampa, South Central Bolivia. Geobios 16:747–753.

Myers, G. S. 1938. Fresh-water fishes and West Indian zoogeography. Smithsonian Rep. 1937:339–364.

———. 1966. Derivation of the freshwater fauna of Central America. Copeia 1966:766–773.

———. 1967. Zoogeographical evidence of the age of the South Atlantic Ocean. Stud. Trop. Oceanog., Inst. Mar. Sci., Univ. Miami 5:614–621.

Nelson, G. J. 1969. Infraorbital bones and their bearing on the phylogeny and geography of osteoglossomorph fishes. Amer. Mus. Novit. 2394:1–37.

Nelson, G. J., and N. Platnick. 1981. *Systematics and Biogeography: Cladistics and Vicariance.* New York: Columbia University Press.

Nelson, J. S. 1984. *Fishes of the World.* Toronto: John Wiley and Sons.

Novacek, M. J., and L. G. Marshall. 1976. Early biogeographic history of ostariophysan fishes. Copeia 1976:1–12.

Oliver, M. K. 1985. Systematics of African cichlid fishes: Determination of the most primitive taxon, and studies of the haplochromines of Lake Malawi (Teleostei: Cichlidae). Ph.D. diss., Yale University.

Parenti, L. R. 1981. A phylogenetic and biogeographic analysis of cyprinodontiform fishes (Telesostei: Atherinomorpha). Bull. Amer. Mus. Nat. Hist. 168:335–557.

———. 1984a. Biogeography of the Andean killifish genus *Orestias* with comments on the species flock concept. Pp. 85–92 *in* A. A. Echelle and I. Kornfield (eds.), *Evolution of Fish Species Flocks*. Orono: University of Maine at Orono Press.

———. 1984b. A taxonomic revision of the Andean killifish genus *Orestias* (Cyprinodontiformes, Cyprinodontidae). Bull. Amer. Mus. Nat. Hist. 178:107–214.

Patterson, C. 1975. The distribution of Mesozoic freshwater fishes. Pp. 156–174 *in* Biogéographie et liaisons intercontinentales au cours du Mésozoïque. Mém. Mus. Natn. Hist. Nat., Paris, n.s. (A) Zool. 88.

———. 1981. The significance of fossils in determining evolutionary relationships. Ann. Rev. Ecol. Syst. 12:195–223.

———. 1983. Aims and methods in biogeography. Pp. 1–28 *in* R. W. Sims, J. H. Price, and P. E. S. Whalley (eds.), *Evolution, Time and Space: The Emergence of the Biosphere*. London: Academic.

———. 1984. *Chanoides,* A marine Eocene otophysan fish (Teleostei: Ostariophysi). J. Vert. Paleontol. 4:430–456.

———. 1990. An overview of the early fossil record of Acanthomorpha. Abstracts of the 1990 Annual Meeting of the American Society of Ichthyologists and Herpetologists, College of Charleston, South Carolina, p. 139.

Pinna, M. C. C. de. 1989. Phylogenetic relationships of loricarioids to other catfish families (Ostariophysi, Siluriformes). Abstracts of the 1989 Annual Meeting of the American Society of Ichthyologists and Herpetologists, San Francisco State University, p. 131.

Regan, C. T. 1911. The classification of the teleostean fishes of the order Ostariophysi. Part 2, Siluroidea. Ann. Mag. Nat. Hist., ser. 8, 8:553–577.

———. 1922. The distribution of the fishes of the order Ostariophysi. Bijd. tot Dierk. 22:203–208.

Roberts, T. R. 1969. Osteology and relationships of characoid fishes, particularly the genera *Hepsetus, Salminus, Hoplias, Ctenolucius* and *Acestrorhynchus*. Proc. Calif. Acad. Sci. 36:391–500.

———. 1973. Interrelationships of ostariophysans. Pp. 373–395 *in* P. H. Greenwood, R. S. Miles, and C. Patterson (eds.), *Interrelationships of Fishes*. London: Academic.

———. 1975. Geographical distribution of African freshwater fishes. J. Linn. Soc. (Zool.) 57:249–319.

Rosen, D. 1974. Phylogeny and zoogeography of Salmoniform fishes and relationships of *Lepidogalaxias salmandroides*. Bull. Amer. Mus. Nat. Hist. 153:267–325.

———. 1976. A vicariance model of Caribbean biogeography. Syst. Zool. 24:431–464.

———. 1978. Vicariant patterns and historical explanation in biogeography. Syst. Zool. 27:159–188.

Rosen, D., and P. H. Greenwood. 1970. Origin of the Weberian apparatus and relationships of the ostariophysan and gonorynchiform fishes. Am. Mus. Novit. 2468:1–49.

———. 1976. A fourth neotropical species of synbranchid eel and the phylogeny and systematics of synbranchiform fishes. Bull. Amer. Mus. Nat. Hist. 157:1–70.

Royero, R. 1988. Morfología de la aleta dorsal en los bagres (Teleostei: Siluriformes), con especial referencia a las familias Americanas. Licenciado thesis, Universidad Central de Venezuela.

Savage, J. M. 1982. The enigma of the Central American herpetofauna. Ann. Missouri Bot. Gard. 69:464–547.

Schaeffer, B. 1947. Cretaceous and Tertiary actinopterygian fishes from Brazil. Bull. Amer. Mus. Nat. Hist. 89:1–40.

———. 1952. The evidence of the fresh-water fishes. Pp. 227–234 *in* E. Mayr (ed.), The problem of land connections across the South Atlantic, with special reference to the Mesozoic. Bull. Am. Mus. Nat. Hist. 99:79–258.

Schaefer, S. A., and G. V. Lauder. 1983. Historical transformation of functional design: Evolutionary morphology of feeding mechanisms in loricarioid catfishes. Syst. Zool. 35:489–508.

Schaefer, S. A., S. H. Weitzman, and H. A. Britzki. 1989. Review of the Neotropical catfish genus *Scoloplax* (Pisces: Loricarioidea:

Scoloplacidae) with comments on reductive characters in phylogenetic analysis. Proc. Acad. Nat. Sci. Phila. 141:181–211.

Schultze, H. P. 1991. Lungfish from the El Molino (late Cretaceous) and Santa Lucia (early Paleocene) formations in southcentral Bolivia. Pp. 441–448 *in* R. Suarez-Soruco (ed.), *Fosiles y Facies de Bolivia,* Vol. 1: *Vertebrados.* Rev. Téc. Yaci. Petrol. Fisc. Bol. 12(3–4):357–718.

Sethi, R. P. 1978. Remarks on paleontology and zoogeography of oviparous cyprinodont fishes (order Atheriniformes Rosen). Prani 1:20–27.

Sigé, B. 1968. Dents de micromammifères et fragments de coquilles d'oeufs de dinosauriens dans la faune vertébrés du Crétacé supérieur de Laguna Umayo (Andes péruviennes). C. R. Acad. Sci., Paris, D, 267:1495–1498.

da Silva Santos, R. 1985. *Laeliichthys ancestralis,* novo gênero e espécie de Osteoglossiformes do Aptiano da Formaçao Areado, Estado de Mina Gerais, Brasil. VIII Cong. Brasil. Paleontol. 1983, MME-DNPM, Ser. Geol. 27, Paleontol./Estrat. 2:151–167.

———. 1987. *Lepidosiren megalos* n. sp. do Terciário do Estado do Acre—Brasil. An. Acad. brasil. Ci. 59:375–384.

Skelton, P. H. 1988. Outgroup relationships of amphiliid catfishes. Abstracts of the 1988 Annual Meeting of the American Society of Ichthyologists and Herpetologists, University of Michigan, Ann Arbor, p. 172.

Stiassny, M. L. J. 1981. The phyletic status of the family Cichlidae (Pisces, Perciformes): A comparative anatomical investigation. Netherlands J. Zool. 31:275–314.

———. 1982. The relationships of the neotropical genus *Cichla* (Perciformes, Cichlidae): A phyletic analysis including some functional considerations. J. Zool. London 197:427–453.

———. 1986. The limits and relationships of the acanthomorph teleosts. J. Zool. London, ser. B, 1:411–460.

———. 1987. Cichlid familial intrarelationships and the placement of the neotropical genus *Cichla* (Perciformes, Labroidei). J. Nat. Hist. 21:1311–1331.

———. 1991. Phylogenetic intrarelationships of the family Cichlidae: An overview. Pp. 1–35 *in* M. H. A. Keenleyside (ed.), *Cichlid Fishes: Behaviour, Ecology and Evolution.* London: Croom Helm.

Stiassny, M. L. J., and J. S. Jensen. 1987. Labroid intrarelationships revisited: Morphological complexity, key innovations, and the study of comparative diversity. Bull. Mus. Comp. Zool. 151:269–319.

Swofford, D. L. 1985. *Phylogenetic Analysis Using Parsimony. Version 2.4 User's Manual.* Champaign: Illinois Natural History Survey.

Taverne, L. 1979. Ostéologie, phylogénèse et systématique des Téléostéens fossiles et actuels du super-ordre des Osteoglossomorphes. Partie 3, Evolution des structures ostéologiques et conclusions générales relatives à la phylogénèse et à la systematique du super-ordre. Addendum. Acad. Roy. Belgique, Mém. Classe des Sci., ser. 2, 43:1–168.

van Couvering, J. A. H. 1982. Fossil cichlid fish of Africa. Spec. Pap. Paleontol. Assoc. 29:1–103.

Vari, R. P. 1979. Anatomy, relationships and classification of the families Citharinidae and Distichodontidae (Pisces, Characoidea). Bull. Brit. Mus. (Nat. Hist.) Zool. 36:261–344.

———. 1983. Phylogenetic relationships of the families Curimatidae, Prochilodontidae, Anostomidae, and Chilodontidae (Pisces: Characiformes). Smithsonian Contr. Zool. 378:1–60.

Walsh, S. J. 1990. A systematic revision of the Neotropical catfish family Ageneiosidae (Teleostei: Ostariophysi: Siluriformes). Ph.D. diss., University of Florida, Gainesville.

Weitzman, S. H. 1962. The osteology of *Brycon meeki,* a generalized characid fish, with an osteological definition of the family. Stanford Ichthyol. Bull. 8:1–77.

Weitzman, S. H., and W. L. Fink. 1983. Relationships of the neon tetras, a group of South American freshwater fishes (Teleostei, Characidae), with comments on the phylogeny of New World characiforms. Bull. Mus. Comp. Zool. 150:339–395.

Wenz, S. 1969. Note sur quelques poissons actinoptérigiens du Crétacé supérieur de Bolivie. Bull. Soc. Géol. France 11:434–438.

Woodburn, W. O., and W. J. Zinsmeister. 1984. The first land mammal from Antarctica and its biogeographic implications. J. Paleontol. 58:913–948.

Chapter 9

Amphibians in Africa and South America: Evolutionary History and Ecological Comparisons

William E. Duellman

I. Introduction

Although amphibians have existed since the mid-Devonian, nearly 350 million years ago, they are represented today by only three orders of Lissamphibia. Representatives of each order occur in Africa and South America, but only the frogs and toads (Order Anura) have high species diversity on those continents. Critical comparisons of the amphibian faunas of Africa and South America are lacking. Laurent (1973) made some general comparisons of the equatorial herpetofaunas of the two continents and subsequently (1979) expanded his earlier review to include the entire African and South American herpetofaunas.

In this chapter I (1) interpret the historical biogeography of African and South American amphibians, especially as it relates to the fragmentation of Gondwanaland, (2) document the adaptive radiation of the different

I thank Alan Channing, Ian Coulson, Robert C. Drewes, and Paul Chippindale for information on African anuran communities; Larry D. Martin, Jean-Luc Perret, John C. Poynton, Margaret M. Stewart, and Elisabeth S. Vrba for calling my attention to pertinent literature; and Erik R. Wild for technical assistance. For their critical comments on all or part of the chapter, I am indebted to Philip S. Humphrey, John D. Lynch, David A. Kizirian, Jean-Luc Perret, John C. Poynton, and Linda Trueb; however, I, alone, am responsible for the analyses and interpretations presented herein. My field studies in South America have been supported at various times by the National Geographic Society, the National Science Foundation, and the Museum of Natural History, University of Kansas. The latter institution and the South African Council for Scientific and Industrial Research partially supported my field studies in Africa.

 ISBN 0-300-05375-4

taxonomic groups on each continent, especially in relation to climatic and vegetational differences, and thereby (3) account for the comparative species diversity of amphibians in Africa and in South America. I focus on the anurans, which comprise 94 percent of the amphibian faunas on the two continents. Salamanders constitute only 1 percent of the amphibian faunas of Africa and South America and seem not to be associated historically with Gondwanaland. The comparatively poorly known caecilians account for about 5 percent of the faunas and are historically a part of Gondwanaland.

II. Materials and Methods

The numbers and distributions of taxa were updated from Frost (1985) through 31 December 1989; the families, genera, and numbers of species in each are listed in the chapter appendix. Data on distributions, habitats, and life histories were drawn from a wide variety of sources—especially Amiet (1972, 1973, 1975, 1978, 1986, and 1989), Amiet and Perret (1969), Drewes (1984), Inger (1968), Lamotte and Lescure (1989), Laurent (1964), Mertens (1938), Passmore and Carruthers (1979), Perret (1966), Poynton (1964), Poynton and Broadley (1985a, 1985b, 1987, 1988, and 1991), Schiøtz (1963, 1967, and 1975), Schmidt and Inger (1959), Stewart (1967), and Tandy and Keith (1972) for Africa; and Cei (1962, 1979, and 1980), Duellman (1979a, 1979c, 1988, 1989b, 1990), Formas (1979), Gallardo (1979), Heyer et al. (1990), Hoogmoed (1979), Lynch (1979), and Rivero-Blanco and Dixon (1979) for South America. In addition to these references, I have relied on data on specimens in various museums and on my extensive field experience in much of South America and more limited field experience in eastern and southern Africa.

The areas of the continents were obtained from *The International Atlas* (Rand McNally 1969). The area for sub-Saharan Africa was calculated from the areas of the countries south of Mauritania, Mali, Niger, Chad, and Sudan; sub-Saharan Africa includes Ethiopia and Somalia. Major vegetation types were obtained by combining similar types of vegetation mapped by UNESCO (1958) for Africa and by Hueck (1972) for South America. Areas of vegetation types were measured from maps by use of a Micro-Plan II image analysis system (Laboratory Computer Systems, Inc., Cambridge, Massachusetts). For purposes of analysis of distribution of amphibians among vegetation types, each species was assigned to one vegetation type; if the species occurs in more than one vegetation type, it was assigned to the type in which it is most widely distributed. The same criteria were used to assign species to microhabitats. The term *family groups* equals families and subfamilies.

Paleogeographic, paleoclimatic, and modern biogeographic data were synthesized from a vast array of literature. Most of that not dealing with amphibians is not cited herein; however, the major sources can be found in compilations edited by Duellman (1979b), Prance (1982), Stehli and Webb (1985), Vuilleumier and Monasterio (1986), Whitmore and Prance (1987), Kingdon (1989), and Woods (1989). Paleogeographic interpretations were based on maps of continental movements provided by Barron et al. (1981), as updated by many authors.

III. Living Groups of Amphibians

In order to place the African and South American amphibians in perspective, it is necessary to review the histories of the living groups of lissamphibians. All three groups existed before Gondwanaland began to break up in the Jurassic.

A. Gymnophiona

The earliest known caecilian is from the Upper Jurassic of North America (Jenkins and Walsh 1990), and the only other gymnophionan fossil is a presumed caeciliaid from the Paleocene of Brazil (Estes and Wake 1972). The Gymnophiona consists of about 160 known species worldwide; caecilians are well represented in southeastern Asia, Africa, and South America. However, the biology and relationships of most caecilians are so poorly known, especially those of the large family Caeciliaidae, that meaningful comparisons cannot be made. Of the six families of caecilians, one is endemic to Africa—Scolecomorphidae, with 2 genera and 5 species (Nussbaum 1985)—and two are endemic to South America—Rhinatrematidae, with 2 genera and 9 species, and Typhlonectidae, with 4 genera and 12 species (Nussbaum and Wilkinson 1989). The apparently paraphyletic family Caeciliaidae is represented in Africa by 6 genera and 17 species and in South America by 11 genera and 56 species; one other genus is restricted to Central America, three are endemic to the Seychelles Islands, and two are endemic to peninsular India (Nussbaum and Wilkinson 1989). The other two families, Ichthyophiidae and Uraeotyphlidae, are endemic to southeastern Asia.

As noted by Savage (1973) and Duellman and Trueb (1986), the historical biogeography of the living families of caecilians is associated with the breakup of Gondwanaland. The primitive rhinatrematids now restricted to northern South America presumably are relicts of a former widespread Gondwanan distribution. The caeciliaids have a classic Gondwanan pattern—South America, Africa, and peninsular India, plus the Seychelles Islands. The scolecomorphids and typhlonectids presumably evolved in situ in Africa and South America, respectively.

B. Caudata

The earliest salamanders are from the upper Jurassic, and this order has an extensive fossil record in Laurasia but no fossils from Gondwanan continents (Estes 1981). Salamanders, with about 360 living species worldwide, are most widespread and diverse in the Northern Hemisphere. The biogeography of salamanders is associated with the history of Laurasia (Duellman and Trueb 1986).

Salamanders are represented in Africa by only three species of salamandrids—*Pleurodeles poireti,* endemic to northern Algeria and Tunisia, and *Pleurodeles waltl* and *Salamandra salamandra,* both of which occur in Europe and northwestern Africa (Frost 1985). Two genera of the primarily Mesoamerican tribe Bolitoglossini of the Plethodontidae enter northwestern South America. *Bolitoglossa* has 47 species in Mesoamerica and 20 species in South America, mostly in Colombia and Ecuador but extending southward to central Bolivia and eastward to northeastern Brazil; *Oedipina* is represented by two species in Chocoan Colombia, one of which also occurs in Central America, where an additional 14 species are found (Frost 1985).

C. Anura

The history of the Anura is complex. The earliest known salientian amphibian is *Triadobatrachus* from the Lower Triassic of Madagascar. The earliest anurans (i.e., true frogs) are *Vieraella* from the early Jurassic of South America and four genera from the late Jurassic—*Notobatrachus* from South America, *Comobatrachus* and *Eobatrachus* from North America, and *Monteschobatrachus* from Europe. Obviously, anurans were widespread in the world prior to the breakup of Pangaea. They generally are separated into the suborders Archaeobatrachia and Neobatrachia (Duellman and Trueb 1986).

The Archaeobatrachia contains only 130 living species, or only slightly more than 3 percent of the living species of anurans. *Notobatrachus* may be the sister group of all other anurans; other Jurassic anurans are too fragmentary to be included in phylogenetic analyses (Trueb, personal communication, 1990). Cannatella (1985) analyzed cladistically the family groups of archaeobatrachians. Two families, Ascaphidae (Recent of North America) and Leiopelmatidae (Recent of New Zealand), were placed as successive sister groups of remaining anurans. Within the Archaeobatrachia, the superfamily Pipoidea contains three families—Rhinophrynidae (late Paleocene to Recent of North America), Palaeobatrachidae (Jurassic-Cretaceous boundary to Pleistocene of Europe; late Cretaceous of North America), and Pipidae (lower Cretaceous of Israel and Upper Cretaceous to Recent in Africa and South America). Thus, it is evident that the historical biogeography of all of these archaeobatrachians was associated with the breakup of Pangaea and the subsequent fragmentation of Gondwanaland. The other families of archaeobatrachians are associated with the history of Laurasia—Bombinatoridae (middle Miocene to Recent of Eurasia), Discoglossidae (Upper Cretaceous of North America; Upper Jurassic to Recent of Europe), Megophryidae (Recent of tropical Asia), Pelobatidae (late Cretaceous to Recent of North America and Eurasia), and Pelodytidae (middle Eocene to Recent of Europe; middle Miocene of North America).

Most living anurans are members of families in the Neobatrachia. Savage (1973) and Duellman and Trueb (1986) emphasized the historical events of the Gondwanan continents in the vicariance biogeography of neobatrachians, for which the fossil record is meager; fossils are known for only 6 of the 17 families. The earliest fossil leptodactylids are from the late Cretaceous of South America, where the family is represented by numerous fossils throughout the Cenozoic and is diverse at present. The earliest bufonids and hylids are from the Paleocene of South America. Bufonids appear in the upper Oligocene in Europe and in the lower Miocene in North America; hylids are present in the Oligocene in North America and in the Miocene of Europe and Australia. The presence of hylids and bufonids in North America and Europe in the mid-Cenozoic suggests that these families might have existed before the breakup of Pangaea; however, it is more likely that they dispersed into northern continents from Gondwanan continents during the late Cretaceous or early Cenozoic. The earliest ranids are from the Oligocene of Europe and the Miocene of North America and northern Africa. The only microhylid fossil is from the Miocene of North America. The Eocene *Indobatrachus* from peninsular India has been assigned tentatively by Lynch (1971) to the Myobatrachidae, which presently is endemic to Australia and New Guinea.

Of the 17 families of living neobatrachians, 4 (Allophrynidae, Brachycephalidae, Pseudidae, and Rhinodermatidae) are endemic to South America, 2 other primarily South American families (Centrolenidae and Dendrobatidae) also have taxa in Mesoamerica, and 1 other family (Leptodactylidae) is also represented in Mesoamerica, North America, and the West Indies. Three families (Arthroleptidae, Heleophrynidae, and Hemisotidae) are endemic to Africa, and one other family (Hyperoliidae) also has taxa on Madagascar and the Seychelles Islands. The Myobatrachidae is endemic to Australia and the Sooglossidae to the Seychelles Islands. The Rhacophoridae is primarily in tropical Asia but has representatives in Madagascar and Africa. The histories of all of these families seem to be related to the fragmentation of Gondwanaland.

Four families of neobatrachians are widespread in both northern and southern continents. The Bufonidae is cosmopolitan in temperate and tropical regions, except for the Australo-Papuan, Madagascan, and Oceanic regions. The distribution of the genus *Bufo* equals that of the family, whereas 8 other genera are endemic to South America, 10 to Africa, and 6 to southeastern Asia, including peninsular India, the Philippines, Borneo, and the Greater Sunda Islands.

The Hylidae contains 33 genera. The paraphyletic genus *Hyla* occurs throughout the range of the family in South, Central, and North America and is the only representative of the family in Eurasia, with two European species entering northern Africa. The family is most diverse in the Neotropics, with 14 genera endemic to South America, 6 genera endemic to Mesoamerica, and 7 others occurring in South America and Mesoamerica. Two genera are endemic to the West Indies and 2 to North America. Three speciose genera of hylids are endemic to the Australo-Papuan region.

The Microhylidae is nearly cosmopolitan in temperate and tropical regions, except for the Palearctic, most of Australia, the West Indies, and most oceanic islands. Two subfamilies of microhylids—Brevicipitinae, with 5 genera, and Phrynomerinae, with 1 genus—are endemic to Africa. The subfamily Melanobatrachinae is represented in Africa by 2 genera and in peninsular India by 1 genus. Subfamilies Cophylinae, with 9 genera, and Scaphiophryninae, with 2 genera, are endemic to Madagascar, and Dyscophinae is represented on Madagascar by 1 genus and in southeastern Asia by another. Subfamilies Asterophryninae, with 7 genera, and Geniophryninae, with 6 genera, are endemic to New Guinea and adjacent archipelagos, except species of two genera of geniophrynines that occur in northern Australia. The Otophryninae is endemic to South America. The remaining microhylid subfamily, Microhylinae, has a distribution unlike that of any other family group of anurans. Eleven genera are endemic to South America, and 4 other genera occur in South and Central America. Two genera are endemic to North America (both enter Central America). The remaining microhylines are in Asia and adjacent archipelagos. Two genera are endemic to India and Sri Lanka, and 8 genera are endemic to southeastern Asia and adjacent archipelagos as far south as Bali.

The Ranidae is demonstrably paraphyletic (Ford 1989). The family as now recognized is cosmopolitan except for southern South America, the West Indies, Australia, and most oceanic islands. The distribution of the genus *Rana* equals the range of the family, and *Rana* is the only genus in the New World and in the Palearctic. Other ranine genera are endemic to sub-Saharan Africa (6) or southeastern Asia and adjacent islands (14). Of the latter, 5 genera occur in southeastern Asia; two of these also occur in peninsular India, and one of these two is represented in the East Indies as far south as the Lesser Sunda Islands. Three genera of ranines are endemic to peninsular India, and 6 genera are endemic to the Asian archipelagos (1 extending to New Guinea and Fiji). One ranine genus, *Tomopterna,* is widely distributed in sub-Saharan Africa, but it also is present on Madagascar and in peninsular India. The subfamily Petropedetinae, with 12 genera, is endemic to sub-Saharan Africa, and the subfamily Mantellinae, with 3 genera, is endemic to Madagascar.

Certainly at least part of the histories of these four wide-ranging families is associated with the fragmentation of Gondwanaland. This is indicated for the Bufonidae by the presence of endemic genera in South America, Africa, and India; if, as suggested by these patterns of distribution, the family was involved in plate movements in the Cretaceous, the absence of bufonids on Madagascar and Australia is an anomaly. The Hylidae was present before the separation of Antarctica-Australia from South America. With the exception of the presence of microhylines in Asia and the endemism of asterophrynines and geniophrynines in New Guinea, the Microhylidae exhibits a classical western Gondwanaland vicariance pattern, with separate subfamilies in South America and Africa, as well as an Indian plate movement, with one subfamily occurring in Africa and peninsular India and three subfamilies in Madagascar. The Ranidae may be an African family, but it seems unlikely that the great radiation of ranines in southeastern Asia and associated archipelagos resulted from differentiation after the collision of the Indian plate with Asia in the Miocene, a scenario suggested by the distribution of *Tomopterna.* Rigorous phylogenetic analyses do not exist for the taxa in these four families; consequently, the current classification may not reflect phylogenetic relationships.

IV. Amphibian Faunas of Africa and South America

The amphibian faunas of Africa and South America contain representatives of 38 family groups (families and subfamilies) composed of 32 family groups of anurans, 2 of salamanders, and 4 of caecilians (table 9.1). The independence of the faunal histories of the two continents is suggested by the presence of only 5 families in common. Two, the anuran Pipidae and gymnophionan Caeciliaidae, are shared primarily only by Africa and South America, but caeciliaids also occur in India and on the Seychelles Islands. Two family groups of anurans, Bufonidae and Raninae, are nearly cosmopolitan except for Australia. A fifth family group, Hylinae, is widespread in the New World and in the Palearctic and Australo-Papuan regions but

Table 9.1. Worldwide distribution of family groups of amphibians that inhabit Africa and South America.

Family Group	Near	Meso	WInd	SAm	Pale	Eth	Mad	India	Orien	Aust
ANURA										
Allophrynidae	–	–	–	+	–	–	–	–	–	–
Arthroleptidae										
Arthroleptinae	–	–	–	–	–	+	–	–	–	–
Astylosterninae	–	–	–	–	–	+	–	–	–	–
Brachycephalidae	–	–	–	+	–	–	–	–	–	–
Bufonidae	+	+	+	+	+	+	–	+	+	–
Centrolenidae	–	+	–	+	–	–	–	–	–	–
Dendrobatidae	–	+	–	+	–	–	–	–	–	–
Discoglossidae	–	–	–	–	+	+[1]	–	–	–	–
Heleophrynidae	–	–	–	–	–	+	–	–	–	–
Hemisotidae	–	–	–	–	–	+	–	–	–	–
Hylidae										
Hemiphractinae	–	+	–	+	–	–	–	–	–	–
Hylinae	+	+	+	+	+	+[1]	–	–	+	–
Phyllomedusinae	–	+	–	+	–	–	–	–	–	–
Hyperoliidae										
Hyperoliinae	–	–	–	–	–	+	+[2]	–	–	–
Kassininae	–	–	–	–	–	+	–	–	–	–
Leptopelinae	–	–	–	–	–	+	–	–	–	–
Leptodactylidae										
Ceratophryinae	–	–	–	+	–	–	–	–	–	–
Hylodinae	–	–	–	+	–	–	–	–	–	–
Leptodactylinae	+	+	+	+	–	–	–	–	–	–
Telmatobiinae	+	+	+	+	–	–	–	–	–	–
Microhylidae										
Brevicipitinae	–	–	–	–	–	+	–	–	–	–
Melanobatrachinae	–	–	–	–	–	+	–	+	–	–
Microhylinae	+	+	–	+	–	–	–	+	+	–
Otophryninae	–	–	–	+	–	–	–	–	–	–
Phrynomerinae	–	–	–	–	–	+	–	–	–	–
Pelobatidae	+	–	–	–	+	+[1]	–	–	–	–
Pipidae	–	–	–	+	–	+	–	–	–	–
Pseudidae	–	–	–	+	–	–	–	–	–	–

Family Group	Near	Meso	WInd	SAm	Pale	Eth	Mad	India	Orien	Aust
Ranidae										
Petropedetinae	–	–	–	–	–	+	–	–	–	–
Raninae	+	+	–	+	+	+	+	+	+	+
Rhacophoridae										
Rhacophorinae	–	–	–	–	–	+	+	+	+	–
Rhinodermatidae	–	–	–	+	–	–	–	–	–	–
CAUDATA										
Plethodontidae										
Plethodontinae	+	+	–	+	+	–	–	–	–	–
Salamandridae	+	–	–	–	+	+[1]	–	–	–	–
GYMNOPHIONA										
Caeciliaidae	–	+	–	+	–	+	–	+[2]	–	–
Rhinatrematidae	–	–	–	+	–	–	–	–	–	–
Scolecomorphidae	–	–	–	–	–	+	–	–	–	–
Typhlonectidae	–	–	–	+	–	–	–	–	–	–

Note: Near, Nearactic; *Meso,* Mesoamerica; *WInd,* West Indies; *SAm,* South America; *Pale,* Palearctic; *Eth,* Ethiopian; *Mad,* Madagascar; *India,* India, *Orien,* Oriental; *Aust,* Australian.
[1]North Africa only.
[2]Also Seychelles Islands.

is represented in Africa by only two European species in North Africa.

Three other Holarctic families are represented by small numbers of species in Africa only north of the Sahara—Discoglossidae, Pelobatidae, and Salamandridae. Thus, the sub-Saharan amphibian fauna consists of only 15 family groups of anurans, 3 of which are shared with South America, and 2 families of caecilians, 1 of which is shared with South America.

Family group endemism is high on both continents. Of the 15 family groups of anurans in sub-Saharan Africa, nine (60%) are endemic. Of the 18 family groups of anurans in South America, 7 (39%) are endemic and 6 others are South American families that range into Central America, so in reality, 13 family groups (72%) are endemic to the Neotropics. One of the 2 families (50%) of African caecilians is endemic to sub-Saharan Africa, whereas 2 of the 3 families (67%) of neotropical caecilians are endemic to South America.

Of the 17 family groups of amphibians occurring in sub-Saharan Africa, 4 are shared with India; 3 each are shared with Madagascar, the Palearctic Region, and the Oriental Region; and 2 are shared with the Seychelles Islands. In contrast, among the

22 family groups of amphibians in South America, 12 are shared with Central America, 7 with North America, and 4 with the West Indies. Other cosmopolitan family groups (e.g., Bufonidae, Raninae) occurring in both South America and Africa are shared with other regions of the world.

V. Historical Relationships of the Amphibians

The data and analyses appropriate for unraveling the historical relationships of African and South American amphibian faunas are extremely limited. Three kinds of data sets have been used to ascertain historical relationships of faunas on the two continents: (1) comparative fossil records, (2) rigorous phylogenetic analyses of taxa, and (3) molecular differences between related taxa. The first of these simply consists of determining what taxa were represented during which geological periods or epochs. The second involves the application of the principles of phylogenetic systematics (Wiley 1981) to character sets; the results can be analyzed further with respect to vicariance biogeography (Wiley 1988). The third data set is derived from immunological distances resulting from microcomplement fixation of serum albumins; times of divergence are calculated from the immunological distances (Maxson and Maxson 1986). Other kinds of biogeographic and evolutionary events (e.g., dispersal) can be inferred from patterns of distribution and speciation, but such inferences are not testable.

A. Fossil Record

The fossil record of amphibians in Africa and South America is poor and does not begin to approach the extensive record for mammals; it is therefore of limited usefulness. The record for South America was summarized by Estes and Báez (1985), but no summary of fossil amphibians in Africa exists. Significant fossils from the two continents are listed in table 9.2. In addition to these, there is an unnamed pipid, possibly from the Cretaceous, which van Dijk (1985a) compared with *Thoraciliacus* from the early Cretaceous of Israel.

The presence of *Vieraella* and *Notobatrachus* in the Jurassic in South America confirms the existence of anurans in Gondwanaland prior to its fragmentation. The presence of the pipid genus *Xenopus* in both Africa and South America during the Cretaceous, as well as another pipid genus, *Saltenia,* in South America in the Cretaceous, indicates widespread distribution and differentiation of pipids in western Gondwanaland before the separation of Africa and South America. The presence of *Bufo* in the Paleocene of South America shows that this genus is old and suggests that it could have been present prior to the separation of the continents in the Cretaceous. Likewise, the presence of a caecilian (*Apodops;* cf. Caeciliaidae) in the Paleocene of South America demonstrates the presence of this group of amphibians in the southern continents and provides evidence that caecilians might have been present there before the breakup of Gondwanaland.

The other fossils are of taxa restricted to either Africa or South America and provide approximate ages of the genera. The earliest known leptodactylids are from the Cretaceous, and the earliest hylids are from the Paleocene. Ranid frogs of the genus *Ptychadena* are restricted presently to forest and

Table 9.2. Fossil record of anurans in Africa and South America.

Age	Africa	South America
Jurassic	—	Leiopelmatidae: *Vieraella,* Notobatrachus**
Cretaceous	Pipidae: *Xenopus*	Leptodactylidae: Unnamed
		Pipidae: *Saltenia,* Xenopus*
Paleocene	—	Bufonidae: *Bufo*
		Hylidae: Unnamed
		Pipidae: *Xenopus*
Eocene	Pipidae: *Eoxenopoides**	Leptodactylidae: *Caudiverbera*
Oligocene	Pipidae: *Hymenochirus, Xenopus*	Leptodactylidae: *Caudiverbera, Eupsophus, Neoprocoela**
Miocene	Pipidae: *Xenopus*	Bufonidae: *Bufo*
	Raninae: *Ptychadena*	Leptodactylidae: *Caudiverbera, Wawelia**
Pliocene	—	Bufonidae: *Bufo*
		Leptodactylidae: *Ceratophrys*
Pleistocene	—	Bufonidae
		Leptodactylidae: *Ceratophrys, Leptodactylus*

*Fossil genera.

savanna habitats in sub-Saharan Africa, but Miocene fossils of *Ptychadena* from Morocco corroborate other evidence for more mesic environments in North Africa in the early Neogene. Likewise, the presence of forest-inhabiting frogs (e.g., *Caudiverbera, Ceratophrys, Eupsophus*) from Argentina in the Eocene-Pliocene indicates more mesic past conditions in South America east of the Andes than exist there now.

B. Phylogenetic Analyses

Historical relationships between African and South American anurans have been supported by cladistic analyses of two groups. Cannatella and Trueb's (1988a, 1988b) rigorous phylogenetic analysis of primarily morphological characters of the living species of pipid frogs shows unequivocally that the family is monophyletic and that *Xenopus* is the sister group to all other pipids (figure 9.1). The 13 species of *Xenopus* live in sub-Saharan Africa today, but fossils are known from the Cretaceous of Africa and South America. A second clade is *Silurana* (de Sá and Hillis 1990), which also occurs in sub-Saharan Africa. The clade consisting of the other genera is well supported and shows a vicariance between *Pseudhymenochirus* and *Hymenochirus* in Africa and *Pipa* in South America. This analysis and the fossil evidence demonstrate that primitive pipids (*Xenopus*) were present on both continents prior to their separation and that *Xenopus* became extinct in South America sometime after the Paleocene. The analysis also sug-

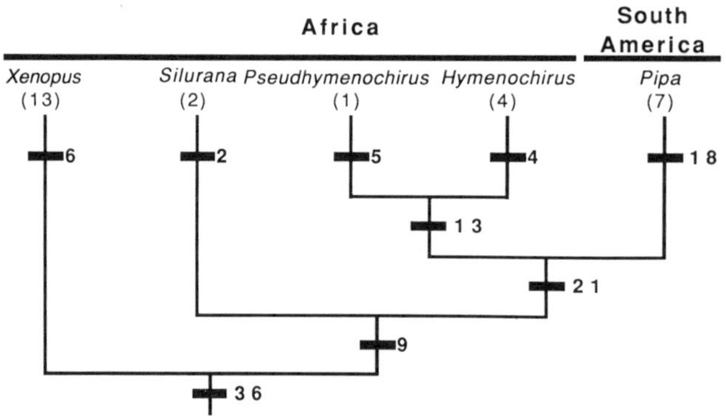

Figure 9.1. Cladogram depicting hypothesized phylogeny of frogs of the family Pipidae, based on Cannatella and Trueb (1988a, 1988b). Numbered bars are synapomorphies for each clade; those on lines to terminal taxa are synapomorphies for the species in those genera. Numbers in parentheses beneath generic names are the numbers of species.

gests that further generic differentiation occurred prior to the middle Cretaceous; only *Hymenochirus* and *Pseudhymenochirus* could have differentiated from each other in Africa after the continents separated.

Using osteological and myological characters in a phylogenetic analysis to determine the interfamilial relationships of dendrobatid frogs, Ford (1989) provided a moderately well corroborated cladogram showing sister-group relationships between the South American Dendrobatidae and the African Arthroleptidae (figure 9.2). Her analysis showed major problems in the classification of ranid frogs. For example, the sister group to the clade consisting of the arthroleptids and dendrobatids is *Dimorphognathus,* a supposed petropedetine ranid in West Africa. Furthermore, petropedetines and ranines do not form monophyletic groups.

This analysis indicates that the vicariance that gave rise to the Arthroleptidae and Dendrobatidae probably occurred at the time of the separation of Africa from South America, and that the differentiation of petropedetine and ranine ranids from one another and of the ranids from arthroleptids + dendrobatids must have occurred even earlier.

C. Immunological Distances

Employing techniques of microcomplement fixation of serum albumins and invoking the evolutionary clock of amino acid changes, Maxson (1981, 1984) substantiated a late Cretaceous separation of African and South American *Bufo.* The presence of *Bufo* in the Paleocene of Brazil helps to support a western Gondwanan distribution of the genus. Maxson's (1981) data further suggest differentiation of *Bufo* and *Schismaderma* in Africa in the Eocene and of the major lineages of African *Bufo* no later than the Miocene. No similar analysis has been performed for South American *Bufo.*

The great diversification of the cosmo-

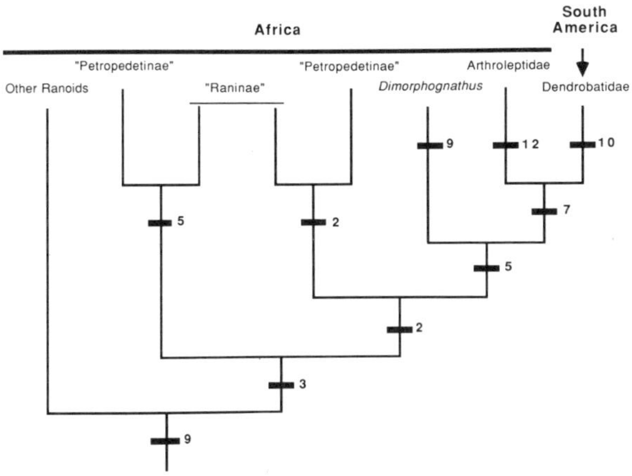

Figure 9.2. Cladogram depicting hypothesized phylogeny of frogs of the families Arthroleptidae and Dendrobatidae and other ranid frogs, based on Ford (1989). Numbered bars are synapomorphies for each clade.

politan genus *Bufo* in both Africa (5 species in North Africa, 2 of which are Eurasian, and 56 species in sub-Saharan Africa) and South America (46 species) and of other genera on both continents—10 genera with 24 species in sub-Saharan Africa and 9 genera with 75 species in South America—cries for phylogenetic analysis. Grandison (1981) based a phylogenetic analysis of African bufonid genera on morphological data and noted that all of the genera and some species of *Bufo* possess some unique derived characters. *Bufo perreti* is also immunologically distant from other *Bufo* (Maxson 1981). A phylogenetic analysis of South American genera showed that all except *Bufo* possess unique derived characters (Cannatella 1986). These analyses suggest that *Bufo* is paraphyletic with respect to the other genera. As emphasized by Cannatella (1986), the South American genera have not been compared critically with African (and Asian) genera, and some of the Neotropical genera may have their closest relatives in Africa (or Asia).

D. Faunal Interchange and Dispersal

The interchange of herpetofaunal elements between adjacent continents increases the diversity of taxa on these continents. When a good fossil record is available, as for many groups of mammals, direction of movements is usually evident. Direction of dispersal can also be inferred from the results of phylogenetic analyses. In many cases, direction of dispersal is evident simply from the distribution of a given species and its congeners. In some instances the direction of such interchanges is not clear. Evidence exists for dispersal between Africa and Madagascar, Africa and Eurasia, South America and Mesoamerica, and South America and the West Indies, but there is no evidence for waif dispersal between Africa and South America.

1. Africa Intercontinental interchanges of African amphibians have occurred with the Iberian Peninsula in southwestern Europe, with southwestern Asia, and with Madagascar. The occurrence in Africa of European species of salamanders (*Pleurodeles waltl* and *Salamandra salamandra*) and anurans (*Alytes obstetricans, Bufo bufo, Bufo viridis, Discoglossus pictus, Hyla arborea, Hyla meridionalis,* and *Rana perezi*) indicate relatively recent invasions from the Iberian Peninsula. *Bufo viridis,* which has a broad distribution across Mediterranean Africa, southwestern Asia, and Europe, may have entered Africa from southwestern Asia. The specific differentiation of *Pleurodeles poireti* and *Pelobates varaldi* suggests that these species were isolated in northwestern Africa from European congeners earlier than the above species (Busack 1986).

The presence of pipid frogs in the lower Cretaceous of the Negev Desert in Israel presents no biogeographic problem (Estes 1975). During the Cretaceous, the Negev Desert area was part of the African continental plate; in the early Cretaceous the equator extended through northern Africa and the Arabian Peninsula, and that region enjoyed humid tropical conditions. Even though the Arabian Peninsula separated from northeastern Africa in the Miocene, no Somalian species of anurans occur in Asia (Lanza 1983). However, Inger (1972) considered *Bufo dodsoni* in northeastern Africa to be the sister species of *B. orientalis* in the southern part of the Arabian Peninsula.

Two family groups of anurans, Hyperoliinae and Raninae, were presumably affected by the separation of Madagascar from Africa. According to Drewes (1984), among hyperoliids, *Tachycnemis* (one species endemic to Seychelles Islands) is the sister group to *Heterixalus* (seven species endemic to Madagascar); together these are the sister group to *Afrixalus + Hyperolius,* both of which are widespread and speciose in sub-Saharan Africa. The ranine genus *Tomopterna* is represented by five species in sub-Saharan Africa, one in Madagascar, and seven in India and adjacent regions. Clarke (1981) concluded on the basis of osteological studies of African ranine frogs that African *Tomopterna* probably were derived from a common ancestor shared with Asian and Madagascan *Tomopterna.* Duellman (1986) argued that these taxa rode the Madagascan-Seychellean-Indian plate from Africa. The only evidence for waif dispersal among these regions is the presence of the ranine *Ptychadena mascariensis* in Madagascar and on the Seychelles and Mascarene islands in the Indian Ocean; the species is widely distributed in sub-Saharan Africa, to which the other 40 species in the genus are confined.

2. South America A major biotic interchange has taken place between South America and Central America; much of this occurred by dispersal after the closure of the Panamanian Portal in the late Cenozoic, but earlier exchanges are known to have occurred via past land connections or intermediate archipelagos. A late Cretaceous or early Cenozoic interchange resulted in diversification in South America of organisms of North American origin and vice versa (Stehli and Webb 1985). The historical interchange of amphibians between South America and Mesoamerica was discussed by Duellman (1979c), Savage (1982), and Vanzolini and Heyer (1985).

As is evident from the list of taxa in table 9.3, many groups of amphibians have been involved in the Mesoamerican–South American interchange. Until recently, most of this interchange was thought to have occurred after the establishment of a land corridor between the continents in the late Cenozoic. However, Savage (1982) provided convincing arguments for the evolution of part of the Neotropical herpetofauna in Mesoamerica during the Cenozoic. His contention is supported by recent phylogenetic analyses.

The principal radiation of bolitoglossine plethodontid salamanders is in Mesoamerica (Wake and Lynch 1976; Wake and Elias 1983). According to dates suggested by genetic differentiation, *Bolitoglossa* entered South America in the Miocene (Hanken and Wake 1982). The occurrence of a species of *Bolitoglossa* and a species of *Oedipina* on Isla Gorgona in the Pacific Ocean, 56 kilometers west of Colombia, of the same species of *Bolitoglossa* on Isla de Colón in the Bay of Panama, and of both species on the mainland is indicative of overwater waif dispersal. These two lines of evidence strongly suggest that *Bolitoglossa* entered South America and underwent a radiation there prior to the establishment of the land bridge between Central America and South America in the late Pliocene (Vanzolini and Heyer 1985).

Frogs of the genus *Rana* are speciose and widely distributed in North America. Some species of the primarily North American *Rana pipiens* group occur in Mesoamerica, where one species ranges as far south as central Panama. The *Rana palmipes* group is primarily Central American, but three species occur in South America. One of these is in the Amazon Basin and two are on the Pacific lowlands; one of the latter extends into Central America. Analysis of morphological and molecular characters by Hillis and de Sá (1988) provided evidence that the three species in South America are the sole members of a clade that presumably differentiated in South America in the Neogene, and that *Rana vaillanti* dispersed from South America into Central America after the Panamanian Portal closed.

Two family groups of anurans, Hylidae (hylines and phyllomedusines) and Telmatobiinae, have undergone extensive radiations in both Mesoamerica and South America. Because both groups are primarily tropical and have the greatest diversity of relatives at the familial and the generic level in South America, they are presumed to be of South American origin. However, the radiations in both family groups in Mesoamerica are indicative of a long history there. Five genera of hylines plus members of several species groups of *Hyla* are endemic to Mesoamerica. The genus *Smilisca,* with six species, two of which enter northwestern South America, is also endemic to Mesoamerica (Duellman 1970). Likewise, two genera of phyllomedusines seem to be of Mesoamerican origin: *Pachymedusa,* which is endemic, and *Agalychnis,* which contains eight species, three of which extend onto the Pacific lowlands of northwestern South America and one of which is restricted to the Amazon Basin (Duellman 1970). Frogs of the telmatobiine genus *Eleutherodactylus* are extremely speciose in the American tropics. Using morphological characters, Lynch (1986) defined a Mesoamerican clade within the genus.

Smaller radiations of two other family groups of South American origin have oc-

Table 9.3. Biogeographical components of the Neotropical amphibian fauna.

South America	Mesoamerica	West Indies
ANURA		
Allophrynidae	Bufonidae (pt)	Bufonidae (pt)
Brachycephalidae	*Atelophryne*	*Peltophryne*
Bufonidae (pt)	*Bufo* (pt)	Hylinae (pt)
Bufo (pt)	*Crepidophryne*	*Calyptahyla*
All other S. Am. genera	Hylinae (pt)	*Hyla* (pt)
Centrolenidae[1]	*Anotheca*	*Osteopilus*
Dendrobatidae[1]	*Hyla* (pt)	Telmatobiinae (pt)
Hemiphractinae[1]	*Plectrohyla*	*Eleutherodactylus* (pt)
Hylinae (pt)	*Pternohyla*	
Hyla (pt)	*Ptychohyla*	
All other S. Am. genera	*Smilisca*[2]	
Phyllomedusinae (pt)	*Triprion*	
Phyllomedusa[1]	Phyllomedusinae (pt)	
Ceratophryinae	*Agalychnis*[2]	
Hylodinae	*Pachymedusa*	
Leptodactylinae[3]		
Telmatobiinae (pt)	Telmatobiinae (pt)	
Eleutherodactylus (pt)	*Eleutherodactylus* (pt)	
All other S. Am. genera	*Hylactophryne*	
Microhylinae (pt)	*Syrrhophus*	
All S. Am. genera[4]	*Tomodactylus*	
Otophryninae	Microhylinae (pt)	
Pipidae[1]	*Gastrophryne*	
Pseudidae	*Hypopachus*	
Rhinodermatidae	Raninae (pt)	
	Rana (pt)	
CAUDATA		
Plethodontinae (pt)	Plethodontinae (pt)	
Bolitoglossa (pt)	*Bolitoglossa* (pt)	
	All other Neotropical genera	
GYMNOPHIONA		
Caeciliaidae (pt)	Caeciliadidae (pt)	
All S. Am. genera[5]	*Dermophis*[2]	
Rhinatrematidae	*Gymnopis*	
Typhlonectidae		

Note: pt indicates part.

[1]Distribution extends into Central America.

[2]Distribution extends into South America.

[3]Distribution of *Leptodactylus* extends into West Indies, and *Pleurodema* into Panama.

[4]Distribution of *Chiasmocleis, Elachistocleis,* and *Relictivomer* extends into Panama, and *Nelsonophryne* into Costa Rica.

[5]Distribution of *Caecilia* and *Oscaecilia* extends into Panama.

curred in Mesoamerica. Within the Bufonidae, the differentiation of species groups of *Bufo* and the presumed derived genera *Atelophryne* and *Crepidophryne* probably occurred in Mesoamerica from early bufonids that entered that region in the Cenozoic. Among the microhylines, *Gastrophryne* and *Hypopachus* apparently differentiated in Mesoamerica during Cenozoic separation from other microhylines in South America.

Other South American groups that have undergone speciation in Mesoamerica include frogs of the genera *Centrolenella, Dendrobates, Phyllobates, Hyla, Leptodactylus,* and *Pipa* and caecilians of the genera *Caecilia* and *Oscaecilia.* These groups possibly dispersed into Mesoamerica after the establishment of the continuous land connection in the late Cenozoic. All are widespread and speciose in South America. Other South American species (e.g., *Bufo alatus, Bufo granulosus, Pleurodema brachyops, Chiasmocleis panamensis, Elachistocleis ovalis, Nelsonophryne aterrimum, Relictivomer pearsei,* and various species of *Hyla, Eleutherodactylus,* and *Leptodactylus*) have extensive ranges in South America and extend only into lower Central America; these species are recent dispersers into Central America.

In contrast to the number of South American groups that have invaded Mesoamerica, relatively few groups of Mesoamerican amphibians have dispersed into South America. With the exception of salamanders of the genus *Bolitoglossa* and frogs of the genus *Rana,* little radiation of Mesoamerican groups has taken place in South America. Only one species of a Mesoamerican phyllomedusine frog genus (*Agalychnis craspedopus*) has differentiated in, and is restricted to, South America. The major contribution of the Mesoamerican amphibian fauna to South America is the continuity of species from lower Central America into the Chocoan Region (humid Pacific versant of Colombia and northern Ecuador). Included in this group are frogs of the genera *Agalychnis, Centrolenella, Eleutherodactylus, Hyla,* and *Smilisca,* salamanders of the genus *Oedipina,* and caecilians of the genus *Dermophis.*

Although the native amphibian fauna of the West Indies is composed of only three families of frogs—Bufonidae, Hylidae, and Leptodactylidae—which are more diverse in South America, there have been some notable radiations in the West Indian frog fauna. Cladistic analyses of morphological (Joglar 1989) and allozymic (Hedges 1989) data suggest multiple invasions of the Greater Antilles by *Eleutherodactylus.* Most of these stocks presumably were transported to their present geographical positions by continental movement from the region that now is Mesoamerica, but members of the *Eleutherodactylus auriculatus* group differentiated in the West Indies from a stock that dispersed northward from northeastern South America through the Lesser Antilles to the Greater Antilles (Hedges 1989). Using morphological characters, Trueb and Tyler (1974) recognized five species of *Hyla* and two endemic genera of hylines in the Greater Antilles; they postulated five separate invasions of the islands from South America. This scenario is being challenged by results of molecular studies by S. B. Hedges and L. R. Maxson (in preparation). Bufonids are represented in the Greater Antilles by the endemic genus *Peltophryne,* with nine species. Pregill (1981) considered this radiation to have resulted from an old invasion; *Peltophryne* shares some characters with the *Bufo granulosus* group in South America, and the two clades may be sister groups. The only other native West Indian frogs are members of the genus *Leptodactylus,* which is widespread in South and Central America. The West Indies have contributed no taxa to the South American fauna.

E. Problematic Groups

Four family groups present systematic and biogeographical problems. None is represented by significant fossils. Comprehensive cladistic analyses have not been performed, nor are immunological data available.

1. Heleophrynidae The four species of *Heleophryne* are confined to streams in the mountains of South Africa. Lynch (1971) considered this genus to be the most primitive of the leptodactyloids (Heleophrynidae, Leptodactylidae, Myobatrachidae). *Heleophryne* apparently is a phylogenetic and distributional relict of a primitive temperate-adapted leptodactyloid stock (Savage 1973). If these assertions are correct, leptodactyloids must have been present in southern Africa and South America prior to the separation of the continents, and the closest relatives of *Heleophryne* should be the temperate-adapted telmatobiine leptodactylids in austral South America.

2. Microhylidae In spite of the lack of fossil evidence, Savage (1973) considered the microhylids to be an ancient group; their extensive Gondwanan distribution and independent radiations in Africa, South America, Madagascar, tropical Asia, and the Australo-Papuan region support this reasoning. The brevicipitine and phrynomerine microhylids are well defined and are restricted to sub-Saharan Africa. Several problems must be solved before a meaningful biogeographic analysis of the family can be attempted:

1. The Otophryninae, represented solely by *Otophryne robusta* in South America, has larval features significantly different from those of other microhylids, has a large external tympanum, and is diurnal. Other microhylids, in contrast, have a small or usually absent tympanum and are nocturnal (Wassersug and Pyburn 1987). *Otophryne* may be the sister group to microhylines.

2. The Microhylinae is represented by 17 genera in the Neotropics and 10 genera in tropical Asia and adjacent archipelagos. Savage (1973:355) hinted that the Oriental and Neotropical microhylines might be evolutionarily independent entities: "My impression is that the two clusters of genera form distinct units that resemble one another because of parallel reduction specializations." Zweifel (1986) concluded that the New World microhylines were monophyletic, and Donnelly and co-workers (1990) noted differences between Oriental and New World microhyline tadpoles. A rigorous phylogenetic analysis of both Old and New World "microhylines" is needed.

3. The Melanobatrachinae may not be a monophyletic group. Without substantiation, Savage (1973) included the Indian *Melanobatrachus* in the Microhylinae and recognized the Hoplophryninae to include the African genera *Hoplophryne* and *Parhoplophryne,* which he considered to be related to the African phrynomerines.

4. The relationships of the Madagascan microhylids are unclear. Savage (1973) and

Duellman (1986) perceived the Madagascan microhylids (Cophylinae, Dyscophinae, and Scaphiophryninae) to have been isolated on Madagascar since its separation from Africa. These subfamilies are endemic to Madagascar, except the Indo-Malayan genus *Calluella,* which Savage (1973) considered to be a primitive Asterophryine, a subfamily otherwise restricted to New Guinea. The picture of Madagascan microhylids is further complicated by the possession of both ranid and microhylid features in the tadpole of the scaphiophrynine *Pseudohemisus granulosum* (Blommers-Schlösser 1975; Wassersug 1984).

3. Ranidae Savage (1973:425) noted: "This family [the Ranidae] is so clearly African in origin, diversification, and radiation that no one has seriously questioned these propositions." Although the family as recognized by Savage has undergone some taxonomic modifications, his statement still is valid. Early ranoid evolution in Africa presumably gave rise to hemisotids, hyperoliids, ranids, and possibly rhacophorids. Within the ranids, there was differentiation of mantellines, petropedetines, and ranines. The Mantellinae, with three genera and 55 species, is endemic to Madagascar; the ancestors of this subfamily may have been affected by the separation of Africa and Madagascar (Duellman 1986) or may have arrived by waif dispersal from Africa (Savage 1973). The Petropedetinae, which is endemic to sub-Saharan Africa, seems to be polyphyletic (Ford 1989). It is not clear if all African ranines are monophyletic or if some African genera may be more closely related to Asian ones (Clarke 1981; Ford 1989). The possible relationships among African and Asiatic ranines imply Cenozoic dispersal through southwestern Asia and northeastern Africa.

4. Rhacophoridae This family is widespread in tropical Asia and adjacent archipelagos where seven genera contain 154 species. Two genera, with 29 species, are endemic to Madagascar, and *Chiromantis,* with 3 species, is endemic to sub-Saharan Africa. Liem (1970) analyzed morphological characters of rhachophorids and concluded that the family originated in Asia, where the primitive *Buergeria* exists; furthermore, because *Chiromantis* shares so many character states with Asiatic rhacophorid genera, he suggested that *Chiromantis* evolved from an Asiatic rhacophorid and reached Africa recently. Liem's conclusions with respect to *Chiromantis* are contrary to those expressed by Laurent (1941, 1943). Savage (1973) suggested that primitive rhacophorids probably were somewhat like *Rhacophorus* (Asia) and *Boophis* (Madagascar) and may have originated in Africa and immigrated to Asia in the early Cenozoic. He also suggested that primitive rhacophorids became extinct in Africa, but the specialized, derived genus *Chiromantis* persists there. Duellman (1986) suggested that rhacophorids may have originated in Africa and were transported on the Madagascar-Seychelles-India plate to Asia, where they underwent a major radiation in the absence of other arboreal frogs. Thus, we are faced with three alternative scenarios—one involving drift and vicariance, one involving dispersal from Africa to Madagascar and Asia, and one involving dispersal from tropical Asia to Madagascar and Africa. The last alternative seems to be the least likely.

Table 9.4. Amphibian diversity among regions, by group.

Region	Family Groups		Genera		Species	
ANURA						
Northern Africa	5	(10%)	6	(2%)	12	(0.3%)
Sub-Saharan Africa	15	(30%)	72	(23%)	606	(16%)
South America	18	(36%)	110	(34%)	1334	(36%)
World	47	(100%)	320	(100%)	3750	(100%)
CAUDATA						
Northern Africa	1	(11%)	1	(1.5%)	3	(0.8%)
World	9	(100%)	64	(100%)	365	(100%)
GYMNOPHIONA						
Sub-Saharan Africa	2	(33%)	9	(26%)	21	(14%)
South America	3	(50%)	16	(47%)	76	(49%)
World	6	(100%)	34	(100%)	154	(100%)

Note: Percentages refer to categories worldwide.

VI. Species Diversity

The numbers of genera and species of amphibians are much greater in South America than in Africa (table 9.4). Of the 5 family groups and 6 genera in North Africa, 2 family groups (Bufonidae and Raninae) are among the 15 family groups in sub-Saharan Africa, and 2 genera (*Bufo* and *Rana*) are among the 72 genera in sub-Saharan Africa. Only *Bufo regularis* occurs in both regions.

Comparison of the numbers of species in sub-Saharan Africa and South America reveals that there are 2.2 times as many species of anurans and 3.6 times as many species of caecilians in South America as in sub-Saharan Africa. This difference is equally striking when the areas of the two continents are compared. Sub-Saharan Africa has about 95 percent of the area of South America, yet the species density of anurans is 2.1 times greater in South America than in Africa (table 9.5). Why are there such striking differences in the numbers of taxa in two continents of approximately the same size that were connected in the past and now lie in about the same range of latitude?

The biotas of Africa and South America have been studied since the mid-1700s. By the beginning of the twentieth century, 158 species of anurans (26% of the present number) were known from sub-Saharan Africa, whereas 373 species (28% of the present number) were known from South America. Species discovery rates were about the same for both continents until 1960, when 461 species were known for Africa and 684 for South America (figure 9.3). After 1960, the species discovery rate remained relatively unchanged in Africa but increased dramatically in South America. Only 2% of the African species were named during the 1980s, whereas 21% of the South Ameri-

Table 9.5. Anuran density by region.

Region	Surface Area (km^2)	Number of Species	Species per 10^6 km^2
Sub-Saharan Africa	16,958,300	606	35.7
South America	17,793,000	1334	74.8

can species were named during that decade. Striking differences also arose in the demography of herpetologists naming species of anurans from Africa and South America. In the 1800s and early 1900s most species were named by taxonomists living in Europe or the United States. Beginning in the 1920s in South America and in the 1940s in Africa, however, significant percentages of the species were being named by residents of Africa and South America. During the 1970s and 1980s, 515 (38.6%) of the currently recognized species of anurans in South America were named; 36.3% of these were named by South Americans. In contrast, during the same 20 years, 102 (16.8%) of the currently recognized species in Africa were named; 42.2% of these were named by Europeans residing in Africa. These data show that even early exploration in South America revealed more species there than in Africa and that the rate of discovery remained about the same until about 30 years ago when numerous South American biologists greatly increased the rate of discovery on that continent.

A probable explanation for the differences in species richness between Africa and South America is the disparity of areas of humid versus subhumid and xeric environments on the two continents (table 9.6). Comparison of the numbers and densities of species of anurans across vegetation types on the two continents reveals some interesting similarities and striking differences (table 9.7). However, these comparisons of numbers of species per unit area can be misleading. For example, the high species densities in montane tropical forest and especially in temperate forest contrast greatly with the low density in tropical savanna–grassland in Africa, but the area covered by the latter type of vegetation is 32 and 139 times greater than that covered by the other two vegetation types, respectively. There are many widespread species in the tropical savanna–grassland, where the number of species at any given site is much greater than at single localities in the other two vegetation types (see Section VIII).

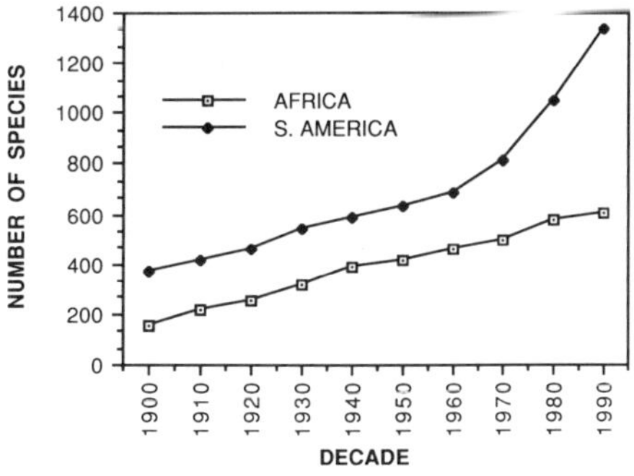

Figure 9.3. Rate of discovery of species of anurans in sub-Saharan Africa and in South America. Data points are numbers of species named that are recognized today.

Table 9.6. Areas of vegetation types in sub-Saharan Africa and South America.

	Sub-Saharan Africa		South America	
Vegetation Type	Area (10^6 km^2)	% Total Area	Area (10^6 km^2)	% Total Area
Lowland tropical rainforest	2.309	16.5	6.365	35.8
Montane tropical rainforest	0.267	1.6	0.760	4.3
Tropical dry forest	4.258	25.0	4.595	25.8
Temperate forest	0.011	0.1	0.446	2.5
Tropical savanna–grassland	8.488	50.0	3.100	17.3
Temperate grassland	0.061	0.4	0.610	4.0
Montane grassland	0.223	1.3	1.540	8.6
Desert	1.341	7.8	0.297	1.7

Note: Total percentages may exceed 100% due to problems in rounding off.

Even though the number of species inhabiting lowland tropical rainforest in South America is more than twice that in Africa, the percentage of the anuran fauna inhabiting that vegetation type is about the same on the two continents, and the species density per area in lowland tropical rainforest is lower in South America than in Africa, where the area covered by lowland rainforest is less than half that of South America. The numbers and percentages of species inhabiting subhumid and xeric habitats (tropical dry forest, savanna, grassland, and desert) are much higher in Africa than in South America; the relatively low density of species in these environments in Africa is a reflection of the large areas encompassed by these types of vegetation. Conversely, the greater numbers of species in montane grasslands (páramo and puna) in South America reflect, at least in part, the much greater area of this type of vegetation in South America compared to Africa.

Montane tropical forests have extremely high densities of species per total area on both continents; in South America the number of species in this type of vegetation is nearly equal to the number of species in lowland tropical rainforest, whereas the number of species in montane forests in Africa is only about one-half of the number of species in lowland rainforest. The much higher densities of species in temperate forest and grassland in Africa, as compared with South America, are a reflection of the small areas of these environments in Africa.

Because of the physiological constraints imposed by problems of water economy, most amphibians are associated with humid environments. Therefore, greater species richness is expected in South America than in Africa because of the greater percentage of South America with humid environments. The high species diversity in lowland rainforest and montane forests in South America is the result of the fragmentation of such

Table 9.7. Numbers, percentages, and densities of species of anurans in different vegetation types in sub-Saharan Africa and South America.

	Sub-Saharan Africa			South America		
Vegetation Type	Species	% Species	Species/ 10^6 km^2	Species	% Species	Species/ 10^6 km^2
Lowland tropical rainforest	229	37.78	99.2	514	38.53	80.0
Montane tropical rainforest	104	17.16	389.5	509	38.16	669.7
Tropical dry forest	60	9.90	14.1	105	7.87	22.0
Temperature forest	6	0.99	545.5	32	2.39	71.7
Tropical savanna–grassland	155	25.57	18.3	21	1.57	6.8
Temperate grassland	24	3.96	393.4	18	1.35	29.5
Montane grassland	16	2.64	71.8	130	9.75	84.4
Desert	12	2.00	8.9	5	0.38	16.8
Total	606	100.00	—	1334	100.00	—

forests. The lowland rainforests are in three regions—the extensive Amazon Basin and the smaller Atlantic coastal forest and Chocoan forest. Likewise, montane forests are separated into numerous regions—eastern and western slopes of the Andes, different highlands in Venezuela and the Guianas, and the Brazilian highlands.

A final reason for the greater species diversity in South America is the influx from Central America of taxa inhabiting humid environments; the greatest concentration of these is on the Pacific versant in northwestern South America. No equivalent dispersal of species into sub-Saharan Africa has occurred.

VII. Ecological Comparisons

Following the analysis of the comparative diversity of anurans on the two continents with respect to major vegetation types, it is possible to view major ecological assemblages of anurans with respect to microhabitat utilization, diets, diel activity, reproductive modes, and ecomorphological guilds of larvae. The following discussion compares anurans in sub-Saharan Africa with those in South America.

A. Microhabitat

Microhabitat utilization is usually associated with morphological adaptations. For purposes of this analysis, African and South American anurans are placed in five groups (table 9.8), which can be viewed as adaptive types.

1. Fossorial Fossorial anurans spend much time underground and have specializations for borrowing. Most fossorial species dig into the ground with the hind feet and thus bury themselves posterior end first; they typically have an enlarged, spadelike,

Table 9.8. Numbers of species of different adaptive types among family groups of anurans.

	Sub-Saharan Africa					South America				
Family Group	Fos	Ter	Arb	Rip	Aqu	Fos	Ter	Arb	Rip	Aqu
Allophrynidae	—	—	—	—	—	0	0	1	0	0
Arthroleptidae										
Arthroleptinae	0	47	0	0	0	—	—	—	—	—
Astylosterninae	0	4	0	21	0	—	—	—	—	—
Brachycephalidae	—	—	—	—	—	0	2	0	0	0
Bufonidae	0	72	0	8	0	0	120	1	0	0
Centrolenidae	—	—	—	—	—	0	0	69	4	0
Dendrobatidae	—	—	—	—	—	0	129	3	0	0
Heleophrynidae	0	0	0	4	0	—	—	—	—	—
Hemisotidae	8	0	0	0	0	—	—	—	—	—
Hylidae										
Hemiphractinae	—	—	—	—	—	0	16	47	0	0
Hylinae	—	—	—	—	—	0	16	259	2	0
Phyllomedusinae	—	—	—	—	—	0	1	35	0	0
Hyperoliidae										
Hyperoliinae	0	12	106	2	0	—	—	—	—	—
Kassininae	0	15	35	0	0	—	—	—	—	—
Leptopelinae	2	14	32	0	0	—	—	—	—	—
Leptodactylidae										
Ceratophryinae	—	—	—	—	—	4	7	0	0	0
Hylodinae	—	—	—	—	—	0	0	0	23	0
Leptodactylinae	—	—	—	—	—	7	106	0	2	2
Telmatobiinae	—	—	—	—	—	8	133	201	39	47
Microhylidae										
Brevicipitinae	12	6	0	0	0	—	—	—	—	—
Melanobatrachinae	0	0	3	0	0	—	—	—	—	—
Microhylinae	—	—	—	—	—	4	30	0	0	0
Otophryninae	—	—	—	—	—	0	1	0	0	0
Phrynomerinae	0	4	0	0	0	—	—	—	—	—
Pipidae	0	0	0	0	20	0	0	0	0	7
Pseudidae	—	—	—	—	—	0	0	0	0	3
Ranidae										
Petropedetinae	0	76	0	10	3	—	—	—	—	—
Raninae	15	23	0	47	2	0	0	0	3	0
Rhacophoridae										
Rhacophorinae	0	0	3	0	0	—	—	—	—	—
Rhinodermatidae	—	—	—	—	—	0	2	0	0	0

Note: Dash indicates no species present.

Fos, fossorial; *Ter*, terrestrial; *Arb*, arboreal; *Rip*, riparian; *Aqu*, aquatic.

keratinized inner metatarsal tubercle. Among South American anurans, this type of fossorial behavior is characteristic of *Chacophrys* and *Lepidobatrachus* (Ceratophryinae), many species of *Pleurodema* (Leptodactylinae), and *Macrogenioglottis* and *Odontophrynus* (Telmatobiinae). In Africa this behavior is found principally in *Breviceps* (Brevicipitinae) and in *Hildebrantia, Pyxicephalus,* and *Tomopterna* (Ranidae), whereas *Hemisus* (Hemisotidae) has a pointed snout and strong forelimbs for burrowing headfirst.

2. *Terrestrial* Many anurans spend most of their lives on the ground but may seek shelter under rocks or logs or in leaf litter on the forest floor. This group includes most bufonids on both continents. In Africa, all arthroleptines and phrynomerines, most petropedetines (e.g., *Cacosternum* and *Phrynobatrachus*), and some kassinines (e.g., *Kassina, Paracassina,* and *Semnodactylus*), brevicipitines (e.g., *Balebreviceps* and *Probreviceps*), and ranines (some *Ptychadena*) are terrestrial. Terrestrial frogs in South America include nearly all dendrobatids; most leptodactylines (e.g., *Adenomera, Physalaemus,* and most *Leptodactylus*), telmatobiines (e.g., *Eupsophus, Phrynopus,* and many *Eleutherodactylus*), and microhylines (e.g., *Chiasmocleis* and *Elachistocleis*); some ceratophryines (*Ceratophrys*), hemiphractines (*Stefania* and some *Gastrotheca*), and hylines (e.g., *Argenteohyla, Phyllodytes,* and some *Ololygon*); and all brachycephalids, otophrynines, and rhinodermatids.

3. *Arboreal* The treefrogs in Africa include all of the melanobatrachines (*Hoplophryne* and *Parhoplophryne*) and rhacophorids (*Chiromantis*) and most hyperoliids, of which *Afrixalus, Hyperolius,* and *Leptopelis*) are the most diverse. In South America, the arboreal component is composed principally of most hylids, nearly all centrolenids, and some telmatobiines (many species of *Eleutherodactylus*).

4. *Riparian* Some frogs are terrestrial but remain at the edges of streams or ponds and seek refuge in the water. In Africa, such riparian frogs include all astylosternines and heleophrynids, many ranines (e.g., *Conraua, Stronglopus,* and many species of *Ptychadena* and *Rana*), and some bufonids (*Nectophryne, Werneria,* and *Wolterstorffina*) and hyperoliines (*Cryptothylax*). The riparian frogs in South America include all hylodines and ranines (only *Rana*), several telmatobiines (*Alsodes, Crossodactylodes, Cycloramphus,* and *Thoropa*), four species of centrolenids, one hyline (*Aplastodiscus*), and two leptodactylines (*Leptodactylus*).

5. *Aquatic* Relatively few species of frogs are aquatic, but two families are strictly aquatic—Pipidae, with 20 species in Africa and 7 in South America, and Pseudidae, with 3 species in South America. Other aquatic African frogs are 3 species of *Petropedetes* and 2 of *Rana*. In South America, the majority of aquatic frogs are telmatobiines (all members of the genera *Atelognathus, Batrachophrynus, Caudiverbera, Somuncuria, Telmatobius,* and *Telmatobufo*); the only others are two monotypic genera of leptodactylines (*Hydrolaetare* and *Limnomedusa*).

Comparison of the numbers and percentages of species characteristic of different microhabitats on the two continents reveals some similarities and some differences (table

9.9). The percentage of fossorial and riparian species in Africa is about three times greater than in South America, where the percentage of arboreal species is nearly twice that in Africa. The percentage of aquatic and terrestrial species is about the same on the two continents.

The discrepancies seem to reflect the greater extent of subhumid and xeric environments in Africa than in South America. Fossorial frogs are more prevalent in subhumid environments than in humid ones. The relatively high number of riparian species in Africa seems to be at least partially correlated with species of ranines (*Ptychadena, Rana,* and *Strongylopus*) that are restricted to riparian situations in otherwise subhumid or xeric environments. However, other groups have highly modified larvae for living in cascading streams and are usually near such streams, many of which are in humid environments. These frogs include the African astylosternines, heleophrynids, and *Petropedetes,* and the South American hylodines and the telmatobiine genera *Alsodes, Crossodactylodes, Cycloramphus,* and *Thoropa.* All of these South American groups are restricted to southeastern Brazil, except *Alsodes,* which inhabits temperate forests in southern Argentina and Chile.

The far greater number and percentage of arboreal species in South America than in Africa may be simply a reflection of the greater area of humid forests in the former. However, the disjunction of three major areas of lowland tropical forest in South America (Amazonian, Chocoan, and Atlantic Coastal) has resulted in distinctive anuran assemblages in each area; Lynch (1979) showed that few species are shared among these forests. In contrast, the relatively small African rainforests are divided into two major areas—the Guinean and the Zairean—which probably were continuous during pluvial periods in the Pleistocene. Although there are species restricted to one or the other of the rainforest areas, many species occur in both.

Another factor in the greater diversity of arboreal frogs in South America may be the abundance of water-holding bromeliads there and the absence of bromeliads in Africa. In South America, many arboreal frogs (centrolenids, hylids, and *Eleutherodac-*

Table 9.9. Adaptive types of anurans in sub-Saharan Africa and South America.

	Sub-Saharan Africa		South America	
Adaptive Type	Species	% Species	Species	% Species
Fossorial	37	6.11	23	1.72
Terrestrial	273	45.05	563	42.20
Arboreal	179	29.53	616	46.18
Riparian	92	15.18	73	5.47
Aquatic	25	4.13	59	4.43
Total	606	100.00	1334	100.00

tylus) use bromeliads for diurnal retreats. In subhumid and xeric areas, many species pass the dry season in bromeliads; some hylids (e.g., *Trachycephalus atlas* and *Gastrotheca fissipes*) flex their casqued, co-ossified heads so as to block the entrance to the bromeliad and thus prevent evaporative water loss. Terrestrial bromeliads also are used by terrestrial anurans (e.g., *Aparasphenodon, Phyllodytes,* and some *Dendrobates, Eleutherodactylus,* and *Gastrotheca*); some of these species and some arboreal *Eleutherodactylus* and *Hyla* deposit their eggs in bromeliads.

Intercontinental comparisons can be refined by viewing microhabitat utilization in different vegetation types (table 9.10). The relative numbers of fossorial frogs are usually higher in all vegetation types in Africa, except that the majority of fossorial frogs in South America are in tropical dry forest. Terrestrial frogs form a greater percentage of the anuran fauna in lowland and montane tropical forests and in temperate and montane grassland in Africa. The percentage of arboreal species in lowland and montane tropical forests in South America is about twice that in Africa. Riparian frogs are more prevalent in all vegetation types in Africa than in South America. The high percentage

Table 9.10. Distribution of anurans in microhabitats in different vegetation types.

		Microhabitats				
Vegetation Type	Region	Fos	Ter	Arb	Rip	Aqu
Lowland tropical rainforest	Africa	0.9	49.7	33.2	9.6	9.6
	S. Amer.	0.4	39.5	57.3	1.4	1.4
Montane tropical forest	Africa	0.0	48.2	18.7	28.6	4.5
	S. Amer.	0.2	31.2	58.2	10.0	0.4
Tropical dry forest	Africa	5.0	38.3	46.7	6.7	3.2
	S. Amer.	15.2	64.7	15.2	2.0	2.9
Temperate forest	Africa	16.7	33.3	0.0	50.0	0.0
	S. Amer.	0.0	50.0	3.1	34.4	12.5
Tropical savanna–grassland	Africa	15.5	33.5	32.9	16.6	1.3
	S. Amer.	9.5	38.2	33.3	9.5	9.5
Temperate grassland	Africa	16.7	50.0	12.5	16.7	4.1
	S. Amer.	11.2	44.4	0.0	0.0	44.4
Montane grassland	Africa	0.0	100.0	0.0	0.0	0.0
	S. Amer.	0.0	73.8	0.8	0.0	25.4
Desert	Africa	15.0	66.7	0.0	8.3	0.0
	S. Amer.	0.0	100.0	0.0	0.0	0.0

Note: Numbers are percentages of species in given vegetation types. *Fos,* fossorial; *Ter,* terrestrial; *Arb,* arboreal; *Rip,* riparian; *Aqu,* aquatic.

of aquatic frogs in temperate forest and in temperate and montane grassland in South America is due to the diversification of one group, the telmatobiine leptodactylids.

B. Diet

Information on diets of anurans is limited to various anecdotal reports and a few detailed analyses—Inger and Marx (1961) and Toft (1982) for Africa, and Duellman (1978, 1989b) and Toft (1980) for South America. Most anurans are sit-and-wait strategists and feed on whatever prey approaches them; the maximum size of the prey depends on the gape of the frog. The diets include mostly insects, with orthopterans, hemipterans, and coleopterans predominating. However, these anurans also eat mollusks, spiders, and annelids, and the larger anurans include small vertebrates in their diets. Some of these, such as the South American hylid *Hemiphractus,* specialize on anurans and large orthopterans. The African ranine *Pyxicephalus adspersus* and South American ceratophryines, *Ceratophrys,* are notorious carnivores; they feed on smaller frogs and small lizards, snakes, and mammals.

Not all sit-and-wait strategists feed on relatively large prey. Most microhylids specialize on ants; the African *Hemisus* and South American *Physalaemus* specialize on termites. Other anurans are active foragers; these species almost always feed on small insects, especially ants, but they also eat termites and small beetles. This foraging strategy is most evident in the diurnal South American dendrobatids that have toxic skin secretions (*Dendrobates, Epipedobates, Minyobates,* and *Phyllobates*), but it also occurs in some South American leptodactylids (e.g., *Edalorhina* and some *Eleutherodactylus*), some African and South American bufonids, and possibly African arthroleptids.

C. Diel Activity

Although most anurans are nocturnal, there are many exceptions. Aquatic anurans seem to be active by day and night, and some riparian anurans are active by day. *Hylodes,* for example, is diurnal along streams in southeastern Brazil. Fossorial anurans appear above ground at night; some species do so only after heavy rains. Arboreal frogs and most terrestrial anurans are generally active only at night. However, among terrestrial anurans in South America, all dendrobatids and bufonids of the genus *Atelopus* are diurnal, as are some cryptically patterned *Bufo* and the leptodactyline *Edalorhina.*

D. Reproductive Modes

Duellman (1986, 1989a) and Duellman and Trueb (1986) defined reproductive modes and summarized their distribution among anurans worldwide. Like most fishes, some members of all three living groups of amphibians deposit eggs in water; this strongly suggests that the production of aquatic eggs and larvae is the primitive (and generalized) reproductive mode in amphibians. Modes that involve the deposition of eggs out of water, development of eggs into nonfeeding larvae, direct development (no free-living larval stage), and incorporation of diverse kinds of parental care are considered to be evolutionarily derived. Duellman and Trueb (1986) recognized 29 reproductive modes in anurans; these can be grouped into five major categories for comparison among family groups of anurans in sub-Saharan Africa and South America (table 9.11).

Table 9.11. Reproductive modes among family groups of anurans.

Family Group	Region	Aquatic Eggs and Larvae	Non-aquatic Eggs; Aquatic Larvae	Foam Nest	Terrestrial Eggs; Direct Development	Eggs or Larvae Carried
Allophrynidae	SA[1]	+	−	−	−	−
Arthroleptidae						
Arthroleptinae	A	+	−	−	+	−
Astylosterninae	A	+	−	−	−	−
Brachycephalidae	SA	−	−	−	+	−
Bufonidae	A, SA	+	+	−	+[2]	−
Centrolenidae	SA	−	+	−	−	−
Dendrobatidae	SA	−	+	−	−	+
Heleophrynidae	A	+	−	−	−	−
Hemisotidae	A	−	+	−	−	+
Hylidae						
Hemiphractinae	SA	−	+	−	−	+[3]
Hylinae	SA	+	+	+	−	−
Phyllomedusinae	SA	−	+	−	−	−
Hyperoliidae						
Hyperoliinae	A	+	+	+[4]	−	−
Kassininae	A	+	+	−	−	−
Leptopelinae	A	−	+	−	−	−
Leptodactylidae						
Ceratophryinae	SA	+	−	−	−	−
Hylodinae	SA	−	+	−	−	−
Leptodactylinae	SA	+	−	+	+	−
Telmatobiinae	SA	+	+	−	+	−
Microhylidae						
Brevicipinae	A	−	−	−	+	−
Melanobatrachinae	A	−	+	−	−	−
Microhylinae	SA	+	+	−	+	−
Otophryninae	SA	+	−	−	−	−
Phrynomerinae	A	+	−	−	−	−
Pipidae	A, SA	+	−	−	−	+[3]
Pseudidae	SA	+	−	−	−	−
Ranidae						
Petropedetinae	A	+	−	−	+	−
Raninae	A, SA	+	−	−	+	−
Rhacophoridae						
Rhacophorinae	A	−	−	+[4]	−	−
Rhinodermatidae	SA	−	+	−	−	+

[1]*SA*, South America; *A*, Sub-Saharan Africa.

[2]Also some African species (*Nectophrynoides*) are ovoviviparous or viviparous.

[3]Direct development (no free-living larvae) in most hemiphractine hylids and some species of *Pipa* (both groups in South America).

[4]Arboreal foam nests.

The generalized mode of aquatic eggs and larvae is common on both continents. Eggs are deposited singly, or in clumps or strings attached to submerged vegetation, or as free masses or surface films in lentic waters. This mode is used in South America by all allophrynids, ceratophryines, pseudids, and ranines, most species of bufonines, hylines, and microhylines, and some species of leptodactylines and telmatobiines. In Africa, the generalized mode is characteristic of all phrynomerines and pipids, most bufonids, and some species of hyperoliines, kassinines, and ranines. In other groups, eggs are deposited in streams; this is characteristic of all astylosternines and heleophrynids, and some arthroleptines, bufonids, petropedetines, and ranines in Africa. African bufonids of the genera *Mertensophryne* and *Stephopaedes* deposit eggs in constrained bodies of water. In South America, eggs are deposited in streams by some bufonids (principally *Atelopus*), hylines, telmatobiines, and *Otophryne*. Some South American hylines deposit eggs in constrained bodies of water—for example, in the axils of bromeliads (*Aparasphenodon, Phyllodytes,* and some *Hyla,* and *Ololygon*) or in tree holes (*Nyctimantis rugiceps* and *Phrynohyas resinfictrix*). Males of some species of *Hyla* (*H. boans, H. faber,* and *H. rosenbergi*) construct shallow basins at the edges of ponds or slow-moving streams, in which the eggs and early larval stages develop.

Many anurans deposit eggs out of water. South American centrolenids, phyllomedusines, and some hylines, as well as some African hyperoliines, kassinines, and petropedetines, deposit eggs on vegetation above ponds or streams into which the larvae fall upon hatching. In South American hylodines and some telmatobiines, and in African leptopelines, eggs are laid on land; upon hatching, the tadpoles either fall into or propel themselves to water. The African melanobatrachines deposit eggs in tree holes, and upon hatching, the larvae develop in the water in these cavities. Other examples of terrestrial eggs and aquatic larvae involve parental care. South American dendrobatids deposit their eggs on land, and adults transport the hatchling tadpoles on their backs to water; in some species of *Dendrobates* the female subsequently provides unfertilized eggs for the larvae to eat. Females of the South American hemiphractines carry the eggs openly on the dorsum or in a dorsal pouch until hatching. In those species in which the eggs hatch as larvae, the tadpoles are deposited in ponds or constrained bodies of water (tree holes or bromeliads). In other species the eggs hatch as froglets. Males of the South American *Rhinoderma* transport larvae in their vocal sacs; one such species deposits the larvae in small ponds, whereas the other species carries the larvae until they metamorphose. The larvae of the African *Hemisus marmoratus* may be transported on the back of the female, who excavates a tunnel from an underground nest chamber to a pond (Bourquin 1985; van Dijk 1985b).

The construction of a foam nest for the eggs seems to protect them from desiccation. In some African hyperoliines, small clutches of eggs are contained in small foamy nests enclosed in leaves of emergent vegetation above ponds; upon hatching, the larvae fall into the water. The African rhacophorid *Chiromantis* constructs large foam nests on trees above water; the exteriors of these nests harden, but the interiors remain moist. Foam nests on the surface of the water or on land at the edge of water are characteristic of most leptodactylines in South America, and the eggs in the terrestrial foam nests of *Adenomera* undergo direct development. The only other South American frog known to construct an aquatic foam nest is *Hyla* cf *rizibiliz* (Haddad et al. 1990).

In addition to those South American hemiphractines in which eggs undergo direct development on the dorsum or in dorsal pouches of the females, the absence of a free-living larval stage is seen in many other frogs on both continents. Direct development of nonaquatic eggs is characteristic of brachycephalids, many telmatobiines (e.g., *Eleutherodactylus, Geobatrachus, Ischnocnema,* and *Phrynopus*), and perhaps some microhylines (e.g., *Myersiella*) in South America. In Africa, direct development of terrestrial eggs presumably occurs in all brevicipitines; it also is characteristic of some arthroleptines (e.g., *Arthroleptis*) and petropedetines (e.g., *Anhydrophryne*). African bufonids of the genus *Nectophrynoides* have internal fertilization; reproductive modes involve production of free-living larvae, direct development of terrestrial eggs, ovoviviparity, and viviparity (Wake 1980).

South American pipids of the genus *Pipa* differ from African pipids in that the fertilized eggs imbed in the back of the female. In some species (e.g., *P. carvalhoi*) the eggs hatch as free-living tadpoles, whereas in *P. pipa* and *P. snethlage* the eggs undergo direct development into miniatures of the adults.

Because the same evolutionarily derived reproductive modes—for example, the occurrence of foam nests, arboreal egg masses, and direct development of terrestrial eggs—appear in unrelated lineages of anurans (Duellman 1989a), I conclude that independent evolution of these modes occurred in different groups of frogs on the continents of Africa and South America. Anurans in South America display a greater variety of modes than those in Africa. Transportation of eggs, or larvae, or both is common in South America and may occur in *Hemisus* in Africa. Ovoviviparity and viviparity are unknown in South American anurans, as are arboreal foam nests.

The diversity of South American taxa having nonaquatic eggs, many of which undergo direct development, seems to be correlated with extent of humid habitats in South America. Lynch (1979) first pointed out the fidelity of these reproductive modes to humid lowland rainforests, where atmospheric humidity is high. Duellman and Trueb (1986) expanded this concept by comparing the diversity of reproductive modes in different environments. As in South America, most African frogs that deposit terrestrial eggs that undergo direct development inhabit humid areas. An exception is *Breviceps,* some species of which inhabit xeric areas, but these frogs deposit their eggs in underground chambers where the soil is relatively moist; furthermore, the attending female may moisten the eggs by secreting water stored in the bladder.

E. Larvae

Because of the biphasic life histories characteristic of most amphibians, larvae and adults are subjected to different selective pressures. This is especially evident in anurans, in which the adults are alike morphologically and ecologically, but the larvae develop in different habitats (e.g., streams versus ponds) and are distinctly different morphologically. In anuran larvae known from Africa and South America, many adaptive types are recognized.

Altig and Johnston (1989) defined ecomorphological guilds of anuran larvae. Their terminology is used in the following discussion, which includes only those guilds and examples from Africa (A) and South America (SA). A primary dichotomy involves larvae that obtain all nutrients for development through metamorphosis from maternal yolk (endotrophic), as contrasted with free-living larvae that obtain nutrients from the environment (exotrophic). Endotrophic larvae are segregated into several types:

Viviparous: Froglet birthed from oviducts; nutrients provided by maternal oviducal materials after exhaustion of yolk—bufonid *Nectophrynoides occidentalis* (A).

Ovoviviparous: Froglet birthed from oviducts; all nutrition provided by yolk—bufonid *Nectophrynoides tornieri* and *N. viviparus* (A).

Paraviviparous: Froglet hatched from egg at eventual site of deposition on body of female—many hemiphractines (*Cryptobatrachus, Hemiphractus, Stefania,* and some *Gastrotheca*), some pipids (*Pipa pipa, P. snethlage*) (all SA). (Hemiphractines of the genera *Flectonotus* and *Fritziana* hatch as advanced nonfeeding tadpoles and complete their development in bromeliads or tree holes.)

Exoviviparous: Larvae transported in vocal sac of male until metamorphosis—rhinodermatid *Rhinoderma* (SA).

Direct development: Froglet hatched from eggs not associated intimately with body of parent—arthroleptine *Arthroleptis* (A), brevicipitine *Breviceps* (A), and many telmatobiines, such as *Eleutherodactylus* (SA).

Nidicolous: Nonfeeding larvae undergo development in or near nest site—bufonid *Nectophrynoides malcolmi* (A), leptodactyline *Adenomera* (SA), microhyline *Syncope* (SA), and telmatobiine *Eupsophus* (SA).

The guilds of exotrophic larvae are divided into two sections, the first of which is composed of those that inhabit lotic systems; some of these inhabit flowing portions of streams and are designated rheophilous or rhyacophilous:

Clasping: Slow-water inhabitants with slightly modified mouths—petropedetine *Natalobatrachus* (A), *Rana palmipes* group (SA).

Adherent: Fast-water inhabitants with multiple rows of labial papillae and teeth—some hylines of the genus *Hyla* (SA).

Suctorial: Fast-water inhabitants that maintain position via oral discs—astylosternine *Trichobatrachus* (A), heleophrynid *Heleophryne* (A), some hylines of the genus *Hyla* (SA), petropedetine *Petropedetes* (A), and ranine *Conraua* (A).

Gastromyzophorous: Fast-water inhabitants with ventral sucker—bufonid *Atelopus* (SA).

Psammonic: Long spiracular tube and reduced mouth—otophrynine *Otophryne* (SA).

Semiterrestrial: Inhabitants of wet rock faces; spend some time out of water—bufonid *Werneria* (A), leptodactyline *Leptodactylus rugulosus* (SA), petropedetine *Nothophryne* (A), ranine *Ptychadena broadleyi* (A), telmatobiine *Thoropa,* and some *Cycloramphus* (SA).

Other inhabitants of lotic systems occur in nonflowing parts of streams and morphologically resemble lentic tadpoles:

Benthic: Inhabit stream bottoms and have depressed bodies and low caudal fins—some dendrobatids of the genus *Colostethus* (SA); leptopeline *Leptopelis* (A); and ranines *Rana angolensis, R. fuscigula,* and *Stronglopus grayii* (all A).

Fossorial: Vermiform burrowers in stream detritus—centrolenids (SA).

Nectonic: Pelagic inhabitants in nonmoving parts of streams—ranine *Strongylopus fasciatus* (A).

Neustonic: Surface feeders with enlarged oral disc directed dorsally—dendrobatid *Colostethus nubicola* (SA), and phylomedusine *Phyllomedusa guttata,* and *P. jandaia* (SA).

The other exotrophic larvae inhabit lentic systems that include bromeliads and tree holes, as well as lakes and ponds:

Arboreal: Inhabitants of bromeliads, tree holes, and other isolated pockets of water—bufonids *Mertensophryne* and *Stephopaedes* (A), some species of dendrobatid *Dendrobates* (SA), hyline *Phrynohyas resinifictrix* (SA), some species of hyline *Ololygon* (SA), leptopeline *Acanthixalus* (A), melanobatrachine *Hoplophryne* (A), and phyllomedusines *Agalychnis calcarifer* and *Phyllomedusa fimbriata* (SA).

Benthic: Bottom inhabitants in lentic water—most bufonids of the genus *Bufo* (A, SA); some dendrobatids of the genera *Colostethus, Dendrobates,* and *Epipedobates* (SA); discoglossid *Alytes obstetricans* (A); hemisotid *Hemisus* (A); some species of hyline genus *Smilisca* (SA); and ranines *Ptychadena* and some *Rana* (A).

Carnivorous: Benthic or pelagic larvae with terminal mouths—ceratophryines *Ceratophrys* and *Leptidobatrachus* (SA), hyline *Hyla marmorata* (SA), pipids *Hymenochirus* (A) and *Pipa carvalhoi* (SA).

Macrophagous: Benthic or pelagic larvae with terminal mouths and oral discs—arthroleptine *Cardioglossa* (A); hylines of the genus *Scarthyla* (SA) and of the *Hyla leucophyllata, H. microcephala,* and *H. parviceps* groups (SA); and kassinine *Afrixalus* (A).

Nectonic: Open water inhabitants with anteroventral mouths—most hylines, including *Osteocephalus* and *Trachycephalus;* many *Hyla* and *Smilisca;* most *Ololygon* and *Phrynohyas* (SA); and most hyperoliine *Hyperolius* (A).

Suspension feeder: Midwater inhabitants with reduced mouthparts—most microhylines (SA), phrynomerine *Phrynomantis* (A), and pipids *Silurana* and *Xenopus* (A).

Suspension-rasper: Midwater inhabitants with full complement of mouth parts—kassinine *Kassina* (A), most phyllomedusines of the genera *Agalychnis* and *Phyllomedusa* (SA).

The distribution of ecomorphological guilds of tadpoles among the family groups of anurans on the two continents clearly shows independent adaptive radiations of larvae on the two continents. Only the lentic-benthic larvae of *Bufo* and the lentic-carnivorous larvae of *Hymenochirus* in Africa and *Pipa* in South America may have existed among west Gondwanan anurans prior to the separation of the continents (see III.C.). The diverse kinds of rheophilous, terrestrial, and arboreal tadpoles evolved independently on the two continents.

VIII. Anuran Communities

A final aspect of the amphibian faunas on the two continents concerns communities of anurans. Numerous studies of anuran communities are available for South America: Crump (1971) for Belém, Brazil; Lescure (1976) for the Rivière Oyapock, French Guiana; Staton and Dixon (1977) for the central Venezuelan llanos; Duellman (1978) for Santa Cecilia, Ecuador; Hoogmoed and Gorzula (1979) for El Manteco, Venezuela; Toft and Duellman (1979) and Schlüter (1984) for Panguana, Peru; Péfaur and Duellman (1980) for the Andes; Weygoldt (1986) for Espírito Santo, Brazil; Gudynas and Rudolf (1987) for Pajas Blancas, Uruguay; Heyer et al. (1990) for Boracéia, Brazil; Rodriguez and Cadle (1990) for Cocha Cashu, Peru; and Zimmerman and Rodrigues (1990) for Manaus, Brazil. Unfortunately, equivalent studies are not available for African anurans. Balinsky (1969), Channing (1976, 1989), and Telford (1982) provided data on breeding communities in southern Africa, and Bowker and Bowker (1979) reported on a breeding community in Kenya; a general idea of community composition can be gained from the works of Schiøtz (1963) on Nigeria and from those of Perret (1966) and Amiet (1972, 1973, 1975, 1978, 1986) on Cameroon.

Sufficient data are available to make some comparisons among communities on the two continents with respect to numbers of taxa, microhabitats, and reproductive modes. The largest communities occur in lowland tropical forests, and the smallest in deserts (table 9.12). A notable discrepancy between Africa and South America exists in the sizes of anuran communities in the different vegetation types. Generally, many more species exist in communities in lowland and montane tropical forests, temperate forests, and tropical grasslands in South America than in Africa. The numbers are about the same in tropical dry forests, deserts, and temperate and montane grasslands on the two continents.

The largest anuran communities in lowland tropical forests on each continent are 48 species at Foulassi, Cameroon (Perret 1966) and 81 species at Santa Cecilia, Ecuador (Duellman 1978). Comparison of these communities shows a proportionally higher utilization of terrestrial and riparian microhabitats at Foulassi and a higher utilization of arboreal microhabitats at Santa Cecilia (table 9.13). There is a greater diversity of reproductive modes, with proportionately fewer species having aquatic eggs and larvae, at Santa Cecilia than at Foulassi. The pattern of higher utilization of terrestrial microhabitats than arboreal ones seems to be characteristic of lowland tropical forests in Africa. For example, at the Sapo National Park in Liberia, P. F. Chippindale (personal communication, 1989) found 25 species of anurans: 15 (60%) terrestrial, 4 (16%) arboreal, 5 (20%) riparian, and 1 (4%) aquatic. The pattern of high utilization of arboreal habitats seems to be characteristic in lowland tropical forests in Amazonian and Chocoan South America (Duellman 1978, 1988, 1990), but there is proportionally higher use of riparian microhabitats (particularly streams) in the Atlantic coastal forests (Heyer et al. 1990; Weygoldt 1986). I have also noted the decline in reproductive modes with terrestrial eggs from

Table 9.12. Numbers of species in anuran communities in different vegetation types in sub-Saharan Africa and South America.

Sub-Saharan Africa				South America		
N	Range	Mean	Vegetation Type	N	Range	Mean
5	20–49	22.8	Lowland tropical rainforest	14	35–81	54.4
2	8–10	9.0	Montane tropical forest	5	20–39	25.4
1	26	26.0	Tropical dry forest	3	22–29	25.3
3	3–5	3.6	Temperate forest	4	5–8	6.5
3	8–13	10.7	Tropical savanna–grassland	2	16–26	21.0
1	9	9.0	Temperate grassland	3	3–17	7.7
1	8	8.0	Montane grassland	6	5–9	7.0
2	1–2	1.5	Desert	3	1	1.0

Table 9.13. Anuran communities in lowland tropical rainforests at Foulassi, Cameroon, and Santa Cecilia, Ecuador.

	Foulassi	Santa Cecilia
N[1]	48	81
Microhabitat		
Terrestrial	22 (45.8%)	24 (19.6%)
Arboreal	17 (35.4%)	51 (62.9%)
Riparian	7 (14.6%)	5 (6.3%)
Aquatic	2 (4.2%)	1 (1.2%)
Reproductive Mode		
Aquatic eggs and larvae	27 (56.3%)	35 (43.2%)
Foam nest, aquatic larvae[2]	1 (2.1%)	9 (11.1%)
Arboreal eggs, aquatic larvae	9 (18.7%)	13 (16.0%)
Terrestrial eggs, aquatic larvae[3]	6 (12.5%)	6 (7.4%)
Terrestrial eggs, direct development	5 (10.4%)	16 (19.8%)
Eggs on female, direct development	—	2 (2.5%)

[1]Number of species.

[2]Foam nests are arboreal at Foulassi.

[3]Larve transported to water on backs of adults at Santa Cecilia.

regions of higher to ones with lower rainfall in the Amazon Basin (Duellman 1988).

Comparison of tropical dry forests on the two continents reveals similar differences in microhabitat utilization. For example, 26 species occur at the Sengwa Wildlife Research Area in Zimbabwe: 2 (7.7%) fossorial, 10 (38.5%) terrestrial, 3 (11.5%) arboreal, 9 (34.6%) riparian, and 2 (7.7%) aquatic (I. Coulson, personal communication, 1989). At Cerro Cora, Paraguay, there are 29 species: 1 (3.5%) fossorial, 13 (44.8%) terrestrial, 13 (44.8%) arboreal, and 2 (6.9%) riparian (Duellman 1988). The patterns are similar for tropical grasslands and savannas on both continents, except that communities are larger in South America. For example, there are 26 species at El Manteco, Venezuela: 1 (3.8%) fossorial, 11 (42.9%) terrestrial, 10 (38.5%) arboreal, 3 (11.5%) riparian, and 1 (3.8%) aquatic (Hoogmoed and Gorzula 1979). Savanna communities in Nigeria are composed of 8–13 species; 11 species are reported at Igbetti, Nigeria: 1 (9.1%) fossorial, 4 (36.4%) terrestrial, 4 (36.4%) arboreal, 1 (9.1%) riparian, and 1 (9.1%) aquatic (Schiøtz 1963).

Comparison of a montane grassland community in the Itombwe Highlands in eastern Zaire (Laurent 1964) with six communities in the northern and central Andes (Péfaur and Duellman 1980; Duellman 1988) reveals that the Zairean community has 8 species (4 terrestrial and 4 riparian), all of which have aquatic larvae in either ponds or streams. The Andean communities each have 5–9 ($\bar{x}$ = 7.0) species. The number of terrestrial species in each Andean community is 4–8 ($\bar{x}$ = 5.83); riparian, 0–1 ($\bar{x}$ = 0.50); and aquatic, 0–2 ($\bar{x}$ = 0.67). Eight reproductive modes are found in the Andean communities: (1) eggs and larvae in ponds, (2) eggs and larvae in streams, (3) eggs and larvae in bromeliads, (4) eggs in aquatic foam nests in ponds, (5) eggs in maternal pouch and larvae in ponds, (6) eggs undergoing direct development in maternal pouch, (7) eggs terrestrial and larvae transported to streams, and (8) terrestrial eggs undergoing direct development.

According to the available data, not only do anuran communities within a given vegetation type in South America usually contain more species than comparable communities in Africa, but they also exhibit notable differences in microhabitat utilization. Generally, proportionately more species in African communities are terrestrial or riparian than in South America, whereas there are proportionately more species in arboreal microhabitats in South America than in Africa. These differences may be correlated with greater arboreal habitat heterogeneity in South America, where there are bromeliads (absent in Africa) and where there is greater floristic diversity in tropical forests than in Africa (Richards 1973). The preponderance of arboreal frogs in South American communities is at least partially the result of the great numbers of species of centrolenids, hylids, and arboreal *Eleutherodactylus* there; these groups account for more than 600 species, a number equal to the entire anuran fauna of Africa, where there are less than 200 arboreal species.

IX. Discussion

The foregoing summary and evaluation of historical and ecological data can be viewed as a first step in a critical comparison of the amphibian faunas of Africa and South America. I take this opportunity to comment on the quality of existing data.

Although recent geophysical work has refined our understanding of the separation of Africa and South America, there is an appalling lack of congruence of geophysical, sedimentary, and biological data regarding the connection of South America with Central America (see discussion in Chapter 2). This is a major hindrance to interpretation of the intercontinental interchange, in spite of the recent major work on the subject by Stehli and Webb (1985). Interpretation of the changing continental stage is clouded by extremely generalized views of climatic changes during the Cenozoic and by controversy over the amount and effects of climatic change in the Quaternary. Van der Hammen and co-workers (1973) and Van der Hammen and Cleef (1986) presented a clear and detailed picture of changing scenes in the northern Andes, but how far can these data be extrapolated to the lowlands, where Colinvaux (1979 and Chapter 16) and Livingstone (1982) dispute sufficient climatic change to result in alternating forest and nonforest biomes? The meager data from the lowlands are inconclusive.

Richards (1973), among others, emphasized the depauperate nature of the African flora. However, Gentry (Chapter 17) noted that although the floras of Africa and South America differ at the continental level, floristic composition and patterns of species richness are extremely similar at the community level on both continents. Can these continental differences and community similarities be interpreted as results of an ancient separation followed by independent adaptive radiations under similar selective pressures? If so, why is South America so much richer floristically than Africa? Is it simply that Africa is a drier continent? These questions are germane to an understanding of the history of the vegetation types and their animal inhabitants on both continents.

The relationships and histories of the major groups of amphibians are obfuscated by inadequate data. The fossil record and cladistic analysis of the archaeobatrachians provide something palatable, but what about the caeciliaids and the other 97 percent of the anurans? Ford's (1989) demonstration of sister-group relationships of African arthroleptines and South American dendrobatids should provide an impetus for rigorous phylogenetic analyses of other groups of neobatrachians. This is especially evident for the African and South American bufonids and microhylids. The limited work on immunological distances suggests that further research may refine our ideas about times of separation of lineages. The historical biogeographic scenario suffers from our inadequate knowledge of phylogenetic relationships and consequent inability to apply the principles of vicariance biogeography or even to evaluate possible dispersals. Four problematic groups are identified, which include 13 percent of the African and South American anuran faunas; until the relationships within these groups are ascertained, any biogeographic suggestions are suspect. Moreover, such groups as the Leptodactylidae (containing 42 percent of the South American frogs) and *Hyla,* which are demonstrably paraphyletic, and others like African Raninae and Petropedetine, which are polyphyletic, cannot be analyzed biogeographically.

The great difference in species diversity between the continents seems to be real; in fact, the species-discovery curves suggest that the discrepancy will grow, especially as more remote regions of South America are explored. The attempt to analyze species diversity with respect to different vegetation types is crude. The vegetation types need to be defined more accurately. Furthermore, the presence of a species in more than one vegetation type must be taken into account. These refinements perhaps exceed the limits of published information, but in order to obtain an accurate assessment of anuran distribution among vegetation types, these kinds of data need to be collected and analyzed.

The ecological comparisons are generalized. Data on many species are inadequate. This dilemma exists because of the differences in methodology and emphasis by different workers. There is a great need for standards of data collecting and manner of reporting data to make realistic comparisons possible. In many cases, association of a species with a microhabitat was made because of similarities in morphology with species of known microhabitats. Of all ecological parameters, diet is the weakest. Even for those species for which diet information is available, the food items generally are recognized solely as broad categories (e.g., ants, orthopterans). Perhaps frogs that feed on ants specialize on certain species. Such refinement in study of food items is beyond the abilities of most amphibian biologists and beyond the realm of interest of systematic entomologists, who might be able to identify fragments of insects.

The reproductive biology of many species is unknown. The categories of reproductive modes defined in Duellman (1985) and the ecomorphological guilds of larvae defined in Altig and Johnston (1989) need to be integrated into a broad synthesis of anuran reproductive biology that incorporates developmental constraints with ecological factors in an evolutionary framework.

Knowledge of anuran communities is pitifully poor, especially in Africa. The "snapshot" approach to community ecology, as used here, was criticized by Duellman and Pianka (1990), who emphasized that community ecology will emerge on a new level of awareness and understanding as the evolutionary time scale is broadened to include historical events, biogeographic histories of the taxa, innate physiological and morphological constraints of the taxa, and climatic histories of the areas. Regarding African and South American anuran communities, at best, climatic histories are known in a general way for some of the areas; the biogeographic histories have been interpreted for few taxa, and next to nothing is known about the physiological and morphological constraints of the taxa. Knowledge of these factors and evidence regarding interspecific competition for limited resources (if any are limited at any time of the year) should provide necessary insights as to whether community organization is stochastic or whether it is deterministic, as implied in the present interpretation.

Appendix

Family groups and genera of amphibians in Africa and South America are listed with the numbers of species recognized in each genus given in parentheses; the numbers in parentheses after family group names are number of genera/number of species). Taxa occurring only in northern Africa are noted by an asterisk (*). These data have been updated from Frost (1985) through 31 December

1989. However, names that have not been associated with any populations have not been included.

Africa

ANURA: Arthroleptidae: Arthroleptinae (2/47): *Arthroleptis* (31), *Cardioglossa* (16); Astylosterninae (5/25): *Astylosternus* (11), *Leptodactylon* (11), *Nyctibates* (1), *Scotobleps* (1), *Trichobatrachus* (1). Bufonidae (11/84): *Bufo* (60), *Capensibufo* (2), *Didynamipus* (1), *Laurentophryne* (1), *Mertensophryne* (2), *Nectophryne* (2), *Nectophrynoides* (8), *Schismaderma* (1), *Stephopaedes* (1), *Werneria* (4), *Wolterstorffina* (2). Discoglossidae* (2/1): *Alytes** (1), *Discoglossus** (1). Heleophrynidae (1/4): *Heleophryne* (4). Hemisotidae (1/8): *Hemisus* (8). Hylidae: Hylinae (1/2): *Hyla* (2). Hyperoliidae: Hyperoliinae (8/120): *Alexteroon* (1), *Arlequinus* (1), *Callixalus* (1), *Chlorolius* (1), *Chrysobatrachus* (1), *Cryptothylax* (2), *Hyperolius* (112), *Nesionixalus* (1); Kassininae (9/50): *Afrixalus* (27), *Kassina* (14), *Kassinula* (1), *Opisthothylax* (1), *Paracassina* (2), *Phlyctimantis* (4), *Semnodactylus* (1); Leptopelinae (2/48): *Acanthixalus* (1), *Leptopelis* (47). Microhylidae: Brevicipitinae (5/18): *Balebreviceps* (1), *Breviceps* (12), *Callulina* (1), *Probreviceps* (3), *Spelaeophryne* (1); Melanobatrachinae (2/3): *Hoplophryne* (2), *Parhoplophryne* (1); Phrynomerinae (1/4): *Phrynomantis* (4). Pelobatidae (1/1): *Pelobates* (1). Pipidae (4/20): *Hymenochirus* (4), *Pseudhymenochirus* (1), *Silurana* (2), *Xenopus* (13). Ranidae: Petropedetinae (12/89): *Anhydrophryne* (1), *Arthroleptella* (2), *Arthroleptides* (2), *Cacosterum* (5), *Dimorphognathus* (1), *Microbatrachella* (1), *Natalobatrachus* (1), *Nothophryne* (1), *Petropedetes* (7), *Phrynobatrachus* (66), *Phrynodon* (1), *Poyntonia* (1); Raninae (8/90): *Aubria* (1), *Conraua* (6), *Hildebrandtia* (7), *Ptychadena* (41), *Pyxicephalus* (2), *Rana* (21), *Strongylopus* (6), *Tomopterna* (6). Rhacophoridae: Rhacophorinae (1/3): *Chiromantis* (3).

CAUDATA: Salamandridae (2/3): *Pleurodeles* (2), *Salamandra* (1).

GYMNOPHIONA: Caeciliaidae (7/16): *Afrocaecilia* (3), *Boulengerula* (1), *Geotrypetes* (3), *Herpele* (2), *Idiocranium* (1), *Schistometopum* (5), *Sylvacaecilia* (1).

South America

ANURA: Allophrynidae (1/1): *Allophryne* (1). Brachycephalidae (2/2): *Brachycephalus* (1), *Psyllophryne* (1). Bufonidae (9/121): *Andinophryne* (3), *Atelopus* (44), *Bufo* (46), *Dendrophryniscus* (3), *Frostius* (1), *Melanophryniscus* (9), *Oreophrynella* (3), *Osornophryne* (5), *Rhamphophryne* (7). Centrolenidae (2/73): *Centrolene* (2), *Centrolenella* (71). Dendrobatidae (5/132): *Colostethus* (78), *Dendrobates* (19), *Epipedobates* (24), *Minyobates* (8), *Phyllobates* (3). Hylidae: Hemiphractinae (6/63): *Cryptobatrachus* (3), *Flectonotus* (2), *Fritziana* (3), *Gastrotheca* (43), *Hemiphractus* (5), *Stefania* (7); Hylinae (14/277): *Aparasphenodon* (2), *Aplastodiscus* (1), *Argenteohyla* (1), *Corythomantis* (1), *Hyla* (182), *Nyctimantis* (1), *Ololygon* (55), *Osteocephalus* (7), *Phrynohyas* (5), *Phyllodytes* (6), *Smilisca* (2), *Scarthyla* (1), *Sphaenorhynchus* (10), *Trachycephalus* (3); Phyllomedusinae (2/36): *Agalychnis* (4), *Phyllomedusa* (32). Leptodactylidae: Ceratophryinae (3/11): *Ceratophrys* (7), *Chacophrys* (1), *Lepidobatrachus* (3); Hylodinae (3/23): *Crossodactylus* (6), *Hylodes* (16), *Megaelosia* (1); Leptodactylinae (12/117): *Adenomera* (7), *Atopophrynus* (1), *Edalorhina* (2), *Hydralaetare* (1), *Leptodactylus* (47), *Limnomedusa* (1), *Lithodytes* (1), *Paratelmatobius* (1), *Physalaemus* (36), *Pleurodema* (12), *Pseudopaludicola* (7), *Vanzolinius* (1); Telmatobiinae (30/428): *Adelphryne* (2), *Alsodes* (11), *Atelognathus* (7), *Barycholos* (2), *Batrachophrynus* (2), *Batrachyla* (3), *Caudiverbera* (1), *Crossodactylodes* (3), *Cycloramphus* (25), *Dischidodactylus* (2), *Eleutherodactylus* (266), *Euparkerella* (1), *Eupsophus* (7), *Geobatrachus* (1), *Holoaden* (2), *Hylarina* (1), *Insuetophrynus* (1), *Ischnocnema* (3), *Macrogenioglottis* (1), *Odontophrynus* (7), *Phyzelaphryne* (1), *Phrynopus* (19), *Phyllonastes* (3), *Proceratophrys* (11), *Scythrophrys* (1), *Somuncuria* (1), *Telmatobius* (33), *Telmatobufo* (3), *Thoropa* (5), *Zachaenus* (3). Microhylidae: Microhylinae (15/34): *Adelastes* (1), *Arcovomer* (1), *Chiasmocleis* (12), *Ctenophryne* (2), *Dasypops* (1), *Dermatonotus* (1), *Elachistocleis* (4), *Hamptophryne* (1), *Hyophryne* (1), *Myersiella* (1), *Nelsonophryne* (2), *Relictovomer* (1), *Stereocyclops* (1), *Synapturanus* (3), *Syncope* (2); Otophryninae: (1/1): *Otophryne* (1). Pipidae (1/7): *Pipa* (7). Pseudidae (2/3): *Lysapsus* (1), *Pseudis* (2). Ranidae: Raninae (1/3): *Rana* (3). Rhinodermatidae (1/2): *Rhinoderma* (2).

CAUDATA: Plethodontidae: Plethodontinae (2/22): *Bolitoglossa* (20), *Oedipina* (2).

GYMNOPHIONA: Caeciliaidae (10/55): *Brasilotyphlus* (1), *Caecilia* (31), *Dermophis* (1), *Lutkenotyphlus* (1), *Microcaecilia* (5), *Mimosiphonops* (1), *Oscaecilia* (7), *Parvicaecilia* (2), *Pseudosiphonops* (1), *Siphonops* (5). Rhinatrematidae (2/9): *Epicrionops* (8), *Rhinatrema* (1). Typhlonectidae (4/12): *Chthonerpeton* (6), *Nectocaecilia* (2), *Potomotyphlus* (1), *Typhlonectes* (3).

Literature Cited

Altig, R., and G. F. Johnston. 1989. Guilds of anuran larvae: Relationships among developmental modes, morphologies, and habitats. Herpetol. Monog. 3:81–109.

Amiet, L.-L. 1972. Notes faunistiques, éthologiques et écologiques sur quelques amphibiens anoures du Cameroun. Ann. Fac. Sci. Cameroun 9:127–153.

———. 1973. Notes faunistiques, éthologiques, et écologiques sur quelques amphibiens anoures du Cameroun (2e série). Ann. Fac. Sci. Cameroun 13:135–161.

———. 1975. Écologie et distribution des amphibiens anoures de la région de Nkongsanba (Cameroun). Ann. Fac. Sci. Yaoundé 20:33–107.

———. 1978. Les amphibiens anoures de la région de Mamfe. Ann. Fac. Sci. Yaoundé 25:189–219.

———. 1986. La batrachofauna sylvicole d'un secteur forestier du Cameroun: La Région de Yaoundé. Mém. Mus. Natn. Hist. Nat., n.s. (A) Zool. 132:29–42.

———. 1989. Quelques aspectes de la biologie des amphibiens anoures du Cameroun. Ann. Biol. 28:73–136.

Amiet, L.-L., and J.-L. Perret. 1969. Contribution à la faune de la région de Yaoundé (Cameroun). II, Amphibiens anoures. Ann. Fac. Sci. Cameroun 3:117–137.

Balinsky, B. I. 1969. Reproductive ecology of the amphibians of the Transvaal highveld. Zool. Africana 4:37–93.

Barron, E. J., C. G. A. Harrison, J. L. Sloan II, and W. W. Hay. 1981. Paleogeography: 180 million years ago to the present. Ecologae Geol. Helveticae 7:443–470.

Blommers-Schlösser, R. 1975. Observations on the larval development of some Malagasy frogs, with notes on their ecology and biology (Anura: Dyscophinae, Scahiophryninae and Cophylinae). Beaufortia 24:7–26.

Bourquin, O. 1985. A note on *Hemisus marmoratum,* compiled from the narrative of D. J. Bourquin. S. Afr. J. Sci. 81:210.

Bowker, R. G., and M. H. Bowker. 1979. Abundance and distribution of anurans in a Kenyan pond. Copeia 1979:278–285.

Busack, S. D. 1986. Biogeographic analysis of the herpetofauna separated by the formation of the Strait of Gibraltar. Nat. Geogr. Res. 2:17–36.

Cannatella, D. C. 1985. A phylogeny of primitive frogs (Archaeobatrachians). Ph.D. diss., University of Kansas.

———. 1986. A new genus of bufonid (Anura) from South America, and phylogenetic relationships of the neotropical genera. Herpetologica 42:197–205.

Cannatella, D. C., and L. Trueb. 1988a. Evolution of pipoid frogs: Intergeneric relationships of the aquatic frog family Pipidae (Anura). Zool. J. Linnean Soc. 94:1–38.

———. 1988b. Evolution of pipoid frogs: Morphology and phylogenetic relationships of *Pseudhymenochirus*. J. Herpetol. 22:439–456.

Cei, J. M. 1962. *Batracios de Chile*. Santiago: Universidad Católica de Chile.

———. 1979. The Patagonian herpetofauna. Pp. 309–339 *in* W. E. Duellman (ed.), *The South American Herpetofauna: Its Origin, Evolution, and Dispersal*. Mus. Nat. Hist. Univ. Kansas Monog. 7. Lawrence: University of Kansas Museum of Natural History.

———. 1980. The amphibians of Argentina. Monit. Zool. Italiano, n.s., Monogr. 2:1–609.

Channing, A. 1976. Life histories of frogs in the Namib Desert. Zool. Africana 11:299–312.

———. 1989. New frog records from the eastern Caprivi strip, South West Africa, Namibia. Madoqua 16:1–4.

Clarke, B. T. 1981. Comparative osteology and evolutionary relationships in the African Raninae (Anura: Ranidae). Monit. Zool. Italiano, n.s., 15 (Suppl.):285–331.

Colinvaux, P. 1979. The ice-age Amazon. Nature 278:399–400.

Crump, M. L. 1971. Quantitative analysis of the ecological distribution of a tropical herpetofauna. Occas. Pap. Mus. Nat. Hist. Univ. Kansas 3:1–62.

de Sá, R. O., and D. M. Hillis. 1990. Phylogenetic relationships of the pipid frogs *Xenopus* and *Silurana:* An integration of ribosomal DNA and morphology. Molec. Biol. Evol. 7:365–376.

Donnelly, M. A., R. O. de Sá, and C. Guyer. 1990. Description of the tadpoles of *Gastrophryne pictiventris* and *Nelsonophryne aterrima* (Anura: Microhylidae), with a review of morphological variation in free-swimming microhylid larvae. Amer. Mus. Novit. 2976:1–19.

Drewes, R. C. 1984. A phylogenetic analysis of the Hyperoliidae (Anura): Treefrogs of Africa, Madagascar, and the Seychelles Islands. Occas. Pap. Calif. Acad. Sci. 139:1–70.

Duellman, W. E. 1970. *The Hylid Frogs of Middle America.* Mus. Nat. Hist. Univ. Kansas Monog. 1. Lawrence: University of Kansas Museum of Natural History.

———. 1978. The biology of an equatorial herpetofauna in Amazonian Ecuador. Mus. Nat. Hist. Univ. Kansas Misc. Publ. 65. Lawrence: University of Kansas Museum of Natural History.

———. 1979a. The herpetofauna of the Andes: Patterns of distribution, origin, differentiation, and present communities. Pp. 371–459 *in* W. E. Duellman (ed.), *The South American Herpetofauna: Its Origin, Evolution, and Dispersal.* Mus. Nat. Hist. Univ. Kansas Monog. 7. Lawrence: University of Kansas Museum of Natural History.

———. (ed.). 1979b. *The South American Herpetofauna: Its Origin, Evolution, and Dispersal.* Mus. Nat. Hist. Univ. Kansas Monog. 7. Lawrence: University of Kansas Museum of Natural History.

———. 1979c. The South American herpetofauna: A panoramic view. Pp. 1–28 *in* W. E. Duellman (ed.), *The South American Herpetofauna: Its Origin, Evolution, and Dispersal.* Mus. Nat. Hist. Univ. Kansas Monog. 7. Lawrence: University of Kansas Museum of Natural History.

———. 1985. Reproductive modes in anuran amphibians: Phylogenetic significance of adaptive strategies. S. Afr. J. Sci. 81:174–178.

———. 1986. Plate tectonics, phylogenetic systematics and vicariance biogeography of anurans: Methodology for unresolved problems. Pp. 59–62 *in* Z. Rocek (ed.), *Studies in Herpetology.* Prague: Charles University.

———. 1988. Patterns of species diversity of anuran amphibians in the American tropics. Ann. Missouri Bot. Gard. 75:79–104.

———. 1989a. Alternative life-history styles in anuran amphibians: Evolutionary and ecological implications. Pp. 101–126 *in* M. N. Bruton (ed.), *Alternative Life-History Styles of Animals.* Dordrecht: Kluwer.

———. 1989b. Tropical herpetofaunal communities: Patterns of community structure in neotropical rainforests. Pp. 61–88 *in* M. L. Harmelin-Vivien and F. Bourlière (eds.), *Vertebrates in Complex Tropical Ecosystems.* New York: Springer-Verlag.

———. 1990. Herpetofaunas in neotropical rainforests: Comparative composition, history, and resource use. Pp. 455–505 *in* A. H. Gentry (ed.). *Four Neotropical Rainforests.* New Haven, Conn.: Yale University Press.

Duellman, W. E., and E. R. Pianka. 1990. Biogeography of nocturnal insectivores: Historical events and ecological filters. Ann. Rev. Ecol. Syst. 21:57–68.

Duellman, W. E., and L. Trueb. 1986. *Biology of Amphibians*. New York: McGraw-Hill.

Estes, R. 1975. Fossil *Xenopus* from the Paleocene of South America and the zoogeography of pipid frogs. Herpetologica 31:263–278.

———. 1981. Gymnophiona, Caudata. Handb. Paläoherpetol. 2:1–115.

Estes, R., and A. Báez. 1985. Herpetofaunas of North and South America during the late Cretaceous and Cenozoic: Evidence for interchange? Pp. 139–197 *in* F. G. Stehli and S. D. Webb (eds.), *The Great American Biotic Interchange*. New York: Plenum.

Estes, R., and M. H. Wake. 1972. The first fossil record of caecilian amphibians. Nature 239:228–231.

Ford, L. S. 1989. The phylogenetic position of poison-dart frogs (Dendrobatidae): Reassessment of the neobatrachian phylogeny with a commentary on complex character systems. Ph.D. diss., University of Kansas.

Formas, J. R. 1979. La herpetofauna de los bosques temperados de Sudamérica. Pp. 341–369 *in* W. E. Duellman (ed.), *The South American Herpetofauna: Its Origin, Evolution, and Dispersal*. Mus. Nat. Hist. Univ. Kansas Monog. 7. Lawrence: University of Kansas Museum of Natural History.

Frost, D. (ed.). 1985. *Amphibian Species of the World*. Lawrence, Kans.: Association of Systematics Collections.

Gallardo, J. M. 1979. Composición, distribución y origen de la herpetofauna chaqueña. Pp. 299–307 *in* W. E. Duellman (ed.), *The South American Herpetofauna: Its Origin, Evolution, and Dispersal*. Mus. Nat. Hist. Univ. Kansas Monog. 7. Lawrence: University of Kansas Museum of Natural History.

Grandison, A. G. C. 1981. Morphology and phylogenetic position of the West African *Didynamipus sjoestedti* Andersson, 1903 (Anura Bufonidae). Monit. Zool. Italiano, n.s., 15 (Suppl.):187–215.

Gudynas, E., and J. C. Rudolf. 1987. La herpetofauna de la localidad costera de "Pajas Blancas" (Uruguay): Lista sistematica comentada y estructura ecologica de la comunidad. Commun. Mus. Ci. PUCRS 46:173–194.

Haddad, C. F. B., J. P. Pombal, Jr., and M. Gordo. 1990. Foam nesting in a hylid frog (Amphibia, Anura). J. Herpetol. 24:225–226.

Hanken, J., and D. B. Wake. 1982. Genetic differentiation among plethodontid salamanders (genus *Bolitoglossa*) in Central and South America: Implications for the South American invasion. Herpetologica 38:272–287.

Hedges, S. B. 1989. Evolution and biogeography of West Indian frogs of the genus *Eleutherodactylus:* Slow-evolving loci and the major groups. Pp. 305–370 *in* C. A. Woods (ed.), *Biogeography of the West Indies*. Gainesville, Fla.: Sandhill Crane.

Heyer, W. R., A. S. Rand, C. A. Gonçalves da Cruz, O. L. Peixoto, and C. E. Nelson. 1990. Frogs of Boracéia. Arq. Zool. São Paulo 31:231–410.

Hillis, D. M., and R. de Sá. 1988. Phylogeny and taxonomy of the *Rana palmipes* group (Salientia: Ranidae). Herpetol. Monog. 2:1–26.

Hoogmoed, M. S. 1979. The herpetofauna of the Guianan region. Pp. 241–279 *in* W. E. Duellman (ed.), *The South American Herpetofauna: Its Origin, Evolution, and Dispersal*. Mus. Nat. Hist. Univ. Kansas Monog. 7. Lawrence: University of Kansas Museum of Natural History.

Hoogmoed, M. S., and S. J. Gorzula. 1979. Checklist of the savanna inhabiting frogs of the El Manteco region with notes on the ecology and the description of a new species of treefrog (Hylidae, Anura). Zool. Meded., Leiden 54:183–216.

Hueck, K. 1972. Mapa de la vegetación de America del Sur, 1–8,000,000. Munich: Paul Seibert Forstliche Forschungsanstalt.

Inger, R. F. 1968. Amphibia. Expl. du Parc National de la Garamba. 52:1–190.

———. 1972. *Bufo* of Eurasia. Pp. 102–118 *in* W. F. Blair (ed.), *Evolution in the Genus* Bufo. Austin: University of Texas Press.

Inger, R. F., and H. Marx. 1961. The food of amphibians. Expl. du Parc National de l'Upemba. Inst. Parcs Nat. Congo Ruanda-Urundi. 64:1–86.

Jenkins, F. S., Jr., and D. M. Walsh. 1990. During the Jurassic caecilians had limbs (abstract). J. Vert. Paleontol. 10 (no. 3, Suppl.):29A.

Joglar, R. L. 1989. Phylogenetic relationships of the West Indian frogs of the genus *Eleutherodactylus:* A morphological analysis. Pp. 371–408 *in* C. A. Woods (ed.), *Biogeography of the West Indies.* Gainesville, Fla.: Sandhill Crane.

Kingdon, J. 1989. *Island Africa.* Princeton, N.J.: Princeton University Press.

Lanza, B. 1983. A list of the Somali amphibians and reptiles. Monit. Zool. Italiana, n.s., 17 (Suppl.):193–247.

Laurent, R. F. 1941. Contribution à l'ostéologie et à la systematique des rhacophorides africains. Rev. Zool. Bot. Africaines 35:85–111.

———. 1943. Contribution à l'ostéologie et à la systematique des rhacophorides non africains. Bull. Mus. Roy. Hist. Nat. Belgique 19(28):1–16.

———. 1964. Adaptive modifications in frogs of an isolated highland fauna in central Africa. Evolution 18:458–467.

———. 1973. A parallel survey of equatorial amphibians and reptiles in Africa and South America. Pp. 259–266 *in* B. J. Meggers, E. S. Ayensu, and W. D. Duckworth (eds.), *Tropical Forest Ecosystems in Africa and South America: A Comparative Review.* Washington, D.C.: Smithsonian Institution Press.

———. 1979. Herpetofaunal relationships between Africa and South America. Pp. 55–71 *in* W. E. Duellman (ed.), *The South American Herpetofauna: Its Origin, Evolution, and Dispersal.* Mus. Nat. Hist. Univ. Kansas Monog. 7. Lawrence: University of Kansas Museum of Natural History.

Lescure, J. 1976. Contribution à l'étude des amphibiens de Guyane Français. 6, Liste préliminaire des anoures. Bull. Mus. Natn. Hist. Nat., Paris (Zool.) 265:475–524.

Liem, S. S. 1970. The morphology, systematics, and evolution of the Old World treefrogs (Rhacophoridae and Hyperoliidae). Fieldiana: Zool. 57:1–145.

Livingstone, D. A. 1982. Quaternary geography of Africa and the refuge theory. Pp. 523–536 *in* G. T. Prance (ed.), *Biological Diversification in the Tropics.* New York: Columbia University Press.

Lynch, J. D. 1971. Evolutionary relationships, osteology, and zoogeography of leptodactyloid frogs. Mus. Nat. Hist. Univ. Kansas Misc. Publ. 53. Lawrence: University of Kansas Museum of Natural History.

———. 1979. The amphibians of the lowland tropical forests. Pp. 189–215 *in* W. E. Duellman (ed.), *The South American Herpetofauna: Its Origin, Evolution, and Dispersal.* Mus. Nat. Hist. Univ. Kansas Monog. 7. Lawrence: University of Kansas Museum of Natural History.

———. 1986. The definition of the Middle American clade of *Eleutherodactylus* based on jaw musculature (Amphibia: Leptodactylidae). Herpetologica 42:248–258.

Maxson, L. R. 1981. Albumin evolution and its phylogenetic implications in African toads of the genus *Bufo.* Herpetologica 37:96–104.

———. 1984. Molecular probes of phylogeny and biogeography in toads of the widespread genus *Bufo.* Molec. Biol. Evol. 1:345–356.

Maxson, R. D., and L. R. Maxson. 1986. Micro-complement fixation: A quantitative estimator of protein evolution. Molec. Biol. Evol. 3:375–388.

Mertens, R. 1938. Herpetologische Ergebnisse einer Reise nach Kamerun. Abh. Senckenberg. Naturf. Ges. 442:1–52.

Nussbaum, R. A. 1985. Systematics of caecilians (Amphibia: Gymnophiona) of the family Scolecomorphidae. Occas. Pap. Mus. Zool. Univ. Michigan 713:1–49.

Nussbaum, R. A., and M. Wilkinson. 1989. On the classification and phylogeny of caecilians (Amphibia: Gymnophiona): A critical review. Herpetol. Monog. 3:1–42.

Passmore, N. I., and V. C. Carruthers. 1979. *South African Frogs.* Johannesburg: Witwatersrand University Press.

Péfaur, J. E., and W. E. Duellman. 1980. Community structure in high Andean herpetofaunas. Trans. Kansas Acad. Sci. 83:45–65.

Perret, J.-L. 1966. Les amphibiens du Cameroun. Zool. Jahrb. Syst. 8:289–464.

Poynton, J. C. 1964. The Amphibia of Southern Africa. Ann. Natal Mus. 17:1–334.

Poynton, J. C., and D. G. Broadley. 1985a. Amphibia Zambesiaca 1: Scolecomorphidae, Pipidae, Microhylidae, Hemisidae, Arthroleptidae. Ann. Natal Mus. 26:503–553.

———. 1985b. Amphibia Zambesiaca 2: Ranidae. Ann. Natal Mus. 27:115–181.

———. 1987. Amphibia Zambesiaca 3: Rhacophoridae and Hyperoliidae. Ann. Natal Mus. 28:161–229.

———. 1988. Amphibia Zambesiaca 4: Bufonidae. Ann. Natal Mus. 29:447–490.

———. 1991. Amphibia Zambesiaca 5: Zoogeography. Ann. Natal Mus. 32:221–277.

Prance, G. T. (ed.). 1982. *Biological Diversification in the Tropics*. New York: Columbia University Press.

Pregill, G. 1981. Cranial morphology and the evolution of West Indian toads (Salientia: Bufonidae): Resurrection of the genus *Peltophryne* Fitzinger. Copeia 1981:273–285.

Rand McNally. 1969. *The International Atlas*. Chicago: Rand McNally.

Richards, P. W. 1973. Africa, the "Odd Man Out." Pp. 21–26 *in* B. J. Meggers, E. S. Ayensu, and W. D. Duckworth (eds.), *Tropical Forest Ecosystems in Africa and South America: A Comparative Review*. Washington, D.C.: Smithsonian Institution Press.

Rivero-Blanco, C., and J. R. Dixon. 1979. Origin and distribution of the herpetofauna of the dry lowland regions of northern South America. Pp. 281–298 *in* W. E. Duellman (ed.), *The South American Herpetofauna: Its Origin, Evolution, and Dispersal*. Mus. Nat. Hist. Univ. Kansas Monog. 7. Lawrence: University of Kansas Museum of Natural History.

Rodriguez, L. B., and J. E. Cadle. 1990. A preliminary overview of the herpetofauna of Cocha Cashu, Manu National Park, Peru. Pp. 410–425 *in* A. H. Gentry (ed.), *Four Neotropical Rainforests*. New Haven, Conn.: Yale University Press.

Savage, J. M. 1973. The geographic distribution of frogs: Patterns and predictions. Pp. 352–445 *in* J. L. Vial (ed.), *Evolutionary Biology of the Anurans: Contemporary Research on Major Problems*. Columbia, Mo.: University of Missouri Press.

———. 1982. The enigma of the Central American herpetofauna: Dispersals or vicariance? Ann. Missouri Bot. Gard. 69:464–547.

Schiøtz, A. 1963. The amphibians of Nigeria. Vidensk. Medd. Dansk. Naturh. Foren. 125:1–92.

———. 1967. The treefrogs (Rhacophoridae) of West Africa. Spol. Zool. Mus. Haun. (Copenhagen) 25:1–346.

———. 1975. *The Treefrogs of Eastern Africa*. Copenhagen: Steenstrupia.

Schlüter, A. 1984. Ökologische Untersuchungen an einem Stillgewasser im tropischen Regenwald von Peru unter besonderer Berucksichtigung der Amphibien. Ph.D. diss., University of Hamburg.

Schmidt, K. P., and R. F. Inger. 1959. Amphibians exclusive of the genera Afrixalus and Hyperolius. Expl. du Parc National d'Upemba. Inst. Parcs Nat. Congo Ruanda-Urundi 56:1–264.

Staton, M. A., and J. R. Dixon. 1977. The herpetofauna of the central llanos of Venezuela: Noteworthy records, a tentative checklist, and ecological notes. J. Herpetol. 11:17–24.

Stehli, F. G., and S. D. Webb (eds.). 1985. *The Great American Biotic Interchange*. New York: Plenum.

Stewart, M. M. 1967. *Amphibians of Malawi*. Albany: SUNY Press.

Tandy, M., and R. Keith. 1972. *Bufo* of Africa. Pp. 119–170 *in* W. F. Blair (ed.), *Evolution in the Genus* Bufo. Austin: University of Texas Press.

Telford, S. R. 1982. Aspects of mate recognition and social behavior in a sub-tropical frog community. Ph.D. diss., University of Witwatersrand.

Toft, C. A. 1980. Feeding ecology of thirteen syntopic species of anurans in a seasonal tropical environment. Oecologia 45:131–141.

———. 1982. Community structure of litter anurans in a tropical forest at Makokou, Gabon: A preliminary analysis in the minor dry season. Rev. Ecol. (Terre Vie) 36:224–232.

Toft, C. A., and W. E. Duellman. 1979. Anurans

of the lower Río Llollapichis, Amazonian Peru: A preliminary analysis of community structure. Herpetologica 35:71–77.

Trueb, L., and M. J. Tyler. 1974. Systematics and evolution of the Greater Antillean hylid frogs. Occas Pap. Mus. Nat. Hist. Univ. Kansas 24:1–60.

UNESCO. 1958. Vegetation map of Africa south of the Tropic of Cancer. Paris: UNESCO.

van der Hammen, T., and A. M. Cleef. 1986. Development of the high Andean páramo flora and vegetation. Pp. 153–201 *in* F. Vuilleumier and M. Monasterio (eds.), *High Altitude Tropical Biogeography.* Oxford: Oxford University Press.

van der Hammen, T., J. H. Werner, and H. Van Dommelen. 1973. Palynological record of the upheaval of the northern Andes: A study of the Pliocene and lower Quaternary of the Colombian eastern cordillera and the early evolution of its biota. Rev. Palaeobot. Palynol. 16:1–122.

van Dijk, D. E. 1985a. An addition to the fossil Anura of southern Africa. S. Afr. J. Sci. 81:207–208.

———. 1985b. *Hemisus marmoratum* adults reported to carry tadpoles. S. Afr. J. Sci. 81:209–210.

Vanzolini, P. E., and W. R. Heyer. 1985. The American herpetofauna and the interchange. Pp. 475–487 *in* F. G. Stehli and S. D. Webb (eds.), *The Great American Biotic Interchange.* New York: Plenum.

Vuilleumier, F., and M. Monasterio (eds.). 1986. *High Altitude Tropical Biogeography.* Oxford: Oxford University Press.

Wake, D. B., and P. Elias. 1983. New genera and a new species of Central American salamanders, with a review of the tropical genera (Amphibia, Caudata, Plethodontidae). Contr. Sci. Los Angeles Co. Mus. Nat. Hist. 345:1–19.

Wake, D. B., and J. F. Lynch. 1976. The distribution, ecology, and evolutionary history of plethodontid salamanders in tropical America. Sci. Bull. Los Angeles Co. Mus. Nat. Hist. 25:1–65.

Wake, M. H. 1980. The reproductive biology of *Nectophrynoides malcolmi* (Amphibia: Bufonidae), with comments on the evolution of reproductive modes in the genus *Nectophrynoides.* Copeia 1980:193–209.

Wassersug, R. J. 1984. The *Pseudohemisus* tadpole: A morphological link between microhylid (Orton Type 2) and ranoid (Orton Type 4) larvae. Herpetologica 40:138–149.

Wassersug, R. J., and W. F. Pyburn. 1987. The biology of the Pe-ret' toad, *Otophryne robusta* (Microhylidae), with a special consideration of its fossorial larva and systematic relationships. Zool. J. Linn. Soc. 91:137–169.

Weygoldt, P. 1986. Beobachtungen zue Ökologie und Biologie von Fröschen an einem neotropischen Bergbach. Zool. Jarhb. Syst. 113:429–454.

Whitmore, T. C., and G. T. Prance (eds.). 1987. *Biogeography and Quaternary History in Tropical America.* Oxford: Clarendon.

Wiley, E. O. 1981. *Phylogenetics: The Theory and Practice of Phylogenetic Systematics.* New York: John Wiley & Sons.

———. 1988. Vicariance biogeography. Ann. Rev. Ecol. Syst. 19:513–542.

Woods, C. A. (ed.). 1989. *Biogeography of the West Indies.* Gainesville, Fla.: Sandhill Crane.

Zimmerman, B. L., and M. T. Rodrigues. 1990. Frogs, snakes, and lizards of the INPA-WWF reserves near Manaus, Brazil. Pp. 426–454 *in* A. H. Gentry (ed.), *Four Neotropical Rainforests.* New Haven, Conn.: Yale University Press.

Zweifel, R. G. 1986. A new genus and species of microhylid frog from the Cerro de la Neblina region of Venezuela and a discussion of relationships among New World microhylid genera. Amer. Mus. Novit. 2863:1–24.

Chapter 10

African–South American Relationships: A Perspective from the Reptilia

Aaron M. Bauer

I. Introduction

In comparison with other vertebrate groups, the taxa of extant reptiles have played a relatively minor role in the formulation and support of biogeographic hypotheses involving Gondwanan connections between Africa and South America. Laurent (1973), in reviewing the equatorial herpetofauna, found little evidence to support a Gondwanan connection between Africa and South America and attributed most shared taxa to over-water dispersals and Holarctic migrations followed by extinctions in the north. Nonetheless, Laurent (1979) later accepted a number of different historical factors, including Gondwanan fragmentation, as causal in the distributions observed in the southern continents. Duellman (1979b), in the same volume as Laurent, postulated geographic origins for the reptiles of South America and, among the taxa discussed herein, considered pelomedusids, "iguanids," and "micrurids" to represent Gondwanan lineages, the last two with their origins in South America. Emydids, trionychids, anguids, and viperids were attributed to Laurasian origins while "boids" were regarded as Pangaean and the majority of the taxa, gekkonids, scincids, amphisbaenids, typhlopids, leptotyphlopids, colubrids, and crocodylids, were considered of uncertain origin. More recently, Cei (1986) has attributed the present distributions of lep-

I thank J. Vindum, W. Lamar, R. McDiarmid, R. Crombie, W. Duellman, A. Wynn, R. Drewes, D. Good, K. de Queiroz, and W. Branch for their contributions to and comments on this chapter. The U.S. National Museum of Natural History and California Academy of Sciences have provided much of the literature used in preparing this chapter.

 ISBN 0-300-05375-4

totyphlopids, boids, amphisbaenians, and iguanids, particularly anoloids, to former Gondwanan connections between Africa and South America.

Laurent's (1979) review is the most recent examination of the relationship between the African and South American herpetofaunas in their entirety. Since the publication of his work, the erection of explicit hypotheses of relationship among many groups and the widespread application of analytical biogeographic methods have modified both the data base and the tools for evaluating this relationship. In this paper I review the current state of knowledge of the phylogenies of those reptile groups occurring in both Africa and South America. The patterns of observed distributions are assessed in light of the phylogenies, and the contribution of reptile systematics to the biogeographic analysis of African–South American relationships is evaluated.

II. Materials and Methods

Discussion is restricted to those reptile groups that are represented by at least some living forms in both Africa and South America. The recent interest in the role of fossils in phylogenetic reconstruction (Gauthier et al. 1988; Donoghue et al. 1989) has highlighted the need for the inclusion of these forms in biogeographic analysis as well. The simple pictures of distribution suggested by modern distribution patterns alone may obscure more complex phenomena needed to explain fossil distributions (Bauer and Vindum 1990). In any biogeographic interpretation, the results of analysis are dependent on the assumption that the taxa involved share a common history. Interpretations based on nonmonophyletic groups may thus seriously alter the outcome of analyses (Springer 1981; Bauer 1988b). I have endeavored to treat fossil taxa when directly pertinent to the phylogenetic hypotheses being examined.

The primary approach to examining the trans-Atlantic relationships of reptiles employed in this review is a survey of the current state of understanding of affinities among those higher taxa having living or fossil representation in both continents under consideration. I have not made interpretations concerning the validity of the phylogenies proposed for most groups but have presented the most recent, well-supported hypotheses available. The literature cited herein dates mostly from the last twenty years, and much of it from the last ten. Older references may be found in Laurent (1979), which serves as a point of departure for this review.

III. Phylogenetic and Biogeographic Relationships of African and South American Reptiles

A. Testudines

Proganochelys, the oldest known turtle, first described from the Late Triassic of Germany, appears to be the sister group of all other turtles (Gaffney and Meeker 1983; Gaffney 1990) and is not assignable to either of the major groups of living turtles, the pleurodires and cryptodires, which together form the Casichelydia (Gaffney 1975a, 1975b, 1984a, 1984b; Gaffney and Meylan 1988). The Pangaean distribution of the group is attested to by the existence of a turtle from the Late Triassic or Early Jurassic of the Orange Free State (Kitching 1980), which has been tentatively identified as the sister taxon of *Proganochelys* (Gaffney 1986).

Pleurodira. Proterochersis robusta, from the Triassic of Germany, exhibits at least one synapomorphy of the Pleurodira (Gaffney 1986; Gaffney and Meylan 1988), but the earliest Gondwanan turtle purported to be a pleurodire is *Notoemys laticentralis* from the Late Jurassic of Argentina (Wood and Freiberg 1977). Living pleurodires and their fossil relatives exclusive of the pre-Cretaceous taxa are placed in the Eupleurodira (Gaffney and Meylan 1988), consisting of two living families. The Chelidae, presently distributed in South America and Australia constitute the sister taxon of the Pelomedusidae, which is distributed in Africa and South America. Gaffney and Meylan (1988) recognized two subfamilies of pelomedusids. Their Pelomedusinae includes the living sister taxa *Pelusios* and *Pelomedusa* and the Cretaceous taxa *Bothremys* and *Taphrosphys*. The living taxa occur throughout sub-Saharan Africa and adjacent islands. The two fossil genera were widely distributed in the Upper Cretaceous and early Tertiary and have been recovered from coastal marine sediments. *Bothremys* occurred in North America as well as Africa, Madagascar, and South America (Broin 1988) and *Taphrosphys* is known from Europe, Africa, North and South America, and perhaps Madagascar (Broin 1980).

The Podocneminae consists of two primary clades, one of which contains all of the living taxa. The monotypic *Erymnochelys* is associated with the western drainages of Madagascar and is the sister taxon of the two living South American genera, *Peltocephalus* and *Podocnemis*. The sister group of the living forms contains the Asiatic (primarily Indian) and Egyptian *Schweboemys*/*Stereogenys* lineage, which ranged from the Eocene to the Pleistocene and consisted of primarily freshwater forms (Wood 1970).

Gaffney and Meylan (1988) did not treat the many fossil pelomedusids that are poorly or incompletely known, due to the difficulty of assessing key character states in this material. A brief summary of this material, derived largely from Broin (1988) is presented below. *Teneremys* has been recorded from the Cretaceous of both Ceará, Brazil, and Niger (Broin 1980, 1988). *Platycheloides* cf. *nyasae,* also from the Lower Cretaceous of Niger, has been stated to show African pelomedusine affinities (Broin 1980, 1988). Other Lower Cretaceous pleurodires, including both pelomedusids and forms related to the chelid lineages, are known from Malawi, north Africa, and Portugal (Broin 1988). An *Erymnochelys*-like form was recorded from Niger (Broin et al. 1974) and fossil podocnemines have been recorded from many sites in central South America (Broin 1988). Modern-type pelomedusines are known from the Tertiary of Africa and highly derived taxa attributed to *Podocnemis* are recorded from several sites in South America, possibly as early as the Late Cretaceous (Wood 1979; Pritchard 1985).

Broin (1980, 1988) provided extensive comments about the historical biogeography of pelomedusids and other pleurodires. By the Late Triassic turtles were present over much of Pangaea. She argued that the Gondwanan affinities of the group were well-supported by fossils which indicated the presence of pleurodires in the Triassic of Argentina and suggested Cretaceous division of the pelomedusid lineage into primarily South American (podocnemine) and African (pelomedusine) lineages. Most extra-Gondwanan pelomedusids, such as *Bothremys,* were apparently marine forms and probably dispersed via shallow coastal water corridors from the south. The arrival of the freshwater *Schweboemys* lineage in India may predate the Africa-India rift (Broin 1988), or may be indicative of secondary contact between the landmasses. The modern distribution of the family was established through subsequent extinctions. "Bothremydids" appear to have died out in the Eocene and the Indian forms in the Pleistocene. *Erymnochelys*-like forms that once occurred on the African mainland and in Europe probably became extinct by the Miocene, in association with changing climatic conditions (Broin 1988).

Pelomedusids are unique among reptile groups in that both phylogenetic analysis and fossil age and locality unambiguously support cladogenesis associated with South American–African rifting in the Cretaceous. An area cladogram derived from pelomedusid relationships, especially if incorporating chelids and fossil forms, would be almost completely congruent with numerous "classical" Gondwanan patterns generated by systematists working with nonvertebrates (see Humphries and Parenti 1986 for examples).

Cryptodira. The only cryptodiran families with living representatives in both Africa and South America are nested well within the upper branches of Gaffney and Meylan's (1988) cladogram for turtles, in the Testudinoidea. The Emydidae (sensu Gaffney and Meylan 1988) is the sister group of the "Bataguridae," a paraphyletic group containing the monophyletic Testudinidae.

Emydids are peripheral to both Africa and South America. Only *Emys orbicularis* occurs in Africa, and it is limited to coastal Morocco, Algeria, and Tunisia. Its arrival in Morocco may date from a Miocene crossing from Spain (Busack 1977). *Trachemys scripta* in northern South America is clearly a recent arrival from North America, perhaps becoming isolated at several localities in Middle and South America during the early Pleistocene (Seidel and Jackson 1990) or even more recently (Legler 1990). Alternative subdivisions of the Emydidae (e.g., Ckhikvadzé 1984 and Gaffney and Meylan 1988) propose no close relationships between the two genera.

Relationships within batagurids are highly problematic (McDowell 1964; Hirayama 1985; Gaffney and Meylan 1988; Capler and Moll 1990; Carr 1990). The only African form is *Mauremys leprosa.* The species is primarily restricted to Mediterranean Africa and the more mesic borders of the Sahara. Records of specimens from tropical west Africa may represent misidentifications of pelomedusids (Le Berre 1989) or may be the result of southward movement along former river courses in the western Sahara (Busack and Ernst 1980). Busack (1986) found a genetic distance of 0.00 between Spanish and Moroccan populations of this species, perhaps suggesting a high migration rate for the species across the narrow gap from Spain, whence it may have migrated in the Miocene (Busack 1977).

The only South American batagurids are referable to *Rhinoclemmys,* with living species in the Choco and a fossil form from the Pleistocene of Brazil (Campos and Broin

1981). This genus appears to be a post-Gondwanan arrival on the continent. However, the picture is somewhat confused by recent findings that the extant South American species are not each other's closest relatives, but rather represent two different historical patterns of colonization (Carr 1990). Other workers have presented evidence that the genus *Rhinoclemmys* is not monophyletic, but rather divided into several clades within a major batagurid subdivision (Hirayama 1985). All recent reviewers of the group (e.g., McDowell 1964; Hirayama 1985) have placed both *Mauremys* and *Rhinoclemmys* in the same major sublineage, the Geoemydinae of Gaffney and Meylan (1988), but this group also contains several Asian genera as well, and Carr (1990) has presented hypotheses of sister-group relationship linking *Rhinoclemmys* to the Indian genus *Melanochelys*. The biogeographic patterns illustrated by batagurids remain obscured by the present confusion over relationships within the group. Nonetheless no direct link between the African and South American taxa is indicated and the histories of the two genera involved on these continents are largely independent.

Patterns of relationship within the Testudinidae are generally well supported (Crumly 1984, 1985; Gaffney and Meylan 1988). The South American tortoises include four mainland species plus one species (or more, depending upon species definition) on the Galapagos Islands. All are members of the genus *Geochelone* sensu lato. This genus, however, was stated by Crumly (1982, 1984) and Coto and Acuña (1986) to be nonmonophyletic, and even the South American forms may represent two distinct lineages (Crumly 1982). The subgenera recognized by Auffenberg (1974) do not appear to be supported by present phylogenetic knowledge. The African testudinids include *Geochelone* as well as a number of other genera. Five of these, *Chersina, Acinixys, Pyxis, Homopus,* and *Psammobates* form a closely related clade of strictly sub-Saharan African and Madagascan distribution. This unit is in turn the sister taxon of *Malacochersus* plus *Testudo.* The former is east African while the latter is broadly distributed around the Mediterranean. *Kinixys* is somewhat more distantly related to the other African genera, with *Geochelone* more distant still. Unlike pelomedusids, fossil representatives of extant testudinid lineages have been recorded primarily from the geographic areas now occupied, although at least some fossil tortoises occurred in high latitudes not now occupied (Estes and Hutchison 1980). The first testudinids appeared in the Eocene and have been recorded from many localities in the Laurasian land masses (Crumly 1984). African tortoises are known from the Fayum Formation of Egypt, but their allocation to modern genera is suspect (Crumly 1984). By the Miocene, however, tortoises putatively representing *Geochelone, Chersina* and *Kinixys* were present (Meylan and Auffenberg 1986) in mainland Africa, although South American testudinid fossils are not known before the Miocene (Auffenberg (1971).

Simpson (1942) regarded the South American testudinids as waif dispersers, arriving from North America during the Miocene. Williams (1950) also proposed a northern origin for the South American *Geochelone,* with essentially the same temporal framework. Bramble (1971) regarded the Old and New World tortoises as nonmonophyletic and derived the South American forms from a Central American ancestor. Crumly (1984) proposed that some or all of the South American *Geochelone* may have arrived in South America via waif dispersal, either from Central or North America or Africa, although he endorsed neither hypothesis preferentially. Pritchard and Trebbau (1984) favored colonization by waif dispersers from Africa, but this was not based on explicit phylogenetic information. More recently, Gaffney and Meylan (1988) have proposed a sister-group relationship between *Geochelone* sensu stricto in Africa and *Chelonoidis* (South American tortoises previously assigned to *Geochelone*) on the basis of shared derived characters.

Evidence points to a Tertiary arrival of tortoises in South America, but the confusion regarding the relationships within *Geochelone* make any biogeographic inferences highly speculative. Nonetheless, there is ample evidence to support the ability of tortoises to disperse across ocean barriers. The peripheral *Geochelone* taxa, those occurring on the Galapagos and on Madagascar, the Seychelles and, formerly, the Mascarene Islands, are generally accepted to have been dispersed over water, as land connections to these islands never existed (Galapagos), or predate the probable origin of the testudinids (Madagascar) (Crumly 1984). The giant tortoises of the Indian Ocean appear to represent two lineages, each with affinities to a different group of Afro-Madagascan *Geochelone* sensu lato (Bour 1984; Crumly 1984). Further, the specific patterns of tortoise distribution in the Indian Ocean (multiple colonizations) have almost certainly been affected by multiple colonizations following vicariantly based regional extinctions (Arnold 1979).

Trionychids are known from a single specimen from the Paleocene of Venezuela (Wood and Patterson 1973) as well as three living genera in Africa. The living genera *Cyclanorbis* and *Cycloderma* appear to form a monophyletic group, but *Trionyx triunguis* is relatively distantly related, sharing its closest affinities with Laurasian taxa (Meylan 1987). No close relationship of the Venezuelan fossil to either African group has so far been suggested.

B. Squamata

1. "Sauria" Lizard taxa shared between Africa and South America include the families Scincidae and Gekkonidae, the two most speciose and widely distributed families in the suborder. Both families are represented by many more African than South American representatives. Amphisbaenids are well represented on both continents, but anguids are very minor components. Higher order relationships including those within the Iguania and the Autarchoglossa are also related to intercontinental biogeographic patterns. The distribution of fossil lizards is generally similar to that of the living forms except that many taxa were formerly represented on northern continents during periods of favorable climatic conditions. By the Late Jurassic most of the major lizard lineages were already well established (Estes 1983a, 1983b), and some of the diversification may have taken place in the Late Triassic, which is poorly represented by terrestrial fossils.

Estes and co-workers (1988) provided the

most recent hypothesis of relationships within the Squamata. They recognized the Iguania as sister taxon of the Scleroglossa, itself consisting of the clades Gekkota and Autarchoglossa. Within the Autarchoglossa the anguimorphan clade is particularly well supported. Patterns within the scincomorphan clade as presented by these authors are not uniformly accepted (see Presch 1988), although the composition of the clade as a whole is basically stable. Within the Scleroglossa the Dibamidae, Amphisbaenia, and Serpentes were regarded by these authors as incertae sedis, but all three may be autarchoglossans. Estes and co-workers (1988, which see for references) suggested that the following, previously proposed hypotheses were the most well documented for the taxa in question: dibamids are allied to acontine scincids, amphisbaenians may be scincomorphs, and snakes are relatives of the Anguimorpha.

Iguania. Frost and Etheridge (1989) provided a phylogenetic analysis of iguania and recognized nine families. Of these the Chamaeleonidae (including those taxa formerly assigned to the Agamidae) occurs in mainland Africa, whereas the Corytophanidae, Hoplocercidae, Iguanidae, Polychridae, and Tropiduridae have South American representatives. The Opluridae (those Madagascan lizards formerly assigned to the Iguanidae) occur on the Indian Ocean islands associated with Africa. Unfortunately Frost and Etheridge (1989), like Etheridge and de Queiroz (1988) and previous authors were unable to provide a fully resolved hypothesis of relationships among the well-supported family group taxa they identified. One interfamilial clade was supported by Frost and Etheridge's (1989) data: Opluridae + Tropiduridae + Phrynosomatidae (perhaps + Polychridae), but several possible topologies within this grouping were obtained.

Williams (1988) suggested affinities of part of the Polychridae (as anoles and leiosaurs) with oplurids and chameleons, and regarded these taxa as Gondwanan in origin, with the (anole + leiosaur)/(oplurid + chameleon) split being associated with the opening of the Atlantic. He further regarded *Polychrus* and the Hoploceridae (as morunasaurs) as closely allied Gondwanan taxa. Williams (1988) accounted for the presence of tropidurids and other South American iguanians through migration from Laurasia. Agamids too were postulated to be Laurasian in origin, invading Africa after the split from South America (see also Moody 1980). Renous (1979), on the basis of highly homoplastic nerve characters, also regarded oplurids and chamaeleonids as closely related. She also allied agamids and New World "iguanids" and developed a complex biogeographic scenario for all lizards, but the nature of the data she employed in the analysis makes all of her conclusions highly suspect.

Iguanians illustrate a primarily Gondwanan distribution, but the extreme southern connection linking oplurids with South American iguanians is difficult to evaluate in light of the most recent (unresolved) phylogeny for the group (Frost and Etheridge 1989). Estes (1983a) presented a biogeographic scenario in which a southern Gondwanan pleurodont group of iguanians was split by the opening of the Atlantic ocean, giving rise to South American iguanids on the one hand and the Madagascan oplurids on the other, with intervening mainland African forms subsequently disappearing, perhaps in relation to climatic shifts associated with the northward movement of that continent (Estes 1983a, 1983b). Estes (1983a) also postulated a northern acrodont group which was vicariantly divided into a primarily African chameleon clade and an Asian (or Asia + India + Australia) agamid clade. Fossil iguanians as a whole are consistent with this reconstruction and, in general, occupy the same regions as living forms.

The overall biogeographic pattern implied by iguanian relationships is compatible with that of pelomedusid turtles. Africa bears a sister-group relationship to (Madagascar + South America). This pattern, however, is likely to be more complex than presented, as the resolution of relationships among the remaining iguanian groups will almost certainly support additional patterns of relationship between Africa and South America, as well as connections to Pangaean landmasses.

Gekkota. Fossil gekkonids have a worldwide distribution but, with the possible exception of a Miocene *Phyllodactylus* from Europe, the subfamilial and generic affinities of the pre-Pleistocene forms are unknown or indeterminable. The most recent treatment of higher order gekkonid systematics is that of Kluge (1987), in an expanded and explicitly cladistic reanalysis of an earlier study (Kluge 1967). Among living forms the Eublepharinae, occurring in North and Central America, Asia, and Africa, are the sister group of the remaining geckos. Grismer (1988) postulated a Laurasian origin for the Eublepharinae and demonstrated that the American *Coleonyx* is only distantly related to the African taxa *Holodactylus* and *Hemitheconyx*. The Diplodactylinae (including the pygopodids), an endemic Pacific radiation, are the sister group of the Gekkoninae (discussed below), which is itself essentially cosmopolitan.

The validity of the large group of gekkonine geckos has recently been challenged by King (1987a), but his criteria for the evaluation of monophyly are incapable of falsifying the naturalness of the clade, which has been accepted by nearly all other workers. Within the Gekkoninae the central Asian genus *Teratoscincus* has been identified as the sister group of all other forms (Kluge 1987), which are in turn divided into several major lineages, each of which is pertinent to the present discussion. On the basis of a single character, the absence of the second ceratobranchial arch, Kluge (1983) erected the tribe Gekkonini. Most of the generic groupings of geckos erected by Russell (1972) on purely morphological grounds remain intact when placed into Kluge's (1983) tribal system. Nonetheless, the single-character division of this huge group of taxa cannot be regarded as inviolate in light of other evidence supporting alternative groupings (Bauer 1990b). Within the remaining gekkonines (= "ptyodactylines") several groups have been recognized as potentially monophyletic.

Joger (1985) and Bauer (1990b) examined the African genera of gekkonines, but neither explicitly demonstrated the monophyly of the groups that they were studying and the resultant hypotheses of relationship may be regarded as tentative at best. Both, however, concurred with Haacke (1976) and Kluge (1983) that the genera *Chondrodactylus, Colopus, Kaokogecko, Rhoptropus,* and *Palmatogecko* represent offshoots of the

Pachydactylus lineage. This union was supported by albumin immunological data (Joger 1985) as well as by uniquely derived morphological features (Russell 1972; Haacke 1976; Kluge 1983; Bauer 1990b). Russell (1972) and Kluge (1983) also included *Tarentola* and *Geckonia* in the *Pachydactylus* group. Among the other African taxa, the grouping (*Narudasia* + *Pristurus* + *Quedenfeldtia* + *Saurodactylus*) may have evolutionary reality (Arnold 1981; Kluge 1983). Another grouping, uniting the Madagascan genera *Ebenavia, Paroedura,* and *Uroplatus* is supported by at least two synapomorphies (Bauer 1990b). For the most part, however, relationships among the African genera of gekkonines remain poorly defined (Bauer and Russell 1989; Bauer 1990b).

No attempt has been made to examine the New World gekkonines as a group, but Kluge (1967, 1969) argued that most showed their closest affinities with African taxa and has indicated that both *Aristelliger* and *Thecadactylus* have African sister taxa (Kluge in Duellman and Pianka 1990). Joger (1985) examined both of these taxa in an immunological study of the African gekkonines and callibrated his dating system by assuming that these forms had separated from the African genera as a result of the opening of the Atlantic Ocean, dated by him at 92 Ma. Consequently, his dating of all events in the evolution of the group was predicated on the vicariant origin of the American taxa. In his consideration of the biogeography of the South American Gekkonidae Vanzolini (1968) noted the similarity of the Brazilian geckos *Bogertia, Briba,* and *Phyllopezus* to *Hemidactylus,* but did not specify explicitly the relationships of these forms to the South American or African members of the last genus. The affinities of *Homonota* and *Gymnodactylus,* two genera that Kluge (1984) regarded as closely allied, at least phenetically, remain particularly obscure. Duellman (1979b) suggested African affinities for the latter, with relatively recent rafting as the mode of transport, but this genus is a South American endemic and Duellman's remarks were based on the incorrect generic assignment of the gecko's putative African ancestors. The most speciose New World geckos are sphaerodactylines. These geckos constitute a New World endemic monophyletic unit generally regarded as a subfamily. The Sphaerodactylinae appears to be the sister group of the ptyodactyline genus *Pristurus* of north Africa and southwest Asia (Kluge 1987). As such it represents the only gekkonid group for which explicit phylogenetic information pertinent to transatlantic relationships exists at the suprageneric level. The probable affinities of *Pristurus* to other padless African geckos (see above) with discrete areas of endemism suggest that further resolution of relationships will be particularly useful biogeographically. The division between these African and American groups has been associated with both Cretaceous Gondwanan fragmentation (Smith et al. 1977) and Tertiary rafting (Kluge 1967), but either explanation remains possible.

Several gekkonine genera are represented in both Africa and South America. Of these only *Phyllodactylus* is moderately speciose on both continents. However, the many constituent species in this genus, which also occur in Asia, Europe, and North America, do not form a natural group (Kluge 1967,

1983, 1987; Russell 1972; Dixon and Anderson 1973; Bauer 1990b). Despite extensive work on the alpha systematics of some representatives of the group (e.g., Dixon 1964; Dixon and Huey 1970), no adequate interpretation of the patterns of relationship has been undertaken. Some preliminary information (Bauer 1990b) suggests that there may be no close connection between the African and South American species. The only hypothesis of origin in South America is that of Vanzolini (1968; Vanzolini and Heyer 1985), who suggested that these geckos arrived over water from the Pacific. This would appear to be supported neither by phylogenetic nor by ecological data.

Three putatively monophyletic genera of gekkonines have distributions spanning the Atlantic. *Lygodactylus* is a chiefly Afro-Madagascan genus with two species in the nonforested regions of South America. Smith and co-workers (1977) regarded the two South American species as representatives of a distinctive genus, *Vanzoia,* and associated them with the basal stock of the sphaerodactylines. Bons and Pasteur (1977) subsequently demonstrated that the supposedly diagnostic features of *Vanzoia* had been misinterpreted and that the American forms are allied to the *Lygodactylus capensis* group in Africa. Although the Smith group (1977) believed that the division of African and South American lygodactyles probably occurred during the Cretaceous and favored a vicariant cladogenetic event, most other authors (Vanzolini 1968; Bailey in Kluge 1969; Bons and Pasteur 1977) postulated more recent over-water dispersal events as the source of *Lygodactylus* in America. Bons and Pasteur (1977), however, regarded the relationship as dating from the early Tertiary and did not entirely rule out the possibility of a tectonically induced division of the *L. capensis* group.

Hemidactylus is represented by many species in Africa, especially in Somalia and Ethiopia. Relatively few taxa occur in South America and the West Indies. Kluge (1969) argued that New World members of the *Hemidactylus mabouia-brookii* complex were prehuman arrivals in the New World. Schwartz (1968) likewise suggested that *Tarentola americana,* the West Indian representative of a primarily North African genus, arrived over water from either Africa or southern Europe prior to human times. Vanzolini (1978) argued an alternate view of essentially the same evidence and supported the human introduction of *H. mabouia* and *H. brookii* into South America, although he recognized two indigenous congeners as well. Other *Hemidactylus* unambiguously represent human introductions into the Neotropics (Kluge 1969). There is significant variation in biochemical features of members of the *H. brookii* complex (R. Crombie, personal communication, 1990) and final determination of the origin of the South American *Hemidactylus* must be postponed until this information has been analyzed.

Kluge (1987) reviewed major trends in gekkonid biogeography, revising his earlier (Kluge 1967) pretectonic hypotheses. Kluge (1987) hypothesized that the breakup of Pangaea approximately 180 Ma was the vicariant event responsible for the divergence of the Laurasian eublepharines and the Gondwanan gekkonine/diplodactyline lineages. The Gondwanan affinities of the non-eublepharine geckos as a whole has been recognized for some time (Cracraft 1974; Bauer 1990a; King 1987b). While subsequent plate movements may have been responsible for some of the subsequent cladogenesis within the geckos (Kluge 1987; Bauer 1990a, 1990b), Kluge (1987) maintained that transatlantic dispersal was probably responsible for the origin of the sphaerodactylines and most of the other genera with transatlantic distributions. Although he provided no estimate of time for the Atlantic crossings, Kluge (1967) had earlier postulated an early Tertiary date for the movement of the sphaerodactylines.

South American gekkonine geckos (including the Sphaerodactylinae) clearly do not share a common origin. The resolution of generic relationships within the group may be particularly valuable for identifying possible repeated patterns of area relationship supportive of vicariance. Although existing evidence is equivocal, I consider it likely that most South American gekkonines, exclusive of *Hemidactylus* and *Lygodactylus,* are descended from ancestors present at the time of the fragmentation of western Gondwanaland.

Autarchoglossa. Estes (1983a) reviewed two biogeographic hypotheses relevant to the relationship of lacertids and teiids (including gymnophthalmids), which were hypothesized as sister taxa by Estes et al. (1988). He favored a Laurasian origin for the group with subsequent dispersal into Africa and South America, respectively. He could not, however, falsify his earlier (Estes 1970) hypothesis that the division of the families was coincident with the rifting of western Gondwanaland. The alternative scincomorph phylogeny of Presch (1988) does not support the sister-group relationship of teiids and lacertids.

Scincids constitute the single largest group of lizards, and like geckos, they have essentially cosmopolitan distributions. Fossils are rare and only a few specimens are known from Africa (Miocene of Morocco) and the Neotropics (Pleistocene of Puerto Rico) (Estes 1983b). Relationships within the Scincidae are based on the work of Greer (1970b, 1970c, 1979, 1990 inter alia). Greer (1970b) recognized the Feylininae, restricted to west Africa and Principe Island, the Acontinae, in southern Africa and Kenya, and two larger subfamilies, the Scincinae, occurring on most major land masses except Australasia and South America, and the Lygosominae, found on all continents. Rieppel (1980, 1984) has questioned the monophyly of the family, but the strongest objections apply to the naturalness of the Scincinae (Estes et al. 1988; Rieppel 1988), which Greer himself (1970b) regarded as primitive and ancestral to the remaining groups. Greer (1970b) accepted an early radiation of scincines in mainland Africa, giving rise to both the acontines and feylinines, as well as the Madagascan and Indian Ocean island scincines (Greer 1970c). Lygosomines were regarded (Greer 1970b) as more recent invaders of Africa from an east Asian or Australian center of initial diversification.

As stated by Greer (1979:354), "There is nothing in the distribution of skinks as a whole that suggests that any skink ever used a Gondwanaland dispersal route." Greer (1970b) postulated that skinks arrived only recently in South America, via either Beringia and North America or over water rafting from Africa. The lygosomine radiation in South America is very small and is limited to representatives of the genus *Mabuya* which, as presently construed, is nonmonophyletic (Greer 1970a, 1976, 1979, 1990). The African relationships of the South American skinks have been noted for some time. Dunn (1936) suggested affinities between Neotropical and west African *Mabuya,* but the only treatment of *Mabuya* as a whole has been the unpublished work of Horton (1973). The patterns of relationship suggested by Horton (1973) were not rigorously evaluated and represent a primarily phenetic approach to the systematics of the group. Horton (1973) regarded the African *Mabuya* as being derived from several major radiations. His "*vittata* group" and "*perroteti-brevicollis* group" are cited as having given rise to the South American taxa, as well as those on Madagascar, the Seychelles, and the Cape Verde Islands. Specifically, he postulated that the most primitive South American *Mabuya* occur in the northeastern part of the continent and that they are derived from west African species which reached Venezuela by waif dispersal in the Miocene. *Mabuya maculata,* endemic to Ilha Fernando do Noronha, Brazil, has been especially closely linked to African taxa (Travassos 1948). Only phenetic similarities have yet been documented, but the closest relatives of this species do appear to be west African (R. Crombie, personal communication, 1990).

Many anguimorphan groups have rich fossil records, but most localities are in the Laurasian land masses (Estes 1983a, 1983b), where most of the diversification of the constituent families appears to have occurred. Only the family Anguidae is represented in both Africa and South America. *Ophisaurus koellikeri* of Morocco and *Anguis fragilis* reported from Algeria and Tunisia are representatives of the subfamily Anguinae, a group with primarily northern affinities (Estes 1983a). In South America there are four species each of *Ophiodes* and *Diploglossus,* both diploglossines. This subfamily is widely distributed in the Caribbean and Central America and the South American species are offshoots of this radiation which is hypothesized to have originated in North America (Estes 1983a). Although recent morphological (Gauthier 1982) and allozyme (Good 1987) analyses differ in their placement of the anguid subfamilies, neither support a sister-group relationship between the Anguinae and Diploglossinae, and the presence of anguids in both continents clearly reflects independent histories of the two subfamilies involved with respect to northern relatives.

Amphisbaenidae. The position of amphisbaenians within the squamata remains obscure (Estes et al. 1988; Rieppel 1988; Gans 1990), although autarchoglossan and perhaps scincomorph affinities are probable (Estes et al. 1988; Böhme 1989). Nonetheless, the distinctiveness of amphisbaenians is reflected by some authors (e.g., Gans 1978) in separate subordinal ranking. No cladistic analyses of any amphisbaenid groups have been conducted, although evidence clearly supports the monophyly of the group as a whole (Gans 1978, 1990; Rieppel 1988). Vanzolini (1951) and Gans (1978) each proposed several living familial (or subfamilial) groupings which, presumably, they regarded as natural units. I here follow Gans (1978) as the best documented of the classification schemes. Two of Gans's families occur in the regions under consideration. His Trogonophidae are limited to north Africa and adjacent southwest Asia, while the Amphisbaenidae sensu stricto are widespread in South America, the West Indies, Africa, and around the Mediterranean Sea. Whatever relationships exist between the continents, they would appear to be at the suprageneric level. The diversity of amphisbaenians on both continents, as well as the existence of fossils from the Miocene of Morocco and the Pleistocene of Bolivia, has been used as evidence to suggest a Gondwanan origin for the group (Duellman 1979b; Laurent 1979). This is supported by the logic that fossorial animals are poor overwater dispersers, but in the absence of phylogenetic information, the relationships among the Old and New World amphisbaenians remain obscure. Gans (1990) has suggested a Mediterranean origin for the Amphisbaenidae, with subsequent migration into the Americas and thence into Africa.

2. Serpentes The phylogeny of snakes remains poorly known and relationships among the familial-level taxa have been reconstructed in several patterns by recent authors (Rieppel 1979; McDowell 1987; Rage 1987; Cadle 1987; Kluge 1991). General agreement exists, however, on the monophyly of the major groups, the Alethinophidia and Scolecophidia (Branch 1986, but see Groombridge 1979). The oldest fossil snakes are known from the early Cretaceous of Algeria (*Lapparentophis defrennei*) (Hoffstetter 1959). This taxon and several late Cretaceous forms have been suggested as sister taxa of the remaining Alethinophidia (Rage 1974, 1987).

Scolecophidians include three families, the Anomalepididae, occurring in Middle and South America, the Typhlopidae (pantropical), and the Leptotyphlopidae in the Americas, continental Africa, and southwest Asia. Relationships among and within the families are obscure, but Branch (1986) reported major differences in hemipenial structure between Old and New World leptotyphlopids. On the other hand, similarities in this and other features have been used to associate the African and mainland South American typhlopids with one another (Roux-Estève 1975; Dixon and Hendricks 1979), and a Gondwanan origin for this family's transatlantic distribution has been put forward (Thomas 1976), although no explicit phylogenetic hypothesis has been presented to support this interpretation.

The Alethinophidia may (Rieppel 1979) or may not (McDowell 1987) consist of two monophyletic subgroups, the Henophidia and Caenophidia. The former group consists of a large group of relatively small families including uropeltids, boids sensu lato, aniliids, and tropidopheids. McDowell (1987) divided these groups into superfamilies. In the Anilioidea (Loxocemidae, Xenopeltidae, Aniliidae, and Uropeltidae) only the aniliids occur in South America, and no groups are represented in Africa, although Duellman (1979b) suggested that aniliids may have reached South America from Laurasia via Africa. McDowell's (1987) Tropidopheoidea contains only the Tropidopheidae, distributed in the Neotropics. The Bolyeroidea includes only the two genera and species of Mascarene "boas." The Booidea of McDowell (1987) incorporates the Pythonidae (Asia, Australia, mainland Africa) and Boidae (boines in the Neotropics, Indian Ocean, Pacific Ocean, erycines in North America, Africa, and Eurasia).

McDowell's (1987) placement of the west African *Calabaria* as a pythonid differs from that of Dowling (1978) and Rieppel (1979), who associate it with *Loxocemus* and *Xenopeltis,* Underwood (1976; Underwood and Stimson 1990), who regarded it as more closely allied to boines, tropidopheids, and bolyerids, and Kluge (1991) who regarded it as the sister taxon of the erycines. Underwood and Stimson (1990) also found support for the monophyly of *Loxocemus* + pythonines, with respect to which they regarded *Xenopeltis* + uropeltids, and "boas" (boids, bolyerines, erycines, tropidopheines, and *Calabaria*) as sequentially more distant outgroups. On this basis they regarded the Asia–Africa/Central American distribution of pythons as relictual Laurasian and the African–South American distribution of boas as Gondwanan in origin. The most recent and comprehensive review of higher order snake phylogeny (Kluge 1991) suggests that the "Henophidia" is a paraphyletic group. This is supported by recent data from tongue morphology (Smith and Mackay 1990). Within the henophidian grade, Kluge (1991) recognized the bolyerines as the sister group of the Caenophidia with tropidopheines, erycines + boines + pythonines, *Loxocemus,* and *Xenopeltis* as sequentially more distant outgroups. Within the boines, Kluge (1991) has presented evidence for the synonymy of the Madagascan genera *Sanzinia* and Acrantophis with the New World *Boa.* He regarded this pattern of relationship as reflective of the same vicariant pattern seen in pelomedusids and in the Iguania.

"Advanced snakes," the Colubroidea, represent a well-supported natural group, but arrangements within the group vary between authors (Rasmussen 1985; McDowell 1986, 1987; Cadle 1982, 1987, 1988). Fossil caenophidians assigned to the genus *Pterosphenus* appear to have occupied South America and Africa, as well as Laurasian landmasses, by the Eocene (Hoffstetter 1958, 1960), but were marine forms. The modern caenophidian lineages, however, are unknown from fossils older than the Oligocene (colubrids—Rage 1974, 1988) and Miocene (elapids and viperids—Hoffstetter 1939). Among the main living lineages, Cadle (1987, 1988) regarded the Viperidae as the sister group of (Elapidae + Atractaspidae + Colubridae). Relationships within vipers are unclear, but the major groupings of viperines and crotalines suggest no close African–South American relationships. On the basis of venom proteins (Detrait and Saint-Girons 1979) the African forms *Bitis, Echis,* and *Cerastes* appear to be more closely related to each other than to *Vipera,* which enters Africa from the Palearctic. *Atheris* (an African endemic) also appears to form a separate lineage. *Causus* has been suggested to be the sister taxon of all of the above-mentioned genera (Cadle 1987). This is supported by its plesiomorphic circulatory anatomy (Groombridge 1984) and by albumin immunological data (Cadle 1982). The affinities of the African genus *Atractaspis* remain unclear with respect to elapids and colubrids (Cadle 1982, 1987). The Elapidae is monophyletic (McCarthy 1985), contrary to evidence for the independent origin of micrurines (coral snakes) suggested by Savitzky (1978) and followed by some workers (e.g., Laurent 1979). Elapids were divided by McDowell (1987) into six subfamilies. Of these the Elapinae includes all of the New World coral snakes while African taxa are placed in the Bungarinae; neither group is demonstrably monophyletic (Cadle 1987). No phylogenetic hypotheses have been erected relating these groups to one another, although a definite, but ancient, shared ancestry has been implied (Cadle and Sarich 1981).

McDowell (1987) presented ten colubrid subfamilies, but most are not demonstrably monophyletic and their relationships to one another are unresolved (Cadle 1987). Previously recognized subfamilial groupings (Dowling 1978; Dowling et al. 1983) were largely revised or discarded. The Boodontinae occurs in mainland Africa and Madagascar. Natricines (sensu McDowell 1987) are a diverse group including numerous African genera, but apparently excluding South American forms. Xenodontines are a New World group. The monophyly of Central and South American clades of xenodontines is supported by immunological data (Cadle 1984a, 1984b), but the naturalness of the entire subfamily is disputable (Cadle 1984c, 1985, 1987; McDowell 1987). The two lineages appear to have been separate since the beginning of the Tertiary and their origin is at present attributed to either southward migrations from North America or a Gondwanan vicariant event (Cadle 1984c, 1985). Cadle (1984c) reviewed previous interpretations of the origin of the South American colubrid fauna. Although most authors proposed southward migration of essentially northern stock, Laurent (1979) favored arrival from Africa over water. Cadle (1984c) regarded the xenodontines as the oldest of the New World colubrids and suggested that their affinities (or those of some Neotropical xenodontines) might lie with African snakes, but no data are currently available to definitively address the issue. Finally, the colubrines occur worldwide and their monophyly is supported by both molecular (Schwaner and Desauer 1982; Dowling et al. 1983; Cadle 1984c) and morphological (McDowell 1987) data. Rasmussen (1985) examined retinal morphology and erected a hypothesis of relationships incorporating all caenophidians, but it is largely incompatible with the other schemes mentioned.

Because of the chaotic state of higher order snake systematics, very few biogeographic scenarios, and no analytical biogeographical treatments have been presented with regard to intercontinental relationships. Henophidians hold the promise of providing some information because they include many small groups with generally well circumscribed ranges. Otherwise there is little explicit evidence from snake phylogeny to support any biogeographic reconstruction.

C. Crocodylia

The phylogeny and distribution of fossil groups have been reviewed recently by Buffetaut (1985b) and Benton and Clark (1988) and will not be discussed in detail here. Nonetheless, several groups do support South American–African connections. Mesosuchians of the genus *Sarcosuchus* from the early Cretaceous show an intimate relationship between the Bahia Basin, Brazil and the northwest African Gadoufaoua fauna (Buffetaut and Taquet 1977), a connection also shared by pelomedusid turtles (see above). *Araripesuchus* illustrates a similar distribution (Buffetaut and Taquet 1979). The subsequent Late Cretaceous and Tertiary history of crocodilians in Africa has been reviewed by Buffetaut (1985b) and Tchernov (1986), and several mesosuchian and eusuchian connections between Africa and South America are suggested by the fossils, although these chiefly represent probable over-water dispersals of saline-adapted forms. The latter author suggested east African affinities to Asiatic forms of *Tomistoma* and *Crocodylus,* but did not highlight South American affinities. The Gavialidae, which is represented by a single Asiatic species, today has fossil representatives from the Eocene of Africa and the Oligocene of South America (Buffetaut 1982, 1985a, 1985b) and is hypothesized to have spread westward over water after the establishment of the Atlantic (Taplin and Grigg 1989). The most recent putative gavialids from Africa are those of the Oligocene, while *Gryposuchus* survived until the Pliocene in South America.

Recent systematic treatments of living groups have relied on morphological (Buffetaut 1978, 1982, 1985a, 1985b; Tarsitano et al. 1989; Frey et al. 1989; Norell 1989), biochemical and DNA (Densmore 1983; Densmore and Desauer 1984; Densmore and Owen 1989), or parasite (Brooks 1979; Brooks and O'Grady 1989) data sets. The primary differences between hypotheses relate to the relative placement of *Gavialis* and *Tomistoma,* and as such are not directly pertinent to the present discussion.

Among South American taxa, all authors (e.g., Benton and Clark 1988; Brooks and O'Grady 1989) recognize the caimans as a monophyletic group. Within this clade *Paleosuchus* has been regarded as the sister group of *Caiman,* with *Caiman* itself paraphyletic, giving rise to *Melanosuchus* (Densmore 1983). Alternatively, a monophyletic *Caiman* has been recognized as the sister taxon of the other two genera (Brooks and O'Grady 1989).

Relationships within *Crocodylus* are unresolved, but *Osteolaemus* appears to be the sister group of all *Crocodylus* (Densmore 1983), of (*Crocodylus* + *Tomistoma*) (Norell in Benton and Clark 1988) or of the African or African + Oceanic *Crocodylus*.

There are few if any known pre-Pliocene fossils of direct ancestry to the living *Crocodylus* and the origin of the living crocodilians as a whole remains obscure (Tarsitano et al. 1989; but see Taplin and Grigg 1989). Although the pan-tropical distribution of *Crocodylus* has been attributed to tectonic events (Brooks and O'Grady 1989), the possibility exists that it is the result of relatively recent dispersal(s). The latter hypothesis is supported by the low levels of protein differentiation between taxa reported by Densmore (1983; Densmore and Owen 1989). Densmore (1983; Densmore and Owen 1989) found that biochemical data supported the distant relationship of *C. cataphractus* to all other congeners. Brooks and O'Grady's (1989) parasite data, however, yielded a cladogram in which this species was most closely allied to the other African species, *C. niloticus*. Fossils assignable to the living African species come from the Plio-Pleistocene (Tchernov 1976) but are uninformative with respect to transatlantic connections.

Taplin and Grigg (1989) presented a biogeographic scenario for crocodilians, based upon the results of Densmore (1983) and Densmore and Owen (1989). On the basis of an Eocene caiman-like alligatorid, *Eocaiman* from Argentina (Langston 1965), they accept a relatively long history of the caiman lineage in South America. They may have originated from North America, where alligators were well established or, less likely

(Taplin and Grigg 1989), arrived from eastern Laurasia via Africa. The absence of resolution within *Crocodylus* (Densmore 1983) has prevented the erection of a detailed biogeographic scenario for the living crocodilians. The occurrence of the genus in the Eocene of Egypt and its spread to most continents, including South America, by the Miocene is problematic as the later fossil record is weak and wide-scale extinctions and recolonizations have been implied (Taplin and Grigg 1989). Sill (1968) supported a northern origin for South American *Crocodylus,* but it is highly likely that Cretaceous or Tertiary links directly to Africa or across a narrow Atlantic played some role in the population of South America by the genus.

IV. Biogeographic Interpretation of Reptilian Phylogenetic Data

Among ten distribution patterns characterizing the herpetofaunas of Africa and South America, Laurent (1979) recognized five associated with shared reptilian fauna. Two of these are interrelated and may be combined. These patterns and their examples as provided by Laurent (1979) are presented in table 10.1. Although Laurent (1979) discussed the Gondwanan affinities of the pelomedusids, perhaps the strongest case for reptilian connections of great age, he did not present any hypothesis for the pattern of cryptodire relations, and with this group, as with crocodilians, he emphasized fossil occurrence as the primary evidence for continental relationships. Thus, while he did not present data pertinent to living crocodiles, Cretaceous and early Tertiary connections were implied by gavialid and mesosuchian distributions.

Some of Laurent's (1979) hypotheses of lizard distributions are now outdated, for they reflected phylogenetic hypotheses that have since been proven false. In particular, the possible relationship between teiids and iguanids has been shown to be false (Estes and Báez 1985; Estes et al. 1988), and the probable paraphyly of the Iguanidae sensu lato renders relatively simple hypotheses about iguanian distribution suspect. The presence of Paleocene gekkonids (Estes 1970) suggested that some gecko taxa, such as *Homonota* and the sphaerodactyline geckos, might also be derived from a Gondwanan ancestor, but Laurent (1979) regarded most other geckos and all of the *Mabuya* as reflecting more recent dispersal into South America.

Among snakes Laurent (1979) regarded leptotyphlopids, typhlopids, and amphisbaenians as reflecting a similar, Gondwanan origin, but this was supported only by the transatlantic distribution of these fossorial groups. He proposed separate origins for the "boid" faunas of the continents, with an indigenous South American stock and the African forms derived from separate invasions of Laurasian forms. This would appear to be contradicted by the demonstration of Madagascan–New World affinities within boines (Kluge 1991). Laurent's (1979) view of colubrid snake distribution was based upon the phylogeny of Underwood (1967), which recognized many groups now accepted as nonmonophyletic. Nonetheless, his hypothesis of several separate over-water dispersals may still be generally valid in explaining the origin of some colubrids. Some views of snake relationships accepted by Laurent (1979) have changed and require corresponding revisions of biogeographic

Table 10.1. Patterns of African–South American reptile distribution after Laurent (1979) and this review, with first fossil occurrences of taxa.

	Laurent (1979)	This Review	
Pattern	Taxa	Taxa	First Fossil Occurrence Africa–South America
Gondwanan or West Gondwanan	Pelomedusidae	Pelomedusidae	Lower Cretaceous/Lower Cretaceous
	Amphisbaenidae	Iguania	Miocene/Upper Cretaceous
	Typhlopidae	Gekkoninae (part)?	Miocene/Paleocene
	Leptotyphlopidae	Gavialidae[1]	Eocene/Oligocene
		Typhlopidae?	no record/no record
		Leptotyphlopidae?	no record/no record
		Amphisbaenidae	Miocene/Pleistocene
		Booidea (part)	Upper Cretaceous/Upper Cretaceous
		Elapidae?	Miocene
		Colubridae (part)?	Miocene/Miocene
South American invaders of Africa (pre–continental separation) followed by African extinction	Iguanidae	—	
African invaders of South America (post–continental separation)	Gavialidae(?)	Colubridae (part)?	see above
	Gekkonidae	Gekkoninae (part)?	see above
	Scincidae	Scincidae	Miocene/no record
	Amphisbaenidae(?)	Testudinidae?	Eocene/Miocene
	Colubrinae(?)	Crocodylidae?	Eocene/Miocene
Holarctic invaders of both Africa and South America	Testudinidae	Bataguridae	[Miocene]/Pleistocene
	Crocodylidae	Emydidae	[Miocene]/[Pleistocene]
	Colubrinae(?)	Anguidae	no record/no record
		Scincomorpha	Miocene/Paleocene
		Booidea (part)	see above

Note: Fossil dates in brackets are estimates (see text for discussion).

[1]Now extinct in both Africa and South America.

interpretations. For example, Savitzky's (1978) hypothesis that micrurine coral snakes were more closely related to some colubrids than to Old World elapids has since been demonstrated to be false (Cadle and Sarich 1981; Cadle 1982, 1987) and Laurent's (1979) dismissal of the problem of elapid distributions must be reevaluated.

Cracraft (1975) regarded four pairs of putative sister taxa as being illustrative of an African–South American late Cretaceous vicariant event: Agamidae-Iguanidae, Lacertidae-Teiidae, Pythoninae-Boinae, and an unspecified division within gekkonine geckos. Pythonines and boines do appear to be sister taxa (Kluge 1991), however, at present there is no phylogenetic evidence directly supporting the sister-group status of the first two of Cracraft's taxon pairs. The split within gekkonines may be an appropriate example, but in the absence of an explicit statement of the taxa involved this cannot be evaluated.

Presch (1983), on the other hand, found no evidence of cladogenetic events of higher lizard categories being illustrative of tectonic events. Instead he proposed that dispersals must have been of prime importance, especially in the populating of the New World. Presch's (1983) analysis was flawed by several factors: (1) the acceptance of hypotheses of relationship now believed to be partially unsupported, (2) the inclusion of widely occurring terminal taxa in the analysis, and (3) the assumption that noncongruence of the geologically derived cladogram and taxon-area cladogram constitutes falsification of the hypothesis he was testing. Further, the null hypothesis that tectonically based cladogenesis would be reflected at the subfamilial or familial level is inappropriate as such hierarchical levels have no basis in nature and are thus not comparable across taxa. Kluge (1987) has subsequently provided a counterexample to Presch's (1983) argument by presenting evidence that Pangaean and Gondwanan rifting were the crucial events in the establishment of the main gekkonid lizard clades.

Complex histories, involving both vicariant and dispersal events, are typical of most land areas, even relatively simple island systems (Murphy 1983; Bauer 1988a). The a priori assumption of the universal explanatory power of any single causal agent is unwarranted and frequently leads to excessive speculation by systematists and biogeographers (Greer 1990). Phylogeny reflects evolutionary history of a geographical nature. However, even groups sharing essentially similar patterns of distribution and subject to the same historical events will be expected to display partially incongruent information with respect to area reconstruction as a result of random events, features of vagility and biological tolerance, or any of innumerable causes (Cracraft 1988).

Both vicariant and dispersal events have been implicated by previous authors in accounting for the distribution of the living and fossil groups common to Africa and South America. While both may be plausible explanations, dispersal events are not testable under generally recognized circumstances. As Kluge (1988, 1991) has noted, they can explain both patterns and nonpatterns. As a result, it is appropriate to attempt to analyze potentially vicariant patterns, capable of being falsified, before resorting to the construction of dispersalist scenarios. The tenets of vicariance biogeography are sufficiently well known to obviate the need for an exhaustive discussion. A common earth history is most parsimoniously invoked as the causal agent in congruent patterns of cladogenesis in multiple lineages within a biota. In the case of the taxa reviewed in this paper, explicit, resolved phylogenies are unavailable for most groups (e.g., *Crocodylus,* gekkonine geckos, amphisbaenians, *Mabuya,* and many snake lineages). Given the poor resolution of relationships among virtually all taxa except turtles (Gaffney 1984a, 1984b, 1988; Gaffney and Meylan 1988) and some snakes (Kluge 1991), it is not possible to effectively employ analytical biographical methods (sensu Humphries and Parenti 1986). Rather, it is necessary to resort to an essentially narrative technique—the reciprocal illumination of well-corroborated geological hypotheses of area relationships by one or more phylogenetic hypotheses of correspondingly distributed biotic groups. Falsification is admittedly not possible under these circumstances (Patterson 1981), however, noncongruence can point to errors in either the geological or biological hypothesis of relationships, or both. Such an approach tends to avoid ad hoc hypotheses of dispersalism and may at least suggest a connection between earth and phyletic patterns which may be tested at some later point when phylogenetic hypotheses have been erected for other biotic elements occurring in the same areas of endemism.

Among the phylogenies discussed above, the turtles, with explicit hypotheses of relationship, provide the greatest opportunity for analytical examination. In particular, the area relationships suggested by pelomedusids and pleurodires in general, as noted by Laurent (1979), would appear to be congruent with many classical Gondwanan taxa and are broadly compatible with geologically based area cladograms. Among lizards, a familial-level phylogeny is available, but the position of many taxa is unresolved. The iguanian phylogeny, to the extent that it is resolved, is broadly compatible with that of pelomedusid turtles and is congruent with other phylogenies that have been used as evidence of a common Gondwanan biota, divided in the west by the opening of the Atlantic Ocean during the Late Cretaceous. The phylogeny of boines also supports this pattern, and more specifically, a close area relationship between Madagascar and South America. This specific pattern is reinforced by the specific examples of *Podocnemis* among the pelomedusids, and the oplurine iguanians among lizards. Other patterns of relationship supporting vicariant Gondwanan cladogenetic events are less well established, and are compatible with either tectonic or dispersal events. Groups relevant to this category include the fossorial amphisbaenids, typhlopids and leptotyphlopids, and one or more groups of geckos. Many of the shared occurrences of reptiles at the family level may in fact represent nonpatterns; this is true in the case of the turtle groups Emydidae and "Bataguridae" and in the anguid lizards. In all three cases the few representatives occurring in both continents share only a distant heritage. Co-occurrence in Africa and South America is the result of unrelated historical events affecting different sublineages. Key groups for which appropriate phylogenies have not been hypothesized or for which available phylogenies informative with respect to African–South American relationships are unresolved or contradictory include colubrids, elapids, and the genera *Crocodylus* and *Geochelone.* In other cases, such as *Mabuya, Lygodactylus,* and *Hemidactylus,* the level of differentiation between Old and New World forms is minimal and Tertiary or later over-water dispersal may be involved.

V. Reptile Diversity and Abundance in Africa and South America

The analysis of continental faunal relationships must take into account both the phylogenetic or historical factors responsible for the presence of the taxa involved and the ecological factors dictating the relative diversity and abundance of organisms within a continent (Duellman 1979b). Intercontinental patterns of reptile diversity and abundance have been reviewed by numerous authors, including Laurent (1973), but a paucity of data regarding precise distributions and habitat use has been, and still is, lacking for all but a tiny fraction of species. Patterns of community structure, as well as morphological convergence between ecologically similar taxa have been examined in the case of African and South American anuran amphibians (Lamotte and Solano 1989; Chapter 9, this volume), but equivalent data for reptiles are fragmentary.

Duellman (1979b) tallied the taxa for the South American herpetofauna and reported 1115 species of reptiles representing 22 families and 203 genera. Of these, 994 species and 114 genera were endemic, with the majority of the remainder shared with Central America or the West Indies. Many species have been described and others relegated to synonymies in the intervening time period (Vanzolini 1986a, 1986b) and the application of phylogenetic systematic techniques to higher levels has even resulted in changes in the families recognized (see table 10.2). No such compilation has ever been published for the African continent, although Welch (1982) presented a checklist for African squamates. Patterns of reptile diversity in Africa and South America as compiled for this review may be gleaned from tables 10.2. to 10.4, which present continentwide taxa totals by family and country-by-country summaries for the major reptile groups. Such compilations can only provide the most general framework for comparisons but may be useful for illustrating at least the broadest general patterns of diversity.

Detailed within-continent distributional patterns are beyond the scope of the present paper, but analyses of this sort are available for South America in Vanzolini and Williams (1970), Vanzolini (1988), Duellman (1979a; 1988), Dixon (1979), Hoogmoed (1979), Rivero-Blanco and Dixon (1979), Gallardo (1979), Cei (1979), and Formas (1979), and on a more restricted scale for Africa in Laurent (1973), Hughes (1983), and Böhme (1985). Within South America such events as the uplift of the Andes in the Miocene, the late Tertiary aridification of the west coast, and the cycling of warm and cool and dry and humid periods in the cis-Andean regions have greatly influenced speciation and diversification (Duellman 1979b). The increasing aridity of the Sahara and Namib, as well as movements of their respective sand seas, have profoundly affected African reptile faunas, and at least some recent speciation events in South African lizards have been attributed to sea level changes and microclimatic changes during glacial-interglacial cycling (Mouton and Oelofsen 1988).

Table 10.2. Numbers of living terrestrial or freshwater genera and species of each reptile family with representatives in Africa, in South America, or in both.

	Africa		South America	
Taxon	Genera	Species	Genera	Species
Testudines				
Pelomedusidae	2 (1)	16 (2)	2	7
Chelidae	—	—	5	19
Kinosternidae	—	—	1	3
Chelydridae	—	—	1	1
Trionychidae	3	5	—	—
Emydidae	1	1	1	1
"Bataguridae"	1	1	1	4
Testudinidae	7 (3)	18 (5)	1	4 (1)
Total Turtles	14 (4)	41 (7)	12	39 (1)
Diapsida				
Archosauromorpha				
Crocodylia				
Crocodylidae	2	3	4	8
Lepidosauromorpha				
"Lacertilia"				
Gekkonidae	28 (9)	183 (75)	14	83 (6)
Agamidae[1]	7	46	—	—
Chamaeleonidae[1]	3 (3)	67 (57)	—	—
Corytophanidae	—	—	2	4
Hoplocercidae	—	—	3	10
Iguanidae	—	—	1 (2)	1 (3)
Opluridac	(2)	(7)	—	—
Polychridae	—	—	9	109
Tropiduridae	—	—	12	259 (7)
Teiidae	—	—	9	35
Gymnophthalmidae	—	—	30	126
Lacertidae	21	91	—	—
Scincidae	22 (12)	174 (64)	1	13
Cordylidae	4	42	—	—
Gerrhosauridae	4 (2)	14 (14)	—	—
Anguidae	1	1	2	8
Varanidae	1	3	—	—
Amphisbaenidae	10	61	5	49
Total Lizards	101 (28)	682 (217)	88 (2)	697 (16)

(continued)

Table 10.2. ***Continued.***

Taxon	Africa		South America	
	Genera	Species	Genera	Species
Serpentes[2]				
Anomalepididae	—	—	4	14
Leptotyphlopidae	2	33	1	38
Typhlopidae	3	46 (8)	1	6
Boidae	1 (2)	4 (3)	5	8
Pythonidae	2	4	—	—
Bolyeriidae	(2)	(2)	—	—
Aniliidae	—	—	1	1
Tropidopheidae	—	—	3	6
Atractaspididae	7	34	—	—
Colubridae	61 (17)	220 (52)	75	412 (3)
Elapidae	10	25	1	34
Viperidae	7	33	6	43
Total Snakes	93 (21)	398 (65)	97	562 (3)
Grand Total	210 (53)	1124 (289)	201 (2)	1306 (20)
	263	1413	203	1326

Sources: Testudines (King and Burke 1989; Iverson 1986; Stubbs 1989; Swingland and Klemens 1989); Crocodylia (King and Burke 1989); South American Elapidae and Viperidae (Campbell and Lamar 1989); Cordylidae (Lang 1991); Amphisbaenidae (Gans 1978); Chamaeleonidae (Klaver and Böhme 1986); Tropiduridae (Rodrigues 1987); Gymnophthalmidae (Harris 1985); Scolecophidia (Hahn 1980); South American Typhlopidae (Dixon and Hendricks 1979); African Typhlopidae (Roux-Estève 1974).

Note: Parenthetical listings for Africa are additional taxa not occurring on the mainland but endemic to Madagascan or Indian Ocean islands, or both. For South America, parenthetical numbers represent Galapagos taxa.

[1]Family group names employed here for iguanian lizards follow the nomenclature of Frost and Etheridge (1989) except that the Agamidae is retained as a group separate from the Chamaeleonidae sensu stricto.

[2]Family group names employed here for snakes follow McDowell (1987).

Table 10.3. South American reptile species by country.

Country	Area (1000 km²)	Chelonia	Crocodylia	Lacertilia	Serpentes	Total
Argentina	2777	12	3	116	84	215
Bolivia	1099	11	5	68	104	188
Brazil	8457	27	6	278	240	561
Chile	757	0	0	71	6	77
Colombia	1139	27	5	205	143	380
Ecuador	284	12 (1)	5	135 (16)	180 (3)	341 (20)
French Guiana	91	10	4	38	70	122
Guyana	215	10	4	53	73	140
Paraguay	407	7	3	33	74	117
Peru	1285	12	5	131	143	291
Suriname	163	9	4	51	75	139
Uruguay	176	5	2	18	31	56
Venezuela	912	19	5	110	140	276
Total	17,828	39 (1)	8	705 (16)	562 (3)	1314 (20)

Sources: Tabulations were obtained primarily from Peters and Donoso-Barros (1986), Peters and Orejas-Miranda (1986), and Vanzolini (1986a, 1986b) and were supplemented by the following literature: General (Medem 1983; Pritchard and Trebbau 1984; Iverson 1986; Dixon 1989; Ernst and Barbour 1989; King and Burke 1989; Swingland and Klemens 1989); Argentina (Cei 1986); Bolivia (Kempff Mercado 1975); Chile (Donoso-Barros 1966); Colombia (Medem 1981; Ayala 1986; Pérez-Santos and Moreno 1988, 1989); Ecuador (Miyata 1982); French Guiana (Chippaux 1986; Gasc 1990; Hoogmoed and Avila-Pires 1991); Suriname (Hoogmoed 1973; Moonen et al. 1979); Uruguay (Achával 1976); Venezuela (Donoso-Barros 1968; Lancini 1970, 1979).
Note: The totals include near shore islands; in the case of Ecuador, the totals for the Galapagos Islands are included parenthetically.

Duellman (1979b) recognized familial diversity as a function of earth history and diversity within such hierarchical groups as a function of habitat diversity. This is certainly true for reptiles in both continents under consideration. In South America by far the greatest diversity of family level taxa occurs in the tropical, chiefly forested regions (Chocó, Caribbean-Orinoco, Amazonia, Guiana Shield, Atlantic Forest, Brazilian Shield) (Duellman 1979b). Vanzolini et al. (1980) listed almost 50 species in 16 families (as recognized in the present review) for the Brazilian Caatingas. Fewer are represented in the Chaco and fewer still in the Pampas. Only "iguanids," teiids, colubroids, and scolecophidians have succeeded in the Andes, while 9 families inhabit the Atacama Desert and 2 (tropidurids and xenodontine colubrids) are found in the cold austral forests.

In Africa diversity is more uniform.

Table 10.4. African reptile species by country.

Country	Area (1000 km²)	Chelonia	Crocodylia	Lacertilia	Serpentes	Total
Algeria	2381	3	1*	44	20	68
Angola	1247	10	3	110	90	213
Benin	112	9	3	31	82	125
Botswana	582	8	1	67	68	144
Burkina Faso	274	3	3	15	49	70
Burundi	28	5	1	22	34	62
Cameroon	475	9	3	64	115	191
Cape Verde	4	0	0	11	0	11
Cent. Afr. Rep.	623	8	2	33	91	134
Comores	2	1	1	18	6	26
Congo	342	9	3	37	81	130
Cote d'Ivoire	322	8	3	30	70	111
Djibouti	23	2	0	38	16	56
Egypt	1002	3	1	46	34	84
Equatorial Guinea	28	6	1	35	63	105
Ethiopia	1237	8	1	57	47	113
Gabon	268	8	3	37	70	118
Gambia	11	4	2	12	26	44
Ghana	239	10	3	41	92	144
Guinea	246	4	2	26	65	97
Guinea Bissau	36	7	2	14	38	61
Kenya	583	10	1	84	98	193
Lesotho	30	0	0	24	24	48
Liberia	111	7	3	30	72	112
Libya	1760	3	0	28	19	50
Madagascar	587	8	1	175	58	243
Malawi	118	8	1	49	51	109
Mali	1240	5	2	23	42	72
Mauritania	1031	4	0	14	20	38
Mascarene Is.	5	1*	0	19*	3	23
Morocco	447	3	0	52	24	79
Mozambique	783	9	1	67	69	146
Namibia	824	13	1	135	83	232
Niger	1267	3	1	29	38	71
Nigeria	924	11	3	47	102	163
Rwanda	26	3	1	26	40	70

Country	Area (1000 km²)	Chelonia	Crocodylia	Lacertilia	Serpentes	Total
Sao Tome and Principe	1	0	0	4	4	8
Senegal	196	8	2	27	43	80
Seychelles	<1	4	1	20	3	27*
Sierra Leone	72	6	3	24	69	102
Somalia	627	5	1	112	77	195
South Africa	1221	16	1	174	102	293
Sudan	2506	8	1	43	63	115
Swaziland	17	5	1	39	56	101
Tanzania	942	11	2	130	121	264
Tchad	1284	4	3	28	31	66
Togo	56	8	3	34	86	131
Tunisia	164	3	0	29	20	52
Uganda	236	8	1	49	90	148
Western Sahara	267	0	0	13	4	17
Zaire	2345	16	3	95	154	268
Zambia	753	8	1	51	62	122
Zimbabwe	391	8	1	70	76	155
Total	30,320	48	3	899	463	1413

Sources: General information for Africa is derived from Welch (1982); additional sources are as follows: General (Hughes 1983; Iverson 1986; Ernst and Barbour 1989; King and Burke 1989; Swingland and Klemens 1989; Le Berre 1989); southern Africa (Broadley 1983; Branch 1988; Branch et al. 1988; Boycott and Bourquin 1988); west Africa (Villiers 1975); Botswana (Auerbach 1987); Burkina Faso (Roman 1980); Burundi (Rosselot 1977); Cape Verde Islands (Schleich 1987); Cameroon (Stucki-Stirn 1979); Central African Republic (Joger 1990); Comores (Blanc 1971); Congo (Trape 1981); Djibouti (Schätti 1989); Egypt (Marx 1968); Equatorial Guinea (Eisentraut 1973); Gambia (Håkansson 1981); Ghana (Hughes 1988); Mascarene Islands (Vinson and Vinson 1969; Owadally 1988); Namibia (Griffin 1987); Botswana (Auerbach 1987); Kenya (Spawls 1978); Mali (Papenfuss 1969; Joger 1981; Schätti 1986); Morocco (S. D. Busack, pers. comm.); Senegal (Cissé and Karns 1979); Seychelles (Cheke 1984; Nussbaum 1984); Sierra Leone (Menzies 1966); Somalia (Lanza 1983); Swaziland (Boycott and Culverwell 1992); Tanzania (Broadley and Howell 1991); Uganda (Pitman 1974); Zambia (Broadley 1971); Zimbabwe (Broadley 1988).
Note: The list is very uneven in its coverage, reflecting the age and comprehensiveness of the most recent sources for each region. Colubrids and lizards in general are especially problematic. In all cases except the Comoros, Mascarene Islands, and Seychelles, totals almost certainly reflect underestimates. Asterisks indicate taxa that have become extinct in historical times.

Comparable familial representation is found in the forested lowlands of west Africa, montane forests of east Africa, and in semiarid southern Africa. The lowest diversity is associated with regions of the Sahara and Namib deserts, where only approximately 7 reptile families actually enter the sand sea regions. Diversity is also reduced in the cool forested regions of coastal South Africa (Branch and Hanekom 1987) and at high elevations in the Drakensberg Mountains, as in Lesotho.

Species richness in South America, like familial diversity is centered in the tropical forests. The lowland forests of South America are home to approximately half of all reptile species in the continent (Dixon 1979). Dixon and Soini (1986) sampled the Iquitos region of Amazonian Peru and in an area of approximately 52,000 km^2 (although most of the records were obtained a short distance from riverine transects) recorded 88 snakes, 42 lizards, 9 turtles, and 4 crocodilians. They reported that the known fauna for the entire Amazon Basin included a total of 117 snake species, 75 lizards, 10 turtles, and 4 crocodilians, for a total of 206 reptiles. Duellman (1978) recorded 6 turtles, 2 crocodilians, 30 lizards, and 53 snakes from a much smaller area (all but 4 of the reptile species were collected in a region of 3 km^2) at Santa Cecilia in Amazonian Ecuador. In the Andes and in the cooler and drier habitats of southern South America, species richness is much reduced and species composition is greatly altered with respect to the tropics. Snakes, which are more speciose than lizards in the Amazonian sites mentioned above, are poorly represented in the south (e.g., 71 lizard species versus only 6 snakes in Chile). Further, there is overall similarity among the lizard faunas of the entire Amazon Basin (Dixon and Soini 1986), whereas in the Andes and desert regions regional lizard endemism is high.

As in South America, patterns of African lizard and snake diversity at the species level are not concordant. Lizard diversity is highest in South Africa and Madagascar, with Namibia, Tanzania, Somalia, and Angola following (table 10.4). Species numbers are also high in Zaire and Kenya. Arid northwest Africa as a unit has the fewest number of lizards, although isolated small or poorly collected nations elsewhere on the continent have also known saurian faunas of fewer than 20 species. Lowland regions of west Africa, which contain the majority of the lowland Guinea Forest Blocks, also have few species (20–40) and support a diversity comparable to that of the Sahara and Sahel. In contrast, snakes are most speciose in Zaire, one of the largest and most ecologically diverse countries on the continent. Tanzania and South Africa also support large numbers of species, as do the west African lowlands. North Africa as a whole does not support a large snake fauna, but the Sahara itself is especially poor in snakes. Turtles and crocodilians account for only a tiny percentage of the reptile fauna of Africa, and their distribution reflects one of two obvious patterns. Testudinids are numerous in arid regions, especially South Africa and Namibia, and are virtually absent from many forested regions. Freshwater turtles and crocodilians are most numerous in association with forested riverine habitats and are most numerous in Zaire and the lowland forests of west Africa.

Hughes (1983) reviewed basic snake faunal patterns in tropical sub-Saharan Africa.

In lowland forests he reported low endemism and a high percentage (approx. 70%) of wide-ranging species distributed throughout the Upper and Lower Guinea Forest Blocks. Hughes (1983) characterized two distinctive savanna faunas, northern and southern, meeting in Kenya. Nearly half of sub-Saharan Africa (8.49 × 10^6 km^2) is tropical savanna, and despite the relative homogeneity of this habitat type, regional endemism, linked apparently to vegetation type, is seen. In contrast, the tropical savannas of South America are relatively small, only about one-third the area of those of Africa, but temperate and montane grasslands, almost absent in Africa, are widespread and support a depauperate but largely endemic reptile fauna.

Endemic montane snake species are numerous in the Ulugurus and Usambaras of Tanzania (see Broadley and Howell 1991) but are a relatively unimportant part of the fauna in the highlands of Ethiopia and Cameroon (Hughes 1983). For chameleons, however, and perhaps for other lizards, these areas may be important centers of speciation and endemism (Böhme and Klaver 1981; Böhme 1985). Indeed, virtually all of the highland areas of Africa support endemic reptile taxa. In South Africa, the Cape Fold Mountains have been especially important as centers of endemism and speciation. Montane forests in South America, which at 0.76 × 10^6 km^2 are triple the area of their African counterparts, also support endemic radiations of several major taxa (Duellman 1979a). In virtually all rocky areas, especially in southern Africa, geographical isolates of rupicolous forms have evolved (Poynton and Broadley 1978). Arenicolous isolates have also formed in association with the Kalahari and Namib sands in the south and the Saharan sands in the north.

The Namib and the equatorial lowlands are not occupied by agamids (Moody 1980), but these areas are a center of diversity for gekkonids and lacertids. The highlands of the Horn of Africa serve as regions of endemism for agamids as well as geckos of the genus *Hemidactylus*. The Cape region of southern Africa is an area of diminution of agamid taxa but has several major radiations of geckos and lacertids, as well as scincids and the bulk of the family Cordylidae. Chameleons have centers of diversity in the east African highlands and in the temperate Cape and Natal coastal belts. Tortoises are also most numerous in the Cape region and the arid regions bordering the Namib Desert.

Many of the differences in gross faunal patterns seen between Africa and South America are attributable to the predominance of more xeric habitats in the former and mesic in the latter. In the lowland forests of South America snake species generally equal or outnumber lizards. The same is true in west Africa, but in South America the forest covers nearly three times the area covered in Africa and thus the number of resident snake taxa is greater in South America. Lizards account for the greatest percentage of the herpetofauna in the drier areas of South America, such as the Atacama Desert and the drier areas of Argentina and all of Chile. Arid and semi-arid areas also support a preponderance of lizards in South Africa, Namibia, and Somalia. Desert habitats in sub-Saharan Africa account for 1.34 × 10^6 km^2 versus only 0.29 × 10^6 km^2 in South America (Chapter 9), contributing to the "over representation" of lizards in the African reptile fauna. This is especially so with

Table 10.5. Nominal gekkonid genera represented in Africa and South America (and associated islands as in tables 10.2–10.4).

Africa	South America
GEKKONINAE	GEKKONINAE
Afroedura (9)	*Bogertia* (1)
Ailuronyx (2)	*Briba* (1)
Chondrodactylus (1)	*Gymnodactylus* (2)
Cnemaspis (13)	*Hemidactylus* (4)
Colopus (1)	*Homonota* (10)
Cyrtodactylus (1)	*Lygodactylus** (2)
Ebenavia (1)	*Phyllodactylus* (21)
Geckolepis (6)	*Phyllopezus* (2)
Geckonia (1)	*Thecadactylus* (1)
Hemidactylus (43)	
Homopholis (6)	
Kaokogecko (1)	
*Lygodactylus** (46)	SPHAERODACTYLINAE
*Microscalabotes** (1)	*Coleodactylus** (5)
*Millotisaurus** (1)	*Gonatodes** (16)
*Narudasia** (1)	*Lepidoblepharis** (15)
Pachydactylus (30)	*Pseudogonatodes** (5)
Palmatogecko (1)	*Sphaerodactylus** (4)
Paragehyra (1)	
Paroedura (9)	
*Phelsuma** (31)	
Phyllodactylus (6)	
*Pristurus** (4)	
Ptenopus (3)	
*Ptyodactylus** (3)	
*Quedenfeldtia** (1)	
*Rhoptropus** (7)	
Saurodactylus (2)	
Stenodactylus (2)	
Tarentola (8)	
Tropiocolotes (2)	
Urocotyledon (4)	
Uroplatus (6)	

Africa		South America
EUBLEPHARINAE		
Hemitheconyx (2)		
Holodactylus (2)		
Total genera	35	14
Total species	258	89

Note: Total number of recognized full species in each genus is provided parenthetically. Asterisks indicate taxa that are primarily diurnal in their activity patterns.

respect to geckos and lacertids. Geckos and skinks, as well as cordylids have been especially successful in association with rocky habitats within deserts. Areas of shifting sands, however, impose stringent morphological and physiological constraints, and in such regions species richness is extremely low (Bauer and Russell 1991).

VI. Comparative Ecology

Many pairs of unrelated taxa have been proposed as ecological and morphological equivalents in South America and Africa (Laurent 1973, 1979). Among reptiles these include iguanids and agamids, teiids and lacertids, and boids and pythonids, as well as species pairs within single families (Laurent 1973). Morphological convergences, however, are not as striking as those seen between Australia and Africa, or the Indian Ocean and southwest Pacific Ocean islands, where similar climatic conditions and resultant vegetational patterns have favored a restricted set of body forms. Nonetheless, gross similarities in body form and habitat do characterize some reptile groups. Gymnophthalmids in South America have been identified as the ecological equivalents of fossorial and semi-fossorial skinks in Africa (Laurent 1973; Dixon and Soini 1986), and the few species of *Ophiodes* in South America may also be included in this category. However, no arid-adapted American equivalents exist to complement the large number of scincids found in southern Africa and the acrodont radiations in Africa in no way approximate the more than 375 "iguanids" in South America.

The gekkonid faunas of the two continents are especially dissimilar. In contrast to the large radiations of nocturnal gekkonids in all parts of Africa, at Cuzco Amazónico, Peru, *Thecadactylus rapicauda* is the only nocturnal lizard present (Duellman 1987). Although Hoogmoed and de Avila-Pires (1989) found that some "diurnal" geckos and gymnophthalmids of the Brazilian leaf litter facultatively forage on moonlit nights, there are few South American counterparts to the African gekkonines (table 10.5). *Gonatodes,* a diurnal, arboreal, or semi-arboreal sphaerodactyline genus, is paralleled by a comparable radiation of *Cnemaspis* and perhaps by some *Lygodactylus* in Africa. *Phyllodactylus* of the Americas finds its match in African species assigned to the same genus and in some of the southern African *Oedura* and *Pachydactylus*. There are no South American equivalents of the large radiations of *Hemidactylus* and *Tarentola* of north Africa or of the "*Pachydactylus* group" of southern Africa, and *Thecadactylus* is essentially the only large arboreal American species, contrasting with a huge Indian Ocean group containing many genera and species of both diurnal (*Phelsuma*) and nocturnal (*Ailuronyx, Uroplatus,* etc.) forms. Perhaps the single most important feature of this continental difference is that geckos appear to be most diverse in association with rupicolous habitats (Russell and Bauer 1989). In South America, the flat lowlands of Amazonia provide little such habitat, and most of the appropriate substrates are high in elevation or latitude and at or beyond the thermal range of most geckos. It remains to be explained why neither mainland Africa nor South America has developed a large, primarily arboreal gekkonid fauna, similar to that of Madagascar, Asia, and the islands of the western Pacific. True psammophily in geckos is rare relative to arboreality and rupicolousness. In South America only three species have become associated with dune dwelling. All are members of the *Phyllodactylus* radiation of the Atacama region. By contrast, the dune-dwelling geckos of Africa include nine species in six genera, two of which, *Palmatogecko* in the Namib and *Geckonia* in coastal northwest Africa, are dune endemics (Bauer and Russell 1991).

Laurent (1973) remarked on the relative abundance of freshwater turtles in South America. Alternatively, the cryptodiran fauna of Africa, especially the testudinids, far exceeds that of the Neotropics. He also drew parallels between the forest lizard faunas and presented an overview of snake equivalents in aquatic, terrestrial, and arboreal habitats in the equatorial regions of the two continents. On the whole, the rainforests of South America would appear to have a richer and certainly more speciose fauna than the corresponding regions of Africa. Nonetheless, the more intensive work in the Americas may indicate that some of the discrepancy is artificial. Other features may relate to the overall area of forested regions in the two continents; the area of South American rainforest (prehuman) was approximately 4.5×10^6 km^2 versus 2.0×10^6 km^2 for Africa (Dixon 1979).

Ecological studies carried out on lizards have tended to emphasize either desert communities or forest communities. As might be expected from the relative proportions of each habitat type in Africa and South America, desert studies have predominated in Africa (Huey and Pianka 1977; Huey et al. 1974; Pianka 1971, 1986; Pianka and Huey 1971), while forest analyses have been conducted primarily in Amazonian South America (Rand and Humphrey 1968; Crump 1971; Duellman 1978, 1987; Dixon and Soini 1986). Nonetheless there exists some information on African forests (Laurent 1973) and on the Atacama (Huey 1979) and Argentinian (Sage 1974) deserts. Andean (Pearson and Ralph 1978) and central Chilean (Jaksic et al. 1980) communities have also been investigated, as have Mediterranean (Bons 1973) and savanna-woodland communities (Simbotwe and Garber 1979; Simbotwe 1984) in Africa.

Lizard species in Amazonia are often widespread (Duellman 1978; Dixon and Soini 1986) and the fauna relatively uniform across large distances. This is less true of snakes, which appear to have more restricted resource requirements than lizards. The reverse seems to be true in Africa, where forest snakes are chiefly widely distributed (Hughes 1983). Polychrids and gymnophthalmids predominate in Amazonia as the primary arboreal and terrestrial lizards respectively (Duellman 1987). Duellman (1987) summarized biomass and abundance data for several Amazon Basin sites, but comparable data are lacking from most other South American habitats and regions and are essentially nonexistent for Africa. Duellman (1987) reported a great discrepancy in species abundance in Amazonian Peru, and a similar trend was noted by Dixon and Soini (1986).

Pianka and his colleagues have extensively studied desert lizard faunas. In Africa these studies have concentrated on the Kalahari (Pianka 1971, 1986; Pianka and Huey 1971), while more limited work has evaluated the ecobiogeographic affinities of the reptiles of arid west central Africa (Papenfuss 1969). The lizards of the Peruvian Atacama Desert (Huey 1979) and the arid regions of Argentina (Sage 1974) have also received attention. Species in desert lizard communities in general appear to be more equitably represented than those in forested regions (see above), with Pianka (1986) reporting the greatest "equitability" in the Kalahari community he examined. Pianka's extensive work (reviewed in Pianka 1986) concentrated on the sandveld regions of the Kalahari. In this relatively simple habitat he reported up to 18 lizard species in sympatry. In rocky mopane savannas in northern Namibia this figure increases by as much as three-fold (Bauer et al. 1993). In the Karoo National Park, South Africa (32,000 ha), Branch and Braack (1989) reported 6 turtles, 35 lizards, and 18 snakes in an area of arid grasslands and plateaus. Of the lizards 8 were geckos of the genus *Pachydactylus* and 7 were representatives of the endemic African family Cordylidae.

VII. Conclusions

The reptile faunas of both Africa and South America are extremely rich and diverse at all levels of hierarchical inclusiveness. Although the two continents share a common history through the Cretaceous, there is relatively little explicit reflection of this in the living reptile families. The majority of reptile families occurring in both continents probably achieved their distributions as a result of vicariance or rafting during the early Tertiary, when the landmasses were still in close proximity to one another, perhaps linked by emergent island groups. Existing phylogenetic information does not currently permit a more definite interpretation of observed distributional patterns. Several cofamilial taxa, such as emydids and anguids, reflect southward migrations of primarily Laurasian groups. Most congeneric transatlantic distributions are likely to reflect more recent colonization of South America by African species, and such patterns are limited to those taxa, geckos and skinks, that are known for their potential vagility. *Phyllodactylus* is probably an exception to the pattern of recent transport, but biogeographic analysis of this group is presently impossible as it is recognizably paraphyletic.

Differences between the faunas of Africa and South America reflect both historical and ecological factors. The predominance of desert and savanna in Africa in contrast to lowland forest, temperate and montane grassland, and high elevation forest in South America is largely responsible for regional patterns of intracontinental species richness. Lizards are favored in the drier habitats of southern, eastern, and northern Africa and are also most plentiful, relative to other reptiles, in the drier regions of South America, such as Chile and coastal Peru and in much of Argentina. Snakes, on the other hand are numerous throughout mesic tropical America and in the lowland forests of west Africa.

Patterns of endemism reflect vegetational boundaries, as for snakes of the African savanna belt, and topographical isolation. Lizard endemism in South America is seen in the large radiations of Andean iguanians and in Africa in the many rupicolous geckos and cordylids. The highest levels of endemism and the greatest species density in Africa can be found in Madagascar, where diverse habitats, dissected terrain, and long isolation have resulted in the retention of groups extinct on the African mainland (*Erymnochelys*) and the radiation of others (e.g., chameleons, *Lygodactylus*).

Although it is not possible to contrast the ecologies of the reptiles in any truly meaningful way, at least some trends are evident. A major difference in the composition of the reptile faunas of the two continents is the absence in South America of ecological equivalents of the many nocturnal gekkonine lizards that are so numerous in Africa. Likewise, Africa has no counterpart to the anoles or *Liolaemus,* both of which represent a bewildering array of diversity and an incredible number of species within a single genus. The study of reptile ecological communities remains in its infancy and many of the most complex and diverse habitats in Africa and South America have yet to be examined.

A true picture of transatlantic historical and ecological patterns of reptile distribution awaits further advances in phylogenetic systematics and basic herpetofaunal community studies. While interest in such studies has continued to grow since the African and South American herpetofaunas were last compared, habitat destruction has proceeded apace and little progress has been made in recruiting indigenous peoples into such areas of pure research. Ambitious cooperative projects that address systematic and ecological problems on a broad scale may be particularly important in the future if the study of biotic patterns, especially those of reptiles, is to expand beyond its present bounds.

Literature Cited

Achával F. 1976. Reptiles. Pp. 26–29 *in* A. Languth (ed.), *Lista de las Especies de Vertebrados del Uruguay.* Montevideo: Museo Nacional de Historia Natural.

Arnold, E. N. 1979. Indian Ocean giant tortoises: Their systematics and island adaptations. Philos. Trans. Roy. Soc. London B 286:127–148.

———. 1981. Estimating phylogenies at low taxonomic levels. Z. f. Zool. Syst. Evolutionsforsch. 19:1–35.

Auerbach, R. D. 1987. *The Amphibians and Reptiles of Botswana.* Gaborone: Mokwepa Consultants.

Auffenberg, W. 1971. A new fossil tortoise, with remarks on the origin of South American tortoises. Copeia 1971:106–117.

———. 1974. Checklist of fossil land tortoises. Bull. Florida St. Mus. 18:121–251.

Ayala, S. C. 1986. Saurios de Colombia: Lista actualista, y distribucion de exemplares Colombianos in los museos. Caldasia 15:71–75.

Bauer, A. M. 1988a. A geological basis for some herpetofaunal disjunctions in the Southwest Pacific, with special reference to Vanuatu. Herpetol. J. 1:259–263.

———. 1988b. Reptiles and the biogeographic interpretation of New Caledonia. Tuatara 30:39–50.

———. 1990a. Phylogenetic systematics and biogeography of the Carphodactylini (Reptilia: Gekkonidae). Bonn. Zool. Monogr. 30:1–220.

———. 1990b. Phylogeny and biogeography of the geckos of southern Africa and the islands of the western Indian Ocean: A preliminary analysis. Pp. 275–284 *in* G. Peters and R. Hutterer (eds.), *Vertebrates in the Tropics*. Bonn: Museum Alexander Koenig.

Bauer, A. M., and A. P. Russell. 1989. A systematic review of the genus *Uroplatus* (Reptilia: Gekkonidae), with notes on its biology. J. Nat. Hist. 23:169–203.

———. 1991. Pedal specialisations in dune-dwelling geckos. J. Arid Environ. 20:43–62.

Bauer, A. M., and J. V. Vindum. 1990. A checklist and key to the herpetofauna of New Caledonia, with remarks on biogeography. Proc. Calif. Acad. Sci. 17:17–44.

Bauer, A. M., W. R. Branch, and W. D. Haacke. 1993. The herpetofauna of Kamanjab and adjacent Damaraland, Namibia. Madoqua 19.

Benton, M. J., and J. M. Clark. 1988. Archosaur phylogeny and the relationships of the Crocodylia. Pp. 295–338 *in* M. J. Benton (ed.), *The Phylogeny and Classification of Tetrapods*, Vol. 1: *Amphibians, Reptiles, Birds*. Oxford: Clarendon Press.

Blanc, C. P. 1971. Les reptiles de Madagascar et des îles voisines. Ann. Univ. Madagascar 8:95–178.

Böhme, W. 1985. Zoogeographical patterns of the lizard fauna of the African subsahara savanna belt, with preliminary description of a new chameleon. Pp. 471–478 *in* K.-L. Schuchmann (ed.), *Proceedings of the International Symposium on African Vertebrates: Systematics, Phylogeny and Evolutionary Ecology*. Bonn: Zoologisches Forschungsinstitut und Museum Alexander Koenig.

———. 1989. Zur systematischen Stellung der Amphisbänen (Reptillia: Squamata), mit besonderer Berücksichtigung der Morphologie des Hemipenis. Z. f. Zool. Syst. Evolutionsforsch. 27:330–337.

Böhme, W., and C. J. J. Klaver. 1981. Zur innerartlichen Gliederung und zur Artgeschichte von *Chamaeleo quadricornis* Tornier, 1899 (Sauria: Chamaeleonidae). Amphib.-Rept. 1:313–328.

Bons, J. 1973. Herpétologie marocaine. II. Origines, évolution et particularités du peuplement herpétologique du Maroc. Bull. Soc. Sci. Nat. Phys. Maroc 53:63–110.

Bons, J., and G. Pasteur. 1977. Solution histologique à un problème de taxinomie herpétologique intéresant les rapports paléobiologiques de l'Amérique du Sud et de l'Afrique. C. R. Acad. Sci., Paris, ser. D 284:2547–2550.

Bour, R. 1984. Les tortues terrestres géantes des îles de l'Océan Indien Occidental: Données géographiques, taxinomiques et phylogénétiques. Studia Palaeocheloniologica 1:17–76.

Boycott, R. C., and O. Bourquin. 1988. *The South African Tortoise Book*. Johannesburg: Southern Book Publishers.

Boycott, R. C., and J. B. Culverwell. 1992. Swaziland herpetofauna: A preliminary synthesis. J. Herpetol. Assoc. Afr. 40:38–41.

Bramble, D. M. 1971. Functional morphology, evolution, and paleoecology of gopher tortoises. Ph.D. diss., University of California, Berkeley.

Branch, W. R. 1986. Hemipenial morphology of African snakes: A taxonomic review. Part 1. Scolecophidia and Boidae. J. Herpetol. 20:285–299.

———. 1988. *Field Guide to the Snakes and Other Reptiles of Southern Africa*. Cape Town: Struik Publishers.

Branch, W. R., and H. H. Braack. 1989. Reptiles and amphibians in the Karoo National Park: A surprising diversity. J. Herpetol. Assoc. Afr. 36:26–35.

Branch, W. R., and N. Hanekom. 1987. The herpetofauna of the Tsitsikamma Coastal and Forest National Parks. Koedoe 30:49–60.

Branch, W. R., E. H. W. Baard, W. D. Haacke, N. Jacobsen, J. C. Poynton, and D. G. Broadley (eds.). 1988. A provisional and annotated checklist of the herpetofauna of southern Africa. J. Herpetol. Assoc. Afr. 34:1–19.

Brattstrom, B. H. 1961. Some new fossil tortoises from western North America with remarks on the zoogeography and paleoecology of tortoises. J. Paleontol. 35:543–560.

Broadley, D. G. 1971. The reptiles and amphibians of Zambia. Puku 7:93–95.

———. 1983. *FitzSimons' Snakes of Southern Africa.* Johannesburg: Delta Books.

———. 1988. A checklist of the reptiles of Zimbabwe, with synoptic keys. Arnoldia Zimbabwe 9:369–430.

Broadley, D. G., and K. M. Howell. 1991. A checklist of the reptiles of Tanzania, with synoptic keys. Syntarsus 1:1–70.

Broin, F. de. 1980. Les tortues de Gadoufaoua (Aptien du Niger); aperçu sur la plaéobiogéographie des Pelomedusidae (Pleurodira). M/Dm. Soc. Géol. France, n.s., 139:39–46.

———. 1988. Les tortues et le Gondwana. Examen des rapports entre le fractionnement du Gondwana et la dispersion géographique des tortues pleurodires à partir du Crétacé. Studia Palaeocheloniologica 2:103–142.

Broin, F. de., E. Buffetaut, J. C. Koeniguer, J. C. Rage, D. Russell, P. Taquet, C. Vergnaud-Grazzini, and S. Wenz. 1974. La faune de vertébrés continentaux du gisement d'In Beceten (Sénonien du Niger). C. R. Acad. Sci., Paris, sér. D 279:469–472.

Brooks, D. R. 1979. Testing hypotheses of evolutionary relationships among parasites: The digeneans of crocodilians. Amer. Zool. 19:1228–1238.

Brooks, D. R., and R. T. O'Grady. 1989. Crocodilians and their helminth parasites: Macroevolutionary considerations. Amer. Zool. 29:873–883.

Buffetaut, E. 1978. Sur l'histoire phylogénétique et biogéographique des Gavialidae (Crocodylia, Eusuchia). C. R. Acad. Sci., Paris, ser. D 287:911–914.

———. 1982. Systematique, origine et evolution des Gavialidae Sud-Americains. Geobios, Mém. Spec. 6:127–140.

———. 1985a. The place of *Gavialis* and *Tomistoma* in eusuchian evolution: A reconciliation of palaeontological and biochemical data. N. Jb. Geol. Paläont. Mh. 1985(12):707–716.

———. 1985b. Zoogeographical history of African crocodiles since the Triassic. Pp. 453–469 *in* K.-L. Schuchmann (ed.), *Proceedings of the International Symposium on African Vertebrates: Systematics, Phylogeny and Evolutionary Ecology.* Bonn: Zoologisches Forschungsinstitut und Museum Alexander Koenig.

Buffetaut, E., and P. Taquet. 1977. The giant crocodilian *Sarcosuchus* in the early Cretaceous of Brazil and Niger. Palaeontol. 20:203–208.

———. 1979. An early Cretaceous terrestrial crocodilian and the opening of the South Atlantic. Nature 280:486–487.

Busack, S. D. 1977. Zoogeography of amphibians and reptiles in Cádiz Province, Spain. Ann. Carnegie Mus. 46:285–316.

———. 1986. Biogeographic analysis of the herpetofauna separated by the formation of the Strait of Gibraltar. Natl. Geogr. Res. 2:17–36.

Busack, S. D., and C. H. Ernst. 1980. Variation in Mediterranean populations of *Mauremys* Gray 1869 (Reptilia, Testudines, Emydidae). Ann. Carnegie Mus. 49:251–264.

Cadle, J. E. 1982. Problems and approaches in the interpretation of the evolutionary history of venomous snakes. Mem. Inst. Butantan 46:255–274.

———. 1984a. Molecular systematics of neotropical xenodontine snakes: I. South American xenodontines. Herpetologica 40:8–20.

———. 1984b. Molecular systematics of neotropical xenodontine snakes: II. Central American xenodontines. Herpetologica 40:21–30.

———. 1984c. Molecular systematics of neotropical xenodontine snakes. III. Overview of xenodontine phylogeny and the history of New World snakes. Copeia 1984:641–652.

———. 1985. The neotropical colubrid snake fauna (Serpentes: Colubridae): Lineage components and biogeography. Syst. Zool. 34:1–20.

———. 1987. The geographic distribution of snakes. Pp. 77–105 *in* R. A. Seigel, J. T. Collins, and S. S. Novak (eds.), *Snakes: Ecology and Evolutionary Biology.* New York: Macmillan.

———. 1988. Phylogenetic relationships among advanced snakes. Univ. Calif. Publ. Zool. 119:1–77.

Cadle, J. E., and V. M. Sarich. 1981. An immunological assessment of the phylogenetic position of New World coral snakes. J. Zool. London 195:157–167.

Campbell, J. A., and W. W. Lamar. 1989. *The Venomous Reptiles of Latin America.* Ithaca, N.Y.: Comstock Publishing.

Campos, D. de A., and F. de Broin. 1981. Tartarugas fosseis do Brasil. Ann. Acad. brasil. Ci. 53:210–211.

Capler, J. M., and E. O. Moll. 1990. Phylogenetic relationships of the *Batagur* complex of tropical Asian turtles. Abstracts, 1990 Joint Annual Meeting of The Herpetologists' League and The Society for the Study of Amphibians and Reptiles, p. 52.

Carr, J. L. 1990. Phylogenetic analysis of the turtle genus *Rhinoclemmys*. Abstracts, 1990 Joint Ann. Meeting of the Herpetol. League and the Soc. for the Study of Amphib. and Rept., p. 52.

Cei, J. M. 1979. The Patagonian herpetofauna. Pp. 309–340 *in* W. E. Duellman (ed.), *The South American Herpetofauna: Its Origin, Evolution and Dispersal.* Mus. Nat. Hist. Univ. Kansas Monogr. 7. Lawrence: University of Kansas Museum of Natural History.

———. 1986. *Reptiles del Centro, Centro-oeste y Sur de la Argentina.* Mus. Reg. Sci. Nat. Torino Monogr. 4. Torino: Museo Regionale di Scienze Naturali.

Cheke, A. S. 1984. Lizards of the Seychelles. Pp. 331–360 *in* D. R. Stoddart (ed.), *Biogeography and Ecology of the Seychelles Islands.* The Hague: W. Junk.

Chippaux, J.-P. 1986. *Les serpentes de la Guyane française.* Paris: Editions de l'ORSTOM.

Cissé, M., and D. R. Karns. 1979. Les Sauriens du Sénegal. Bull. Inst. Fond. Afr. Noire, ser. A, 40:144–211.

Ckhikvadzé, V. M. 1984. Classification des tortues de la famille des Emydidae et leurs liens phylogénétiques avec d'autres familles. Studia Palaeocheloniologica 1:105–113.

Coto R., and A. and R. A. Acuña M. 1986. Filogenia de *Geochelone costaricensis* y la familia Testudinidae (Reptilia: Testudines) en el continente americano. Rev. Biol. Trop. 34:199–208.

Cracraft, J. 1974. Continental drift and vertebrate distribution. Ann. Rev. Ecol. Syst. 5:215–261.

———. 1975. Mesozoic dispersal of terrestrial faunas around the southern end of the world. Mém. Mus. Natn. Hist. Nat., n.s. (A) Zool. 88:29–54.

———. 1988. Deep-history biogeography: Retrieving the historical pattern of evolving continental biotas. Syst. Zool. 37:221–236.

Crumly, C. R. 1982. A cladistic analysis of *Geochelone* using cranial osteology. J. Herpetol. 16:215–234.

———. 1984. The evolution of land tortoises (Family Testudinidae). Ph.D. diss., Rutgers University, Newark, N.J.

———. 1985. A hypothesis for the relationship of land tortoise genera (Family Testudinidae). Studia Palaeocheloniologica 1:115–124.

Crump, M. L. 1971. Quantitative analysis of the ecological distribution of a tropical herpetofauna. Occas. Pap. Mus. Nat. Hist. Univ. Kansas 3:1–62.

Densmore, L. D., III. 1983. Biochemical and immunological systematics of the order Crocodilia. Pp. 397–465 *in* M. K. Hecht, B. Wallace, and G. H. Prance (eds.), *Evolutionary Biology,* vol. 16. New York: Plenum.

Densmore, L. D., III, and H. C. Dessauer. 1984. Low levels of protein divergence detected between *Gavialis* and *Tomistoma:* Evidence for crocodilian monophyly. Comp. Biochem. Physiol. 77B:715–720.

Densmore, L. D., III, and R. D. Owen. 1989. Molecular systematics of the Order Crocodilia. Amer. Zool. 29:831–841.

Detrait, J., and H. Saint-Girons. 1979. Communautés antigéniques des venins et systématique des Viperidae. Bijd. tot Dierk. 49:71–80.

Dixon, J. R. 1964. The systematics and distribution of lizards of the genus *Phyllodactylus* in North and Central America. N. Mexico St. Univ. Res. Cent. Sci. Bull. 64, no. 1:iv + 139 pp.

———. 1979. Origin and distribution of reptiles in lowland tropical rainforests of South America. Pp. 217–240 *in* W. E. Duellman (ed.), *The South American Herpetofauna: Its Origin, Evolution and Dispersal.* Mus. Nat. Hist.

Univ. Kansas Monogr. 7. Lawrence: University of Kansas Museum of Natural History.

———. 1989. A key and checklist to the neotropical snake genus *Liophis* with country lists and maps. Smithsonian Herpetol. Inform. Serv. 79:1–40.

Dixon, J. R., and S. C. Anderson. 1973. A new genus and species of gecko (Sauria: Gekkonidae) from Iran and Iraq. Bull. So. Calif. Acad. Sci. 72:155–160.

Dixon, J. R., and F. S. Hendricks. 1979. The wormsnakes (Family Typhlopidae) of the Neotropics, exclusive of the Antilles. Zool. Verhand. Leiden 173:1–39.

Dixon, J. R., and R. B. Huey. 1970. Systematics of the lizards of the gekkonid genus *Phyllodactylus* of mainland South America. Los Angeles Co. Mus. Contrib. Sci. 192:1–78.

Dixon, J. R., and P. Soini. 1986. *The Reptiles of the Upper Amazon Basin, Iquitos Region, Peru*. Milwaukee: Milwaukee Public Museum.

Donoghue, M. J., J. A. Doyle, J. Gauthier, A. G. Kluge, and T. Rowe. 1989. The importance of fossils in phylogeny reconstruction. Ann. Rev. Ecol. Syst. 20:431–460.

Donoso-Barros, R. 1966. *Reptiles de Chile*. Santiago: Ediciones de la Universidad de Chile.

———. 1968. The lizards of Venezuela (check list and key). Carib. J. Sci. 8:105–122.

Dowling, H. G. 1978. Colubridae. Pp. 112.1–112d.4 *in* H. G. Dowling and W. E. Duellman (eds.), *Systematic Herpetology: A Synopsis of Families and Higher Categories*. New York: HISS Publications.

Dowling, H. G., R. Highton, G. C. Maha, and L. R. Maxson. 1983. Biochemical evaluation of colubrid snake phylogeny. J. Zool. London 201:309–329.

Duellman, W. E. 1978. The biology of an equatorial herpetofauna in Amazonian Ecuador. Misc. Publ. Mus. Nat. Hist. Univ. Kansas 65:1–352.

———. 1979a. The herpetofauna of the Andes: Patterns of distribution, origin, differentiation and present communities. Pp. 371–460 *in* W. E. Duellman (ed.), *The South American Herpetofauna: Its Origin, Evolution and Dispersal*. Mus. Nat. Hist. Univ. Kansas Monogr. 7. Lawrence: University of Kansas Museum of Natural History.

———. 1979b. The South American herpetofauna: A panoramic view. Pp. 1–28 *in* W. E. Duellman (ed.), *The South American Herpetofauna: Its Origin, Evolution and Dispersal*. Mus. Nat. Hist. Univ. Kansas Monogr. 7. Lawrence: University of Kansas Museum of Natural History.

———. 1987. Lizards in an Amazonian rain forest community: Resource utilization and abundance. Natl. Geogr. Res. 3:489–500.

———. 1988. Patterns of species diversity in anuran amphibians in the American tropics. Ann. Missouri Bot. Gard. 75:79–104.

Duellman, W. E., and E. R. Pianka. 1990. Biogeography of nocturnal insectivores: Historical events and ecological filters. Ann. Rev. Ecol. Syst. 21:57–68.

Dunn, E. R. 1936. Notes on American mabuyas. Proc. Acad. Nat. Sci. Phila. 87:533–557.

Eisentraut, M. 1973. Die Wirbeltierfauna von Fernando Poo und Westkamerun. Bonn. Zool. Monogr. 3:1–428 + 5 pls.

Ernst, C. H., and R. W. Barbour. 1989. *Turtles of the World*. Washington, D.C.: Smithsonian Institution Press.

Estes, R. 1970. Origin of the North American lower vertebrate fauna: An inquiry into the fossil record. Forma et Functio 3:139–163.

———. 1983a. The fossil record and early distribution of lizards. Pp. 365–398 *in* A. Rhodin and K. Miyata (eds.), *Studies in Herpetology and Evolutionary Biology: Essays in Honor of Ernest Edward Williams*. Cambridge, Mass.: Harvard University Press.

———. 1983b. *Handbuch der Paläoherpetologie*. Teil 10A, *Sauria terrestria, Amphisbaenia*. Stuttgart: Gustav Fischer Verlag.

Estes, R., and A. M. Báez. 1985. Herpetofaunas of North and South America during the Late Cretaceous and Cenozoic: Evidence for interchange? Pp. 139–197 *in* F. G. Stehli and S. D. Webb (eds.), *The Great American Biotic Interchange*. New York: Plenum.

Estes, R., and J. H. Hutchison. 1980. Eocene lower vertebrates from Ellesmere Island, Canadian Arctic Archipelago. Palaeogeogr. Palaeoclimatol. Palaeoecol. 30:325–347.

Estes, R., K. de Queiroz, and J. Gauthier. 1988. Phylogenetic relationships within Squamata. Pp. 119–281 *in* R. Estes and G. Pregill (eds.), *Phylogenetic Relationships of the Lizard Families: Essays Commemorating Charles L. Camp.* Stanford, Calif.: Stanford University Press.

Etheridge, R., and K. de Queiroz. 1988. A phylogeny of Iguanidae. Pp. 283–367 *in* R. Estes and G. Pregill (eds.), *Phylogenetic Relationships of the Lizard Families: Essays Commemorating Charles L. Camp.* Stanford, Calif.: Stanford University Press.

Formas, J. R. 1979. La herpetofauna de los bosques temperados de Sudamerica. Pp. 341–370 *in* W. E. Duellman (ed.), *The South American Herpetofauna: Its Origin, Evolution and Dispersal.* Mus. Nat. Hist. Univ. Kansas Monogr. 7. Lawrence: University of Kansas Museum of Natural History.

Frey, E., J. Riess, and S. F. Tarsitano. 1989. The axial tail musculature of recent crocodiles and its phyletic implications. Amer. Zool. 29:857–862.

Frost, D. R., and R. Etheridge. 1989. A phylogenetic analysis and taxonomy of Iguanian lizards (Reptilia: Squamata). Misc. Publ. Univ. Kansas Mus. Nat. Hist. 81:1–65.

Gaffney, E. S. 1975a. A phylogeny and classification of the higher categories of turtles. Bull. Amer. Mus. Nat. Hist. 155:387–436.

———. 1975b. A taxonomic revision of the Jurassic turtles *Portlandemys* and *Plesiochelys.* Amer. Mus. Novit. 2574.1–19.

———. 1984a. Historical analysis of theories of chelonian relationship. Syst. Zool. 33:283–301.

———. 1984b. Progress towards a natural hierarchy of turtles. Studia Palaeocheloniologica 1:125–131.

———. 1986. Triassic and Early Jurassic turtles. Pp. 183–187 *in* K. Padian (ed.), *The Beginning of the Age of Dinosaurs.* Cambridge: Cambridge University Press.

———. 1988. A cladogram of the pleurodiran turtles. Acta Zool. Cracov. 31:487–492.

———. 1990. The comparative osteology of the Triassic turtle *Proganochelys.* Bull. Amer. Mus. Nat. Hist. 194:1–263.

Gaffney, E. S., and L. J. Meeker. 1983. Skull morphology of the oldest turtles: A preliminary description of *Proganochelys quenstedti.* J. Vert. Paleontol. 3:25–28.

Gaffney, E. S., and P. A. Meylan. 1988. A phylogeny of turtles. Pp. 157–219 *in* M. J. Benton (ed.), *The Phylogeny and Classification of Tetrapods,* Vol. 1: *Amphibians, Reptiles, Birds.* Oxford: Clarendon.

Gallardo, J. M. 1979. Composición, distribución y origen de la herpetofauna Chaqueña. Pp. 299–308 *in* W. E. Duellman (ed.), *The South American Herpetofauna: Its Origin, Evolution and Dispersal.* Mus. Nat. Hist. Univ. Kansas Monogr. 7. Lawrence: University of Kansas Museum of Natural History.

Gans, C. 1978. The characteristics and affinities of the Amphisbaenia. Trans. Zool. Soc. London 34:347–416.

———. 1990. Patterns in amphisbaenian biogeography: A preliminary analysis. Pp. 133–143 *in* G. Peters and R. Hutterer (eds.), *Vertebrates in the Tropics.* Bonn: Museum Alexander Koenig.

Gasc, J.-P. 1990. *Les lézards de Guyane.* Paris: Editions Raymond Chabaud.

Gauthier, J. A. 1982. Fossil xenosaurid and anguid lizards from the early Eocene Wasatchian Formation, southeast Wyoming, and a revision of the Anguioidea. Contr. Geol., Univ. Wyoming 21:7–54.

Gauthier, J. A., A. G. Kluge, and T. Rowe. 1988. Amniote phylogeny and the importance of fossils. Cladistics 4:105–209.

Good, D. A. 1987. An allozyme analysis of anguid subfamilial relationships (Lacertilia: Anguidae). Copeia 1987:696–701.

Greer, A. E. 1970a. The relationships of the skinks referred to the genus *Dasia.* Breviora 348:1–30.

———. 1970b. A subfamilial classification of scincid lizards. Bull. Mus. Comp. Zool. 139:151–184.

———. 1970c. The systematics and evolution of the subsaharan Africa, Seychelles, and Mauritius scincid lizards. Bull. Mus. Comp. Zool. 140:1–23.

———. 1976. On the evolution of the giant Cape

Verde scincid lizard *Macroscincus coctei.* J. Nat. Hist. 10:691–712.

———. 1979. A phylogenetic subdivision of Australian skinks. Rec. Aust. Mus. 32:339–371.

———. 1990. *Biology and Evolution of Australian Lizards.* Chipping Norton, Eng.: Surrey Beatty and Sons.

Griffin, M. 1987. *Checklist of Amphibians, Reptiles and Mammals Which Are Known or Expected to Occur in S.W.A./Namibia.* Windhoek: Department of Nature Conservation.

Grismer, L. L. 1988. Phylogeny, taxonomy, classification, and biogeography of eublepharid geckos. Pp. 369–469 *in* R. Estes and G. Pregill (eds.), *Phylogenetic Relationships of the Lizard Families: Essays Commemorating Charles L. Camp.* Stanford, Calif.: Stanford University Press.

Groombridge, B. C. 1979. Variations in morphology of the superficial palate of henophidian snakes and some possible systematic implications. J. Nat. Hist. 13:447–475.

———. 1984. The facial carotid artery in snakes (Reptilia: Serpentes): Variations and possible cladistic significance. Amphib.-Rept. 5:145–155.

Haacke, W. D. 1976. The burrowing geckos of Southern Africa, 5 (Reptilia: Gekkonidae). Ann. Transvaal Mus. 30:71–89.

Hahn, D. E. 1980. Liste der rezenten Amphibien und Reptilien, Anomalepididae, Leptotyphlopidae, Typhlopidae. Das Tierreich 101:i–xii + 1–93.

Håkansson, N. T. 1981. An annotated checklist of reptiles known to occur in the Gambia. J. Herpetol. 15:155–161.

Harris, D. M. 1985. Infralingual plicae: Support for Boulenger's Teiidae (Sauria). Copeia 1985:560–565.

Hirayama, R. 1985. Cladistic analysis of batagurine turtles (Batagurinae: Emydidae: Testudinoidea): A preliminary result. Studia Palaeocheloniologica 1:141–157.

Hoffstetter, R. 1939. Contribution à l'étude des elapides actuels et fossiles et de l'ostéologie des ophidiens. Arch. Mus. Hist. Nat. Lyon 15(3):1–82.

———. 1958. Un serpent marin du genre *Pterosphenus* (*P. sheppardi* nov. sp.) dans l'Éocène supérieur de l'Equateur (Amérique du Sud). Bull. Soc. Géol. France 8:45–50.

———. 1959. Un serpent terrestre dans le Crétacé inférieur du Sahara. Bull. Soc. Géol. France, ser. 7, 1:897–902.

———. 1960. Presence de *Pterosphenus* (serpent Paleophide) dans l'Éocène supérieur du bord occidental du Desert Libyque. C. R. Somm. Soc. Géol. France 1960(2):41–42.

Hoogmoed, M. S. 1973. *Notes on the Herpetofauna of Surinam IV: The Lizards and Amphisbaenians of Surinam.* The Hague: W. Junk.

———. 1979. The herpetofauna of the Guianan region. Pp. 241–280 *in* W. E. Duellman (ed.), *The South American Herpetofauna: Its Origin, Evolution and Dispersal.* Mus. Nat. Hist. Univ. Kansas Monogr. 7. Lawrence: University of Kansas Museum of Natural History.

Hoogmoed, M. S., and T. C. S. de Avila-Pires. 1989. Observations on the nocturnal activity of lizards in a marshy area in Serra do Navio, Brazil. Trop. Zool. 2:165–173.

———. 1991. Annotated checklist of the herpetofauna of Petit Saut, Sinnamary River, French Guiana. Zool. Meded. Leiden 65:53–88.

Horton, D. 1973. Evolution in the genus *Mabuya* (Lacertilia, Scincidae). Ph.D. diss., University of New England, Armidale, Australia.

Huey, R. B. 1979. Parapatry and niche complementarity of Peruvian desert geckos (*Phyllodactylus*): The ambiguous role of competition. Oecologia 38:249–259.

Huey, R. B., and E. R. Pianka. 1977. Patterns of niche overlap among broadly sympatric versus narrowly sympatric Kalahari lizards (Scincidae: Mabuya). Ecology 58:119–128.

Huey, R. B., E. R. Pianka, M. E. Egan, and L. W. Coons. 1974. Ecological shifts in sympatry: Kalahari fossorial lizards (*Typhlosaurus*). Ecology 55:304–316.

Hughes, B. 1983. African snake faunas. Bonn. Zool. Beitr. 34:311–365.

———. 1988. Herpetology in Ghana (West Africa). Brit. Herpetol. Soc. Bull. 25:29–38.

Humphries, C. J., and L. R. Parenti. 1986. *Cladistic Biogeography.* Oxford: Clarendon.

Iverson, J. B. 1986. *A Checklist with Distribution*

Maps of the Turtles of the World. Richmond, Ind.: J. B. Iverson.

Jaksic, F. M., H. Nuñez, and F. P. Ojeda. 1980. Body proportions, microhabitat selection, and adaptive radiation of *Liolaemus* lizards in central Chile. Oecologia 45:178–181.

Joger, U. 1981. Zur Herpetofaunistik Westafrikas. Bonn. Zool. Beitr. 32:297–340.

———. 1985. The African gekkonine radiation: Preliminary phylogenetic results, based on quantitative immunological comparisons of serum albumins. Pp. 479–494 *in* K.-L. Schuchmann (ed.), *Proceedings of the International Symposium on African Vertebrates: Systematics, Phylogeny and Evolutionary Ecology.* Bonn: Zoologisches Forschungsinstitut und Museum Alexander Koenig.

———. 1990. The herpetofauna of the Central African Republic, with description of a new species of *Rhinotyphlops* (Serpentes: Typhlopidae). Pp. 85–102 *in* G. Peters and R. Hutterer (eds.), *Vertebrates in the Tropics*. Bonn: Museum Alexander Koenig.

Kempff Mercado, N. 1975. *Ofidios de Bolivia.* La Paz: Academia Nacional de Ciencias de Bolivia.

King, F. W., and R. L. Burke (eds.). 1989. *Crocodilian, Tuatara, and Turtle Species of the World: A Taxonomic and Geographic Reference*. Washington, D.C.: Association of Systematics Collections.

King, M. 1987a. Monophyleticism and polyphyleticism in the Gekkonidae: A chromosomal perspective. Aust. J. Zool. 35:641–654.

———. 1987b. Origin of the Gekkonidae: Chromosomal and albumin evolution suggests Gondwanaland. Search 18:252–254.

Kitching, J. 1980. [no title]. Soc. Vert. Paleontol. News Bull. 120:37.

Klaver, C., and W. Böhme. 1986. Phylogeny and classification of the Chamaeleonidae (Sauria) with special reference to hemipenis morphology. Bonn. Zool. Monogr. 22:1–64.

Kluge, A. G. 1964. A revision of the South American gekkonid lizard genus *Homonota* Gray. Amer. Mus. Novit. 2193:1–41.

———. 1967. Higher taxonomic categories of gekkonid lizards and their evolution. Bull. Amer. Mus. Nat. Hist. 135:1–59.

———. 1969. The evolution and geographical origin of the New World *Hemidactylus mabouia-brooki* complex (Gekkonidae, Sauria). Misc. Publ. Mus. Zool., Univ. Michigan 138:1–78.

———. 1983. Cladistic relationships among gekkonid lizards. Copeia 1983:465–475.

———. 1987. Cladistic relationships in the Gekkonoidea (Squamata, Sauria). Misc. Publ. Mus. Zool., Univ. Michigan 173:i–iv + 1–54.

———. 1988. Parsimony in vicariance biogeography: A quantitative method and a Great Antillean example. Syst. Zool. 37:315–328.

———. 1991. Boine snake phylogeny and research cycles. Misc. Pub. Mus. Zool., Univ. Michigan 178:1–58.

Lamotte, M., and H. Solano. 1989. Amphibiens anoures de la savane de Calabozo (Venezuela) et des savanes de l'ouest Africain: Convergences et differences. Bull. Soc. Zool. France 114:27–46.

Lancini, A. R. 1970. *Los ofidios*. Caracas: Oficina Central de Information.

———. 1979. *Serpientes de Venezuela.* Caracas: Gráficas Armitano.

Lang, M. 1991. Generic relationships within Cordyliformes (Reptilia: Squamata). Bull. Inst. Roy. Sci. Nat. Belg., Biol. 61:1–68.

Langston, W., Jr. 1965. Fossil crocodilians from Colombia and the Cenozoic history of the Crocodilia in South America. Univ. Calif. Publ. Geol. Sci. 52:1–152.

Lanza, B. 1983. A list of the Somali amphibians and reptiles. Monit. Zool. Italiano, n.s., 18 (Suppl.):193–247.

Laurent, R. F. 1973. A parallel survey of equatorial amphibians and reptiles in Africa and South America. Pp. 259–266 *in* B. J. Meggers, E. S. Ayensu, and W. D. Duckworth (eds.), *Tropical Forest Ecosystems in Africa and South America: A Comparative Review.* Washington, D.C.: Smithsonian Institution Press.

———. 1979. Herpetofaunal relationships between Africa and South America. Pp. 55–71 *in* W. E. Duellman (ed.), *The South American Herpetofauna: Its Origin, Evolution and Dispersal*. Mus. Nat. Hist. Univ. Kansas

Monogr. 7. Lawrence: University of Kansas Museum of Natural History.

Le Berre, M. 1989. *Faune du Sahara,* Vol. 1: *Poissons, Amphibiens, Reptiles.* Paris: R. Chabaud-Lechevalier.

Legler, J. M. 1990. The genus *Pseudemys* in Mesoamerica: Taxonomy, distribution, and origins. Pp. 82–105 *in* J. W. Gibbons (ed.), *Life History and Ecology of the Slider Turtle.* Washington, D.C.: Smithsonian Institution Press.

McCarthy, C. J. 1985. Monophyly of elapid snakes (Serpentes: Elapidae): An assessment of the evidence. Zool. J. Linn. Soc. 83:79–93.

McDowell, S. B. 1964. Partition of the genus *Clemmys* and related problems in the taxonomy of aquatic Testudinidae. Proc. Zool. Soc. London 143:239–279.

———. 1986. The architecture of the corner of the mouth of colubroid snakes. J. Herpetol. 20:353–407.

———. 1987. Snake systematics. Pp. 3–50 *in* R. A. Seigel, J. T. Collins, and S. S. Novak (eds.), *Snakes: Ecology and Evolutionary Biology.* New York: Macmillan.

Marx, H. 1968. Checklist of the reptiles and amphibians of Egypt. Spec. Publ. U.S. Naval Med. Res. Unit No. 3, Cairo.

Medem, F. 1981. *Los Crocodylia de Sur America,* Vol. 1: *Los Crocodylia de Colombia.* Bogotá: Colciencias.

———. 1983. *Los Crocodylia de Sur America,* Vol. 2: *Venezuela-Trinidad-Tobago-Guyana-Suriname-Guyana Francesa-Ecuador-Perú-Bolivia-Brasil-Paraguay-Argentina-Uruguay.* Bogotá: Colciencias.

Menzies, J. I. 1966. The snakes of Sierra Leone. Copeia 1966:169–179.

Meylan, P. A. 1987. The phylogenetic relationships of soft-shelled turtles (Family Trionychidae). Bull. Amer. Mus. Nat. Hist. 186:1–101.

Meylan, P. A., and W. Auffenberg. 1986. New land tortoises (Testudines: Testudinidae) from the Miocene of Africa. Zool. J. Linn. Soc. 86:279–307.

Mittermeier, R. A. 1972. Zoogeography of fossil and living turtles. Aust. Nat. Hist. 17:265–269.

Miyata, K. 1982. A checklist of the amphibians and reptiles of Ecuador with a bibliography of Ecuadorian herpetology. Smithsonian Herpetol. Inform. Serv. 54:1–70.

Moody, S. M. 1980. Phylogenetic and historical biogeographical relationships of the genera in the Family Agamidae (Reptilia: Lacertilia). Ph.D. diss., University of Michigan, Ann Arbor.

Moonen, J., W. Eriks, and K. van Deursen. 1979. *Surinaamse Slangen in Kleur.* Paramaribo: C. Kersten.

Mouton, P. le F. N., and B. W. Oelofsen. 1988. A model explaining patterns of geographic character variation in *Cordylus cordylus* (Reptilia; Cordylidae) in the south-western Cape, South Africa. S. Afr. J. Zool. 23:20–31.

Murphy, R. 1983. The reptiles: origins and evolution. Pp. 130–158 *in* T. J. Case and M. L. Cody (eds.), *Island Biogeography in the Sea of Cortez.* Berkeley: University of California Press.

Norell, M. A. 1989. The higher level relationships of the extant Crocodylia. J. Herpetol. 23:325–335.

Nussbaum, R. A. 1984. Snakes of the Seychelles. Pp. 361–377 *in* D. R. Stoddart (ed.), *Biogeography and Ecology of the Seychelles Islands.* The Hague: W. Junk.

Owadally, A. W. 1988. Herpetology in Mauritius: A history of extinction, future hope for conservation. Bull. Brit. Herp. Soc. 23:11–20.

Papenfuss, T. J. 1969. Preliminary analysis of the reptiles of arid central west Africa. Wasmann J. Biol. 27:249–325.

Patterson, C. 1981. Methods of paleobiogeography. Pp. 446–489 *in* G. J. Nelson and D. E. Rosen (eds.), *Vicariance Biogeography: A Critique.* New York: Columbia University Press.

Pearson, O. P., and C. P. Ralph. 1978. The diversity and abundance of vertebrates along an altitudinal gradient in Peru. Mem. Mus. Hist. Nat. "Javier Prado" 18:1–97.

Pérez-Santos, C., and A. G. Moreno. 1988. *Ofi-*

dios de Colombia. Mus. Reg. Sci. Nat. Torino Monogr. 6. Torino: Museo Regionale di Scienze Naturali.

———. 1989. Addenda y corrigenda al libro "Ofidios de Colombia." Boll. Mus. Reg. Sci. Nat. Torino 7:1–17.

Peters, J. A., and R. Donoso-Barros. 1986. *Catalogue of the Neotropical Squamata,* Part 1: *Snakes.* Washington, D.C.: Smithsonian Institution Press.

Peters, J. A., R. Donoso-Barros, and B. Orejas-Míranda. 1986. *Catalogue of the Neotropical Squamata,* Part 2: *Lizards and Amphisbaenians.* Washington, D.C.: Smithsonian Institution Press.

Pianka, E. R. 1971. Lizard species density in the Kalahari Desert. Ecology 52:1024–1029.

———. 1986. *Ecology and Natural History of Desert Lizards.* Princeton: Princeton University Press.

Pianka, E. R., and R. B. Huey. 1971. Comparative ecology, niche segregation, and resource utilization among gekkonid lizards in the southern Kalahari. Copeia 1971:691–701.

Pitman, C. R. S. 1974. *A Guide to the Snakes of Uganda,* rev. ed. Codicote: Wheldon & Wesley.

Poynton, J. C., and D. G. Broadley. 1978. The herpetofauna. Pp. 925–948 *in* M. J. A. Werger (ed.), *Biogeography and Ecology of Southern Africa.* The Hague: W. Junk.

Presch, W. 1983. The family-subfamily taxon in lizards: Effects of plate tectonics on biogeographic patterns. Pp. 191–198 *in* N. I. Platnick and V. A. Funk (eds.), *Advances in Cladistics,* Vol. 2. New York: Columbia University Press.

———. 1988. Phylogenetic relationships of the Scincomorpha. Pp. 471–492 *in* R. Estes and G. Pregill (eds.), *Phylogenetic Relationships of the Lizard Families: Essays Commemorating Charles L. Camp.* Stanford, Calif.: Stanford University Press.

Pritchard, P. C. H. 1985. Evolution and zoogeography of South American turtles. Studia Palaeocheloniologica 1:225–233.

Pritchard, P. C. H., and P. Trebbau. 1984. *The Turtles of Venezuela.* Soc. Stud. Amphib. Rept., Contrib. Herpetol. 2. Oxford, Ohio: Society for the Study of Amphibians and Reptiles.

Rage, J.-C. 1974. Les serpents des phosphorites du Quercy. Palaeovertebrata 6:274–303.

———. 1987. Fossil history. Pp. 51–76 *in* R. A. Seigel, J. T. Collins, and S. S. Novak (eds.), *Snakes, Ecology and Evolutionary Biology.* New York: Macmillan.

———. 1988. The oldest known colubrid snakes: The state of the art. Acta Zool. Cracov. 31:457–473.

Rand, A. S., and S. S. Humphrey. 1968. Interspecific competition in the tropical rain forest: Ecological distribution among lizards at Belém, Pará. Proc. U.S. Natl. Mus. 125:1–17.

Rasmussen, J. B. 1985. A re-evaluation of the systematics of the African rear-fanged snakes of Bogert's groups XII–XVI, including a discussion of some evolutionary trends within Caenophidia. Pp. 531–548 *in* K.-L. Schuchmann (ed.), *Proceedings of the International Symposium on African Vertebrates, Systematics, Phylogeny and Evolutionary Ecology.* Bonn: Zoologisches Forschungsinstitut und Museum Alexander Koenig.

Renous, S. 1979. Application des principes cladistiques à la phylogénèse et la biogéographie des lacertiliens. Gegenbaurs Morphol. Jb. 125:376–432.

Rieppel, O. 1979. A cladistic classification of primitive snakes based on skull structure. Z. f. Zool. Syst. Evolutionsforsch. 17:140–150.

———. 1980. The trigeminal jaw adductor musculature of *Tupinambis,* with comments on the phylogenetic relationship of the Teiidae. Zool. J. Linn. Soc. 69:1–29.

———. 1984. The cranial morphology of the fossorial lizard genus *Dibamus* with a consideration of its phylogenetic relationships. J. Zool. London 204:289–327.

———. 1988. The classification of the Squamata. Pp. 261–293 *in* M. J. Benton (ed.), *The Phylogeny and Classification of the Tetrapods,* Vol. 1: *Amphibians, Reptiles, Birds.* Oxford: Clarendon.

Rivero-Blanco, C., and J. R. Dixon. 1979. Origin and dispersal of the herpetofauna of the dry lowland regions of northern South America. Pp. 281–298 *in* W. E. Duellman (ed.), *The South American Herpetofauna: Its Origin, Evolution and Dispersal*. Mus. Nat. Hist. Univ. Kansas Monogr. 7. Lawrence: University of Kansas Museum of Natural History.

Rodrigues, M. T. 1987. Sistemática, ecologia e zoogeografia dos *Tropidurus* do grupo *torquatus* ao sul do Rio Amazonas (Sauria, Iguanidae). Arq. Zool. São Paulo 31:105–230.

Roman, B. 1980. *Serpents de Haute-Volta*. Ouagadougou: C.N.R.S.T.

Rosselot, B. 1977. Les serpentes dangereux du Burundi. Issoudun, France: République Française Ministère de la Coopération.

Roux-Estève, R. 1974. Révision systématique des Typhlopidae d'Afrique Reptilia-Serpentes. Mém. Mus. Natn. Hist. Nat., n.s. (A) Zool. 87:1–313.

———. 1975. Recherches sur la morphologie, la biogéographie et la phylogenie des Typhlopidae d'Afrique. Bull. Inst. Fond. Afr. Noire 36:428–508.

Russell, A. P. 1972. The foot of gekkonid lizards: A study in comparative and functional anatomy. Ph.D. diss., University of London.

Russell, A. P., and A. M. Bauer. 1989. The morphology of the digits of the golden gecko, *Calodactylodes aureus*, and its implications for the occupation of rupicolous habitats. Amphib.-Rept. 10:125–140.

Sage, R. D. 1974. The structure of lizard faunas: Comparative biologies of lizards in two Argentina deserts. Ph.D. diss., University of Texas.

Savitzky, A. H. 1978. The origin of the New World proteroglyphous snakes and its bearing on the study of venom delivery systems in snakes. Ph.D. diss., University of Kansas.

Schätti, B. 1986. Herpetologische Ausbeute einer Sammelreise nach Mali (Amphibia, Reptilia). Rev. Suisse Zool. 93:765–778.

———. 1989. Amphibien und Reptilien aus der Arabischen Republik Jemen und Djibouti. Rev. Suisse Zool. 96:905–937.

Schleich, H.-H. 1987. Herpetofauna Caboverdiana. Spixiana Suppl. 12:1–75.

Schwaner, T. D., and H. C. Dessauer. 1982. Comparative immunodiffusion survey of snake transferrins focused on the relationships of the natricines. Copeia 1982:541–549.

Schwartz, A. 1968. Geographic variation in the New World gekkonid lizard *Tarentola americana* Gray. Proc. Biol. Soc. Wash. 81:123–142.

Seidel, M. E., and D. R. Jackson. 1990. Evolution and fossil relationships of slider turtles. Pp. 68–73 *in* J. W. Gibbons (ed.), *Life History and Ecology of the Slider Turtle*. Washington, D.C.: Smithsonian Institution Press.

Sill, W. D. 1968. The zoogeography of the Crocodylia. Copeia 1968:76–88.

Simbotwe, M. P. 1984. Resource partitioning in a woodland reptile community of the Kafue Flats, Zambia. Afr. J. Ecol. 22:281–288.

Simbotwe, M. P., and S. D. Garber. 1979. Feeding habits of lizards in the genera *Mabuya, Agama, Ichnotropis* and *Lygodactylus* in Zambia, Africa. Trans. Kansas Acad. Sci. 85:55–59.

Simpson, G. G. 1942. A Miocene tortoise from Patagonia. Amer. Mus. Novit. 1209:1–6.

Smith, H. M., R. L. Martin, and T. A. Swain. 1977. A new genus and two new species of South American geckos (Reptilia: Lacertilia). Pap. Avul. Zool. São Paulo 30:195–213.

Smith, K. K., and K. A. Mackay. 1990. The morphology of the intrinsic tongue musculature in snakes (Reptilia, Ophidia): Functional and phylogenetic implications. J. Morph. 205:307–324.

Spawls, S. 1978. A checklist of the snakes of Kenya. J. E. Afr. Nat. Hist. Soc. Natl. Mus. 31(167):1–18.

Springer, V. G. 1981. Comments on Solem's land-snail biogeography, with an hypothesized explanation of the distribution of the Endodontidae. Pp. 225–230 *in* G. J. Nelson and D. E. Rosen (eds.), *Vicariance Biogeography: A Critique*. New York: Columbia University Press.

Stubbs, D. (comp.). 1989. *Tortoises and Fresh-*

water Turtles, an Action Plan for Their Conservation. Gland, Switzerland: IUCN/SSC Tortoise and Freshwater Turtle Specialist Group.

Stucki-Stirn, M. C. 1979. *Snake Report 721*. Teuffenthal, Switzerland: Herpeto-Verlag.

Swingland, I. R., and M. W. Klemens (eds.). 1989. The Conservation Biology of Tortoises. Occas. Pap. IUCN Species Survival Commission 5:i–iv + 1–204.

Taplin, L. E., and G. C. Grigg. 1989. Historical zoogeography of the eusuchian crocodilians: A physiological perspective. Amer. Zool. 29:885–901.

Tarsitano, S. F., E. Frey, and J. Riess. 1989. The evolution of the Crocodilia: A conflict between morphological and biochemical data. Amer. Zool. 29:843–856.

Tchernov, E. 1976. Crocodylidae from Pliocene/Pleistocene formations of the Rudolf basin. Pp. 370–378 *in* Y. Coppens, F. C. Howell, G. L. Isaac, and R. E. F. Leakey (eds.), *Earliest Man and Environments in the Lake Rudolf Basin*. Chicago: University of Chicago Press.

———. 1986. Evolution of the crocodiles in east and north Africa. Cah. Paléontol., Trav. Paléontol. Est-Afr. Éditions du CNRS, Paris.

Thomas, R. 1976. Systematics of Antillean snakes of the genus *Typhlops* (Serpentes: Typhlopidae). Ph.D. diss., Louisiana State University.

Trape, J.-F. 1981. Liste des serpents de la Republique Populaire du Congo Paper, Office Rech. Sci. Techn. Outre-Mer, Centre de Brazzaville.

Travassos, H. 1948. Nota sobre a "*Mabuya*" da Ilha Fernando de Noronha (Squamata, Scincidae). Rev. Brasil. Biol. 8:201–208.

Underwood, G. 1967. *A Contribution to the Classification of Snakes*. London: British Museum (Natural History).

———. 1976. A systematic analysis of boid snakes. Pp. 151–175 *in* A. d'A. Bellairs and C. B. Cox (eds.), *Morphology and Biology of Reptiles*. London: Academic Press.

Underwood, G., and A. F. Stimson. 1990. A classification of pythons (Serpentes, Pythoninae). J. Zool. London 221:565–603.

Vanzolini, P. E. 1951. A systematic arrangement of the family Amphisbaenidae (Sauria). Herpetologica 7:113–123.

———. 1968. Geography of the South American Gekkonidae (Sauria). Arq. Zool. São Paulo 17:85–112.

———. 1978. On South American *Hemidactylus* (Sauria, Gekkonidae). Pap. Avulsos Zool. 31:307–343.

———. 1986a. Addenda and corrigenda to Part I: Snakes. 26 pp. *in* J. A. Peters and R. Donoso-Barros, *Catalogue of the Neotropical Squamata,* Part 1: *Snakes*. Washington, D.C.: Smithsonian Institution Press.

———. 1986b. Addenda and corrigenda to Part 2: Lizards and amphisbaenians. 25 pp. *in* J. A. Peters and B. Orejas-Miranda, *Catalogue of the Neotropical Squamata,* Part 2: Lizards and Amphisbaenians. Washington, D.C.: Smithsonian Institution Press.

———. 1988. Distributional patterns of South American lizards. Pp. 317–342 *in* P. E. Vanzolini and W. R. Heyer. *Proceedings of a Workshop on Neotropical Distribution Patterns*. Rio de Janeiro: Academia Brasileira de Ciências.

Vanzolini, P. E., and W. R. Heyer. 1985. The American herpetofauna and the interchange. Pp. 475–487 *in* F. G. Stehli and S. D. Webb (eds.), *The Great American Biotic Interchange*. New York: Plenum.

Vanzolini, P. E., and E. E. Williams. 1970. South American anoles: The geographic differentiation and evolution of the *Anolis chrysolepis* species group (Sauria, Iguanidae). Arq. Zool. São Paulo 19:1–124.

Vanzolini, P. E., A. M. A. Ramos-Costa, and L. J. Vitt. 1980. *Repteis das Caatingas*. Rio de Janeiro: Academia Brasileira Ciències.

Villiers, A. 1975. *Les serpents de l'Ouest Africain,* 3d ed. Dakar: Université de Dakar et Institut Fondamental d'Afrique Noire.

Vinson, J., and J.-M. Vinson. 1969. The saurian fauna of the Mascarene Islands. Mauritius Inst. Bull. 6:203–320.

Welch, K. R. G. 1982. *Herpetology of Africa: A Checklist and Bibliography of the Orders Amphisbaenia, Sauria and Serpentes*. Malabar, Fla.: Robert F. Krieger.

Williams, E. E. 1950. Variation and selection in the cervical central articulations of living turtles. Bull. Amer. Mus. Nat. Hist. 94:509–561.

———. 1988. A new look at the Iguania. Pp. 429–488 *in* P. E. Vanzolini and W. R. Heyer (eds.), *Proceedings of a Workshop on Neotropical Distribution Patterns*. Rio de Janeiro: Academia Brasileira de Ciências.

Wood, R. C. 1970. A review of the fossil Pelomedusidae (Testudines, Pleurodira) of Asia. Breviora 357:1–24.

———. 1979. Studies of the systematics, evolution, and ecology of pleurodiran turtles in South America. Natl. Geogr. Soc. Res. Rept., 1970 Projects:591–599.

Wood, R. C., and M. A. Freiberg. 1977. Redescription of *Notoemys laticentralis,* the oldest fossil turtle from South America. Acta Geol. Lilloana 13:187–204.

Wood, R. C., and B. Patterson. 1973. A fossil trionychid turtle from South America. Breviora 415:1–10.

Chapter 11

Avian Biological Relationships between Africa and South America

François Vuilleumier and Allison V. Andors

I. Introduction

The avifaunas of Africa and South America appear distinct on first inspection. In African rainforests, one observes bulbuls (Pycnonotidae), turacos (Musophagidae), and Old World flycatchers (Muscicapidae), whereas in South America common birds are antbirds (Formicariidae), woodcreepers (Dendrocolaptidae), and toucans (Ramphastidae). In savannas, similarly, African taxa are unrelated to South American ones. But hawks (*Buteo*), lapwings (*Vanellus*), nightjars (*Caprimulgus*), pipits (*Anthus*), and thrushes (*Turdus*) occur on both continents.

The dual aspect of dissimilarity and resemblance is what led avian zoogeographers to delimit faunal regions in the first place. Although Sclater (1858) long ago carried out intercontinental comparisons, detailed African–South American studies became popular only later. For example, Keast (1972a) identified the ecomorphological equivalents of South American flycatchers (Tyrannidae) in unrelated African families,

We thank Peter Goldblatt and his colleagues for their invitation to attend the 37th Annual Systematics Symposium at the Missouri Botanical Garden, and for support that made our attendance possible. H. M. F. Alvarenga, W. J. Bock, L. M. Chiappe, G. S. Keith, and S. L. Olson offered constructive suggestions or allowed us access to information that expedited our research considerably. J. M. Williams kindly assisted in typing the manuscript. We acknowledge the intellectual inspiration provided by the writings of the late J. P. Chapin and R. E. Moreau, and of B. P. Hall and A. Keast for Recent birds, and of the late P. Brodkorb, J. Cracraft, S. L. Olson, and P. V. Rich for fossils. Financial support was provided by the Department of Ornithology of the American Museum of Natural History.

 ISBN 0-300-05375-4

Amadon (1973) compared the taxonomic composition of Amazon and Congo (Zaire) rainforest birds, and Houston (1985) contrasted the behavior of Afrotropical and Neotropical vultures. Some workers tried more synthetic comparisons of Recent birds (Keast 1972b, 1988) or of fossils (Cracraft 1973a; Rich 1979).

In this report we review Recent and fossil birds of Africa and South America. To the best of our knowledge, this chapter is the first to compare the fossil birds of the two continents since Cracraft (1973a) and Rich (1979). Throughout the chapter a dagger (†) indicates extinct avian taxa.

II. The Recent African and South American Avifaunas

An African ornithologist visiting South America will notice that all waterbird families and many other nonpasserine families are shared by the two continents. Whereas many passerine birds (numbers of species and of individuals) in South American forests are suboscines, in other vegetation types one finds respectable numbers of both species and individuals of oscines. Wherever the South American passerine avifauna is dominated by Tyrannidae or Furnariidae, however, the nonpasserines will be familiar to the African ornithologist, but the passerines will not.

A visitor to Africa who is knowledgeable about South American but not Eurasian birds will be overwhelmed by the novelty of larks, chats, warblers, weavers, or starlings. The same visitor will note the abundance of birds in open habitats and the apparent poverty of birds in rainforests. In open habitats hawks, doves, and finches will make the South American visitor feel at home, but forest birds will be largely unfamiliar.

The South American avifauna has about the same number of families, but almost twice as many species and more than twice as many genera as the African one (table 11.1). Although the proportion of nonpasserines to passerines is nearly equal in both continents, there are twice as many nonpasserine species, and almost twice as many passerine species in South America. Of the passerines more are suboscines than oscines, whereas only 6 suboscine species occur in Africa (Chapin 1932:370–371): *Pitta* (2 species, Pittidae), *Pseudocalyptomena graueri,* and *Smithornis* (3 species, Eurylaimidae).

A. Taxonomic Comparisons

1. Taxa Unique to Africa Hall (1972:589) listed 11 families or subfamilies, and Cracraft (1976) gave 8 families, as endemic to Africa: Struthionidae (1 genus and species; not listed by Hall 1972), Balaenicipitidae (1 genus and species), Scopidae (1 genus and species), Sagittariidae (1 genus and species), Musophagidae (22 species in 5 genera; 18 species according to Moreau 1966 and Hall 1972), Coliidae (6 species in 1 genus), Phoeniculidae (7 species in 1 genus), and Promeropidae (2 species in 1 genus).

Cracraft (1976) included several subfamilies as endemic: Buphaginae (Sturnidae, 2 species in 1 genus), Malaconotinae (Laniidae, about 39 species in 6 genera; not included by Hall 1972), Prionopinae (Laniidae, 9 species in 1 genus; also included by Hall 1972 at the subfamily rank), Picathartinae (Muscicapinae, 2 species in 1 genus; included as a family by Hall 1972), Bubalornithinae (Ploceidae, 2 species in 2 genera; not included by Hall 1972), and Viduinae

Table 11.1. Aspects of the African and South American avifaunas.

Component of Avifauna	Africa	South America
Number of families	75	79
Number of endemic families	10	10
Number of genera	359	835
Number of species	1498	2671
Number of nonpasserine species (% of total species)	575 (38)	1065 (40)
Number of suboscine passerine species (% of total species)	6 (0.4)	963 (36)
Number of oscine passerine species (% of total species)	917 (61)	643 (24)

Sources: Dorst 1976; Cracraft 1976; Keast 1972b; Moreau 1966; Meyer de Schauensee 1982.

(Estrildidae, 9 species in 3 genera; not included by Hall 1972). Both Hall (1972) and Fry (1983) included the Numididae (7–8 species in 3–4 genera) in the endemic families, but neither included the ostrich (Struthionidae).

Cracraft (1976:44) suggested that several of the endemic African families (Struthionidae, Sagittariidae, Musophagidae, Coliidae, and Phoeniculidae) are relict. Mourer-Chauviré (1982) showed that the Sagittariidae and Coliidae were once much more widespread in the Old World and are now relictual in Africa. To this list can be added the Numididae (Numidinae), if the European (Mlíkovský 1989) and Asian (Olson 1974) Eocene occurrences of the problematic genus †*Telecrex* have been correctly allocated to this family or subfamily.

2. *Taxa Unique to South America* Olrog (1969) listed 10 families as "restricted to South America": Rheidae (2–3 species in 2 genera), Anhimidae (3 species in 2 genera), Opisthocomidae (1 genus and species), Psophiidae (3 species in 1 genus), Thinocoridae (4 species in 2 genera), Cariamidae (2 species in 2 genera), Steatornithidae (1 genus and species), Conopophagidae (8–9 species in 1–2 genera), Phytotomidae (3 species in 1 genus), and Catamblyrhynchidae (1 genus and species). To these 10 families, Cracraft (1976) added the Rhinocryptidae (about 30 species in 12 genera), which also occur in Central America and as fossils in the West Indies (Olson and Kurochkin 1987), but which have their major radiation in South America.

As stated by Olrog (1969:849), "These [10] exclusively South American families include only one or two genera each with few species, and are in no instance dominant families, but ought to be regarded as relicts of ancient faunas." Three of these 10 families (Rheidae, Opisthocomidae, Cariamidae) have prolonged fossil records in South America, and 2 (Cariamidae *sensu lato,* Mourer-Chauviré 1983; Steatornithidae, Olson 1987) have an extra–South American record. If the Rhinocryptidae are added to this list, then they, too, are relictual (Feduccia and Olson 1982). Thus the relictual nature of the few South American endemic or near endemic families parallels the situation of the few African endemic families or subfamilies.

3. Taxa Common to Africa and South America Amadon (1973) gave a list of taxa common to the rainforests of the Congo and Amazon basins, Fry (1983) of families common to savannas of Africa and South America, and Dorst and Vuilleumier (1986) of taxa found in the alpine zones of the two continents. No author seems to have given a list of *all* shared taxa.

About 40 avian families are shared between Africa and South America: Podicipedidae, Pelecanidae, Phalacrocoracidae, Anhingidae, Heliornithidae, Ardeidae, Ciconiidae, Threskiornithidae, Phoenicopteridae, Anatidae, Accipitridae, Falconidae, Phasianidae, Rallidae, Jacanidae, Rostratulidae, Recurvirostridae, Charadriidae, Scolopacidae, Laridae, Rhynchopidae, Columbidae, Psittacidae, Cuculidae, Strigidae, Tytonidae, Caprimulgidae, Apodidae, Trogonidae, Alcedinidae, Capitonidae, Picidae, Alaudidae, Hirundinidae, Corvidae, Motacillidae, Turdidae, Sylviidae, Fringillidae, and Emberizidae.

Africa has 75 families and South America has 79 (Dorst 1976), thus 53 percent of African and 51 percent of South American families are shared. Table 11.2 lists 10 of the 40 shared families and gives numbers of species for the two continents. Most families have more species in South America than in Africa, but in 4 families (Alaudidae, Hirundinidae, Turdidae, and Motacillidae), the reverse is true. In one case (Hirundinidae) South America has an impressive radiation, but for the other families the discrepancies are spectacular. Thus, Africa has many species of larks, whereas only one species occurs in South America. Africa has a great diversity of Turdidae, whereas the South American thrush-like birds are much less diverse. Similarly, the Motacillidae are diverse in Africa but are represented only by *Anthus* in South America.

About 58 genera are shared between Africa and South America (Moreau 1966:112 mentioned only 38): *Podiceps, Pelecanus, Phalacrocorax, Anhinga, Botaurus, Ixobrychus, Nycticorax, Bubulcus, Butorides, Egretta, Casmerodius, Ardea, Ciconia, Plegadis, Phoenicopterus, Elanus, Circus, Accipiter, Buteo, Spizaetus, Falco, Pandion, Dendrocygna, Sarkidiornis, Anas, Netta, Oxyura, Rallus, Porzana, Porphyrio, Gallinula, Fulica, Haematopus, Himantopus, Recurvirostra, Burhinus, Vanellus, Charadrius, Gallinago, Larus, Sterna, Rhynchops, Columba, Tyto, Otus, Lophostrix, Athene, Bubo, Strix, Glaucidium, Asio, Caprimulgus, Chaetura, Megaceryle, Picoides, Hirundo, Anthus,* and *Turdus*. Africa has a total of 359 genera, and South America 835 (Dorst 1976), hence 16 percent of African and 7 percent of South American genera are shared.

The 19 species shared between Africa and South America (table 11.3; Winterbottom 1965; Moreau 1966; Olrog 1969) represent 1.3 percent of the 1498 species in Africa and 0.7 percent of the 2671 species in South America. These are low percentages.

Some of these shared species have dispersed from one continent to the other very recently. Proof is given by the spread of the Cattle Egret *Bubulcus ibis* from Eurafrica

Table 11.2. Species numbers in 10 families of birds shared by Africa and South America.

Family	Number of Species	
	Africa	South America
Nonpasserines		
Anatidae (ducks, geese)	18	36
Accipitridae (hawks)	47	53
Columbidae (pigeons, doves)	28	47
Psittacidae (parrots)	19	111
Picidae (woodpeckers)	29	82
Passerines (nonoscines)		
Alaudidae (larks)	50	1
Hirundinidae (swallows)	31	20
Corvidae (crows, jays)	9	14
Turdidae (chats, thrushes)	94	32
Motacillidae (pipits)	24	7

Sources: Dorst 1976; Moreau 1966; Meyer de Schauensee 1982.

to South America, and from there eventually to most parts of the world (Handtke and Mauersberger 1977). The time span of this series of dispersal events is less than a century. Moreau (1966) supposed that *Casmerodius albus, Dendrocygna bicolor, D. arcuata, Netta erythrophthalma,* and *Sarkidiornis melanotos* made direct crossings but did not suggest the direction of these events.

Of the 15 species discussed by Winterbottom (1965:209) "all except the three ducks [*Dendrocygna bicolor, D. viduata,* and *Netta erythrophthalma*] and the Cattle Egret *Ardeola* [= *Bubulcus ibis*] made their way independently into the two continents from the Holarctic." Because the distribution of *Dendrocygna viduata* in Africa goes "at least as far west as the Gambia," Winterbottom (1965:210) believed it "plausible to conjecture that it spread from Africa to South America by the same route as the Cattle Egret." Winterbottom (1965:210) also believed *Dendrocygna bicolor* and *Larus cirrocephalus* to have spread from Africa to South America because "the [tropical] trade winds, with their easterly component, would help an east-west passage but hinder a flight in the opposite direction." For *Netta erythrophthalma,* by contrast, Winterbottom (1965) suggested that the dispersal was from South America to Africa. Winterbottom (1965) claimed that because *N. erythrophthalma* is of more southerly distribution than the two mentioned above, west to east dispersal would have been assisted by westerly winds south of the tropics. However, the South American range of *N. erythrophthalma* is tropical (Meyer de Schauensee 1982), so an east to west dispersal is just as likely.

Table 11.3. Bird species shared by Africa and South America.

Species (family)	W	M	O
Podiceps nigricollis (Podicipedidae)	–	–	–
Casmerodius albus (Ardeidae)	+	+	+
Butorides striatus (Ardeidae)	+	–	–
Bubulcus ibis (Ardeidae)	+	+	+
Nycticorax nycticorax (Ardeidae)	+	–	+
Plegadis falcinellus (Threskiornithidae)	+	–	–
Phoenicopterus ruber (Phoenicopteridae)	+	–	–
Dendrocygna bicolor (Anatidae)	+	+	+
Dendrocygna viduata (Anatidae)	+	+	+
Netta erythrophthalma (Anatidae)	+	+	+
Sarkidiornis melanotos (Anatidae)	–	+	+
Pandion haliaetus (Pandionidae)	+	+	–
Falco peregrinus (Falconidae)	+	+	+
Gallinula chloropus (Rallidae)	+	+	+
Haematopus ostralegus (Haematopodidae)	–	–	+
Gallinago gallinago (Scolopacidae)	–	–	+
Himantopus himantopus (Recurvirostridae)	+	–	–
Tyto alba (Tytonidae)	+	+	+
Larus cirrocephalus (Laridae)	+	–	+

Note: W = Winterbottom 1965; M = Moreau 1966; O = Olrog 1969. A plus (+) means presence, a minus (–) means absence from the work indicated.

B. Examples of Taxa Common to Africa and South America

1. Ratites The large, flightless ratites (ostrich, rheas, emu, cassowaries, kiwis) occur disjunctly in Africa, South America, Australia, New Guinea, and New Zealand, respectively. Two groups are of interest here, the ostrich (Struthionidae) and rheas (Rheidae). The ostrich is now widespread in sub-Saharan Africa but lived until recently in North Africa and the Middle East. The rheas occur in western, central, and southern South America.

The ostrich and the rheas are usually considered to be related, but Bock and Bühler (1990:34) were "unable to accept the monophyly of the Ratitae or that the Rheidae and the Struthionidae are sister groups." If the monophyly of these two families is in doubt, then the two groups may not indicate biogeographic relationships between Africa and South America. The fossil evidence for ostrich and rhea origins and dispersal is summarized later.

2. Ardeidae Many herons (Ardeidae) are shared between Africa and South America. We discuss here the tiger herons. "Each area [Congo and Amazon] has a species of tiger-heron, probably closely related though placed in separate genera" (Amadon 1973:271). Earlier Chapin (1932:377) noted that the African *Tigriornis,* "though resembling *Tigrisoma* of South America, is plainly closer to *Zonerodius* of New Guinea."

Bock (1956) reviewed all taxa of tiger herons, including another South American genus, *Zebrilus.* In his phylogenetic scheme, Bock (1956) treated *Tigriornis, Zonerodius,* and *Tigrisoma* as equally interrelated and *Zebrilus* as more distant. Whether there is enough evidence to resolve the trichotomy, and to discover whether the African and South American genera are sister taxa, we do not know.

We agree with Bock (1956:24) that "the tiger herons are close to the ancestral stock of the Ardeidae" and that their distribution "is justly called relict." The African and South American tiger herons appear to indicate an ancient relationship between these continents, but the timing of the early steps of this relationship is difficult to pinpoint at present.

3. Vanellus Bock (1958) argued that the 19 genera of lapwings should all be included in *Vanellus.* The plover-like lapwings live in steppes, marshes, and grasslands worldwide with the exception of North America. Of the 24 species recognized by Bock (1958), 3 occur in South America and 12 in Africa. The 3 South American species are Andean *resplendens,* lowland temperate *chilensis,* and lowland tropical *cayanus. V. chilensis* and *V. resplendens* are sister species. *V. cayanus* is not closely related to them. Bock (1958) suggested that the 3 South American species represented two separate invasions from Africa but was not able to identify their African relatives.

4. Tytonidae and Strigidae About equal numbers of Tytonidae and Strigidae occur on both continents: 26 species in 10 genera in Africa (Voous 1966) and 30 species in 12 genera in South America (Meyer de Schauensee 1982).

The genera *Ciccaba* and *Strix* are of interest for African–South American relationships, depending on their taxonomic treatment. *Strix* is usually considered shared by the two continents, whereas *Ciccaba* is unique to South America. However, Voous (1964:472) pointed out that Croizat "has used the distribution pattern of the *Ciccaba* owls as a basic example of his 'across the Atlantic' type of distribution." Voous (1964:476) argued that the South American species of *Strix,* a genus closely related to *Ciccaba,* "occupy a position in between a strigine and a ciccabine condition." He believed that *Strix* and *Ciccaba* owls were probably indistinguishable taxonomically. Voous added (1964:476) that "any attempt at a zoogeographic interpretation of the distribution of these owls seems premature," a conclusion that would lead "to severe repercussions on Croizat's zoogeographic discussion."

Lophostrix is another pertinent genus. One African species, *lettii,* placed formerly in the monotypic African genus *Jubula,* is included in *Lophostrix* (Chapin 1939).

Lophostrix cristata is widely distributed in South America. Erard (1989) found African *lettii* and South American *cristata* very similar ecologically. Voous (1966:504) was not convinced of the wisdom of treating *lettii* and *cristata* as members of the genus *Lophostrix* and wrote that "it is most likely that the superficial resemblance of these tropical African and American 'eared' forest owls has mainly resulted from a life under similar ecological conditions."

5. Piciformes The Piciformes comprise woodpeckers (Picidae), barbets (Capitonidae), toucans (Ramphastidae), and honeyguides (Indicatoridae). Woodpeckers and barbets occur in Africa and South America, whereas honeyguides (Africa and Asia) do not occur in South America, and toucans (Neotropical) do not occur in Africa. Short (1970, 1985) reviewed the African–South American relationships of piciforms.

Ten genera of woodpeckers are found in South America, and 6 in Africa. Only *Picoides* is shared. This sort of comparison, suggesting a low level of relationship, is misleading. The only African species of *Picoides* is related to Eurasian congeners. The 2 closely interrelated South American species of *Picoides* are related to North American congeners. The key to understanding African–South American woodpecker relationships does not lie in further study of *Picoides* but in an analysis of the genera *Campethera, Geocolaptes,* and *Dendropicos,* which constitute the major African radiation. Short (1970) has argued that these 3 genera (campetherines) are related to the American colaptines, a group including *Colaptes* (7 species), *Piculus* (9), and *Veniliornis* (12).

Three genera of barbets live in South America and 7 in Africa. None are shared. South American taxa inhabit forests, but African ones live in open woodlands and scrublands. Reflecting this ecological dichotomy, speciation in South American barbets has been correlated with forest refugia, but in African ones the correlation has been with alternating episodes of isolation and dispersal among woodlands and savannas. Unlike the situation in woodpeckers, however, Short (1985) thinks that speculations on the taxonomic relationships between African and South American barbets are premature. Sibley and Ahlquist (1985) have nevertheless interpreted the Capitonidae as paraphyletic on the basis of DNA-DNA hybridization. The biogeographic significance of this hypothesis is interpreted in view of fossil evidence in a later section.

C. Ecological Comparisons

1. Species Diversity Table 11.4 compares species diversity in Africa and South America. There are two exceptions to the main pattern of greater diversity in South America than in Africa. The first is that the South African fynbos and its South American equivalent, the Chilean matorral, have about the same numbers of species (28 and 31, respectively). The equivalence of species numbers in Mediterranean climate sites is not, however, surprising in view of Blondel and co-workers' (1984) finding that species numbers were about the same in matorral of Chile, California, and Provence (38, 31, and 31). The second exception, that tropical African savannas have 708 species versus only 521 in South America, is discussed later.

The figures of table 11.4 substantiate

Table 11.4. Species richness of birds in Africa and South America.

	Numbers of species		
	Africa	South America	Source
Total number of species	1498	2671	1
Lowland rainforest	266	946	2
Congo and Amazon rainforest	212	592	3
Lowland forest plots of 2 km^2	175	227	4
Lowland forest plots of 15 ha	112	131	5
Montane forest	120	400	6
Tropical savannas	708	521	7
Mediterranean scrub	138	128	8
Alpine grassland and moorland	53	166	9

Sources: 1 = Dorst 1976; 2 = Keast 1988; 3 = Amadon 1973; 4 = Erard 1989; 5 = Pearson 1977 (Peru in South America); 6 = Keast 1988; 7 = Fry 1983; 8 = Broekhuysen 1966 and Vuilleumier 1991; 9 = Dorst and Vuilleumier 1986.

Keast's (1988:347) conclusion that "basic differences in the continental environments underlie the prominence of savannah-adapted forms in Africa and rain forest adapted ones in South America." South America has more species of birds than Africa, but Africa has more savanna species than South America.

2. *Lowland Rainforest* Amadon (1973) estimated diversity at 212 species for the lowland rainforest of the Congo and 592 for that of the Amazon. Because this biome covers 1,500,000 km^2 in Africa and 5,000,000 km^2 in South America, it is not surprising that fewer species of birds occur in Africa. The question is whether, given a patch of lowland rainforest of equal size and ecological diversity on the two continents, there are fewer species in Africa. The answer seems to be yes.

Pearson (1977) compared lowland forest avifaunas in South America (Ecuador, Peru, Bolivia) and in Africa (Gabon). He (1977:232) "attempted to observe all birds 25 m on either side of [a circular path 2.5–3 km in length] . . . giving a total area of approximately 15 ha" and "chose the plots to be as representative of the general forest in each area as possible." Sixty-two percent of the potential species were observed in Gabon, 61 percent in Peru, and 63 percent in Ecuador, but only 53 percent in Bolivia. Because of uneven sample sizes, the appropriate African–South American comparison is between Gabon and Peru.

Pearson's (1977) table 5 gives a total of 214 species for Peru and 154 for Gabon. Eliminating the rare species (Pearson, 1977:238), the numbers become 131 species for Peru and 112 for Gabon. Pearson (1977) attributes the greater richness in South America to two kinds of factors, historical and ecological. Pearson did not discuss historical factors. The ecological ones include fruiting phenology, seasonal variation in insect availability, habitat heterogeneity, and influence of nonavian competitors (more mammalian insectivores and frugivores in Africa).

Karr (1976) mistnetted undergrowth birds in lowland rainforests of Africa (Liberia) and Panama (not in South America but geographically and faunistically close). About the same number of hours were spent in each area, and the plots were of about 2.44 ha in each area. Karr found fewer species in the African plot (101) than in the Panamanian one (117).

Karr (1976:466) found "more insectivore-frugivore species in the Panama forest (10 vs. 1)" and no carrion eaters in the African forest. He found that "Panama has more bark-gleaners (5 vs. 1), more hover-gleaners (15 vs. 7), and more salliers (3 vs. 0)" (Karr 1976:467). Karr (1976), like Pearson (1977), hypothesized historical effects to explain these differences. Relying on refuge reconstructions for South America and Africa, Karr suggested that more opportunities for speciation had existed in South America. He also invoked longer lasting human influences in Africa than in South America: "Perhaps African forests are what New World forests will become after a long history of human exploitation; that is, as forests of the Neotropics are restricted to smaller and smaller refuges, specialized genera and species will disappear leaving a subset of the original fauna of lower diversity and composed of more opportunistic species" (Karr 1976:474). Finally, again like Pearson (1977), Karr suggested that nonavian competitors influence species diversity.

The most detailed comparisons between African and South American lowland rainforest birds are those of Erard (1989), who studied avian communities on 2 km^2 plots in Gabon and French Guiana for long periods of time. More species occur in South America (277) than in Africa (175). Equal numbers of raptors occur at both sites, but many more species of ground foragers (18 vs. 8) are found in South America. South America has more nectarivores (13 species or 6 percent vs. 1 species or 1 percent) and more frugivores (66 vs. 20 species, or 29 percent vs. 11 percent), but Africa has more species feeding on arthropods (132 species or 75 percent vs. 12 species or 56 percent). Africa has more canopy foragers than South America (51 percent of African species vs. 40 percent of South American ones).

Erard (1989) gave several explanations to account for similarities and differences between the avifaunas of the two areas. One is the difference in the structure of the forests themselves, especially that "the Guiana forest studied appears to be structurally more diverse than that of North-East Gabon" (Erard 1989:116). Another involves evolutionary differences having to do with the long-term histories of these avifaunas. South American forests have more species because there has been more faunal enrichment through speciation there than in Africa.

3. *Tropical Savannas* A total of 708 species occur in African versus only 521 in South American savannas (Fry 1983), a ratio of 1.36. Why should African savannas have more species? Keast (1972b) suggested that 50 percent of sub-Saharan Africa, or

4,000,000 square miles, is covered with woodland, open forest, savanna, grassland, and low-scrub steppe and that 60 percent of South America, or 4,200,000 square miles, is covered with similar types of vegetation. Thus the areas of open vegetation types including savannas are about the same on both continents and we should expect about equal avian diversities. However, various authors differ in their definition of savanna. If we take into account only Keast's (1972b) "woodland and open forest" because this vegetation type corresponds more closely to what some authors (e.g., Moreau 1966) have called savanna, then the areas are as follows: 2,480,000 square miles for Africa versus only 1,540,000 for South America, a ratio of about 1.61 and comparable to the avian ratio. The greater area of savanna in Africa helps explain the greater number of bird species there.

Avian families with high species diversity in African savannas (Fry 1983) are: Ploceidae (61 species), Sylviidae (56), Estrildidae (36), Accipitridae (33), Alaudidae (31), Turdidae (31), Laniidae (27), Muscicapidae (26), Nectariniidae (23), Sturnidae (23), Capitonidae (22), and Phasianidae (22). These birds include arboreal and ground seedeaters, insectivores, birds with mixed diets, predators, nectarivores, and berry eaters. This diversity underlines the variety of ecological opportunities available to African savanna-dwelling birds.

In South America the most speciose families (Fry 1983) are: Tyrannidae (72 species), Fringillidae (65), Furnariidae (27), Icteridae (25), Psittacidae (25), Trochilidae (24), Thraupidae (23), and Accipitridae (21). These birds include seedeaters, insectivores, nectarivores, frugivores, predators, and mixed-diet species; ground foragers and arboreal feeders are both well represented. Ecologically, South American savanna faunas are as diverse as the African ones.

Fry (1983) also showed that 61 percent of South American savanna birds were insectivores (vs. 51 percent in Africa), but that 23 percent of African birds were seedeaters (vs. 16 percent in South America). There are more carrion feeders in African savannas (14 percent vs. 8 percent) but more nectarivores in South America (5 percent vs. 3 percent). About equal percentages of fruit eaters are found in both avifaunas (11 percent in South America vs. 10 percent in Africa).

As in rainforest avifaunas, both short-term ecological factors and long-term historical ones are at play in savanna avifaunas. More speciation may have taken place in African than in South American savannas, thus accounting for more species enrichment. But the differences in habitat heterogeneity have not been studied sufficiently to assess their role in ecological diversification.

4. Mediterranean Scrublands We discuss the avifaunas of Mediterranean-climate areas in South Africa (fynbos) and Chile (matorral). Broekhuysen (1966) recorded 138 species of birds in fynbos and divided his list into 8 categories based on frequency of occurrence: vagrants (26 species); rare (60); occasional (18); fairly frequent (14); regular (10); common (5); abundant (3); dominant (2). Eliminating the 26 vagrant and the 60 rare species, one is left with the 52 species making up the core fynbos avifauna. The 5 abundant and dominant species of the fynbos (Broekhuysen 1966) are *Onychognathus morio, Promerops cafer, Cossypha caffra, Nectarinia violacea,* and *Prinia maculosa.*

The fynbos avifauna includes widespread species (*Elanus caeruleus, Aquila verreauxi*), localized species (*Geocolaptes olivaceus, Erythropygia coryphaeus*), and endemics (*Francolinus capensis, Chaetops frenatus, Bradypterus victorini, Promerops cafer, Nectarinia violacea, Serinus leucopterus*).

Vuilleumier (1991) compiled a list of 128 matorral bird species in Chile. Assuming that the African (Broekhuysen 1966) and Chilean (Vuilleumier 1991) lists are comparable, the fynbos has 10 more species (138) than the matorral (128). The most abundant matorral species are *Diuca diuca, Anairetes parulus, Troglodytes aedon, Mimus thenca, Zonotrichia capensis, Asthenes humicola,* and *Elaenia albiceps*. The matorral avifauna (Vuilleumier 1991) includes widespread species (*Elanus leucurus, Parabuteo unicinctus*), localized species (*Cyanoliseus patagonus, Phytotoma rara*), and endemics (*Nothoprocta perdicaria, Mimus thenca, Scelorchilus albicollis*), much as in the fynbos.

Cody and Mooney (1978) divided the avifaunas of Mediterranean scrublands of Africa and South America into 9 guilds. Several guilds have equal or about equal numbers of species on both continents: foliage insectivores (5 species in Africa, 4 in South America), sallying flycatchers (2 each), nectarivores (3 vs. 2), trunk/bark foragers (1 vs. 2), aerial feeders (1 each), and crepuscular insectivores (1 each). South America has more ground foragers than Africa (12 vs. 8), and more raptors and scavengers (5 vs. 3), but Africa has more seed/fruit eaters (4 vs. 2). South America has a higher density of birds (about 16 pairs vs. about 9 pairs per ha). Africa has a lower density of foliage insectivores. Cody and Mooney (1978:287) point out an important difference between Africa and South America: "The abundant sugarbird *Promerops,* which has no New World equivalent in morphology (it is very large), introduced an element of discord [in the otherwise similar nectarivore guild organization] that must be interpreted relative to the copious nectar production of the huge *Protea* blooms, which themselves have no counterparts in the New World vegetation."

Local ecological peculiarities contribute to the intercontinental differences observed. Thus in South Africa, the *Protea* fynbos provides a suitable environment for *Promerops, Onychognathus,* and *Serinus*. In Chile, the "well-developed herb layer in its matorral . . . contributes food . . . for [its] birds and lizards" (Cody and Mooney 1978:315–316). "The effect is a greater diversity and density of Chilean ground-foraging birds . . . and a shifting down of the foraging zones of the foliage insectivores" (Cody and Mooney 1978:316).

Blondel et al. (1984:198) suggested that in the Chilean matorral "five species [out of 31] have distribution patterns that can probably be ascribed to Pleistocene speciation events." They comprise 4 ground foragers (*Nothoprocta perdicaria, Pteroptochos megapodius, Scelorchilus albicollis, Mimus thenca*) and 1 foliage insectivore (*Thripophaga* [= *Asthenes*] *humicola*). Among the South African fynbos birds, we suggest that Pleistocene effects have played a role in the distribution of a ground forager (*Francolinus capensis;* Hall 1963; Winterbottom 1968), a foliage insectivore (*Bradypterus victorini;* Winterbottom 1968), a nectarivore (*Nectarinia violacea;* Winterbottom 1968), and 3 seed/fruit eaters (*Serinus* spp.).

More complete study of the fossil avifauna from Langebaanweg (early Pliocene; Rich 1980b) is likely to shed light on the past history of the fynbos avifauna (Olson 1985a:68, discussion of fossil ibis †*Geronticus apelex*).

5. *Alpine Zone* The Alpine zone biome has a much smaller area in Africa (afroalpine, about 6,000 km^2) than in South America (páramo, puna, about 185,000 km^2) (Dorst and Vuilleumier 1986). We summarize the conclusions of an intercontinental comparison (Dorst and Vuilleumier 1986).

"The Andean páramo and puna avifauna consists of 166 species in 84 genera and 32 families . . . and the afroalpine avifauna of Ethiopia and East Africa contains 53 species in 38 genera and 23 families" (Dorst and Vuilleumier 1986:123). Shared genera are landbirds (*Buteo, Falco, Bubo, Anthus, Turdus*) and waterbirds (*Anas, Rallus, Vanellus, Gallinago*). No species is shared.

Dorst and Vuilleumier (1986) divided these faunas into 18 "trophic or foraging guilds." About equal species richness was noted in the predator guild and in the medium to large marsh insectivore/omnivore guild (6 species in each), but for other guilds the patterns differ. Thus Africa has more scavengers (5 vs. 3 species) and more aerial insectivores (6 vs. 3), but South America has more large ground seedeaters and insectivores (7 vs. 3), small ground insectivores (22 vs. 8), small bush insectivores (12 vs. 4), nectarivores/insectivores (8 vs. 4), small bush/ground seedeaters (16 vs. 5), and large swimming vegetarians (9 vs. 2). Some guilds are absent from Africa, for instance the large swimming piscivores (4 species of Podicipedidae and 1 Phalacrocoracidae in the Andes), and the small insectivores living in rushes around lakes and lagoons (3 species in the Andes). Dorst and Vuilleumier (1986:131) remarked that "the discrepancy between Africa and the Andes in water birds of [the piscivore guild] clearly reflects the lack of suitable habitat in Africa." But the lakeshore vegetation of the Ethiopian highlands is not inhabited by bird species like the Andean rush specialists *Tachuris rubrigastra, Phleocryptes melanops,* and *Agelaius thilius.*

Other than an area effect (more species in the Andes with larger surface area), reasons for the differences between Africa and South America include ecological effects, human interference (sheep grazing and long-term occupation of the puna by complex civilizations), and long-term evolutionary factors. Since ecological factors appear to be of equal significance in both areas, and human interference has been negligible in Africa, the major factors that have impinged upon diversity in these high altitude biomes are probably historical in nature. Dorst and Vuilleumier (1986) showed that species density is greater in Africa. "When the numbers of species per 1000 km^2 (a measure of relative species packing) in Africa and the Andes are compared, it can be seen that Africa has between 9 and 11 species per 1000 km^2 of high altitude biome, whereas the Andes have only between 1 and 3. In other words, Africa has between 3 and 11 times more species per unit area of afroalpine biome than the Andes have of páramo or puna biome" (Dorst and Vuilleumier 1986:136).

Dorst and Vuilleumier (1986:141) concluded that "some speciation occurred in Africa, but not very much when compared to the Andes. Indeed, there is ample evidence that speciation has been very active in the latter mountains, since at least one-third of all species are members of superspecies, most if not all of which are the result of Pleistocene speciation events."

6. Ecological Comparisons: General Remarks Table 11.5 summarizes some generalizations that emerge from comparisons of diversity at the biome level:

1. Lowland tropical forests have fewer taxa in Africa than in South America. An exception is that canopy foragers are more diverse in Africa.

2. African savannas have greater diversity, a pattern that runs contrary to the overall pattern of greater diversity in South America.

3. African fynbos has more seedeaters and fruit eaters than the South American matorral.

4. The afroalpine has more scavengers and aerial insectivores than the Andean páramo and puna and about as many predators.

5. Africa has more scavengers, about equal numbers of predators, perhaps more seedeaters, and about equal diversity of sallying flycatchers.

D. Trophic Comparisons

1. Predators Thiollay (1986) published an overview of raptorial birds in tropical lowland rainforests, based largely on his own field work. Many more species occur in South America (27–30) than in Africa (8–12). Only 2 genera, *Accipiter* (4 species in Africa, 3 in South America) and *Spizaetus* (1 species in Africa, 2 in South America), are shared.

About half these species belong to the guild of active vertebrate predators (4–7 species in Africa, vs. 8–14 in South America). These vertebrate predators fall into four size classes: small species (< 400 g), medium species (400–900 g), small eagles (900–1800 g), and large eagles (2500–5000 g). The greater diversity in South America is due to the diversity of forest Falconidae (*Micrastur*) there. Thus lowland rainforests of South America have up to 9 sympatric Falconidae, whereas the family is lacking in similar African habitats.

Thiollay discusses other raptorial guilds including slow canopy hunters, sallying foragers, wasp-nest specialists (migratory *Pernis apivorus* in Africa; resident *Daptrius americanus* in South America), fast-flying bird or bat hunters, reptile specialists (up to 4 species of *Leucopternis* in South America vs. 1 in Africa in *Dryotriorchis*), and the convergent tree crown feeders *Polyboroides* (Africa) and *Geranospiza* (South America), which forage into cavities or epiphytes and raid nests.

2. Scavengers The major scavenging birds are vultures: Accipitridae (Aegypiinae or Gypaetinae) in Africa and Cathartidae in South America. Other birds with vulturine habits have evolved independently (falconid caracaras; ciconiid marabous, *Leptotilos*).

Rich (1983) summarized the fossil history of vultures. The New World Vultures (now restricted to the Americas) occurred as fossils in the New World since the early Eocene, and in the Old World since the mid-Eocene. They disappeared from the Old World by the late Oligocene–early Miocene. The Old World vultures have a New World record going back to the early Miocene. The present-day South American Cathartidae

Table 11.5. A comparison of diversity patterns in four biomes of Africa and South America.

	Diversity	
	Africa	South America
Tropical lowland rainforest		
Species richness	lower	higher
*Canopy foragers	more	fewer
Sallying flycatchers	fewer	more
*Predators	equal	equal
Carrion feeders	none	yes
Fruit eaters	fewer	more
Ecological effects	important	important
Historical effects (speciation)	important	important
Tropical lowland savanna		
*Species richness	higher	lower
*Sallying flycatchers	equal	equal
*Predators	equal	equal
*Carrion feeders	equal	equal
*Seedeaters	more	fewer
Ecological effects	important	important
Historical effects (speciation)	important	important
Mediterranean shrubland		
*Species richness	higher	lower
*Sallying flycatchers	equal	equal
*Foliage insectivores	equal	equal
*Seedeaters and fruit eaters	more	fewer
Predators	fewer	more
Carrion feeders	fewer	more
Ecological effects	important	important
Historical effects (speciation)	important	important
Alpine habitats		
Species richness	lower	higher
Bush insectivores	fewer	more
Bush/ground seed eaters	fewer	more
*Aerial insectivores	more	fewer
*Predators	equal	equal
*Carrion eaters	more	fewer
Ecological effects	important	important
Historical effects (speciation)	minor	important

Note: An asterisk denotes deviation from the usual pattern with less diversity in Africa.

(*Vultur, Coragyps, Sarcoramphus, Cathartes*) have a fossil record in the New World only, going back to the Oligocene (Alvarenga 1985a). Today they occur in forest and nonforest habitats throughout South America. Recent Old World aegypiine vultures (*Gyps, Torgos, Aegypius, Neophron, Necrosyrtes, Gypaetus, Trigonoceps*) have a Pleistocene fossil record in the Old World only.

Houston (1985:856) stated that "despite their different ancestry, the two groups of modern vultures are extremely similar in superficial appearance. They are textbook examples of convergent evolution. Both groups have a similar diversity of species, and share many aspects of plumage pattern and a common dependence on soaring flight. In contrast, the habitats occupied by those two vulture groups are markedly different in one respect. The Old World vultures are confined to open habitats such as savannas, grasslands, and semideserts, and none of these species is found in any of the forested areas of Africa or Asia." Houston (1985) argued that the high diversity of forest-dwelling cathartids, which find their carrion prey by smell, is correlated with the high density of mammals in South American forests. By contrast, the low mammal density in African forests "has been insufficient to enable any scavenging birds to specialize on it" (Houston 1985:863).

3. Flycatchers Keast (1972a) divided the South American Tyrannidae into nine ecological categories (guilds) and identified their ecomorphological counterparts in Africa. He concluded that 14 families or subfamilies of African birds, including about 120 genera and 540 species, can be considered counterparts to the Tyrannidae. Six of these families or subfamilies (Muscicapinae, Sylviinae, Turdinae, Eurylaimidae, Paridae, Dicruridae) are more analogous to the Tyrannidae than the others. Keast (1972a:269) suggested that "in contrast to the continued evolution in isolation enjoyed by the tyrannids, and most of the other South American groups, Africa has received continued waves of colonists so that the avifauna has been built up by a series of strata superimposed one upon the other. As each new family became established there must have been a period of intense competition and a resorting and redivision of adaptive zones and niches, until stability was again reached. Since the new colonists had themselves evolved in a large land mass they were, presumably, highly efficient and competitive. The result would have been that each of the individual families of birds [in Africa] tended to become restricted, or confined, to the narrow zone where it was optimally efficient."

4. Nectarivores Characteristic nectarivores are the Trochilidae (nonpasserines) in the New World and the Nectariniidae (passerines) in the Old World. Wolf and Gill (1986) compared the adaptations of these two groups in the high Andes and East African mountains. Many more hummingbirds occur in the Andes than sunbirds in East African mountains (78 vs. 15 species). In spite of this difference, Wolf and Gill (1986) found that local communities (= assemblages of species living in a particular study site) were organized similarly. The major constraint on community organization is the temporal availability of nectar as an energy supply, which in turn depends on the phenology of flowering of the nectar-producing plants.

One difference between hummingbirds and sunbirds lies in their pair bond: sunbirds are monogamous during the breeding season, whereas hummingbirds are not.

III. The Fossil Avifaunas of Africa and South America

The avian fossil record in Africa and South America largely postdates the final separation of those continents in the Cretaceous. This generalization applies equally to the insular avifaunas of the Canary Islands (Rothe 1964; Báez 1992) and of Madagascar (Keith 1980), the oldest elements of which are, respectively, Mio-Pliocene and Quaternary. In continental South America and Africa, the oldest fossils are, respectively, an Early Cretaceous (late Aptian–Albian) feather from Brazil (Martins Neto and Kellner 1988; Chapter 14, this volume) and a Middle Eocene pseudodontorn (†Pelagornithidae), †*Gigantornis eaglesomei,* from Nigeria (Andrews 1916). Bird-like ichnofossils from the Late Triassic–Early Jurassic of southern Africa (†*Trisauropodiscus* spp., †*Masitisisauropus* spp., †*Masitisisauropezus* spp., †*Masitisisauropodiscus* spp.; Ellenberger 1970, 1972, 1974, Olsen and Galton 1984) and from the Early Jurassic of Morocco (unnamed taxon; Lockley et al. 1992: figs. 13 and 15d) antedate the opening of the Atlantic, but the avian affinities of these, discussed recently by Lockley and co-workers (1992), cannot be regarded as being well established. Putative bird tracks have also been reported from the Late Cretaceous (Maastrichtian) of Morocco (unnamed taxa; Ambroggi and Lapparent 1954) and Argentina (†*Yacoraitichnus avus* and unnamed taxon; Alonso and Marquillas 1986, Leonardi 1987: pl. XVIII, C). Tables 11.6 and 11.7 summarize the times of appearance and stratigraphic distribution of avian families (excepting ichnotaxa) in Africa and South America.

A. The Insufficiences of the Avian Fossil Record

The fossil records of Africa and South America are poorer than those of North America and Europe. Simpson (1980a) compiled statistics on worldwide avian family diversity for each period and epoch of the late Mesozoic and Cenozoic. When similar statistics are assembled for Africa and South America (table 11.6), the relative impoverishment of the avian fossil records of these continents can be appreciated. Jurassic birds have not been recorded from either of the two landmasses, which were then contiguous, nor have Cretaceous, Paleocene, or Oligocene birds been recovered in Africa. In South America, Cretaceous collections from Brazil, Chile, and Argentina (Martins Neto and Kellner 1988; Chiappe 1991a, 1991b; Olson 1992), Paleocene collections from Itaboraí, Brazil (Alvarenga 1983, 1985b; Sibley and Ahlquist 1990:701), and Eocene collections from Argentina (Tonni 1980) account for 10 first appearances in 100 Myr, a figure that is dwarfed by Simpson's global figure of 56 for the same interval. In Africa, Eocene collections from Nigeria (Andrews 1916) and from the Fayum region, Egypt (Rasmussen et al. 1987) yield 15 first appearances in 21 Myr, an impressive figure when compared to the world total of 41 for this epoch. Oligocene and Miocene first appearances for South America (10 and 17), and Miocene first ap-

Table 11.6. Data on first appearances of families of birds in the fossil record, based on Simpson (1980a) and table 11.7.

Geologic Period or Epoch	Number of First Known Appearances			Percentage Extinct before the Holocene			Percentage Surviving into the Holocene		
(Duration in Myr)	SA	Af	Wo	SA	Af	Wo	SA	Af	Wo
Quaternary (1.8)	39	3	92	0	0	3	100	100	97
Pliocene (3.7)	3	37	9	0	0	0	100	100	100
Miocene (18.4)	17	9	24	12	11	25	88	89	75
Oligocene (12.9)	10	0	18	10	—	39	90	—	61
Eocene (21.2)	4	15	41	75	20	32	25	80	68
Paleocene (8.6)	3	0	3	67	—	67	33	—	33
Cretaceous (77.5)	3	0	12	100	—	75	0	—	25
Jurassic (64)	0	0	1	—	—	100	—	—	0

Note: Durations of periods and epochs are based on Kent and Gradstein (1985) for the Jurassic and Cretaceous, Berggren et al. (1985) for the Tertiary, and Savage and Russell (1983) for the Quaternary. SA = South America; Af = Africa; Wo = world.

pearances for Africa (9), also compare favorably to world totals (Oligocene: 18; Miocene: 24), due to significant collections from Brazil (Oligocene), Argentina (Oligocene and Miocene), Peru (Miocene), Morocco (Miocene), Tunisia (Miocene), and Kenya (Miocene). African Pliocene deposits, notably those at Langebaanweg, South Africa (Rich 1980b; Hendey 1981) and Olduvai, Tanzania (Brodkorb 1985), record the first Afrotropical appearances of 37 families, which greatly outnumber families appearing in South America (3) and globally (9) at this time. The low world total for the Pliocene seems due to the comparative recency and shortness of this epoch (3.7 Myr) rather than to a dearth of fossils. The numerous Quaternary appearances in South America (39) and the world (92) indicate a bias toward preservation or discovery in the youngest sediments. The low number of Quaternary appearances in Africa (3) is an artifact of the extensive Pliocene record, which contains many families that survive to the present. High survivorship of Plio-Pleistocene families is also evident in the statistics for South America and the world.

B. Avifaunal Comparisons

The fossil avifaunas of Africa and South America contain 64 and 73 families, respectively, compared to the slightly more diverse Recent avifaunas (75 and 79 families, respectively). One extinct (†Pelagornithidae) and 35 extant families (Spheniscidae, Podicipedidae, Diomedeidae, Procellariidae, Hydrobatidae, Pelecanoididae, possibly Pelecanidae, Sulidae, Phalacrocoracidae, Anhingidae, Ardeidae, Ciconiidae, Threskiornithidae, Phoenicopteridae, Anatidae, Accipitridae, Falconidae, Phasianidae, Rallidae, Jacanidae, Recurvirostridae, Burhinidae, Charadriidae, Scolopacidae, Laridae, Columbidae, Psittacidae, Cuculidae, Tytonidae, Strigidae, Apodidae, Alcedinidae, Picidae, Hirundinidae, Corvidae) have a fossil record on both continents. Only 10 of these shared families (†Pelagornithidae, possibly Spheniscidae, Ardeidae, Ciconiidae, Phoenicopteridae, Anatidae, Accipitridae, Rallidae, Jacanidae, Cuculidae) have a Paleogene record on either continent, the rest being confined to the Neogene. Most families in the latter category have African or South American records restricted to the Quaternary. All 35 extant shared families are widespread, occurring in Africa and South America and on other continents (table 11.8). The only extinct shared family, the †Pelagornithidae, were wide-ranging pelagic birds with a Tertiary record in Eurasia, Africa, North and South America, Antarctica, New Zealand, and Japan. Not surprisingly, a large proportion of the shared families are waterbirds with high potential for overwater dispersal. The neospecies shared by Africa and South America (table 11.3) belong to families that are present in the fossil records of both continents (exceptions: the nearly cosmopolitan Osprey *Pandion haliaetus*, Pandionidae, and the widespread Oystercatcher *Haematopus ostralegus*, Haematopodidae, with no fossil records in South America).

Africa and South America each possess a few seemingly endemic paleofamilies that supplement the list of endemic neofamilies mentioned earlier: the African †Xenerodiopidae (Egypt) and †Aepyornithidae (Madagascar) and the South American †Patagopterygidae, †Opisthodactylidae, †Onychopterygidae, †Cunampaiidae, and †Cladornithidae. Africa possesses 1 (Scopidae), and South America possesses 2 or 3 (Rheidae, Opisthocomidae, Psophiidae?), endemic neofamilies with similarly endemic paleodistributions.

The fossil records of several higher-level taxa presently confined to Africa or South America indicate more extensive past distributions. Characteristically African taxa (Struthionidae, Balaenicipitidae, Numidinae [Phasianidae], Musophagidae, Coliidae, Phoeniculidae, Sagittariidae) and quintessentially South American families (Cariamidae, †Phorusrhacidae, Steatornithidae, Nyctibiidae) are present in the Tertiary of Eurasia (Mourer-Chauviré 1982, 1988; Olson 1985d; Peters 1987). The †Phorusrhacidae and the presently Neotropical Anhi-

Table 11.7. Stratigraphic distribution of avian taxa in South America and Africa.

	South America							Africa[1]						
Taxon	K	P	E	O	M	Pl	Q	K	P	E	O	M	Pl	Q
†Enantiornithes	+	–	–	–	–	–	–	–	–	–	–	–	–	–
†Patagopterygidae	+	–	–	–	–	–	–	–	–	–	–	–	–	–
†Aepyornithidae	–	–	–	–	–	–	–	–	–	?	–	?	–	+
Struthionidae	–	–	–	–	–	–	–	–	–	–	–	+	+	+
†Opisthodactylidae	–	+	–	–	+	–	–	–	–	–	–	–	–	–
Rheidae	–	–	–	–	–	+	+	–	–	–	–	–	–	–
Tinamidae	–	–	–	–	+	+	+	–	–	–	–	–	–	–
Spheniscidae	–	–	–	?– –	+	+	–	–	–	–	–	+	+	+
Gaviidae	+	–	–	–	–	–	–	–	–	–	–	–	–	–
Podicipedidae	–	–	–	–	–	–	+	–	–	–	–	–	+	+
Diomedeidae	–	–	–	–	+	–	–	–	–	–	–	–	+	+
Procellariidae	–	–	?– –	+	?	+	+	–	–	–	–	–	+	+
Hydrobatidae	–	–	–	–	–	–	+	–	–	–	–	–	+	–
Pelecanoididae	–	–	–	–	–	–	+	–	–	–	–	–	+	–
†Pelagornithidae	–	–	–	–	+	+	–	–	–	+	–	–	–	–
Pelecanidae	–	–	–	–	+	–	–	–	–	–	–	–	+	+
Sulidae	–	–	–	–	+	+	+	–	–	–	–	–	+	+
Phalacrocoracidae	–	–	–	–	+	+	+	–	–	+	–	+	+	+
Anhingidae	–	–	–	–	+	–	–	–	–	–	–	+	+	+
†Xenerodiopidae	–	–	–	–	–	–	–	–	–	+	–	–	–	–
Ardeidae	–	–	–	–	–	–	+	–	–	+	–	+	+	?
Balaenicipitidae	–	–	–	–	–	–	–	–	–	+	–	+	+	–
Scopidae	–	–	–	–	–	–	–	–	–	–	–	–	+	–
Ciconiidae	–	–	–	?	–	–	+	–	–	+	–	+	+	?
Threskiornithidae	–	–	–	–	+	–	+	–	–	–	–	–	+	+
Phoenicopteridae	–	–	–	?	?– –	+	+	–	–	+	–	+	+	+
Anhimidae	–	–	–	–	–	–	+	–	–	–	–	–	–	–
†Presbyornithidae	–	–	+	–	–	–	–	–	–	–	–	–	–	–
Anatidae	–	–	–	+	+	–	+	–	–	+[2]	–	+	+	+
Cathartidae	–	–	–	+	–	+	+	–	–	–	–	–	–	–
†Teratornithidae	–	–	–	–	+	+	+	–	–	–	–	–	–	–
Pandionidae	–	–	–	–	–	–	–	–	–	+	–	–	–	–
Accipitridae	–	–	–	+	+	–	+	–	–	+	–	+	+	+
Falconidae	–	–	–	–	+	–	+	–	–	–	–	–	+	+

	South America							Africa[1]						
Taxon	K	P	E	O	M	Pl	Q	K	P	E	O	M	Pl	Q
Cracidae	–	–	–	–	+	–	+	–	–	–	–	–	–	–
Phasianidae	–	–	–	–	–	–	+	–	–	–	–	+	+	+
†Onychopterygidae	–	–	+	–	–	–	–	–	–	–	–	–	–	–
Opisthocomidae	–	–	–	–	+	–	–	–	–	–	–	–	–	–
Turnicidae	–	–	–	–	–	–	–	–	–	–	–	–	+	+
Gruidae	–	–	–	–	–	–	–	–	–	+	–	–	+	+
Aramidae	–	–	–	+	–	–	+	–	–	–	–	–	–	–
Psophiidae	–	–	–	–	?[3]	–	–	–	–	–	–	–	–	–
Rallidae	–	–	–	+	–	–	+	–	–	+	–	–	+	+
†Cunampaiidae	–	–	+	–	–	–	–	–	–	–	–	–	–	–
Cariamidae	–	–	–	+	+	+	+	–	–	–	–	–	–	–
†Phorusrhacidae s.l.	–	+	+	+	+	+	+	–	–	–	–	–	–	–
Otididae	–	–	–	–	–	–	–	–	–	–	–	+	+	+
Jacanidae	–	–	–	–	–	–	+	–	–	+	–	–	+	–
Rostratulidae	–	–	–	–	–	–	–	–	–	–	–	–	+	–
Haematopodidae	–	–	–	–	–	–	–	–	–	–	–	–	–	+
Recurvirostridae	–	–	–	–	–	–	+	–	–	–	–	–	+	–
Burhinidae	–	–	–	–	–	–	+	–	–	–	–	–	–	+
Glareolidae	–	–	–	–	–	–	–	–	–	–	–	–	+	–
Charadriidae	–	–	–	–	?	–	+	–	–	–	–	–	+	+
Scolopacidae	–	–	–	–	+	–	+	–	–	–	–	–	+	+
Thinocoridae	–	–	–	–	–	?	+	–	–	–	–	–	–	–
Laridae	–	–	–	–	–	+	+	–	–	–	–	–	+	+
Pteroclididae	–	–	–	–	–	–	–	–	–	–	–	–	+	–
Columbidae	–	–	–	–	–	–	+	–	–	–	–	–	+	+
Psittacidae	–	–	–	–	–	–	+	–	–	–	–	–	+	+
Musophagidae	–	–	–	–	–	–	–	–	–	+	–	+	–	–
Cuculidae	–	+	–	–	–	–	+	–	–	–	–	–	?	+
†Cladornithidae	–	–	–	+	–	–	–	–	–	–	–	–	–	–
†Zygodactylidae	–	–	–	–	–	–	–	–	–	–	–	?[4]	–	–
Tytonidae	–	–	–	–	–	–	+	–	–	–	–	+	+	+
Strigidae	–	–	–	–	+	–	+	–	–	–	–	–	+	+
Steatornithidae	–	–	–	–	–	–	+	–	–	–	–	–	–	–
Nyctibiidae	–	–	–	–	–	–	+	–	–	–	–	–	–	–

(continued)

Table 11.7. ***Continued.***

Taxon	South America							Africa[1]						
	K	P	E	O	M	Pl	Q	K	P	E	O	M	Pl	Q
Caprimulgidae	–	–	–	–	–	–	+	–	–	–	–	–	–	–
Apodidae	–	–	–	–	–	–	+	–	–	–	–	–	+	?
Trochilidae	–	–	–	–	–	–	+	–	–	–	–	–	–	–
Coliidae	–	–	–	–	–	–	–	–	–	–	–	–	+	–
Trogonidae	–	–	–	–	–	–	+	–	–	–	–	–	–	–
Alcedinidae	–	–	–	–	–	–	+	–	–	–	–	–	+	–
Momotidae	–	–	–	–	–	–	+	–	–	–	–	–	–	–
Coraciidae	–	–	–	–	–	–	–	–	–	–	–	–	+	–
Upupidae	–	–	–	–	–	–	–	–	–	–	–	–	+	–
Bucerotidae	–	–	–	–	–	–	–	–	–	–	–	+	–	–
Bucconidae	–	–	–	–	–	–	+	–	–	–	–	–	–	–
Indicatoridae	–	–	–	–	–	–	–	–	–	–	–	–	+	–
Ramphastidae	–	–	–	–	–	–	+	–	–	–	–	–	–	–
Picidae	–	–	–	–	–	–	+	–	–	–	–	–	?––	+
Passeriformes	–	–	–	–	+[5]	–	–	–	–	–	–	–	–	–
Dendrocolaptidae	–	–	–	–	–	–	+	–	–	–	–	–	–	–
Furnariidae	–	–	–	–	–	–	+	–	–	–	–	–	–	–
Formicariidae	–	–	–	–	–	–	+	–	–	–	–	–	–	–
Hirundinidae	–	–	–	–	–	–	+	–	–	–	–	–	+	+
Motacillidae	–	–	–	–	–	–	–	–	–	–	–	–	–	+
Troglodytidae	–	–	–	–	–	–	+	–	–	–	–	–	–	–
Mimidae	–	–	–	–	–	–	+	–	–	–	–	–	–	–
Muscicapidae	–	–	–	–	–	–	+[6]	–	–	–	–	?	?––	+
Sylviidae	–	–	–	–	–	–	–	–	–	–	–	–	?	+
Nectariniidae	–	–	–	–	–	–	–	–	–	–	–	–	?––	+
Emberizidae	–	–	–	–	–	–	+	–	–	–	–	–	–	–
Vireonidae	–	–	–	–	–	–	+	–	–	–	–	–	–	–
Icteridae	–	–	–	–	–	–	+	–	–	–	–	–	–	–
Fringillidae	–	–	–	–	–	–	–	–	–	–	–	–	?––	+
Estrildidae	–	–	–	–	–	–	–	–	–	–	–	–	+	+
Ploceidae	–	–	–	–	–	–	–	–	–	–	–	–	?	+
Sturnidae	–	–	–	–	–	–	–	–	–	–	–	–	+	+
Corvidae	–	–	–	–	–	–	+	–	–	–	–	–	?	+

Sources: Alvarenga 1982, 1983, 1985a, 1985b, 1988; Alvarenga and Bonaparte 1992; Andrews 1916; Báez 1992; Balouet 1981; Brodkorb 1978, 1980, 1985; Brodkorb and Mourer-Chauviré 1982,

1984a, 1984b; Brunet 1961, 1971; Campbell and Frailey 1985; Campbell and Tonni 1980; Chiappe 1991a, 1991b, 1991c; Cuello 1988; Haarhoff 1988; Harrison 1974, 1980; Harrison and Walker 1976a, 1976b, 1976c, 1979, 1982; Hendey 1978, 1981; Hill and Walker 1979; Lambrecht 1930; Marshall et al. 1984; Muizon and DeVries 1985; Nasif 1988; Olson 1975, 1981, 1984, 1985a, 1985b, 1985c, 1985d, 1992; Olson and Eller 1989; Olson and Feduccia 1980a, 1980b; Rasmussen et al. 1987, 1992; Rich 1972, 1974 (and works cited therein), 1979, 1980a, 1980b; Rich and Haarhoff 1985; Rich and Walker 1983; Rothe 1964; Sibley and Ahlquist 1990; Simpson 1971, 1973, 1975, 1979a, 1979b; Steadman 1986; Tambussi 1987; Tonni 1980; Tonni and Tambussi 1986, 1988; Ubilla et al. 1990; Vuilleumier 1985b, 1985c.

Notes: K = Cretaceous; P = Paleocene; E = Eocene; O = Oligocene; M = Miocene; Pl = Pliocene; Q = Quaternary.

[1]We assign the avifauna from the Jebel Qatrani Formation, Fayum Province, Egypt, a Late Eocene (Priabonian) age, following Van Couvering and Harris (1991). This fauna was previously dated as early Oligocene (Olson and Rasmussen 1986; Rasmussen et al. 1987). For present purposes we assume that Olduvai Bed I = Pliocene and Bed II = Pleistocene (Savage and Russell 1983; Hay 1990). Our tabulation of the rich fossil avifauna from Olduvai Gorge, Tanzania, is based on preliminary information (e.g., Rich 1974; Harrison 1980; Brodkorb 1985). A comprehensive review of this avifauna is in preparation (D. G. Matthiesen, Ph.D. candidate at University of Florida).

[2]Anatidae sp. from the Eocene Jebel Qatrani Formation, Fayum Province, Egypt, in the American Museum of Natural History (AMNH 5082).

[3]†*Anisolornis excavatus* Ameghino from the Santacrucian of Argentina resembles the Psophiidae (Olson 1985d:165).

[4]"*Jynx* cf. *torquilla*" Brunet 1961, a problematic zygodactyl bird from the middle Miocene of Beni Mellal, Morocco, seems "to have some similarities to the [European] Zygodactylidae" (Olson 1985d:122).

[5]Probable suboscine passerine from the lower-middle Miocene Pinturas Formation, Santa Cruz Province, Argentina (Chiappe 1991c).

[6]*Turdus* sp. (Winge 1888:49).

Table 11.8. Global distribution of avian taxa with fossil records in South America or Africa.

	Biogeographical Realm							
Taxon	Pa	Ne	Af	In	Oc	Au	An	Nt
†Enantiornithes	+	+	–	–	–	+	–	+
†Patagopterygidae	–	–	–	–	–	–	–	+
Palaeognathae	+*	+	+*	+	+*	+*	+*	+*
†Aepyornithidae	?	–	+	–	–	–	–	–
Struthionidae	+*	–	+*	+	–	–	–	–
†Opisthodactylidae	–	–	–	–	–	–	–	+
Rheidae	–	–	–	–	–	–	–	+*
Tinamidae	–	*	–	–	–	–	–	+*
Spheniscidae	–	–	+*	–	–	+*	+*	+*
Gaviidae	+*	+*	–	–	–	–	–	+
Podicipedidae	+*	+*	+*	*	*	+*	*	+*
Diomedeidae	*	*	+*	*	+*	+*	*	+*
Procellariidae	+*	+*	+*	*	+*	+*	+*	+*
Hydrobatidae	*	+*	+*	*	+*	+*	*	+*
Pelecanoididae	–	–	+	–	–	+*	*	+*
†Pelagornithidae	+	+	+	–	–	–	+	+
Pelecanidae	+*	+*	+*	+*	*	+*	+*	?*
Sulidae	+*	+*	+*	*	+*	+*	*	+*
Phalacrocoracidae	+*	+*	+*	?*	*	+*	*	+*
Anhingidae	+*	+*	+*	?*	*	+*	*	+*
†Xenerodiopidae	+	–	–	–	–	–	–	–
Ardeidae	+*	+*	+*	*	*	+*	*	+*
Balaenicipitidae	+	–	+*	–	–	–	–	–
Scopidae	–	–	+*	–	–	–	–	–
Ciconiidae	+*	+*	+*	+*	*	+*	–	+*
Threskiornithidae	+*	+*	+*	*	+*	+*	*	+*
Phoenicopteridae	+*	+*	+*	*	–	+	–	+*
Anhimidae	–	+	–	–	–	–	–	+*
†Presbyornithidae	+	+	–	–	–	–	–	+
Anatidae	+*	+*	+*	*	+*	+*	+*	+*
Cathartidae	+	+*	–	–	–	–	–	+*
†Teratornithidae	–	+	–	–	–	–	–	+
Pandionidae	+*	+	*	*	*	*	–	*
Accipitridae	+*	+*	+*	*	+*	+*	+*	+*
Falconidae	+*	+*	+*	*	+*	+*	*	+*
Cracidae	?	+*	–	–	–	–	–	+*

Taxon	Biogeographical Realm							
	Pa	Ne	Af	In	Oc	Au	An[1]	Nt
Phasianidae	+*	+*	+*	+*	+*	+*	+*	+*
†Onychopterygidae	–	–	–	–	–	–	–	+
Opisthocomidae	–	–	–	–	–	–	–	+*
Turnicidae	*	–	+*	*	+*	+*	–	–
Gruidae	+*	+*	+*	+*	*	+*	–	*
Aramidae	–	+*	–	–	–	–	–	+*
Psophiidae	–	–	–	–	–	–	–	?*
Rallidae	+*	+*	+*	*	+*	+*	+*	+*
†Cunampaiidae	–	–	–	–	–	–	–	+
Cariamidae s.l.	+	+	–	–	–	–	–	+*
†Phorusrhacidae s.l.	+	+	–	–	–	–	+	+
Otididae	+*	–	+*	*	*	+*	–	–
Jacanidae	+*	+*	+*	*	*	*	–	+*
Rostratulidae	*	–	+*	*	–	*	–	*
Haematopodidae	*	+*	+*	*	*	*	+*	*
Recurvirostridae	+*	+*	+*	*	*	*	+*	+*
Burhinidae	+*	+	+*	*	*	+*	–	+*
Glareolidae	+*	+	+*	*	*	*	–	–
Charadriidae	?*	*	+*	*	+*	+*	*	+*
Scolopacidae	+*	+*	+*	*	+*	+*	*	+*
Thinocoridae	–	–	–	–	–	–	–	+*
Laridae	+*	+*	+*	*	+*	+*	+*	+*
Pteroclididae	+*	–	+*	*	–	–	–	–
Columbidae	+*	+*	+*	*	+*	+*	+*	+*
Psittacidae	+*	+*	+*	*	+*	+*	+*	+*
Musophagidae	+	–	+*	–	–	–	–	–
Cuculidae	+*	+*	+*	*	+*	+*	*	+*
†Cladornithidae	–	–	–	–	–	–	–	+
†Zygodactylidae	+	–	–	–	–	–	–	–
Tytonidae	+*	+*	+*	*	+*	+*	+*	+*
Strigidae	+*	+*	+*	+*	+*	+*	+*	+*
Steatornithidae	+	+	–	–	–	–	–	+*
Nyctibiidae	–	–	–	–	–	–	–	+*
Caprimulgidae	+*	+*	*	*	*	*	–	+*
Apodidae	+*	+*	+*	*	+*	+*	*	+*
Trochilidae	–	*	–	–	–	–	–	+*
Coliidae	+	–	+*	–	–	–	–	–

(*continued*)

Table 11.8. *Continued.*

Taxon	Biogeographical Realm							
	Pa	Ne	Af	In	Oc	Au	An[1]	Nt
Trogonidae	+*	*	*	*	–	–	–	+*
Alcedinidae	+*	+*	+*	*	+*	+*	*	+*
Momotidae	+	–	–	–	–	–	–	+*
Coraciidae	+*	+	+*	*	*	*	+*	–
Upupidae s.str.	+*	–	+*	*	+	–	–	–
Bucerotidae	+	–	*	*	*	–	–	–
Bucconidae	–	–	–	–	–	–	–	+*
Indicatoridae	–	–	+*	*	–	–	–	–
Ramphastidae	–	–	–	–	–	–	–	+*
Picidae	+*	+*	+*	*	–	–	–	+*
Dendrocolaptidae	–	–	–	–	–	–	–	+*
Furnariidae	–	–	–	–	–	–	*	+*
Formicariidae	–	–	–	–	–	–	–	+*
Hirundinidae	+*	+*	+*	*	*	*	*	+*
Motacillidae	+*	*	+*	*	*	*	*	*
Troglodytidae	+*	+*	–	*	–	–	*	+*
Mimidae	–	+*	–	–	–	–	–	+*
Muscicapidae[2]	+*	+*	+*	*	+*	*	*	+*
Sylviidae[2]	+*	+*	+*	*	+*	*	*	–
Nectariniidae	*	–	+*	*	*	*	–	–
Emberizidae	+*	+*	*	*	–	–	*	+*
Vireonidae	–	+*	–	–	–	–	–	+*
Icteridae	–	+*	–	–	–	–	*	+*
Fringillidae	+*	+*	+*	*	+*	–	–	*
Estrildidae	*	–	+*	*	*	*	–	–
Ploceidae	+*	+	+*	*	–	–	–	–
Sturnidae	+*	–	+*	*	+*	*	–	–
Corvidae	+*	+*	+*	*	+*	*	+?	+*

Sources: Table 11.7; Balouet and Olson 1989; Case et al. 1987; Olson and James 1982; Peters et al. 1931–1987; Rich and Van Tets 1984; Steadman 1988, 1989; Steadman and Kirch 1990; Steadman and Olson 1985.

Note: Fossil or subfossil (+) and Recent (*) occurrences are recorded separately. The geographical units adopted are Udvardy's (1987: 187, figure 1) biogeographical realms. In this system, South America is classified as Neotropical, North Africa is classified as Palaearctic, and sub-Saharan Africa is classified as Afrotropical based on floristic, faunistic, and ecosystemic considerations. Pa = Palaearctic; Ne = Nearctic; Af = Afrotropical; In = Indomalayan; Oc = Oceanian; Au = Australian; An = Antarctic; Nt = Neotropical.

[1]Antarctic includes New Zealand and Falkland Is.

[2]*Sensu* Sibley and Monroe 1990.

midae, Cracidae, and Steatornithidae occur also in the Tertiary of North America (Olson 1985d, 1987; Houde 1992, personal communication, 1988). Altogether, 24 family- or higher-level taxa with a South American (but not African) fossil record are found outside of South America, and 23 or 24 families with an African (but not South American) record occur outside of Africa, either as fossils or as living populations (table 11.8). Thus, the avian fossil records of Africa and South America register some endemism and much dispersal (immigration and emigration to adjacent regions), at least during the Neogene and Quaternary intervals for which the record is best.

C. Vicariant Patterns

A phylogenetic analysis of African and South American birds that would refine our knowledge of African–American zoogeographical relationships is beyond the purview of this study. However, one such analysis, based on DNA comparisons, has already been presented (Sibley and Ahlquist 1990). Sibley and Ahlquist correlate the rhea-ostrich and Neotropical cuckoo–Old World cuckoo divergences with the opening of the Atlantic between Africa and South America about 90 Ma. Although they believe that other groups, including barbets (discussed earlier), trogons, and the sungrebe-finfoot clade, may indicate an African–South American connection, these divergences evidently occurred too late to have been affected by long-term plate movements. In the following sections we discuss each of the aforementioned African–American divergences in turn.

1. Ostrich-Rhea Divergence Ostriches (Struthionidae) first appear in the Middle or Upper Miocene of Kenya (*Struthio* sp.; Harrison 1980) and in the Upper Miocene of Turkey (†*S. brachydactylus*) and Republic of Moldova (†*S. orlovi*). Their Miocene, Pliocene, and Quaternary record (Olson 1985d) is restricted to Africa and Eurasia. The living *S. camelus* is confined to Africa.

Rheas (†Opisthodactylidae and Rheidae) have been reported from the Upper Paleocene (†*Diogenornis fragilis*) of Brazil; from the Lower Miocene (†*Opisthodactylus patagonicus*), Miocene or Pliocene (unnamed tracks; Leonardi 1987: pl. XX, A), Upper Pliocene (†*Heterorhea dabbenei*), and Quaternary (†*Rhea anchorenensis,* †*R. pampeana,* †*R. azarae, R. americana, Pterocnemia* sp.) of Argentina; and from the Quaternary of Uruguay and Brazil (*R. americana;* Cuello 1988). A relationship between the †Opisthodactylidae (†*Diogenornis,* †*Opisthodactylus*) and the Rheidae (†*Heterorhea, Rhea, Pterocnemia*) has been hypothesized by Patterson and Kraglievich (1960). Tonni (1980) has pointed out that "fossil rheas have been found within the present area of distribution of the family [Rheidae]." The two extant species of Rheidae (*R. americana, Pterocnemia pennata*) are confined to South America.

The oldest rhea (Paleocene) and the oldest ostrich (Miocene) postdate the Cretaceous separation of South America from Africa. Thus, known rhea and ostrich fossils provide only a minimum date for the hypothesized rhea-ostrich divergence. We therefore turn to the distribution of putative relatives for evidence concerning the zoogeographic affinities of these taxa.

Martin (1983) transfers †*Remiornis minor* (Lemoine), a large ground bird from the late Paleocene of France, from the neognathous †Gastornithidae to the palaeognathous ratites. Although he does not refer †*Remiornis* to a specific family, this taxon is noteworthy because of its Northern Hemisphere distribution and its age, which rivals that of †*Diogenornis*. †*Remiornis* and †*Diogenornis* demonstrate the presence of ratites in the Northern and Southern hemispheres by the early Tertiary.

Peters (1988) believes that †*Palaeotis weigelti*, a small flightless palaeognath from the Middle Eocene of Germany previously placed in the Otididae (Lambrecht 1928) and Struthionidae (Houde and Haubold 1987), may be related to rheas. The presence of rhea-like forms in South America (†*Diogenornis*) and Europe (†*Palaeotis*) during the early Tertiary suggests that this group was once more widespread and that their present distribution in South America is relictual.

Houde (1986, 1988) traces the ancestry of ostriches to the "*Lithornis*-cohort" or †Lithornithidae, a group of palaeognathous carinates (†*Lithornis*, †*Paracathartes*, †*Pseudocrypturus*) from the Paleocene and Eocene of North America and Europe. He considers ostriches to have evolved in the Northern Hemisphere long after the breakup of Gondwanaland and to have emigrated from Europe to Africa some time during the early or middle Tertiary. He identifies †*Palaeotis* as a palaeognathous, tridactylous ostrich that offers further evidence for the origin of the Struthionidae in Eurasia (Houde and Haubold 1987).

A Northern Hemisphere origin for ostriches is also implied by Olson (1985d), who identifies the †Geranoididae (Eocene of North America), †Eogruidae (Eocene-Oligocene of central Asia), and †Ergilornithidae (Oligocene-Pliocene of Eurasia) as representatives of a neognathous lineage that progressively evolved flightlessness, didactyly, and palaeognathy, ultimately giving rise to flightless, two-toed, palaeognathous Struthionidae in the Oligocene or Miocene (Olson 1985d). Cracraft (1973b) interprets as convergence the didactylous condition of ergilornithids and struthionids. Like Sibley and Ahlquist, he posits a common, Southern Hemisphere origin for ostriches and rheas (Cracraft 1974).

If ostriches arose from either the geranoidid-eogruid-ergilornithid or lithornithid lines, then struthionids may have originated in the Tertiary of the Northern Hemisphere where their supposed predecessors, ergilornithids and †*Palaeotis*, have been recovered. If, however, ostriches diverged from a common ancestor shared with rheas, then struthionids may have arisen in the Cretaceous of the Southern Hemisphere and subsequently dispersed to Eurasia. Available evidence does not permit us to choose between these hypotheses.

The vicariance hypothesis of ostrich-rhea distribution requires that the progenitors of these groups were isolated in the Southern Hemisphere as the result of geotectonic events and that the birds' powers of dispersal were minimal (Houde 1986). Both premises of this hypothesis seem violated by the fossil record of ostriches, rheas, and their putative relatives, which suggests vagility and a protracted Northern Hemisphere history during the Tertiary.

2. *Old World Cuckoo–New World Cuckoo Divergence* Sibley and Ahlquist (1990) recognize six families of cuculiforms, of which three (Crotophagidae, Neomorphidae, Opisthocomidae) are Neotropical, one (Coc-

cyzidae) is Neotropical and Nearctic, and two (Centropodidae, Cuculidae) are Old World in distribution. They consider the Coccyzidae to be more closely related to the Old World cuckoos than to the strictly Neotropical cuckoos, and they interpret the Hoatzin (*Opisthocomus hoazin,* Opisthocomidae) of South America as a highly modified cuckoo that is more closely related to roadrunners (Neomorphidae) and anis and Guira Cuckoos (Crotophagidae) than to any other living birds. The apparently plesiomorphic, anisodactyl foot of the Hoatzin (Bock 1992) distinguishes it from other cuculiforms, which have zygodactyl toes (Sibley and Ahlquist 1990:365). Until the phylogenetic significance of zygodactyly in cuckoos has been evaluated, we consider it reasonable to transfer the Hoatzin to Aves Incertae Sedis and to recognize only a single heterogeneous cuckoo family (Cuculidae) in place of the remaining five cuculiform families recognized by Sibley and Ahlquist.

The Neotropical fossil record of the Cuculidae is limited to one undescribed Paleocene occurrence in Brazil (Sibley and Ahlquist 1990:370, 701) and to several Quaternary occurrences of neospecies in Brazil, Mexico, and the West Indies (Cuello 1988). The Afrotropical record of this family is sparser, with an unnamed cuculid in the Plio-Pleistocene of Olduvai, Tanzania (Rich 1974; Brodkorb 1985) and one named paleospecies (†*Coua primaeva*) in the Pleistocene of Madagascar (Brodkorb 1971). Other Southern Hemisphere records of cuculids include one undescribed paleospecies in the Pleistocene of Australia (Rich and Van Tets 1984) and one neospecies in the Late Quaternary of New Caledonia (Balouet and Olson 1989).

Brodkorb (1971) and Olson (1985d) recognize the following Northern Hemisphere paleospecies of Cuculidae: †*Dynamopterus velox* (Eo-Oligocene, France), †*Neococcyx mccorquodalei* (lower Oligocene, Saskatchewan), †*Cursoricoccyx geraldinae* (lower Miocene, Colorado), an unnamed large cuckoo (lower Pliocene, North Carolina), and †*Geococcyx conklingi* (Pleistocene, Mexico, New Mexico). Brodkorb (1971) lists Pleistocene occurrences of cuculid neospecies in Israel, Italy, Czechoslovakia, Hungary, and the United States.

To summarize, the oldest known cuculid is from the Paleocene of the Neotropics. The next oldest cuculids are Eocene or Oligocene in age and are northern (Palaearctic and Nearctic) in distribution. Miocene and Pliocene cuculids are also northern (Nearctic). Quaternary cuculids are found in both the Northern and Southern hemispheres, in the Nearctic, Neotropical, Palaearctic, Afrotropical, Australian, and Oceanian Realms. Recent cuculids are cosmopolitan (table 11.8).

These data indicate that the Cuculidae have had a long Tertiary history in Laurasia, notwithstanding their early appearance in South America. The close relationship of the Coccyzinae to Old World cuckoos suggests that the progenitors of this subfamily spread to the New World via a northern route, reaching South America by the Paleocene (60 Ma; Sibley and Ahlquist 1990:701). Prior to this dispersal event, progenitors of crotophagine and neomorphine cuckoos may have been isolated in South America by the rifting of Africa–South America. Thus, by the time coccyzines arrived in South America, crotophagines and neomorphines with transatlantic affinities to centropodines and cuculines may already have been well-established. If they were, then older (Cretaceous) cuculid fossils with African relationships may yet be discovered in South America.

3. Barbet, Trogon, and Sungrebe-Finfoot Divergences Barbets (Capitonidae *auct.*; Megalaimidae, Lybiidae, and Ramphastidae: Capitoninae of Sibley and Ahlquist 1990) have been recovered from the Miocene of Germany (†*Capitonides europaeus*, †*C. protractus*, †*Capitonides* sp.), France, and Florida (Olson 1985d). All known fossils are northern in distribution, whereas modern representatives are Neotropical, Afrotropical, and Indomalayan.

According to Sibley and Ahlquist (1986:87), "The DNA indicates that the branching between the Old World [Megalaimidae, Lybiidae] and the New World barbets [Ramphastidae: Capitoninae] took place [during the Early Eocene] about 55 million years ago. The toucans [Ramphastidae: Ramphastinae] branched from the New World barbet lineage more recently, [during the Early Oligocene] about 30 million years ago." The oldest barbets (†*Capitonides* spp. from Bavaria), dated as early Miocene (early Burdigalian; Olson 1985d), postdate the Old World barbet–New World barbet divergence by roughly 38 Myr.

Trogons (Trogonidae) occur in the early Oligocene of Switzerland (unnamed specimen) and early Miocene of France (†*Paratrogon gallicus;* Olson 1985d), and in the Pleistocene of the Dominican Republic (*Temnotrogon roseigaster*) and Brazil (*Trogon surrucura;* Mourer-Chauviré 1980). The possibly related †Archaeotrogonidae, which lack the heterodactyl foot structure of modern trogons, are known from four paleospecies in the late Eocene to late Oligocene (†*Archaeotrogon venustus*) and Oligocene (†*A. zitteli*, †*A. cayluxensis*, †*A. hoffstetteri*) of France (Mourer-Chauviré 1980). Recent trogons are Neotropical, Afrotropical, and Indomalayan.

The New World trogons diverged from the African trogons in the early Miocene, approximately 17 Ma, and from the Asian trogons in the middle Miocene, about 15 Ma, assuming a passerine calibration factor of $\triangle T_{50}H$ 1.0 = 2.3 Myr (Sibley and Ahlquist 1990:703). If we assume an ostrich-rhea calibration factor of $\triangle T_{50}H$ 1.0 = 4.7 Myr (Sibley and Ahlquist 1990:703), the divergences were both early Oligocene, about 35 Ma and 31 Ma, respectively. Thus, the oldest fossil trogon could be either older than, or roughly contemporaneous with, the postulated divergence between extant Old World and New World trogons.

The Heliornithidae have no published fossil record (Olson 1985d). The modern Sungrebe (*Heliornis fulica*) is Neotropical and the finfoots are respectively Afrotropical (*Podica senegalensis*) and Indomalayan (*Heliopais personata*). Transatlantic relationships between *Heliornis* and the African and Asian finfoots are suspected, but divergences within this group have not been firmly dated (Sibley and Ahlquist 1990).

To summarize, the Old World barbet–New World barbet and Old World trogon–New World trogon divergences are placed in the Cenozoic, subsequent to the opening of the Atlantic. Thus, transatlantic crossings between Africa and South America cannot be invoked to explain these relationships (Sibley and Ahlquist 1990). The northern distribution of fossil trogons (including archaeotrogonids) and barbets suggests the possibility of northerly dispersal of these groups between the Old and New Worlds via North America. European fossil trogonids and archaeotrogonids may be a product of one such (Oligocene) dispersal episode.

IV. Relative Isolation of Africa versus South America

The oldest diverse avifauna from Africa, that of the Late Eocene Jebel Qatrani Formation of Fayum Province, Egypt (Van Couvering and Harris 1991), contains at least 19 species in 14 families, including the oldest fossil records of the Musophagidae, Pandionidae, Jacanidae, and Balaenicipitidae. Other families represented are supposed †Aepyornithidae, Anatidae (personal observation), Accipitridae, Rallidae, Gruidae, Phoenicopteridae, †Xenerodiopidae, Ardeidae, Ciconiidae, and Phalacrocoracidae. The Fayum assemblage is similar to one that would be expected along a tropical swampy river in central Africa today (Olson and Rasmussen 1986; Rasmussen et al. 1987).

The late Miocene fauna of Beni-Mellal in Morocco contains putative Phasianidae, †Zygodactylidae ("*Jynx* cf. *torquilla*"), Tytonidae, Bucerotidae, and Muscicapidae (Brunet 1961, 1971), of which only the bucerotid, †*Bucorvus brailloni,* seems to have been positively identified (Olson 1985d). As pointed out by Brunet (1971) and Olson (1985d:137), this record of ground hornbills is situated well north of the modern range of *Bucorvus* in sub-Saharan Africa.

The late Miocene Beglia Formation of Tunisia has produced five species in the Struthionidae, Phalacrocoracidae, Anhingidae, Balaenicipitidae, and Ciconiidae. The birds of this assemblage indicate the former proximity of savanna, papyrus swamp, and freshwater stream habitats. The Tunisian ostrich resembles Eurasian and African *Struthio camelus;* the stork resembles †*Leptoptilos falconeri* from the early Pliocene of India; and the anhinga is probably conspecific with †*Anhinga pannonica* from the early Pliocene of Hungary (Rich 1972, 1974). The balaenicipitid, †*Paludavis richae,* may be conspecific with a contemporaneous whalehead in the Siwalik Series of Pakistan (Harrison and Walker 1982).

The aforementioned fossil avifaunas from Egypt, Morocco, and Tunisia indicate that North Africa was more humid during the Late Eocene, Miocene, and early Pliocene than it is at present. The Tunisian avifauna suggests that some dispersal was possible between Africa and Eurasia in the late Tertiary (Rich 1972). Prior to the formation of the Strait of Gibraltar in the late Miocene, exchange of terrestrial faunas evidently took place between Morocco and Iberia (Busack 1986). In the opinion of Astibia and coworkers (1990:465), "Late Cretaceous Europe (or at least its Franco-Iberian part) had closer faunal links with the southern continents than with either western North America or mainland Asia, which suggests interchange between Europe and Africa sometime during the Cretaceous."

In the New World, faunal exchange between North America and insular South America became possible in the late Pliocene (Late Blancan–Chapadmalalan), after about 3 Ma, when the Bolivar geosyncline was uplifted to form the Isthmus of Panama (Marshall et al. 1979). The ensuing interchange of terrestrial biotas between the Americas was extensive and has been well documented for mammals (Webb 1976; Simpson 1980b) and adequately so for birds (Vuilleumier 1985b). Prior to the Great American Faunal Interchange, from 65 Ma to about 2 Ma, South America was continuously or episodically an island continent and its avifauna was isolated from that of the Northern Hemisphere (Chiappe 1991b). During the Cretaceous, the Antarctic region comprised a source area for South American and African tetrapods (Molnar 1989). A land bridge may have connected South America with Antarctica as early as the latest Cretaceous and as late as the Eocene, when polydolopid marsupials (Woodburne and Zinsmeister 1982) and phorusrhacid ground birds (Case et al. 1987) are known to have been present on both continents. In the late Miocene and early Pliocene, a change from warm, subtropical to cool, temperate marine conditions in the Cape region of southwest Africa and along the western coast of South America was accompanied by establishment of similar seabird faunas in South Africa (Olson 1983) and Peru (Muizon and DeVries 1985). Olson (1983:401) suggests that many of these seabird taxa (Spheniscidae, Procellariidae, Pelecanoididae) originated in southerly latitudes and that "some of the same species of seabirds that moved into South Africa also moved up the west coast of South America concurrently."

In general, it appears that African avifaunas have had a long (late Mesozoic and Cenozoic) history of exchange with those of Eurasia, whereas South American avifaunas have been characterized by prolonged isolation and relatively recent (Late Cenozoic) interchange with those of North America. The appreciable resemblance between African and South American fossil and Recent avifaunas, documented earlier, can be explained by differential dispersal across barriers (the dispersal hypothesis), as in the case of waterbirds, or by formation of geographic barriers to gene exchange (the vicariance hypothesis), as in the case of ostriches, rheas, and perhaps cuckoos. Most of the faunal resemblance seems to be of Cenozoic origin and to be adequately explained by the dispersal hypothesis.

Literature Cited

Alonso, R. N., and R. A. Marquillas. 1986. Nueva localidad con huellas de dinosaurios y primer hallazgo de huellas de aves en la Formación Yacoraite (Maastrichtiano) del norte argentino. Actas IV Congr. Argent. Paleontol. Bioestrat., Mendoza 2:33–41.

Alvarenga, H. M. F. 1982. Uma gigantesca ave fóssil do Cenozóico brasileiro: *Physornis brasiliensis* sp. n. An. Acad. brasil. Ci. 54:697–712.

———. 1983. Uma ave Ratitae do Paleoceno brasileiro: Bacia calcária de Itaboraí, Estado do Rio de Janeiro, Brasil. Bol. Museu Nacional, n.s., 41:1–11.

———. 1985a. Notas sobre os Cathartidae (Aves) e descrição de um novo gênero do Cenozóico brasileiro. An. Acad. brasil. Ci. 57:349–357.

———. 1985b. Um novo Psilopteridae (Aves: Gruiformes) dos sedimentos Terciários de Itaboraí, Rio de Janeiro, Brasil. VIII Congr. brasil. Paleont., Sér. Geol. No. 27, Paleont./ Estratig. No. 2:17–20.

———. 1988. Ave fóssil (Gruiformes: Rallidae) dos folhelhos de bacia de Taubaté, Estado de São Paulo, Brasil. An. Acad. brasil. Ci. 60:321–328.

Alvarenga, H. M. F., and J. F. Bonaparte. 1992. A new flightless landbird from the Cretaceous of Patagonia. Pp. 51–64 *in* K. E. Campbell,

Jr. (ed.), *Papers in Avian Paleontology Honoring Pierce Brodkorb*. Science Series No. 36. Los Angeles: Natural History Museum of Los Angeles County.

Amadon, D. 1973. Birds of the Congo and Amazon forests: A comparison. Pp. 267–277 *in* B. J. Meggers, E. S. Ayensu, and W. D. Duckworth (eds.), *Tropical Forest Ecosystems in Africa and South America: A Comparative Review.* Washington, D.C.: Smithsonian Institution Press.

Ambroggi, R., and A. F. de Lapparent. 1954. Les empreintes de pas fossiles du Maestrichtien d'Agadir. Notes et Mém. Serv. Géol. Maroc 10(122): 43–63.

Andrews, C. W. 1916. Note on the sternum of a large carinate bird from the (?) Eocene of southern Nigeria. Proc. Zool. Soc. London 1916:519–524.

Astibia, H., E. Buffetaut, A. D. Buscalioni, H. Cappetta, C. Corral, R. Estes, F. Garcia-Garmilla, J. J. Jaeger, E. Jimenez-Fuentes, J. Le Loeuff, J. M. Mazin, X. Orue-Etxebarria, J. Pereda-Suberbiola, J. E. Powell, J. C. Rage, J. Rodriguez-Lazaro, J. L. Sanz, and H. Tong. 1990. The fossil vertebrates from Laño (Basque Country, Spain); new evidence on the composition and affinities of the Late Cretaceous continental faunas of Europe. Terra Nova 2:460–466.

Báez, M. 1992. Zoogeography and evolution of the avifauna of the Canary Islands. Pp. 425–431 *in* K. E. Campbell, Jr. (ed.), *Papers in Avian Paleontology Honoring Pierce Brodkorb*. Science Series No. 36. Los Angeles: Natural History Museum of Los Angeles County.

Balouet, J.-C. 1981. *Zeltornis ginsburgi* n. g., n. sp. (Ardeidae, Aves), héron géant du Miocène inférieur du Djebel Zelten (Libye). Comptes Rendus de l'Académie des Sciences Paris, 2d ser., 293:235–239.

Balouet, J. C., and S. L. Olson. 1989. Fossil birds from late Quaternary deposits in New Caledonia. Smithsonian Contr. Zool. No. 469:i–iv + 1–38.

Berggren, W. A., D. V. Kent, J. J. Flynn, and J. A. Van Couvering. 1985. Cenozoic geochronology. Bull. Geol. Soc. Amer. 96: 1407–1418.

Blondel, J., F. Vuilleumier, L. F. Marcus, and E. Terouanne. 1984. Is there ecomorphological convergence among mediterranean bird communities of Chile, California, and France? Evol. Biol. 18:141–213.

Bock, W. J. 1956. A generic review of the family Ardeidae (Aves). Amer. Mus. Novit. 1779:1–49.

———. 1958. A generic review of the plovers (Charadriinae, Aves). Bull. Mus. Comp. Zool. 118:27–97.

———. 1992. Methodology in avian macrosystematics. Pp. 53–72 *in* J. F. Monk (ed.), *Bulletin of the British Ornithologists' Club, Centenary Supplement* 112A.

Bock, W. J., and P. Bühler. 1990. The evolution and biogeographical history of the palaeognathous birds. Pp. 31–36 *in* R. van den Elzen, K.-L. Schuchmann, and K. Schmidt-Koenig (eds.), *Proceedings of the International Centennial Meeting of the Deutsche Ornithologen-Gesellschaft, Current Topics in Avian Biology.* Bonn: Verlag der Deutschen Ornithologen-Gesellschaft.

Brodkorb, P. 1971. Catalogue of fossil birds: Part 4 (Columbiformes through Piciformes). Bull. Florida St. Mus. Biol. Sci. 15:163–266.

———. 1978. Catalogue of fossil birds: Part 5 (Passeriformes). Bull. Florida St. Mus. Biol. Sci. 23:139–228.

———. 1980. A new fossil heron (Aves: Ardeidae) from the Omo Basin of Ethiopia, with remarks on the position of some other species assigned to the Ardeidae. Contr. Sci. Los Angeles Co. Mus. Nat. Hist. 330:87–92.

———. 1985. Preliminary report on Pliocene/Pleistocene birds of East Africa. Acta XVIII Congr. Internat. Ornithol. Moscow, 1982, Vol. 1:174–177.

Brodkorb, P., and C. Mourer-Chauviré. 1982. Fossil anhingas (Aves: Anhingidae) from early man sites of Hadar and Omo (Ethiopia) and Olduvai Gorge (Tanzania). Geobios 15:505–515.

———. 1984a. Fossil owls from early man sites

of Olduvai Gorge, Tanzania. Ostrich 55:17–27.

———. 1984b. A new species of cormorant (Aves: Phalacrocoracidae) from the Pleistocene of Olduvai Gorge, Tanzania. Geobios 17:331–337.

Broekhuysen, G. J. 1966. The avifauna of the Cape "Protea-Heath Macchia" habitat in South Africa. Proc. 2d Pan-African Ornithol. Congr., pp. 323–334.

Brunet, J. 1961. Le gisement de vertébrés miocènes de Beni Mellal (Maroc): Part 5, Oiseaux. Notes et Mém. Serv. Géol. Maroc 20 (155):105–108.

———. 1971. Oiseaux miocènes de Beni-Mellal (Maroc); un complément à leur étude. Notes et Mém. Serv. Géol. Maroc 31(237):109–111.

Busack, S. D. 1986. Biogeographic analysis of the herpetofauna separated by the formation of the Strait of Gibraltar. Natl. Geogr. Res. 2:17–36.

Campbell, K. E., Jr., and C. D. Frailey. 1985. Paleontological investigations in southeastern Peru. Natl. Geogr. Soc. Res. Rep. 18:189–199.

Campbell, K. E., Jr., and E. P. Tonni. 1980. A new genus of teratorn from the Huayquerian of Argentina (Aves: Teratornithidae). Contr. Sci. Los Angeles Co. Mus. Nat. Hist. 330:59–68.

Case, J. A., M. O. Woodburne, and D. S. Chaney. 1987. A gigantic phororhacoid(?) bird from Antarctica. J. Paleontol. 61:1280–1284.

Chapin, J. P. 1932. The birds of the Belgian Congo, part I. Bull. Amer. Mus. Nat. Hist. 65:1–391.

———. 1939. The birds of the Belgian Congo, part II. Bull. Amer. Mus. Nat. Hist. 75:1–632.

Chiappe, L. M. 1991a. Cretaceous avian remains from Patagonia shed new light on the early radiation of birds. Alcheringa 15:333–338.

———. 1991b. Cretaceous birds of Latin America. Cretaceous Res. 12:55–63.

———. 1991c. Fossil birds from the Miocene Pinturas Formation of southern Argentina. (Abstract.) J. Vert. Paleontol. 11 (Suppl. 3):21A–22A.

Cody, M. L., and M. A. Mooney. 1978. Convergence versus nonconvergence in mediterranean-climate ecosystems. Annu. Rev. Ecol. Syst. 9:265–321.

Cracraft, J. 1973a. Continental drift, paleoclimatology, and the evolution and biogeography of birds. J. Zool. 169:455–545.

———. 1973b. Systematics and evolution of the Gruiformes (Class Aves). 3. Phylogeny of the Suborder Grues. Bull. Amer. Mus. Nat. Hist. 151:1–128.

———. 1974. Phylogeny and evolution of the ratite birds. Ibis 116:494–521.

———. 1976. Avian evolution on southern continents: Influences of palaeogeography and palaeoclimatology. Proc. 16th Internat. Ornithol. Congr., pp. 40–52.

Cuello, J. P. 1988. Lista de las aves fósiles de la región neotropical y de las islas antillanas. Paula-Coutiana 2:3–79.

Dorst, J. 1976. Historical factors influencing the richness and diversity of the South American avifauna. Proc. 16th Internat. Ornithol. Congr., pp. 17–35.

Dorst, J., and F. Vuilleumier. 1986. Convergences in bird communities at high altitudes in the tropics (especially the Andes and Africa) and at temperate latitudes (Tibet). Pp. 120–149 *in* F. Vuilleumier and M. Monasterio (eds.), *High Altitude Tropical Biogeography.* New York: Oxford University Press.

Ellenberger, P. 1970. Les niveaux paléontologiques de première apparition des Mammifères primordiaux en Afrique du Sud et leur ichnologie. Establissement [sic] de zones stratigraphiques détaillées dans le Stormberg du Lesotho (Afrique du Sud) (Trias Supérieur à Jurassique). Pp. 343–370 *in* International Union of Geological Sciences, Commission on Stratigraphy, Sub-Commission on Gondwana Stratigraphy and Palaeontology, *Proceedings and Papers, Second Gondwana Symposium, South Africa, 1970.* Pretoria: Council for Scientific and Industrial Research.

———. 1972. Contribution à la classification des

pistes de vertébrés du Trias: Les types du Stormberg d'Afrique du Sud (I). Palaeovertebrata, Mém. Extra., 1972:1–152.
———. 1974. Contribution à la classification des pistes de vertébrés du Trias: Les types du Stormberg d'Afrique du Sud (II: Le Stormberg Supérieur, I. Le biome de la zone B/1 ou niveau de Moyeni: ses biocénoses). Palaeovertebrata, Mém. Extra., 1974:i–ii, 1–202.
Erard, C. 1989. Bird community structure in two rainforests: Africa (Gabon) and South America (French Guiana): A comparison. Pp. 89–122 *in* M. L. Harmelin-Vivien and F. Bourlière (eds.), *Vertebrates in Complex Tropical Systems*. New York: Springer-Verlag.
Feduccia, A., and S. L. Olson. 1982. Morphological similarities between the Menurae and the Rhinocryptidae, relict passerine birds of the Southern Hemisphere. Smithsonian Contr. Zool. 366:1–22.
Fry, C. H. 1983. Birds in savannah ecosystems. Pp. 337–357 *in* F. Bourlière (ed.), *Tropical Savannahs*. Amsterdam: Elsevier.
Haarhoff, P. J. 1988. A new fossil stork (Aves, Ciconiidae) from the late Tertiary of Langebaanweg, South Africa. Ann. S. Afr. Mus. 97:297–313.
Hall, B. P. 1963. The francolins, a study in speciation. Bull. Brit. Mus. (Nat. Hist.) Zool. 10:107–204.
———. 1972. Causal ornithogeography of Africa. Proc. 15th Internat. Ornithol. Congr., pp. 585–593.
Handtke, K., and G. Mauersberger. 1977. Die Ausbreitung der Kuhreihers *Bubulcus ibis* (L.). Mitt. Zool. Mus. Berlin 53 (Suppl.), Annal. Ornithol.:1.
Harrison, C. J. O. 1974. A re-examination of material of the extinct marabou stork, *Leptoptilos falconeri:* With descriptions of some new species. Bull. Brit. Ornithol. Club 94:42–49.
———. 1980. Fossil birds from Afrotropical Africa in the collection of the British Museum (Natural History). Ostrich 51:92–98.
Harrison, C. J. O., and C. A. Walker. 1976a. Cranial material of Oligocene and Miocene flamingos: With a description of a new species from Africa. Bull. Brit. Mus. (Nat. Hist.) Geol. 27:305–314.
———. 1976b. A new fossil pelican from Olduvai. Bull. Brit. Mus. (Nat. Hist.) Geol. 27:315–320.
———. 1976c. A review of the bony-toothed birds (Odontopterygiformes): With descriptions of some new species. Tertiary Res. Special Pap. No. 2:1–62.
———. 1979. A Recent and an extinct cormorant species from the Middle Pleistocene of Tanzania. Ostrich 50:182–183.
———. 1982. Fossil birds from the Upper Miocene of northern Pakistan. Tertiary Res. 4:53–69.
Hay, R. L. 1990. Olduvai Gorge; a case history in the interpretation of hominid paleoenvironments in East Africa. Geol. Soc. Amer. Spec. Pap. 242:23–37.
Hendey, Q. B. 1978. Preliminary report on the Miocene vertebrates from Arrisdrift, South West Africa. Ann. S. Afr. Mus. 76:1–41.
———. 1981. Palaeoecology of the late Tertiary fossil occurrences in 'E' Quarry, Langebaanweg, South Africa, and a reinterpretation of their geological context. Ann. S. Afr. Mus. 84:1–104.
Hill, A., and A. Walker. 1979. A fossil marabou (Aves: Ciconiidae) from the Miocene Ngorora Formation, Baringo District, Kenya. Netherlands J. Zool. 29:215–220.
Houde, P. 1986. Ostrich ancestors found in the Northern Hemisphere suggest new hypothesis of ratite origins. Nature 324:563–565.
———. 1988. Paleognathous birds from the early Tertiary of the Northern Hemisphere. Publ. Nuttall Ornithol. Club, No. 22:1–148.
———. 1992. The oldest known anseriform: An early Eocene "screamer" from Wyoming (Anhimidae). Abstracts, 110th Stated Meeting of the American Ornithologists' Union, Ames, Iowa, 24–28 June 1992: no. 96 [p. 36].
Houde, P., and H. Haubold. 1987. *Palaeotis weigelti* restudied: A small Middle Eocene ostrich (Aves: Struthioniformes). Palaeovertebrata 17:27–42.
Houston, D. C. 1985. Evolutionary ecology of

Afrotropical and Neotropical vultures in forests. Pp. 856–864 *in* P. A. Buckley, M. S. Foster, E. S. Morton, R. S. Ridgely, and F. G. Buckley (eds.), *Neotropical Ornithology,* Ornithol. Monogr. Amer. Ornithol. Union No. 36. Washington, D.C.: American Ornithologists' Union.

Karr, J. R. 1976. Within and between habitat avian diversity in African and Neotropical lowland habitats. Ecol. Monogr. 46:457–481.

Keast, A. 1972a. Ecological opportunities and dominant families, as illustrated by the Neotropical Tyrannidae (Aves). Evol. Biol. 5:229–277.

———. 1972b. Faunal elements and evolutionary patterns: Africa, South America, and Australia. Proc. 15th Internat. Ornithol. Congr., pp. 594–622.

———. 1988. Physical geography relative to ornithogeography: Africa compared to South America and Australia. Proc. 6th Pan-Afr. Ornithol. Congr., pp. 347–374.

Keith, S. 1980. Origins of the avifauna of the Malagasy Region. Proc. 4th Pan-Afr. Ornithol. Congr., pp. 99–108.

Kent, D. V., and F. M. Gradstein. 1985. A Cretaceous and Jurassic geochronology. Bull. Geol. Soc. Amer. 96:1419–1427.

Lambrecht, K. 1928. *Palaeotis weigelti* n.g. n.sp., eine fossile Trappe aus der mitteleozänen Braunkohle des Geiseltales. Jb. hallesch. Verband., n.s., 7:1–11.

———. 1930. Studien über fossile Riesenvögel. Geologica Hungarica, Series Palaeontologica, Fasc. 7:1–37.

Leonardi, G. (ed.). 1987. Glossary and manual of tetrapod footprint palaeoichnology. Brasília: Departamento Nacional da Produção Mineral.

Lockley, M. G., S. Y. Yang, M. Matsukawa, F. Fleming, and S. K. Lim. 1992. The track record of Mesozoic birds: Evidence and implications. Philos. Trans. Roy. Soc. London B 336:113–134.

Marshall, L. G., R. F. Butler, R. E. Drake, G. H. Curtis, and R. H. Tedford. 1979. Calibration of the great American interchange. Science 204:272–279.

Marshall, L. G., A. Berta, R. Hoffstetter, R. Pascual, O. A. Reig, M. Bombin, and A. Mones. 1984. Mammals and stratigraphy: Geochronology of the continental mammal-bearing Quaternary of South America. Palaeovertebrata, Mém. Extra., 1984:1–76.

Martin, L. D. 1983. The origin and early radiation of birds. Pp. 291–338 *in* A. H. Brush and G. A. Clark, Jr. (eds.), *Perspectives in Ornithology, Essays Presented for the Centennial of the American Ornithologists' Union.* Cambridge: Cambridge University Press.

Martins Neto, R. G., and A. W. A. Kellner. 1988. Primeiro registro de pena na formação Santana (Cretáceo Inferior), bacia do Araripe, Nordeste do Brasil. An. Acad. brasil. Ci. 60:61–68.

Meyer de Schauensee, R. 1982. *A Guide to the Birds of South America.* Philadelphia: Intercollegiate Press, Pan-American Section, International Council for Bird Preservation.

Mlíkovský, J. 1989. A new guineafowl (Aves: Phasianidae) from the late Eocene of France. Ann. Naturhist. Mus. Wien 90A:63–66.

Molnar, R. E. 1989. Terrestrial tetrapods in Cretaceous Antarctica. Pp. 131–140 *in* J. A. Crame (ed.), *Origins and Evolution of the Antarctic Biota,* Geol. Soc. Special Publ. No. 47. Oxford: Alden.

Moreau, R. E. 1966. *The Bird Faunas of Africa and Its Islands.* New York: Academic.

Mourer-Chauviré, C. 1980. The Archaeotrogonidae of the Eocene and Oligocene Phosphorites du Quercy (France). Contr. Sci. Los Angeles County Mus. Nat. Hist. 330:17–31.

———. 1982. Les oiseaux fossiles des Phosphorites du Quercy (Éocène supérieur à Oligocène supérieur): Implications paléobiogéographiques. Geobios, Mém. Spéc. 6:413–426.

———. 1983. Les Gruiformes (Aves) des Phosphorites du Quercy (France). 1. Sous-ordre Cariamae (Cariamidae et Phorusrhacidae). Systématique et biostratigraphie. Palaeovertebrata 13:83–143.

———. 1988. Les Caprimulgiformes et les Coraciiformes de l'Éocène et de l'Oligocène des phosphorites du Quercy et description de deux genres nouveaux de Podargidae et Nyctibiidae. Acta XIX Congr. Internat. Ornithol., Vol. 2:2047–2055.

Muizon, C. de, and T. J. DeVries. 1985. Geology and paleontology of late Cenozoic marine deposits in the Sacaco area (Peru). Geol. Rundschau 74:547–563.

Nasif, N. L. 1988. Primer registro de flamencos (Phoenicopteridae) del Terciario superior, Valle del Cajón (Provincia de Catamarca, Argentina). Ameghiniana 25:169–173.

Olrog, C. C. 1969. Birds of South America. Pp. 849–878 *in* E. J. Fittkau, J. Illies, H. Klinge, G. H. Schwabe, and H. Sioli (eds.), *Biogeography and Ecology in South America,* Vol. 2. The Hague: W. Junk.

Olsen, P. E., and P. M. Galton. 1984. A review of the reptile and amphibian assemblages from the Stormberg of southern Africa, with special emphasis on the footprints and the age of the Stormberg. Palaeontol. Afr. 25:87–110.

Olson, S. L. 1974. *Telecrex* restudied: A small Eocene guineafowl. Wilson Bull. 86:246–250.

———. 1975. An evaluation of the supposed anhinga of Mauritius. Auk 92:374–376.

———. 1981. The generic allocation of *Ibis pagana* Milne-Edwards, with a review of fossil ibises (Aves: Threskiornithidae). J. Vert. Paleontol. 1:165–170.

———. 1983. Fossil seabirds and changing marine environments in the late Tertiary of South Africa. S. Afr. J. Sci. 79:399–402.

———. 1984. A hamerkop from the early Pliocene of South Africa (Aves: Scopidae). Proc. Biol. Soc. Washington 97:736–740.

———. 1985a. Early Pliocene ibises (Aves, Plataleidae) from south-western Cape Province, South Africa. Ann. S. Afr. Mus. 97:57–69.

———. 1985b. An early Pliocene marine avifauna from Duinefontein, Cape Province, South Africa. Ann. S. Afr. Mus. 95:147–164.

———. 1985c. Early Pliocene Procellariiformes (Aves) from Langebaanweg, south-western Cape Province, South Africa. Ann. S. Afr. Mus. 95:123–145.

———. 1985d. The fossil record of birds. Pp. 79–252 *in* D. S. Farner, J. R. King, and K. C. Parkes (eds.), *Avian Biology,* Vol. 8. Orlando, Fla.: Academic.

———. 1987. An Early Eocene oilbird from the Green River Formation of Wyoming (Caprimulgiformes: Steatornithidae). Docum. Lab. Géol. Lyon, No. 99:57–69.

———. 1992. *Neogaeornis wetzeli* Lambrecht, a Cretaceous loon from Chile (Aves: Gaviidae). J. Vert. Paleontol. 12(1):122–124.

Olson, S. L., and K. G. Eller. 1989. A new species of painted snipe (Charadriiformes: Rostratulidae) from the Early Pliocene at Langebaanweg, Southwestern Cape Province, South Africa. Ostrich 60:118–121.

Olson, S. L., and A. Feduccia. 1980a. *Presbyornis* and the origin of the Anseriformes (Aves: Charadriomorphae). Smithsonian Contr. Zool. 323:1–24.

———. 1980b. Relationships and evolution of flamingos (Aves: Phoenicopteridae). Smithsonian Contr. Zool. 316:1–73.

Olson, S. L., and H. F. James. 1982. Prodromus of the fossil avifauna of the Hawaiian Islands. Smithsonian Contr. Zool. 365:i–vi + 1–59.

Olson, S. L., and E. N. Kurochkin. 1987. Fossil evidence of a tapaculo in the Quaternary of Cuba (Aves: Passeriformes: Scytalopodidae). Proc. Biol. Soc. Washington 100:353–357.

Olson, S. L., and D. T. Rasmussen. 1986. Paleoenvironment of the earliest hominoids: New evidence from the Oligocene avifauna of Egypt. Science 233:1202–1204.

Patterson, B., and J. L. Kraglievich. 1960. Sistemática y nomenclatura de las aves fororracoídeas del Plioceno argentino. Publ. Mus. Mun. Cienc. Nat. Tradicional Mar del Plata 1:1–51.

Pearson, D. L. 1977. A pantropical comparison of bird community structure of six lowland forest sites. Condor 79:232–244.

Peters, D. S. 1987. Ein "Phorusrhacide" aus dem Mittel-Eozän von Messel (Aves: Gruiformes: Cariamae). Docum. Lab. Géol. Lyon, No. 99:71–87.

———. 1988. Ein vollständiges Exemplar von *Palaeotis weigelti* (Aves, Palaeognathae).

Cour. Forsch.-Inst. Senckenberg 107:223–233.

Peters, J. L., et al. 1931–1987. Check-list of birds of the world, Vols. 1–16. Cambridge: Harvard University Press.

Rasmussen, D. T., and R. F. Kay. 1992. A Miocene anhinga from Colombia, and comments on the zoogeographic relationships of South America's Tertiary avifauna. Pp. 225–230 *in* K. E. Campbell, Jr. (ed.), *Papers in Avian Paleontology Honoring Pierce Brodkorb*. Science Series No. 36. Los Angeles: Natural History Museum of Los Angeles County.

Rasmussen, D. T., S. L. Olson, and E. L. Simons. 1987. Fossil birds from the Oligocene Jebel Qatrani Formation, Fayum Province, Egypt. Smithsonian Contrib. Paleobiol. 62:1–20.

Rich, P. V. 1972. A fossil avifauna from the upper Miocene Beglia Formation of Tunisia. Notes Serv. Géol., Tunis, 35, Trav. géol. tunisienne No. 5 (1971), Formation Beglia, Fascicule 1:29–66.

———. 1974. Significance of the Tertiary avifaunas from Africa (with emphasis on a mid to late Miocene avifauna from southern Tunisia). Ann. Geol. Surv. Egypt 4:167–210.

———. 1979. Fossil birds of Old Gondwanaland: A comment on drifting continents and their passengers. Pp. 321–332 *in* J. Gray and A. J. Boucot (eds.), *Historical Biogeography, Plate Tectonics, and the Changing Environment*. Eugene: Oregon State University Press.

———. 1980a. *"New World Vultures" with Old World Affinities?: A Review of Fossil and Recent Gypaetinae of both the Old and the New World*. Contributions to Vertebrate Evolution, Vol. 5. Basel: S. Karger.

———. 1980b. Preliminary report on the fossil avian remains from Late Tertiary sediments at Langebaanweg (Cape Province), South Africa. S. Afr. J. Sci. 76:166–170.

———. 1983. The fossil history of vultures: A world perspective. Pp. 3–25 *in* S. R. Wilbur and J. A. Jackson (eds.), *Vulture Biology and Management*. Berkeley: University of California Press.

Rich, P. V., and P. J. Haarhoff. 1985. Early Pliocene Coliidae (Aves, Coliiformes) from Langebaanweg, South Africa. Ostrich 56:20–41.

Rich, P. V., and G. Van Tets. 1984. What fossil birds contribute towards an understanding of origin and development of the Australian avifauna. Pp. 421–446 *in* M. Archer and G. Clayton (eds.), *Vertebrate Zoogeography and Evolution in Australasia (Animals in Space and Time)*. Carlisle and Victoria Park, West. Aust.: Hesperian Press.

Rich, P. V., and C. A. Walker. 1983. A new genus of Miocene flamingo from East Africa. Ostrich 54:95–104.

Rothe, P. 1964. Fossile Strausseneier auf Lanzarote. Natur und Museum 94:175–187.

Savage, D. E., and D. E. Russell. 1983. *Mammalian Paleofaunas of the World*. London: Addison-Wesley.

Sclater, P. L. 1858. On the general geographical distribution of the members of the class Aves. J. Proc. Linn. Soc. Zool. 2:130–145.

Short, L. L. 1970. The affinity of African with Neotropical woodpeckers. Ostrich 8 (Suppl.): 35–40.

———. 1985. Neotropical-Afrotropical barbet and woodpecker radiations: A comparison. Pp. 559–574 *in* P. A. Buckley, M. S. Foster, E. S. Morton, R. S. Ridgely, and F. G. Buckley (eds.), *Neotropical Ornithology*. Ornithol. Monogr. Amer. Ornithol. Union, No. 36. Washington, D.C.: American Ornithologists' Union.

Sibley, C. G., and J. E. Ahlquist. 1985. The relationships of some groups of African birds, based on comparisons of the genetic material, DNA. Proc. Intern. Symp. African Vertebrates, Bonn, 1985, pp. 115–165.

———. 1986. Reconstructing bird phylogeny by comparing DNA's. Scient. Amer. 254:82–92.

———. 1990. *Phylogeny and Classification of Birds: A Study in Molecular Evolution*. New Haven and London: Yale University Press.

Sibley, C. G., and B. L. Monroe, Jr. 1990. *Distribution and Taxonomy of Birds of the World*. New Haven and London: Yale University Press.

Simpson, G. G. 1971. Fossil penguin from the late Cenozoic of South Africa. Science 171: 1144–1145.

———. 1973. Tertiary penguins (Sphenisciformes, Sphenicidae) from Ysterplaats, Cape Town, South Africa. S. Afr. J. Sci. 69:342–344.

———. 1975. Notes on variation in penguins and on fossil penguins from the Pliocene of Langebaanweg, Cape Province, South Africa. Ann. S. Afr. Mus. 69:59–72.

———. 1979a. A new genus of late Tertiary penguin from Langebaanweg, South Africa. Ann. S. Afr. Mus. 78:1–9.

———. 1979b. Tertiary penguins from the Duinefontein site, Cape Province, South Africa. Ann. S. Afr. Mus. 79:1–7.

———. 1980a. Fossil birds and evolution. Contr. Sci. Los Angeles Co. Mus. Nat. Hist. 330:3–8.

———. 1980b. *Splendid Isolation: The Curious History of South American Mammals.* New Haven and London: Yale University Press.

Steadman, D. W. 1986. Holocene vertebrate fossils from Isla Floreana, Galápagos. Smithsonian Contr. Zool. 413:1–103.

———. 1988. Fossil birds and biogeography in Polynesia. Acta XIX Congr. Internat. Ornithol., Vol. 2:1526–1534.

———. 1989. New species and records of birds (Aves: Megapodiidae, Columbidae) from an archeological site on Lifuka, Tonga. Proc. Biol. Soc. Washington 102:537–552.

Steadman, D. W., and P. V. Kirch. 1990. Prehistoric extinction of birds on Mangaia, Cook Islands, Polynesia. Proc. Nat. Acad. Sci. U.S.A. 87:9605–9609.

Steadman, D. W., and S. L. Olson. 1985. Bird remains from an archaeological site on Henderson Island, South Pacific: Man-caused extinctions on an "uninhabited" island. Proc. Nat. Acad. Sci. U.S.A. 82:6191–6195.

Tambussi, C. P. 1987. Catálogo crítico de los Tinamidae (Aves: Tinamiformes) fósiles de la República Argentina. Ameghiniana 24:241–244.

Thiollay, J.-M. 1986. Diversité spécifique et écologie comparée des Falconiformes de forêts tropicales sur trois continents. Mém. Mus. Natn. Hist. Nat., n.s. (A) Zool. 132:67–77.

Tonni, E. P. 1980. The present state of knowledge of the Cenozoic birds of Argentina. Contr. Sci. Los Angeles Co. Mus. Nat. Hist. 330:105–114.

Tonni, E. P., and C. P. Tambussi. 1986. Las aves del Cenozoico de la República Argentina. Actas IV Congr. Argent. Paleontol. Bioestrat., Mendoza 2:131–142.

———. 1988. Un nuevo Psilopterinae (Aves: Ralliformes) del Mioceno tardío de la Provincia de Buenos Aires, República Argentina. Ameghiniana 25:155–160.

Ubilla, M., D. Perea, C. Tambussi, and E. P. Tonni. 1990. Primer registro fósil de Phoenicopteridae (Aves: Charadriiformes) para el Uruguay (Mio-Plioceno). An. Acad. brasil. Ci. 62:61–68.

Udvardy, M. D. F. 1987. The biogeographical realm Antarctica: A proposal. J. Royal Soc. New Zealand 17:187–194.

Van Couvering, J. A., and J. A. Harris. 1991. Late Eocene age of Fayum mammal faunas. J. Human Evol. 21:241–260.

Voous, K. H. 1964. Wood owls of the genera *Strix* and *Ciccaba.* Zool. Meded. Leiden 39:471–478.

———. 1966. The distribution of owls in Africa in relation to general zoogeographical problems. Ostrich 6 (Suppl.):499–506.

Vuilleumier, F. 1984. Faunal turnover and development of fossil avifaunas in South America. Evolution 38:1384–1396.

———. 1985a. Forest birds of Patagonia: Ecological geography, speciation, endemism, and faunal history. Pp. 255–304 *in* P. A. Buckley, M. S. Foster, E. S. Morton, R. S. Ridgely and F. G. Buckley (eds.), *Neotropical Ornithology.* Ornithol. Monogr. Amer. Ornithol. Union, No. 36. Washington, D.C.: American Ornithologists' Union.

———. 1985b. Fossil and Recent avifaunas and the Interamerican Interchange. Pp. 387–424 *in* F. G. Stehli and S. D. Webb (eds.), *The Great American Biotic Interchange.* New York: Plenum.

———. 1985c. Fossil evidence on the development of South American avifaunas. Acta XVIII Congr. Internat. Ornithol., pp. 348–357.

———. 1991. Invasions in the Mediterranean avifaunas of California and Chile. Pp. 323–354 *in* R. Groves and F. di Castri (eds.), *Biogeography of Mediterranean Invasions*. Cambridge: Cambridge University Press.

Webb, S. D. 1976. Mammalian faunal dynamics of the great American interchange. Paleobiology 2:220–234.

Winge, O. 1888. Fugle fra Knoglehuler i Brasilien. E Museo Lundii 1(2):1–54.

Winterbottom, J. M. 1965. Avifaunal relationships between the Neotropical and Ethiopian regions. Hornero 10:209–214.

———. 1968. Remarks on the possible origin of the avifauna of the South West Cape. Ibis 110:91–93.

Wolf, L. L., and F. B. Gill. 1986. Physiological and ecological adaptations of high montane sunbirds and hummingbirds. Pp. 103–119 *in* F. Vuilleumier and M. Monasterio (eds.), *High Altitude Tropical Biogeography.* New York: Oxford University Press.

Woodburne, M. O., and W. J. Zinsmeister. 1982. Fossil land mammal from Antarctica. Science 218:284–286.

Chapter 12

Evolution of the Neotropical Cenozoic Land Mammal Fauna in Its Geochronologic, Stratigraphic, and Tectonic Context

Larry G. Marshall and Thierry Sempere

I. Introduction

Three features of the Cenozoic record of South American land mammals render it ideal for macroevolutionary studies. First, the record is remarkably complete, particularly in the south (i.e., Argentina and Bolivia), and multiple faunas are known for every geological epoch. Second, South America had no direct land connections with other continents during most of Cenozoic time and this resulted in the development of a peculiar autochthonous fauna. And third, most major changes in taxonomic composition or "stages of evolution" of land mammal faunas can be correlated with deformations or uplifts of the Andes Cordillera, Caribbean plate history, sea level lowstands, and "immigration events" (Pascual and Odreman Rivas 1973; Pascual 1984; Baez 1986; Marshall and Cifelli 1990; Legarreta et al. 1990; Pascual and Ortíz-Jaureguizar 1990).

In this chapter we explore the possibility that geological processes and associated crustal movements influenced tempo and mode aspects in the macroevolutionary his-

Aspects of this study were supported by funds from the National Geographic Society (2467-82, 2908-84, 3381-86); Gordon Barbour Fund, Department of Geological and Geophysical Sciences, Princeton University; National Science Foundation (BSR-8614133, EAR-8804423, INT-8814059); two PRA Fellowships from the Organization of American States (F 10072); the Missouri Botanical Garden, St. Louis; and UR 1H of ORSTOM (Institut Français de Recherche Scientifique pour le Développment en Coopération). Helpful information was provided by Jean Dejax and Denise Sigogneau-Russell, and the chapter has been improved thanks to comments by Andrew Tomlinson and Eric Nelson.

 ISBN 0-300-05375-4

tory of Cenozoic land mammal faunas in South America. We attempt to identify these apparent interrelationships by evaluating the land mammal record in its geochronologic, stratigraphic, and tectonic context.

A review of the poorly known record of Mesozoic mammals in South America is given in Appendix A. Abbreviations used in this chapter are: *c*, approximately; *ka*, thousands of years ago; *Ma*, megaannum, or millions of years ago, a point in time; *Myr*, millions of years, a duration of time; SALMA, South American Land Mammal Age. The geological time scale follows Berggren et al. (1985).

II. The Age of Mammals in South America

The record of Cenozoic land mammals in South America is sufficiently complete that geologists and vertebrate paleontologists have long used it to subdivide geologic time. The occurrence of unique associations of taxa that are inferred to have existed during a restricted interval of time resulted in the recognition of discrete faunal-time units called South American Land Mammal Ages (SALMAS). These ages were established on the basis of knowledge of the stage of evolution of the taxa; on their time of first or last appearance, or both, in the fossil record; and on changing faunal associations through time (Marshall 1985).

Beginning in 1975, multidisciplinary programs involving magnetostratigraphic, isotopic (Ar/Ar, K-Ar, fission-track), and refined biostratigraphic studies of mammal-bearing rocks have permitted calibration of some SALMAS with the geologic time scale. As a result of these studies, during the past 15 years, many modifications have been made in age or usage of SALMAS as conceived by earlier workers (i.e., Pascual and Odreman Rivas 1971; Patterson and Pascual 1972).

The composition and chronology of SALMAS used in this study (figure 12.1) follow Marshall (1985) as supplemented by Marshall, Drake et al. (1986), Marshall, Cifelli et al. (1986), Marshall (1989), and Marshall and Salinas (1990a); several additional emendations are proposed and documented below.

Three sets of SALMAS can be recognized based on faunal content. The oldest set, Tiupampian-Divisaderan SALMAS, includes Paleocene, Eocene, and early Oligocene faunas (figure 12.1). These ages are composed of Stratum 1 faunas of G. G. Simpson (1980), which include groups that were in South America at the beginning of the Age of Mammals or were dispersants from North America during latest Cretaceous or early Paleocene time. Included are marsupials and members of the placental orders Leptictida, Pantodonta, Condylarthra, Xenarthra, Litopterna, Notoungulata, Pyrotheria, Trigonostylopoidea and Xenungulata.

The second set, Deseadan-Montehermosan SALMAS, includes late Oligocene–middle Pliocene faunas. These faunas are characterized by the appearance of caviomorph Rodentia and platyrrhine Primates (Stratum 2 or "Old Island Hoppers" of G. G. Simpson 1980), which were dispersants from Caribbean Island stocks and first appear in late Oligocene (Deseadan) time, and procyonids ("New Island Hoppers" of G. G. Simpson 1980), which were dispersants from North or Central America and first appear in late Miocene (Huayquerian) time.

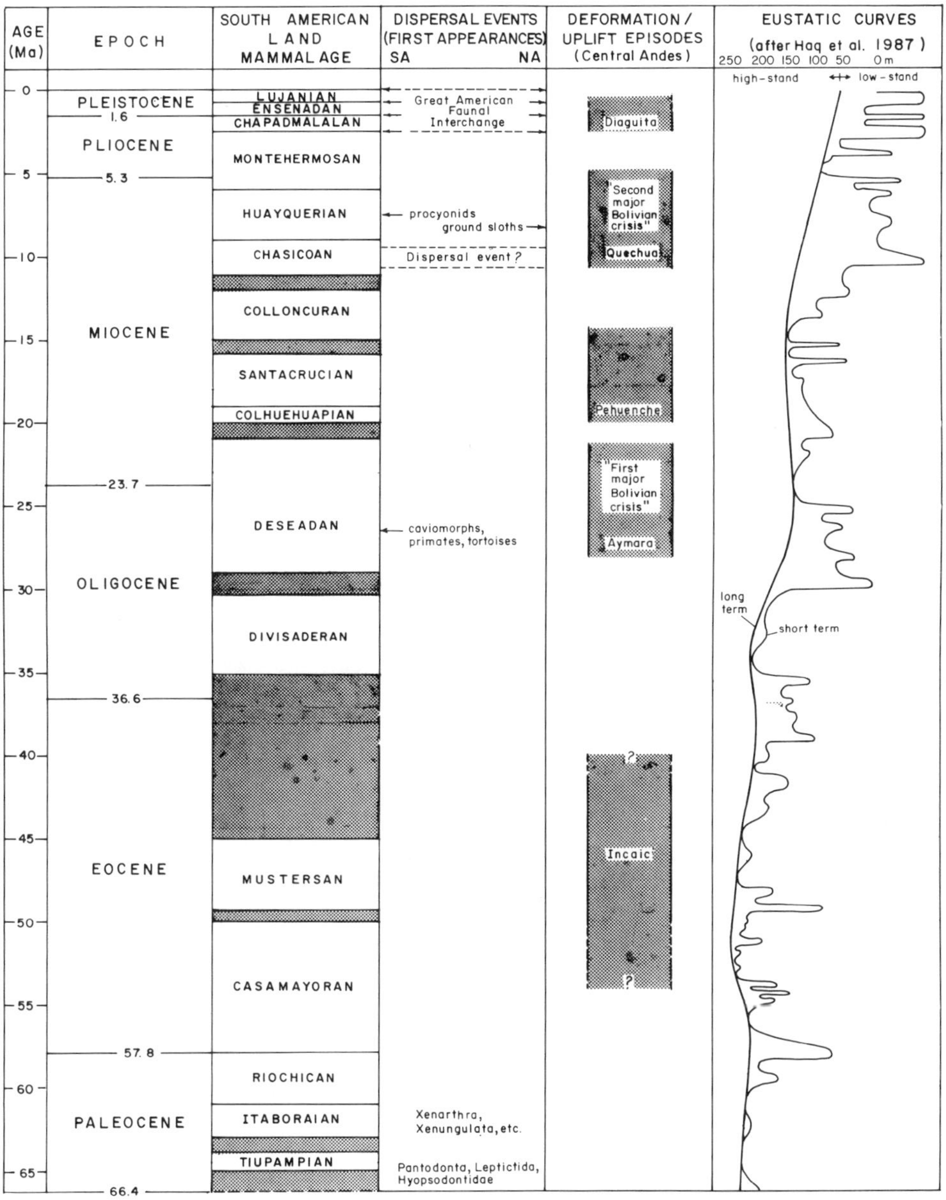

Figure 12.1. Chronology of South American Land Mammal Ages, dispersal events, deformation/uplift episodes, and sea levels during Cenozoic time ("Aymara" episode *sensu* Sébrier et al. 1988, but see Sempere 1991 for a critique of its validity). Geologic time scale follows Berggren et al. (1985).

The youngest set of late Pliocene–Pleistocene faunas—Chapadmalalan-Lujanian SALMAS—includes members of Soricidae, Leporidae, Heteromyidae, Sciuridae, Cricetidae, Gomphotheriidae, additional Procyonidae, Mustelidae, Canidae, Ursidae, Felidae, Tapiridae, Equidae, Tayassuidae, Camelidae, Cervidae, and Hominidae (Stratum 3 of G. G. Simpson 1980), which walked to South from North America as participants in the Great American Faunal Interchange (GAFI) following the emergence of the Panamanian Land Bridge about 2.5 Ma.

III. Geologic Factors

Plate tectonics serves as a common basis for the two primary geologic factors that influenced the evolution of the South American land mammal fauna. First, the history of the interaction between South America and the adjacent Caribbean Plate, which was periodically land-connected to North America, highlights the opportunities that may have existed for land mammal immigration to South America from the north, permitting allochthonous additions to the South American fauna. Second, the fact that the Andean Cordillera was built during the Cenozoic on the leading (western) edge of South America by processes related to plate convergence and subduction directly leads to the idea that this presently high mountain range produced a rain shadow as soon as it had reached a sufficient elevation. According to latitudes and wind paleocirculations, such a rain shadow created so-called stress environments, where extinction and speciation rates were accelerated and from where new successful species or complete faunas could disperse to other regions of South America. Because of these ecological changes, the building of the Andean Cordillera apparently induced or monitored most aspects of the autochthonous evolution of the South American land mammal fauna.

A. Caribbean Tectonics and Its Relation to Faunal Immigration Events

As illustrated by the three sets of SALMAS, only three principal immigration events occurred since latest Cretaceous time. The first—probably including several subevents from latest Cretaceous to early Paleocene—and the third—the GAFI, from late Pliocene to Recent—are of outstanding importance, because they involved reciprocal interchanges between North and South America. In contrast, the second immigration event, in the late Oligocene, which involved only caviomorph Rodentia and platyrrhine Primates (Old Island Hoppers), appears as a one-way (nonreciprocal) event. Geological data support inferences from the paleontological data that the first and third immigration events were due to the establishment of landbridges between these continents through tectonic and magmatic processes and eustatic sea level changes, and that the second event was likely related to the formation of a land connection only with the Caribbean Islands and their endemic faunal stocks.

Thus, only one primary immigration event, albeit of noteworthy consequence, occurred between the submergence of the Paleocene landbridge and the emergence of the Pliocene one. It has long been recognized that, as a result of these historical dispersal opportunities, the evolution of the South American land mammal fauna was almost totally autochthonous during most of Cenozoic time (the "Splendid Isolation" of G. G. Simpson 1980).

As documented below, the mammals in the first set of SALMA faunas are, at least in part, related to late Cretaceous–early Paleocene faunas of North America. Rage (1978, 1981, 1986) listed several nonmammalian vertebrate groups that very likely participated in one or more late Cretaceous interchanges between North and South America (see also Bonaparte 1984, 1986a), including dinosaurs (Avisauridae, Ceratopsidae, Hadrosauridae, Titanosauridae), snakes (Anilioidea, Boidea), and possibly lizards (Teiidae) and shore-dwelling turtles (Bothremydidae). Other possible participants are listed by Gayet and co-workers (1991, 1992). To explain some aspects of this terrestrial interchange, Rage (1978) proposed that a landbridge connecting the Americas existed in the late Cretaceous, although such a conclusion was not sustained by geodynamic data at the time he raised this possibility. Recent studies (Pindell and Dewey 1982; Beck 1986; Pindell et al. 1988; Stephan et al. 1990), however, document consistent and complementary geological data that corroborate the paleontological evidence for a late Cretaceous–Paleocene landbridge (figure 12.2A).

Although the origin of the Caribbean Plate remains poorly known, it appears that it was initially located south of and in connection with nuclear Central America and that it began its east-northeastward migration in Campanian time, shortly after its development as a separate plate (Pindell et al. 1988; Pindell and Barrett 1990). The Greater Antilles and the now-submerged Aves Ridge were in the late Cretaceous–early Paleocene a subaerial volcanic arc on the northeastern edge of the Caribbean plate (figure 12.2A). This arc is likely to have provided the pathway for the first immigration event or events. Although another volcanic arc existed on the southwestern edge of this plate (presently Panama and Costa Rica), it probably did not serve as a land connection between the Americas because of its discontinuous character (figure 12.2A). Collisions and accretions of Caribbean continental and oceanic terranes against southern Yucatan and northwestern South America are known to have occurred in the late Senonian–early Paleocene (Anderson et al. 1985; Bourgois et al. 1987; Chevalier et al. 1988; Pindell et al. 1988), and such tectonic processes must have played some role in the uplift and emergence of the Aves Ridge in latest Cretaceous time. The landbridge provided by the Aves Ridge was probably strongly dependent on the magmatic and tectonic activity of the volcanic arc and on sea level changes. Magmatic-tectonic quiescence or marine high-stand periods, or both, would have promoted the sinking of part of the isthmus below sea level. In view of the probable fragility of the Aves Ridge connection, it seems likely that faunal interchange occurred in several pulses until the landbridge finally disappeared in late Paleocene time, as suggested by faunal data. This pulse-hypothesis allows for opportunities for endemic evolution of first-wave immigrants to South America and for subsequent dispersal of some of these stocks back to North America during a later pulse (Marshall and Muizon 1988).

The eastern and western edges of the Caribbean Plate can be visualized as two arms extending from North America and reaching

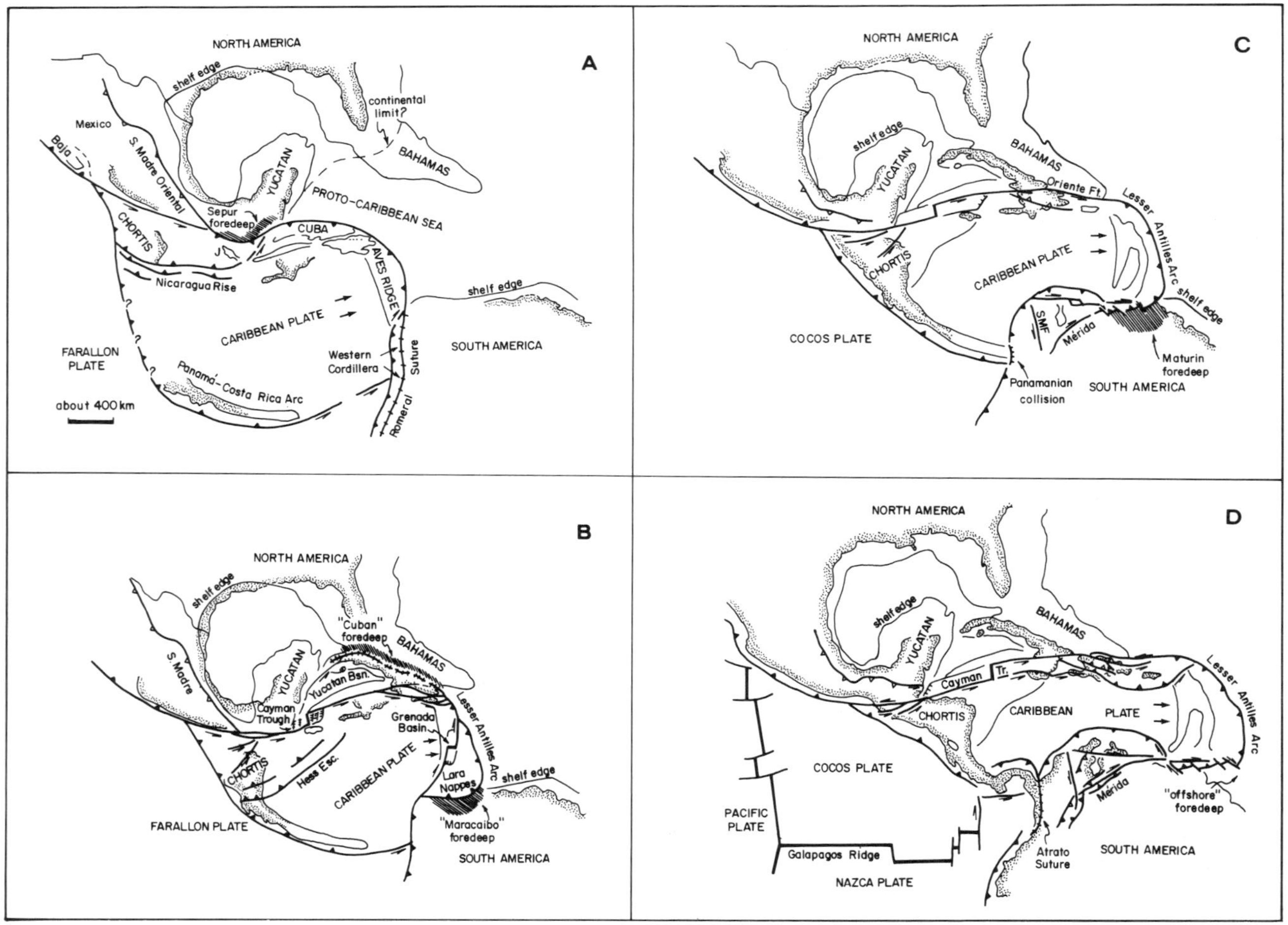

Figure 12.2. Configuration of Caribbean area during the late Cretaceous–early Paleocene (*A*), the middle Eocene (*B*), the early Miocene (*C*), and today (*D*), illustrating potential routes for dispersal of North American or Caribbean Island faunal stocks. Modified from Pindell et al. (1988).

out for South America (figures 12.2B and 12.2C). Today, the western arm (Central America and the Panamanian isthmus) is fully connected to South America, whereas the eastern arm is fragmented into numerous islands stretching from Florida to the southernmost Lesser Antilles (figure 12.2D). Present and restored land configurations and repartitions along the southwestern and northeastern edges of the Caribbean region show a clear diminution in land and island surfaces from north to south owing to the transition from the Greater to the Lesser Antilles and to the fact that Central America grew subaerially southward from North America (Coney 1982). The longitudinal asymmetry of these two southern land or island appendixes of North America has an important consequence: both arms, even where and when formed by island arcs, could easily have received faunal stocks from North America, especially during times of local tectonic uplift or volcanic buildup and global sea level low-stands; in contrast, it would have been much more difficult for South American stocks to migrate northward, because of the permanence of marine barriers just north of the continental mass. These geologic features help to explain the nonreciprocal nature of the late Oligocene immigration event.

Complete land connection between the Americas existed via the northeastern arm during the late Cretaceous–early Paleocene and via the southwestern arm during the Pliocene-Quaternary. But between these epochs (i.e., during Eocene to Miocene times), the Caribbean Plate continued moving east-northeastward relative to South America, and the paleogeographic relation of North and South America was of the type represented in figure 12.3. The immigration to South America of caviomorph Rodentia and platyrrhine Primates, of which the first South American remains are in rocks of late Oligocene (Deseadan) age, therefore occurred when the paleogeography of the Caribbean–South American interface was dominated by the two island arcs (the western and eastern arms).

It seems quite unlikely that these rodents and monkeys could have been rafted directly from Africa to South America (see discussion in Patterson and Wood 1982), because these continents were at that time at least 2000 kilometers apart. It is therefore probable that the caviomorphs and platyrrhines came to South America from nearby islands somewhere on the Caribbean plate, and that these stocks had come to such islands from either North America or Africa at a prior time (late Paleocene? or Eocene?).

We thus favor the following scenario for immigration of the Old Island Hoppers to South America:

1. Islands along the western and eastern edges of the Caribbean Plate were colonized by North American faunas every time interchange with the northern continent and between islands was geologically possible. In the late Paleocene or Eocene, stocks of caviomorph Rodentia and/or platyrrhine Primates came to one or more Caribbean Islands as waif (over water) dispersants from North America or Africa.
2. According to the tectonic and magmatic history of each island arc and to sea level changes, endemic evolution of the island faunas was triggered by vicariance and by geographic isolation.

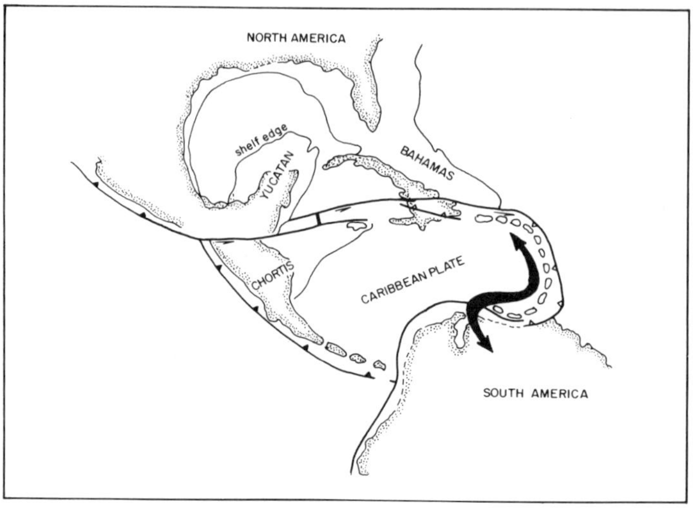

Figure 12.3. Configuration of Caribbean area during late Oligocene–early Miocene, showing potential island areas where caviomorph Rodentia and platyrrhine Primates may have existed at the time of their dispersal to South America in early Deseadan (c. 27 Ma) time. The eastern island arc is favored as the source of these immigrant stocks. Modified from Pindell et al. (1988).

3. Because of the east-northeastward motion of the Caribbean Plate relative to South America, some islands at the southern tips of both arms were tangentially accreted to the South American continent (Speed 1985). This does not necessarily imply that such accreted islands were then land-connected to South America (as exemplified by present-day Margarita Island off the coast of Venezuela).
4. Some accreted islands were populated by endemic Old Island Hoppers that had experienced radiations and autochthonous evolution in their isolated environments, even after accretion if the island mass remained offshore.
5. Dispersal to South America of the endemic stocks from offshore accreted islands would have been optimal at: (a) times of major sea level lows, which would have considerably minimized or eliminated marine barriers; (b) times of major uplift of the leading edge of South America, as due to the similar effects of acceleration of plate convergence and subsequent magmatic reactivation; or (c) a time when both of these factors were operative.

The earliest known Old Island Hoppers are of late Oligocene (c. 27 Ma) age. It is noteworthy that the major marine low-stand registered in the Eocene-Miocene interval occurred at the early Oligocene–late Oligocene boundary and was followed by a period of low relative sea level (between 30 and 25 Ma according to Haq et al. 1987, and between 31 and 28 Ma according to Macharé et al. 1988), that onset of widespread magmatic activity in the northern Andes occurred in the late Oligocene (Aspden et al. 1987), and that intense tectonic and magmatic activity coevally developed in the central Andes starting about 27 Ma (Sempere et al., "Tectonic crisis," 1990). Such a chronological coincidence supports a scenario that the Old Island Hoppers took advantage of the late-Oligocene sea level lows and coeval uplift of the northern Andes to obtain a foothold in South America. They came from accreted islands off Colombia and/or Venezuela that had been initially populated by stocks, possibly already endemic, from North America or Africa.

This scenario incorporates numerous geologic events that are securely known to have occurred: the east-northeastward displacement of the Caribbean Plate relative to South America and tangential accretion of the southern tips of the western and eastern Caribbean island arcs to the South American continent; and late Oligocene regressions and Andean uplift, which diminished marine distances between islands and the continent. Awareness of these important geological factors constrains paleontological interpretations, particularly with regard to the place and time of immigration of the "Old Island Hoppers."

Sea level low-stand seems to have been a necessary coeval condition for faunal immigration to South America. Consider that although an allochthonous terrane, which probably belonged to the western edge of the Caribbean Plate, was accreted to western Colombia in middle Miocene time (Restrepo and Toussaint 1988), no immigrant fauna is known from this period. No major sea level low is known in the Miocene prior to 10.5 Ma (Haq et al. 1987), and this time period correlates with initiation of the second major tectonic crisis ("Quechua") in the central (and northern?) Andes (Sempere et al. 1989, 1990a) and approximates the arrival of procyonids (which first appear at 7.5–7.0 Ma). Accretion of allochthonous blocks would have provided a nearby source for immigrant mammals, but actual immigration was triggered by global sea level low-stand or regional uplift.

The formation of the present Panamanian isthmus is easy to understand in light of Caribbean Plate motion and geometry. Since the Caribbean Plate was created in early Campanian time (Pindell and Barrett 1990; Pindell et al. 1988), it has behaved as a somewhat plastic mass forced between the rigid North and South American continents. Its relative east-northeastward movement led to the formation of an isthmus when the eastern edge of the plate intruded into the passage between the southern Yucatan block and the northwestern tip of Colombia (figure 12.2A). As the Caribbean Plate continued eastward, this isthmus disintegrated, and North and South America were again separated. Paleontological data indicate that this occurred in the late Paleocene (Gayet et al. 1991, 1992). The next opportunity for formation of a new landbridge occurred in the late Pliocene during the squeezing of the western edge of the Caribbean Plate in the Yucatan–northwest Colombia passage, tectonic uplift of westernmost Colombia (McCourt et al. 1984), and a coeval sea level low-stand (Haq et al. 1987).

B. Andean Uplift and South American Land Mammal Autochthonous Evolution

The building of the Andean Cordillera on the leading (western) edge of South America is the most prominent geologic feature of the Cenozoic history of that continent. Although compression and transcurrent tectonics had probably built a low-elevation proto-cordillera in the late Jurassic and early Cretaceous, at least along some parts of the Pacific margin (Davidson and Vicente 1973; Sempere In press), true continuous compressive conditions were established only in the late Cretaceous and have subsequently dominated the tectonic evolution of the active margin (Marocco et al. 1987; Sempere In press). Compression of the margin has generated deformations in specific areas, which, depending on their age and position in the orogen, formed thin-skinned fold-thrust belts or thick-skinned zones of tilted blocks. Extensional conditions developed locally within the orogen.

Most Andean deformation during the Cenozoic resulted in overall crustal shortening of the margin, and therefore in crustal thickening and uplift (Jordan et al. 1983). Uplift of the central Andes was probably enhanced by an asthenospheric wedge beneath the Cordillera root (Isacks 1988). Sedimentary basins developed in conjunction with deformation. Though some of the basins are of extensional origin and others are related to wrench-faulting, most are foreland basins generated by local or regional shortening and related tectonic loading (Sempere 1991). The infillings of the Andean foreland basins provide an excellent record of the tectonic history of the related source-areas, because these infillings underwent contemporaneous deformation and uplift. The mere existence of thick-filled foreland basins directly testifies to a practically continuous deformation in the uplifted source-areas.

Chronology of deformation in the Andes is a crucial and debated issue. Literature on Andean geology typically uses an implicit catalogue of diastrophic phases (reviewed by Moya and Salfity [1982]; Appendix B). The earliest of these were named in Peru by Steinmann (1929) who recognized three angular unconformities that he interpreted as recording deformational episodes and to which he gave the now-classic names (from oldest to youngest) Peruvian, Incaic, and Quechua phases (figure 12.4; Appendix B). Subsequently, and especially in the past two decades, these names have been widely used throughout the Andes, and other names have been added in Argentina and Bolivia (i.e., Yrigoyen 1976, 1979; Sébrier et al. 1988;

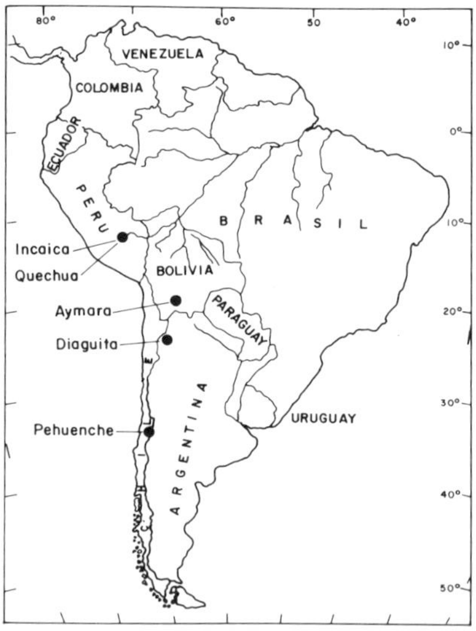

Figure 12.4. Areas to which terminology for the five principal deformation/uplift phases of the Andean orogeny was originally applied. After Steinmann (1929), Yrigoyen (1976, 1979), Sébrier et al. (1988).

figures 12.4 and Appendix B). As a result, the Andean literature now contains a plethora of named phases, which may at times appear confusing. The named diastrophic phases have been used in reference to magmatic uplift or subsidence pulses, directly observable deformational structures (compressional or extensional), deformations inferred from angular unconformities (frequently interpreted as tectonic pulses), disconformities, or simple facies changes. Correlation between distinct diastrophic phenomena chiefly relies on apparent or inferred rough synchroneity, which can unfortunately result in ready use of the same name for phenomena that may not be geographically, genetically and even chronologically related.

Furthermore, precise datings of Steinmann's Incaic and Quechua unconformities have shown that in Peru alone there are at least two "Incaic" (McKee and Noble 1989; Noble et al. 1990) and four to five "Quechua" (Lavenu and Marocco 1984; Sébrier et al. 1988) compressional phases (Appendix B). As noted by Marocco and co-workers (1987), this increase in numbers of recognized phases suggests that compression may well have been practically continuous at the scale of the Andes. Sempere and co-workers ("Tectonic crisis," 1990) demonstrated that the first major Andean-age deformation in Bolivia could not be correlated with the late Eocene Incaic phase of central Peru, as was previously believed, because it developed over about an 8 Myr span in the late Oligocene and early Miocene. This knowledge thus favors the notion that Andean compressive deformations developed over appreciable time spans and did *not* occur as short-lived pulses. In summary, development of continuously filling foreland basins during several epochs and in several areas of the plate margin indicate virtually continuous tectonic subsidence (i.e., continuous tectonic loading and hence continuous deformation in the uplifted source-areas; Sempere et al. 1989), suggesting that each associated deformational episode lasted a substantial amount of time (figure 12.1).

The only way to resolve the confusing intricacy of present nomenclature and correlations of Andean diastrophic phases is to document in each case *what* occurred, and *when* and *where* it occurred. Secure correlations are possible only when they are based on firm chronostratigraphic data that incorporate information on radioisotopic (K-Ar, Ar/Ar, fission-track) dates, magnetostratigraphy, and biostratigraphy. Discrimination among structural, magmatic, and sedimentary phenomena is needed for precise interpretation of objective regional data. Subjective long-distance correlations of events of assumed coeval tectonic origin based on rough synchroneity should be avoided. We emphasize that secure knowledge of the timing of phases—not their names—is the important issue. Liberal application of phase names to uncalibrated deformations, which is now done widely in South America, gives a false impression of knowledge of geological processes that can lead to erroneous interpretations. Furthermore, the unconformities used to identify these phases may have both tectonic and eustatic components.

If we favor long-lasting deformational (uplift) regimes, it appears that the quiescence periods that separate deformational episodes are intervals of relatively short duration, in contrast with the traditional view that the compressional pulses are the short intervals. Specifically, the onsets of deformational episodes in the Andes appear related to notable modification of the modalities of subduction (velocity or direction of convergence, subduction angle, etc.) to which the active margin reacted by progressive compressional failure (Jordan et al. 1983; Jordan and Alonso 1987; Isacks 1988; Sempere et al., "Tectonic crisis," 1990), and to episodic or punctuated diastrophic phenomena associated with collision of various features on the oceanic plates with the forearc (Nelson and Forsythe 1989). Fluctuations in spreading and convergence rates related to the global events also partly controlled Andean regional tectonism and magmatism, because both are linked to plate convergence and subduction.

The intense deformational and magmatic episode that initiated in the late Oligocene ("first major Bolivian crisis," figure 12.1) seems to have been widespread along the Andes. It was probably triggered by the breakup of the Farellon Plate into the Cocos and Nazca plates, and by the subsequent increase and change in direction in convergence velocity (Pardo-Casas and Molnar 1987; Pilger 1984; Jordan and Alonso 1987; Sempere et al., "Tectonic crisis," 1990; Sempere et al. In press). Because of the noteworthy intensity of this tectonic episode, the Andean Cordillera began at that time to be a prominent relief feature on the Pacific margin of South America. No appreciable reliefs were known prior to this tectosedimentary episode in what is now the high western region of Bolivia (Sempere et al., "Altiplano," 1990). The late Oligocene–early Miocene tectonic episode thus represents a significant turning point in Andean building history.

One outstanding ecological effect of the Andean buildup was the creation of a rain barrier which appeared, probably diachronously, when the Cordillera had locally reached an elevation sufficient to concentrate precipitation from dominant moisture-rich winds (figure 12.5). In northern Chile and probably throughout the central Andes, changes in land mammal faunas suggest that a rain shadow first appeared in association with the late Oligocene–early Miocene tectonic episode in Deseadan time; it was not well established, however, until middle Miocene time (Alpers and Brimhall 1988). The effect of the rain shadow on land mammal evolution in southern South America is discussed below.

IV. The Cenozoic Land Mammal Record in South America

A. Set 1: Tiupampian-Divisaderan

Tiupampian The earliest known Cenozoic faunas are assigned to the Tiupampian SALMA (*sensu* Marshall 1989). Included are the rich fauna described by Marshall and Muizon (1988), from what Sempere and Marshall (In press) and Gayet and coworkers (1991) regard as the Santa Lucía Formation at Tiupampa in southcentral Bolivia (the El Molino Formation of previous workers), and the depauperate fauna described by Sigé (1972), from the Umayo Formation at Laguna Umayo in southern Peru (the Vilquechico Formation of previous workers; Jaillard et al. In press). The age of these local faunas is debated, and they have been referred to the late Maastrichtian (i.e.,

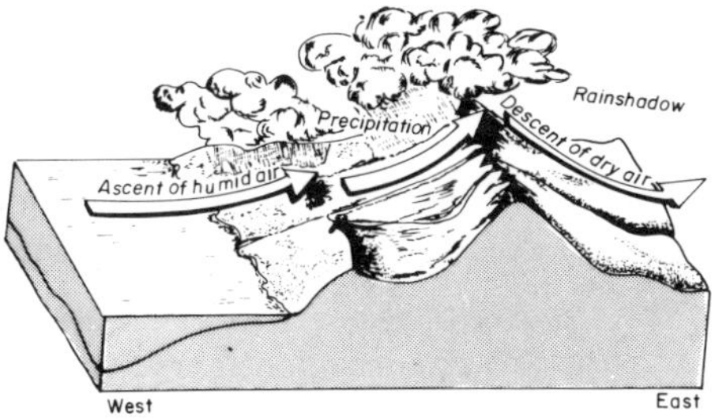

Figure 12.5. West to east transect of Andes, showing mechanism for development of a rain shadow south of 30° S. The same mechanism occurs, in the opposite direction, for northeasterly proceeding moisture-laden winds north of this latitude. The rain shadow intensified as the elevation of the Andes increased.

Sigé 1972; Marshall and Muizon 1988; Marshall 1989) and the early, but not earliest, Paleocene (i.e., Van Valen 1988; Ortiz-Jaureguizar and Pascual 1989). The fauna at Tiupampa occurs stratigraphically above the upper member of the El Molino Formation, which is regarded as early Paleocene, whereas the lower and basal middle members of the El Molino are regarded as latest Campanian-Maastrichtian (Gayet et al. 1991; Sempere and Marshall In press). Accordingly, Tiupampian is here placed in the early, but not earliest, Paleocene, as suggested by the majority of available paleontologic and stratigraphic data (figure 12.1).

The Tiupampa local fauna includes several groups of marsupials (Peradectidae, Microbiotheriidae, Didelphidae, Caroloameghiniidae, Hathliacynidae; *sensu* Marshall et al. 1989) and members of the placental orders Pantodonta, Leptictida, Condylarthra (Hyopsodontidae, Didolodontidae) and a possible Notoungulata (Henricosborniidae or Oldfieldthomasiidae). A peradectid, a microbiotheriid or pediomyid, a possible didelphid marsupial, a condylarth (didolodontid), and a possible notoungulate (Perutheriidae) are recorded from Laguna Umayo (Sigé 1972; Marshall, Hoffstetter, and Pascual 1983; Marshall and Muizon 1988; Gayet et al. 1991). A polydolopoid marsupial (*Epidolops* sp.) is now known from a new Tiupampian fauna of the Santa Lucía Formation at Estancia Blanco Rancho in Bolivia (Gayet et al. 1991).

The orders Leptictida and Pantodonta and a family of Condylarthra (Hyopsodontidae) are known in South America only in Tiupampian faunas; four other placental orders—Xenarthra, Litopterna, Trigonostylopoidea, and Xenungulata—are first known in Itaboraian (middle Paleocene) faunas. This suggests that a faunal hiatus, which apparently represents a missing SALMA, exists between Tiupampian and Itaboraian time (figure 12.1). This hiatus provides the time necessary for Xenarthra, Litopterna, Trigonostylopoidea and Xenungulata to appear and, in some cases, to diversify. This view is based on the assumption that the absence of these groups in known Tiupampian faunas is real and that they were absent elsewhere in South America in coeval faunas (Marshall 1989).

Another probable Tiupampian fauna is known from the lower part of the 195-meter-thick Mealla Formation in the southern part of the Andean basin of northwest Argentina. Pascual and co-workers (1978) described specimens of *Simpsonotus praecursor* and *S. major* (Henricosborniidae, Notoungulata) that were collected 18, 30, and 43 meters above the unconformable contact with the Maastrichtian-age Yacoraite Formation (*sensu* Marquillas and Salfity 1989). This contact marks a chronological hiatus in southern Bolivia and probably also in northwest Argentina. Pascual and co-workers (1978:366) regarded *Simpsonotus* (which they assigned a Riochican age) as the most primitive known South American notoungulate, although *Perutherium altiplanense* at Laguna Umayo (see Marshall, Hoffstetter, and Pascual 1983) and an indeterminate molar of Henricosborniidae or Oldfieldthomasiidae at Tiupampa are also among the earliest records of this order.

The Yacoraite and overlying Mealla formations of northwestern Argentina are regarded as stratigraphic and temporal equivalents of the El Molino and lower Santa Lucía formations, respectively, in southern Bolivia (Gayet et al. 1991). Marquillas and Salfity (1989) and Pascual et al. (1981) assigned the Mealla Formation an early, but not earliest, Paleocene age—the basal part is regarded as temporally equivalent to the late Danian age marine Salamanca Formation of the San Jorge basin in southern Argentina. Jaillard and Sempere (1989) suggested that the K/T boundary was close to the unconformity between the Yacoraite and Mealla formations, although it is now thought that the K/T boundary may lie *in* this unconformity because it is an erosional hiatus. Ortiz-Jaureguizar and Pascual (1989) equate the Tiupampian with the Salamancan marine stage, which is consistent with Marquillas and Salfity's (1989) age assignment for the Mealla Formation and the chronology adopted here.

In summary, our knowledge of Tiupampian faunas is restricted to three localities within the Andean basin of northern Argentina, southern Bolivia, and southern Peru. However, a possible fauna of this age is also reported from the Banco Negro Inferior in the base of the Río Chico Formation at Punta Peligro in the San Jorge basin of southern Argentina. Known taxa include the possible multituberculate Sudamericidae and the condylarths Arctocyonidae and/or Mioclaenidae (Pascual and Ortiz-Jaureguizar 1991).

Itaboraian This SALMA was originally proposed by Paula Couto (1952) for the rich fauna from fissure fillings in the limestone São José de Itaborai Formation near Rio de Janeiro, Brazil. The Itaboraian age was subsequently rejected and later workers regarded this fauna as Riochican (i.e., late Paleocene; Patterson and Pascual 1972). Based on a geochronologic study of the Río Chico Formation in Chubut Province, southern Argentina (Marshall et al. 1981), and on the co-occurrence of the marsupial *Epidolops* (Polydolopidae) and placental *Carodnia* (Xenungulata), Marshall (1985:57) resurrected the Itaboraian and redefined it to include the Itaborai fauna of Brazil and faunas from the lower Río Chico Formation at Punta Peligro and Las Flores near the contact of the continental Río Chico and conformably underlying marine Salamanca formations in the San Jorge basin of southern Argentina, which were of demonstrated or apparent middle Paleocene age (about 63.5–61.0 Ma). This arrangement was rejected by Soria (1987) and Van Valen (1988) who reverted to including so-called Itaboraian faunas in a broadly defined Riochican age. Marshall's arrangement was subsequently accepted by Ortiz-Jaureguizar and Pascual (1989), and a modified version was used by Legarreta and co-workers (1990), although the scheme proposed by the latter is in places inconsistent with the available geochronologic data published by Marshall and associates (1981). Pascual and Ortiz-Jaureguizar (1991) recently suggested that the fauna from the Banco Negro Inferior (BNI) may be of Tiupampian age, while that from Las Flores, which occurs 60 meters above the BNI, they regard as Itaboraian (along with the *Kibenikhoria* fauna from Cañadon Hondo and the *Carodnia* level from Bajo de la Patagonia). However, it is now recognized that the Itaborai fauna from Brazil may represent an assemblage that spans middle Paleocene and early Eocene time (Gayet et al. 1991); consequently, this age should probably be abandoned. Completion of a magnetostratigraphic-radioisotopic study of the El Molino–Santa Lucía sequence in Bolivia will provide the needed geochronologic control to resolve these debated issues (Butler, Marshall, Sempere, and Swisher In prep.).

Riochican This SALMA is based on faunas from the upper part of the Río Chico Formation in the San Jorge basin of southern Argentina (Marshall, Hoffstetter, and Pascual 1983:10–16). As noted by Marshall (1985:57), these rocks and faunas encompass late Paleocene time (about 61.0–57.8 Ma). The Riochican age probably includes the Río Loro Formation in Tucumán Province, northwest Argentina, which yielded *Eoastrapostylops riolorense* (Eoastrapostylopidae, Trigonostylopoidea; Soria and Powell 1982), but not the Mealla Formation (*contra* Ortiz-Jaureguizar and Pascual 1989).

Casamayoran The Casamayoran SALMA (conventionally early Eocene) is based on knowledge of faunas from multiple localities in Chubut and northern Santa Cruz provinces, southern Argentina (for a review see Marshall, Hoffstetter, and Pascual 1983:18–20). The biostratigraphy of these rocks and faunas is discussed by Cifelli (1985).

The only major fauna outside of Patagonia that is referred to this age is from the Lumbrera Formation in Jujuy and Salta provinces, northwest Argentina. The taxa described or recorded to date include: marsupials (*Bonapartherium hinakusijum, Prepidolops didelphoides, P. molinae, Patene simpsoni*), Trigonostylopoidea (Astrapotheriidae, *Albertogaudrya carahuasensis*), Notoungulata (Notostylopidae, *Boreastylops lumbrerensis;* Isotemnidae, *Pampatemnus infernalis* and *P. deuteros;* Oldfieldthomasiidae, *Colbertia lumbrerensis*) and Litopterna (Adianthidae, *Indalecia grandensis*) (Carvajal et al. 1977; Pascual 1980a, 1980b; Vucetich 1980; Pascual et al. 1981; Bond 1981; Vucetich and Bond 1982; Bond and Vucetich 1983, 1985; Goin, Palma, et al. 1986). An early Eocene age for the Lumbrera Formation is corroborated by magnetostratigraphic data (Butler and Marshall In prep.).

Mustersan This conventionally middle Eocene SALMA is based on a few local faunas in the San Jorge basin of Chubut Province, southern Argentina (Marshall, Hoffstetter, and Pascual 1983:20–21).

Mammals of Mustersan age (noted by Alonso et al. 1988) are also known from the base of the Pozuelos Formation on the southern Puna in northwest Argentina. Taxa from the two localities include: marsupials (*Reigia punae, Prepidolops alonsoi*), Xenarthra (Dasypodidae, cf. *Astegotherium* or *Pros-*

tegotherium), Condylarthra (Didolodontidae, aff. *Ernestokokenia nitida*), Notoungulata (Oldfieldthomasiidae, Interatheriidae, Archaeohyracidae, Notohippidae, Isotemnidae) and Pyrotheria (Pyrotheriidae, cf. *Propyrotherium*) (Pascual 1983; Bond 1984; Alonso et al. 1988).

Divisaderan A taxonomically small fauna consisting of only nine genera was described by Simpson and co-workers (1962) from the Divisadero Largo Formation in Mendoza Province, west-central Argentina. These authors regarded the fauna as approximately early Deseadan or latest pre-Deseadan (late Eocene). Pascual and co-workers (1965) recognized this fauna as representing a new SALMA, the Divisaderan, which they regarded as late Eocene and placed in the final part of the faunal hiatus between Mustersan and Deseadan. Subsequent workers placed the Divisaderan in the early (Marshall, Cifelli, et al. 1986) and middle (MacFadden et al. 1985) Oligocene. The latter age assignment was based on some reputed faunal links between Divisaderan and early Deseadan faunas, and geochronologic data which showed that Deseadan faunas were considerably younger than previously envisioned (see below). Based on these data, Divisaderan was "pulled" into the middle part of the Oligocene by upward placement of Deseadan.

A multivariate study of Cenozoic land mammal families (Ortiz-Jaureguizar 1986) indicates that the Divisadero Largo fauna is more closely (but still distantly) affiliated to Mustersan than to Deseadan, while studies of genera and families using different techniques show that this fauna is taxonomically intermediate between Mustersan and Deseadan (Marshall and Cifelli 1990).

Recently, Charrier and co-workers (1990) and Wyss and co-workers (1990) reported discovery of a fauna in what may correspond to the lower part of the Coya-Machali Formation at Termas del Flaco in central Chile. The fauna, collected in association with volcanic rocks that yielded K-Ar (whole rock) ages of 35 to 31 Ma, includes: Marsupialia (Groeberiidae, gen. et sp. nov.; Argyrolagidae, gen. et sp. nov.; Polydolopidae, *Polydolops* sp. nov.), Edentata (Dasypodidae, inc. sed.; Tardigrada, *Pseudoglyptodon* sp. nov.), and Notoungulata (Notostylopidae, gen. et sp. nov.; Interatheriidae, three gen. et sp. nov.; Archaeohyracidae, *Pseudhyrax* sp. nov., gen. et sp. nov.; Isotemnidae, *Periphragnis* sp., *Rhyphodon* sp.; Notohippidae, gen. et sp. nov.). This fauna appears to be of Divisaderan aspect, and the associated K-Ar dates (35 to 31 Ma) agree with the consensus that this age spans from about 36 to about 30 Ma (Marshall, Cifelli, et al. 1986; Marshall and Cifelli 1990); it is accordingly, but tentatively, assigned to this SALMA (figure 12.1).

Discussion Multiple taxonomic similarities (10 examples, see below) link late Cretaceous and Paleocene faunas in North America with those of Tiupampian and Itaboraian age in South America. These links document the occurrence of one or more dispersal events (in whatever direction) between the Americas at the beginning of the Cenozoic or in latest Cretaceous time. Among marsupials, for example, *Peradectes* is known in Tiupampian faunas and in early Paleocene faunas of North America; caroloameghiniids

(*Glasbius*) in the late Cretaceous of North America and Tiupampian (*Roberthoffstetteria*) of South America; Didelphidae in the early Eocene of North America and Tiupampian of South America; and Pediomyidae, a sister group of Microbiotheriidae (first record in Tiupampian of South America), in the late Cretaceous of North America (Marshall and Muizon 1988; Marshall et al. 1989). Among placentals, a Leptictida (cf. *Cimolestes*) is known from Tiupampa and *Cimolestes* occurs in the late Cretaceous and early Paleocene of North America; Pantodonta (*Alcidedorbignya*) at Tiupampa are first known in rocks of middle Paleocene age in North America; mioclaenine hyopsodontid Condylarthra (*Tiuclaneus, Molinodus*) at Tiupampa are first recorded in early Paleocene (Puercan) faunas of North America; Xenarthra (Dasypodidae) first appear in Itaboraian faunas in South America and in the late Paleocene (late Tiffanian) in North America; and Dinocerata, which first appear in the late Paleocene (late Tiffanian) of North America, are regarded as a sister-group of Xenungulata (*Carodnia*), which appear in the Itaboraian of South America (Gingerich 1985; Schoch and Lucas 1985; Marshall and Muizon 1988). Arctostylopidae, which appear in the late Paleocene (late Tiffanian) of North America, were long regarded as notoungulates, which are known from the Tiupampian of South America (Gingerich 1985; Marshall and Muizon 1988), although this relationship has recently been abandoned and this family is now placed in a separate order, Arctostylopida (Cifelli et al. 1989). In addition, phororhacoid ground birds, which in South America are first recorded in Itaboraian faunas of Brazil (Alvarenga 1985), may have participated in this interchange (Mourier-Chauviré 1982).

The currently available geologic data support the existence of a landbridge between the Americas in late Cretaceous or early Paleocene time that accounts for these faunal similarities. Furthermore, sharp marine regressions occurred at about 68 and 63 Ma (Haq et al. 1987), and would have favored dispersal.

The only SALMAS for which geochronologic (magnetostratigraphic or isotopic) data are available and for which calibration with the geologic time scale is reasonably certain are Riochican and Divisaderan (in Chile). Calibration of the other SALMAS (Tiupampian, Itaboraian, Casamayoran, and Mustersan) is based on "stage of evolution" of the faunas or on known relationships of the rocks—that is, there is a conformable contact between rocks containing Riochican and Casamayoran faunas; therefore the top of the former and base of the latter are assumed to be time-equivalent).

Of all mammal-bearing rocks of early Cenozoic age in the San Jorge basin, those containing Mustersan faunas "are the most restricted in known extent and usually appear as local channel deposits, frequently incised" into rocks of Casamayoran age. "Uplift and erosion were evidently increasingly active, because the Musters was not deposited or was entirely removed in many localities and the Deseado often rests directly on the Casamayor" (G. G. Simpson 1940:660). This erosional disconformity may be due to a eustatic sea level low-stand, which would have caused erosion and channeling of Mustersan into Casamayoran. Filling of the Mustersan channels would have originated during subsequent rising (sea) base level. The sharp marine regression registered at about 49 Ma by Haq and co-workers (1987) may equate with the Mustersan-Casamayoran stratigraphic boundary (figure 12.1).

Casamayoran and Mustersan faunas were apparently coeval with development of the Incaic deformation (*sensu* Steinmann 1929; Charrier and Malumián 1975; Noble et al. 1979; Jordan and Alonso 1987), which generated in western Bolivia a wide foreland basin (Sempere et al. 1989; Sempere et al., "Altiplano," 1990; Sempere et al., "Tectonic crisis," 1990). At least in central Peru, where the name Incaic was originally applied (figure 12.4), one pulse of this deformation episode is well dated as upper middle Eocene (c. 43 Ma), and another one seems older than 55 Ma (Noble et al. 1990; Appendix B). In southern Chile and Argentina (where it is called the "Segunda Fase del Primer Movimiento Andino" by Groeber [1929]), deformation occurred between 50–42 Ma (Ramos 1989:900). In the southern Bolivian part of the Andean basin, the initiation of this deformation is apparently indicated by a local angular unconformity, possibly about 51–50 Ma, which separates the Cayara from the underlying El Molino, Santa Lucía, and Impora formations (Marocco et al. 1987; Sempere 1990; Marshall and Sempere 1991). The onset of the first pulse of the Incaic deformation correlates with a series of sharp marine regressions (Haq et al. 1987; figure 12.1).

B. Set 2: Deseadan-Montehermosan

Deseadan The type locality of this SALMA is along the Río Deseado in northeast Santa Cruz Province southern Argentina, and the majority of well known faunas of this age in Argentina are from the San Jorge basin in Chubut Province. Other important faunas are from the Salla-Luribay basin in Bolivia and the Taubate basin in Brazil (Marshall, Hoffstetter, and Pascual 1983).

Deseadan was long regarded as early Oligocene (Patterson and Pascual 1972), but recent geochronologic studies of rocks and faunas at two localities conclusively demonstrate that this SALMA encompasses late Oligocene–early Miocene time.

MacFadden, Campbell, and co-workers (1985), after examining magnetostratigraphic and isotope data, report that the Salla Beds of Bolivia span from about 28.5 to 24.0 Ma. Hayashida and Danhara (1985) report fission-track dates (27.2 ± 1.6, 25.0 ± 1.5, 26.1 ± 1.9, 24.0 ± 1.5 Ma) that complement this age assignment. Naeser and co-workers (1987) obtained additional K-Ar (28.0 ± 0.9, 27.2 ± 0.9, 27.9 ± 0.9, 25.1 ± 0.7 Ma) and fission-track (34.5 ± 4.0, 24.2 ± 3.6, 23.5 ± 2.2 Ma) dates and conclude that the principal fossil horizons range from 27 to 24 Ma, and the complete section ranges from about 28 to 22 Ma. Sempere and co-workers ("Tectonic crisis," 1990) reinterpret the geochronologic data and suggest that the section may be 3.0–4.5 Myr younger (i.e., 24.5–21.0 Ma; the principal fossiliferous level being c. 23.5 Ma) than favored by MacFadden and co-workers (1985). The correlation of Sempere et al. ("Tectonic crisis," 1990) is followed here because it agrees with conclusions reached independently by McRae (1989) and Marshall.

Marshall, Cifelli, and co-workers (1986) report 10 K-Ar dates on basalts and tuffs associated with the Deseadan fauna at Scarritt

Pocket in the San Jorge basin and conclude that it ranges from about 23.4 to 21.0 Ma. The base of Deseadan time is constrained by a series of K-Ar dates on basalts overlying faunas at the Gran Barranca and Valle Hermoso that yielded "best ages" of 28.8 and 28.7 Ma, respectively (Marshall, Cifelli, et al. 1986). In addition, restudy of field notes and photos of the Pico Truncado locality by Marshall indicates that the Deseadan fauna is in a slump-block of the cliff from below the basalt dated at 27.6 Ma and is not associated with the 33.6 Ma basalt as originally believed (i.e., Marshall et al. 1977; Marshall, Cifelli, et al. 1986). Collectively, these dates suggest that Deseadan ranges from late Oligocene to earliest Miocene—about 29.0 to 21.0 Ma.

Support for a basal age of about 29 Ma for known Deseadan relies on the following facts and observations. In Bolivia an unconformity occurs below the Luribay Conglomerate, which conformably underlies the Salla Beds. This unconformity is about 27 Ma and represents initiation of what is called the "first major tectonic crisis" in Bolivia (Sempere et al., "Tectonic crisis," 1990; Appendix B). In central Argentina, this unconformity is apparently present at the base of the Rodados Lustrosos of Gorroño et al. (1979) and of the Santa María Conglomerates of Schiller (1912), (Sempere et al. In Press). In southern Mendoza Province the Rodados Lustrosos conformably underlies sediments that contain a Deseadan fauna ("Agua de la Piedra Formation-sedimentario" = "Complejo vulcano" of Gorroño et al. [1979]). The Rodados Lustrosos (= "Conglomerado Violáceo" of Kokogián and Mancilla [1989: Lámina 9]) unconformably overlie the Divisadero Largo Formation in central Mendoza Province, the middle member of which yielded the Divisaderan SALMA fauna of Pascual and co-workers (1965). The stratigraphic equivalence of the Rodados Lustrosos and Conglomerado Violáceo is supported by associated fossil land mammal faunas (Gorroño et al. 1979).

The initiation of deposition of the Santa María Conglomerates is therefore not coeval with the Incaic deformation, as concluded by Ramos (1988). If the fauna dated 35 to 31 Ma from what may correspond to the basal part of the Coya-Machali Formation in central Chile is indeed Divisaderan, then this age assignment is consistent for both the Divisaderan-Deseadan boundary and correlation with the initiation of the first major Bolivian crisis. These data further suggest that a small hiatus, here interpreted to be about 2 Myr (figure 12.1), occurs between Divisaderan and Deseadan time. This hiatus is believed to correspond to the termination of deposition of continental sediments during a sea level low-stand along the South American Pacific coast between 31 and 28 Ma (Macharé et al. 1988:34). The first major Bolivian crisis continued to about 19 Ma (Sempere et al., "Tectonic crisis," 1990), while in central Argentina and adjacent Chile this crisis ends prior to deposition of the Farellones Formation volcanics at about 20 Ma (Vergara et al. 1988; Godoy and Palma 1990; Sempere et al. In press).

Deseadan faunas are conventionally identified by the stratigraphic range of *Pyrotherium* and the first record in South America of caviomorph Rodentia and platyrrhine Primates. Only one genus of primate, *Branisella,* is known from a level at Salla that Sempere et al. ("Tectonic crisis," 1990) suggest is about 23.5 Ma, but MacFadden (1990) place at 26 Ma. Deseadan age caviomorph rodents, however, are remarkably diverse and are classified by Patterson and Wood (1982) into seven families and 14 genera. There are no rodents in the *Pyrotherium*-bearing Deseadan at the Gran Barranca and Valle Hermoso in the San Jorge basin, which are capped by basalts dated about 29 Ma, and the single rodent specimen reported by Marshall, Cifelli, and co-workers (1986: 942) from Pico Truncado below the 27.6 Ma basalt needs to be verified. There are no caviomorph rodents known from the Divisaderan age Divisadero Largo Formation in Argentina or from the lower Coya-Machali Formation (35–31 Ma) in Chile. The earliest uncontested caviomorph rodents in South America occur at Scarritt Pocket and throughout the fossiliferous part of the Salla section (MacFadden et al. 1985) and are therefore younger than 27 Ma.

The diversity of caviomorph rodents at their time of first appearance in South America has been interpreted to indicate that they arrived before Deseadan time. "Assuming the arrival of but a single immigrant stock, some time is needed to permit radiation to account for their documented diversity in Deseadan time. It has been hypothesized that caviomorphs arrived during the paleontological hiatus between Mustersan (middle Eocene) and Deseadan time. Thus, they were long thought to have arrived in the late Eocene, although mammal faunas of this age are not known. Patterson and Wood (1982) recently suggested that the Deseadan age caviomorph diversity is so great that a middle Eocene arrival is more reasonable. The important point is that all authorities on South American fossil caviomorphs agree upon their pre-Deseadan arrival. Thus, caviomorphs cannot conceptually nor operationally be used as an immigrant datum to define the beginning of Deseadan time" (Marshall, Cifelli, et al. 1986:945).

We propose a new model for the paleobiogeographic history of caviomorph rodents that is consistent with available geologic, geochronologic, and paleontologic data, previous paleontologic models, and current views on Caribbean plate tectonics, but which is at odds with the above scenario.

1. Caviomorphs and primates were waif dispersants, from Africa or North America, which colonized one or more large Caribbean islands during late Paleocene or Eocene time. This aspect of the model agrees with the Eocene time of arrival proposed by earlier workers. The availability of multiple large islands in close proximity permitted opportunities for vicariance and diversification. Dispersal of these stocks to South America was in part prevented by a marine seaway and possibly further impeded by the then-westward drainage of the Amazon River, which shifted from the Pacific to the Atlantic Ocean in middle Miocene time because of the increasing rate of Andean buildup (Kulm et al. 1981; Nuttall 1990:figure 453).

2. Southward dispersal of caviomorphs to Bolivia and Argentina was coeval with the onset in the central Andes of a major tectonic episode dated about 27 Ma in Bolivia (Sempere et al., "Tectonic crisis," 1990; Sempere 1991). The central Andes became an important mountain range at this time (Marshall and Sempere 1991). This date also coincides with activation of a magmatic arc 250–450 kilometers east of the present Peru-Chile trench axis (c. 26 Ma; Alpers and

Brimhall 1988), initiation of a volcanic arc in the western Andes (27–26 Ma; Jordan and Alonso 1987), breakup of the Farellon plate and formation of the Nazca plate, and increase in convergence rates (Rapela and Kay 1988). The tectonic crisis produced large intermontane basins in what is now the Altiplano. In one of these, the Salla-Luribay basin, the Luribay Conglomerate began to accumulate about 25.5 Ma, and the overlying Deseadan Salla Beds between 24.5 and 21.0 Ma (Sempere et al., "Tectonic crisis," 1990). These orogenic movements apparently occurred throughout the Andes (Sébrier et al. 1988:96), and the mountain ranges of Colombia and Venezuela were "bodily uplifted" (Harrington 1962:1804) as a result of a tectonic pulse about 24 Ma (Duque-Caro 1976). This apparently correlates with the Proto-Andean Orogeny of Van Houten and James (1984:figure 5). "The Ecuadorian 'sea portal' was finally closed and the Colombian-Peruvian foredeep lost its precarious connection with the Pacific waters, receiving only continental sediments" (Harrington 1962:1804). This uplift phase correlates with a series of sharp marine regressions between 30 and 25.5 Ma (Haq et al. 1987; see also Duque-Caro 1976).

This aspect of the model explains why caviomorphs are taxonomically diverse at their time of first appearance in the lower part of the Salla Beds but are absent in earliest Deseadan faunas at the Gran Barranca and Valle Hermoso (i.e., from below basalts dated c. 29 Ma) and in faunas of Divisaderan age (35–31 Ma). It also explains their phylogenetic distinction from myomorph rodents and provides a model that supports the work of Graur and co-workers (1991), who recognize caviomorphs as a distinct order. Along with the caviomorphs came platyrrhine primates, first known at Salla, and land tortoises, first called *?Chelonoidis* sp. (Testudinidae, Chelonia), at Quebrada Saguayo in the Petaca Formation of central Bolivia—a temporal equivalent of the Salla Beds (Sanjinés and Jiménez 1975; Sempere et al., "Tectonic crisis," 1990)—and at Cabeza Blanca in southern Argentina (Broin 1991). This model implies that the arrival of caviomorphs to continental South America is an important late Oligocene immigrant datum event. It is also possible that at this time stocks of sloths dispersed from South America to islands in the Greater Antilles (Ford 1990).

These and other data suggest a twofold subdivision of Deseadan time. First is Deseadan of more than 27 Ma, with *Pyrotherium* but no caviomorphs (i.e., Gran Barranca, Valle Hermoso); second is Deseadan between about 27 and 21 Ma, with both *Pyrotherium* and caviomorphs (i.e., Salla Beds and most known Deseadan local faunas) (figure 12.1).

In early Deseadan time, the Andes began to increase significantly in height. South of 40° S latitude, ecologies along at least their eastern flank changed from warm closed-woodland habitats favoring mammals with browsing adaptations like low-crowned cheek teeth to drier and cooler savanna-woodland habitats favoring those with grazing adaptations like high-crowned cheek teeth (Cifelli 1985). True grassland habitats came into existence at this time, and grazing opportunities were seized by many groups of mammals. The magnitude of these new opportunities is reflected in the fact that hypsodonty (ever-growing high-crowned cheek teeth) evolved at least six times among ungulates during Deseadan time (Patterson and Pascual 1972; Cifelli 1985). Although the chronology and sequence of these changes among Deseadan local faunas has yet to be securely documented, it is clear that this SALMA records an important turning point in the evolution of the South American land mammal faunas.

Deseadan faunas thus lived during the development of a major tectonic episode, which is well documented in Bolivia between about 27 and 19 Ma (Sempere et al., "Tectonic crisis," 1990) and probably affected most, if not all, of the Pacific margin of South America (Sempere et al. In press). Coeval (26–17 Ma) marine sediments were deposited in forearc basins from southern Chile to northern Colombia (Macharé et al. 1988), where they unconformably overlie the 31–28 Ma depositional hiatus noted above. This sea level high-stand correlates with increase in convergence rate between the Nazca and South American plates (and higher sea level resulting from increase in volume of spreading oceanic ridges) and associated global rise in oceanic temperatures (Macharé et al. 1988). Because the onset of this tectosedimentary episode marks the beginning of intense mountain building along the active margin, the evolution of Deseadan and younger faunas was increasingly influenced by the ecological changes induced by the growing Andes.

Colhuehuapian This SALMA is based on a few local faunas from the San Jorge basin in Chubut Province, southern Argentina (Marshall, Hoffstetter, and Pascual 1983:24). A probable sea level low-stand in this area occurred between Deseadan and Colhuehuapian time, and the San Jorge basin "was widely and deeply eroded" (G. G. Simpson 1940:661). Colhuehuapian deposits probably originated when the global base sea level rose again at 20–19 Ma (Haq et al. 1987).

Colhuehuapian occurs within the 21–19 Ma hiatus between Deseadan and Santacrucian. Because it shows closer taxonomic similarity with Santacrucian than with Deseadan faunas, it was regarded as possibly also ranging from about 19–18 Ma (Marshall 1985; Marshall, Drake, et al. 1986; Marshall, Cifelli, et al. 1986). This assignment was regarded as tentative because no secure geochronologic data were yet available for rocks and faunas of that age.

Pascual and co-workers (1965:177) further questioned the validity of recognizing a Colhuehuapian age distinct from the Santacrucian because, as noted by G. G. Simpson (1940:665), the "mammals are recognizably different, but the average difference is so slight that they suggest closely successive stages of the same general fauna." In addition, it is conventionally recognized that a distinct step in evolution exists between known Deseadan and known Colhuehuapian faunas, and for this reason it has long been believed that there exists a "hiatus" in the faunal record between these ages (Pascual and Odreman Rivas 1971, 1973). This hiatus appears to be real, and it correlates with a significant sea level low-stand around 21 Ma (Haq et al. 1987). Bellosi (1990) observed a hiatus in the marine sequence in the San Jorge basin and correlated it with this low-stand and the hiatus between Deseadan and Colhuehuapian. We conclude from current geochronological data on Deseadan and Santacrucian rocks that this hiatus is of lesser duration than envisioned by earlier workers.

Although these controversies cannot be

resolved at this time, we feel that Colhuehuapian faunas in general are indeed transitional between Deseadan and Santacrucian. This affinity is championed by Ortiz-Jaureguizar (1986) who, following taxonomic and adaptive similarities, placed Deseadan, Colhuehuapian and Santacrucian SALMA faunas in a Ciclo Faunístico Patagónico. In this chapter we place reliance on these faunal observations and on the sea level data of Haq and coworkers (1987), and place Colhuehuapian between 20 and 19 Ma (figure 12.1).

Santacrucian This SALMA is based on faunas from the Santa Cruz Formation in the Magallanes or Austral basin of southern Argentina and adjacent Chile (Marshall 1976; Marshall, Hoffstetter, and Pascual 1983:28; Marshall and Salinas 1990a, 1990b). Geochronologic (magnetostratigraphic, isotopic, fission-track) studies indicate that these rocks and faunas are about 18–15 Ma (Marshall, Cifelli, et al. 1986; Marshall and Salinas 1990a). However, a date of about 19 Ma on a tuff from the base of this formation, near Lago Argentino (Marshall and Salinas 1990a), suggests that the lower boundary of this SALMA could be slightly older than previously believed. Colhuehuapian could be 20–19 Ma, as suggested by marine transgression beginning about 20 Ma (Haq et al. 1987). This view is adopted here.

Santacrucian faunas are also known from various localities of the Río Frias Formation in the Rio Mayo Embayment along the western edge of the San Jorge basin (Marshall and Salinas 1990a). These faunas were earlier included in the Friasian SALMA, but study of the marsupials and a K-Ar date of about 17 Ma from the type Friasian SALMA locality along the Alto Río Cisnes, Chile, indicates that these rocks and faunas are temporally equivalent to the Santa Cruz Formation (Marshall 1990; Marshall and Salinas 1990a). This view is supported by knowledge that the Santa Cruz Formation is mapped in the central and eastern parts of the San Jorge basin (Bellosi 1990:376; Fitzgerald et al. 1990), while the western part of this basin contains what is called the Río Frias Formation (Marshall and Salinas 1990a: Appendix A). Furthermore, the marine deposits at Comodoro Rivadavia, which are called the Chenque Formation, appear to be coeval with early Miocene marine beds in the Magallanes basin known as the Centinela Formation (Bellosi 1990:figure 1).

Deposition of the Santa Cruz and Río Frias formations began with onset of an uplift episode about 19 Ma. In the Magallanes basin, this deformation caused the regression of the Centinela Formation epicontinental sea (Ramos 1989:899; Marshall and Salinas 1990a), and in the San Jorge basin, regression of the Chenque Formation epicontinental sea (Marshall and Salinas 1990a; Bellosi 1990). The resulting emersion was apparently not abrupt, because the contact between the marine and continental formations in both basins is conformable. Coeval marine regressions occurred along the entire Pacific coast of South America, suggesting that this tectonic episode might have occurred throughout the Andes (Macharé et al. 1988).

Onset of this orogenic episode seems to be coeval with what is called the Pehuenche deformation defined in the region of Aconcagua along the central Argentina–Chile border (Yrigoyen 1976, 1979; see also Malumián and Ramos 1984; figure 12.5; Appendix B). In this area the Pehuenche deformation began about 20 Ma, as dated by uplift of an associated granitoid (Ramos and Cingolani 1989). Coeval deformation phases are also recognized in northwestern (Jordan and Alonso 1987; Diaz and Miserendino 1988; Miserendino and Diaz 1988) and southern (Ramos 1989; Marshall and Salinas 1990a) Argentina.

Colloncuran As recently emended by Marshall and Salinas (1990a), the name of the SALMA between Santacrucian and Chasicoan is Colloncuran (*sensu* Groeber 1929) and not Friasian, which is a Santacrucian equivalent. The type fauna from the Miembro Tobáceo Las Bayas of the Collón Curá Formation at Pilcaniyeu Viejo in Río Negro Province, Argentina (Bondesio, Rabassa, et al. 1980) occurs above an ignimbrite that has K-Ar dates ranging between 15.4 and 14.0 Ma (Marshall et al. 1977; Marshall and Salinas 1990a). It is the most taxonomically diverse local fauna classically included in the Friasian as conceived by earlier workers.

The Collón Curá Formation to the south of San Carlos de Bariloche in Río Negro Province, Argentina, disconformably (in most places) overlies the continental Ñorquinco and Rincón de Cholila formations are bly overlies the partially marine Rincón de Cholila Formation (Cazau 1972). The Ñorquinco Rincón de Cholila formations are placed by most recent workers in the Ñirihuau Formation (Spalletti 1983; Cazau et al. 1989), which is divided into three members. The upper member (*sensu* Cazau et al. 1989:313) has isotopic dates of 17.1, 16.7, and 16.4 Ma from its base; the underlying middle member has dates of 22.0, 21.6, and 19.7 Ma (Cazau et al. 1989:317). Isotopic dates are not available for the lower member, although a few fossil fish have been interpreted to indicate that it is of Oligocene age (Pascual et al. 1984). If this age assignment is correct, then the unconformity between the Ñirihuau and underlying Ventana formations recognized by Cazau and co-workers (1989) may be coeval with that at about 27 Ma, below the Rodados Lustrosos. (The Ventana has an average age of 27.5 Ma, based on 18 K-Ar dates; Cazau et al. 1989: 316.) This interpretation is consistent with the age and stratigraphic relationships of rocks in the Ñirihuau basin as proposed by Uliana and Biddle (1987).

Pascual and Odreman Rivas (1973:306) report the occurrence of so-called Friasian age mammals in the Ñorquinco Formation, which, in the present context, appears to equate with the upper member of the Ñirihuau Formation. This faunal evidence, along with the reported isotopic dates, can be interpreted in two ways. First, that the mammals from the upper member will prove to be Santacrucian in age upon restudy; or second, that if the mammals prove to be Colloncuran in age, it *may* be necessary to place the Santacrucian-Colloncuran boundary at about 16.0 Ma. This second interpretation does not conflict with the most reliable dates on Santacrucian rocks, all of which are older than 17 Ma (Marshall, Drake, et al. 1986; Marshall and Salinas 1990a). Furthermore, the disconformity between the Ñirihuau and Collón Curá formations was apparently coeval with initiation of emplacement of the Reloncaví granitoids about 16 Ma (Rapela and Kay 1988). These data collectively suggest that the Santacrucian-Colloncuran boundary is about 16 Ma and that these ages are, at least in Patagonia, separated by a small hiatus (figure 12.1).

It has long been recognized that continued uplift of the Andes Cordillera created a rain shadow along its eastern flank by blocking moisture-laden Pacific winds (Patterson and Pascual 1972). These winds now exist only south of about 30° S latitude, whereas northeast prevailing moisture-rich winds predominate north of this latitude. This rain shadow initiated the desiccation of Patagonia and resulted in the appearance of drier forest habitats and more extensive grasslands (Pascual and Ortíz-Jaureguizar 1990). These changes represent an accentuation of trends begun in Deseadan time. Although this uplift began to accelerate about 19 Ma, the southern Andes did not become high enough to produce a rain shadow effect until somewhat later (2000–3000 meters are required for this effect; see Alpers and Brimhall 1988:1653). The rain shadow is first evident in Friasian faunas (*sensu* Patterson and Pascual 1972; Pascual et al. 1965) south of 46° S latitude about 15 Ma (Marshall and Salinas 1990a), which presently approximates the Santacrucian-Colloncuran boundary (figure 12.1).

Because of striking variation in topography of the southern Andes, the effect of this rain shadow would not have been uniform. The Andes north of 46°30′ S average 2000 meters lower in minimum height than the Andes further south (i.e., south of Lago General Carrera–Buenos Aires; Ramos 1989:900). These elevation differences existed in early Miocene time and monitored deposition of foreland molasse deposits along their eastern flank. South of 46°30′ S, the areally extensive Santa Cruz Formation attains a thickness of 1500 meters, but to the north, the disjunct Río Frias Formation is only about 250 meters (Ramos 1989; Marshall and Salinas 1990a). The greater thickness of sediments in the Magallanes basin is due to major tectonic subsidence, a feature not evident in the Río Mayo Embayment. The rain shadow would thus have been most evident south of 46°30′ S and have influenced faunas in the Magallanes basin (Santa Cruz Formation); to the north it was less pronounced and had a relatively smaller effect on faunas in the Río Mayo Embayment (Río Frias Formation) of the San Jorge basin and northward to about 40° S, which in the early Miocene was at or near sea level (Ramos 1982; Macharé et al. 1988). North of 40° S, the Andes were then apparently much higher (Ramos 1988). Different ecologies almost certainly existed in these adjacent basins, and this may help to explain long-observed differences in taxonomic composition of their coeval faunas (Marshall and Salinas 1990a).

Concurrently, a rain shadow developed along the coastal areas of northern Chile (north of 33° S) and Peru as they blocked northeast prevailing moisture-laden winds from the Amazon basin (Alpers and Brimhall 1988; Kalin Arroyo et al. 1988; Marshall and Sempere 1991). It is estimated that the central Andes near 22° S latitude attained at least half their present elevation by about 15 Ma, at which time hyperarid conditions were established along their western flank in northern Chile and southern Peru (Alpers and Brimhall 1988). In the central Andes, "major topographic relief *first* developed during the middle and late Miocene" (Jordan and Gardeweg 1989).

Independent studies of the central and southern Andes are thus complementary in suggesting coeval uplift and onset of major rain shadow–induced ecological changes about 15 Ma. Furthermore, the climatic changes were apparently accentuated by the contemporaneous development of the Antarctic ice cap, which produced a temperature decrease in coastal waters supplied by the ancestral Humboldt Current (derived from cold polar waters), an increase in upwelling intensity, and onset of high-latitude temperate climates between 15 and 13 Ma (Alpers and Brimhall 1988). A small but sharp regression about 14 Ma (Haq et al. 1987) may signify a drop in ocean level owing to onset of buildup of the Antarctic ice cap (figure 12.1).

A Colloncuran fauna is also known from Quebrada Honda in southernmost Bolivia. These mammal-bearing rocks were dated between 13 and 12 Ma (MacFadden et al. 1990). The fauna was regarded as Santacrucian by Frailey (1988), although the geochronologic data of the MacFadden group support a Colloncuran (Friasian) age as originally suggested by Hoffstetter (1977).

An even more spectacular fauna, which is at least in part of Colloncuran age, is known from the Villavieja Formation of the Honda Group in the La Venta badlands of Colombia. Stirton (1953) originally assigned this fauna to the Friasian (= Colloncuran), although Kay et al. (1987) report that work in progress suggests it is "more similar to Santacrucian than Friasian assemblages of Chile and Argentina, as they are currently known."

The available geochronologic data on these rocks are as follows: a 115-meter-thick magnetostratigraphic section of predominantly normal polarity was correlated with Epoch 15 (15.2–13.6 Ma; Hayashida 1984); fission-track dates of 14.6 ± 1.1, 15.7 ± 1.1 and 16.1 ± 0.9 Ma are reported by Takemura and Danhara (1986) from the Monkey Unit in the lower Honda Group; an Ar/Ar date of circa 13.5 Ma is reported from the uppermost Honda Group (Flynn et al. 1989); and K-Ar dates of 16.1–14.6 Ma are reported from the lower part of this group (van der Wiel 1990). These data suggest that the upper part of these beds are of Colloncuran age, although the lower part may be temporally equivalent with high-latitude Santacrucian faunas in Argentina and Chile (Flynn et al. 1989).

Takai and co-workers (1988) illustrated a tooth of a caenolestid marsupial, which probably came from the Monkey Unit. It is clearly a member of the Palaeothentinae and is possibly referable to *Palaeothentes* (C. Villarroel, personal communication, 1989). This fossil lends further support for Santacrucian affinities, although as-yet undescribed specimens of *Palaeothentes* are reported from the Collón Curá Formation (Colloncuran SALMA) of Argentina (see Marshall 1990 and references therein).

The taxonomic and temporal differences among the high-latitude Santacrucian and Colloncuran faunas of Argentina and Chile (53°–40° S), the medial-latitude Colloncuran age but Santacrucian aspect fauna of Quebrada Honda, Bolivia (c. 22° S), and the low-latitude Colloncuran age (at least in part) but Santacrucian aspect La Venta fauna of Colombia (5° N) may be due to heterochrony, which is "temporal overlap of assemblages assigned to successive, presumed non-overlapping ages, or assemblages assigned to the same . . . LMA being time transgressive or not precisely time-equivalent" (Flynn et al. 1984). The Santacrucian SALMA fauna was apparently time transgressive from south to north. The rain shadow effect starting circa

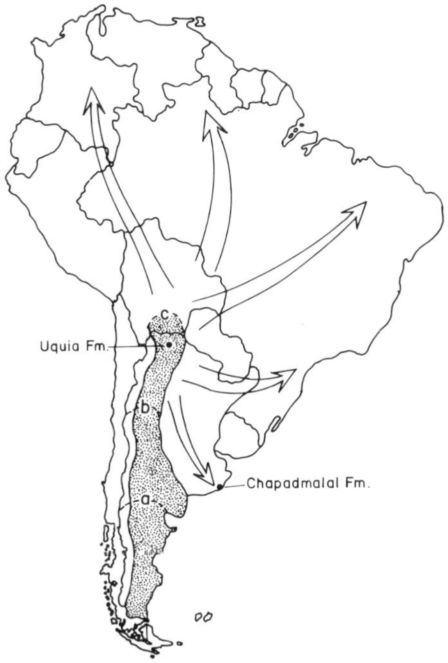

Figure 12.6. Proposed extent of "stress environment" (stippled area) during early Miocene (*a*), middle and late Miocene (*b*), and Pliocene (*c*) time. The western edge coincides with the eastern edge of the Andes, and the eastern edge approximates the westernmost occurrence of the loessoid sediments of the late Pliocene–Pleistocene Pampean Formation (Marshall et al. 1984:figure 4). Arrows indicate dispersal of faunas out of the stress environment.

15 Ma along the eastern flank of the southern Andes can be regarded as having generated a "stress environment" south of 30° S latitude that favored species originations and promoted rapid faunal change (figure 12.6). The rain shadow–induced changes in Santacrucian faunas (19–16 Ma) are evident in Colloncuran faunas (15–12 Ma) of the Collón Curá Formation. However, north and east of the stress environment (i.e., where the moisture-laden prevailing winds came from the east) faunas of Santacrucian aspect survived longer, and in Bolivia and Colombia appear in rocks that are temporally equivalent, in total or in part, to Colloncuran faunas in southern Argentina. The concurrent onset of a distinct south (temperate) to north (tropical) temperature gradient was also a contributing element and was possibly accentuated by about a 20° northward shift of the equator at the end of Santacrucian time (Pickford 1990a). Refined geochronologic and systematic studies of rocks and faunas are needed to test this heterochrony model.

Chasicoan This SALMA is based on the fauna from the Vivero Member of the Arroyo Chasicó Formation in southwest Buenos Aires Province, Argentina (Bondesio, Laza, et al. 1980). A few additional local faunas in Argentina are known but are not taxonomically diverse (Marshall, Hoffstetter, and Pascual 1983:33). No specific geochronologic data are available on Chasicoan age rocks and faunas, although for reasons discussed by Marshall (1985:63), this SALMA was believed to span 12–9 Ma. This age estimate is supported by a date of circa 11 Ma on a tuff from the Saladillo Formation and a date of 10.7 ± 1.7 Ma from a possible stratigraphic equivalent, the El Morterito Formation in northwest Argentina (Strecker et al. 1989; Malizia et al. 1990). These rock units underlie the Chiquimil Formation, which contains a Chasicoan fauna in its lower part (Pascual and Odreman Rivas 1973:319). In addition, Ramos (1988) equated initiation of the Quechua orogeny with an unconformity between the Chasicoan age La Pilona beds (noted by Pascual and Odreman Rivas 1973) and older beds in the valley of Uspallata and Cacheuta area in western Argentina. These data suggest that the base of Chasicoan time may be near 11 Ma and that a hiatus exists between Chasicoan and Colloncuran. This hiatus *might* correspond, at least in part, to the extensive epicontinental Paranense sea which is represented in the Chaco-Paraná basin of Argentina by the Parana Formation (Bertels 1979; Pascual 1984; Fernández 1989) and in the Subandean belt of Bolivia by the Yecua Formation (Sempere 1990), and to the Río-negrense sea in the Colorado basin of Argentina, which has an associated K-Ar data of 9.14 Ma (Zinsmeister et al. 1981).

Huayquerian The type fauna of this SALMA is from the Huayquerias Formation near San Carlos, Mendoza Province, west-central Argentina (Marshall, Hoffstetter, and Pascual 1983:35). A tuff from the top of this formation yielded a K-Ar date of 5.8 Ma (Marshall et al. 1985), and a magnetostratigraphic section with associated K-Ar dates from the Huayquerian age Araucanense (= Andalhuala) Formation (Garcia et al. 1984) at Puerto de Corral Quemado in Catamarca Province, northwest Argentina, shows that these rocks and faunas range from 7.5 to 6.4 Ma (Butler et al. 1984).

Faunas of this age are well known over much of the central and northern Argentina, extending from the eastern flank of the Andes to near the Atlantic coast (Marshall, Hoffstetter, and Pascual 1983). Outside of Argentina, Huayquerian age faunas are abundant only in Bolivia (Marshall and Sempere 1991), especially in the northern Altiplano, where they have been dated between 9 and 8 Ma (Marshall et al. 1992). Based on these data, Huayquerian is believed to span 9.0–6.0 Ma (Marshall 1985; figure 12.1).

Rocks of Huayquerian age were deposited in association with a major deformational episode that began approximately 10 Ma and is called the Quechua tectosedimentary episode (based on Steinmann 1929) in northwest Argentina (see Jordan and Alonso 1987; Díaz and Miserendino 1988; Miserendino and Díaz 1988; Jordan and Gardeweg 1989). The deformational episode is dated 9.5–8.0 Ma on the Puna (Coira et al. 1982: 321) and correlates with the "F3 and F4 tectonic phases" (c. 10–7 Ma) of Sébrier and co-workers (1988) in the central Andes, with the "second major tectonic crisis" starting about 11–10 Ma in Bolivia (Sempere et al. 1989; Sempere et al., "Altiplano," 1990), and with the first Neogene tectonic movements and basin subsidence in the Sierras Pampeanas of northwest Argentina (Malizia et al. 1990). Maximum uplift and deformation of the southern Andes occurred about this same time (c. 9 Ma; Ramos 1989:899).

This tectosedimentary episode, starting circa 10 Ma, led to the last major uplift of the Andes. It now appears that uplift since the early Miocene resulted in about a 3500-meter height increase of at least part of the central Andes (Tosdal et al. 1984). The rain shadow effects intensified, and by Montehermosan time (6.0–2.5 Ma), true desert habitats were established in northwest Argentina (Solbrig 1976; Marshall and Patterson 1981), extensive grasslands in the Pampas of Argentina, and dry, low-temperature, high montane habitats on the Altiplano of Bolivia (B. Simpson 1983; Marshall and Cifelli 1990; Pascual and Ortiz-Jaureguizar 1990 and references therein). The stress environment extended northward as far as southern Bolivia (figure 12.6), and aridity in intermontane basins of northwest Argentina intensified as a result of blockage of moisture-laden easterly winds by the high Sierras Pampeanas (Strecker et al. 1989: 527). In northwest Argentina, uplift was renewed about 7 Ma and became pronounced between 4.0 and 3.4 Ma (Strecker et al. 1989).

Huayquerian faunas are characterized by the first South American record of Procyonidae (raccoons and their allies). The earliest known fossil of this group is of the genus *Cyonasua* from a level of the Araucanense (= Andalhuala) Formation in Catamarca Province, northwest Argentina, and is dated at 7.5–7.0 Ma (Butler et al. 1984; Marshall 1985). Procyonids were waif dispersants from Central or North America; their arrival correlates with the reciprocal waif dispersal of ground sloths from South to North America, where they are first securely known below a tuff dated at 8.19±0.16 Ma (Marshall 1985 and references therein). This apparently reciprocal dispersal event may have occurred during the major marine regression at approximately 10 Ma (Haq et al. 1987) that was coeval with partial emergence of the isthmian region (Duque-Caro, 1989; figure 12.1).

Montehermosan Taxonomically diverse faunas of this age are known from the Tunuyan Formation near San Carlos in Mendoza Province, west-central Argentina, which occurs above a tuff in the top of the Huayquerias Formation (see above) dated 5.8 Ma. They are also known from the Corral Quemado Formation in Catamarca Province, northwest Argentina, for which magnetostratigraphic and associated K-Ar data range from 6.4 to 3.5 Ma (Butler et al. 1984). Numerous faunas on the northern Altiplano of Bolivia occur between tuffs securely dated 5.4 and 2.8 Ma (Marshall et al. 1992). These geochronologic studies demonstrate that Montehermosan spans about 6.0–2.5 Ma.

The type locality of the Montehermosan SALMA is at Monte Hermoso along the Atlantic coast near Bahia Blanca, Buenos Aires Province, Argentina. No geochronologic studies have been made on these rocks, although evidence on stage of evolution indicates that the fauna falls within latest Montehermosan time (i.e., 3.5–2.5 Ma; Marshall 1985:65).

The earliest known cricetid rodents in South America (*Auliscomys formosus, Bolomys bonapartei;* Reig 1978) are reported from the "Hermosense típico" stratigraphic unit (*sensu* Bonaparte 1960) at Monte Hermoso. This group is more diverse in Chapadmalalan and later faunas. The time and means of cricetid dispersal to South America is a subject of continuing controversy, and three divergent models have been proposed to explain their paleobiogeographic history.

In the first model, cricetids were among the first wave of North American immigrants to walk across the Panamanian landbridge following its emergence about 2.5 Ma. Their notable taxonomic diversity is explained by a prior or subsequent radiation (Patterson and Pascual 1972). This model dictates two possibilities: that the type fauna of the Montehermosan SALMA is about 2.5 Ma (younger than conventionally envisioned) and is temporally equivalent to early Chapadmalalan faunas, or that the cricetids from Monte Hermoso may not have been collected in situ from the Hermosense típico stratigraphic unit but were derived from overlying Chapadmalalan beds that occur at the same locality (see below).

According to the second model, sylvan ancestral forms of cricetids arrived as waif dispersants from Central America in late Miocene time, during the Messinian marine low-stand; diversified into pastoral, grassland forms in savanna-grassland areas of northwest South America; and dispersed southward to Argentina through a savanna-grassland corridor along the eastern foothills of the Andes, which formed during the first glacial advance in the Pliocene (c. 3.5 Ma). This model is based on the known fossil record, on well-documented ecologic changes during the Pliocene and Pleistocene, and on the belief that the Hermosense típico is of latest Montehermosan age and that the cricetid fossils were derived in situ from this level (Marshall 1979, 1985).

To account for their present diversity and the fact that the earliest known fossils are specialized pastoral forms, the third model proposes that cricetids were waif dispersants in the early Miocene, experienced cladogenesis in high-altitude areas of the northern and central Andes, and dispersed throughout South America only after the onset of glaciations about 3.5 Ma (Reig 1986).

There are some problems with the third model. The high-altitude Altiplano of Bolivia and Puna of northern Argentina and adjacent Chile are identified by Reig as the primary area of cricetid diversification in the central Andes (figure 12.7). However, this high-altitude environment is of recent origin, and developed in conjunction with the Quechua orogenic uplift starting circa 10 Ma (Allmendinger 1986; Marshall et al. 1992). Thus, the high Altiplano-Puna did not exist

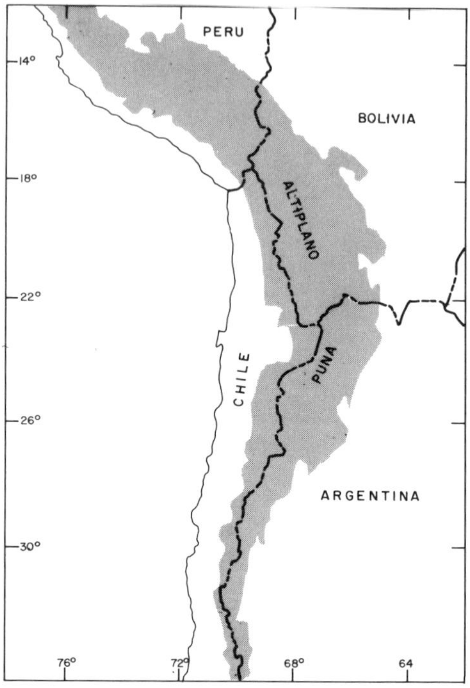

Figure 12.7. Location and distribution (in stipple) of the high-elevation Altiplano-Puna in west-central South America. After figure 1 in Isacks (1988).

at the hypothesized early Miocene time of cricetid arrival. The basic foundation of the Altiplano did exist beginning in the late Oligocene, but it was high only after the late Miocene.

There is no late Cenozoic fossil record of land mammals on the Argentina-Chile Puna, and although there is a remarkable record on the adjacent Bolivian Altiplano, from which some small body size taxa are known, it includes no cricetids (Hoffstetter 1986; Marshall, Hoffstetter, and Pascual 1983; Marshall et al. 1984; Marshall and Sempere 1991). The hypothesized early Miocene–Pliocene presence of cricetids in these areas is therefore unsupported by the available fossil record.

The Altiplano-Puna was the site of major volcanic activity, especially after approximately 11 Ma (Sempere et al., "Altiplano," 1990). In northern Chile alone there are more than 470 volcanoes documented for the Plio-Pleistocene, and for the Altiplano-Puna of Bolivia, Argentina, and Chile there were possibly more than 5000 volcanoes during the late Miocene-Pleistocene (J. A. Naranjo, personal communication, 1991). In the volcanic province between 19°30′ S and 22°30′ S, there were peaks of volcanic activity between 12 and 9 Ma and between 6 and 3 Ma, and the volume of lava and ignimbrites extruded during these periods was on the order of 1900 km^3 and 5250 km^3, respectively (Baker and Francis 1978). Ignimbrite-ash sheets are common, and some covered extensive areas. The Ataña Ignimbrite of the Chilean Puna, for example, was formed by four major eruptions that were emplaced in rapid succession about 4.1 ± 0.4 Ma and probably covered an area of 7700 km^2 (Gardeweg and Ramírez 1987). On the northern Altiplano, the extensive Toba 76–Cota Cota Tuff sheet (5.4 Ma), which might be related to collapse of the Soledad Caldera and eruption of the Soledad Tuffs (Redwood 1987), may have covered 5000 km^2, and the Pérez Ignimbrite–Ayo Ayo Tuff–Chijini Tuff sheet (c. 2.8 Ma) covered more than 20,000 km^2 (Marshall et al. 1992). These sheets and associated deformations separate early Huayquerian from Montehermosan faunas, and Montehermosan from Ensenadan-Lujanian faunas (Marshall et al. 1992). Hoffstetter (1986) has shown that the changes between these successive low diversity faunas are characterized by regional extinction and subsequent immigration from lowland areas.

These geological and paleontological data collectively suggest that during the late Miocene and Pliocene, recurrent and widespread ash sheets resulted in the repeated elimination of biotas over large areas of the Altiplano-Puna, and that subsequent faunas were derived from immigration of select high-altitude, cold-tolerant stocks from surrounding, mostly lowland, areas (Hoffstetter 1986). It therefore appears unlikely that the Altiplano-Puna would have provided opportunity for cricetid diversification as envisioned by Reig; on the contrary, it was probably an area of repeated regional extinctions and historically low faunal diversity maintained by immigrant stocks from more stable populations at lower elevations. There are no paleontological data to support the notion that the Altiplano-Puna had an endemic fauna or that it was the site of significant in situ evolution.

A parallel to the faunal history of the Altiplano discussed above is provided by its flora. "Across biomes, diversity varies predictably in the same way as productivity. Diversity decreases from the tropics to the poles and from the bases to the summits of mountains. . . . As the equator-pole gradient increased during the Tertiary and the Andes progressively rose, we would therefore expect that there was a gradual development of a gradient in productivity from low to high latitudes and from low to high elevations. Correlated with these trends in productivity, we would expect a progressive decrease in numbers of species in high latitudes and high elevations from that present in the same terrestrial localities in the Eocene. In effect, the development of global thermal zonality and increased elevations both had the same effect, the reduction in primary productivity and elimination of species (reduction in diversity)" (B. Simpson 1983:114).

In the Altiplano-Puna in the central Andes, selection was strong, and "the demands made of species adapted to life in the lowland humid tropics were beyond the limits of variation present in most of them. As conditions became more and more disparate from the lowland tropical conditions, fewer and fewer species of the original lowland vegetation survived. . . . Most of the original inhabitants were replaced by successful alien colonists. The most successful of these immigrants came from areas ecologically similar to the new kinds of habitats being created. . . . While the central Andes probably did not become extremely dry until very late in the Tertiary, pockets of arid habitats developed relatively early in interandean valleys. . . . These valleys were undoubtedly of importance in providing migration routes for some of the taxa preadapted to arid conditions that eventually colonized the dry regions of the Altiplano. . . . Among these were open habitats in various parts of the Neotropics and the interandean dry regions. . . . The relative importance of these source areas depends on the distance from the source to some extent, but primarily to the degree of ecological similarity of the source area" (B. Simpson 1983:114–116, 121).

Simpson (1983:116, 117) concludes that many of the successful immigrants "would have belonged to groups that ultimately became cosmopolitan because of their colonization abilities." Among these were "weedy cosmopolitan elements," which by analogy are represented by the known large body size Pleistocene mammals on the Altiplano (i.e., taxa with wide geographic and altitudinal distributions: *Glyptodon, Megatherium, Glossotherium, Scelidodon, Macrauchenia, Onohippidium, Cuvieronius*—see Hoffstetter 1986). Because the Altiplano is an extensive flat area, "species distributions are usually determined by climatic, rather than geographic factors. . . . Some autochthonous speciation would have occurred, but not on the scale of that [in the northern Andes]."

Discussion Caviomorph Rodentia and platyrrhine Primates (Old Island Hoppers) were derived from Caribbean Island stocks that obtained a foothold in South America at the beginning of the late Oligocene (c. 27 Ma). The appearance of these groups represents an important "immigrant datum event" in what is here interpreted to be early, but not basal, Deseadan.

Procyonids (New Island Hoppers) are first recorded in South America in faunas of Huayquerian age (7.5–7.0 Ma). An optimal opportunity for the dispersal of procyonid stocks from North or Central America would have been enhanced near 10–11 Ma by a sea level low-stand and/or initiation of a second major Andean tectonic episode (figure 12.1).

We favor the view that cricetid rodents were probably the first group to disperse from North to South America following initial emergence of the Panamanian isthmus about 2.5 Ma and that they were participants in the Great American Faunal Interchange. This view is based on the observation that the earliest taxa reported from the Hermosense típico were never demonstrated to have been collected in situ, and it appears likely that they may have been derived from overlying sediments of Chapadmalalan age. Given that the overall stratigraphic thickness of fossiliferous beds at Monte Hermoso is only 7 meters (Marshall 1985:65), this appears to be a distinct possibility. The only way to substantiate this prediction is by additional controlled collecting in the Monte Hermoso area.

C. Set 3: Chapadmalalan-Lujanian

Chapadmalalan-Uquian These SALMAS are discussed together because, for reasons documented below, we regard them as temporal equivalents.

Chapadmalalan is based almost exclusively on the fauna from the Chapadmalal Formation between Mar del Plata and Miramar along the Atlantic coast, Buenos Aires Province, Argentina. The authoritative study of the fauna is by L. Kraglievich (1934); the geology was described by his son J. L. Kraglievich (1952), who distinguished 15 levels, from oldest to youngest, I–XV. The only other nearby fauna assigned to this age is that from the Miembro de Limolitas Claras of Bonaparte (1960), which overlies the Hermosense tipico at Monte Hermoso and which Vignati (1925) earlier regarded as Chapadmalense (see Marshall, Hoffstetter, and Pascual 1983:37). A mustelid (*Conepatus*) and a tayassuid (*Platygonus*) were recovered from level III of the Chapadmalal Formation, while four species of cricetid rodents (*Akodon kermacki, Dankemys simpsoni, Graomys dorae, Reithrodon auritus;* Reig 1978) are known from other levels. These taxa represent the first record in the Chapadmalal area of North American immigrants that walked to South America across the Panamanian landbridge. Their appearance represents an important immigrant datum event for South American faunas and, with cricetids, marks the initiation of the GAFI. The date of this dispersal event based on evidence from elsewhere in South America and from the reciprocal side of this event in North America will predictably constrain the lower age of Chapadmalalan age rocks and faunas.

Early participants in the GAFI are also known in Jujuy Province, northwest Argentina, about 1900 kilometers north-northwest of the Chapadmalal Formation. Here, remains of two tayassuids (*Platygonus, ?Mylohyus*) and an equid (*Hippidion*) were recovered from the lower part of the Uquía Formation at Esquina Blanca. A magnetostratigraphic section with associated K-Ar dates indicates that this formation ranges from 2.5 to 1.5 Ma (Marshall et al. 1982). The lowest level that yielded fossil mammals (including the equid and tayassuids) is in the basal part of the Matuyama Chron and is therefore circa 2.5–2.4 Ma. These data suggest that the Chapadmalal Formation may be of similar age. In this context it is of historical interest to note that L. Kraglievich (1930, 1934) had regarded the fauna from the lower level at Uquía (his Uquiense) as similar to Chapadmalalan faunas.

In North America, the earliest South American participants in the GAFI—a porcupine, *Erethizon;* a capybara, *Neochoerus;* a glyptodont, *Glyptotherium;* armadillos, *Holmesina, Dasypus;* a ground sloth, *Glossotherium;* and a large ground bird, *Titanis*—are known from multiple well-dated units and occur in the latest part of the Gauss Chron (Marshall 1985; Webb and Barnosky 1989). The level within the Gauss Chron is not surely known, although it was long interpreted to be between 2.8 and 2.6 Ma, which resulted in assignment of a 2.8 Ma age to the base of the Chapadmalal Formation (Marshall 1985). This age, 2.8 Ma, does correlate with the onset of climate deterioration and faunal changes elsewhere in the world (Vrba et al. 1989; Pickford 1990b) or with general climatic-environmental change in South America (Mörner 1990). Reconsideration of these data now suggest that the earliest GAFI dispersants are from a level corresponding to 2.5±0.1 Ma (Webb and Barnosky 1989:416) at the Gauss-Matuyama boundary (Opdyke 1990). This age is consistent with the 2.5–2.4 age at Uquía, and it corresponds to the dispersal of microtine rodents into North America from Eurasia and of *Equus* to Eurasia from North America. At this time, worldwide avenues for mammalian dispersal apparently became available, and the initiation of the GAFI was simply part of this global event (Marshall 1985 and references therein; Webb and Barnosky 1989).

New and better-constrained evidence of changes in Caribbean and Pacific foraminifera indicates that initial emergence of the landbridge occurred around 2.4 Ma, and final separation of oceanic realms at 1.8 Ma (Keller et al. 1989). Initial emergence at approximately 2.4 Ma coincides with a major global cooling event, onset of glaciation in both hemispheres, increased temperature gradients in the mid- to high-latitude Atlantic, a drop in sea level, culmination of a trend in climatic deterioration begun near 2.8 Ma, and a widespread erosional hiatus (Keller et al. 1989:107; see also Webb and Barnosky 1989:418; Vrba et al. 1989; Pickford 1990a; Ruocco 1990).

It has long been recognized that taxa participating in the GAFI were tolerant of, or specifically adapted to, savanna-woodland habitats (Webb 1978). Furthermore, the extensive savanna habitats that served as dispersal corridors existed only during times of glacial advance (Webb 1978; Marshall 1979). The first well-calibrated glacial chronology of southern Argentina, based on magnetostratigraphic study of moraines, shows that a major glacial advance (and hence "savanna maxima") occurred around 2.3 Ma, which is close to the Gauss-Matuyama boundary (Mörner and Sylwan 1989).

This dispersal opportunity also correlates with a major marine regression (Haq et al. 1987) and, in a broader sense, with the Diaguita deformation in northwestern Argentina (3±1 Ma; Jordan and Alonso 1987), where tectonic motions occurred between 2.5 and 0.6 Ma (Strecker et al. 1989). It also correlates with the Q4 or F6 deformation on the northern Altiplano of Bolivia (2.8–1.6 Ma; Sébrier et al. 1988; Marshall et al. 1992). The extinction event identified by Marshall and Cifelli (1990) in Chapadmalalan (*s.s.*) time is apparently related to these geologic-climatic changes and onset of the GAFI. Furthermore, restudy using multivariate techniques of L. Kraglievich's (1934) data, upon which the distinction of Chapadmalalan is based, shows that this fauna groups more closely to Uquian faunas, not to Montehermosan faunas as originally concluded (Prado and Ortiz-Jaureguizar 1989). The Chapadmalalan and Uquian faunas are in dire need of systematic revision, for previous age studies were based on poorly constrained long-distance correlations using outdated taxonomic lists, rather than on study of the fossils themselves (Zárate and Fasano 1989).

There nevertheless exists overwhelming evidence to indicate that Chapadmalalan and Uquian (Uquiense) faunas are temporal equivalents and that the basal age of both is circa 2.5 Ma, based on the co-occurrence of GAFI participants. This equivalence possibility was suggested earlier by Marshall (1985). Recognition of this equivalence requires nomenclatural changes in the SALMA sequence. The name Chapadmalalan dates from Ameghino 1908 and has priority over the name Uquian, which dates from L. Kraglievich 1930 and 1934. Chapadmalalan should therefore be used in place of Uquian as presently conceived (Marshall et al. 1984; Marshall 1985; figure 12.1). This change is made on the premise that first appearances—immigration events—are more useful than last appearances in defining SALMA boundaries (e.g., Woodburne 1977).

This new interpretation of age for the Chapadmalal Formation further permits reevaluation of the magnetostratigraphic section of these rocks provided by Orgeira and Valencio (1984). These workers show that the majority of the formation is of reversed polarity, which we interpret to represent the base of the Matuyama Chron. Our view supports an age equivalence for the Chapadmalal and basal Uquía formations. In addition, Orgeira and Valencio (1984) show that the Uquian age Barranca de Los Lobos and Vorohué formations (*sensu* Marshall et al. 1984) are predominantly of normal polarity, which we interpret as equivalent to the Olduvai Event in the Matuyama Chron. These interpretations are consistent with all other data reviewed here, although they are considerably different from the conclusions reached by Orgeira and Valencio (1984). Ruocco (1990), after conducting a magnetostratigraphic study of the Barranca de Los Lobos, Vorohué, and San Andrés formations, concluded that they are older than 1.89 Ma and younger than 2.9 Ma, although the latter age is poorly constrained.

We propose a possible explanation for the long-observed "older stage of evolution" of faunas from the Chapadmalal Formation relative to the "younger stage" of those from the lower part of the Uquía Formation. The faunas from Uquía are in the stress environment produced by the intensified rain shadow and predictably are changing faster relative to faunas from the Chapadmalal Formation, which are well outside of this environment (figure 12.6). The observed taxonomic differences can be attributed to heterochrony, because the faunas from the Chapadmalal Formation have long been regarded as closer to Montehermosan faunas, while those from the Uquía Formation show closer affinity to those from the Barranca de Los Lobos, Vorohué, and San Andrés formations that overlie the Chapadmalal Formation (Marshall et al. 1984; Marshall 1985 and references therein). The stress environment in Argentina thus had a west to east heterochrony component that apparently had a lag time of several hundred thousand years, and the north to south continent-wide component for Santacrucian-Colloncuran faunas had an apparent lag time of several million years.

Ensenadan The type fauna of this SALMA comes from the Ensenada Formation at Ensenada along the Río de La Plata, Buenos Aires Province, Argentina. Magnetostratigraphic studies of the Ensenada Formation in the city of Buenos Aires show that it correlates with the Matuyama Chron and therefore ranges from 2.41 to 0.7 Ma (Valencio and Orgeira 1983; Nabel and Spiegelman 1988). Unfortunately, only a few fossil mammals were associated with these magnetostratigraphic sections, and the relative position of larger collections from Ensenada are not known. The data also demonstrate that the lower part of the Ensenada Formation is in the upper part of the Gauss Chron, which is 2.84–2.41 Ma. This means it is temporally equivalent to at least some Chapadmalalan age rocks and faunas elsewhere in Argentina. There are, however, no data that demonstrate that Ensenadan faunas occur in the lower part of the Ensenada Formation, although Orgeira and Valencio (1984) erroneously regarded both rocks and SALMA faunas as equivalents.

Ruocco (1990) provides a magnetostratigraphic study of the Ensenadan age Miramar Formation (*sensu* Marshall et al. 1984) in Buenos Aires Province and concludes that it is bracketed somewhere between 1.89 and 0.97 Ma.

A detailed geochronologic (magnetostratigraphic, isotopic) study of the approximately 250-meter-thick Tarija Formation in southern Bolivia shows that it spans 1.0–0.7 Ma (MacFadden et al. 1983). The stage of evolution of this exceedingly rich fauna suggests that it is principally of middle or late Ensenadan age. These data, along with those on the Ensenada Formation of Argentina, are concordant in placing an upper age limit for Ensenadan time at 0.7 Ma and a lower boundary older than 1.0 Ma. For reasons discussed by Marshall (1985), placement of the Chapadmalalan-Ensenadan boundary near 1.5 Ma still appears reasonable and is consistent with correlation of the normal-polarity Barranca de Los Lobos and Vorohué formations with the Olduvai Event. The 1.5 Ma boundary correlates with a sharp marine regression (Haq et al. 1987) and dispersal event (see below; figure 12.1).

Lujanian The type fauna comes from the Luján Formation to the west of Buenos Aires, Argentina (Marshall et al. 1984:29). A recent magnetostratigraphic study of the Bonarense sedimentary rocks in the city of Buenos Aires (early Lujanian age; Marshall et al. 1984) indicates that they are in the Brunhes Normal epoch and hence younger than 0.7 Ma (Valencio and Orgeira 1983). MacFadden et al. (1983) obtained identical results on a magnetostratigraphic study of the Lujanian-bearing part of the Ñuapua Formation in southeast Bolivia. Ruocco (1990) reported that magnetostratigraphic study of the Lujanian age Arroyo Seco and Santa Isabel formations in Buenos Aires Province indicates they are "younger than 1.0 Ma, maybe even 0.6 Ma."

Lujanian thus extends from about 0.7 Ma to about 10 ka (Ochsenius 1985). Both the lower and upper boundary correlate with sharp marine regressions (Haq et al. 1987; figure 12.1).

Tonni and co-workers (1987) recognized a new SALMA, the Centinelan, between the Ensenadan and Lujanian. Its validity has yet to be firmly established, however, and these authors did not specify how this new age differs from the Bonarense, which is a conceptual equivalent (see Marshall et al. 1984:29).

The end of Lujanian time in South America is marked by the extinction of most large body size mammals (i.e., larger than 100 kg), which are commonly referred to as megafauna (Ochsenius, 1985). The peak of these extinctions occurred between 12 and 10 ka (Marshall et al. 1984; Ochsenius 1985) and resulted in the disappearance of 8 of the 37 families (or 21%) recorded in Lujanian faunas—Glyptodontidae, Megalonychidae, Megatheriidae, Mylodontidae, Macraucheniidae, Toxodontidae, Equidae, and Gomphotheriidae. This is the only significant extinction event seen in an analysis of South American land mammal family patterns through Cenozoic time (Marshall and Cifelli 1990). Furthermore, 56 of the 153 recorded land mammal genera (37%) went extinct; 54 of these (96%) were megafauna (Marshall and Cifelli 1990). The cause of these extinctions is not known with certainty, although they have been attributed to overhunting by early humans, climatic change, and a combination of the two (Ochsenius 1985; Hoffstetter 1986).

Discussion Chapadmalalan-Lujanian faunas were affected by two interrelated phenomena that influenced or determined their taxonomic composition: glaciations and associated climatic changes, and the Great American Faunal Interchange.

Glaciations: Today, savannas have a disjunct and predominantly three-part distribution throughout South America, occurring over much of Uruguay and east-central Argentina, over much of eastern Bolivia and central Brazil, and in parts of northeast Colombia, Venezuela, and the Guianas (figure 12.8c and d). They are presently separated by extensive warm-wet subtropical and tropical biomes. Many species of mammals have populations in these disjunct savanna "islands" but not in the intervening tropical forest areas. These and other data demonstrate that the savanna islands were interconnected in the recent past (Webb 1978; Ochsenius 1985; Marshall 1979, 1985).

During times of glacial maxima (advance), the subtropical and tropical biomes experienced cold-dry climates, resulting in their contraction and in the concurrent expansion and interconnection of savanna biomes (Kalin Arroyo et al. 1988; figures 12.8 and 12.9). A savanna corridor formed along the eastern flank of the Andes and provided a north-south dispersal route for savanna biotas (Webb 1978; figure 12.8). This cold-dry, uniformly nonforested savanna corridor linked the Pampas of Argentina with the Llanos of Colombia and continued north across the Panamanian isthmus as far as Florida (Webb 1978; Marshall 1979, 1985). In the region of Panama, tropical forests persisted in lowland areas (Bush and Colinvaux 1990; Piperno et al. 1990), so the savanna corridor apparently occupied the extensive continental shelf areas that were exposed during significant coeval sea level low-stands (Haq et al. 1987). This "savanna maxima" model explains both dispersal and speciation opportunities during the past 2.5 Myr (Marshall 1985).

The glacial sequence in South America has yet to be fully documented and well dated although Mörner and Sylwan (1989) published a magnetostratigraphic study of 15 separate terminal moraines near Lago Buenos Aires in southern Argentina. They grouped the moraines into five major glaciation complexes: Initioglacial I, circa 2.3 Ma; Initioglacial II, circa 1.2 Ma; Daniglacial, circa 0.7 Ma; Gothiglacial, circa 0.3? Ma; and a two-part Finiglacial, one < 120 ka,

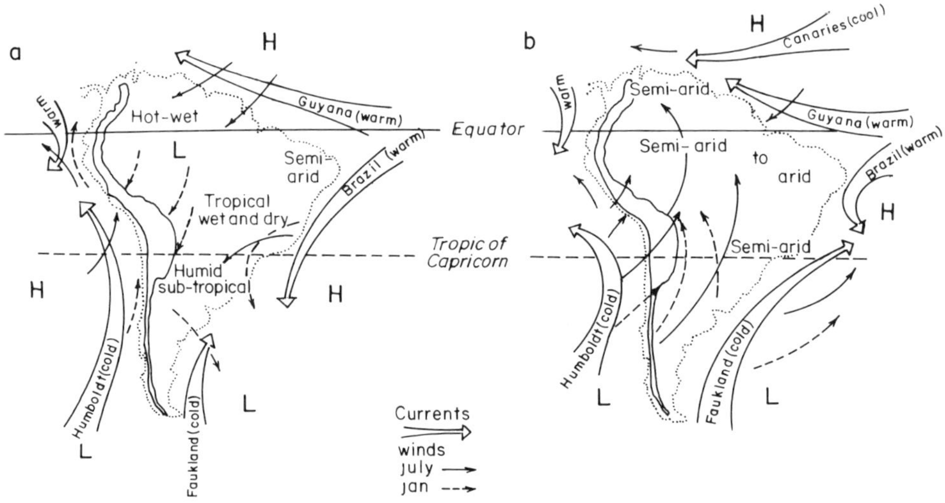

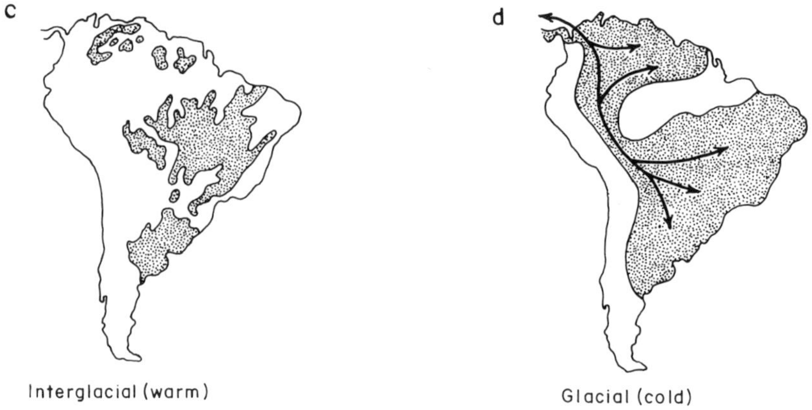

Figure 12.8. Probable interglacial (*a*) and glacial (*b*) climates for South America (*H*, high pressure area, *L*, low pressure area). After figure 5 in Kalin Arroyo et al. (1988), as modified from Damuth and Fairbridge (1970). Approximate distribution of savannas (stippled areas) during interglacials, as today (*c*), and during glacial maxima (*d*). Modified from figure 1 in Marshall (1979).

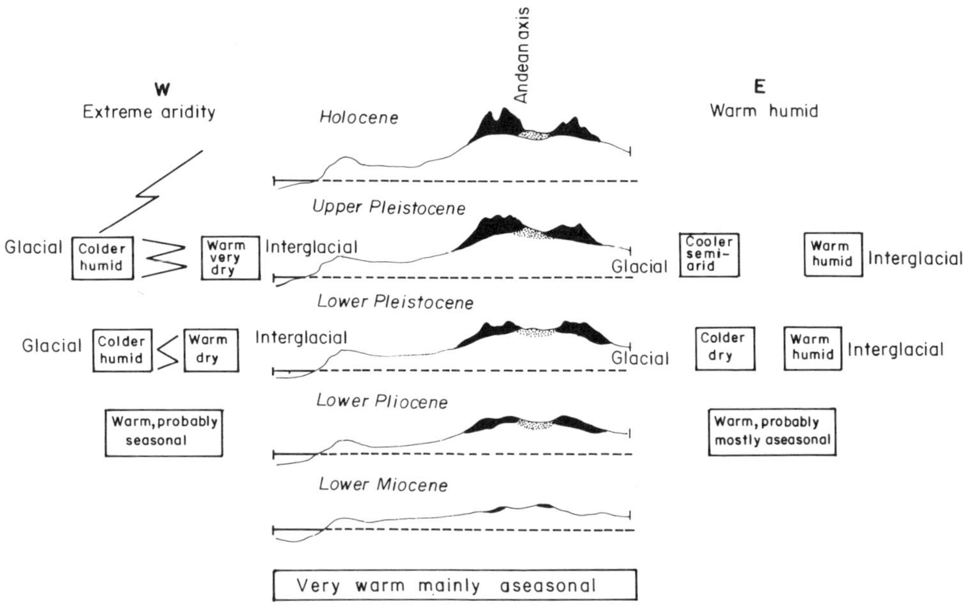

Figure 12.9. West to east cross section of central Andes, showing climatic tendencies during five successive periods, from early Miocene to Holocene. Solid black = volcanic rocks, stippled area = Altiplano. After figure 4 in Kalin Arroyo et al. (1988). Note that the model of Andean orogeny used here is based on the view that successive uplift occurred in the same area, although the model used in figure 12.10 is now preferred.

another > 14 ka. The four youngest of these may correlate with the Calvario, Kaluyo, Sorata, and Choqueyapu (I and II) glaciations recognized by Servant and Fontes (1978) and Lavenu et al. (1984) on the Altiplano of Bolivia. There is also evidence for a glacial advance in Bolivia in the Pliocene, below the 2.8 Ma Chijini tuff (Clapperton 1979; Marshall et al. 1992), and another in southern Argentina in the late Miocene–early Pliocene (Mörner and Sylwan 1989). The important point is that there were at least five and possibly six major glacial advances and savanna maxima during Chapadmalalan-Lujanian time. We therefore predict, based on a model of this time interval, that a minimum of nine distinct faunal types (five glacial, four interglacial) will be represented, and these have yet to be identified and calibrated.

Great American Faunal Interchange: All of the taxa that participated in the GAFI were tolerant of, or specifically adapted to, savanna-woodland habitats (Webb 1978). Large-scale dispersal of these taxa was thus possible only during times of glacial advance and the contemporaneous presence of a savanna corridor from Argentina to Florida. That opportunities permitting dispersal of savanna taxa existed only about five times during the past 2.5 Myr explains the long-observed fact that the interchange occurred in a series of successional and reciprocal waves or pulses (Webb 1976; Hoffstetter 1986). The chronology of these savanna maxima dispersal waves proposed by Marshall (1985) requires reconsideration in light of the well-calibrated moraine sequence in Argentina. It now appears that there were major dispersal events at the beginning of Chapadmalalan (2.5–2.3 Ma), Ensenadan (c. 1.5–1.2 Ma), and Lujanian (0.7 Ma) time, possibly one during Lujanian time (c. 0.3? Ma), and one at the end of Lujanian time (14–12 ka) that apparently permitted entry of humans into the Neotropics. All of these events correlate with major marine regressions (Haq et al. 1987; figure 12.1).

V. Conclusions

The dominating Andean Cordillera is a relatively new feature for the South American landscape. The central Andes, for example, became a notable mountain belt, albeit of limited altitude, only about 27 Ma; they attained about 50 percent of their present height by circa 15 Ma and the remaining 50 percent subsequent to early Miocene time.

Land mammal fossils have been recovered from sedimentary basins that developed in a variety of tectonic contexts. Description of these contexts aids in the understanding of some aspects of the faunas. There is still no consensus, however, for the style of tectonic evolution of the Andean Cordillera. Early models envisioned the Andean Cenozoic tectonic regime as characterized by long tensional periods, short synchronous compressional pulses, and generally high-angle faulting (e.g., Sébrier et al. 1988), whereas newer models favor dominance of progressive crustal shortening, thrust propagation, and foreland basin evolution (e.g., Jordan and Alonso 1987; Sempere et al., "Tectonic crisis," 1990; Marshall and Sempere, 1991). A shared basic premise in these divergent models is that the Andes developed by a stepwise series of eastward migrating magmatic arc systems (figure 12.10) and that uplift is primarily related to subduction geometry and plate convergence velocity along the western edge of the South American Plate (Coira et al. 1982; Jordan and Alonso 1987; Alpers and Brimhall 1988; Sempere et al., "Tectonic crisis," 1990). The eastward-stepping depocenters were deformed and served as source areas for the next younger basins to the east (Williams et al. 1990). Knowing this enables us to understand the distribution of mammal-bearing sediments in space and time (i.e., intermontane Andean basins formed only after c. 27 Ma, etc.).

As summarized in figure 12.1, the tectonic episodes correlate fairly well with SALMAS and their boundaries (i.e., Deseadan with the "first major Bolivian crisis" of Sempere et al., "Tectonic crisis," 1990; Santacrucian with the Pehuenche deformation [called "Quechua" in the southern Andes]; Huayquerian with the Quechua in the central Andes; and Chapadmalalan with the Diaguita in the central Andes). These correlations suggest that faunal changes are induced by geological processes and events.

It has also long been recognized that deformation/uplift (tectosedimentary) episodes have a marine regression counterpart (e.g., Harrington 1962). This feature is supported by the tectonic data summarized in this paper and the eustatic curves provided by Haq et al. (1987; figure 12.1). However, transgression and regression phases in the

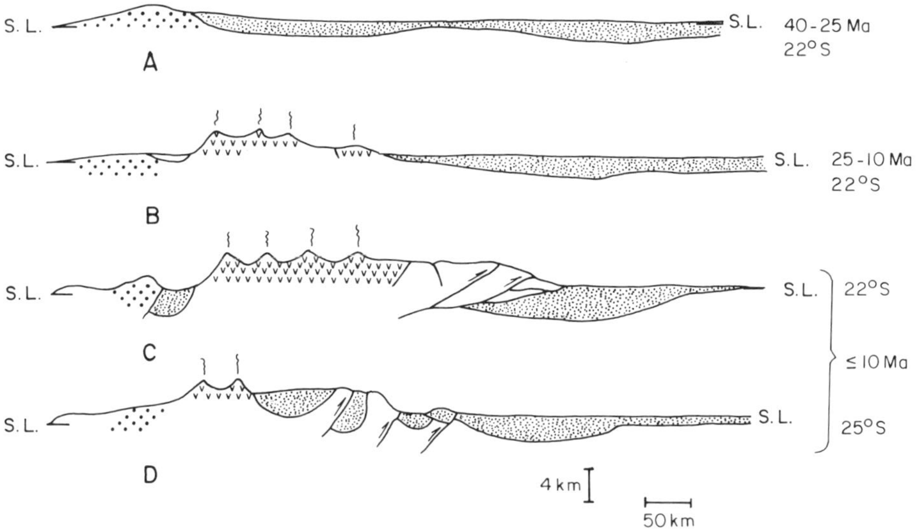

Figure 12.10. Generalized west (left) to east (right) cross section of central Andes, showing eastward migrating magmatic arc systems and magnitude of uplift during three periods of time. Stippled = sedimentary deposits; vvv = volcanic rocks; dotted = ancient high arc area. After figure 7 in Jordan and Alonso (1987).

Cenozoic of South America may result from eustatic or tectonic processes or both. Consequently, forced correlations of eustatic cycles with continental events can lead to erroneous conclusions (e.g., those of Legarreta et al. 1990).

During periods of uplift and accompanying marine regression, "dispersal opportunities" for continental faunas were optimized by: (a) decrease in width of marine barriers separating continental plates; (b) regression of epicontinental seas, permitting free interchange of previously isolated or areally restricted faunas; and (c) newly elevated land areas serving as bridges between previously separated areas. It is therefore not surprising that all important immigration events correlate with uplift and marine regression events: for example, the arrival of caviomorph rodents and primates in early Deseadan time; the arrival of procyonids in Huayquerian time; and the beginning of the GAFI in Chapadmalalan time. When such immigration events correlate with times of uplift-induced environmental change, it is difficult, if not impossible, to identify the separate influences of each on concurrent faunal turnover (Webb 1984; Marshall and Cifelli 1990).

Marine epicontinental seas on the South American craton apparently played an important role in regulating aspects of land mammal biochronology and evolution during Cenozoic time by controlling the land surface available to the terrestrial faunas. Such regulation is evident, for example, in the distribution of early Miocene marine epicontinental seas in southern South America (figure 12.11). Colhuehuapian faunas were apparently coeval with a rise in sea level between 20 and 19 Ma, and rocks of this age were either deposited along the shoreline of these epicontinental seas (e.g., the type Colhuehuapian fauna, labeled *A* in figure 12.11) or covered by the transgression (as at Gaiman, labeled *B*). Early Santacrucian faunas were either deposited along these same shorelines (as that from the Pinturas Formation, labeled *C* in figure 12.11) or near the shoreline but above the marine beds, when regression began about 19 Ma (e.g., at Karaiken, labeled *D*). The dune deposits noted by Bown and Larriestra (1990) in the Pinturas Formation (*C*) are readily understood in this paleogeographic context, because they represent coastal dunes and not inland desert dunes. Dune deposits in the continental Santa Cruz Formation from along the Atlantic coast, noted by Bown and Larriestra (1990), probably represent coastal dunes that accompanied this marine regression, as indicated by the sequence-stratigraphy observations of Legarreta et al. (1990). More importantly, the data illustrated in figure 12.11 clearly indicate that the classic Santacrucian SALMA fauna, which is known from the Santa Cruz Formation in the Magallanes basin, *must* have developed from Colhuehuapian stocks that immigrated there from the north following the marine regression, and that this Santacrucian fauna did not evolve from in situ stocks. The Magallanes basin, which received these immigrant stocks, was previously uninhabited, was in a temperate area, and probably provided opportunities for rapid evolution. These factors can only be appreciated when the distribution of continental and marine deposits are viewed in temporal and areal perspective.

The rain shadow resulting from the increased elevation of the Andes during Deseadan time intensified during late Cenozoic time (B. Simpson 1983). The rain shadow area along the eastern flank of the southern Andes was a stress environment that extended progressively northward to the central Andes (figure 12.6). It initiated and monitored rapid changes in land mammal faunas, while contemporaneous faunas from areas outside the stress environment show slower rates of change. SALMA faunas of

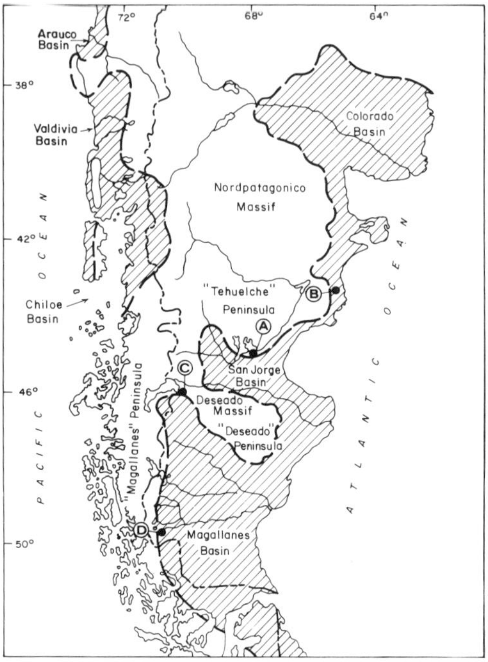

Figure 12.11. Distribution of marine epicontinental seas during early Miocene time in Patagonia. The rich Santacrucian SALMA faunas (c. 19–15 Ma) from the Magallanes Basin were almost certainly derived from immigration of entire faunas from the land-positive areas between 38° and 45° S. *A*, type Colhuehuapian SALMA locality; *B*, Gaiman (Colhuehuapian); *C*, Pinturas Formation (early Santacrucian); *D*, Karaiken (early Santacrucian). Modified from Ramos (1982, 1989) and Bellosi (1990).

Miocene and Pliocene age were therefore apparently, at least in part, diachronous. A continent-wide south-to-north heterochrony component, as suggested by Santacrucian-Colloncuran faunas, apparently had a lag time of several million years, and a west-to-east heterochrony component, as suggested by Chapadmalalan faunas, had a lag time of several hundred thousand years. This apparent heterochrony has yet to be securely documented and calibrated, although present evidence suggests that it could have significant ramifications for understanding aspects of land mammal macroevolutionary processes during Miocene and Pliocene time in South America. Furthermore, this heterochrony model for Argentina implies that the classic and well-known Argentine faunas that occur extensively in this stress environment may uniformly represent a stage of evolution that is advanced relative to coeval faunas from elsewhere on the continent. The stress environment not only resulted from the rain shadow but was influenced by its location in temperate South America, by prevailing wind patterns, oceanic currents, building of the Antarctic ice cap, and possibly by a circa 20° northward shift of the equator between middle Miocene and Pliocene time (Pickford 1990a).

Pleistocene faunas in particular were influenced by pronounced habitat and altitude heterogeneity. The high-altitude Andean faunas of Colombia, Ecuador, Peru, and Bolivia were compositionally distinct from Pampean faunas located below about 2000 meters in Argentina, Uruguay, and Brazil (Hoffstetter 1986). However, there was remarkably little endemism in the Andean faunas, which were for the most part composed of taxa that also occurred at lower elevations. Altitude and associated climatic differences thus served as a filter that restricted many taxa to more equable habitats at lower elevations. These features explain the historically low diversity of faunas on the Bolivian Altiplano, which were probably maintained by repeated immigration of filtered stocks from lower elevations (Hoffstetter 1986).

The available data suggest a remarkable correlation and apparent interrelation between geological events and induced environmental changes (uplift of Andes, formation of rain shadow, interaction of Caribbean and South American plates, giant volcanic ash-flows, marine regressions, etc.) and changes in the land mammal fauna (immigration events, community reorganization, SALMA boundaries, etc.). We therefore conclude that important aspects of the macroevolutionary history of Cenozoic land mammal faunas in South America were initiated and/or monitored by geological processes. We realize, however, that many ideas and concepts presented in this chapter are provocative and need to be tested.

Note added while in press: Subsequent to the completion and submission of this paper in 1990, Flynn et al. (1991) reported a single rodent tooth from the Divisaderan fauna (c. 35 to 31 Ma) in Chile. The precise stratigraphic level and age of this tooth are not designated, although it is regarded as pre-Deseadan. This discovery, at face value, warrants modification of some dispersal models presented in this chapter. We caution, however, that the models are tectonically sound and that slight modifications of them still permit their viability while incorporating this new discovery.

Appendix A

A brief summary of the poorly known record of Mesozoic mammals in South America.

The earliest known evidence of mammals is numerous trackways from the "Matildense" or "Complejo Porfírico" at Estancia Laguna Manantiales in northern Santa Cruz Province, southern Argentina. These trackways were interpreted to be of a jumping pantothere mammal, named *Ameghinichnus patagonicus* by Casamiquela (1961), ancestral to marsupials and possibly placentals (see Casamiquela 1975). The age of these rocks is uncertain, although Casamiquela (1975) suggested that they are middle or late Jurassic.

Unnamed and unstudied trackways of possible mammals are also reported from the Caiuá Formation of the Bauru Group in northwestern Paraná State, Brazil (Leonardi 1977). The age of the Caiuá Formation is not well established, although Soares et al. (1980) suggest that it is late early Cretaceous (Aptian or Albian).

Using partial jaws and dentitions, Carvalho and Campós (1988) described *Candidodon itapecuruense* from the Albian-Cenomanian age Itapecuru Formation in the Parnaiba basin, northern Brazil. They regarded this taxon as a triconodont mammal. However, study of the original specimens by Marshall revealed that it is definitely not a mammal, although its exact identity is not certain; Bertini and co-workers (In press) suggest that it may be a crocodile of the family Atoposauridae.

Fossil remains of Mesozoic mammals are securely known from only five localities in South America.

1. The oldest of these is in the lower part of the early Cretaceous (Hauterivian age) La Amarga Formation in southeastern Neuquén Province, southern Argentina. A single species of pantothere, *Vincelestes neuquenianus* Bonaparte (1986b), is represented by nearly complete skulls and skeletons.

2. A partial right mandible of what is interpreted to be a possible placental was recovered from the lower part of the Adamantina Formation at Loc. 99 near Santo Anastacio in the southwestern part of São Paulo State, Brazil (Bertini et al., In press). The placental affinity is based on the presence of what appears to be four premolars. Based on knowledge of the fossil vertebrates (reviewed by Bertini et al., In press), the Adamantina Formation is regarded as Senonian, possibly Cenomanian to Campanian in age (Soares et al. 1980).

3. Mourier and co-workers (1986) report and figure a single trigon of a therian (either marsupial or placental) that was collected from rocks of late Cretaceous (upper Santonian to Campanian) age near Bagua in north-central Peru.

4. A single edentulous right mandible of a mammal is reported from the late Cretaceous (Campanian-Maastrichtian) age Anacleto Member of the Río Colorado Formation in Río Negro Province, northern Patagonia, Argentina (Goin, Carlini, and Pascual 1986). It was tentatively identified as a marsupial by the Goin group, but because of its fragmentary nature and lack of diagnostic marsupial characters, it is now regarded as Mammalia *incertae sedis* (Marshall and Muizon 1988).

5. The only taxonomically diverse mammal fauna of undisputed Mesozoic age is from the Campanian (*sensu lato*) age Los Alamitos Formation in southeast Río Negro Province, southern Argentina. The taxa, reviewed by Bonaparte (1990), include 14 species of non-tribosphenic mammals: Eotheria (Triconodontidae, *Austrotriconodon mckennai*), Allotheria (Multituberculata, Ferugliotheriidae, *Ferugliotherium windhauseni*), Theria (Symmetrodonta, Bondesiidae, *Bondesius ferox;* family indet., *Casamiquelia rionegrina;* ?Spalacotheriidae, *Brandonia intermedia;* Barbenreniidae, *Barberenia araujoae, Quirogatherium major*), Eupantotheria (Dryolestoidea, Dryolestidae, *Groebertherium stipanicici, G. novasi, Leonardus cuspidatus;* Mesungulatidae, *Mesungulatum houssayi;* Reigitheriidae, *Reigitherium bunodonta*) and ?Paratheria (Gondwanatheria, Gondwanatheriidae, *Gondwanatherium patagonicum, Vucetichia gracilis*).

The paleobiogeography and evolutionary history of South American Cretaceous mammals is discussed at length by Bonaparte (1990). Comparison of South American taxa with groups in Africa is hampered by the fact that only two sparse faunas of Cretaceous age are presently known on the African continent. One is from rocks (Couches Rouges) regarded as early Cretaceous (possibly Berriasian) age in the High Atlas Mountains of Morocco. The mammals are represented by Eotheria, Allotheria and Theria, including three Symmetrodonta (*Thereuodon dahmanii, Atlasodon monbaroni, Microderson laaroussii*), a Droylestoidea (*Donodon prescriptoris*), an amphilestid triconodont, a multituberculate, and a Pappotherida (*Tribotherium africanum*) (Sigogneau-Russell 1989, 1991a, 1991b, 1991c; Sigogneau-Russell, Monbaron, and Russell 1988; Sigogneau-Russell, Monbaron, and de Kaenel 1990). The other is from the Koum Basin in northern Cameroon, where triconodonts and pantotheres were recovered (Jacobs et al. 1988; Brunet et al. 1988).

Although symmetrodonts, dryolestid pantotheres, and triconodonts are known in Africa and South America, there are no genera of Mesozoic mammals common to both, although *Thereuodon* of Africa and *Bondesius* of South America are very similar. The available mammal data do not support the existence of faunal links during Cretaceous time but suggest considerable endemism, at least in South America (Bonaparte 1990).

Appendix B

Terminology for deformation/uplift phase or episode names used by various authors for the southern and central Andes and their correlation with the geologic time scale of Berggren et al. (1985).

Age (Ma)	Epochs (Berggren et al. 1985)	Deformation, Present Age, Original Authorship	Groeber (1929, 1951)	Malumián and Ramos (1984)	Ramos (1989)	Ramos (1988)	Pascual (1984)	Pascual and Ortiz-J (1990)	Charrier and Malumian (1975)
1	Pleistocene								
	___________			- Diaguita -		- Diaguitic -			
2									
		- Diaguita -	- Mov. T-IV -				- Diaguita -	- Diaguita -	
3	Pliocene	(Yrigoyen	(2° Pref.						
		1976)	III M.)						
4									
5	___________								
6			-(3° F. II M.)						
7									
8									
9					- deforma -				
				- Quechuic -	tion				
10		- Quechua -	- Mov. T- III -		(no name)	- Quechua -			- Quechua -
		(Steinmann							
11		1929)							
12									
			-(2° F. II M.)						
13									
14									
15							- Quechua -	- Quechua -	
	Miocene								
16									
17									
18					- Quechua -				
19		- Pehuenche -	- Mov. T-II -	- Pehuenchic -					
		(Yrigoyen	(2° F. I M.)						
20		1976)							
21									
22									

Salfity et al. (1984)	Jordan and Alonso (1987)	Diaz and Miserendino (1988)	Sébrier et al. (1988)	Soler and Bonhomme (1988)	Lavenu et al. (1989)	McKee and Noble (1989)	Noble et al. (1990)	Frutos (1990)	Sempere et al. (1990a)
		- Diaguita -	- F6 -		- Quechua - 4				
- Diaguita -									
	- Diaguita -								
		FASE III							
				- Quechua - 3					
		- Quechua - Principal				- Quechua - III		- Quechuan - II	
- Quechua - Principal			- F5 -						
		FASE II			- Quechua - 3				- Second - major tectonic crisis
						- Quechua - II			
	- Quechua -	- Quechua -	- F4 -	- Quechua - 2			- Quechua - II		
- Quechua - Inicial		- Quechua - Inicial						Quechuan I	
					- Quechua - 2				
			- F3 -	- Quechua - 1					
		FASE I							
						- Quechua - I			

(*continued*)

Appendix B. ***Continued.***

Age (Ma)	Epochs (Berggren et al. 1985)	Deformation, Present Age, Original Authorship	Groeber (1929, 1951)	Malumián and Ramos (1984)	Ramos (1989)	Ramos (1988)	Pascual (1984)	Pascual and Ortiz-J (1990)	Charrier and Malumian (1975)
23									
24									
25							- Pehuenche -		
26									
27		- Aymara - (Sébrier et al. 1988)	- Mov. T-I - (in part)						
28									
29									
30								- Pehuenche -	
	Oligocene								
31									
32									
33									
34									
35									
36								- Inca -	
									- Incaica -
37									
38							- Inca -		
39									
40									
41									
42				- Incaic -	- Incaic -				
43		?							
44									

Salfity et al. (1984)	Jordan and Alonso (1987)	Diaz and Miserendino (1988)	Sébrier et al. (1988)	Soler and Bonhomme (1988)	Lavenu et al. (1989)	McKee and Noble (1989)	Noble et al. (1990)	Frutos (1990)	Sempere et al. (1990a)
- Pehuenche -		-							
		- Pehuenche -							- First -
									major
	- Pehuenche -								tectonic
									crisis
							- Quechua -		
							I		
			- F2 -						
			Aymara						
					- Quechua -				
					1				
								Incaic II	
		- Incaica -		- Incaic -					
	- Incaic -		- Fl -						
						- Incaic -	- Incaic		
						II	II		

(*continued*)

Appendix B. ***Continued.***

Age (Ma)	Epochs (Berggren et al. 1985)	Deformation, Present Age, Original Authorship	Groeber (1929, 1951)	Malumián and Ramos (1984)	Ramos (1989)	Ramos (1988)	Pascual (1984)	Pascual and Ortiz-J (1990)	Charrier and Malumian (1975)
45									
46									
47									
	Eocene								
48									
49		- Incaica -	- Mov. T-I -						
		(Steinmann	(in part)						
50		1929)	(1° F. I M.)						
51									
52									
53									
54									
55		?							
56									
57									
58									
59							- Laramico -		
60									
61									
62									
	Paleocene								- Laramica -
63									
64								- Laramic -	
65									
66									

Salfity et al. (1984)	Jordan and Alonso (1987)	Diaz and Miserendino (1988)	Sébrier et al. (1988)	Soler and Bonhomme (1988)	Lavenu et al. (1989)	McKee and Noble (1989)	Noble et al. (1990)	Frutos (1990)	Sempere et al. (1990a)
- Incaica								Incaic I	
							? - Incaic - I ?		
						? - Incaic - I ?			
								Laramian	

Literature Cited

Allmendinger, R. W. 1986. Tectonic development, southeastern border of the Puna Plateau, northwestern Argentine Andes. Geol. Soc. Amer. Bull. 97:1070–1082.

Alonso, R. N., W. D. Berman, M. Bond, A. A. Carlini, R. Pascual, and M. A. Reguero. 1988. Vertebrados Paleogenos de la Puna Austral: Sus aportes a la evolución biogeográfica. V Jorn. Argent. Paleontol. Vert., La Plata, pp. 38–39.

Alpers, C. N., and G. H. Brimhall. 1988. Middle Miocene climatic change in the Atacama Desert, northern Chile: Evidence from supergene mineralization at La Escondida. Geol. Soc. Amer. Bull. 100:1640–1656.

Alvarenga, H. 1985. Un novo Psilopteridae (Aves: Gruiformes) dos sedimentos Terciarios de Itaboraí, Rio de Janeiro, Brasil. VIII Congr. Brasil. Paleontol. 1983 MME-DNPM, Ser. Geol. 27, Paleontol./Estrat. 2:17–20.

Ameghino, F. 1908. Las formaciones sedimentarias de la región litoral de Mar del Plata y Chapadmalal. Anales Mus. Nac. Buenos Aires (3), 10:343–428.

Anderson, T. H., R. J. Erdlac, Jr., and M. A. Sandstrom. 1985. Late-Cretaceous allochthonous and post-Cretaceous strike-slip displacement along the Cuilco-Chixoy-Polochic fault, Guatemala. Tectonics 4:453–475.

Aspden, J. A., W. J. McCourt, and M. Brook. 1987. Geometrical control of subduction-related magmatism: The Mesozoic and Cenozoic plutonic history of Western Colombia. J. Geol. Soc. London 144:893–905.

Baez, A. M. 1986. El registro Terciario de los anuros en territorio Argentino: Una revaluación. Actas IV Congr. Argent. Paleontol. Bioestrat., Mendoza 2:107–118.

Baker, M. C. W., and P. W. Francis. 1978. Upper Cenozoic volcanism in the Central Andes—ages and volumes. Earth Planet. Sci. Letts. 41:175–187.

Beck, C. 1986. Collision caraïbe, dérive andine, et évolution géodynamique mésozoïque-cénozoïque des Caraïbes. Rev. Géol. Dynam. Géogr. Phys. 27: 163–182.

Bellosi, E. S. 1990. Discontinuidades en la sedimentación litoral "Patagoniana" de la Cuenca San Jorge (Terciario Medio). Actas III Reun. Argent. Sedimentol., San Juan, 20–24 May 1990, pp. 372–377.

Berggren, W. A., D. V. Kent, J. J. Flynn, and J. A. Van Couvering. 1985. Cenozoic geochronology. Geol. Soc. Amer. Bull. 96:1407–1418.

Bertels, A. 1979. Paleobiogeografía de los foraminíferos del Cretácico Superior y Cenozoico de América del Sur. Ameghiniana 16(3–4):273–356.

Bertini, R. J., L. G. Marshall, M. Gayet, and P. M. Brito. In press. Vertebrate fauna from the Adamantina and Marília formations (Upper Bauru Group), Late Cretaceous, Brazil. Neues Jahrbuch für Geologie und Paleontologie, Abhandlungen, Stuttgart.

Bonaparte, J. F. 1960. La sucesión estratigráfica de Monte Hermoso (Prov. de Buenos Aires). Acta Geol. Lilloana 3:273–287.

———. 1984. El intercambio faunístico de vertebrados continentales entre América del Sur y del Norte a fines del Cretácico. Mem. III Congr. Latinoamer. Paleontol., Mexico, pp. 438–450.

———. 1986a. History of the terrestrial Cretaceous vertebrates of Gondwana. Actas IV Congr. Argent. Paleontol. Bioestratigr., Mendoza 2:63–95.

———. 1986b. Sobre *Mesungulatum houssayi* y nuevos mamíferos cretácicos de Patagonia, Argentina. Actas IV Congr. Argent. Paleontol. Bioestratigr., Mendoza 2:48–61.

———. 1990. New late Cretaceous mammals from the Los Alamitos Formation, northern Patagonia. Natl. Geogr. Res. 6(1):63–93.

Bond, M. 1981. Un nuevo Oldfieldthomasiidae (Mammalia, Notoungulata) del Eoceno inferior (Fm. Lumbrera, Grupo Salta) del NW Argentina. Anais II Congr. Latinoamer. Paleontol. 2:521–536.

———. 1984. Nuevos mamíferos de la Fm. Pozuelos y Casa Grande (Eógeno) del NW Argentino. I Jorn. Argent. Paleontol. Vert., Resumenes, p. 21.

Bond, M., and M. G. Vucetich. 1983. *Indalecia grandensis* gen. et sp. nov. del Eoceno temprano del noroeste Argentino, tipo de una nueva subfamilia de los Adianthidae (Mammalia, Litopterna). Rev. Asoc. Geol. Argent. 38:107–117.

———. 1985. Adiciones al conocimiento de *Colbertia* (Mammalia, Notoungulata). II Jorn. Argent. Paleontol. Vert., p. 4.

Bondesio, P., J. H. Laza, G. J. Scillato Yane, E. P. Tonni, and M. G. Vucetich. 1980. Estado actual del conocimiento de los vertebrados de la Formación Arroyo Chasicó (Plioceno Temprano) de la Provincia de Buenos Aires. Actas II Congr. Argent. Paleontol. Bioestratigr. y I Congr. Latinoamer. Paleontol. 3:101–127.

Bondesio, P., J. Rabassa, R. Pascual, M. G. Vucetich, and G. J. Scillato-Yane. 1980. La formación Collón-Curá de Pilcaniyeu Viejo y sus alrededores (Río Negro, Argentina): Su antigüedad y las condiciones ambientales según su distribución, su litogénesis y sus vertebrados. Actas II Congr. Argent. Paleontol. Bioestratigra. y I Congr. Latinoamer. Paleontol. 3:85–99.

Bourgois, J., J. F. Toussaint, H. Gonzalez, J. Azema, B. Calle, A. Desmet, L. A. Murcia, A. P. Acevedo, E. Parra, and J. Tournon. 1987. Geological history of the Cretaceous ophiolitic complexes of northwestern South America (Colombian Andes). Tectonophysics 143:307–327.

Bown, T. M., and C. N. Larriestra. 1990. Sedimentary paleoenvironments of fossil platyrrhine localities, Miocene Pinturas Formation, Santa Cruz Province, Argentina. J. Human Evol. 19:87–119.

Broin, F. de. 1991. Fossil turtles from Bolivia. Pp. 509–527 *in* R. Suarez (ed.), *Fosiles y Facies de Bolivia.* Rev. Téc. Yaci. Petro. Fisc. Bol. 12(3–4).

Brunet, M., L. Jacobs, J. Congleton, Y. Coppens, J. Dejax, L. J. Flynn, J. Hell, Y. Jehenne, G. Mouchelin, and D. Pilbeam. 1988. Première decouverte, d'un fragment de mandibule de mammifère dans le Crétacé inférieur d'Afrique (Cameroun, Bassin de Koum). Compt. Rend. Acad. Sci. Paris, sér. 2, 307: 1675–1680.

Bush, M. B., and P. A. Colinvaux. 1990. A pollen record of a complete glacial cycle from lowland Panama. J. Veg. Sci. 1:105–118.

Butler, R. F., L. G. Marshall, R. E. Drake, and G. H. Curtis. 1984. Magnetic polarity stratigraphy and 40K–40Ar dating of Late Miocene and Early Pliocene continental deposits, Catamarca Province, NW Argentina. J. Geol. 92:623–636.

Carvajal, E., R. Pascual, R. Pinedo, J. A. Salfity, and M. G. Vucetich. 1977. Un nuevo mamífero de la Formación Lumbrera (Grupo Salta) de la comarca de Carahuasi (Salta, Argentina): Edad y correlaciones. Publ. Mus. Mun. Cienc. Nat. "Lorenzo Scaglia," Mar del Plata 2(7):148–163.

Carvalho, I. de S., and D. de A. Campos. 1988. Um mamífero Triconodonte do Cretáceo Inferior do Maranhão Brasil. An. Acad. brasil. Ci. 60(4):437–446.

Casamiquela, R. M. 1961. Sobre la presencia de un mamífero en el primer elenco (Icnológico) de vertebrados de Jurásico de la Patagonia (Noticia). Physis 22:225–233.

———. 1975. Sobre la significación de *Ameghinichnus patagonicus,* un mamífero brincador del Jurásico medio de Santa Cruz (Patagonia). Actas I Congr. Argent. Paleontol. Bioestratigr., Tucumán 2:71–85.

Cazau, L. 1972. Cuenca del Ñirihuau-Norquinco-Cushamen. Pp. 727–740 *in* A. F. Leanza (ed.), *Geología Regional Argentina.* Córdoba, Argentina: Academia Nacional de Ciencias.

Cazau, L., D. Mancini, J. Cangini, and L. Spalletti. 1989. Cuenca de Ñirihuau. Pp. 299–318 *in* G. Chebli and L. Spalletti (eds.), *Cuencas Sedimentarias Argentinas.* Serie Correlación Geológica no. 6. San Miguel de Tucumán: Universidad Nacional de Tucumán.

Charrier, R. 1973. Interruptions of spreading and the compressive tectonic phases of the Meridional Andes. Earth Planet. Sci. Letts. 20(2): 242–249.

Charrier, R., and N. Malumián. 1975. Orogenesis y epeirogénesis en la Región Austral de

América del Sur durante el Mesozoico y el Cenozoico. Rev. Asoc. Geol. Argent. 30(2): 193–207.

Charrier, R., and J. C. Vicente. 1972. Liminary and Geosynclinal Andes: Major orogenic phases and synchronical evolution of the Central and Austral sectors of the southern Andes. Conf. Sobre Probl. Tierra Solída, Buenos Aires 2:451–470.

Charrier, R., A. R. Wyss, M. A. Norell, J. J. Flynn, M. J. Novacek, M. C. McKenna, C. C. Swisher III, D. Frassinetti, and P. Salinas. 1990. Hallazgo de mamíferos fósiles del Terciario inferior en el sector de Termas del Flaco, Cordillera Principal, Chile Central: Implicaciones paleontológicas, estratigráficas y tectónicas. Actas II Simposio sobre el Terciario de Chile, Concepción, December 1990, 1:73–84.

Chevalier, Y., J. F. Stephan, J. R. Darboux, M. Gravelle, H. Bellon, A. Bellizzia, and R. Blanchet. 1988. Obduction et collision pré-Tertiaire dans les zones internes de la Chaîne Caraïbe vénézuélienne, sur le transect Ile de Margarita–Péninsula d'Araya. Compt. Rend. Acad. Sci. Paris, sér. 2, 307:1925–1932.

Cifelli, R. L. 1985. Biostratigraphy of the Casamayoran, Early Eocene, of Patagonia. Amer. Mus. Novit. 2820:1–26.

Cifelli, R. L., C. R. Schaff, and M. C. McKenna. 1989. The relationships of the Arctostylopidae (Mammalia): New data and interpretation. Bull. Mus. Comp. Zool. 152(1):1–44.

Clapperton, C. M. 1979. Glaciation in Bolivia before 3.27 Myr. Nature 277:375–377.

Coira, B., J. Davidson, C. Mpodozis, and V. Ramos. 1982. Tectonic and magnetic evolution of the Andes of northern Argentina and Chile. Earth-Sci. Rev. 18:303–332.

Coney, P. J. 1982. Plate tectonic constraints on the biogeography of Middle America and the Caribbean region. Ann. Missouri Bot. Gard. 69:432–443.

Damuth, J. E., and R. W. Fairbridge. 1970. Equatorial Atlantic deep-sea arkosic sands and ice-age aridity in tropical South America. Geol. Soc. Amer. Bull. 81:189–206.

Davidson, J., and J. C. Vicente. 1973. Características paleogeográficas y estructurales del área fronteriza de las nacientes del Teno (Chile) y Santa Elena (Argentina) (Cordillera Principal, 35° a 35°15′ de latitud Sur). V Congr. Geol. Argent., Buenos Aires 5:11–55.

Díaz, J. I., and A. Miserendino. 1988. El ámbito deposicional y tectónico del Grupo Payogastilla (Provincia de Salta, Republica Argentina). V Congr. Geol. Chileno, Santiago, 2:C87-C103.

Duque-Caro, H. 1976. Características estratigráficas y sedimentarias del Terciario marino de Colombia. II Congr. Geol. Latinoamer. 1973, Caracas, pp. 945–964.

———. 1989. Cenozoic paleogeography in NW South America: Major drainage pattern implications. Pacific Science Association, IV Inter. Congr., Valparaíso-Viña del Mar, Chile, 7–10 August. The Pacific, Bridge or Barrier? Abstracts, p. 202.

Fernandez, C. 1989. La cuenca Chaco-Paranaense Argentina: Sus tendencias evolutivas y algunas posibilidades exploratorias. Bol. Info. Petrol. Año 6(18):1–17.

Fitzgerald, M. G., R. M. Mitchum Jr., M. A. Uliana, and K. T. Biddle. 1990. Evolution of the San Jorge Basin, Argentina. Bull. Amer. Assoc. Petrol. Geol. 74:879–920.

Flynn, J. J., B. J. MacFadden, and M. C. McKenna. 1984. Land-mammal ages, faunal heterochrony, and temporal resolution in Cenozoic terrestrial sequences. J. Geol. 92:687–705.

Flynn, J. J., L. G. Marshall, J. Guerrero, and P. Salinas. 1989. Geochronology of Middle Miocene ("Friasian" Land Mammal Age) faunas from Chile and Colombia. Abstracts to Annual Meeting of the Geological Society of America, St. Louis, Missouri. Stratigraphy 3:A133.

Flynn, J. J., M. A. Norell, C. C. Swisher III, and A. R. Wyss. 1991. Pre-Deseadan, post-Mustersan mammals from Central Chile: An update. J. Vert. Paleontol., Abstracts, 11 (suppl. to no. 3):29A.

Ford, S. 1990. Locomotor adaptations of fossil platyrrhines. J. Human Evol. 19:41–174.

Frailey, C. D. 1988. *The Miocene Vertebrates of Quebrada Honda, Bolivia.* Part 2, *Edentata.* Occasional Papers 123. Lawrence, Kans.:

University of Kansas, Museum of Natural History.

Garcia, A. V., M. H. Carrion, and M. Omil. 1984. El perfil tipo de la formación Yasyamayo (Grupo Santa María), Departamento Tafi, Provincia de Tucumán. Actas IX Congr. Geol. Argent. (S. C. de Bariloche) 4:174–189.

Gardeweg, M., and C. F. Ramírez. 1987. La Pacana caldera and the Atana Ignimbrite: A major ash-flow and resurgent caldera complex in the Andes of northern Chile. Bull. Volcanol. 49:547–566.

Gayet, M., L. G. Marshall, and T. Sempere. 1991. The Mesozoic and Paleocene vertebrates of Bolivia and their stratigraphic context: A review. Pp. 393–433 *in* R. Suarez (ed.), *Fosiles y Facies de Bolivia.* Rev. Téc. Yaci. Petro. Fisc. Bol. 12(3–4):393–433.

Gayet, M., J. C. Rage, T. Sempere, and P. Y. Gagnier. 1992. Modalités des échanges de vertébrés continentaux entre l'Amérique du Nord et l'Amérique du Sud au Crétacé supérieur et au Paléocène. Bull. Soc. Geol. France 163:781–791.

Gingerich, P. D. 1985. South American mammals in the Paleocene of North America. Pp. 123–137 *in* F. G. Stehli and S. D. Webb (eds.), *The Great American Biotic Interchange*. New York: Plenum.

Godoy, E., and W. Palma. 1990. El corrimiento del Fierro y su propagación como plegamiento en el alto río Maipo, Andes de Chile Central: Un evento oligoceno a mioceno inferior. Actas II Simposio sobre el Terciario de Chile, Concepción, December 1990, 1:97–104.

Goin, F. J., A. A. Carlini, and R. Pascual. 1986. Un probable marsupial del Cretácico Tardío del norte de la Patagonia, Argentina. Actas IV Congr. Argent. Paleontol. Bioestratigr., Mendoza 2:43–47.

Goin, F. J., R. M. Palma, R. Pascual, and J. E. Powell. 1986. Persistencia de un primitivo Borhyaenidae (Mammalia, Marsupialia) en el Eoceno temprano de Salta (Fm. Lumbrera, Argentina): Aspectos geológicos y paleoambientales relacionados. Ameghiniana 23:47–56.

Gorroño, P., R. Pascual, and R. Pombo. 1979. Hallazgo de mamíferos Eogenos en el sur de Mendoza: Su implicancia en las dataciones de los "Rodados Lustrosos" y del primer episodio orogénico del Terciario en esa región. Actas VII Congr. Geol. Argent., Neuquén, 9–15 April 1978, 2:475–487.

Graur, D., A. H. Winston, and Li Wen-Hsiung. 1991. Is the guinea-pig a rodent? Nature 351:649–652.

Groeber, P. 1929. Líneas fundamentales de la geología del Neuquén, sur de Mendoza y regiones adyacentes. Bol. Direc. Gen. Miner. Geol. Hidrol. 58:1–110.

———. 1951. La Alta Cordillera entre las latitudes 34° y 29°30′. Rev. Cienc. Geol. Mus. Cienc. Nat. "Bernardino Rivadavia" 5:235–352.

Haq, B. U., J. Hardenbol, and P. R. Vail. 1987. Chronology of fluctuating sea levels since the Triassic. Science 235:1156–1166.

Harrington, H. J. 1962. Paleogeographic development of South America. Bull. Amer. Assoc. Petrol. Geol. 46(10):1773–1814.

Hayashida, A. 1984. Paleomagnetic study of the Miocene continental deposits in La Venta badlands, Colombia. Kyoto University Overseas Research Reports of New World Monkeys 4:77–83.

Hayashida, A., and T. Danhara. 1985. Fission-track dating of the volcanic ash layers in continental deposits of Salla, Bolivia: Reexaminations. Kyoto University Overseas Research Reports of New World Monkeys 5:39–42.

Hoffstetter, R. 1977. Un gisement de mammifères miocènes à Quebrada Honda (Sud-Bolivien). Compt. Rend. Hebdomadaires Séanc. Acad. Sci., sér. D, 284:1517–1520.

———. 1986. High Andean mammalian faunas during the Plio-Pleistocene. Pp. 218–245 *in* F. Vuilleumier and M. Monasterio (eds.), *High Altitude Tropical Biogeography.* New York: Oxford University Press.

Isacks, B. L. 1988. Uplift of the Altiplano-Puna plateau and bending of the Bolivian Orocline. J. Geophys. Res. 93:3211–3231.

Jacobs, L. L., J. D. Congleton, M. Brunet, J. Dejax, L. J. Flynn, J. V. Hell, G. Mouchelin, and D. Pilbeam. 1988. Mammal teeth from the Cretaceous of Africa. Nature 336:158–160.

Jaillard, E., and T. Sempere. 1989. Cretaceous sequence stratigraphy of Peru and Bolivia. Pp. 1–27 *in* Proceedings, Simposios sobre el Cretácico de América Latina (IGCP 242/GSGP). Part A, *Eventos y registro sedimentario del Cretácico de América Latina,* L. Spalletti (ed.). Centro de Investigaciones Geologicas, Universidad Nacional de La Plata.

Jaillard, E., H. Cappetta, P. Ellenberger, M. Feist, N. Grambast-Fessard, J. P. LeFranc, and B. Sigé. In press. The late Cretaceous Vilquechico Formation, South Peru: Sedimentology, paleontology, biostratigraphy, correlations. Cretaceous Res.

Jordan, T. E., and R. N. Alonso. 1987. Cenozoic stratigraphy and basin tectonics of the Andes Mountains, 20°–28° South Latitude. Bull. Amer. Assoc. Petrol. Geol. 71(1):49–64.

Jordan, T. E., and M. Gardeweg. 1989. Tectonic evolution of the Late Cenozoic Central Andes (20°–33° S). Pp. 193–207 *in* Z. Ben-Avraham (ed.), *The Evolution of the Pacific Ocean Margins.* New York: Oxford University Press.

Jordan, T. E., B. Isacks, R. Allmendinger, J. Brewer, V. A. Ramos, and C. Ando. 1983. Andean tectonics related to geometry of subduction plates. Geol. Soc. Amer. Bull. 94(3):341–361.

Kalin Arroyo, M. T., F. A. Squeo, J. J. Armesto, and C. Villagran. 1988. Effects of aridity on plant diversity in the northern Chilean Andes: Results of a natural experiment. Ann. Missouri Bot. Gard. 75:55–78.

Kay, R. F., R. H. Madden, J. M. Plavcan, R. L. Cifelli, and J. G. Diaz. 1987. *Stirtonia victoriae,* a new species of Miocene Colombian primate. J. Human Evol. 16:173–196.

Keller, G., C. E. Zenker, and S. M. Stone. 1989. Late Neogene history of the Pacific-Caribbean gateway. J. S. Amer. Earth Sci. 2:73–108.

Kokogián, D. A., and O. Mancilla. 1989. Análisis estratigráfico secuencial de la Cuenca Cuyana. Pp. 169–201 *in* G. A. Chebli and L. A. Spalletti (eds.), *Cuencas Sedimentarias Argentinas.* Serie Correlación Geológica 6. San Miguel de Tucumán: Universidad Nacional de Tucumán.

Kraglievich, J. L. 1952. El perfil geológico de Chapadmalal y Miramar, Provincia de Buenos Aires. Rev. Mus. Mun. Cienc. Nat. Trad. Mar del Plata 1:8–37.

Kraglievich, L. 1930. La Formación Friaseana del río Frías, río Fénix, Laguna Blanca, etcétera, y su fauna de mamíferos. Physis 10:127–161.

———. 1934. *La antigüedad pliocena de las faunas de Monte Hermoso y Chapadmalal: Deducidas de su comparación con las que le procedieron y sucedieron.* Montevideo: Imprenta El Siglo Ilustrado.

Kulm, L. D., H. Schrader, J. M. Resig, T. M. Thornburg, A. Masias, and L. Johnson. 1981. *Late Cenozoic carbonates on the Peru continental margin: Lithostratigraphy, biostratigraphy, and tectonic history.* Memoir 154. Boulder, Colo.: Geological Society of America.

Lavenu, A., and R. Marocco. 1984. Sédimentation continentale et tectonique d'une chaîne liée à une zone de subduction: L'exemple des Andes Centrales (Pérou-Bolivie) pendant le Tertiaire. Bull. Cent. Rech. Explor.-Prod. Elf-Aquitaine, 8(1):57–70.

Lavenu, A., M. Fornari, and M. Sébrier. 1984. Existence de deux nouveaux épisodes lacustres quaternaires dans l'Altiplano péruvo-bolivien. Cahiers ORSTOM, Sér. Geol., 14:103–114.

Legarreta, L., M. A. Uliana, and M. A. Torres. 1990. Secuencias deposicionales cenozoicas de Patagonia central: sus relaciones con las asociaciones de mamíferos terrestres y episodios marinos epicontinentales. Evaluación preliminar. Actas II Simposio sobre el Terciario de Chile, Concepción, December 1990, 1:135–176.

Leonardi, G. 1977. Two new ichnofaunas (vertebrates and invertebrates) in the eolian Cretaceous sandstones of the Caiuá Formation in northwestern Paraná. Actas I Simp. Geol. Region. Soc. Brasil. Geol., São Paulo 1:112–128.

McCourt, W. J., J. A. Aspden, and M. Brook. 1984. New geological and geochronological data from the Colombian Andes: Continental growth by multiple accretion. J. Geol. Soc. London 141:831–845.

MacFadden, B. J. 1990. Chronology of Cenozoic primate localities in South America. J. Human Evol. 19(1–2):7–21.

MacFadden, B. J., O. Siles, P. Zeitler, N. M. Johnson, and K. E. Campbell, Jr. 1983. Magnetic polarity stratigraphy of the Middle Pleistocene (Ensenadan) Tarija Formation of southern Bolivia. Quat. Res. 19:172–187.

MacFadden, B. J., K. E. Campbell, R. L. Cifelli, O. Siles, N. M. Johnson, C. W. Naeser, and P. K. Zeitler. 1985. Magnetic polarity stratigraphy and mammalian fauna of the Deseadan (Late Oligocene–Early Miocene) Salla Beds of northern Bolivia. J. Geol. 93(3):223–250.

MacFadden, B. J., A. Anaya, H. Perez, C. W. Naeser, P. K. Zeitler, and K. E. Campbell. 1990. Late Cenozoic paleomagnetism and chronology of Andean basins of Bolivia, with comments on possible orocline bending. J. Geol. 98:541–555.

Macharé, J., T. Devries, J. Barron, and E. Fourtanier. 1988. Oligo-Miocene transgression along the Pacific margin of South America: New paleontological and geological evidence from the Pisco Basin (Peru). Géodynamique 3(1–2):25–37.

McKee, E. H., and D. C. Noble. 1989. Cenozoic tectonic events, magmatic pulses, and base- and precious-metal mineralization in the central Andes. Pp. 189–194 *in* G. E. Ericksen, M. T. Cañas Pinochet, and J. A. Reinemund (eds.), *Geology of the Andes and Its Relation to Hydrocarbon and Mineral Resources*. Earth Science Series 11. Houston: Circum-Pacific Council for Energy and Mineral Resources.

McRae, L. E. 1989. Chronostratigraphic variability in fluvial sequences as revealed by paleomagnetic isochrons: Examples in Miocene strata from a central Andean intermontane basin (Salla, Bolivia) and the northwest Himalayan foreland (lower Siwalik Group, Chinji Formation, Pakistan). Ph.D. diss., Dartmouth College.

Malizia, D. C., M. Strecker, R. Herbst, and M. Carrion. 1990. La formación El Morterito (Neogeno de Sierras Pampeanas): Características sedimentológicas y paleontológicas. Actas, III Reunión Argentina de Sedimentológia, San Juan, 20–24 May 1990, pp. 175–180.

Malumián, N., and V. A. Ramos. 1984. Magmatic intervals, transgression-regression cycles and oceanic events in the Cretaceous and Tertiary of southern South America. Earth Planet. Sci. Letts. 67:228–237.

Marocco, R., T. Sempere, M. Cirbián, and J. Oller. 1987. Mise en évidence d'une déformation paléocène en Bolivie du Sud: Sa place dans l'évolution géodynamique des Andes Centrales. Compt. Rend. Acad. Sci. Paris, sér. 2, 304(18):1139–1143.

Marquillas, R. A., and J. A. Salfity. 1989. Dispersion regional de los miembros de la Formación Yacoraite (Cretacico Superior) en el Norte Argentino. Pp. A252-A272 *in* Proceedings, Simposios sobre el Cretácico de América Latina (IGCP 242/GSGP). Part A, *Eventos y registro sedimentario del Cretácico de América Latina,* L. Spalletti (ed.). Centro de Investigaciones Geologicas, Universidad Nacional de La Plata.

Marshall, L. G. 1976. Fossil localities for Santacrucian (Early Miocene) mammals, Santa Cruz Province, southern Patagonia, Argentina. J. Paleontol. 50:1129–1142.

———. 1979. A model for paleobiogeography of South American cricetine rodents. Paleobiology 5:126–132.

———. 1985. Geochronology and land-mammal biochronology of the transamerican faunal interchange. Pp. 49–85 *in* F. G. Stehli and S. D. Webb (eds.), *The Great American Biotic Interchange*. New York: Plenum.

———. 1989. The K-T boundary in South America: On which side is Tiupampa? Natl. Geogr. Res. 5(3):268–270.

———. 1990. Fossil Marsupialia from the type Friasian Land Mammal Age (Miocene), Alto Río Cisnes, Aisen, Chile. Rev. Geol. Chile 17(1):19–55.

Marshall, L. G., and R. Cifelli. 1990. Analysis of changing diversity patterns in Cenozoic Land

Mammal Age faunas, South America. Palaeovertebrata 19(4):169–210.

Marshall, L. G., and C. de Muizon. 1988. The dawn of the Age of Mammals in South America. Nat. Geogr. Res. 4(1):23–55.

Marshall, L. G., and B. Patterson. 1981. Geology and geochronology of the mammal-bearing Tertiary of the Valle de Santa María and Río Corral Quemado, Catamarca Province, Argentina. Fieldiana: Geol. 9:1–80.

Marshall, L. G., and P. Salinas. 1990a. Stratigraphy of the Río Frias Formation (Miocene) along the Alto Río Cisnes, Aisén, Chile. Rev. Geol. Chile 17(1):57–87.

———. 1990b. Vertebrados continentales del Mioceno Temprano de Magallanes, sur de Chile. Anales Inst. Patagonia 19(1):27–38.

Marshall, L. G., and T. Sempere. 1991. The Eocene to Pleistocene vertebrates of Bolivia and their stratigraphic context: A review. Pp. 631–652 *in* R. Suarez (ed.), *Fosiles y Facies de Bolivia.* Rev. Téc. Yaci. Petro. Fisc. Bol. 12(3–4):631–652.

Marshall, L. G., R. Pascual, G. H. Curtis, and R. E. Drake. 1977. South American geochronology: Radiometric time scale for middle to late Tertiary mammal-bearing horizons, Patagonia. Science 195:1325–1328.

Marshall, L. G., R. F. Butler, R. E. Drake, and G. H. Curtis. 1981. Calibration of the beginning of the Age of Mammals in Patagonia. Science 212:43–45.

———. 1982. Geochronology of type Uquian (Late Cenozoic) land mammal age, Argentina. Science 216:986–989.

Marshall, L. G., R. Hoffstetter, and R. Pascual. 1983. Mammals and stratigraphy: Geochronology of the continental mammal-bearing Tertiary of South America. Palaeovertebrata Mém. Ext. 1983:1–93.

Marshall, L. G., C. de Muizon, and B. Sige. 1983. *Perutherium altiplanense,* un notogulé du Crétacé Supérieur du Pérou. Palaeovertebrata 13:145–155.

Marshall, L. G., A. Berta, R. Hoffstetter, R. Pascual, O. A. Reig, M. Bombin, and A. Mones. 1984. Mammals and stratigraphy: Geochronology of the continental mammal-bearing Quaternary of South America. Palaeovertebrata, Mém. Ext. 1984:1–76.

Marshall, L. G., R. E. Drake, and G. H. Curtis. 1985. 40K–40Ar age calibration of Late Miocene–Pliocene mammal-bearing Huayquerias and Tunuyan Formations, Mendoza Province, Argentina. J. Paleontol. 60:448–457.

Marshall, L. G., R. L. Cifelli, R. E. Drake, and G. H. Curtis. 1986. Vertebrate paleontology, geology and geochronology of the Tapera de Lopéz and Scarritt Pocket, Chubut Province, Argentina. J. Paleontol. 60(4):920–951.

Marshall, L. G., R. E. Drake, G. H. Curtis, R. F. Butler, K. M. Flanagan, and C. W. Naeser. 1986. Geochronology of type Santacrucian (Middle Tertiary) Land Mammal Age, Patagonia, Argentina. J. Geol. 94(4):449–457.

Marshall, L. G., J. A. Case, and M. O. Woodburne. 1989. Phylogenetic relationships of the families of marsupials. Current Mammalogy 2:433–502.

Marshall, L. G., C. C. Swisher III, A. Lavenu, R. Hoffstetter, and G. H. Curtis. 1992. Geochronology of the mammal-bearing Late Cenozoic on the northern Altiplano, Bolivia. J. S. Amer. Earth Sci. 5(1):1–19.

Miserendino, A., and J. I. Díaz. 1988. Nuevas consideraciones sobre la edad y correlaciones del Grupo Payogastilla (Terciario Superior, Prov. de Salta, Rep. Argentina). V Congr. Geol. Chileno, Santiago, 2:C69-C86.

Mörner, N.-A., and C. Sylwan. 1989. Magnetostratigraphy of the Patagonian moraine sequence at Lago Buenos Aires. J. S. Amer. Earth Sci. 2(4):385–389.

Mourer-Chauviré, C. 1982. Les oiseaux fossiles des Phosphorites du Quercy (Eocène supérieur a Oligocène supérieur): Implications paléobiogéographiques. Geobios 6 (Mémoire Spécial): 413–426.

Mourier, T., E. Jaillard, G. Laubacher, C. Noblet, A. Pardo, B. Sigé, and P. Taquet. 1986. Découverte de restes dinosauriens et mammalien d'âge Crétacé supérieur à la base des couches rouges du synclinal de Bagua (Andes nord-péruviennes): Aspects stratigraphiques, sédimentologiques et paléogéographiques concernant la régression fini-crétacée. Bull. Soc. Géol. France, sér. 8, 2:171–175.

Moya, M. C., and J. A. Salfity. 1982. Los ciclos magmáticos en el noroeste Argentino. Actas V Congr. Latinoamer. Geol. Argent. 3:523–536.

Nabel, P., and A. Spiegelman. 1988. Caracterización sedimentaria y paleomagnética de una sección del Pampeano, en el subsuelo de la Ciudad de Buenos Aires. Rev. Asoc. Geol. Argent. 43(2):224–230.

Naeser, C. W., E. H. McKee, N. M. Johnson, and B. J. MacFadden. 1987. Confirmation of a Late Oligocene–Early Miocene age of the Deseadan Salla Beds of Bolivia. J. Geol. 95:825–828.

Nelson, E. P., and R. D. Forsythe. 1989. Ridge collision at convergent margins: Implications for Archean and post-Archean crustal growth. Tectonophysics 161:307–315.

Noble, D. C., E. H. McKee, and F. Mégard. 1979. Early Tertiary "Incaic" tectonism, uplift, and volcanic activity, Andes of central Peru. Geol. Soc. Amer. Bull. 90:903–907.

Noble, D. C., E. H. McKee, T. Mourier, and F. Mégard. 1990. Cenozoic stratigraphy magmatic activity, compressive deformation, and uplift in northern Peru. Geol. Soc. Amer. Bull. 102:1105–1113.

Nuttall, C. P. 1990. A review of the Tertiary nonmarine molluscan faunas of the Pebasian and other inland basins of north-western South America. Bull. Brit. Mus. (Nat. Hist.) 45(2): 165–371.

Ochsenius, C. 1985. Pleniglacial desertization, large-animal mass extinction and Pleistocene-Holocene boundary in South America. Rev. Geogr. Norte Grande 12:35–47.

Opdyke, N. D. 1990. Magnetic stratigraphy of Cenozoic terrestrial sedments and mammalian dispersal. J. Geol. 98:621–637.

Orgeira, M. J., and D. A. Valencio. 1984. Estudio paleomagnético de los sedimentos asignados al Cenozoico Tardío aflorantes en La Barranca de Lobos, Provincia de Buenos Aires. Actas IX Congr. Geol. Argent. (S. C. de Bariloche, 1984) 4:162–173.

Ortiz-Jaureguizar, E. 1986. Evolución de las comunidades de mamíferos cenozoicos sudamericanos: Un estudio basado en técnicas de análisis multivariado. IV Congr. Argent. Paleontol. Bioestratigr., Mendoza 2:191–207.

Ortiz-Jaureguizar, E., and R. Pascual. 1989. South American land-mammal faunas during the Cretaceous-Tertiary transition: Evolutionary biogeography. Pp. 231–252 *in* Proceedings, Simposios sobre el Cretácico de América Latina (IGCP 242/GSGP). Part A, *Eventos y registro sedimentario del Cretácico de América Latina,* L. Spalletti (ed.). Centro de Investigaciones Geologicas, Universidad de La Plata.

Pardo-Casas, F., and P. Molnar. 1987. Relative motion of the Nazca (Farallon) and South American plates since Late Cretaceous time. Tectonics 6:233–248.

Pascual, R. 1980a. Nuevos y singulares tipos ecológicos de marsupiales extinguidos de América del Sur (Paleoceno tardío o Eoceno temprano) del Noroeste Argentino. Actas II Congr. Argent. Paleontol. Bioestratigr. y I Congr. Latinoamer. Paleontol. 2:151–173.

———. 1980b. Prepidolopidae, nueva familia de Marsupialia Didelphoidea del Eoceno Sudamericano. Ameghiniana 17(3):216–242.

———. 1983. Novedosos marsupiales Paleogenos de la Formación Pozuelos (Grupo Pastos Grandes) de la Puna, Salta, Argentina. Ameghiniana 20(3–4): 265–280.

———. 1984. La sucesión de las Edades-mamífero, de los climas y del diastrofismo sudamericanos durante el Cenozoico: Fenómenos concurrentes. Anales Acad. Nac. Cienc. Exact. Físic. Nat. 36:15–37.

Pascual, R., and O. E. Odreman Rivas. 1971. Evolución de las comunidades de los vertebrados del Terciario argentino: Los aspectos paleozoogeográficos y paleoclimáticos relacionados. Ameghiniana 8:372–412.

———. 1973. Las unidades estratigráficas del Terciario portadores de mamíferos: Su distribución y sus relaciones con los acontecimientos diastróficos. Actas V Congr. Geol. Argent. 3:293–338.

Pascual, R., and E. Ortiz-Jaureguizar. 1990. Evolving climates and mammal faunas in Cenozoic South America. J. Human Evol. 19(1–2):23–60.

———. 1991. El ciclo faunístico Cochabambiano (Paleoceno Temprano): Su incidencia en la historia biogeográfica de los mamíferos Sudamericanos. Pp. 559–574 *in* R. Suárez (ed.), *Fosiles y Facies de Bolivia.* Rev. Téc. Yaci. Petro. Fisc. Bol. 12(3–4).

Pascual, R., E. J. Ortega Hinojosa, D. Gondar, and E. Tonni. 1965. Las edades del Cenozoico mamalífero de la Argentina, con especial atención a aquellas del territorio Bonaerense. Anales Com. Invest. Cient., Buenos Aires 6:165–193.

Pascual, R., M. G. Vucetich, and J. Fernández. 1978. Los primeros mamíferos (Notoungulata, Henricosborniidae) de la Formación Mealla (Grupo Salta, Subgrupo Santa Barbara): Sus implicancias filogenéticas, taxonómicas y cronológicas. Ameghiniana 15(3–4):366–390.

Pascual, R., M. Bond and M. G. Vucetich. 1981. El subgrupo Santa Barbara (Grupo Salta) y sus vertebrados: Cronología, paleoambientes y paleobiogeografía. Actas VIII Congr. Geol. Argent. 3:743–758.

Pascual, R., P. Bondesio, M. G. Vucetich, G. J. Scillato-Yane, M. Bond, and E. P. Tonni. 1984. Vertebrados fósiles cenozoicos. In *Geología y Recursos Naturales de la Provincia de Río Negro.* Actas IX Congr. Geol. Argent., Relatorio 2:439–461.

Patterson, B., and R. Pascual. 1972. The fossil mammal fauna of South America. Pp. 247–309 *in* A. Keast, F. C. Erk and B. Glass (eds.), *Evolution, Mammals and Southern Continents.* Albany, N.Y.: State University of New York Press.

Patterson, B., and A. E. Wood. 1982. Rodents from the Deseadan Oligocene of Bolivia. Bull. Mus. Comp. Zool. 149:371–543.

Paula Couto, C. de. 1952. Fossil mammals from the beginning of the Cenozoic in Brazil: Condylarthra, Litopterna, Xenungulata, and Astrapotheria. Bull. Amer. Mus. Nat. Hist. 99:355–394.

Pickford, M. 1990a. Dynamics of Old World biogeographic realms during the Neogene: Implications for biostratigraphy. Pp. 413–442 *in* E. H. Lindsay et al. (eds.), *European Neogene Mammal Chronology.* New York: Plenum.

———. 1990b. Uplift of the roof of Africa and its bearing on the evolution of mankind. J. Human Evol. 5(1):1–20.

Pilger, R. H. 1984. Cenozoic plate kinematics, subduction and magmatism: South American Andes. J. Geol. Soc. London 141:793–802.

Pindell, J. L., and S. F. Barrett. 1990. Geologic evolution of the Caribbean: A plate-tectonic perspective. Pp. 405–432 *in* J. E. Case and G. Dengo (eds.), *The Geology of North America,* Vol. H: *The Caribbean Region.* A Decade of North America Geology Series. Boulder, Colo.: Geological Society of America.

Pindell, J. L., and J. F. Dewey. 1982. Permo-Triassic reconstruction of western Pangea and the evolution of the Gulf of Mexico/Caribbean region. Tectonics 1:179–212.

Pindell, J. L., S. C. Cande, W. C. Pitman III, D. B. Rowley, J. F. Dewey, J. Labrecque, and W. Haxby. 1988. A plate-kinematic framework for models of Caribbean evolution. Tectonophysics 155:121–138.

Piperno, D. R., M. B. Bush, and P. A. Colinvaux. 1990. Paleoenvironments and human occupation in late-glacial Panama. Quat. Res. 33:108–116.

Prado, J. L., and E. Ortiz-Jaureguizar. 1989. La edad-mamífero Chapadmalalense: Su validez en base a la comparación de su fauna con la de las edades que le precedieron y sucedieron. VI Jorn. Paleontol. Vert., Resumenes, pp. 35–38.

Rage, J. C. 1978. Une connexion continentale entre Amérique du Nord et Amérique du Sud au Crétacé supérieur? L'exemple des vertébrés continentaux. Compt. Rend. Somm. Séanc. Soc. Géol. France 6:281–285.

———. 1981. Les continents Péri-atlantiques au Crétacé Supérieur: Migrations des faunes continentales et problèmes paléogéographiques. Cretaceous Res. 2:65–84.

———. 1986. South American/North American terrestrial interchanges in the latest Cretaceous: Short comments on Brett-Surman and Paul (1985), with additional data. J. Vert. Paleontol. 6:382–383.

Ramos, V. A. 1982. Las ingresiones Pacíficas del Terciario en el norte de la Patagonia (Argentina). III Congr. Geol. Chileno, Concepción, pp. A262-A288.

———. 1988. The tectonics of the Central Andes; 30° to 33° S latitude. Geol. Soc. Am. Bull. 218 (Special Paper): 31–54.

———. 1989. Andean foothills structures in northern Magallanes Basin, Argentina. Bull. Amer. Assoc. Petrol. Geol. 73(7):887–903.

Ramos, V. A., D. Pérez, and M. B. Aguirre-nodiorita Matienzo: intrusivo mioceno de la Alta Cordillera de Mendoza. Rev. Asoc. Geol. Argent. 43(3):404–408.

Ramos, V. A., D. Perez, and M. B. Aguirre-Urreta. 1990. Geología del filo de Zurbriggen, Aconcagua, Mendoza. Actas X Prim. Congr. Geol. Argent., San Juan, 1990, 2:361–369.

Rapela, C. W., and S. M. Kay. 1988. Late Paleozoic to Recent magmatic evolution of northern Patagonia. Episodes 11(3):175–182.

Redwood, S. D. 1987. The Soledad Caldera, Bolivia: A Miocene caldera with associated epithermal Au-Ag-Cu-Pb-Zn mineralization. Geol. Soc. Amer. Bull. 99:395–404.

Reig, O. A. 1978. Roedores cricétidos del plioceno superior de la Provincia de Buenos Aires (Argentina). Pub. Mus. Mun. Cienc. Nat. "Lorenzo Scaglia," Mar del Plata 2:164–190.

———. 1986. Diversity patterns and differentiation of high Andean rodents. Pp. 405–439, *in* F. Vuilleumier and M. Monasterio (eds.), *High Altitude Tropical Biogeography.* New York: Oxford University Press.

Restrepo, J. J., and J. F. Toussaint. 1988. Terranes and continental accretion in the Colombian Andes. Episodes 11:189–193.

Ruocco, M. 1990. Paleomagnetic analysis of continental deposits of the last 3 Ma from Argentina: Magnetostratigraphic and fine structures of reversals. Paleogeophysics and Geodynamics, Department of Geology and Geochemistry, University of Stockholm, Sweden, 100 p.

Salfity, J. A., S. A. Gorustovich, M. C. Moya, and R. Amenguel. 1984. Marco tectónico de la sedimentación y efusividad cenozoicas en la Puna Argentina. Actas IX Congr. Geol. Argent. (S. C. Bariloche) 1:539–554.

Sanjinés, G., and F. Jiménez. 1975. Comunicación preliminar acerca de la presencia de fósiles vertebrados en la Formación Petaca del área de Santa Cruz. Rev. Téc. Yaci. Petro. Fisc. Bol. 4(3):147–158.

Schiller, W. 1912. La Alta Cordillera de San Juan y Mendoza y parte de la provincia de San Juan. Anales Minist. Agric. Sec. Geol. Mineralog. Miner. 7(5):1–68.

Schoch, R. M., and S. G. Lucas. 1985. The phylogeny and classification of the Dinocerata (Mammalia, Eutheria). Bull. Geol. Inst. Univ. Uppsala, n.s., 11:31–57.

Sébrier, M., A. Lavenu, M. Fornari and J.-P. Soulas. 1988. Tectonics and uplift in Central Andes (Peru, Bolivia and Northern Chile) from Eocene to Present. Géodynamique 3(1–2):85–106.

Sempere, T. 1990. Cuadros estratigráficos de Bolivia: Propuestas nuevas. Rev. Téc. Yaci. Petro. Fisc. Bol. 11(2–3):215–227.

———. 1991. Cenozoic tectonic "phases" in Bolivia: Some needed clarifications. Actas VI Congr. Geol. Chileno, Viña del Mar 1:877–881.

———. In press. Kimmeridgian? to Paleocene tectonic evolution of Bolivia. *In* J. A. Salfity (ed.), *Cretaceous Tectonics in the Andes.* Earth Evolution Sciences Monograph Series. Wiesbaden: Vieweg.

Sempere, T., and L. G. Marshall. In press. Relocation of the Tiupampa vertebrates in the late Cretaceous-Paleocene stratigraphy of Bolivia: Toward a solution to a paleontological debate? J. S. Amer. Earth Sci.

Sempere, T., G. Herail, J. Oller, and P. Baby. 1989. Geologic structure and tectonic history of the Bolivian orocline. Abstracts of the 28th Inter. Geol. Congr., Washington, D.C. 3:72–73.

Sempere, T., G. Hérail, J. Oller, P. Baby, L. Barrios, and R. Marocco. 1990a. The Altiplano: A province of intermontane foreland basins related to crustal shortening in the Bolivian orocline area. Symposium International

"Géodynamique Andine," Grenoble, France, 15–17 May 1990. Pp. 167–170 *in Résumés des Communications*. Paris: Editions de l'Orstom.

Sempere, T., G. Hérail, J. Oller, and M. G. Bonhomme. 1990b. Late Oligocene-Early Miocene major tectonic crisis and related basins in Bolivia. Geology 18:946–949.

Sempere, T., L. G. Marshall, S. Rivano, and E. Godoy. In press. Late Oligocene–early Miocene compressional tectosedimentary episode and associated land-mammal faunas in the Andes of Central Chile and adjacent Argentina (32°–37° S). Tectonophysics.

Servant, M. and J. C. Fontes. 1978. Les lacs quaternaires des hauts plateaux des Andes boliviennes. Cahiers ORSTOM, Sér. Géol. 10(1):9–23.

Sigé, B. 1972. La faunule de mammifères du Crétacé supérieur de Laguna Umayo (Andes péruviennes). Bull. Mus. Natn. Hist. Nat. no. 99, Sciences de la Terre, 19:375–409.

Sigogneau-Russell, D. 1989. Découverte du premier Symmétrodonte (Mammalia) du continent africain. Compt. Rend. Acad. Sci. Paris, sér. 2, 309:921–926.

———. 1991a. Découverte du premier Mammifère tribosphénique du Mésozoïque africain. Compt. Rend. Hebdomadaires Séanc. Acad. Scienc. Paris, sér. 2, 313:1635–1640.

———. 1991b. First evidence of Multituberculata (Mammalia) in the Mesozoic of Africa. Neues Jahrbuch für Geologie und Palaeozoogie, mh., Abt. B, 2:120–126.

———. 1991c. Nouveaux Mammifères thériens du Crétacé inférieur du Maroc. Compt. Rend. Hebdomadaires Séanc. Acad. Sci. Paris, Paléontol., sér. 2, 313:279–285.

Sigogneau-Russell, D., M. Monbaron, and D. E. Russell. 1988. Découverte de Mammifères dans le Mésozoïque moyen d'Afrique. Compt. Rend. Acad. Scienc. Paris, sér. 2, 307:1045–1050.

Sigogneau-Russell, D., M. Monbaron, and E. de Kaenel. 1990. Nouvelles données sur le gisement à Mammifères mésozoïques du Haut-Atlas marocain. Geobios 23(4):461–483.

Simpson, B. 1983. An historical phytogeography of the high Andean flora. Rev. Chilena Hist. Nat. 56(2):109–122.

Simpson, G. G. 1940. Review of the mammal-bearing Tertiary of South America. Proc. Amer. Phil. Soc. 83:649–710.

———. 1980. *Splendid Isolation: The Curious History of South American Mammals*. New Haven: Yale University Press.

Simpson, G. G., J. L. Minoprio, and B. Patterson. 1962. The mammalian fauna of the Divisadero Largo Formation, Mendoza, Argentina. Bull. Mus. Comp. Zool. 127:237–293.

Soares, P. C., P. M. B. Landim, V. J. Fulfaro, and A. F. Sobreiro Neto. 1980. Ensaio de caracterizacão estratigráfica do Cretáceo no Estado de São Paulo: Grupo Baurú. Rev. Bras. Geoc. 10(3):177–185.

Solbrig, O. T. 1976. The origin and floristic affinities of the South American temperate desert and semidesert regions. Pp. 9–49 *in* D. W. Goodall (ed.), *Evolution of Desert Biota*. Austin: University of Texas Press.

Soria, M. F. 1987. Estudios sobre los Astrapotheria (Mammalia) del Paleoceno y Eoceno. Parte I, Descripción de *Eoastrapostylops riolorense* Soria y Powell, 1982. Ameghiniana 24:21–34.

Soria, M. F., and J. E. Powell. 1982. Un primitivo Astrapotheria (Mammalia) y la edad de la Formación Río Loro, Provincia de Tucumán, República Argentina. Ameghiniana 18:155–168.

Spalletti, L. A. 1983. Paleogeografía de la Formación Ñirihuau y sus equivalentes en la región occidental de Neuquén, Río Negro y Chubut. Rev. Asoc. Geol. Argent. 38(3–4):454–468.

Speed, R. C. 1985. Cenozoic collision of the Lesser Antilles arc and continental South America and the origin of the El Pilar fault. Tectonics 4:41–69.

Steinmann, G. 1929. *Geologie von Peru*. Heidelberg: Carl Winters Universitäts Buchhandlung.

Stephan, J.-F., B. Mercier de Lépinay, E. Calais, M. Tardy, C. Beck, J.-C. Carfantan, J.-L. Olivet, J.-M. Vila, P. Bouysse, A. Mauffret,

J. Bourgois, J.-M. Thery, J. Tournon, R. Blanchet, and J. Dercourt. 1990. Paleogeodynamic maps of the Caribbean: 14 steps from Lias to Present. Bull. Soc. Géol. France (8)6:915–919.

Stirton, R. A. 1953. Vertebrate paleontology and continental stratigraphy in Colombia. Geol. Soc. Amer. Bull. 64:603–622.

Strecker, M., P. Cerveny, A. Bloom, and D. Malizia. 1989. Late Cenozoic tectonism and landscape development in the foreland of the Andes: Northern Sierras Pampeanas (26°–28° S), Argentina. Tectonics 8(3):517–534.

Takai, M., T. Setoguchi, C. Villarroel, N. Shigehara, and A. L. Rosenberger. 1988. Preliminary report of small mammal fossils from the La Venta Fauna, South America. Kyoto University Overseas Research Reports of New World Monkeys 6:11–14.

Takemura, K., and T. Danhara. 1986. Fission-track dating of the upper part of Miocene Honda Group in La Venta badlands, Colombia. Kyoto University Overseas Research Reports of New World Monkeys 5:31–37.

Tonni, E. P., D. Berman, F. Fidalgo, F. J. Goin, and J. C. Quiroga. 1987. La fauna local Centinela del Mar y una nueva unidad mamífero para el Pleistoceno Tardío de la Provincia de Buenos Aires (Argentina). X Congr. Geol. Argent., San Miguel de Tucumán 3:175–177.

Tosdal, R. M., A. H. Clark, and E. Farrar. 1984. Cenozoic polyphase landscape and tectonic evolution of the Cordillera Occidental, southern-most Peru. Geol. Soc. Amer. Bull. 95: 1318–1332.

Uliana, M. A., and K. T. Biddle. 1987. *Permian to Late Cenozoic Evolution of Northern Patagonia: Main Tectonic Events, Magmatic Activity, and Depositional Trends.* Geophysical Monograph 40. Washington, D.C.: American Geophysical Union.

Valencio, D. A., and M. J. Orgeira. 1983. La magnetoestratigrafía del Ensenadense y Bonaerense de la Ciudad de Buenos Aires, Parte II. Rev. Asoc. Geol. Argent. 38(1):24–33.

van der Wiel, A. M. 1990. Uplift age of the Garzon Massif (Eastern Cordillera, S. Colombia) in relation to the infill of the adjacent S. Neiva Basin. Symposium International "Géodynamique Andine," Grenoble, France, 15–17 May 1990. Pp. 217–218 *in Résumés des Communications.* Paris: Editions de l'Orstom.

Van Houten, F. B. and H. E. James. 1984. Late Cenozoic delta complex in southwestern Maracaibo Basin, northeastern Colombia. Geol. Soc. Amer. Memoir 162:325–332.

Van Valen, L. M. 1988. *Paleocene dinosaurs or Cretaceous Ungulates in South America.* Evol. Monogr. 10.

Vergara, M., R. Charrier, F. Munizaga, S. Rivano, P. Sepúlveda, P. Thiele, and R. E. Drake. 1988. Miocene vulcanism in the Central Chilean Andes (31°30′ S–34°35′ S). J. S. Amer. Earth Sci. 1(2):199–209.

Vignati, M. A. 1925. La geología de Monte Hermoso. Physis 8:126–127.

Vrba, E. S., G. H. Denton, and M. L. Prentice. 1989. Climatic influences on early hominid behavior. OSSA, Int. J. Skeletal Res. 14:127–156.

Vucetich, M. G. 1980. Un nuevo Notostylopidae (Mammalia, Notoungulata) proveniente de la Formación Lumbrera (Grupo Salta) del noroeste Argentino. Ameghiniana 17(4):363–372.

Vucetich, M. G., and M. Bond. 1982. Los primeros Isotemnidae (Mammalia, Notoungulata) registrados en la Formación Lumbrera (Grupo Salta), del noroeste Argentino. Ameghiniana 19(1–2):7–18.

Webb, S. D. 1976. Mammalian faunal dynamics of the Great American Interchange. Paleobiology 2:216–234.

———. 1978. A history of savanna vertebrates in the New World. Part II, South America and the Great American Interchange. Ann. Rev. Ecol. Syst. 9:393–426.

———. 1984. On two kinds of rapid faunal turnover. Pp. 417–436 *in* W. A. Berggren and J. A. Van Couvering (eds.), *Catastrophes and Earth History: The New Uniformitarianism.* Princeton, N.J.: Princeton University Press.

Webb, S. D., and A. D. Barnosky. 1989. Faunal dynamics of Pleistocene mammals. Ann. Rev. Earth Planet. Sci. 17:413–438.

Williams, G. D., P. Turner, S. Flint, I. Stimpson, E. J. Jolley, and A. J. Hartley. 1990. Andean basin dynamics in northern Chile. Symposium International "Géodynamique Andine," Grenoble, France, 15–17 May 1990. P. 231 *in Résumés des Communications*. Paris: Editions de l'Orstom.

Woodburne, M. O. 1977. Definition and characterization of mammalian chronostratigraphy. J. Paleontol. 21:220–234.

Wyss, A. R., M. A. Norell, J. J. Flynn, M. J. Novacek, R. Charrier, M. C. McKenna, C. C. Swisher III, D. Frassinetti, P. Salinas, and M. Jin. 1990. A new early Tertiary mammal fauna from Central Chile: Implications for Andean stratigraphy and tectonics. J. Vert. Paleontol. 10(4):518–522.

Yrigoyen, M. R. 1976. Observaciones geológicas alrededor del Aconcagua. Actas I Congr. Geol. Chileno, Santiago 1(A):169–190.

———. 1979. Cordillera Principal. II Simposio de Geol. Reg. Argent., Córdoba, Academia Nacional de Ciencias 1: 651–694.

Zárate, M. A., and J. L. Fasano. 1989. The Plio-Pleistocene record of the central eastern Pampas, Buenos Aires Province, Argentina: The Chapadmalal case study. Palaeogeogr. Palaeoclimatol. Palaeoecol. 72:27–52.

Zinsmeister, W. J., L. G. Marshall, R. E. Drake, and G. H. Curtis. 1981. First radioisotope (potassium-argon) age of marine Neogene Rionegro Beds in northeastern Patagonia, Argentina. Science 212:440.

Chapter 13

Mammal Evolution in the African Neogene and a New Look at the Great American Interchange

Elisabeth S. Vrba

I. Introduction

Empirical information on macroevolution has increased dramatically in the past few decades. Studies on how species' origins, extinctions, and diversities are distributed in genealogies, over time, and over geography have boomed. Most interpretations of such compound patterns have relied heavily on particular and local rather than general causes. For instance, tectonic and biotic factors unique to America have traditionally been invoked to explain the waves of mammalian extinctions and speciations associated with the Great American Faunal Interchange (Marshall et al. 1982). Similarly, prominent aspects of mammalian macroevolution in Africa have often been attributed to such local factors as rifting and volcanism (Denys et al. 1986; Pickford 1990) or to interspecies competition (e.g., Ewer 1973; Walker 1984; Wolpoff 1988). In general, among the many interpretations of major evolutionary patterns spawned by the increase in appropriate data, particular and local explanations have proved popular. Some have even claimed that macroevolutionary patterns do not have any consistent order but that all are random clusters of causally independent microevolutionary events (e.g., Hofmann 1989).

A contrasting view holds that there are

I greatly enjoyed attending this most successful conference, and I thank Peter Goldblatt and Peter Raven for inviting me to it. I am grateful to Jim Brown, Larry Marshall, and David Webb for helpful comments on the arguments in this chapter. David Hodgins has been a great help on several aspects of this research; and Susan Hochgraf prepared most of the illustrations. This chapter was partly researched while holding National Science Foundation grant BSR-8907673.

 ISBN 0-300-05375-4

some general rules, even laws, of macroevolution that have operated repeatedly and systematically across genealogy, time, and geography (Vrba 1990). In my view, a major challenge of our subject is to disentangle the causal roles of general and unique factors. Comparison of African and South American histories, particularly during the better known later Cenozoic, provides a good opportunity for taking up this challenge.

Africa and South America have had very different biotic, biogeographic, and tectonic histories during the Neogene. Their phylogenetic compositions are dissimilar today (e.g., large mammals, figure 13.1) and were much more so during the Neogene. Whereas Africa had early connections and extensive biotic exchanges with northern landmasses (Adams et al. 1983; Rogl and Steiniger 1983; Thomas 1985), South America remained largely isolated from North America, except for episodic filter bridges, until the late Neogene emergence of the Panama isthmus (Chapter 2). Whereas African topography and climate were transformed by phases of eastern uplift and rifting (Baker et al. 1982; Pickford 1990), South America was following a different (and presumably independent) tectonic schedule of Andean uplift along its western margin with associated local climatic changes (Chapter 12). To the extent that such region-specific physical and biotic factors influenced macroevolution, the major changes in the two continental records are expected to be independent and to differ especially in chronology (Vrba 1985c).

Yet Africa and South America also share potential causal factors. At least since the

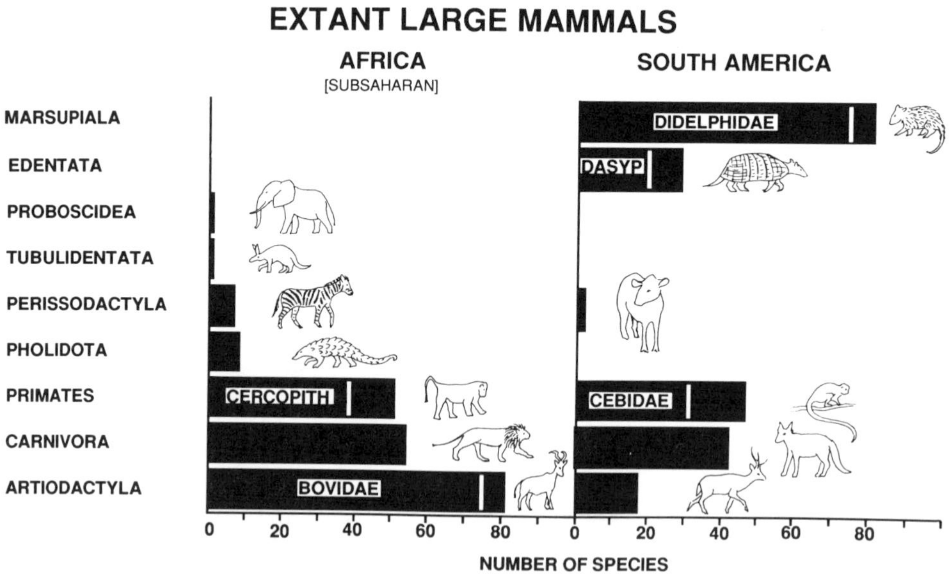

Figure 13.1. Species diversities in orders of extant large mammals in sub-Saharan Africa and South America. Data from Honacki et al. (1982).

early Miocene, each, unlike other continents, has had a huge tropical landmass. Global climatic change was potentially an extrinsic common cause of evolution on both continents. To the extent that climate acted as a general cause, there should be consistency between the continents in timing and in ecological implications of evolutionary responses. I will also hypothesize that some general causes are intrinsic to the mammalian biotas of both Africa and South America in spite of their different phylogenetic compositions.

Taken together, the seven hypotheses I explore here form part of a compatible theory. I refer to it as the "habitat theory" for brief reference (discussed in Vrba 1992) because it emphasizes how species relate to their habitats and how they respond to habitat alteration initiated by physical change. It offers an alternative to the competition paradigm, namely to the traditional emphasis on interspecies competition as a cause of extinctions and on "empty niches" as akin to magnets that attract species' origins (reviews in Benton 1983; Martin and Klein 1984; Vrba 1985c). The Great American Interchange of mammals is often cited as the best paleontological example of the competition paradigm (Benton 1983). I therefore use it, in comparison with the contemporaneous African record, to test the habitat theory.

The earlier Neogene evidence, from Africa and South America and of global climate, is less well dated and abundant than that for the Plio-Pleistocene. First, I document only briefly some major episodes of African mammal turnover for the Neogene as a whole. Next, I discuss global climatic changes in relation to the habitat theory. Finally, I use the better Plio-Pleistocene fossil record to test some hypotheses of the habitat theory and to suggest future testing of others.

II. Major Features of the African Neogene Mammal Record

A. Early to Middle Miocene

During the Oligocene only 15 families are recorded, but during the Rusingan (early Miocene, 24–17 Ma) 42 mammal families first appeared in Africa (figure 13.2). Nearly half were large herbivores. Some have argued (e.g., Janis 1982) that early Miocene faunal changes in Africa represent a general spread of woodland and savanna vegetation types, consistent with the global palynological evidence of a progressive post-Eocene retreat of the tropical forest belt to the lower latitudes (Leopold 1969). I know of no good African palynological evidence across the Oligocene-Miocene boundary to support this.

The graph of generic originations (figure 13.3) singles out the latest Rusingan period, 19–17 million years ago. This was a time of major faunal interchange between Eurasia and Afro-Arabia (Neogene Dispersal Phase 1 of Thomas 1985; several contributions in Lindsay et al. 1989) due to land connections that temporarily closed the eastern Tethys Sea 19 million years ago (Rogl and Steiniger 1983; Adams et al. 1983). Thomas (1985) noted that at this time many more taxa dispersed into Africa (including taxa of Carnivora, Perissodactyla, Artiodactyla, Lagomorpha, Insectivora, and Rodentia) than from Africa northward into Eurasia (including Proboscidea in the Proboscidean Datum Event, which commenced in Europe 19–18 million years ago; Tassy 1989). I sug-

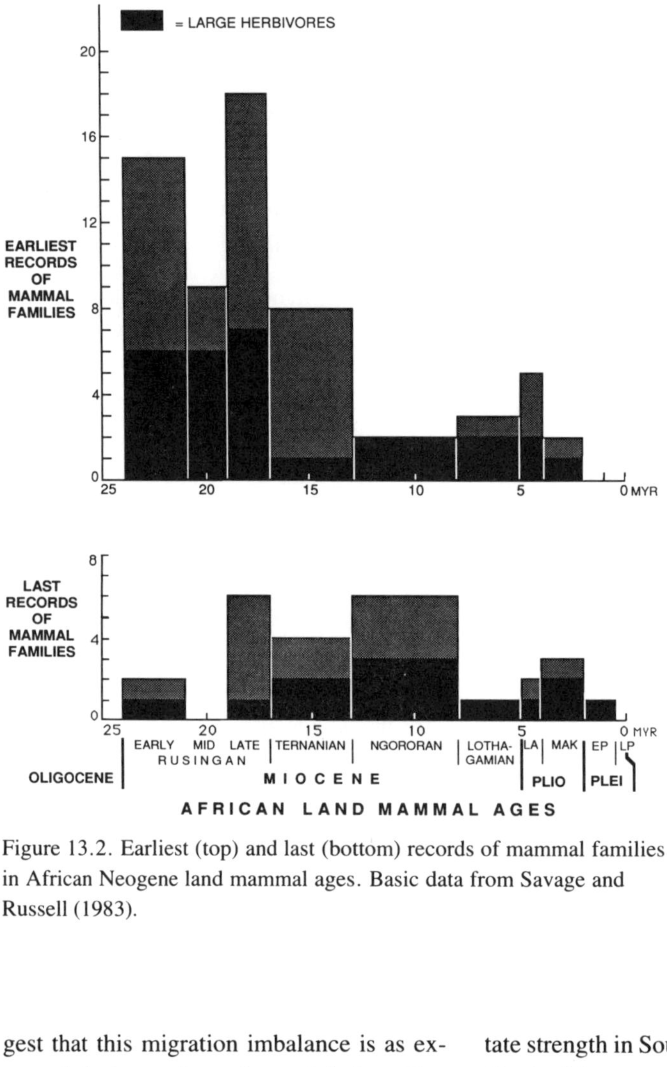

Figure 13.2. Earliest (top) and last (bottom) records of mammal families in African Neogene land mammal ages. Basic data from Savage and Russell (1983).

gest that this migration imbalance is as expected during a time of net global cooling (see figure 13.5), when more high-latitude forms should move toward the equator than in the reverse direction (further discussed below). Thus, most genera in the late Rusingan peak of figure 13.3 are Eurasian immigrants.

The major contrast between African and South American large mammals today (figure 13.1) is that between bovid domination of African savannas and marsupial and edentate strength in South America. Bovidae typify the Neogene expansion of the African savanna biomes. Their relatively abundant and well-dated fossil record can resolve the timing of some of the gross features of figures 13.1 and 13.2. Bovid tribal durations (Vrba, 1985a) in figure 13.4 show that Bovidae first appeared in sub-Saharan Africa near 18 Ma, again reflecting the wave of originations at this time. The bovids also exemplify a second Miocene period of interchange with Eurasia, and of local extinction and origina-

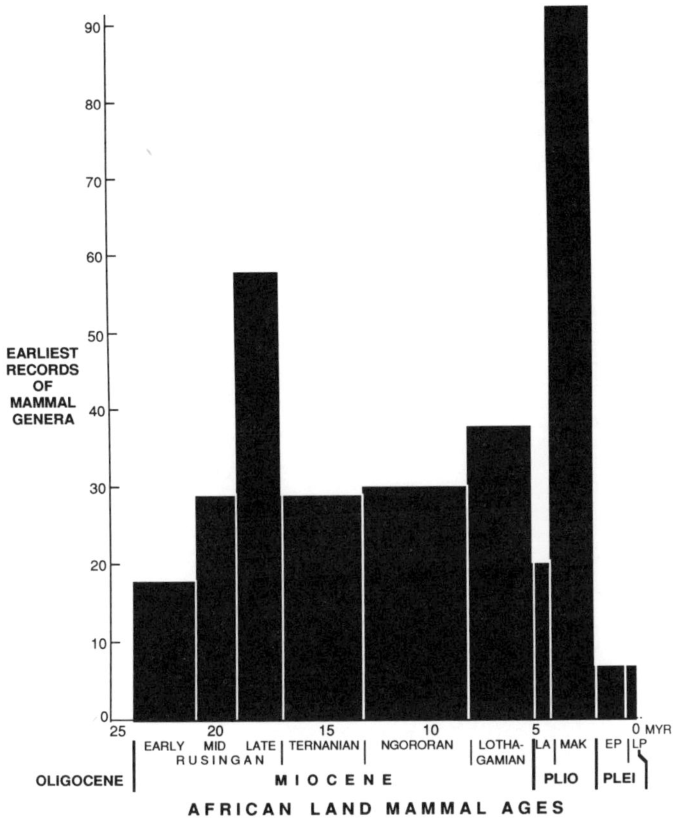

Figure 13.3. Earliest records of mammal genera in African Neogene land mammal ages. Basic data from Savage and Russell (1983).

tion, about 15–14 Ma (figure 13.4), when several taxa, including Caprini (goats and allies) and Antilopini (gazelles), first entered Africa from the north (Neogene Dispersal Phase 2, 15 ± 1 Ma, according to Thomas 1985). After global warming, accompanied by the Langhian marine transgression about 16.8 Ma, there followed a period of land aridification, ocean cooling, and regression during which brief land connections again closed the eastern Mediterranean to allow this second major Neogene dispersal phase (Rogl and Steiniger 1983; Thomas 1985).

B. Late Miocene

A third dramatic macroevolutionary phase, in the late Miocene, 8–5 Ma, is especially well represented by the Bovidae (figure 13.4). Of the 14 bovid tribes currently recognized worldwide, no fewer than 8 appeared for the last or first time in sub-Saharan Africa over this period (Vrba 1985a): one became extinct in Africa (Boselaphini), at least 2 were Eurasian immigrants (Bovini and Ovibovini, and possibly Reduncini), and at least a further 4 I interpret as African originations. The abundant survivors of most of these 7

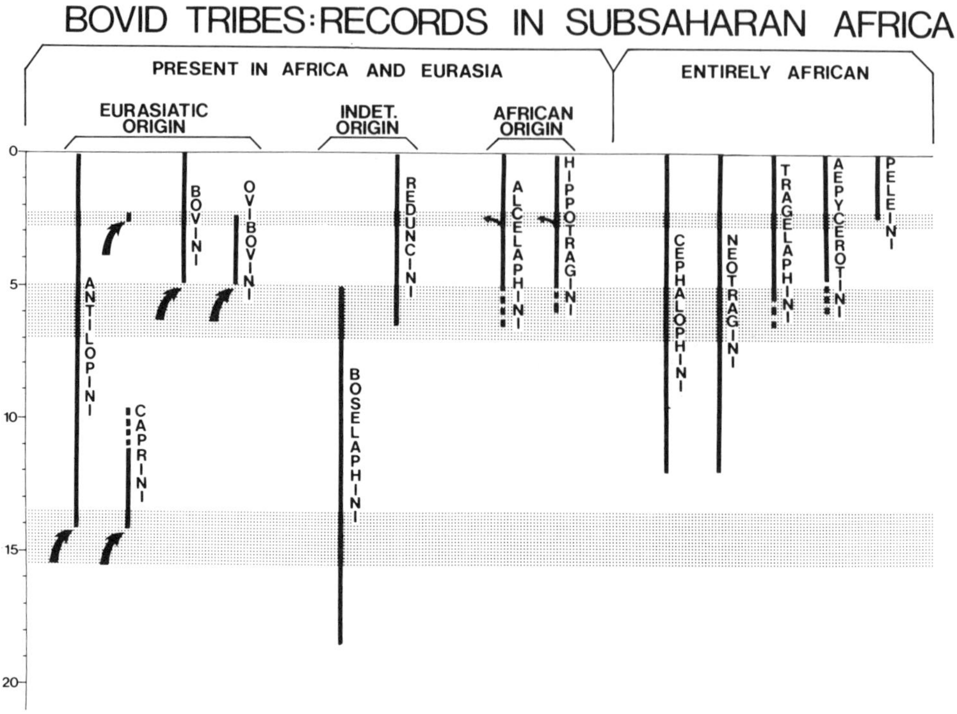

Figure 13.4. Records of bovid tribes in sub-Saharan Africa. Times of marked turnover are shaded. Large arrows indicate immigrants from Eurasia; small arrows indicate emigration to Eurasia. The only bovid tribe that has never been recorded in Africa is Rupicaprini. Adapted from Vrba (1985c).

first-appearing tribes are today exclusive (or nearly so) grazers with a marked preference for open habitats. Four multiplied especially rapidly in Africa. Together, these advents strongly suggest a general increase in the ratio of grass cover to wood cover. This inference of climatic change from bovid evolution is consistent with African palynological evidence (Van Zinderen Bakker 1986; Coetzee 1986). The 6 antelope tribes that have survived in Africa since this late Miocene origin are today the typical and dominant large mammals of African savannas. This led me to suggest that this period marked the establishment of the extensive modern African savanna biota of which the Hominidae, incidentally, are also "founder members" (Vrba 1985b).

Data sets from around the world indicate that 8–5 Ma was a period of global net cooling and changes to more openly vegetated environments and a time of major biotic turnover. Late Miocene evidence for Antarctic ice volume changes and Alaskan glaciation, multiple marine regressions, strong cooling of oceanic surface waters, and the Mediterranean salinity crisis are reviewed in Vrba and co-workers (1989). Evidence of major terrestrial responses includes findings (Quade et al. 1989; Cerling, personal communication, 1991) that C4 plants (mainly grasses) took over dominance from C3 plants fairly abruptly (7.5–7.0 and toward 5 Ma) in Pakistan, and more gradually (9–5 Ma) in East African sequences. This vegetation reconstruction for Pakistan is supported by the presence and nature of mammal turnover between 7.4–7.0 Ma (Barry et al. 1985; Flynn and Jacobs 1982). Marked faunal and floral turnover at this time is also reported from other areas on Earth; and Tedford's (1985:262) comments for Australia are representative of events on several continents: "The most striking aspect of the Neogene record of fossil mammals in Australia is the faunal turnover that takes place between Miocene and Pliocene times. This event manifests itself in terms of changes in taxonomic composition, mean body-size and adaptation."

C. Tectonism in Relation to Neogene African Faunal Change

Much has been written about Neogene African tectonism and postulated local effects on climate, flora, and fauna. According to Baker and co-workers (1982), starting in the late Eocene there were three major episodes of warping and uplift, accompanied by volcanism and faulting in East Africa; the latest of these episodes occurred in the late Pliocene. Unfortunately, the direct dating (independently of biostratigraphy) of such episodes seems very imprecise. Estimates using various lines of evidence differ widely: the same onset of rift-related climatic changes in East Africa was considered to have occurred at about 7, 12, and 10 Ma, respectively, in Coppens (1982, 1984, 1986, in Pickford 1990).

Pickford (1990) proposed that three major Neogene phases of mountain uplift in the Western Rift causally influenced East African biotic turnover. He considered that these tectonic episodes modified regional climates mainly by xerification, intensified by global climatic cooling. He suggests that "from about 5 m.y. to lower Pleistocene times, global [climatic] changes were unidirectional. [He used the smoothed oxygen-isotope curve based on benthic foraminifera from Miller and Fairbanks 1985.] Uplift of the roof of Africa in contrast was episodic during this period, and it is postulated that this episodic uplift caused equally punctuated changes in the mammal faunas of East Africa" (1990:14). The argument is problematic. First, detailed analyses of the deep-sea oxygen-isotope data for the relevant time period do in fact show large-scale episodic fluctuations in spite of an overall cooling trend (see figures 13.5, 13.6, 13.7, and 13.16, and associated references). Second, Pickford (1990) suggested that the three tectonic events took place about 8, 3, and 0.5 Ma. But he based these dates *on the fossil mammal record* itself. For the time period surrounding the event that occurred 8 Ma, as reviewed above, there is evidence from around the world for climatic change with terrestrial xerification and biotic turnover. The same holds for the event that occurred 3 Ma, as will be discussed below. Testing what causal component was contributed by tectonic events (of uncertain date) to local biotic turnover that is chronologically and ecologically consistent with documented global change and biotic turnover elsewhere, becomes very difficult. The hypothesis that global climate contributed is falsifiable, but the hypothesis that tectonism did so is probably not in this case.

III. Global Climatic Changes
A. Short-Term Climatic Cycles

Over the past 65 million years, global climatic changes based on the foraminiferan record (figure 13.5) show large fluctuations in a net cooling trend. For such parts of the record where the data allow magnification of the curve in figure 13.5, smaller excursions in the form of periodic cycles become evident. For instance (figure 13.6), over the past 600,000 years, glacials and interglacials alternated regularly, associated with summer sunshine cycles. It is now generally accepted (e.g., Berger et al. 1984) that earth's paleoclimate has cycled periodically between global cooling and warming (although not always accompanied by waxing and waning of polar ice) due to forcing by cyclic astronomical changes (figure 13.6 top). Three Milankovitch cycles have been documented: one with a periodicity of about 100,000 years, the big excursions in figure 13.6, and two shorter cycles with periodicities of about 40,000 and 23,000 years.

Plio-Pleistocene evidence shows that the cycles were accompanied not only by large-scale expansion and retreat of ice at the poles (Denton 1985; Hays et al. 1976; Shackleton et al. 1984), but also by major climatic and vegetational changes in the land tropics (review in Rind and Peteet 1985). African evidence ranging from the East African low latitudes (Livingstone 1975; Flenley 1979; Hamilton 1982) all the way to the South African Cape (Vogel 1985) shows that the last glacial maximum was substantially colder than today (by 5–6° C) and more arid, with a greater reduction in forests and more

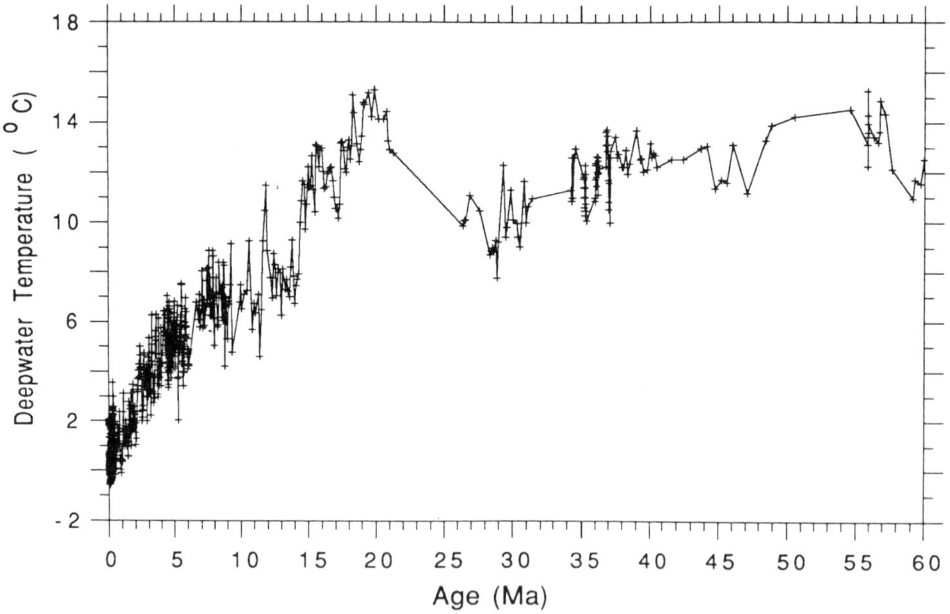

Figure 13.5. Cenozoic deep-water temperature inferred from $\delta^{18}O$ composites in Prentice and Matthews (1991:figure 1); based on the difference between the benthics and planktonics.

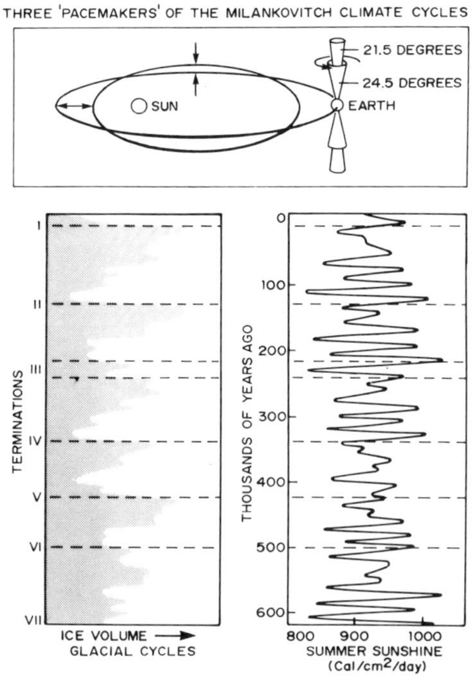

Figure 13.6. Milankovitch cycles, 23,000, 41,000, and 100,000 years long, result, respectively, from changes in orientation and tilt of earth's spin axis and orbital eccentricity (top); from changes in volume of earth's ice sheets (left); and from the effect on the intensity of summer sunshine at high northern latitudes (right). Adapted from Broecker and Denton (1990).

widespread grasslands. During the Holocene, warmer and wetter conditions with higher wood to grass ratios than at present are reported from many African areas, especially 9000–6000 years ago (Vogel 1985). Comparable data are known from the earlier African record, as I discuss below. Similar correlations are evident from Pleistocene tropical strata in Asia (e.g., Loffler 1975; Walker and Flenley 1979), Hawaii (Porter 1979), and America (Bradbury et al. 1981; Prance 1982; Deevey in Lewin 1984; Leyden 1984). Hooghiemstra (1984) found that a high-resolution pollen record covering the past 3.5 million years from the high plain of Bogota, Colombia, correlated well with the deep-sea record of Milankovitch cycles.

B. Longer-Term and Larger Climatic Changes

Since they have astronomical causes, these cycles in some form are expected to have accompanied the entire history of life (see Olsen 1986 for the Triassic; Park and Herbert 1987 for the Cretaceous). So far we have data only for small parts of the fossil record. These show that over longer periods (1 to several million years apart) there were major changes, apparently due to tectonism, in the mean and mode of these cycles and in polar ice volume. This is exemplified by the past 3.5 million years of climatic history (figure 13.7). The left curve shows two major intensifications of the cyclic cold extremes. The earlier one, about 2.5 Ma, may be due to changes in ocean-atmosphere circulation patterns that were precipitated by Panama closure (Maier-Reimer et al. 1990). At this time the Arctic became extensively glaciated (Shackleton et al. 1984) with global cooling. Vrba et al. (1989) reviewed many studies on major climatic changes that occurred near 2.5 Ma around the world, including terrestrial evidence from areas as far-flung as Holland, China, New Zealand, South America, and Africa. The other cooling intensification, near 0.8 Ma, may have been causally influenced by the uplift of western North America and of the Himalayas and Tibetan Plateau (Ruddiman et al.

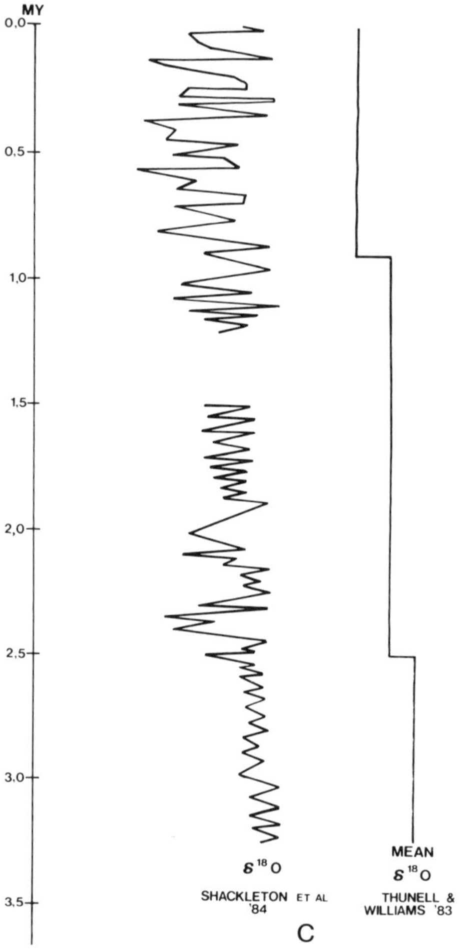

Figure 13.7. Left: Climatic curve, 3.5–0 million years (glacial periods to left, interglacial periods to right), based on deep-sea oxygen isotope data from a North Atlantic site, adapted from Shackleton et al. (1984). Right: Schematic summary of mean oxygen-isotope values from deep-sea sites in the Pacific, Caribbean, Mediterranean, and the ocean off northwestern Africa. Adapted from Thunell and Williams (1983).

1986). The curve on the right (figure 13.7) confirms that these two mean "cooling steps" represent global phenomena. Ruddiman and Raymo (1988) investigated the strength and timing of cyclic effects on the deep-sea foraminiferan record over the past 3.5 million years. They concluded that between about 3.5 and 2.4 Ma the effects were minimal, from 2.4 Ma until 0.8–0.7 Ma the 41,000-year cycle dominated with moderate amplitude, and since then the high-amplitude 100,000-year cycle gained dominant power.

IV. The Habitat Theory

Subparts of the habitat theory have had a long history within the theories of others, who include Lyell (1842) on initiation of species' origins and extinctions by geological changes, Gulick (1873) on the role of habitats and vicariance in differentiation, Mayr (1963) on the importance of allopatry in speciation, Croizat and co-workers (1974) on vicariance biogeography, Eldredge and Gould (1972) on punctuated equilibria, Coope (1979) on constant habitats of species in drifting geographic distributions, Paterson (1982) on the relationship between species' habitats and their fertilization systems, and Haffer (1982) on refuge theory. Certain orientations (especially some extreme ones) within each of the cited arguments are not required by the habitat theory. The parts that are used in the habitat theory are put together in a new way. Some of the component hypotheses and evidence are variously discussed in Vrba (1980, 1985c, 1987a, 1988, 1993; Greenacre and Vrba 1984), and the habitat theory as a whole is presented in Vrba (1992).

A. Habitats and Habitat Specificity

A habitat is a place that contains the resources necessary for an organism or species to live. A given species' habitat may include places that are unoccupied by that species. This follows Davis (1960:538): "The blood stream of man is suitable for malaria parasites and constitutes a habitat for them before infection takes place." In analogy with Hutchinson's (1957) distinction, a species' fundamental habitat includes all the places, plus necessary resources, in which a species *can* live, and the realized habitat at a given time includes all those in which it *does* live. The concept is time specific: with climatic changes a species' habitat patches in particular areas may shift, shrink, disappear, reappear, and expand through time.

All species have limited habitat specificities. Although species can be flexible in the range of habitats in which they can live, no species can live in all life-supporting environments on earth. This is an axiom for the habitat theory. I will focus here on the habitat specificities of terrestrial mammals.

The limits with respect to variables such as temperature, rainfall, substrate, food, and vegetation cover of modern species' habitat specificities can be estimated and quantified (Nix 1986; N. Caithness, personal communication, discussed in Vrba 1992); and the relative importance of particular habitat variables can be assessed statistically (Greenacre and Vrba 1984). Estimated ranges of habitat variables for modern species can be used together with cladograms based on independent characters to test hypotheses of long-term habitat fidelity, that is, that habitat specificities are heritable properties of clades through millions of years (Vrba 1992). I have found considerable support for such habitat fidelity (e.g., Vrba 1987a).

B. Climatic Effects on Species Habitats and Distributions

Relative to the habitat adaptations of most extant species, at least some of the known Milankovitch climatic changes were large. And the less frequent changes in cyclic mean and mode must have been even more strongly felt by the biota.

Figure 13.8 shows some possible consequences for species through one climatic cycle. Species may respond by undergoing no net change at all: they may persist without branching, and with no (or negligible) distribution change (*a*). Species may respond by one of three kinds of change that are represented in all macroevolutionary hypotheses, as they are in the hypotheses of the habitat theory. The first one is passive: they may respond only by vicariance or shifting of their distributions, without branching or extinction (*b, c*). Then there are two kinds of active responses with noticeable macroevolutionary results: Species can undergo vicariance to the limit, namely become extinct (*d*). Or vicariance can result in speciation (*e, f, g*).

All distribution movements (shifting, expansion, and shrinkage by vicariance, *b–g*) result in species turnover in particular areas. I will refer collectively to all such distribution movements as *distribution drift*. Turnover always involves distribution drift; but turnover in a given area may occur through one of three kinds of processes: distribution drift on its own (immigration or emigration), speciation, or extinction (real, not merely local). The processes that bring about allopa-

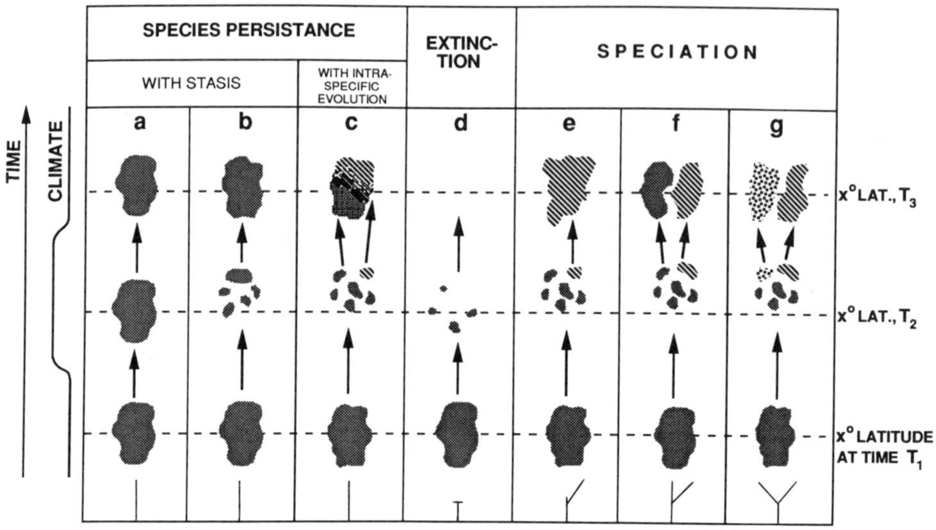

Figure 13.8. The three categories of responses of species and their distributions to environmental change are summarized in this hypothetical diagram: *a, b, c,* species persistance, with or without vicariance and latitudinal shifting of the distribution; *d,* extinction; *e, f, g,* speciation. The climatic excursions indicated can represent, for instance, cooling to the left and warming to the right.

try of populations—by vicariance or by dispersal over a barrier—are a crucial element in the habitat theory. Under the present view, a phase of allopatry is the necessary precursor to both speciation and extinction, allopatry mostly results from vicariance, and allopatry by vicariance is always brought about by physical environmental change. I have argued that even most dispersal over barriers is initiated by physical environmental change (Vrba 1993). Let us first consider the passive response of distribution drift with simple unbranching persistence, that is, without either speciation or extinction (figure 13.8 *b, c*).

C. Distribution Drift of Persistent Species

Climatic and vegetation changes, which were of large scale at least for some known periods, swept back and forth over continents with a periodicity about 1/20th of the duration time of the average species of terrestrial mammals (about 2 million years, from observation of the African record). This raises a hypothesis on how the species survived as long as they did.

> *Hypothesis 1:* Most known species survived many climatic cycles through the passive response of geographic shifting and vicariance of their distributions while maintaining habitat fidelity.

Many complex and linked organismal characters are involved in the habitat specificities of species. This basic set of coadapted habitat adaptations is expected to resist breaking apart and to show stability through long time.

A species' minimum duration in the fossil record is estimated based on how long a particular morphology remains relatively unchanged and distinct in the record. Two components of Hypothesis 1 need testing: that most species really did undergo distribution drift over their known durations, and that these species maintained habitat fidelity with respect to particular habitat components.

The incidence of distribution drift, say in mammals, can in principle be tested statistically at the level of communities or for particular clades. One needs independent evidence of climatic changes associated with the relevant fossil strata, such as of isotopic changes, snowline migrations, lake-level fluctuations, pollen spectra, or invertebrate turnover. One direct test would be to sample the stratigraphic column in a particular locality that did fluctuate climatically over a given time interval or, better still, in several localities that are strategically placed over geography. Figure 13.9 shows diagrammatically what might occur through time in a given tropical African area if the hypothesis holds. Let us say that in the left area in figure 13.9 we have a rich and complete fossiliferous sequence. (The sedimentary regime would have had to be rapid enough to have buried successive samples of large mammals for this example.) Then we should see five faunal and floral fossil assemblages, from three biomes, superposed on each other in

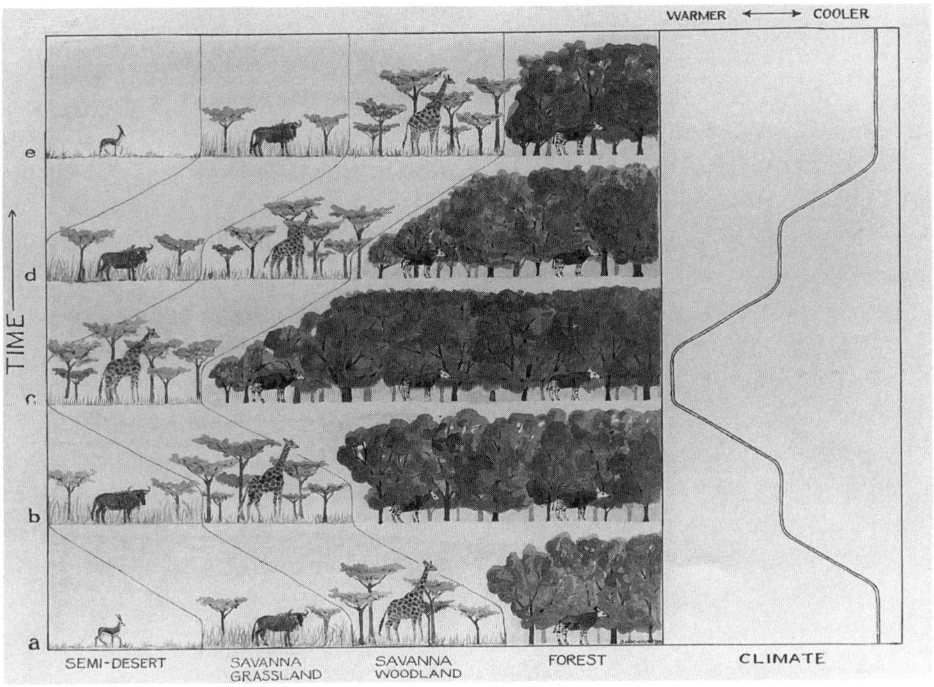

Figure 13.9. Distribution drift exemplified by a transect through an African area and through time. This hypothetical diagram shows distributions and habitats of species in four adjacent biomes drifting over geography (left) as climate changes (right). The horizontal axis for the biome transect in this example can represent equatorial to higher latitudes from right to left.

the same vertical sequence, from (starting at the surface) desert, savanna grassland, savanna woodland, again savanna grassland, and finally desert again at the base. In this example, some eurybiomic taxa, like the Eland *Taurotragus oryx,* would survive throughout. But, if a significantly greater proportion of all taxa is present at some times and absent at others, in consistent association with biome type, this would corroborate the hypothesis.

Many late Pleistocene studies, and even some studies involving longer periods, already support this hypothesis. There is impressive evidence for plants from all continents, such as Huntley and Webb's (1989) data for North American tree populations in the late Quaternary, and Hooghiemstra's (1984) for tropical South America 3.5–0 Ma. In Northwest Africa, Dupont and Hooghiemstra (1989) recorded periodically fluctuating percentages of pollen of Chenopodiaceae, Amaranthaceae, Poaceae, Cyperaceae, and savanna elements through the past 700 ka. They concluded that the Saharan-Sahelian biome boundary had shifted repeatedly over ~14–23° latitude (up to ~2550 km or ~1600 miles). Coope (1979; Coope and Brophy 1972) showed that most late Cenozoic beetle taxa maintained long-term morphological stasis, while their distributions drifted through huge latitudinal distances. For the Southern Indian Ocean, Howard (1985) demonstrated that planktonic foraminiferan communities drifted faithfully northward during each glacial cycle, and southward during each interglacial cycle, over the past 600 ka. The distances covered were ~3–8° latitude; that is, up to nearly 900 km (~560 miles).

According to some of these results (e.g., Howard, 1985), large proportions of the same two communities drifted over geography together. But we should expect that to some extent the *ecological associations of taxa* were different in the past (even in the Quaternary) from those today. One reason is that species may have become extinct, and new ones originated, since the time of the ancient community. A second reason is hypothetically depicted in figure 13.10, in which two extant species, *A* and *B,* are compared with respect to three habitat variables. *A* and *B* differ but overlap in tolerance of temperature, rainfall, and substrate. *A* is more generalized with respect to mean annual rainfall and *B* with respect to substrate. As a result, *A* and *B* are allopatric today, but were partly sympatric during a past colder period. In general, each species has a particular tolerance range for each habitat component that may or may not be identical to that of other species, and a *combination* of such ranges for all its requirements that is unique. Due to the complex interactions of factors like topography with climatic regimes, the associations of different environmental variables underwent "recombination" during past times, and some recombination among taxa in communities necessarily followed suit. A third reason for shuffling of community elements at different times is that taxa differ in how quickly their distributions can spread in response to climatic change.

In agreement with these expectations, evidence exists that extant species with Pleistocene fossil records did occur in past taxonomic associations different from their modern ones (e.g., Huntley and Webb 1989 for North American trees; Bush and Colinvaux 1990 for Central American forest com-

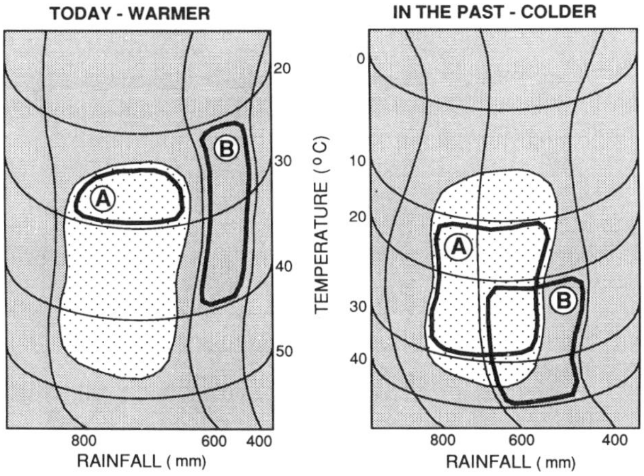

HABITAT VARIABLES	TOLERANCE RANGE	
	SPECIES A	SPECIES B
TEMPERATURE	10° - 30°C	20° - 40°C
RAINFALL	400 - 800 mm	400 - 600 mm
SUBSTRATE	ONLY	AND

Figure 13.10. Hypothetical illustration of how two species, *A* and *B*, with overlapping tolerances for three habitat variables, can have disjunct distributions today, yet sympatric ones in the past.

munities; Coope 1979 and Coope and Brophy 1972 for beetles; Sutcliffe 1985 for European mammals). For this reason, vicariated biome islands are unlikely to be true community "refugia" in the strict sense of providing refuge for an association of taxa that is identical to that of the more widespread parent community (Bush and Colinvaux 1990). I use the term *refugium* in the less strict sense of a biome refugium; for instance, a forest refugium surrounded by more open vegetation might have preserved the characteristic forest vegetation physiognomy, although its detailed taxonomic composition differed from that of the more widespread parent community.

It is difficult to find continuous sequences of high resolution that record mammalian taxa through successive Milankovitch cycles. But one can employ comparative tests effectively, particularly with respect to Pleistocene distributions. We know that some past interglacial times were warmer than today, as was the exceptionally warm early Holocene period of the current interglacial time. We also know that the height of the last glacial period was considerably colder in many terrestrial areas. The hypothesis predicts that the northern distribution limits of most modern taxa were farther north during the warmer periods, and farther south during glacial periods. Abundant data corroborate the prediction. Sutcliffe (1985) gave many instances from Europe. For instance, hippopotamids are today tropical. During the previous interglacial period, about 125,000 years ago, *Hippopotamus amphibius* was present in Britain. During past glacial times many species that today inhabit the Arctic tundra, like the collared lemming, *Dicrostonyx torquatus,* occurred in southern Europe. About the muskox, *Ovibos moschatus,* Sutcliffe (1985:97) wrote: "At present an inhabitant of the high Arctic, where it can survive north of 80° latitude, the former occurrence of this tundra animal in central France, 30° south of its present range strikingly illustrates the magnitude of the vegetational and faunal displacements that must have accompanied the last major glacial advance in Europe." Mammalian distribution drift of comparable magnitudes is evident from America (e.g., Martin and Sneed 1989). Some evidence from the African fossil record will be mentioned below.

D. Temporal Biases in Turnover: The Turnover-Pulse Hypothesis

Recall that turnover can result from distribution drift alone (only emigration or immigration without macroevolutionary change), or from speciation or extinction. The turnover-pulse hypothesis (figure 13.11) was originally stated as follows (Vrba 1985c:232):

> *Hypothesis 2:* Speciation does not occur unless forced (initiated) by changes in the physical environment. Similarly, forcing by the physical environment is required to produce extinctions and most migration events. Thus, most lineage turnover in the history of life has occurred in pulses, nearly (geologically) synchronous *across diverse phylogenies,* and in synchrony with changes in the physical environment. Most turnover-pulses are small peaks involving few lineages and/or restricted

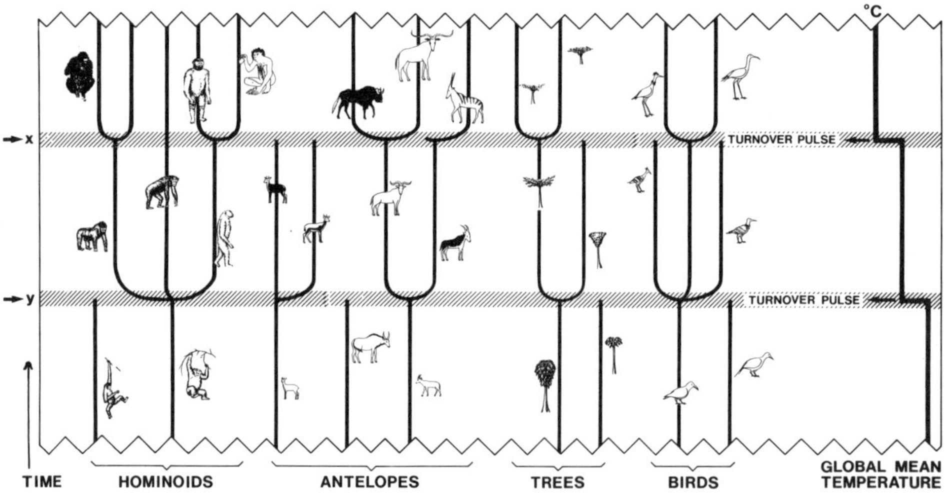

Figure 13.11. Turnover-pulses: In this hypothetical diagram, major climatic changes at times *y* and *x* initiated turnover pulses, i.e., coincident speciations and extinctions in groups as different as hominoids, antelopes, trees, and birds.

geographic areas. Some are massive and of global extent.

Where this hypothesis holds, biotic interactions like predation or competition, occurring on their own in the absence of physical changes, are not sufficient to cause speciation or extinction. Physical changes are needed to force the system off balance. They in turn cause a host of biotic changes. Whereas physical changes are needed to initiate species turnover, they are not sufficient, as shown by the prevalence of distribution drift (figure 13.9) without significant morphological evolution.

The more traditional views (reviewed in Vrba 1985c) predict a more even temporal distribution of turnover events. That is, they predict a constant "background noise" of turnover against the time axis, showing gentle-to-steeper peaks in response to marked physical changes. The most extreme turnover response to physical events, mass extinction, is generally acknowledged.

The extreme statements in Hypothesis 2 are difficult to test. Paleontological recovery of species' actual first and last appearances is beset by taphonomic problems (Behrensmeyer and Hill 1980). Even if there were real turnover-pulses sufficiently spaced out in time (see Hypothesis 3, below), the turnover events recorded in most fossil successions might be expected to present a "smeared-out," gradual pattern against time (Signor and Lipps 1982). One possible approach is to use molecular rates of change (by testing for the "molecular clock," not by assuming it) together with an exceptionally good fossil record to obtain estimates of divergence times (Vrba in litt.). This is what my colleagues R. DeSalle, J. Gatesy, R. Vaisnys, and I are currently doing to test for speciation pulses. We are using a combination of cladograms based on fossil-plus-Recent morphology and on mtDNA sequences of African antelopes, together with information on molecular evolutionary rates and on paleoeclimate. Because our data will not have the resolution to test the general statement of the hypothesis, we will test a more restricted version:

> *Hypothesis 3:* Major global climatic changes, which in the climatic record occurred from one to several million years apart, accounted for the vast majority of speciation events.

This predicts, first, that in many good data sets turnover-pulses should be statistically distinguishable. If this proposition were true, we might expect greater than 80 percent of all branching events in our bovid sample to be bunched temporally at a few time levels. A second prediction is that at each such time there should be independent evidence of climatic change that is widespread over the known geographic ranges of the relevant taxa. A distribution of divergence times that is indistinguishable from random would contradict the first prediction; and one that is uncorrelated with major climatic changes would contradict the second.

E. Ecological Biases in Turnover Rates: Specialists and Generalists

Although related species and clades may share habitat characters (Vrba 1992), they often differ heritably in breadth of habitat specificity. That is, they differ in the breadth of the range of particular resources that they require in order to survive.

This subject is often discussed in terms of generalists and specialists with respect to some resource category. Hypotheses about differences in particular cases need to be tested (e.g., Vrba 1987a). For instance, euryphagy and stenophagy refer, respectively, to broad and narrow tolerance of food resources, that is, to the greater or lesser variety of foods or food species that the organisms eat. A particularly important contrast is that between *eurybiomic* and *stenobiomic* species (Vrba 1987a). The term *biome* has mainly been applied to distinctive land communities, although it is also used to describe marine communities. Biomes of land mammals are characterized by gross vegetational physiognomy, precisely the habitat variable that emerged as important to African antelopes (see correspondence analysis of census data, Greenacre and Vrba 1984). Thus a stenobiomic species is restricted to a particular biome, implying in the terrestrial case a narrow range of vegetation physiognomy, whereas a eurybiomic species can use resources in more than one biome.

The concepts of generalist and specialist make sense only when specified in relation to particular resources, because each species (and each organism within it) commonly has a mosaic of specializations and generalizations, although one category usually predominates. Of primary importance is not so much the specialist-generalist contrast, but the distribution of resource patches across the range of environmental changes that occurred during the history of a phylogeny. For example, most ecologists would describe the aardvark, *Orycteropus afer,* as stenophagous (it feeds exclusively on ants and termites), as a substrate specialist (it needs sandy or clay soil for digging its burrows), and as stenophotic (it is nocturnal) (e.g., Smithers 1985). Yet the aardvark is eurybiomic: its specialized "resource patches" range from semi-desert to dense, moist woodland across Africa. The historical environment of the aardvark lineage was cycling precisely within this environmental spectrum; thus, one might expect the aardvark resource patches to have remained relatively undisturbed over time. Nevertheless, in most herbivores, euryphagy is associated with eurybiomy, and stenophagy with stenobiomy (Vrba 1987a).

Arguments that link breadth of resource use to speciation and extinction rates are not new (e.g., Eldredge 1979; Eldredge and Cracraft 1980; Rensch 1959; Simpson 1953; Stebbins 1950; Vrba 1980). All basically suggest that lineages of specialists should be less speciose than lineages of generalists. My arguments (since Vrba 1980, especially 1987a) on this differ in crucial respects from others. I stress the historical range of biomic changes in the larger environment of a clade and how species' resource patches are distributed across that range, subdivision of the concept of resource use with a focus on eurybiomy and stenobiomy, and selection pressures from physically initiated environmental changes as the important causal agents rather than biotic interactions.

F. The Resource-Use Hypothesis

I propose that breadth of resource use relates to speciation and extinction rates as follows:

> *Hypothesis 4:* a. Clades of species whose resources have tended to persist as environmental extremes came and went during their histories had low incidence of vicariance, speciation, and extinction.

Such eurybiomic lineages include generalists that can flexibly use different resources in different environments and specialists whose specialist patches persisted in alternative environments.

b. Clades of species whose resources have tended to disappear during one of the recurrent environmental extremes during their histories had high incidence of vicariance, speciation, and extinction. This category includes clades of specialists whose resources are confined to a narrow biome range.

I tested a part of this hypothesis by comparing speciation rates of 20 monophyletic groups of mammals with several ecological variables of their modern survivors, including feeding behavior and vegetation cover dependence. The results supported Hypothesis 4 (Vrba 1987a). Stenobiomic and eurybiomic lineages accorded with the prediction. Grazers *and* browsers showed consistently higher speciation rates than did herbivores in which each organism both grazes and browses. This suggests that *both* biomic extremes of the astronomical cycles were active causal agents of speciation in African savanna mammals. Myrmecophages, large carnivores, and omnivores all had low speciation rates, as expected.

G. Geographic Biases in Turnover Rates: Topography and Latitude

For a given set of taxa and a given time interval, can we identify areas on Earth that have had significantly higher turnover rates than others?

One such hypothesis involves topographic diversity, which refers to the number of "patches" of high and low elevation and of different substrates per area. It generally implies climatic diversity, and on a land surface vegetational diversity (figure 13.12).

Hypothesis 5: Of two large areas of similar size, both subject over the same time to Milankovitch cyclic extremes that remain habitable for organisms, the area that is more diverse in topography will have higher incidences of selection pressures and vicariance per species.

The prediction is that higher rates of speciation, extinction, and distribution drift will occur in the topographically more diverse area than in the one that is less so. The subprediction of particular interest is that, *ceteris paribus,* topographically diverse areas (figure 13.12, left) are more likely to be "factories" of species diversity due to the constant vicariation of species' habitats and distributions. To test this, one needs mea-

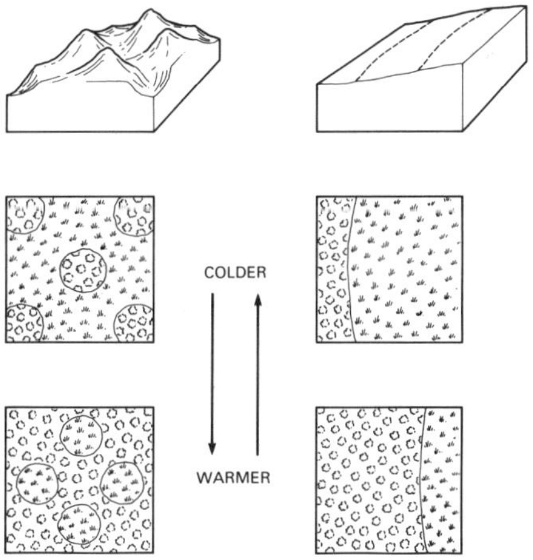

Figure 13.12. As the climate changes, areas of high topographic diversity (left) are expected to undergo higher rates of vicariance of species' habitats and distributions than areas of low topographic diversity (right).

sures of topographic and climatic diversity per unit area (generated, for example, by the methods of Nix 1986, N. Caithness, personal communication, 1990) for comparison with cladistic estimates of speciation rates.

Latitudinal asymmetries in turnover rates are also predicted under the habitat theory. Let us explore one hypothesis that applies to a world with strong thermal contrasts, namely with an ice cap on one or both pole(s). Figure 13.13 shows a hypothetical continent cycling between warming and cooling. Compare the fate of the grassland and woodland biomes with the fate of the equatorial forest biome. The grassland and woodland biomes mainly drift between north and south. The species specific for these higher-latitude biomes are riding along on these "habitat plates," that drift back and forth rapidly (in geological terms) over the

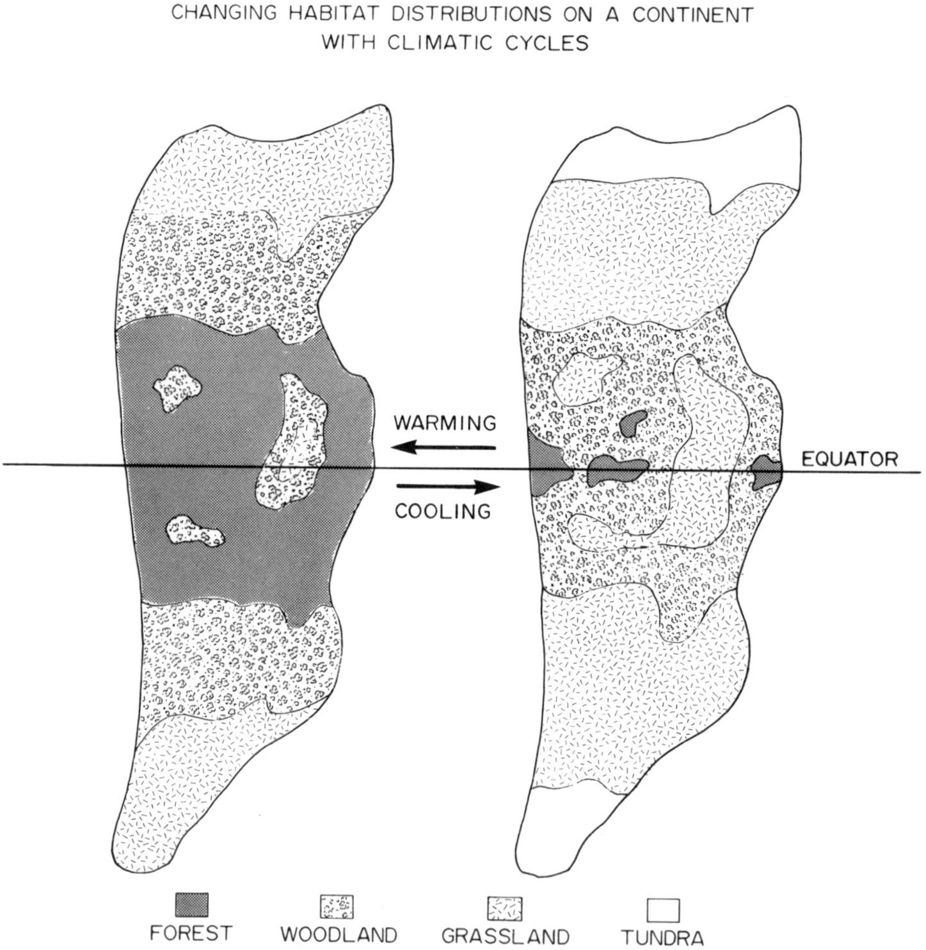

Figure 13.13. Illustration of habitat distributions changing with climatic cycles on a hypothetical continent.

tectonic plates that drift more slowly beneath. Due to local topographic heterogeneities (figure 13.12), these species will, of course, experience the pressures of environmental change. For instance, as the grassland biomes drift toward the equator upon cooling, they are expected to leave behind at high latitudes vicariant grassland islands surrounded by tundra. But as they move toward the equator, they have space into which to spread and may connect right across the hemispheres. These biomes and their species do not suffer much net loss of area at either stroke of the climatic cycle.

In contrast, the forest biome during cooling is much reduced in area (as if drawn into a fissure circling the equator) and more vicariated. Thus, *ceteris paribus,* we may expect different macroevolutionary histories at lower and higher latitudes.

> *Hypothesis 6:* During periods of strong latitudinal thermal contrasts and little change in Milankovitch amplitude, biomes closest to the equator should in general have higher speciation and extinction rates than biomes at adjacent, higher latitudes.

Vicariance of species' habitats and distributions accompanies nearly all lineage turnover. Hypothesis 6 implies that vicariance between *habitable* alternatives is highest in the tropics on the kind of world prevalent at least since the Miocene, namely one with an ice cap at one or both poles. During such intervals the high latitudes are unlikely to be centers of biotic vicariance: although alternative environments come and go at high latitudes, in synchrony with climatic cycles, only the interglacial alternative offers extensive areas that are habitable for organisms. The glacial extreme causes mainly extinctions, not bursts of speciation. I suggested (Vrba 1985c) that during ice-free phases such as the Cretaceous (Barron et al. 1981) the reverse might be predicted, that is, more vicariance and evolution at high than at low latitudes.

I stress that higher extinction *and speciation rates* at low latitudes are predicted under conditions of *little change in amplitude of the Milankovitch cycles* for the following reason. Recall that there were times of major change in the mean and mode of these cycles. At such times (two major cooling phases are shown in figure 13.7) the following is suggested.

> *Hypothesis 7:* At the first major intensification of the cold extreme of the Milankovitch cycle, after a longer period of roughly constant cyclic amplitude, a massive preponderance of extinctions over speciations occurs at low latitudes. Conversely, at such intensification of the warm cyclic extreme, more extinctions than speciations occur at high latitudes.

V. The Habitat Theory Applied to Late Pliocene America

South and North America were isolated from each other until about 3–2.5 Ma when the Panama landbridge emerged. This resulted in mammal migrations in both directions (figure 13.14, adapted from Marshall et al. 1982), especially close to 2.5 million years—the Great American Faunal Interchange (appendix). (South American paleontological data are from Marshall 1985; Marshall et al. 1979; Marshall et al. 1982; Chapter 12; Patterson and Pascual 1972;

Figure 13.14. Mammal taxa involved in the Great American Interchange across the Panama landbridge in the late Pliocene. From south to north: 1. Glyptodontidae; 2. Myrmecophagidae; 3. Megatheriidae; 4. Erethizontidae; 5. Hydrochoeridae; 6. Didelphidae; 7. Mylodontidae; 8. Dasypodidae; 9. Toxodontidae. From north to south: 10. Felidae; 11. Soricidae; 12. Cricetidae; 13. Camelidae; 14. Canidae; 15. Mustelidae; 16. Tapiridae; 17. Leporidae; 18. Sciuridae; 19. Tayassuidae; 20. Procyonidae; 21. Ursidae; 22. Gomphotheriidae; 23. Equidae; 24. Cervidae. Adapted from Marshall et al. (1982).

Simpson 1980; Webb 1969, 1976, 1985; and Webb and Marshall 1982; see also contributions in Stehli and Webb 1985.) I use the term *post-landbridge* to mean the period from about 2.5–1.5 Ma; for South America, Marshall and Sempere (Chapter 12) subsume this period under the Chapadmalalan Land Mammal Age (which includes the Uquian of earlier writings); in North America (Webb 1985) this period includes the late Blancan and earlier Irvingtonian.

A. Three Imbalances between Northern and Southern Mammal Histories

Figure 13.15 illustrates three major differences between the fates of northern and southern mammal faunas after landbridge formation (Marshall et al. 1982).

First, more northern genera moved south than vice versa (*a,* figure 13.15; appendix). The explanation of Marshall and co-workers (1982) for this draws on MacArthur and Wilson's (1967) equilibrium theory, which predicts that more migrants should emigrate from the area with the larger source pool diversity of taxa and greater surface area. In the present case, generic diversity in pre-landbridge North American was indeed greater than in South America, and the area of North America is larger (24×10^6 km^2) than that of South America (18×10^6 km^2). Nevertheless, I will suggest that a different explanation may be the correct one.

The second difference occurred in South America (*d,* figure 13.15). Here from pre- to post-landbridge, the generic extinction rates of South American natives increased much more than those of North American immigrants, although those too increased. In contrast, in North America from pre- to post-landbridge, the generic extinction rates stayed roughly constant for both northern natives and southern immigrants (*b,* figure 13.15). Traditional explanations for the South American extinction imbalance (e.g., Simpson 1950, 1980; see also more recently Webb 1976 for ungulates) invoked replacement of adaptively inferior South American natives through competitive exclusion by the North American invaders.

That both the natives and the immigrants in South America experienced higher extinction rates (*d,* figure 13.15) is explained by Marshall and co-workers (1982) by supersaturation above the previous equilibrium diversity: basing themselves again on MacArthur and Wilson (1967), they argued that the post-landbridge generic increase in South America was greater than that continent can carry, resulting in disproportionate extinctions.

The third difference is perhaps the most interesting: in South America the generic origination rate of North American immigrants was greater than that of the South American incumbents (*e,* figure 13.15; appendix). In contrast, in North America (*c,* figure 13.15) the corresponding origination rates remained roughly constant (in both cases lower than the extinction rates). Furthermore, the generic origination rate of North American immigrants in the south was also greater than that of the North American natives that stayed behind in the north. Early explanations invoked the competitive superiority of North American invaders, which radiated into empty niches vacated by competitively excluded South American taxa. More recently, Marshall and co-workers (1982) concluded that this difference re-

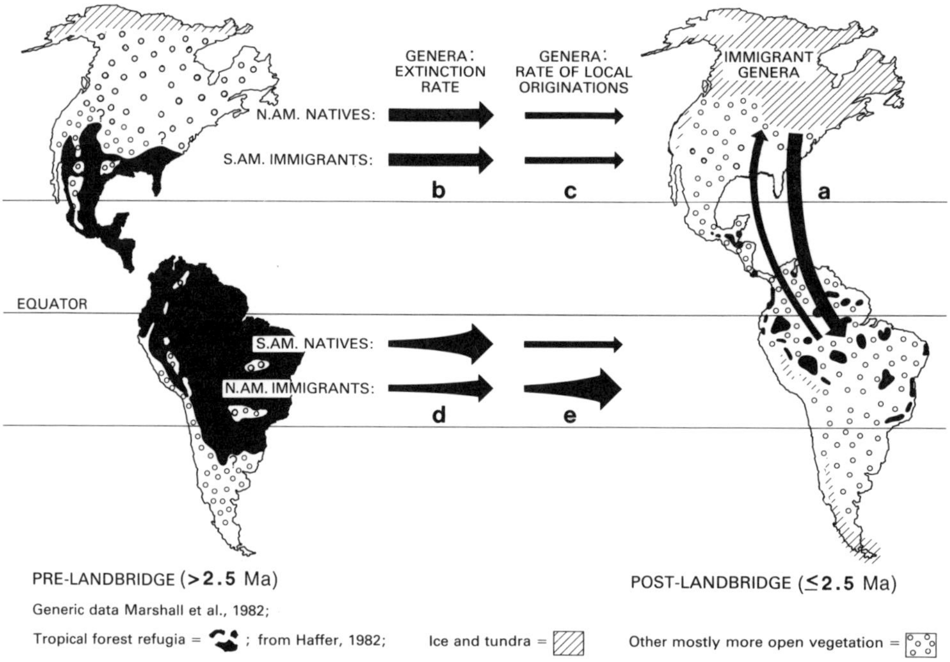

Figure 13.15. Comparison of North and South American mammal faunas at the generic level, from pre- and post-landbridge times: *a,* migration rates; *b* and *d,* extinction rates; *c* and *e,* origination rates. Arrow widths represent relative rate magnitudes (after Marshall et al. 1982), and in *b–e* whether rates remained constant or increased from pre- to post-landbridge times. Hypothetical extent of forest and dense woodland is indicated by solid pattern, of more open vegetation by small circles; the reconstructed extent of tropical forests at about 2.5 Ma, post-landbridge, follows Haffer's (1982) hypothesis for the last glacial age.

mains unexplained and is not expected from MacArthur and Wilson's (1967) theory.

B. Global Climatic Context of the Interchange

Let us now look at this evidence from the point of view of the habitat theory. The global climatic record for the past 6 million years in figure 13.16 includes information on ice volume and temperature fluctuations. The oxygen-isotope record is based on benthic foraminifers and is a composite from many areas. Some interpret it to reflect global temperature changes. (But note that there is an unresolved debate on whether the planktonic record from particular areas, or the benthic record, or a combination of the two, more closely reflects temperature change [Vrba et al. 1989]. I am here assuming that the benthic record does.)

The period from about 3.5 Ma to 2.5 Ma, when the Panama landbridge emerged, seems to have been one of sustained net cooling (see also the record of Shackleton et al. 1984 in figure 13.7). The refrigeration event close to 2.5 Ma was unprecedented in

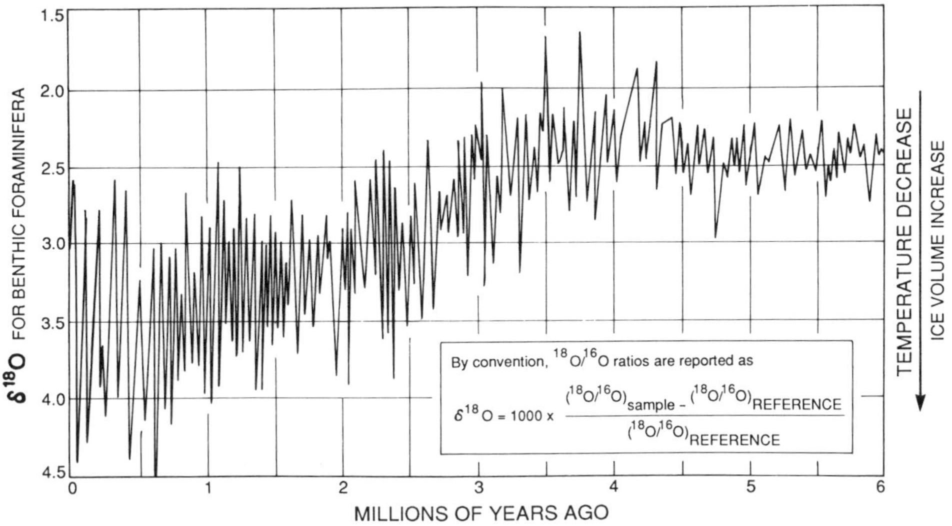

Figure 13.16. Global climatic change indicated by a composite oxygen-isotope curve for benthic foraminifers from many marine localities. The word *reference* in the formula refers to the standard for comparing oxygen-isotope ratios among calcite samples: it is PDB, a belemnite from the Pee Dee Formation, South Carolina. Adapted from Vrba et al. (1989).

the entire Cenozoic (compare with figure 13.5); and after about 2.5 million years, the average of the climatic cycles became even colder.

The period before (3–4 Ma) may be inferred to have been considerably warmer than the current interglacial cycle. Thus, it is realistic (and testable) to hypothesize for pre-landbridge time a greater extension of tropical forest and dense woodland biomes than today. This is roughly indicated in figure 13.15 by solid shading for forest and dense woodland, and circles for open woodland and grassland savanna. Especially in South America, because of its mainly low-latitude position, forest-woodland biomes should have dominated with relatively little open woodland and grassland savannas.

Earth cooled at least as much with the severe cooling change toward 2.5 Ma as it did during the last glacial period (Shackleton et al. 1984), and tropical forests are expected to have shrunk as much. This is represented in figure 13.15 by Haffer's reconstruction (1982, based on several papers in Prance 1982) of American forest refugia 18,000 years ago. But a contrasting view should be noted here: Bush and Colinvaux (1990:105) argued that "the wet highlands of Panama were never refugia for tropical rain forest taxa at any time during the Quaternary, rather rain forest species existed in unfamiliar communities in the Panamanian lowlands." My hypothesis on early Pliocene glacial environments does not require a diminution of rainforest as extreme as in Haffer (1982), but only that at least woodlands with open patches had some continuity through the isthmus.

Thus, in the Americas this period was

marked by *two* major sets of physical changes: tectonism culminating in the final emergence of the landbridge and net global climatic cooling, unprecedented in the Cenozoic. Previous explanations focused on the tectonic change and on biotic interactions following faunal interchange. In contrast, the habitat theory focuses on both physical changes but downplays the role of biotic interactions as initiating causes of speciations and extinctions.

American rainforest fossil assemblages are very rare (Marshall et al. 1982). This is the expected result of the well-known taphonomic bias against bone fossilization in rainforests, relative to that in dense woodland, open woodland, or grassland (Behrensmeyer and Hill 1980). I will briefly refer to some implications of this bias as I apply the habitat theory to the observed imbalances in migration, extinction, and origination rates (figure 13.15).

C. The Habitat Theory of the Migration Imbalance

The few biome generalists from *both* sides are expected to spread their ranges upon continental joining, because their fundamental habitats are likely to continue on the other side. In this respect the predictions for south and north are the same. However, over a period of global *cooling,* the distributions of the numerous vegetation specialists, whether on forest or on grassland savanna or on some narrow range in between, are predicted mostly to move with their habitats *toward the equator.* Simply put, from the north, biome generalists, dense-vegetation specialists, *and* the savanna specialists from a relatively huge pre-landbridge savanna biome (figure 13.15) should have a net vector of southward migration near 2.5 Ma. In contrast, from the south, only the biome generalists are expected to move north, except for the few savanna specialists from the relatively small pre-landbridge savanna biomes in South America.

Thus, the habitat theory precisely predicts the more numerous northern immigrants to South America (figure 13.15); and this prediction remains robust even if the taphonomic bias against fossil forms restricted to rainforest had operated. Under this view, in this particular case, gross geographic areas and taxic diversities of source pools are irrelevant. Total areas and taxic diversities are only expected to make a difference at much smaller spatial scales than represented by these two continents. Under the habitat theory, both these gross variables are subdivided into ecological components, and it is the distinction between these components that makes the difference.

This view predicts that *most* immigrants into the north (figure 13.14; appendix) should have been relative biome generalists, and vice versa. My reading of the literature, so far, suggests that a more detailed study might indeed find this to be so: several of the immigrants from the south, first recorded in the North American late Blancan (near 2.5 Ma), have living relatives that are generalists with respect to biomes and food (e.g., MacDonald 1984). For example, the single living species of capybara, *Hydrochaeris hydrochaeris,* today has a wide distribution in South America and Panama. Although always found near water, it is known from open grassland to rainforest habitats. Living armadillos, Dasypodidae, of which forms like *Dasypus* and *Kraglievichia* are first recorded in the North American late Blancan, today mostly eat a broad range of insects, fungi, tubers, and carrion and live in habitats from arid desert to rainforest. Some extinct edentates, like the giant ground sloths that entered the north at the same time (e.g., the mylodontid *Glossotherium* that entered in the late Blancan and the megatheriid *Eremotherium* that did so in the early Irvingtonian—other sloths had already entered North America during the late Miocene, probably by dispersal along island arcs, appendix) and glyptodonts are thought to be unspecialized herbivores by several of the analysts of the Great American Faunal Interchange (e.g., Webb and Marshall 1982).

In contrast, I suggest that the more numerous northern taxa, like cricetids, camelids, cervids, and equids that first appear in the south near 2.5 Ma (Chapadmalalan), may well be found to include a significant majority of specialists on open biomes. (Note that this does not require that northern emigrants were *all* open biome specialists, nor even that the generalists (peccaries and some carnivore taxa are likely candidates) were very rare, only that the specialist proportion of all migrants from the north was sufficiently higher than that of migrants from the south.) Webb (1978) did argue that many northern emigrants to the south were tracking savanna corridors through the isthmus. He did, however, suggest the same for southern emigrants.

D. The Habitat Theory of the Extinction Imbalance

According to Hypothesis 7, during a time of exceptional cooling, the extinction rates for specialists on tropical forest and dense woodland should be highest, because their habitats are disappearing to a greater extent, whereas specialists on open woodlands and grasslands are likely to have suitable habitats to move into. Thus, as most dense-vegetation specialists before the interchange were South American, the highest post-landbridge extinction rate for South American natives (*d,* figure 13.15) is as predicted by the habitat theory. In fact, as 2.5 Ma was a time of exceptional cooling especially relative to previous mild climates, the South American taxa that did survive are expected to include a disproportionate number of biome generalists.

Any taphonomic bias against forms restricted to rainforest is expected to have decreased the number of recovered fossil taxa particularly among South American natives. That is, if one could cancel this bias, the true post-landbridge extinction rate of South American natives is expected to be even higher (absolutely, and relative to that of North American immigrants) than it appears to be at present (*d,* figure 13.15).

The important point is that the North American immigrants need not be in any way adaptively superior to produce the observed extinction imbalance. They only need to include a large proportion of open woodland and savanna genera to survive in greater numbers. Interspecies competition need not be invoked at all.

For this view to survive, an ecological analysis of the interchange taxa (figure 13.14) near 2.5 Ma would have to demonstrate that *most* South American extinctions in the late Pliocene and early Pleistocene really were of specialists on disappearing forest and dense woodland resources.

E. The Habitat Theory of the Origination Imbalance

North American immigrants had much higher generic origination rates than did the South American native survivors (*e*, figure 13.15).

First, I will consider this under the assumption that there is a component of the observed difference that requires a nontaphonomic explanation: taphonomic bias against forest forms may have disproportionately diminished the originations from South American natives that were recovered in the fossil record (and thus the recovered totals seem to favor northern immigrants more than was really the case), but this bias was insufficient to create the imbalance we see (*e*, figure 13.15) from balance (or from a reversed imbalance) in real life. That is, North American immigrants really did have a significantly higher origination rate. One possible explanation in this case is the expectation, already mentioned, that South American taxa that did survive this massive climatic change included a disproportionate number of biomic generalists, with predicted low speciation rates (Hypothesis 4). This is indeed observed in African mammals over this time period (Vrba 1987a). Again, a preliminary survey of the South American survivors suggests that this might be found true upon ecological and genealogical analysis. For instance, all modern species in *Didelphis* as well as several other living opossums seem to be generalists with respect to vegetation cover, food, and other habitat variables.

This hypothesis also predicts that the large diversifications in South America (figure 13.14; appendix), from northern immigrant ancestors, involved relatively more habitat specialists. The largest radiations occurred in cricetine rodents, proboscideans, perissodactyls, artiodactyls, and carnivorans. The multiplication of cricetid species is the most remarkable of all. A comparison of the extant small mammals of South America and Africa (figure 13.17) shows that African cricetids, although known in Africa for about 18 million years, attained less than one-third of the species richness of the South American cricetids, first recorded only about 2.5 Ma ago according to Marshall and Sempere (Chapter 12; see earlier estimates in previous writings, e.g., 3.5–2.8 Ma in Marshall et al. 1985). South American cricetines multiplied into about 40 genera and more than 200 species. This lineage multiplication would still be remarkable even if they arrived earlier in the Pliocene by waif transport or entered via several ancestral migrants, as variously suggested (e.g., Reig 1978; Hershkovitz 1972; provided, of course, that the number of independent migrants was not too high).

South American cricetids include numerous specialist grazers with habitats at high elevations, prime candidates for frequent habitat vicariance (figure 13.12). The earliest-known cricetid records are specialized grazing forms (Marshall 1985). I suggest that, if this group is analyzed genealogically and ecologically, this radiation may well be found to support the hypothesis that biome specialization promotes high speciation rates (Hypothesis 4), especially in tropical areas (Hypothesis 6) of high topographic diversity (Hypothesis 5). The difference in cricetid diversification between Africa and South America may be accounted for by the fact that South America today has about seven times as much *montane* grassland as Africa, with a further three times as much

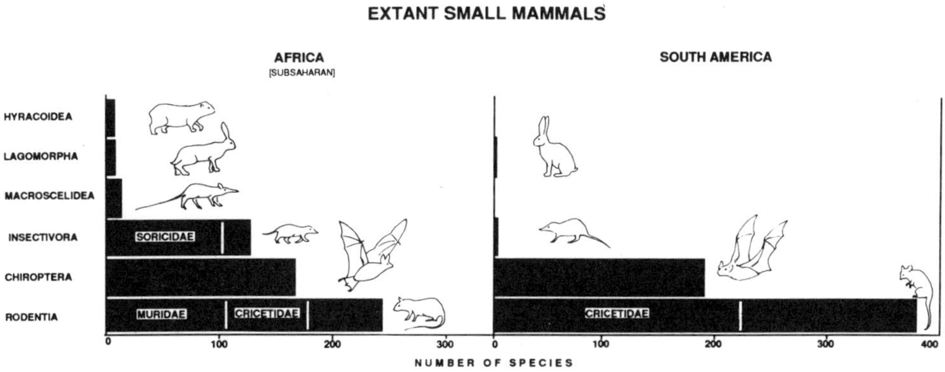

Figure 13.17. Species diversities in orders of extant small mammals in sub-Saharan Africa and South America. Data from Honacki et al. (1982).

montane area that is presently forested but may not always have been so in the Plio-Pleistocene (Chapter 9).

A different possibility is that taphonomic bias resulted in serious underrepresentation of forms restricted to forest relative to a real situation as follows: pre-landbridge mammal diversity may have been higher in South than in North America (*contra* Marshall et al. 1982; as is the case today, Honacki et al. 1982) with a large proportion of southern forest specialists (and a smaller segment from Central and North America) unseen by paleontologists. Although many southern forest specialists became extinct near 2.5 Ma, a large number may have survived whose remains were not recovered. From these southern survivors may have originated many new species that, again due to taphonomic processes, were not recovered. There may have been an origination imbalance favoring South America, or none at all.

In the case of this scenario, there would still remain the observed imbalance (*e*, figure 13.15) in taxa recovered from vegetation zones other than forest. The previous explanation might still apply to that.

VI. Comparison with Late Pliocene Africa

Imagine that one could "rerun the experiment" of Panama closure, but with both continents displaced southward so that the equator is near Washington, D.C. Then these unequal migratory and evolutionary results are predicted to be exactly reversed: more immigrants from south to north, more extinctions of North American natives, and greater radiation in North America of South American immigrants.

That test is impossible. But we *can* apply one involving comparison with Africa. Remember, under the habitat theory, climate and habitat specificities play a preeminent role. This is unlike the traditional explanations, in which interspecific competition (and its effects) due to interchange are preeminent. In Africa, climate and habitat specificities must have played a similar role as in South America, although there was no tropical interchange due to continental joining. Thus, four sets of observations must apply to Africa as well if the hypotheses for the American phenomenon are valid:

First, in Africa during the late Pliocene there would have had to be large-scale climatic and vegetation changes resulting from global cooling. Second, there should be evidence of substantial migration of species mainly from northern but also from southern areas toward the equator near 2.5 Ma. Third, there should be evidence of large-scale extinction of tropical African "native" forms, more than at higher latitudes. Fourth, there should be evidence of massive diversification of savanna specialists in the African tropics from the late Pliocene onward. The needed detailed studies have not yet been done. But the preliminary African evidence already suggests support for all four predictions. I will briefly mention two sets of mammal data.

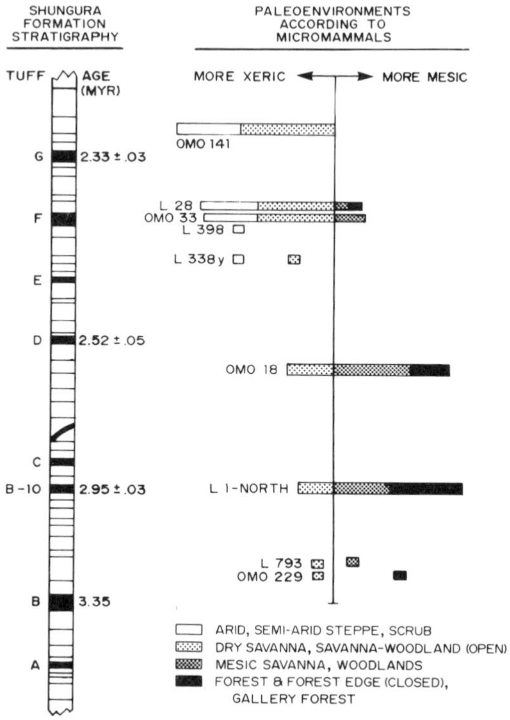

Figure 13.18. Relative micromammal frequencies indicate paleoenvironmental change in Late Pliocene strata of the Shungura Formation, Ethiopia. Tuffs A–G form lower boundaries of members A–G, respectively. The references OMO141, L28, etc., identify particular fossil samples and excavation localities. Adapted from Wesselman (1985:figure 1).

A. Micromammal Evidence

Wesselman (1985) reported changes in the frequencies of micromammal taxa (rodents, insectivores, bats, and other small mammals) in categories from mesic-closed to open-xeric habitats from the Shungura Formation, near the Ethiopian equator (figure 13.18). These results show a strong shift, somewhere between just more than 2.5 Ma and about 2.4 Ma, toward species indicative of more open, arid, and seasonally cold environments. Near 2.9 Ma Wesselman (1985) records mammals today restricted to closed-canopy rainforest, such as Demidoff's bushbaby, *Galago demidovii,* and several species of bats. In contrast, near 2.5 Ma in the same stratigraphic sequence, there appear forms that are today adapted to semi-desert or desert, such as a *Jaculus* species, a gerboa, and a close relative of the naked molerat, *Heterocephalus glaber.*

According to Wesselman (1985), near the African equator, over the time period of the Great American Faunal Interchange, the climate cooled and aridified markedly, taxa immigrated from high-latitude grasslands and steppes, and local extinctions and originations occurred. Pollen data for the same stratigraphic sequence also indicate a change from more wooded to more open conditions over the same period (Bonnefille 1976, 1983).

B. Bovid Evidence

Figure 13.19 is a stratigraphic plot of all sub-Saharan antelope species (Bovidae) over the time 3.3–1.0 Ma. Briefly, the taxa with first global records near 2.5 Ma bear the unmistakable stamp of the spread of grassland and aridification. Examples are taxa of the open-grassland Alcelaphini like the genus of giant hartebeests, *Megalotragus*; a new species of springbuck, *Antidorcas* sp.; and the genus of giant buffaloes, *Pelorovis,* which throughout its subsequent record is associated with cold open environments (Vrba 1987b).

Based on cladistic analyses, the ancestors of the lineages I mentioned so far were African tropical endemics. But several other taxa appearing near 2.5 Ma seem to be immigrants from Eurasia (marked I in figure 13.19), some more certainly, like *Antilope,* the genus of the modern Indian blackbuck, and others possibly like the earliest fossils of the genus *Oryx*. Other taxa also appear first in Africa near 2.5 Ma, like *Equus*. Incidentally, fossils of *Homo* are also first recorded at about 2.5 Ma and stone tools (widely seen as evidence of the hominine behavioral phenotype) first appear near 2.5 Ma (review in Vrba 1988).

The data in figure 13.19 include only antelopes. Evidence is mounting that many different groups of animals and plants were undergoing turnover as well close to 2.5 Ma. In fact, this part of the African record, and indeed of the global record, may upon additional study be found to be a massive turnover-pulse, namely a concentration across different phylogenies of migrations, extinctions, and speciations coincident with global climatic cooling. But this remains speculative at present. So far, I can report that a statistically significant number of first appearances of sub-Saharan Bovidae (figure 13.19) fall within a time interval of about 200,000 years (Vrba 1988).

The evidence from micromammals (figure 13.18) and from bovid records that have been physically dated (some of the left records in figure 13.19; the southern endemic record, on the right, has been dated only by biochronology) for the event 2.5 Ma comes from East Africa. Could this East African environmental change have been largely a local phenomenon (Pickford 1990), mostly (or even entirely) accounted for by uplift in the rift valley? I think that the evidence that at least a major part of the East African change was connected to the global event of 2.5 Ma is very good for four reasons:

1. The magnitude of the first extensive Arctic glaciation at this time (Shackleton et al. 1984), together with the evidence from the last glacial period cited above, leads one to expect an impact in continental areas, including in the African tropics.

2. The nature of the change observed in East Africa, namely the relative spread of grassland and aridification and the arrival of Eurasian immigrants, is precisely as predicted for global cooling from the analogy of the last glacial period.

3. The consistency in the timing and in the nature of changes observed in terrestrial areas as far-flung as South America (Hooghiemstra 1984), Europe (Van der Hammen et al. 1971), Africa, and other continents and in the oceans strongly suggests a major and global cause.

4. There are data sets that directly link the African climate to evidence from deep-sea cores. For instance, Janacek and Ruddiman (1987) report an increase in terriginous sediment in Atlantic marine cores off West Africa, far from the East African rift, indicating an opening of African vegeta-

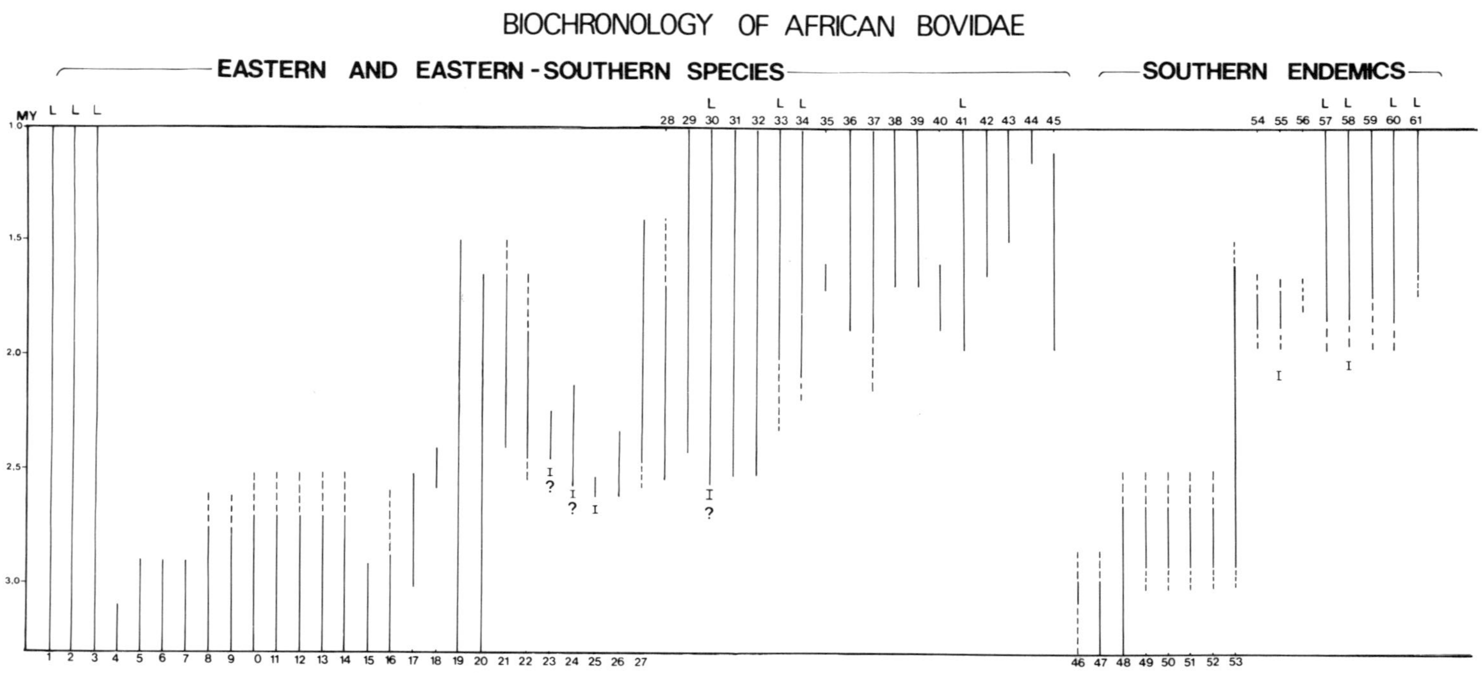

Figure 13.19. Biochronology of sub-Saharan African Bovidae from 3.3–1.0 Ma. Each line represents a species, except 2, 24, 30, and 48, which may each refer to more than one species. L = living; I = immigrant from Eurasia. Adapted from Vrba (1988).

tion and an increase in wind-borne dust, at 2.5 Ma.

VII. Concluding Comments

During the Neogene, major phases of mammalian turnover occurred near 18, between 15 and 14, between 8 and 5, and near 2.5 Ma. For each of these time periods there is evidence for faunal interchange with Eurasia and for local extinctions of some groups followed by multiplication in Africa of other local or immigrant groups.

Such turnover episodes have been reported throughout the fossil record. In particular, fossil case histories in which some major organismal groups declined as others radiated are common. Some of these are widely accepted as evidence for the competition paradigm, namely for the importance of interspecies competition as a cause of extinctions and of "empty niches" as causative in attracting species' origins (Benton 1983). Well-known examples are the "replacement" of dinosaurs by mammals, of perissodactyls by artiodactyls, and of marsupials by placentals during the Great American Faunal Interchange.

I question this traditional emphasis on the competition paradigm in macroevolutionary explanations. I am especially skeptical of the initiating role of "empty niches" in speciation and phylogenetic radiation. If empty niches are taken as synonymous with unused resource space (and disregarding for the moment that "empty niche" is a contradiction in terms, e.g., Lewontin 1978), they are certainly everywhere and were always there when evolutionary radiations occurred. For instance, there are (and were) nearly empty niches for organisms with cellulose-digesting enzymes and organisms that can fix atmospheric nitrogen (Begon et al. 1990), as well as empty niches for snakes that eat grass (Lewontin 1978), and so on. This ubiquitousness of empty niches is precisely one reason why I find the concept empty.

I accept that there is evidence for interspecies competition. Indeed, many historic extinctions on islands seem to have been caused by purely biotic factors: competitive exclusion of native species occurred in the absence of notable physical changes, after human introductions of invading species (Brown 1989; Diamond 1984). Nevertheless, I doubt the importance during life's history of such purely biotic initiation of true (not merely local) extinctions.

Instead of expanding on reasons for such doubts, I have explored a different approach, the habitat theory which includes several mutually consistent hypotheses. It emphasizes how species relate to their habitats, and how they respond to habitat alteration initiated by physical change. The Great American Faunal Interchange of mammals is often cited as the very best paleontological example of the competition paradigm (Benton 1983). Therefore I have compared it with the contemporaneous African record to test aspects of the habitat theory. I have suggested some additional tests, the needed evidence for which has not yet been gathered; until these tests are performed, much of what I have written here remains speculative. Nevertheless, I conclude that the habitat theory has passed my preliminary tests very well and that it may well turn out to explain the major features of the Great American Faunal Interchange more consistently than does the competition paradigm.

Appendix

Mammalian genera in the Great American Faunal Interchange (with avian genus *Titanis*): Earliest records of immigrants.

Arrivals in North America from South America—Hemphillian to Irvingtonian.

Order and Family	Genus	Time of Arrival
HEMPHILLIAN (about 9.5–5.0 Ma)		
Edentata		
Megalonychidae (ground sloths)	*Pliometanastes**	8.0 Ma
Mylodontidae (ground sloths)	*Thinobadistes**	6.7 Ma
BLANCAN (about 5.0–1.8 Ma)		
Aves		
Phorusrhachidae (large terrestrial predator)	*Titanis**	Late Blancan
Edentata		
Glyptodontidae (glyptodonts)	*Glyptotherium**	Late Blancan
Dasypodidae (armadillos)	*Kraglievichia**	Late Blancan
	*Dasypus**	Late Blancan
Mylodontidae (ground sloths)	*Glossotherium**	Late Blancan
Rodentia		
Erethizontidae (porcupines)	*Erethizon**	Late Blancan
Hydrochoeridae (capybaras)	*Neochoerus**	Late Blancan
Sirenia		
Trichechidae (manatees)	*Trichechus**	Late Blancan
IRVINGTONIAN (about 1.8 Ma–Late Pleistocene)		
Edentata		
Megatheriidae (ground sloths)	*Nothrotheriops**	Early Irvingtonian
	*Eremotherium**	Early Irvingtonian
Myrmecophagidae (giant anteaters)	*Myrmecophaga**	Early Irvingtonian
Dasypodidae (armadillos)	*Pampatherium*	Early Irvingtonian
Rodentia		
Hydrochoeridae (capybaras)	*Hydrochoerus**	Early Irvingtonian
Marsupialia		
Didelphidae (opossums)	*Didelphis**	Early Irvingtonian

Arrivals in South America from North America—Huayquerian to Ensenadan.

Order and Family	Genus	Time of Arrival
HUAYQUERIAN (about 9.0–6.0 Ma)		
Carnivora		
Procyonidae (raccoons, etc.)	*Cyonasua**	Late Huayquerian?
MONTEHERMOSAN (about 6.0–2.5 Ma)		
Carnivora		
Procyonidae (raccoons, etc.)	*Chapalmalania*	Late Montehermosan
Rodentia		
Cricetidae	*Auliscomys**	Late Montehermosan
	Bolomys	Late Montehermosan

Order and Family	Genus	Time of Arrival
CHAPADMALALAN (about 2.5–1.5 Ma)		
Carnivora		
Mustelidae (skunks, etc.)	*Conepatus**	Early Chapadmalalan
	*Galera**	Early Chapadmalalan
	Galictis	"Uquian"
	Stipanicicia	"Uquian"
Ursidae (bears)	*Arctodus (Arctotherium)**	"Uquian"
Canidae (dogs and allies)	*Protocyon*	"Uquian"
	Dusicyon	"Uquian"
Felidae (cats and allies)	*Smilodon (Smilodontidion)**	"Uquian"
Rodentia		
Cricetidae	*Akodon*	Early Chapadmalalan
	Dankomys	Early Chapadmalalan
	Graomys	Early Chapadmalalan
	Reithrodon	Early Chapadmalalan
	Oryzomys	Early Chapadmalalan
	Zygodontomys	Early Chapadmalalan
	*Calomys**	"Uquian"
	Euneomys	"Uquian"
	Scapteromys	"Uquian"
Artiodactyla		
Tayassuidae (peccaries)	*Argyrohyus**?	
	*Platygonus**	"Uquian"
Camelidae (camels and allies)	*Lama**	"Uquian"
	*Hemiauchenia**	"Uquian"
Cervidae	*Antifer*?	"Uquian"
	Habromeryx?	"Uquian"
	Hippocamelus	"Uquian"
	Ozotoceros	"Uquian"
Perissodactyla		
Equidae (horses)	*Hippidion**	"Uquian"
	Onohippidium	"Uquian"
Tapiridae (tapirs)	*Tapirus**	"Uquian"
Proboscidea		
Gomphotheriidae	*Stegomastodon**?	"Uquian"
ENSENADAN (about 2.4–0.7 Ma)		
Carnivora		
Procyonidae (raccoons, etc.)	*Nasua (Brachynasua?)**	Ensenadan
Canidae (dogs and allies)	*Cerdocyon*	Ensenadan
	*Canis (Theriodictis)**	Ensenadan
Mustelidae (skunks and allies)	*Lyncodon*	Ensenadan
	*Lutra**	Ensenadan
Felidae (cats and allies)	*Felis* (Puma)*	Ensenadan
	Felis (Jaguarius)*	Ensenadan

(continued)

Appendix. *Continued.*

Order and Family	Genus	Time of Arrival
Rodentia		
Cricetidae	*Necromys*	Ensenadan
	Ptyssophorus	Ensenadan
Artiodactyla		
Tayassuidae (peccaries)	*Catagonus*	Ensenadan
	*Tayassu (Dicotyles)**	Ensenadan
Camelidae (camels and allies)	*Vicugna*	Ensenadan
	*Paleolama**	Ensenadan
Perissodactyla		
Equidae (horses)	*Equus**	Ensenadan

Sources: Marshall 1985; Marshall et al. 1979, 1982; Patterson and Pascual 1972; Webb 1985; Webb and Marshall 1982.
Note: Genera marked with an asterisk are inferred to be "primary immigrants" (immigrants that came directly from the other continent); others are "secondary immigrant" genera (genera that evolved from primary immigrants, *sensu* Marshall et al. 1982). Chronology mainly after Marshall and Sempere (Chapter 12), who include in the Chapadmalalan, dated about 2.5–1.5 Ma, the Uquian of previous writings (genera denoted here by "Uquian").

Literature Cited

Adams, C. G., A. W. Gentry, and P. J. Whybrow. 1983. Dating the terminal Tethyan event. Utrecht Micropalaeontol. Bull. 30:273–298.

Baker, B. H., P. A. Mohr, and L. A. J. Williams. 1982. Geology of the Eastern Rift System of Africa. Pp. 242–276 *in* A. M. Quennell (ed.), *Rift Valleys Afro-Arabian*. Stroudsburg, Penn.: Hutchinson Ross.

Barron, E. J., S. L. Thompson, and S. H. Schneider. 1981. An ice-free Cretaceous? Results from climate model simulations. Science 212:281–283.

Barry, J. C., N. M. Johnson, S. M. Raza, and L. L. Jacobs. 1985. Neogene mammalian faunal change in southern Asia: Correlations with climatic, tectonic, and eustatic events. Geology 13:637–640.

Begon, M., J. L. Harper, and C. R. Townsend. 1990. Ecology: Individuals, populations and communities. Oxford: Blackwell Scientific Publications.

Behrensmeyer, A. K., and A. P. Hill (eds.). 1980. *Fossils in the Making*. Chicago: University of Chicago Press.

Benton, M. J. 1983. Large-scale replacements in the history of life. Nature 302:16–17.

Berger, A., J. Imbrie, J. Hays, G. Kukla, and B. Saltzman (eds.). 1984. *Milankovitch and Climate, Part 1*. Dordrecht, Neth.: Reidel.

Bonnefille, R. 1976. Palynological evidence for an important change in the vegetation of the Omo Basin between 2.5 and 2 million years ago. Pp. 421–432 *in* Y. Coppens, F. C. Howell, G. L. Isaac, and R. E. F. Leakey (eds.), *Earliest Man and Environments in the Lake Rudolph Basin*. Chicago: University of Chicago Press.

———. 1983. Evidence for a cooler and drier climate in the Ethiopian uplands towards 2.5 million years ago. Nature 303:487–491.

Bradbury, J. P., et al. 1981. Late Quaternary environmental history of Lake Valencia, Venezuela. Science 214:1299–1305.

Broecker, W. S., and G. H. Denton. 1990. What drives glacial ages? Scient. Amer. 262:48–56.

Brown, J. H. 1989. Patterns, modes and extents of invasions by vertebrates. Pp. 85–109 *in* J. A. Drake (ed.), *Biological Invasions: A Global Perspective*. New York: John Wiley & Sons.

Bush, M. B., and P. Colinvaux. 1990. A pollen record of a complete glacial cycle from lowland Panama. J. Veg. Sci. 1:105–118.

Coetzee, J. J. 1986. Palynological evidence for

major vegetation and climatic change in the Miocene and Pliocene of the south-western Cape. S. Afr. J. Sci. 82:71–72.

Coope, C. R. 1979. Late Cenozoic fossil Coleptera; evolution, biogeography, ecology. Ann. Rev. Ecol. Syst. 10:247–267.

Coope, G. R., and J. A. Brophy. 1972. Late glacial environmental changes indicated by a coleopteran succession from North Wales. Boreas 1:97–142.

Croizat, L., G. Nelson, and D. E. Rosen. 1974. Centers of origin and related concepts. Syst. Zool. 23:265–287.

Davis, J. H. 1960. Notes and comment: Proposals concerning the concept of habitat and a classification of types. Ecology 41:537–541.

Denton, G. H. 1985. Did the Antarctic ice sheet influence Late Cainozoic climate and evolution in the southern hemisphere? S. Afr. J. Sci. 81:224–229.

Denys, C., J. Chorowicz, and J. J. Tiercelin. 1986. Tectonic and environmental control on rodent diversity in the Plio-Pleistocene sediments of the African Rift System. Pp. 381–389 *in* R. W. Renault, I. Reid, and J. J. Tiercelin (eds.), *Geological Society Special Publication No. 25*. London: Blackwell Scientific Publications.

Diamond, J. M. 1984. Historic extinction: A rosetta stone for understanding prehistoric extinctions. Pp. 824–866 *in* P. S. Martin and R. G. Klein (eds.), *Quaternary Extinctions*. Tucson: University of Arizona Press.

Dupont, L. M., and H. Hooghiemstra. 1989. Variations of the African trade wind regime during the last 140,000 years: Changes in pollen flux evidenced by marine sediment records. Pp. 733–770 *in* M. Leinen and M. Sarnthein (eds.), *Paleocolimatology and Paleometeorology: Modern and Past Patterns of Global Atmospheric Transport*. Dordecht, Neth.: Kluwer.

Eldredge, N. 1979. Alternative approaches to evolutionary theory. Bull. Carnegie Mus. Nat. Hist. 13:7–19.

Eldredge, N., and J. Cracraft. 1980. *Phylogenetic Patterns and the Evolutionary Process*. New York: Columbia University Press.

Eldredge, N., and S. J. Gould. 1972. Punctuated equilibria: An alternative to phyletic gradualism. Pp. 82–115 *in* T. J. M. Schopf (ed.), *Models in Paleobiology*. San Francisco: W. H. Freeman.

Ewer, R. F. 1973. *The Carnivores*. London: Weidenfeld and Nicholson.

Flenley, J. R. 1979. *The Equatorial Rain Forest, a Geological History*. London: Butterworths.

Flynn, L. J., and L. L. Jacobs. 1982. Effects of changing environments on Siwalik rodent faunas of northern Pakistan. Palaeogeogr., Palaeoclimatol., Palaeoecol. 38:129–138.

Greenacre, M. J., and E. S. Vrba. 1984. A correspondence analysis of biological census data. Ecology 65:984–997.

Gulick, J. T. 1873. On diversity and evolution under one set of external conditions. J. Linn. Soc. (Zool.) 11:496–505.

Haffer, J. 1982. General aspects of the refuge theory. Pp. 6–24 *in* G. G. Prance (ed.), *Biological Diversification in the Tropics*. New York: Columbia University Press.

Hamilton, A. C. 1982. *Environmental History of East Africa*. New York: Academic Press.

Hays, J. D., J. Imbrie, and N. J. Shackleton. 1976. Variations in the earth's orbit: Pacemaker of the ice ages. Science 194:1121–1132.

Hershkovitz, P. 1972. The Recent mammals of the Neotropical Region: A zoogeographic and ecological review. Pp. 311–431 *in* A. Keast, F. C. Erk, and B. Glass (eds.), *Evolution, Mammals and Southern Continents*. Albany: SUNY Press.

Hoffman, A. 1989. *Arguments on Evolution*. Oxford: Oxford University Press.

Honacki, J. H., K. E. Kinman, and J. W. Koeppl (eds.). 1982. *Mammal Species of the World*. New York: Allen Press.

Hooghiemstra, H. 1984. *Vegetational and Climatic History of the High Plain of Bogotá, Colombia: A Continuous Record of the Last 3.5 Million Years*. Vaduz, Liechtenstein: A. R. Gantner Verlag.

Howard, W. R. 1985. Late Quaternary Southern Indian Ocean circulation. S. Afr. J. Sci. 81:253–254.

Huntley, B., and T. Webb III. 1989. Migration: Species' response to climatic variations caused by changes in the earth's orbit. J. Biogeogr. 16:5–19.

Hutchinson, G. E. 1957. Concluding remarks, Cold Spring Harbor Symposium on Quantitative Biology 22:415–427.

Janacek, T. R., and W. F. Ruddiman. 1987. Plio-Pleistocene sedimentation in the south equatorial Atlantic divergence. P. 193 *in XII INQUA Research, Ottawa, Canada, Program and Abstracts*. Ottawa: National Research Council of Canada.

Janis, C. 1982. Evolution of horns in ungulates: Ecology and paleoecology. Biol. Rev. 57: 261–318.

Leopold, E. B. 1969. Late Cenozoic palynology. Pp. 377–438 *in* R. S. Tschudy and W. Leuthold (eds.), *Aspects of Palynology.* New York: Wiley Interscience.

Lewin, R. 1984. Fragile forests implied by Pleistocene pollen. Science 226:36–37.

Lewontin, R. C. 1978. Adaptation. Scient. Amer. 239:212–230.

Leyden, B. W. 1984. Guatemalan forest synthesis after Pleistocene aridity. Proc. Nat. Acad. Sci. U.S.A. 81:48–56.

Lindsay, E. H., V. Fahlbusch, and P. Mein (eds.). 1989. *European Neogene Mammal Chronology.* New York: Plenum.

Livingstone, D. A. 1975. Late Quaternary climatic change in Africa. Ann. Rev. Ecol. Syst. 6:249–280.

Loffler, E. 1975. Beobachtungen zur periglazialen Hohenstufe in den Hochgebirgen von Papua New Guinea. Erdkunde 29:285–292.

Lyell, C. 1842. *Principles of Geology.* Boston: Hilliard Gray.

MacArthur, R. H., and E. O. Wilson. 1967. *The Theory of Island Biogeography.* Princeton: Princeton University Press.

MacDonald, D. (ed.). 1984. *The Encyclopedia of Mammals.* Vols. 1 and 2. London: Allen and Unwin.

Maier-Reimer, E., U. Mikolajewicz, and T. Crowley. 1990. Ocean general circulation model sensitivity experiment with an open Central American isthmus. Paleoceanography 5(3):349–366.

Marshall, L. G. 1985. Geochronology and land-mammal biochronology of the Transamerican faunal interchange. Pp. 49–85 *in* F. G. Stehli and S. D. Webb (eds.), *The Great American Biotic Interchange*. New York: Plenum.

Marshall, L. G., R. F. Butler, R. E. Drake, G. H. Curtis, and R. H. Tedford. 1979. Calibration of the Great American Interchange. Science 204:272–279.

Marshall, L. G., S. D. Webb, J. J. Sepkoski, and D. M. Raup. 1982. Mammalian evolution and the Great American Interchange. Science 215:1351–1357.

Martin, P. S., and R. G. Klein. 1984. *Quaternary Extinctions*. Tucson: University of Arizona Press.

Martin, R. A., and J. M. Sneed. 1989. Late Pleistocene records of caribou and elk from Georgia and Alabama. Georgia J. Sci. 47: 117–122.

Mayr, E. 1963. *Animal Speciation and Evolution.* Cambridge: Harvard University Press.

Miller, K. G., and R. G. Fairbanks. 1985. Cainozoic $\delta^{18}O$ record of climate and sea level. S. Afr. J. Sci. 81:248–249.

Nix, H. 1986. A biogeographic analysis of Australian elapid snakes. Pp. 4–15 *in* R. Longmore (ed.), *Atlas of Elapid Snakes of Australia.* Canberra: Australian Government Printing Service.

Olsen, P. E. 1986. A 40-million old lake record of early Mesozoic orbital climate forcing. Science 234:842–844.

Park, J., and T. D. Herbert. 1987. Hunting for paleoclimatic periodicities in a geologic time series with an uncertain time scale. J. Geophys. Res. 92:14,027–14,040.

Paterson, H. E. H. 1982. Perspective on speciation by reinforcement. S. Afr. J. Sci. 78:53–57.

Patterson, B., and R. Pascual. 1972. The fossil mammal fauna of South America. Pp. 247–309 *in* A. Keast, F. C. Erk and B. Glass (eds.), *Evolution, Mammals, and Southern Continents.* Albany: SUNY Press.

Pickford, M. 1990. Uplift of the roof of Africa and its bearing on the evolution of mankind. Human Evol. 5:1–20.

Porter, S. C. 1979. Hawaiian glacial ages. Quat. Res. 12:161–187.

Prance, G. T. (ed.). 1982. *Biological Diversification in the Tropics*. New York: Columbia University Press.

Prentice, M. L., and R. K. Matthews. 1991. Tertiary ice sheet dynamics: The snow gun hypothesis. J. Geophys. Res. 96:6811–6827.

Quade, J., T. E. Cerling, and J. R. Bowman. 1989. Development of Asian monsoon revealed by marked ecological shift during the latest Miocene in northern Pakistan. Nature 342:163–166.

Reig, O. A. 1978. Roedores cricetidos del plioceno superior de la Provincia Buenos Aires (Argentina). Publ. Mus. Mun. Cienc. Nat. "Lorenzo Scaglia," Mar del Plata 2:164–190.

Rensch, B. 1959. *Evolution above the Species Level*. London: Methuen.

Rind, D., and D. Peteet. 1985. Terrestrial conditions at the last glacial maximum and CLIMAP sea-surface temperature estimates: Are they consistent? Quat. Res. 24:1–22.

Rogl, F., and F. F. Steiniger. 1983. Neogene Paratethys, Mediterranean and Indo-pacific seaways. Pp. 171–200 *in* P. Brenchley (ed.), *Fossils and Climate*. New York: John Wiley & Sons.

Ruddiman, W. F., and M. E. Raymo. 1988. Northern hemisphere climate regimes during the last three million years: Possible tectonic connections. Philos. Trans. Roy. Soc. London, B Ser., 318:411–430.

Ruddiman, W. F., M. Raymo, and A. McIntyre. 1986. Matuyama 41,000-year cycles: North Atlantic Ocean and northern hemisphere ice sheets. Earth Planet. Sci. Letts. 80:117–129.

Savage, D. E., and D. E. Russell. 1983. *Mammalian Paleofaunas of the World*. London: Addison-Wesley.

Shackleton, N. J., et al. 1984. Oxygen isotope calibration of the onset of ice-rafting and history of glaciation in the North Atlantic region. Nature 307:620–623.

Signor, P. W., and Lipps, J. H. 1982. Sampling bias, gradual extinction patterns and catastrophes in the fossil record. Geol. Soc. Amer. Spec. Pap. 190:291–296.

Simpson, G. G. 1950. History of the fauna of Latin America. Pp. 369–408 *in* G. A. Baitsell (ed.), *Science in Progress,* 7th ser. New Haven: Yale University Press.

———. 1953. *The Major Features of Evolution*. New York: Columbia University Press.

———. 1980. *Splendid Isolation: The Curious History of South American Mammals*. New Haven: Yale University Press.

Smithers, R. H. N. 1985. *The Mammals of the Southern African Subregion*. Pretoria: University of Pretoria.

Stebbins, L. 1950. *Variation and Evolution in Plants*. New York: Columbia University Press.

Stehli, F. G., and S. D. Webb (eds.). 1985. *The Great American Biotic Interchange*. New York: Plenum.

Sutcliffe, A. J. 1985. *On the Track of Ice Age Mammals*. Cambridge: Harvard University Press.

Tassy, P. 1989. The "Proboscidean Datum Event": How many proboscideans and how many events? Pp. 237–252 *in* E. H. Lindsay, V. Fahlbusch, and P. Mein (eds.), *European Neogene Mammal Chronology.* New York: Plenum.

Tedford, R. H. 1985. Late Miocene turnover of the Australian mammal fauna. S. Afr. J. Sci. 81:262–263.

Thomas, H. 1985. The early and middle Miocene land connection of the Afro-Arabian Plate and Asia: A major event for hominoid dispersal? Pp. 42–50 *in* E. Delson (ed.), *Ancestors: The Hard Evidence*. New York: Alan R. Liss.

Thunell, R. C., and D. F. Williams. 1983. The stepwise development of Pliocene-Pleistocene paleoclimate and paleoceanographic conditions in the Mediterranean: Oxygen isotope studies of DSDP sites 125 and 132. Pp. 111–112 *in* J. E. Meulenkamp (ed.), *Reconstruction of Marine Paleoenvironments*. Utrecht Micropaleontological Bulletin, 30.

Van der Hammen, T., T. A. Wijmstra, and W. H. Zagwijn. 1971. The flora record of the Late Cenozoic of Europe. Pp. 391–424 *in* K. K. Turekian (ed.), *The Late Cenozoic Glacial Ages*. New Haven: Yale University Press.

Van Zinderen Bakker, E. M. 1986. African climates and palaeoenvironments since Messinian times. S. Afr. J. Sci. 82:70–71.

Vogel, J. C. 1985. Southern Africa at 18,000 years before present. S. Afr. J. Sci. 81:250–251.

Vrba, E. S. 1980. Evolution, species and fossils: How does life evolve? S. Afr. J. Sci. 76:61–84.

———. 1985a. African Bovidae: Evolutionary events since the Miocene. S. Afr. J. Sci. 81:263–266.

———. 1985b. Ecological and adaptive changes associated with early hominid evolution. Pp. 63–71 *in* E. Delson (ed.), *Ancestors: The Hard Evidence*. New York: Alan R. Liss.

———. 1985c. Environment and evolution: Alternative causes of the temporal distribution of evolutionary events. S. Afr. J. Sci. 81:229–236.

———. 1987a. Ecology in relation to speciation rates: Some case histories of Miocene-Recent mammal clades. Evol. Ecol. 1:283–300.

———. 1987b. A revision of the Bovini (Bovidae) and a preliminary revised checklist of Bovidae from Makapansgat. Palaeontol. Afr. 26:33–46.

———. 1988. Late Pliocene climatic events and hominid evolution. Pp. 405–426 *in* F. E. Grine (ed.), *The Evolutionary History of Robust Australopithecines*. New York: Aldine.

———. 1990. Comments on macroevolutionary characters and processes. Review of "Arguments on Evolution: A Paleontologist's Perspective" by A. Hoffman. Trends in Ecol. and Evol. 5:235–236.

———. 1992. Mammals as a key to evolutionary theory. J. Mammal. 73:1–28.

———. 1993. Turnover-pulses, the Red Queen, and related topics. Amer. J. Sci. 293-A:418–452.

Vrba, E. S., G. H. Denton, and M. L. Prentice. 1989. Climatic influence on early hominid behavior. OSSA, Int. J. Skeletal Res. 14:127–156.

Walker, A. 1984. Extinction in hominid evolution. Pp. 119–152 *in* N. Nitecki (ed.), *Extinction*. Chicago: University of Chicago Press.

Walker, D., and J. Flenley. 1979. Late Quaternary vegetational history of the Enga Providence of Upland Papua New Guinea. Philos. Trans. Roy. Soc. London 286:265–344.

Webb, S. D. 1969. Extinction-origination equilibria in Late Cenozoic land mammals of North America. Evolution 23:688–702.

———. 1976. *Pleistocene Mammals of Florida*. Gainesville: University of Florida Press.

———. 1978. A history of savanna vertebrates in the New World. Part II: South America and the Great Interchange. Ann. Rev. Ecol. Syst. 9:393–426.

———. 1985. Main pathways of mammalian diversification in North America. Pp. 201–217 *in* F. G. Stehli and S. D. Webb (eds.), *The Great American Biotic Interchange*. New York: Plenum.

Webb, S. D., and L. G. Marshall. 1982. Historical biogeography of recent South American land mammals. Pp. 39–52 *in* M. A. Mares and H. H. Genoways (eds.), *Mammalian Biology in South America: A Symposium Held at Pymatuning Laboratory of Ecology, University of Pittsburgh*. Pittsburgh: University of Pittsburgh Press.

Wesselman, H. B. 1985. Fossil micromammals as indicators of climatic change about 2.4 million years ago in the Omo Valley, Ethiopia. S. Afr. J. Sci. 81:260–261.

Whewell, W. 1847. *The Philosophy of the Inductive Sciences*, 2 vols., 2d ed. London: J. W. Parker.

Wolpoff, M. H. 1988. Divergence between early hominid lineages. Pp. 485–497 *in* F. E. Grine (ed.), *Evolutionary History of the Robust Australopithecines*. New York: Aldine.

Part IV

Ecosystems,
Communities,
and Climates

Chapter 14

Tectonics, the Santana Lagerstätten, and the Implications for Late Gondwanan Biogeography

John G. Maisey

I. Introduction

Tectonic, sedimentological, stratigraphic, and paleontological evidence suggest that the separation of South America and Africa toward the end of the Lower Cretaceous was not a simple, short-lived event, but had its origins much earlier in the Mesozoic. It is not clear just when this separation commenced, nor when it was completed. Pitman and co-workers (Chapter 2) regard it as having begun in the Valanginian, about 135 Ma, based on data in Larson and Ladd (1973) and Rabinowitz and LaBrecque (1979). This initial phase of rifting probably terminated about 106 Ma (Rabinowitz and LaBrecque 1979). The impact such a protracted geological event had on the biota that was resident in the affected regions is largely unknown. It is usually inferred to have had great influence on present-day distributions, as a major vicariant event. In my view, it is questionable whether a tectonic process spanning virtually 100 million years can be viewed as a causal factor in biogeographical analyses. It is likely that other factors of far shorter temporal duration, including climatic changes and eustatic fluctuations, had a more immediate impact upon the biota. Although such phenomena may themselves be

Many thanks are extended to Malcolm McKenna, Dave Grimaldi, and an anonymous reviewer for reading the manuscript and pointing out various errors and omissions. Any that remain are mine alone. None of this compilation would have been possible without the generous support of Dr. Herbert Axelrod, a true philanthropist and patron of the arts and natural sciences. My thanks go also to many friends and colleagues in Brazil, especially to Diogenes de Almeida Campos, Chief Paleontologist at the Departamento Nacional da Producão Mineral, Rio de Janeiro.

 ISBN 0-300-05375-4

linked to tectonic processes on a larger scale, it may be inappropriate to invoke a polyphased tectonic event such as the opening of the South Atlantic (or even the equatorial Atlantic) as a "cause" of any observed biogeographical distribution pattern. In this chapter, I will address these issues from an essentially paleontological position, for which I make no apologies. As my data base, I will examine the rich and taxonomically diverse fossil assemblages of the Santana Formation (Albian to late Aptian) of northeastern Brazil.

The majority of contributors to this volume are neontologists whose areas of research involve Recent taxa having a distribution restricted at some level to Africa and South America. Such distributions are frequently alluded to as "Gondwanan," although in a disproportionately high number of cases the taxa in question have either a poor fossil record or none at all. In such cases it is by no means certain that the modern distribution actually reflects a late Mesozoic vicariant event, as the presence of a formerly continuous ancestral population has not been demonstrated in what was previously a continuous tract of continent known today as Western Gondwanaland.

Interpretation of Recent "Gondwanan" distributions is made even more difficult by the complexity of tectonic and related sedimentological events during the breakup of South America and Africa (see Chapter 2). This separation "event" has now been documented with some precision (e.g., Simpson 1977; Rabinowitz and LaBrecque 1979; Reyment and Dingle 1987; Szatmari et al. 1987; Popoff 1988). Separation of South America from Africa cannot be explained by simple rotational drifting. Popoff (1988) contends that the "South American Plate" consists of at least three subplates, whose rifting-drifting histories were sequential and overlapping, instead of synchronous. According to that model, during the last phases of rifting (before final separation occurred between northeastern Brazil and western Africa) some rotational movement of South America may have occurred about an axis situated in the Ceará region (Szatmari et al. 1987), although this model is not universally accepted. As advocated, however, such rotation would produce northerly compression between the two continents, possibly resulting in ephemeral secondary connections (islands or landbridges, perhaps), which could have been maintained for a significant length of time into the Late Cretaceous, well after the development of oceanic crust between northeastern Brazil and West Africa in the Middle Cretaceous. Even after separation was completed, and a narrow equatorial ocean floor existed between Africa and South America, Szatmari and co-workers (1987) have alluded to compressional events (perhaps related to collision and rebound of Africa and Europe). These apparently closed the south end of the Benue Trough and buckled the equatorial Atlantic floor along an east-west axis during the Late Cretaceous and well into the Tertiary, and at times led to exposure and erosion of sediments along this axis. Added to this is the possibility of secondary (Tertiary) connections via island systems developed at the midoceanic ridge within the equatorial Atlantic (for example, along islands formed during the Cretaceous and early Tertiary along the Ascension Fracture Zone). Thus there are plenty of opportunities for vicariant and dispersal events which significantly postdate the Cretaceous separation of South America and Africa and could produce a pseudo-Gondwanan distribution pattern. The cautionary lesson to be learned is not to judge present (or past) distribution patterns from first impressions (McKenna 1981). Present-day "Gondwanan" taxa do not necessarily range back in time to the later Mesozoic. Of course there are taxa with an ancestry that can be traced back to prerifting times, but as the following analysis of a "good" late Gondwanan biota will reveal, these are few and far between. My personal belief (albeit biased by my exposure to the fossil assemblages in the Lower Cretaceous Santana Formation from northeastern Brazil) is that in many cases Recent "Gondwanan" taxa simply were not there in the Cretaceous. Having said that, a number of biogeographically interesting taxa are certainly recognizable in these fossil assemblages. Given the complexity of geological events during and after this time, however, caution should be exercised in interpreting the age of a vicariant event for "Gondwanan" taxa not known to have Mesozoic Gondwanan representatives.

II. The Santana Formation

For several years I have been compiling data about the Santana Formation, in the Araripe basin of the Chapada do Araripe in northeastern Brazil (Maisey 1991). This deposit is perhaps known best as one of the world's most prolific sources of fossil vertebrates, especially well-preserved fishes. The term *Lagerstätten* (fossil concentrations) has been applied to such deposits (Seilacher et al. 1985) that have yielded a diverse paleobiota. In fact, the Santana Formation is unique in representing not one but two such Lagerstätten. A lower lacustrine sequence (Crato Member) has yielded an unparalleled assemblage of terrestrial and freshwater insects and other arthropods; some important vertebrate fossils including fishes, a frog, and a bird feather; and a number of interesting terrestrial plant macrofossils. The Crato Member seems closest to the "Green River type" Lagerstätten category of Allison and Briggs (1991). Higher in the sequence is a series of sandstones, limestones, clays, and shales (Romualdo Member); large calcareous concretions within the clays and shales have produced many magnificent specimens of fossil fishes, reptiles, and plants indicative of a mixed brackish environment that suffered periodic stress resulting in major mortality events. Allison and Briggs (1991) have equated this occurrence with their "Orsten type" Lagerstätten. In terms of preservation, taxonomic diversity, and biotic representation and in biogeographical significance, therefore, the Santana Formation Lagerstätten provide an opportunity to examine two different aspects of the Early Cretaceous biota from a biogeographically important region at a supposedly critical moment in its tectonic and biogeographical history, during the final phases of rifting and early drifting within the equatorial Atlantic domain.

The Chapada do Araripe is located in northeastern Brazil, mainly in the south of Ceará but extending into the neighboring states of Pernambuco and Piauí. The Chapada (plateau) reaches altitudes of 600–900 m and extends some 200 km east-west and 50 km north-south. It has long been recognized as a source of beautifully preserved fossil vertebrates (Agassiz 1841, 1844; Jordan 1923; Silva Santos and Valença 1968; Maisey 1991) and is also an important source of fossil insects (Grimaldi and Maisey 1990; Grimaldi in Maisey 1991).

The Araripe Basin is a tectonic half-graben with a northerly border fault system formed by part of the Patos Lineament. Sedimentologically, the Mesozoic (mainly Lower Cretaceous) sequence it contains more or less agrees with the tripartite basin-filling model developed by Schlische and Olsen (1990; see also Olsen 1991), with a lower fluvial part (represented by the Missão Velha Formation), a deep lacustrine middle part (represented by the Crato Member), and a shallow lacustrine or fluvial upper part (represented by the Romualdo Member and Exu Formation). The stratigraphic sequence within the Santana Formation is certainly more complex than this, with varying degress of maritime (?) influence from time to time including an evaporitic episode (represented by the middle Ipubi Member) and a marine carbonate level near the top of the Romualdo Member. There continues to be controversy surrounding the stratigraphic succession of the Araripe basin (Small 1913; Beurlen 1962, 1963, 1971; Braun 1966; Moraes et al. 1963; Moraes et al. 1976; Lima 1979; Silva 1986a), but it can be summarized as consisting of a basal series of Lower Cretaceous or Upper Jurassic shales and sandstones (Missão Velha formation of most authors), followed by interbedded shales, laminated carbonates, gypsites, marls, and shales with carbonate concretions, siltites, and limestones (Santana Formation), and overlain by fluviatile arenites (Exu Formation). Palynological data published recently suggest a Late Aptian or Early Albian age for the Crato Member, and Middle to Late Albian age for the Romualdo Member of the Santana Formation (Pons et al. 1990).

The paleontological diversity represented by the Crato and Romualdo Lagerstätten has been well documented (Maisey 1991). The taxa found to date are listed in the appendix (excluding some synonymized taxa), with a notation concerning which of the Lagerstätten the fossils are from. Within each Lagerstätten more than one original fauna may be represented (there is clear evidence of mixing originally discrete terrestrial and aquatic faunas, for example, and there may be more subtle and as-yet undetected mixing also). For some records (e.g., molluscs) the published data do not clearly indicate precise stratigraphic occurrences or which assemblage they are from, and in a few cases (e.g., *Cladocyclus ferus*) the provenance is uncertain (in other cases even the taxonomic validity is questionable; where data are insufficient such dubious taxa are omitted). It is immediately apparent even from a cursory examination of these listings that the fossil assemblages from the Crato and Romualdo Lagerstätten are profoundly different, with little faunal overlap. What is not apparent is that within the Romualdo Member there seem to be at least three distinctive assemblages of uncertain environmental provenance (from accumulated field and laboratory examination of the concretions) and another more typically marine assemblage higher in the succession that has yielded numerous molluscs plus echinoids, isolated fish bones.

III. Biogeographical Remarks

It is immediately obvious from these data that the Santana Formation provides a unique sample of the rich and highly diverse late Gondwanan flora and fauna of this biogeographically critical region. Several of the taxa represented here are of immediate interest because they shed light on the early history of some extant groups with a Gondwanan distribution. Others seem more cosmopolitan, especially at family level. The majority of genera are known at present only from the Araripe basin, but some have wider distribution. The absence of data from other localities undoubtedly reflects preservational or collecting bias, which severely restricts the scope of any biogeographical analysis. In several instances, however, the fossil record suggests western Gondwanan endemism among some now-extinct groups.

Wiley's (1976:99) phylogenetic and biogeographic investigations of fossil and recent gars (Actinopterygii, Lepisosteidae) predicted that "the vicariant sister group of gars may be found among the fossil fauna of South America." The recent discovery of *Obaichthys,* a cladistically primitive gar from the Santana Formation (Wenz and Brito 1992) supports that contention and provides some support for a Gondwanan origin of this group, although its post-Mesozoic history is confined to North America and Europe. Wiley (1976) also set a minimum age for the living genera *Lepisosteus* and *Atractosteus* as Late Cretaceous. *Obaichthys* is potentially important for the information it will provide concerning the phylogenetic position and minimum age of gars within neopterygian fishes. Gars briefly survived in Africa and South America following their separation but apparently became extinct on both continents during the later Cretaceous.

Gondwanan amiids are rare; only three records are known, and two of these are from Brazil (*Calamopleurus* from the Santana Formation and *Amiidarum* from near Bahia). An Eocene amiid from Mali (being studied at the Natural History Museum, London) seems close to *Calamopleurus.* At least one group of amiids therefore migrated into Gondwana before it broke apart during the Early Cretaceous. These fishes survived this geologic event in Africa but became extinct there since the Eocene; we do not know their demise in South America after Lower Cretaceous times. It would be interesting to know whether *Amiidarum* is also allied to *Calamopleurus* or represents another lineage of gondwanan amiids.

The aspidorhynchid *Vinctifer* has been reported from several Lower Cretaceous (Aptian and Albian) localities in other parts of Brazil, notably from the Riachuelo and Muribeca formations (Sergipe-Alagoas basin; Silva Santos 1981, 1985) and from the Marizal Formation of the Tucano basin (Silva Santos 1972). *Vinctifer* has yet to be unequivocally demonstrated from African Cretaceous deposits, but Taverne (1969) has described a headless "*Belonostomus*" from the Lower Cretaceous of Equatorial Guinea that could in fact be *Vinctifer.* Recently there has been an attempt at intrabasinal correlation in Brazil, using *Vinctifer* as a marker fossil (Brito and Campos 1982; Brito 1984).

The dentition of *Vinctifer* is reduced in comparison with *Aspidorhynchus* and *Belonostomus.* There is a well-developed gill-raker basket in *Vinctifer,* suggesting it was a filter-feeder. Interestingly another toothless aspidorhynchid ("*Belonostomus*" *sweeti,* Etheridge and Woodward) is known from the Lower Cretaceous of Queensland; according to Molnar (personal communication, 1991) this form may also have been a filter-feeder. Perhaps there existed a specialized Early Cretaceous group of Gondwanan filter-feeding aspidorhynchids. *Vinctifer* also occurs in the Aptian of Venezuela (Maisey In prep.).

Cladocyclus is a predaceous halecostome fish known from the Lower Cretaceous of the Araripe, Sergipe, and Recôncavo basins of Brazil. In West Africa it is probably represented by "*Proportheus*" *kameruni*. *Cladocyclus* is morphologically close to *Chiromystus* (= *Itaparica*) from the Lower Cretaceous of Brazil (Silva Santos 1986), and to *Chirocentrites* from the later Cretaceous of Africa (Taverne 1986). Together, *Cladocyclus, Chiromystus,* and *Chirocentrites* probably form a monophyletic group of western Gondwanan ichthyodectiforms that did not survive the Cretaceous (Maisey 1991).

The generalized gonorynchiform genera *Dastilbe* and *Tharrhias* have been cited previously in discussions of Gondwanan biodiversity (e.g., Patterson 1975, 1984). Although *Dastilbe* has been formally recognized only in Brazil (where it is very widespread), it has been suggested that *Parachanos* from West Africa is identical (Silva Santos 1947; Taverne 1981; Patterson 1984). Taverne (1981) even went so far as to suggest that nominal species of *Dastilbe,* plus *Parachanos aethiopicus,* might represent separate populations of a single species; their predrift geographical relationship is certainly close. *Tharrhias* is known so far only from Brazil. It resembles *Dastilbe* closely in many morphological features, but the two genera are regarded as distinct. Although both occur in the Santana Formation, they are restricted to separate assemblages (*Dastilbe* to the Crato Member, *Tharrhias* to the Romualdo). From available field observations, these taxa are not sympatric, and are seemingly confined to different Lagerstätten, although this is not apparent from previous works that simply list the fossil taxa recovered from the Santana Formation as a whole. Gonorynchiforms are believed to be the sister group of otophysan fishes, including carps, minnows, tetras, catfishes, knifefishes, etc., many of whose present-day distributions are evocative of a western Gondwanan vicariant history (see Chapter 8). Upper Cretaceous characiform and siluriform otophysans have been reported from Brazil recently (Gayet and Brito 1989); Tertiary characiforms are also known (Schaeffer 1947). The absence of Lower Cretaceous records for these groups stands in contrast to their supposed Gondwanan origins based on Recent distributions.

Both coelacanth genera from the Santana Formation (*Mawsonia, Axelrodichthys*) are now reported from Africa. *Mawsonia* is widespread in both continents, with Neocomian records from Bahia (Mawson and Woodward 1907; Woodward 1908; Carvalho 1982) and Zaire (Casier 1961), Aptian examples from Araripe (Maisey 1986) and Niger (Wenz 1975), and Albian occurrences in Egypt and Morocco (Weiler 1935; Tabaste 1963) and perhaps from Algeria and Niger also (Wenz 1980). *Axelrodichthys* is presently described only from Araripe in Brazil (Maisey 1986) but has been reported recently from the Lower Cretaceous of Niger (Gee 1988). *Mawsonia* and *Axelrodichthys* are closely related, apparently forming a sister group to *Chinlea* from the Triassic of North America. Cladistically these coelacanths seem to be distant from the living *Latimeria.* Quite possibly *Mawsonia* and *Ax-*

elrodichthys are western Gondwanan endemics; *Mawsonia* seems initially to have survived the separation of South America from Africa, but to have become extinct before the Upper Cretaceous.

Pelomedusid turtles present a fascinating biogeographical conundrum. Early Cretaceous fossils occur in South America and Africa, and today the group is found in both continents and Madagascar, suggesting a vicariant history perhaps dating back prior to the separation of South America from Africa. Greater biogeographical complexity is nevertheless indicated by younger (Upper Cretaceous and Tertiary) pelomedusid fossils, which are presently known from every continent except Antarctica and Australia, and which include marine forms. Although Recent pelomedusids have a distribution that is strongly evocative of Gondwanan vicariance, pelomedusid phylogeny is still poorly resolved (Gaffney 1977, 1988) with many unplaced extinct taxa. It may be that the group is too large to be sensitive even to a major event like the separation of South America and Africa. With improved phylogenetic data, however, it is possible that some included taxa will be shown to have a vicariant history, possibly related to the breakup of western Gondwanaland. The present-day distribution pattern of pelomedusids should be treated with caution by biogeographers, however, as it is clearly a relic of an earlier, more cosmopolitan post-Gondwanan distribution.

The Santana Formation crocodilians are also of biogeographical interest. Two taxa have been described, the "notosuchid" *Araripesuchus* (Price 1959) and the "trematochampsid" *Caririsuchus* (Kellner 1987). *Araripesuchus* has been reported from the Lower Cretaceous of Niger (Buffetaut 1981). Other notosuchids, including forms with a distinctive mammal-like dentition, occur in the Early Cretaceous of Brazil and Malawi but have not yet been discovered in the Santana Formation. Supposed trematochampsids are recorded from several Cretaceous sites in South America and Africa. An earlier Cretaceous (Hauterivian) example is known from Patagonia (*Armargasuchus;* Chiappe 1989). Other trematochampsid records are considerably younger, including *Trematochampsa* from the Senonian of Niger and the Campanian of Madagascar (Buffetaut 1974; Buffetaut and Taquet 1979) and *Itasuchus* from the Maastrichtian of Brazil (Price 1955; Buffetaut 1985). Gasparini and co-workers (1991) are doubtful that *Hasuchus, Caririsuchus,* and *Amargasuchus* are trematochampsids. At best, all we can say is that trematochampsids may have been present in western Gondwanaland prior to its fragmentation and went on to survive in both South America and Africa during the Upper Cretaceous. There is disagreement (reviewed by Gasparini et al. 1991) over the phylogenetic relationships of trematochampsids and other "mesosuchians," making it difficult to discern any clear biogeographic pattern among these taxa.

It is premature to remark on Gondwanan pterosaur biogeography, except to note the incredible diversity of morphologically bizarre taxa from the Santana Formation. It seems impossible that such distinctive forms will not eventually be recognized elsewhere, particularly in Africa. Wellnhofer (1985, 1987) has referred some of the Santana pterosaurs to families known from the Cenomanian of England (Ornithocheiridae, Criorhynchidae), although this is disputed by others (e.g., Kellner 1989). Tapejarid pterosaurs may be related to azhdarchids from the Cretaceous of Asia (Kellner, personal communication, 1991).

The insect fossil record from the Crato Member of the Santana Formation is spectacular. More than 15 orders of insects (all living) are represented, and many of the taxa are interesting biogeographically.

Numerous Ephemeroptera are present (both as alates and nymphs), including the oldest fossil representatives of the families Oligoneuridae, Euthyplociidae, and possibly the Potamanthidae (McCafferty 1990). The Oligoneuridae and Euthyplociidae are pantropical today, except for some highly specialized genera. A widespread Gondwanan distribution, followed by a vicariant event toward the end of the Mesozoic, is consistent with the fossil and Recent records. According to McCafferty's (1990) phylogenetic hypothesis, the Heptageniidae (sister group of the Oligoneuridae) should be of Lower Cretaceous age as well, inferring even greater antiquity for the superfamily Heptagenioidea, to at least the Jurassic.

The occurrence of a potamanthid is unexpected, because the group has no previous fossil record and is today only holarctic and oriental in distribution. Its modern distribution may therefore be a relicit one. The presence of Ephemeridae is somewhat unexpected; this family is poorly represented in the Neotropics by a few forms today, with probably nearctic affinities. Siphloneuridae do not occur today in tropical South America or Africa, but are common in holarctic and transarctic areas; according to McCafferty (1990), their occurrence is therefore not unexpected here during the past. The extinct Mesozoic family Hexagenitidae was previously known only from Laurasia.

Among the Odonata from the Crato Member there are specimens apparently belonging to the superfamily Coenagrioidea, representing the first known Early Cretaceous coenagrioids and the first fossil members of the superfamily from South America. Gomphaeschnine anisopterans (strongly represented here) are otherwise known (living or fossil) only from the northern hemisphere. Another significant discovery is a near-mature macromiid nymph (distinctive for its long antennae and some other features). It is the first fossil macromiid and is also the first macromiid record outside of Africa.

The fossil isopteran *Meiatermes* represents only the fourth known Cretaceous termite genus, and the only Gondwanan one. It represents the first fossil record of the Hodotermitidae from the tropics (there are six Recent species).

The presence of fossil raphidiopterans (snake flies) in the insect assemblage is unexpected, because all Recent representatives and all previous reliable fossil records of the group are confined entirely to the northern hemisphere (there is a dubious fossil record from Chile; for further discussion, see Oswald 1990:157). The family Baissopteridae was probably broadly distributed throughout Laurasia and northern Gondwanaland during the Upper Jurassic and Lower Cretaceous; Oswald (1990:157) interpreted the Brazilian fossil occurrence as "a component of this ancestral fauna whose fossils have been displaced westward by the fragmentation of Gondwanaland."

Western Gondwanaland may have been

the primary area of evolution for many angiospermous plants during the Early Cretaceous, especially monocotyledonous groups, and it may represent the site of angiosperm origins (Raven and Axelrod 1974). Modern angiosperm distributions and Cretaceous palynological data are essentially compatible with that hypothesis (Doyle et al. 1982). Early Cretaceous angiosperm leaves are reported from Patagonia, suggesting rapid dispersal from a putative western Gondwanan center of origin (Romero and Archangelsky 1986). That view is now threatened by the description of Triassic angiosperm-like pollen from the eastern United States (Cornet 1989) and the discovery of undescribed parallel-veined angiosperm-like macrofossils from the Newark Series (Cornet, personal communication, 1991). The diverse paleoflora of the Santana Formation is dominated by ferns, cheirolepidiaceous gymnosperms, and ephedroids, but some angiosperm-like leaves have also been noted (Crane in Maisey 1991). As yet, no laminar, reticulate-veined leaves like those of extant dicotyledonous plants have been found in the Santana Formation. These occur in the Patagonian material, perhaps reflecting different climatic conditions from the presumably more arid Santana paleoenvironment. Reticulate-veined angiosperms were certainly well established elsewhere in Gondwanaland during Santana Formation times. The presence of an essentially modern dipteridacean fern (Babinsky, personal communication, 1991) is interesting; it is the first Cretaceous record for Brazil but is apparently the last for South America (Archangelsky, personal communication, 1991). This group was cosmopolitan during the Mesozoic but is today restricted to Asia (China). The example from Araripe is morphologically closer than other Mesozoic examples to modern dipteridaceans, suggesting that the latter represent a relict Pangaean distribution.

IV. Tectonic and Geographic Relationships

Pitman and co-workers (Chapter 2) have reviewed current hypotheses concerning the breakup of South America and Africa. The rich fossil assemblages of the Santana Formation provide an unparalleled opportunity to investigate the impact of this tectonic event on late Gondwanan biota.

Beginning with the development of successor troughs and intraplate domes and sags during the Mid to Late Triassic, crustal extension and associated faulting and vulcanicity continued through the Late Jurassic. During the Early Cretaceous, large-scale drifting occurred, accompanied by epicontinental rifting and sagging in both South America and Africa. Separation commenced earlier in the south than in the north. According to Szatmari and co-workers (1987) rotational drifting initially occurred around a pole located in the northeasterly part of Brazil, producing a fanlike South Atlantic basin that tapered northward. Rabinowitz and LaBrecque (1979) estimated that this initial rifting between South America and Africa lasted almost 30 million years, from the middle Valanginian to the middle Albian.

A succession of prerift, corift and synrift tectonic phases has been recognized in each of three tectonic-sedimentary domains (Popoff 1988). The corresponding tectonic phases of Popoff's domains are not of equivalent ages; those farther south (in the "Austral Domain," extending northward to approximately the level of the Porto Alegre and Walvis-Mombasa discontinuities) predate their equivalent phases of the "Tropical Domain" (extending from these discontinuities northward to about the level of the Pernambuco and Ngaoundere lineaments), which in turn predate their equivalent tectonic phases in the more northerly "Equatorial Domain." The most significant aspects of this model (at least, for biogeographical purposes) are that (a) continental separation is polyphased within each domain and (b) separation events are diachronous from domain to domain; in Popoff's (1988:409) words, current data "are contrary to the earlier views of rigid plates in the context of continental reconstructions." At approximately 20-million-year intervals, rifting-drifting was apparently propagated from south to north within successive domains, with differential deformation between domains being accommodated by northeast-southwest-trending transfer fault systems. Dewey and Burke (1974) suggested an analogous situation in Africa, with temporary extension between two plates along the Benue rift zone (see Chapter 2).

At the local level, there is further evidence for more complex tectonic-sedimentological histories within Brazilian Cretaceous marginal basins than was previously supposed. For example, the Tucano and Jatobá basins of northeast Brazil were originally interpreted as a simple system, connected since the beginning of rifting, and infilled by a fluvio-deltaic-lacustrine system that prograded from south to north. They have now been reinterpreted as a series of half-grabens, each with a disjunct depocenter, each having evolved separately, and each isolated (like the larger tectonic realms) by transfer fault systems. Magnavita and Cupertino (1988) concluded that these fault systems are controlled by basement anisotropy and that the Tucano and Jatobá basins may reflect underlying crustal stretching in a northwest-southeast direction, leading to the establishment of an intracontinental rift that was aborted during the final stages of separation between Africa and South America.

The tectonic setting of the Araripe basin is critical. In all probability, its development is also controlled in part by the stretching of anisotropic basement, and this basin perhaps represents a northerly extension (or even the limit) of this failed Tucano-Jatobá intracontinental rift system. The Araripe basin is separated from the Tucano-Jatobá complex by a major tectonic feature, the Pernambuco lineament, and is itself delimited on its northern side by the Patos lineament. The Araripe basin cannot be regarded as a tectonically and sedimentologically isolated feature, as there are important structural and early (Jurassic–Early Cretaceous) depositional similarities with the Tucano-Jatobá complex, and in particular with the more northerly Jatobá sub-basin. It is tempting to interpret the Araripe basin as lying at a failed triple junction, between the Patos and Pernambuco lineaments and the Tucano-Jatobá rift system.

The question is whether the series of major tectonic events that led to the separation of South America and Africa can be shown to have directly affected the subsequent evolutionary and biogeographical history of taxa. We are only just beginning to tie what is known about the biota represented by the Santana Formation assemblages into a broader tectonic and biogeographic picture. Available data are often too general to give detailed biogeographic resolution, a situation that can be remedied only by controlled sampling and analysis (something that has been sadly lacking in the past). We can nevertheless discern a speciose and diverse Aptian-Albian biota in northeastern Brazil that contains many apparently endemic taxa. The fossil record provides a glimpse at late Gondwanan biodiversity at a critical time in its history as the equatorial Atlantic seaway became established. We can observe the impact this may have had on some groups of organisms, albeit somewhat generally. Groups such as pelomedusid turtles seem too large and diverse to display any sensitivity to such a major geological event. Other groups with a more convincing Gondwanan Cretaceous history may indeed have been affected by tectonic events. Our present state of knowledge does not permit us to resolve the scale of the geological events suspected to be responsible, however. Potentially important local vicariant events (e.g., resulting from intra- or interbasinal tectonism) might easily be overinterpreted as "Gondwanan breakup" events. Post-Gondwanan dispersals, brought about by secondary ephemeral geological events, may also confuse the issue.

V. Conclusions

Several observations may be made concerning the biogeographical significance of the biota represented by the Santana Formation:

1. The majority of fossil taxa are biogeographically uninformative, given present states of knowledge, because at generic level they seem endemic (perhaps falsely, reflecting a strong positive preservational bias at this locality), whereas at a more general (e.g., familial) level many of the taxa are cosmopolitan (e.g., most fishes, insects).

2. Some of the fossil taxa are biogeographically informative concerning Recent descendants (e.g., gars, gonorynchiform fishes, families of Ephemeroptera, Odonata). The known Lower Cretaceous distribution may indeed be Gondwanan, but we should still be cautious in accepting that the assumedly vicariant distribution of the descendant taxa is the product of an Early Cretaceous event; it may be tied in to later (Cretaceous or Tertiary) known geological events.

3. There are additionally several Gondwanan fossil taxa that now belong to extinct lineages, whose distribution may be a product of an early Cretaceous vicariant event. Examples include *Calamopleurus*-like amiids, *Vinctifer*-like aspidorhynchids, the *Cladocyclus-Chiromystus-Chirocentrites* group, the *Mawsonia-Axelrodichthys* group, plus notosuchid and trematochampsid crocodilians. In general, such examples are more readily identifiable than is the case for extant groups with Gondwanan distributions.

4. Some groups, notably pelomedusid turtles, have Recent and Early Cretaceous distributions consistent with a Gondwanan vicariant model, but their intervening fossil record is not in accord with that hypothesis.

5. Many Recent taxa with a modern distribution pattern that is evocative of Gondwanan ancestry are absent from the Early Cretaceous records. "Absence of evidence is not evidence of absence," but it is not evidence of presence, either! In certain cases, the fossil record suggests a much later bio-

geographical event. Geological data are not inconsistent with this possibility, and several postrifting-postdrifting opportunities for vicariance or dispersal apparently existed within the region, well into the Tertiary. *By simply lumping together all Recent "Gondwanan" taxa, we run the risk of creating the biogeographical equivalent of a paraphyletic group in need of further resolution into several individual vicariant events within one region.*

6. The responses of various taxa to events that occurred during the separation of South America and Africa are different. For example, several vertebrate groups survived the event on one or both continents for only a short while, becoming extinct before the Tertiary. Most insect families survived to the present day, however. It would be interesting to understand the factors responsible for such discrepancies; presumably ecological, climatic, and geographic factors are involved. These factors may have been influenced by the break-up and separation of South America and Africa, especially after the equatorial Atlantic seaway was fully developed. But that is a far cry from invoking a vicariant "Gondwanan breakup event" as the culprit.

Appendix

Fossils from the Santana Formation

Taxa	Horizon
FISHES	
Chondrichthyes	
Elasmobranchii	
Hybodontidae	
Tribodus limae	R
Rhinobatidae	
"*Rhinobatos*" *beurleni*	R
Osteichthyes	
Actinopterygii	
Semionotidae	
Araripelepidotes temnurus	R, C
Lepisosteidae	
Obaichthys decoratus	R
Pycnodontiformes	
Neoproscinetes penalvai	R
Temanja palma	R
Amiidae	
Calamopleurus cylindricus	R
Ionoscopidae	
Oshunia brevis	R
Oshunia new sp.	R
Aspidorhynchidae	
Vinctifer comptoni	R
Ichthyodectiformes	
Cladocyclus gardneri	R, C
Cladocyclus ferus	?
Araripichthyidae	
Araripichthys castilhoi	R
Albuloidei	
Brannerion latum	R
Brannerion new sp.	R
Paraelops cearensis	R
Pachyrhizodontidae	
Rhacolepis buccalis	R
Notelopidae	
Notelops brama	R
Notelops new sp.	R
Euteleostei incertae sedis	
Santanichthys diasii	R
Gonorynchiformes	
Dastilbe elongatus	C
Dastilbe crandalli	C
Tharrhias araripis	R
Clupeiformes incertae sedis	
Clupeiform new g. & sp.	R
Sarcopterygii	
Coelacanthidae	
Axelrodichthys araripensis	R
Mawsonia cf. *gigas*	R
TETRAPODS	
Lissamphibia	
indet. anuran	C

Taxa	Horizon
Testudines	
Pelomedusidae	
Araripemys barretoi	R
pelomedusid, new g. & sp.	R
indet.	C
Crocodylia	
Notosuchidae	
Araripesuchus gomesii	R
Trematochampsidae	
Caririsuchus camposi	R
Pterosauria	
indet.	C
? Ornithocheiridae	
Araripesaurus castilhoi	R
Brasileodactylus araripensis	R
Santanadactylus brasilensis	R
Santanadactylus araripensis	R
Anhangueridae	
Anhanguera blittersdorffi	R
Anhanguera santanae	R
? Criorhynchidae	
Tropeognathus mesembrinus	R
Family indet.	
Araripedactylus dehmi	R
Cearadactylus atrox	R
Tapejaridae	
Tapejara wellnhoferi	R
Tupuxuara longicristatus	R
Dinosauria	
indet.	R
Aves	
indet. (feather)	C
INSECTS (all from Crato Member)	
Ephemeroptera	
indet. 1	
indet. 2	
Hexagenitidae	
Protoligoneuria limai	
Siphloneuridae	
Siphgondwanus occidentalis	
indet. 1	
indet. 2	
indet. 3	
Oligoneuriidae	
Colocrus indivicum	
Ephemeridae	
Australiphemera revelata	
Microphemera neotropica	
Potamanthidae	
indet.	
Leptophlebiidae	
indet. 1	
indet. 2	
indet. 3	
indet. ephemeroid 1	
indet. ephemeroid 2	
Odonata	
Zygoptera	
Pseudostigmatidae	
Euarchistigma atrophium	
Protoneuridae	
Eoprotoneura hyperstigma	
Anisoptera	
Gomphidae	
Cordulagomphus tuberculatus	
C. fenestratus	
C. santanensis	
Gomphaeschnaoides obliqua	
Aeschnidiidae	
Wightonia araripina	
Pseudomacromiidae	
Pseudomacromia sensibilis	
Dermaptera	
Cretolabia cearae	
Phasmatodea	
Cretophasmatidae	
Cretophasma araripensis	
Isoptera	
Hododermitidae	
Meiatermes araripina	
Homoptera	
Jascopidae	
Paracarsonus aphrodoides	
Platyjassites inflatifrons	
Platyjassites sp.	
Cicadellidae	
Proerrhomus rugosus	
Proerrhomus sp. A	
Proerrhomus sp. B	
Ovojassus concavifer	
O. minor	

(*continued*)

Appendix. ***Continued.***

Taxa	Horizon
Hallex xestocephalus	
H. gongrogony	
H. brevipes	
H. laticeps	
H. gracilior	
H. sp.	
Cicadoidea	
Cicadoprosbolidae	
Architettix compacta	
Cercopoidea	
Cercopionidae	
Cercopion reticulata	
Aleyrodoidea	
Boreoscytidae ?	
Megaleurodes megocellata	
Fulgoroidea	
Achilidae	
Acixiites immodesta	
A. costalis	
Fennahia cretacea	
Cixiidae ?	
indet.	
Lalacidae	
Protodelphax miles	
P. macroceps	
P. rhinion	
P. chamus	
P. sp.	
Ancorale flaccidum	
A. aschemon	
A. sp.	
Kinnarocixius quassus	
K. sp.	
Lalax mutabilis	
Lalax sp.	
Patulopes setosa	
P. myndoides	
P. sp.	
Carpopodus difficilis	
C. sp. A	
C. sp. B	
Psestocixius fuscus	
P. delphax	
Vulcanoia membranosa	
V. apicalis	
V. acuceps	
V. sp.	
Heteroptera	
Belostomatidae	
indet.	
Pentatomorpha	
indet.	
Gerromorpha (?)	
indet.	
Naucoridae	
indet.	
Corixidae	
indet.	
Peloridiomorpha	
indet.	
Saldidae	
indet.	
Reduviidae	
indet.	
Cimicomorpha	
indet.	
Blattodea	
Blattidae	
Mesoblattinopsis schneideri	
Raphidiomimidae	
indet.	
Orthoptera	
indet.	
Caelifera	
indet.	
Ensifera	
Eneopteridae	
Araripegryllus camposae	
A. nanus	
A. marianoi	
A. femininus	
A. spinosus	
A. serrilhatus	
A. megacephalus	
Santanagryllus hesselae	
Brontogryllus excelsus	
Gryllidae	
Caririgryllus elongatus	
C. pilosus	
C. arthaudi	
C. mesae	
Cratogryllus pentagonalis	

Taxa	Horizon
C. guimaraesae	
Cearagryllidae	
Cearagryllus monstruosus	
C. robustus	
C. gorochovi	
C. perforatorius	
C. poliacanthus	
C. microcephalus	
Myrmecophilidae	
Araripemyrmecophilops gracilis	
Gryllotalpidae	
Palaeoscapteriscops cretacea	
Archaeogryllotalpoides ornatus	
Tridactylidae	
indet.	
Tettigonioidea	
Prophalangopsidae	
Kevania araripensis	
Cretohaglopsis santanaensis	
Phasmomimella araripensis	
Elcanidae	
Cratoelcana zessini	
C. damianii	
Acridoidea	
Archaeopneumoridae	
Archaeopneumoridae cretacea	
Bouretidae	
Bouretia elegans	
Neuroptera	
Chrysopidae	
Limaia conspicua	
Araripechrysa magnifica	
Caririchrysa criptovenata	
C. confusa	
Myrmeleontidae	
Caldasia cretacea	
Blittersdorffia pleoneura	
B. volkheimeri	
Araripeneura regia	
A. gracilis	
Pseudomyphes araripensis	
Caririneura microcephala	
Ascalaphidae	
Cratopteryx robertosantosi	
Nymphidae	
Babinskaia pulchra	
B. formosa	
Neha maculata	
Alloneura acuminata	
Nemopteridae	
Roesleria exotica	
Megalopteryx audax	
M.? robusta	
Rhaphidioptera	
Baissopteridae	
Baissoptera brasiliensis	
Coleoptera	
Pyrochroidae	
Cretaceimelittommoides cearensis	
Dytiscidae	
indet.	
Staphylinidae	
indet.	
Passalidae	
indet.	
Nitidulidae	
? indet.	
Curculionoidea	
indet.	
Carabidae	
indet.	
Scarabaeidae	
indet.	
Diptera	
Tipuloidea	
indet. Tipulidae	
Bibionoidea	
indet. Bibionidae	
Sciaroidea	
indet.	
Mycetophilidae	
indet. Macrocerinae	
Chironomoidea	
indet. Mesotendipedidae	
Asiloidea	
Araripogon axelrodi	
Hymenoptera	
Symphyta	
Anaxyelidae	
Prosyntexis gouleti	
Apocrita	
Ephialtitidae	
Karataus kourios	

(*continued*)

Appendix. ***Continued.***

Taxa	Horizon
Proctotrupidae	
Proctoprocto asodes	
Mesoserphidae	
indet.	
Aculeata	
Tiphiidae	
Architiphia rasnitsyni	
Rhopalosomatidae	
Mesorhopalosoma cearae	
Sphecidae	
Cretosphex parvus	
C. magnus	
indet. sphecid	
indet. Aculeata	
Formicidae	
Cariridris bipetiolata	
Lepidoptera	
Micropterygidae	
Parasabatinica caldasae	
indet.	
Undopterix caririensis	
Gracilepterix pulchra	
Trichoptera	
indet.	
OTHER ARTHROPODS	
Arachnida	
Araneae	
g. & sp. indet.	C
Scorpionida	
Araripescorpius ligabuei	C
Crustacea	
Ostracoda	
Pattersoncypris micropapillosa	R
Copepoda	
Kabatarina pattersoni	R
Branchiopoda	
indet. anostracan	R
indet. conchostracan	C
Malacostraca	
Natantia	
indet. palaemonids	R, C
Brachyura	
Araripecarcinus ferreirai	R

Taxa	Horizon
MOLLUSCS	
Gastropoda	
Turritellidae	
Craginia araripensis	
Gymnentone romualdoi	
Turritella lucianoi	
Turritella sp.	
Mesalia sp.	
Glauconia sp.	
Naticidae	
Natica sp.	
Polinices sp.	
Lunatia sp.	
Ampullina sp.	
Epitoniidae	
Epitonium smalli	
Scalidae	
Turriscala sp.	
Scala sp.	
Cerithiidae	
Cerithium sp. cf. *C. austinense*	
Hemicerithium? sp.	
Aporrhaidae	
Aporrhais sp.	
Neritidae	
Neritoma sp.	
Pelecypoda	
Pteriidae	
Eupteria sp. cf. *E. flicki*	
Pteria sp. A	
Pteria sp. B	
Isognomonidae	
Aguileria sp.	
Anomiidae	
Anomia sp.	
Mytilidae	
Brachidontes sp.	
Crassatellidae	
Crassatella sp.	
Ostreidae	
Ostraea sp.	
Corbulidae	
Corbula sp. A	
Corbula sp. B	
Veneridae	
Venus sp.	

Taxa		Horizon
Unionidae		
indet.		
Cyrenidae		
indet.		
ECHINODERMATA (from marine horizon high in Romualdo Member)		
Echinoidea		
Clypeidae		
	Pygurus tinocoi	
Faujasidae		
	Pygidiolampas araripensis	
PLANTS		
	Pteridophyta indet.	C
	Dipteridaceae indet.	C
	Brachyphyllum obesum	R
	B. castilhoi	R
	indet.	C
	Araucarites vulcanoi	C
	Podozamites lanceolatus	
	Choffatia francheti	
	Nymphaeites choffati	
	indet. gnetales	C
	indet. angiosperm	C

Note: R = Romualdo Member; C = Crato Member. Some identifications of plants are tentative in view of ongoing investigations; see Duarte 1985, 1989. Palynological records are incomplete; for lists see Lima 1978, 1979, 1980; Doyle et al. 1982; Pons el al. 1990.

Literature Cited

Agassiz, L. 1841. On the fossil fishes found by Mr. Gardner in the Province of Ceará, in the north of Brazil. Edin. New Phil. J. 30:82–84.

———. 1844. Sur quelques poissons fossiles du Brésil. Compt. Rend. Acad. Sci. Paris 18:1007–1015.

Allison, P. A., and E. G. Briggs. 1991. The taphonomy of soft-bodied animals. Pp. 120–140 *in* S. T. Donovan (ed.), *The Process of Fossilization*. London: Belhaven Press.

Beurlen, K. 1962. A geologia da Chapada do Araripe. An. Acad. brasil. Ci. 34(3):365–370.

———. 1963. Geologia e estratigrafia da Chapada do Araripe. XVII Congr. brasil. Geol., Recife, pp. 1–47.

———. 1971. As condições ecológicas e faciológicas da formação Santana na Chapada do Araripe (Nordeste do Brasil). An. Acad. brasil. Ci. 43 (Suppl.):411–415.

Braun, O. P. G. 1966. Estratigrafia dos sedimentos da parte interior da regiao Nordeste do Brasil (Bacias de Tucano—Jatoba, Mirandiba e Araripe). D.N.P.M./Div. Geol. Miner., Bol., No. 236:1–15.

Brito, I. M. 1984. The Upper Lower Cretaceous in Brazil, its divisions and boundaries. An. Acad. brasil. Ci. 56:287–293.

Brito, I. M., and D. A. Campos. 1982. O Cretáceo no Brasil. An. Acad. brasil. Ci. 54:197–218.

Buffetaut, E. 1974. *Trematochapsa taqueti,* un crocodilien nouveau du Sénonien inférieur du Niger. Compt. Rend. Acad. Sci. Paris 279:1749–1752.

———. 1981. Die biogeographische Geschichte der Krokodilier, mit Beschreibung einer neuen Art, *Araripesuchus wegeneri*. Geol. Rundschau 70(2):611–624.

———. 1985. Présence de Trematochampsidae (Crocodylia, Mesosuchia) dans le Crétacé supérieur du Brésil. Implications paléobiogéographiques. Compt. Rend. Acad. Sci. Paris, 2d ser. 301:1221–1224.

Buffetaut, E., and P. Taquet. 1979. An early Cretaceous terrestrial crocodilian and the opening of the South Atlantic. Nature 280:486–487.

Carvalho, M. S. S. de. 1982. Ó gênero *Mawsonia* na ictiofáunula do Cretáceo do Estado da Bahia. An. Acad. brasil. Ci. 54(3):519–539.

Casier, E. 1961. Materiaux pour la fauna ichthyologique Eocretacique du Congo. Ann. Mus. Roy. Afr. Centr., Tervuren, ser. 8, Sci., géol. 39:1–96.

Chiappe, L. M. 1989. A new trematochampsid crocodile from the Early Cretaceous of northwestern Patagonia, Argentina and its palaeobiogeographical and phylogenetic implications. Cretaceous Res. 9:379–389.

Cornet, B. 1989. Late Triassic angiosperm-like pollen from the Richmond Rift Basin of Virginia, USA. Palaeontographica Abt. B, 213:37–87.

Dewey, J. F., and K. Burke. 1974. Hot spots and continental breakup: Implications for collisional orogeny. Geol. 2:57–60.

Doyle, J. A., S. Jardiné, and A. Doernenkamp. 1982. *Afropollis*, a new genus of early angiosperm pollen, with notes on the Cretaceous palynostratigraphy and paleoenvironments of northern Gondwana. Bull. Cent. Rech. Explor.-Prod. Elf-Aquitaine 6(1):39–117.

Duarte, L. 1985. Vegetais fosseis da Chapada do Araripe. An. VIII Congr. Brasil. Paleontol., Rio de Janeiro, pp. 557–568.

———. 1989. Remains of the Lower Cretaceous plants from north east of Brazil. An. XI Congr. Brasil. Paleontol., Curitiba, pp. 219–233.

Gaffney, E. S. 1977. The side-necked turtle family Chelidae: A theory of relationships using shared derived characters. Amer. Mus. Novit. 2620:1–28.

———. 1988. A cladogram of the pleurodiran turtles. Acta Zool. Cracov. 31(15):487–492.

Gasparini, Z., L. M. Chiappe, and M. Fernandez. 1991. A new Senonian peirosaurid (Crocodylomorpha) from Argentina and a synopsis of the South American Cretaceous crocodilians. J. Vert. Paleontol. 11(3):316–333.

Gayet, M., and P. M. Brito. 1989. Ichthyofaune nouvelle du Crétacé supérieur du Groupe Baurn (États de Sâo Paulo et Minas Gerais, Brésil). Geobios No. 22(b):841–847.

Gee, H. 1988. Cretaceous unity and diversity. Nature 332:487.

Grimaldi, D. A., and J. G. Maisey. 1990. Introduction. Pp. 5–14 *in* D. A. Grimaldi (ed.), *Insects from the Santana Formation, Lower Cretaceous of Brazil*. Bull. Amer. Mus. Nat. Hist. No. 195.

Jordan, D. S. 1923. Peixes cretaceos do Ceará e Piauhy. Monogr. Serv. geol. min. Brasil, Rio de Janeiro 3:1–97.

Kellner, A. W. A. 1987. Ocorrência de um novo crocodiliano no Cretáceo Inferior da Bacia do Araripe, Nordeste do Brasil. An. Acad. brasil. Ci. 59(3):219–232.

———. 1989. A new edentate pterosaur of the Lower Cretaceous from the Araripe Basin, northeast Brazil. An. Acad. brasil. Ci. 61(4): 439–446.

Larson, R. L., and J. W. Ladd. 1973. Evidence for the opening of the South Atlantic in the Early Cretaceous. Nature 246:210–212.

Lima, M. R. de. 1978. Palinologia da Formação Santana (Cretáceo do Nordeste do Brasil). I. Introdução geológica e descrição sistemática dos esporos da Subturma Azonotriletes. Ameghiniana 15:333–365.

———. 1979. Paleontologia da Formação Exu, Cretáceo do Nordeste Brasil): Estagio Actual de Conhecimentos. An. Acad. brasil. Ci. 51:3.

———. 1980. Palinologia da Formação Santana (Cretáceo do Nordeste do Brasil). III. Descrição sistemática dos esporos da Turma Plicates (Subturma Costates). Ameghiniana 17:15–47.

McCafferty, W. P. 1990. Ephemeroptera. Pp. 20–50 *in* D. A. Grimaldi (ed.), *Insects from the Santana Formation, Lower Cretaceous of Brazil*. Bull. Amer. Mus. Nat. Hist. No. 195.

McKenna, M. C. 1981. Early history and biogeography of South America's extinct land mammals. Pp. 43–77 *in* R. L. Ciochon and A. B. Chiarelli (eds.), *Evolutionary Biology of the New World: Monkeys and Continental Drift*. New York: Plenum.

Magnavita, L. P., and J. A. Cupertino. 1988. A new approach to the geological configuration of the Lower Cretaceous Tucano and Jatoba basins, northeastern Brazil. Rev. bras. Geoc. 18(2):222–230.

Maisey, J. G. 1986. Coelacanths from the Lower Cretaceous of Brazil. Amer. Mus. Novit. 2866:1–30.

———. 1991. (ed.). *Santana Fossils: An Illustrated Atlas*. Neptune, N.J.: TFH Publications.

Mawson, J., and A. S. Woodward. 1907. On the Cretaceous Formation of Bahia (Brazil), and on vertebrate fossils collected therein. Quart. J. Geol. Soc., London, 63:128–139.

Moraes, J. F. S., F. C. Barros, and E. Ramos. 1963. *Reconhecimento Fotogeológico da Região Nord-este do Brasil.* D.N.P.M. Fôlha Crato SB240, Esc. 1:250,000. Rio de Janeiro: Div. Fom. Prod. Mineral.

Moraes, J. F. S., J. da Silva Santos, and J. de C. Masarenhas. 1976. *Projeto Santana. Relatório Final da Etapa I,* vol. 1. Rio de Janeiro: C.P.R.M./D.M.P.M.

Olsen, P. E. 1991. Tectonic, climatic, and biotic modulation of lacustrine ecosystems: Examples from Newark Supergroup of Eastern North America. Pp. 209–224 *in* B. J. Katz (ed.), *Lacustrine Basin Exploration: Case Studies and Modern Analogs.* A.A.P.G. Mem. 50. Tulsa: American Association of Petroleum Geologists.

Oswald, J. D., 1990. Raphidioptera. Pp. 154–163 *in* D. A. Grimaldi (ed.), *Insects from the Santana Formation, Lower Cretaceous of Brazil.* Bull. Amer. Mus. Nat. Hist. No. 195.

Patterson, C. 1975. The distribution of Mesozoic freshwater fishes. Mém. Mus. Natn. Hist. Nat., n.s. (A) Zool. 88:156–174.

———. 1984. Family Chanidae and other teleostean fishes as living fossils. Pp. 132–139 *in* N. Eldredge and S. M. Stanley (eds.), *Living Fossils.* New York: Springer Verlag.

Pons, D., P. Y. Berthou, and D. A. Campos. 1990. Quelques observations sur la palynologie de l'Aptien Supérieur et de l'Albien du bassin d'Araripe (NE du Brésil). Actas 1 Simpósio bacia do Araripe 1990:241–252.

Popoff, M. 1988. Du Gondwana à l'Atlantique sud: Les connexions du fossé de las Bénoué avec les bassins du Nord-Est brésilien jusqu'a l'overture du golfe de Guinée au Crétacé inférieur. Pp. 409–431 *in* J. Sougy and J. Rodgers (eds.), *The West African Connection.* J. Afr. Earth Sci., spec. publ., 7(2).

Price, L. I. 1955. Novos crocodilideos dos arenitos da Série Bauru, Cretáceo do Estado de Minas Gerais. An. Acad. brasil. Ci. 27(4): 487–498.

———. 1959. Sôbre um crocodilideo notosúquio do Cretácico Brasilèiro. Bol. Div. Geol. Mineral. 188:1–155.

Rabinowitz, P. D., and J. LaBrecque. 1979. The Mesozoic South Atlantic Ocean and evolution of its continental margin. J. Geophys. Res. 84:5973–6002.

Raven, P. H., and D. I. Axelrod. 1974. Angiosperm biogeography and past continental movements. Ann. Missouri Bot. Gard. 61: 539–673.

Reyment, R. A., and R. V. Dingle. 1987. Palaeography of Africa during the Cretaceous Period. Palaeogeogr. Palaeoclimatol. Palaeoecol. 59:93–116.

Romero, E. J., and S. Archangelsky. 1986. Early Cretaceous angiosperm leaves from southern South America. Nature 234:1580–1582.

Schaeffer, B. 1947. Cretaceous and Tertiary actinopterygian fishes from Brazil. Bull. Amer. Mus. Nat. Hist. 89(1):5–39.

Schlische, R. W., and P. E. Olsen, 1990. Quantitative filling model for continental extensional basins with applications to early Mesozoic rifts of eastern North America. J. Geol. 98(2):135–155.

Seilacher, A., W. E. Reif, and F. Westphal. 1985. Sedimentological, ecological and temporal patterns of fossil Lagerstätten. Philos. Trans. Roy. Soc. London, B 311:5–23.

Silva, M. A M. da. 1986a. Lower Cretaceous sedimentary sequences in the Araripe basin, northeastern Brazil: A revision. Rev. Bras. Geoc. 16(3):311–319.

———. 1986b. Lower Cretaceous unconformity truncating evaporite-carbonate sequence, Araripe basin, northeastern Brazil. Rev. Bras. Geoc. 16(3):301–310.

Silva Santos, R. da. 1947. Uma rediscrição de *Dastilbe elongatus,* com algumas considerações sobre o genero *Dastilbe.* D.G.M./ D.N.P.M., Notas Prelim. Estudos, 42:1–7.

———. 1972. Peixes da Formação Marizal, Estado da Bahia. Ph.D. diss., University of São Paulo.

———. 1981. Sobre a occurência de uma fáunula de peixes na Formação Riachuelo, estado de Sergipe. An. Acad. brasil. Ci. 53(1):203 (summary).

———. 1985. Sobre a presença do *Vinctifer* Jordan (Pisces, Aspidorhynchiformes) na Formação Muribeca, Estado de Alagoas. VIII Congr. brasil. Paleont., Rio de Janeiro, 1983, M.M.E.-D.N.P.M., Ser. Geol., No. 27, Paleontol. Estrat. 2:147–150.

———. 1986. Posição taxinômica do *Cladocyclus woodwardi* (Silva Santos) do Cretáceo do Brasil. An. Acad. brasil. Ci. 58(2):229–231.

Silva Santos, R. da., and L. G. Valença. 1968. A Formação Santana e sua paleoictiofauna. An. Acad. brasil. Ci. 40(3):339–360; 40(4):491–497.

Simpson, E. S. W. 1977. Evolution of the South Atlantic. Geol. Soc. S. Afr., Alex L. du Toit Mem. Lect., Annex. 80(15):1–15.

Small, H. 1913. Geologia e suprimento de água subterrânea no Ceará e parte do Piaui. Insp. Obras Contra Secas., Ser. Geol. 25:1–180.

Szatmari, P., J. B. L. Françolin, O. Zanotto, and S. Wolff. 1987. Evolução tectônica da margem equatorial Brasileira. Rev. bras. Geoc. 17(2):180–188.

Tabaste, N. 1963. Etude de restes de Poissons du Crétacé Saharien. Mem. IFAN 68:437–485.

Taverne, L. 1969. Sur la présence d'un Aspidorhynchidae (Pisces Holostéens, ordre des Aspidorhynchiformes) dans les terrains éocrétaciques de la Guinée Equatoriale. Rev. Zool. Bot. Afr. 79:3–4.

———. 1981. Ostéologie et position systématique d'*Aethalionopsis robustus* (Pisces, Teleostei) du Crétacé inférieur de Bernissart (Belgique) et considerations sur les affinities des Gonorhynchiformes. Bull. Acad. Roy Belg. Cl. Sci., Sér. 5, 67:958–982.

———. 1986. Ostélogie et affinités systématiques de *Chirocentrites vexillifer* du Crétacé Supérieur de la Mésogée Eurafricaine: Considerations sur la phylogénie des Ichthyodectiformes, poissons téléostéens du Jurassique et du Crétacé. Ann. Roy. Soc. Zool. Belg. 116(1):33–54.

Weiler, W. 1935. Ergebnisse der Forschungsreisen Prof. E. Stromers in den Wüsten Agyptens. II. Wirbeltierreste de Baharije-Stufe, 16 Neue Untersuchungen an den Fischresten. Abh. Bayer. Acad. Wissen. Math.-Nat. Abt., n.s., 32:57.

Wellnhofer, P. 1985. Neue pterosaurier aus der Santanaformation (Apt.) der Chapada do Araripe, Brasilien. Palaeontographia, Abt. A, 187:105–182.

———. 1987. New crested pterosaurs from the Lower Cretaceous of Brazil. Mitt. Bayer. Staatsslg. Paläont. Hist. Geol. 27:175–186.

Wenz, S. 1975. Un nouveau Coelacanthidé du Crétacé inférieur du Niger, remarques sur la fusion des os dermiques. Colloq. Internat. C.N.R.S., Probl. Actu. Paléont.-Evol. Vert., 218 (for 1973):175–190.

———. 1980. A propos du genre *Mawsonia*, Coelacanthe géant du Crétacé inférieur d'Afrique et du Brésil. Mem. Soc. Geol. Fr., n.s., 139:187–190.

Wenz, S., and P. M. Brito. 1992. Première découverte de Lepisosteidae (Pisces, Actinopterygii) dans le Crétacé inférieur de la Chapada do Araripe (N-E du Brésil). Conséquences sur la phylogénie des Ginglymodi. C. R. Acad. Sci., Paris, 314(II):1519–1525.

Wiley, E. O. 1976. The phylogeny and biogeography of fossil and recent gars (Actinopterygii: Lepisosteidae). Mus. Nat. Hist. Univ. Kansas Misc. Publ. 64. Lawrence: University of Kansas Museum of Natural History.

Woodward, A. S. 1908. Cretaceous fishes from Brazil. Quart. J. Geol. Soc., London 64:358–362.

Chapter 15

Evolution of African Climate

Daniel A. Livingstone

I. Introduction

Much of our understanding of the evolution of continental climates comes from lake sediment or from historical records, tree rings, bogs, or glacier ice. The latter sources of information are less generally available in Africa, which is, however, splendidly provided with diverse and ancient lakes. Long-term information about the past climates of Africa comes very largely from lake sediments, with a smaller amount from bones, land-forms, and cave deposits. Although the nilometers with which Egyptians recorded water level on the Nile provide what is probably the longest run of historical climatic data in the world, historical records of climate are few and short for most of Africa.

Lakes Tanganyika, Malawi, Turkana, and Mobutu are among the oldest in the world. Seismic investigation shows that Tanganyika contains several thousand meters of sediment (Rosendahl 1988; Rosendahl and Livingstone 1983). Making allowance for dewatering and applying the late-Quaternary rate of sedimentation to this thickness suggests that it may be well more than 20 million

Research leading to this chapter was supported by grants from NSF, ORSTOM and CNRS and by a Guggenheim Fellowship. The chapter itself was prepared without support. My ideas about the evolution of tropical climate have been influenced by discussion and correspondence with many colleagues, especially, in recent years, Margaret Bolick, Raymonde Bonnefille, David Burney, Joel Casanova, Paul Colinvaux, Kurt Haberyan, Thomas Johnson, Anne-Marie Lezine, Joseph Maitima, Jean Maley, Syukuro Manabe, Henry Osmaston, Patricia Palmer, Robert Peet, Jonathan Richardson, Nicholas Shackleton, Curt Stager, and Michael Talbot. Errors that have survived these discussions are my own.

 ISBN 0-300-05375-4

years old. The late-Quaternary sediment is highly fossiliferous and displays climatically driven changes in phytoplankton and surrounding vegetation. The rate of sedimentation is high and the temporal resolution about one year.

Lake Malawi is of comparable age (Ebinger et al. 1984; Scholz and Rosendahl 1989), but the shallower basins of Mobutu and Turkana, though very old, have probably not held water continuously. Lake Kivu seems to have been wet since its formation during the Pliocene. Meteoritic Bosumtwi in Ghana, volcanic Barombi Mbo in Cameroon, and probably other lakes in both East and West Africa are of the order of a million years old. Many other basins should provide a proxy climatic record for at least 100,000 years. Most published lacustrine records from Africa cover less than 40,000 years.

There is a splendid series of rift lakes in eastern Africa between 15° N and 16° S latitude and a smaller series along the Cameroon rift in the west. Elsewhere lakes are much rarer. We cannot use in Africa the approach of glaciated North America and Europe, where widespread lakes permit continental-scale mapping of past climate. To reconstruct African climate we must have a sound understanding of climate dynamics and use African data, where they exist, to test that understanding and to serve as control points on our reconstructions.

African lakes present us with wonderful possibilities for paleoecological and biogeographic research. They seem to contain a sedimentary record comparable in microfossil richness and resolvability to the best postglacial records of North America and Europe but cover a span of time that is more than a thousand times as long. Even the small part of that record we have already consulted has produced results directly pertinent to the subject of this volume. For example, both Jean Maley and I have independently discovered that pollen of the kapok tree and the oil palm occurs in West African lake deposits back to the early Holocene. *Ceiba pentandra* and *Elaeis guineensis* might still be post-Columbian introductions from Africa to the New World, but they cannot be post-Columbian introductions to Africa from America.

II. Evolution of Climate

Recent decades have seen many exciting advances in climate dynamics. Much new information has come from the ocean, where piston samplers raise cores representing several to many ice ages. Planktonic foraminifera in these cores have been combined with paleomagnetic data to create a worldwide $^{18}O/^{16}O$ stratigraphy.

Interpretation of the isotopes in climatic terms is less certain than stratigraphic correlation. Initially, changes in isotopic composition were attributed to sea surface temperature, suggesting that the tropical ocean had cooled by 6° C during each ice age (Emiliani 1966). Later much of the isotope change was attributed to ^{16}O becoming sequestered in glacier ice (Shackleton 1967). Neither the total mass of glacier ice nor its isotopic composition is known with great accuracy, but the glacier effect is believed to be large; only a small part of the isotopic change in tropical cores is attributed to temperature, probably differences of 1–2° C.

These revised estimates of world ocean temperature come from an analysis of planktonic foraminiferal assemblages (Imbrie and

Kipp 1971). The principal components are extracted from modern assemblages and used to construct transfer functions relating the composition of the assemblages to the sea surface temperature where they live. These transfer functions are then applied to older assemblages to estimate past temperature.

This method has been used very successfully to map past global sea surface temperature (CLIMAP 1981). The surface temperature of the temperate oceans has changed greatly (6° C or more). The ice-age Gulf Stream crossed the Atlantic at the latitude of Portugal rather than Norway and ice-age pack-ice hugged the coast of western Europe. This result is consistent with continental paleoclimatic data showing that ice-age Europe was a polar desert.

Even more satisfying has been power spectrum analysis of down-core data, which shows peaks corresponding to the three major Milankovich periods of 20,000, 40,000, and 100,000 years in the motion of the earth around the sun (Imbrie and Imbrie 1980). This robust result does not depend on the exact interpretation of the isotope record: both high glacier volume and low sea surface temperature are indices of glacial climate.

It was once believed that glaciation was restricted to the Pleistocene, but there is now good evidence for significant though less extensive continental glaciation far back into the Cenozoic. Although they have not always produced such cold climates as they do in our epoch, Milankovich irregularities in the earth's orbit are probably as old as life on our planet. They may produce ice ages only when continents and oceans lie in certain especially sensitive configurations, but Milankovich climatic periodicity is not rare.

Bradley (1929) showed that the Eocene Green River Formation of Wyoming contains rhythmically (perhaps annually?) banded sediments with variations in thickness that correspond nicely to the three principal Milankovich periodicities. Similar rhythmites are known from the Triassic lake beds of eastern North America (Olsen 1986). The climate of the world, and of Africa with it, has probably varied with periods of 20,000, 40,000, and 100,000 years throughout the evolutionary history of the groups of organisms under consideration in this volume.

There remains a problem of causality, because Milankovich periodicities affect the latitudinal distribution of solar radiation over the earth's surface, not the total amount of radiation reaching the earth. One pole should sometimes be warm while the other is cool. The global pattern of marine temperature change confirms earlier continental findings that the two poles are actually in phase (Heusser and Flint 1977; Porter 1981). After several false starts involving changes in global reflectivity and sea level, it now seems that greenhouse gases synchronize the two hemispheres. Changes in carbon dioxide content have been measured in glacial ice cores (Neftel et al. 1982; Oeschger et al. 1984), and the potential of these varying levels to lower the ice-age temperature of the world has been modelled (Broccoli and Manabe 1987). It is not certain how the composition of the atmosphere is controlled by Milankovich changes, but probably the causal connection lies in circulation and planktonic productivity where deep water forms in the Norwegian Sea or the Southern Ocean.

Longer-period changes in atmospheric

composition may spring from tectonic changes in the global mean rate of silicate weathering. At least some climatic changes over centuries or millennia, such as the Allerod oscillation at the end of the last ice age, are too fast to be linear responses to Milankovich changes in the geographic distribution of solar radiant heating. Perhaps ocean circulation and the partitioning of carbon dioxide between atmosphere and ocean respond chaotically to Milankovich forcing, and some of the climatic change is apparently periodic (Broecker et al. 1990). Climatic changes coincident with El Niño and the Southern Oscillation are almost certainly driven by rapid changes in ocean circulation, perhaps through a feedback loop like that described by Ramanathan and Collins (1991).

Although the grand pattern of glacial-interglacial oscillations is synchronized globally by greenhouse gases, a significant part of regional climatic change is due more directly to Milankovich changes in the distribution of solar radiation. Such changes may modulate both the West African monsoon and the Asian monsoon, a dominant influence on the climate of eastern Africa.

Electronic computation has permitted the construction of equilibrium models of the global circulation of the atmosphere and the ocean. Dynamic models would require much more computer power, but even the equilibrium models are both impressive and useful. They contain no arbitrary fine-tuning constants to make them match reality. Simulations of the modern climate that rest on such first principles as the measured distribution of solar radiation and the physical properties of the atmosphere produce an output that is strikingly similar to climatic maps drawn from weather records (Manabe 1967; Manabe and Bryan 1969; Manabe and Broccoli 1985, 1990).

Using as boundary conditions in these models the foraminifera-based sea surface temperature field, the calculated global distribution of solar energy, and the atmospheric carbon dioxide measured in ice-core analyses, one gets estimates of ice-age climatic conditions. For North America or Europe, the result accords with the geologic evidence of past climate (Kutzbach and Wright 1985). We seem to be approaching a reliable understanding of the climate dynamics of the earth.

There are, however, weak links in the chain of evidence that leads to this satisfying conclusion. Past sea surface temperature estimates rest on correlations between planktonic foraminiferal assemblages and the water temperature where they live. Such correlations can lead to very fruitful hypotheses but do not demonstrate causality. Before great weight is placed on such correlations, they ought to be subjected to rigorous experimental test. Correlations between the distribution of organisms and the temperature of their environment should be tested both by experimental investigation of their thermal tolerance and by competition experiments, because the ranges of organisms and the composition of biotic communities are matters of ecology as well as physiology. At present a dangerously large part of our understanding of the past climate of the world rests on the untested hypothesis embedded in a transfer function.

Planktonic foraminifera are very difficult to raise through several generations in the laboratory. Perhaps their competition dynamics might be investigated with transplant experiments using enclosures at sea. For the present we have no alternative to the transfer functions, but until they have been subjected to rigorous experimental test we should not forget that they represent untested correlations.

Most microfossil evidence of past climate suffers from the same logical problem. The standard procedure of pollen analysts is identical to that of paleoceanographers, from whom they learned it. It is buttressed to some extent by physiological experiments with controlled temperature growth chambers, and occasionally by transplant experiments. Dahl (1951), for example, explicitly tested the hypothesis that some late-glacial vascular plants are reliable low-temperature indicators. He transplanted their living descendants from high Norwegian mountains to altitudes near sea level, where he found that they could not live to reproduce, even in a weeded garden. Ecological tests to substantiate our correlation-based ideas about the effect of temperature on forest composition would be very difficult because of the size and long life of forest trees.

III. The Climate of Africa

Over the broad sweep of Cenozoic time all continents have continued the crustal drift they began during the Mesozoic, but Africa has drifted less than any other continent. It does not lie exactly in its Cretaceous position and attitude, but drift is less severe here than elsewhere (Briggs 1987).

This might suggest that African Cenozoic climate has been relatively stable. Actually, even during the Quaternary, African climate underwent very significant change. During the last ice age the glaciers of East African mountains descended more than a thousand meters below their present limits (Osmaston 1975, 1989). During the early Holocene there were extensive lakes and rivers even in parts of the Sahara that are arid today (Petit-Maire 1988, 1989). During the late Pleistocene many lakes, including some very large ones, were dry or much shallower than now (Kendall 1969; Richardson et al. 1978; Haberyan and Hecky 1987). In both East and West Africa (Kendall 1969; Talbot et al. 1984; Maley, Livingstone et al. 1990; Maley and Brenac 1987) the humid forest was much reduced in density or extent. All of these are changes severe enough and extensive enough to have serious biogeographic consequences.

There have been spirited attempts to work out the history of such climatic parameters as cloud cover, raindrop size (Maley 1982), wind direction (Hooghiemstra 1988, 1989), and wind intensity (Talbot 1981; Talbot and Johannessen, personal communication, 1991) but such data are few. The past pressure field and the wind velocities resulting from that field would be invaluable, but the second is difficult and the first is impossible to reconstruct from fossil indicators. Most paleoclimatologists investigate only temperature and humidity, and in the most general terms. It is easier to conclude from fossil evidence that a past climate was warm or cool than to tell how warm or how cool it was. It is difficult to distinguish changes in mean temperature from changes in temperature range or the incidence of extremes.

A. Moisture

Indicators of past humidity seldom separate the effects of rainfall and evaporation; they tell us little about the frequency of short-term departures from mean conditions, but they are our least ambiguous indicators of past climate. High lake levels indicate humid conditions, although they do not tell whether increased rainfall or decreased evaporation produced them. Two lakes of very different shape and catchment area may respond differently to the same climatic change (Hutchinson 1957:chapter 4) but in general, the lakes of a region tend to go up and down together. When dated changes of lake level are plotted on maps, they present a clear synoptic picture of changing humidity through time. This has been done for Africa (Street-Perrott and Harrison 1984; Street-Perrott et al. 1989) and shows that the last Pleistocene ice age was generally dry over most of the continent, with wetter conditions being characteristic of the Holocene. Most exceptions lie in North and South Africa, which are influenced by temperate westerlies.

There is, however, considerable regional variation in the details of humidity change. In southern Africa the latter half of the Holocene tends to be wetter, whereas elsewhere the early half is wetter. The southern pattern of a more humid late Holocene seems to include Madagascar (Burney 1987) and Lake Malawi (Thomas Johnson and George DeBusk, personal communications, 1991) but not all of the continent between their latitudes and southern Africa.

Moist conditions are revealed by incised strandlines around lake basins; dry conditions are more evident in sediment cores. Sometimes low lake levels are reflected by clastic materials deposited in a lake (Haberyan and Hecky 1987), by unconformities in the sediment (Kendall 1969; Talbot et al. 1984), or by the ostracod assemblage (Richardson and Dussinger 1986). Other indicators demonstrate a high concentration of dissolved salts in the water rather than a drop in level. Diatoms record concentration changes (Richardson et al. 1978; Haberyan 1987; Haberyan and Hecky 1987). Autochthonous minerals such as carbonates and zeolites also indicate times of high concentration (Kendall 1969; Talbot and Kelts 1986; Richardson and Richardson 1972; Stoffers and Holdship 1976). Hydrogen Index and carbon isotopic composition have the advantage of being more generally applicable and possibly more sensitive (Talbot and Livingstone 1989).

As a group, these sedimentary indicators of low water level or high concentration of dissolved salts provide a more continuous record of climatic change than do raised strandlines. Sediments have not yet been searched systematically in Africa for short-term periodicities or other changes in the spectrum of climatic change, although such lakes as Crescent Island Crater of Naivasha in Kenya or Bosumtwi in Ghana contain a very detailed record of concentration.

Sedimentary studies are too few for continentwide plotting of the results. They demonstrate at a few localities a pattern of continuous short-term climatic variability, whereas the strandline data give an impression of stasis for hundreds or thousands of years at a time. This apparent difference is due to the characteristic response time of the different climatic indicators. Brief occupations of a particular elevation by lake level do not incise or construct strandlines, but even a single season in which carbonates are precipitated may show up as a sharp band in the sedimentary record.

Harvey (1976), from his data at Lake Mobutu in Uganda, estimated that the dry late Pleistocene had a rainfall only 70 percent of modern. A suggestion by Clark (1967) that the rainfall of Africa has varied between 50 percent and 150 percent of the modern level still seems reasonable for many parts of the continent, although not for the central parts of the Sahara. Some of the former Saharan lakes might be attributed to aquifers nourished by distant humid tropical mountains, but that is not likely to be the explanation for extensive early Holocene lakes in the central part of the desert (Petit-Maire 1988).

Colinvaux (Chapter 16) casts doubt on the current belief that the climate of South America was dry during the last ice age. I agree with him that the extension of this conclusion from peripheral parts of the continent where it is supported by stratigraphic evidence (see, for example, papers cited by Livingstone and van der Hammen 1978, in addition to those cited by Colinvaux) has been something of an act of faith. It it not supported by his own research in tropical lowland localities of either Central or South America.

For Africa, however, the evidence for widespread ice-age aridity is incontrovertible. The papers of Street-Perrott and her collaborators (Street-Perrott and Harrison 1984; Street-Perrot et al. 1989) summarize the lacustrine data, but there is an enormous body of supporting evidence from pollen as well. One might reasonably regard the belief that the last ice age had been dry in tropical Africa to be the best-established of those currently held about the climatic history of the continent. The only contradictory evidence seems to come from humid lowland Cameroon—and, of course, from temperate latitudes, where the belt of westerlies could be depressed into what are now very arid lands (see references cited in Livingstone 1975).

B. Temperature

The most reliable estimates of past temperature from Africa are those of Osmaston (1975, 1989) based on glacial geomorphology. From many different glaciers of different shape and different elevation he was able to estimate changes in firn line, a much more reliable climatic indicator than glacier terminus. Assuming that there was no change in snowfall, and that the glacial retreat since the height of the last ice age was due to temperature alone, Osmaston concluded that the climate of the last ice age was some 5° C cooler than now. If one accepts Harvey's (1976) estimate of the rainfall increase since that time and applies that increase to Osmaston's calculations, the temperature changes have been 7–8° C.

Similar estimates of late Quaternary temperature change have been made for other tropical mountains, but some workers would rather attribute this change to a higher lapse rate rather than to general cooling. In Africa, as in South America (Chapter 16), late Pleistocene temperatures seems to have been low all the way to sea level.

For example, Moreau (1963) pointed out that the bird faunas of Mt. Cameroon and the East African mountains were very similar, and that this similarity could be due to a decline in temperature sufficient for montane birds to spread across the intervening lowland. An apparently simpler explanation, that the birds flew directly across Africa, fails because most East African montane birds have not been able to make the much shorter direct flight to Ethiopia. Ethiopia is separated from East Africa by a lowland belt so arid that it would require an even greater change of climate than the lowlands between East Africa and Cameroon to support a continuous range of montane birds.

A similar biogeographic pattern is displayed by several other groups of organisms, particularly the grasses (Clayton 1976). Mountaintop grasses are of different species in the eastern and western African highlands, but those at intervening altitudes are held in common by Mt. Cameroon and by Ruwenzori, Mt. Kenya, Kilimanjaro, and the other highland areas in East Africa. The simplest explanation for this biogeographic pattern is that there was a time when the climate was cool enough for temperate grasses to spread over most of the intervening lowland.

Because tropical African mountains rise so sharply from the lowland plateau, a moderate decrease in temperature would do little to reduce the distance between the East and West African ranges of their grasses. A temperature decrease of some 10° C seems to be called for. It is clear, on the same basis, that the temperature has not fallen by as much as 16° C: that would permit range confluence of the mountaintop species not shared by both mountain areas.

Grasses are especially reliable indicators of temperature because the mountain species are mostly members of the temperate subfamily Pooideae (Livingstone and Clayton 1980). They use the Calvin photosynthetic pathway, whereas most grasses growing in the warm tropical lowlands use the Hatch-Slack pathway. The biochemistry, physiology, and ecology of their dependence on low temperatures are well understood, so we are on surer ground with them than we are with most pollen grains or planktonic foraminifera.

Grass pollen grains are so similar that only maize, a post-Columbian introduction, can be distinguished reliably from the many hundreds of species in the modern flora of Africa. The cuticle of grass leaves, however, is highly diagnostic, and it is especially easy to separate the cool-temperate pooids from other grasses. As little as a single siliceous phytolith may suffice to establish the presence of pooid grasses in a fossil deposit. It is especially significant that Talbot and co-workers (1984) report up to 40 percent of pooid cuticles among Pleistocene and early Holocene fossil grasses at Lake Bosumtwi in Ghana. This low-altitude lake is separated by a long distance from any existing pooids. The nearest known is *Poa annua,* an introduced weed from a settler's lawn on the lower slopes of Mt. Cameroon. To extend the range of that species to Bosumtwi would require only 2.4° C of cooling, but I have compared some of the fossil cuticles from Bosumtwi with *Poa annua* and they are quite different. If the fossils at Bosumtwi come instead from some native pooid that grows higher on Mt. Cameroon, such as species of *Agrostis* and *Streblochaete* which resemble them more closely, their presence would indicate a late-Pleistocene cooling of at least 6.7° C.

Talbot and co-workers (1984) also report from the Pleistocene levels at Bosumtwi a single cuticle fragment of *Arundinaria alpina,* the mountain bamboo. This suggests a temperature at least 8.7° C cooler than modern.

The range of pooid grasses is limited by nighttime temperature during the season of active growth. The stratigraphic evidence of the former occurrence of pooid grasses near sea level would require only a cool season, if we were dealing with a strongly seasonal climate, not a climate in which the entire year was cool. Bosumtwi, however, lies in the humid tropics. The annual range of mean temperature is only 3.2° C at nearby Kumasi (Lebedev 1970).

Three different sorts of evidence, glacial geomorphology, biogeographic ranges, and microfossils, agree in suggesting that the late Pleistocene was substantially cooler than the late Holocene, perhaps by 6–10° C. This is puzzling, because paleoceanographers believe that the temperature of the tropical ocean was at most a few degrees cooler during the last ice age. General circulation models do not permit such a discrepancy between the temperature of land and ocean during an ice age. Even when they incorporate interactive cloudiness (Broccoli and Manabe 1987), if the reconstructed sea surface temperature field is a boundary condition, they do not generate a continental temperature field such as the geologic evidence suggests for tropical Africa.

Possibly the land is less constrained than the ocean by the cloud-mediated feedback that limits sea surface temperature (Ramanathan and Collins 1991). Another possible explanation may lie embedded in the logical structure of the marine temperature reconstructions. All evidence suggests that the sea surface temperature has fluctuated with Milankovich periodicities. During each ice age, the plankton of most latitudes could simply migrate south to find its thermal desideratum. An equatorial plankton, however, would have no place to go, for the pelagic environment does not afford even the microclimatic refuges equatorial plants and animals find on land in times of climatic adversity. Perhaps evolution of an equatorial plankton that could live only at interglacial temperatures is prevented by recurring ice ages. No living plankton assemblage may require temperatures as high as those of the modern equatorial ocean. Without a high-temperature fauna to register such temperatures, transfer functions may not display the full range of glacial-interglacial equatorial temperature change.

It is possible, though unlikely, that all three kinds of African continental temperature indicators are misleading or that the global circulation models are somehow deficient.

In the absence of direct African information about the climate of remote times, African archaeologists, biogeographers, and evolutionists are sometimes tempted to reason from the marine, ice core, or temperate glacial records of climatic change in distant places. The discrepancy between marine and African continental records of temperature change is a clear warning that this is not a safe procedure. If we are to understand the evolution of the African biota in terms of climate, we must have an African climatic record with which to compare the record of African fossils.

C. Vegetational Change

African pollen, like other paleoclimatic indicators, reveals a continent of restless change. In some parts of Africa the vegetation consists so largely of grasses and of weakly represented zoophilous plants that pollen changes are muted and difficult to interpret (Livingstone 1971). Even in such regions, though, careful analysis reveals (Meadows and Meadows 1988; Scott 1976; Sugden and Meadows 1989) significant changes in late-Quaternary vegetation. Where anemophilous trees are abundant, vegetational change is easier to identify. There is little difference in this respect between a pollen diagram covering the last few tens of thousands of years in highland Kenya (Bakker 1964; Coetzee 1967) or Uganda (Livingstone 1967) and one from the temperate zone. Even at lower elevations pollen diagrams can be clear and convincing (Kendall 1969), provided that they include abundant trees such as *Podocarpus,* Moraceae, *Olea,* and *Celtis* that shed their pollen broadcast on the wind.

At such sensitive sites it is abundantly clear that African vegetation has changed enormously over the past 20,000 years. At tropical latitudes forest trees were much less important in the late Pleistocene than they have been in the Holocene. This is consistent with other climatic indicators of late-Pleistocene drought. At high elevations near tree line, late-Pleistocene coolness would have the same effect.

Many African palynologists, following the early lead of Bakker (1964), perceive in their diagrams indications that the late Pleistocene was some 5–8° C cooler than the Holocene. This is a reasonable conclusion, given all the other evidence for such a temperature change, but I am less confident than most in accepting it as a conclusion required by the pollen evidence (Livingstone 1967). The appearance of *Olea hochstetteri* type pollen at Bosumtwi (Maley and Livingstone 1983) seems to provide a case where pollen requires a substantial lowering of temperature. *Podocarpus,* which is often taken as an indicator of low temperature, can grow at low elevations (Eggeling and Dale 1951; Maley, Caballe et al. 1990) if moisture or cloud cover is abundant.

Few plants in the tropical African flora are restricted to the low temperatures of high elevations, and most of those few do not produce abundant and easily recognizable pollen. We should pay closer attention to the pollen of normally temperate families such as the Chenopodiaceae, Caryophyllaceae, and Rosaceae. Many of their few representatives in tropical Africa grow only at higher elevations, and the number of African species is so small as to favor precise pollen identifications. As with grasses, fossils other than pollen, such as fruits, seeds, and leaves, would make identifications more precise.

Ever since Kendall (1969) showed that rainforest trees were much reduced during the late Pleistocene we have needed to know just how much they were reduced. The forest taxa must have persisted somewhere, because no substantial part of the modern flora could evolve in so short a time as 12,000 years. Maps (Hamilton 1976) depicting refuge areas of Pleistocene forest are not well controlled by fossil data. In the absence of pollen diagrams from central parts of existing blocks of tropical forest, it is not possible to know the extent of forest reduction, nor the ranges of the forest taxa in late Pleistocene and earlier glacial times.

There is substantial reduction in the percentages of rainforest taxa in the Pleistocene levels of published pollen diagrams, but these are all from peripheral parts of the presently forested area. Possibly in the Zairean or Guinean forest core a vegetation persisted that was similar in composition to modern forest. Complete diagrams have not been published from Bosumtwi in Ghana and Barombi Mbo in Cameroon, but the available data (Maley 1987; Maley and Livingstone 1983) indicate that the Pleistocene vegetation included many genera that grow in the pollen catchment areas of those lakes today, together with a few taxa, notably *Olea,* now restricted to higher altitudes. Together with these forest taxa there was a higher percentage of grass in the pollen rain, suggesting a more open vegetation. Possibly this more open vegetation was a wooded grassland, but it was not very similar to any wooded grassland that grows between these lakes and the Sahara today. At Barombi Mbo in Cameroon, however, the late Pleistocene was poor in grasses and rich in forest taxa (Maley and Brenac 1987; Maley, Caballe et al. 1990; Maley, Livingstone et al. 1990). This seems to have been a refuge for forest vegetation, not just the forest flora, during the late Pleistocene. It is not easy to understand the climatic basis for this apparently wet oasis when most of equatorial Africa was dry. Although this is a unique finding for Africa, Colinvaux (Chapter 16) has found something very like it in several lowland tropical sites in the New World.

Palynologists are accustomed to plants that react individualistically to climatic change and are not surprised when African Pleistocene vegetation consists of taxa that no longer grow together. It is probably instructive that so many forest taxa persisted where they grow today despite climatic differences. Most spectacularly in the Cape, but in many other small areas in Africa, one finds many endemic species. It is improbable that these areas of high endemism escaped the general climatic changes of late-Quaternary Africa, and equally improbable that the endemics have returned to them recently from some distant and as yet undiscovered refuge where climate did not change. Probably the endemics, as elsewhere the forest taxa, persisted more or less in place with smaller populations in a vegetation diluted by species now growing only at higher elevations or in drier places.

Pollen analysis gives us some indication of the kinds of phytogeographic change that have affected Africa during the last 20,000 years. Early claims of European trees on isolated massifs of the central Sahara during the late Pleistocene seem less credible since Cour and Duzer (1976) demonstrated modern long-distance dispersal of temperate pollen to the central Sahara. There is good evidence, however, of substantial range extensions from the tropical side of the Sahara deep into what is now desert (Ritchie 1987). Bonnefille and co-workers (1990) have shown that the Restionaceae, a family with its main distribution in Southern Africa, were important in the vegetation of Burundi in the late Pleistocene. This does not represent a great shift in range, however, because one species of the family grows in neighboring Zaire today.

A regional flora very different from the modern one comes from the late Miocene of highland Ethiopia, where Yemane and co-workers 1987) found an assemblage of pollen grains including several from lowland rainforest plants. The site has been elevated by tectonic activity since deposition of the fossils, which represent plants that do not grow in the cool highlands today.

African plants have changed their relative abundances in the vegetation of much, probably all, of the continent. They have moved up and down mountains, and even considerable distances out over the surrounding lowlands; they have moved into habitats that are now too dry or too wet for them. There is, however, surprisingly little fossil evidence of Quaternary change in regional floras. The existence of temperate taxa on African mountains, even on the equator, shows that there have been extensive migrations, and the close similarity of some equatorial mountaintop taxa to their temperate-zone relatives suggests changes that are rather recent. Outside the Sahara, however, there is little stratigraphic evidence of very great shifts of range. A palynologist who makes identifications by comparing all fossils only with the taxa of the appropriate modern regional flora is not likely to make many misidentifications for that reason, but neither is such a worker likely to be the first to discover great shifts in range if they actually occurred.

IV. Refuge Theory and Biotic Diversity

Only a few decades ago most biologists seemed to attribute the high species diversity of the humid tropical regions to a climate that had not changed since the Cretaceous. Now most seem to attribute the high species diversity, at least of the Amazon basin, to a Quaternary climate punctuated by periods of such extreme aridity that organisms requiring moist conditions were restricted to small humid refuges where they were able to evolve rapidly by genetic drift. In Africa, at least, it has not been easy to understand where and how the constituent species of rain forest could have survived these periods of ice-age aridity.

With respect to vascular plants, at least, it seems most likely that the rainforest species persisted over a wide area in which their populations were smaller than they are today and restricted to sites of particularly favorable micro-climate. This is what Colinvaux (Chapter 16) believes to have been the case in North America as well as the Neotropics. Although I agree with him in general, such widespread local persistence does not seem inconsistent with the presence in some places of fairly extensive blocks of mesic vegetation sheltering fairly large subsets of the constituent species of modern forest. In North America, Delcourt and Delcourt (1977) seem to have identified such a block in Louisiana. Maley, Livingstone, and co-workers (1990) seem to have identified such a block in humid lowland Cameroon. The African data are still too few to tell us if the Cameroonian block was only local, or if it extended to Gabon or even through the Zairean cuvette to the moist highlands along the Western Rift. I would interpret the South American data as evidence not of a humid ice age, but of a dry ice age that was still sufficiently humid to sustain the modern forest taxa of vascular plants throughout the humid lowlands.

Viewed in those terms, the difference between rainforest species richness in Africa and South America seems only to be expected. In both continents the ice ages were both cooler and dryer than the Holocene. Because most of the Quaternary was closer to an ice age than to the climatic regime of modern times, that means that most of the time the climates of both Africa and South America were drier and cooler than they are now.

Exactly how much cooler we do not know, but the terrestrial evidence is easy to reconcile with a glacial-interglacial temperature difference between 5° and 10° C. How much drier we cannot say either, but in rough terms, rainfall may have been reduced by 25 percent in South America, by 50 percent in Africa. The different between these estimates is consistent with the relative importance of monsoon rain on the two continents, and with the likely reduction in monsoonal flows.

Such a reduction in temperature and humidity is likely to have restricted most rainforest species to altitudes of only a few hundred meters where the present rainfall is perhaps two meters or more. The total area of such land is enormously greater in South America than in Africa, and one would expect that the number of species would be very much greater as well. This would follow from the argument developed by Preston (1962), regardless of what mechanisms of speciation were involved in the evolution of the rainforest assemblage.

The paleoecological data seem to provide a satisfactory explanation for the difference in species diversity between Africa and South America. They do not, as yet, do much to explain the details of similarity and contrast between the floras of the two continents, nor do they shed much light on the patterns of endemism within each continent.

V. Conclusion

A review that centers on the late Quaternary is unsatisfactory for many biogeographic purposes, but one must be constrained by the availability of evidence. To what extent did the late-Quaternary pattern prevail during earlier times? It is commonly assumed that the alternation of dry glacial periods with wet interglacials is general, and not limited to the last few tens of thousands of years. The best evidence that this is so comes from marine cores (Hooghiemstra 1989; Sarnthein et al. 1981), which contain dust, diatoms, and pollen blown off the African continent into a marine stratigraphic context that covers a reassuring length of time. Similar evidence is starting to come in from caves (Brook et al. 1990).

The seismic stratigraphy of Lake Malawi and Lake Tanganyika contains evidence of an erosional unconformity at a depth of about 50 meters in the sediment (Scholz and Rosendahl 1988). The simplest explanation for this feature, which seems to occur at the same depth in both lakes, is that it represents a time some 100,000 years ago when lake level fell several hundred meters and remained low long enough for substantial erosion of the exposed lake beds to occur. Such a dry period does not correspond to any known feature in the climatic history of the rest of the world. This seems to suggest that local correlatives of known glacial-interglacial oscillations have not been the only significant climatic changes in Africa.

It is sometimes assumed that the broad sweep of climatic change in Africa, the modulation of the Milankovich pattern that makes some parts of Cenozoic time more glacial than others, has been the same as it was in better-known parts of the world. This is a reasonable assumption and receives some support from the African continental record (Bonnefille 1983; Bonnefille and Riolet 1980). The safety of this assumption depends on how much order there is in the universe. It will be very much safer when it is supported by even one continuous African continental record covering several million years, and when some nonclimatic explanation can be found for the seismic stratigraphy of Lakes Malawi and Tanganyika.

It is clear from Colinvaux's contribution (Chapter 16) that the late-Quaternary histories of Africa and South America have been different. Africa has been affected more by late-Pleistocene drought, and the lowland forest has been reduced more in extent and density, than has been the case in South America. Perhaps Africa is so much drier a continent that any reduction in its already low rainfall has a more noticeable effect.

Africa differs in another way: it extends far into the north temperate zone. Much West African rainfall comes from the monsoon winds that blow off the Atlantic into the Saharan summer low-pressure area. Much of the rainfall of eastern Africa is also monsoonal, although here the driving force is low pressure over Central Asia. Both monsoons were much weaker during the ice ages, which may have deprived Africa of much of its rain. South America is less monsoonal and more affected by occasional equatorward surges of polar air. Such surges would be even more frequent during an ice age and may have provided South America with a more dependable moisture supply.

If the two most similar tropical continents in the world can be so different, we should be careful about applying to either of them climatic conclusions derived from the glacial stratigraphy of distant temperate continents or the isotope and foraminiferal stratigraphy of the deep sea.

Literature Cited

Bakker, E. M. van Zinderen. 1964. A pollen diagram from equatorial Africa (Cherangani, Kenya). Geol. en Mijnbouw 43:123–128.

Bonnefille, R. 1983. Evidence for a cooler and drier climate in the Ethiopian uplands towards 2.5 Myr ago. Nature 303:487–491.

Bonnefille, R., and G. Riollet. 1980. Palynologie, végétation et climats de Bed I et Bed II à Olduvai, Tanzanie. Acta 8th Panafr. Prehist. and Quat. Congr., pp. 123–127.

Bonnefille, R., A. C. Hamilton, H. P. Linder, and G. Riolet. 1990. 30,000-year-old fossil Restionaceae pollen from Central Equatorial Africa and its biogeographical significance. J. Biogeogr. 17:307–314.

Bradley, W. H. 1929. The varves and climate of the Green River Epoch. U.S. Geol. Surv. Prof. Pap. 158E:85–110.

Briggs, J. C. 1987. *Biogeography and Plate Tectonics*. Amsterdam: Elsevier.

Broccoli, A. J., and S. Manabe. 1987. The influence of continental ice, atmospheric CO_2, and land albedo on the climate of the last glacial maximum. Clim. Dyna. 1:87–99.

Broecker, W. S., G. Bond, M. Klas, G. Bonani, and W. Wolfli. 1990. A salt oscillator in the glacial Atlantic? 1. The concept. Paleoceanography 5:469–477.

Brook, G. A., D. A. Burney, and J. B. Cowart. 1990. Paleoenvironmental data for Ituri, Zaire, from sediments in Matupi Cave, Mt. Hoyo. Pp. 1–29 *in* N. T. Boaz (ed.), *Evolution of Environments and Hominidae in the African Western Rift Valley.* Virginia Mus. Nat. Hist. Mem. 1.

Burney, D. A. 1987. Late Holocene vegetational change in central Madagascar. Quat. Res. 28:130–274.

Clark, J. D. 1967. *The Atlas of African Prehistory.* Chicago: University of Chicago Press.

Clayton, W. D. 1976. The chorology of African mountain grasses. Kew Bull. 31:273–288.

CLIMAP Project Members. 1981. Seasonal reconstructions of the Earth's surface at the last glacial maximum. Geol. Soc. Amer., Map and Chart Ser. MC-36.

Coetzee, J. A. 1967. Pollen analytical studies in East and Southern Africa. Palaeoecol. Afr. 3:1–146.

Cour, P., and D. Duzer. 1976. Persistance d'un climat hyperaride au Sahara Central et méridional au cours de l'Holocène. Rev. Géogr. Phys. Géol. Dynam. 18:175–198.

Dahl, E. 1951. On the relation between summer temperature and the distribution of alpine vascular plants in the lowlands of Fennoscandia. Oikos 3:22–52.

Delcourt, P. A., and H. R. Delcourt. 1977. The Tunica Hills, Louisiana-Mississippi: Late glacial locality for spruce and deciduous forest species. Quat. Res. 7:218–237.

Ebinger, C. J., M. J. Crow, B. R. Rosendahl, D. A. Livingstone, and J. LeFournier. 1984. Structural evolution of Lake Malawi, Africa. Nature 308:627–629.

Eggeling, W. J., and I. R. Dale. 1951. *The Indigenous Trees of the Uganda Protectorate.* Entebbe: Government of Uganda.

Emiliani, C. 1966. Paleotemperature analysis of Caribbean cores P6304-8 and P6304-9 and a generalized temperature curve for the past 425,000 years. J. Geol. 74:109–126.

Haberyan, K. A. 1987. Fossil diatoms and the history of Lake Rukwa, Tanzania. Freshwater Biol. 17:429–436.

Haberyan, K. A., and R. E. Hecky. 1987. The late Pleistocene and Holocene stratigraphy and paleolimnology of Lakes Kivu and Tanganyika. Palaeogeogr. Palaeoclimatol. Palaeoecol. 61:169–197.

Hamilton, A. 1976. The significance of patterns of distribution shown by forest plants and animals in tropical Africa for the reconstruction of Upper Pleistocene palaeoenvironments: A review. Palaeoecol. Afr. 9:63–92.

Harvey, T. J. 1976. The paleolimnology of Lake Mobutu Sese Seko, Uganda-Zaire: The last 28,000 years. Ph.D. diss., Duke University.

Heusser, C. J., and R. F. Flint. 1977. Quaternary glaciations and environments of northern Isla Chiloe, Chile. Geol. 5:305–308.

Hooghiemstra, H. 1988. Changes of major wind belts and vegetation zones in NW Africa 20,000–5000 yr B.P., as deduced from a marine pollen record near Cap Blanc. Rev. Palaeobot. Palynol. 55:101–140.

———. 1989. Variations of the NW African trade wind regime during the last 140,000 years: Changes in pollen flux evidenced by marine sediments records. Pp. 733–770 *in* M. Leinen et al. (eds.), *Proceedings NATO Advanced Research Workshop on Paleoclimatology and Paleometeorology: Modern and Past Patterns of Global Atmospheric Transport,* NATO ASI Ser. Dordrecht: Reidel.

Hutchinson, G. E. 1957. *A Treatise on Limnology,* Vol. 1: *Geography, Physics and Chemistry.* New York: John Wiley & Sons.

Imbrie, J., and J. Z. Imbrie. 1980. Modelling the climatic response to orbital variations. Science 207:943–953.

Imbrie, J., and N. G. Kipp. 1971. A new micropaleontological method for quantitative paleoclimatology: Application to a Late Pleistocene Caribbean core. Pp. 71–181 *in* K. K. Turekian (ed.), *The Late Cenozoic Glacial Ages.* New Haven: Yale University Press.

Kendall, R. L. 1969. An ecological history of the Lake Victoria Basin. Ecol. Monogr. 39:121–176.

Kutzbach, J. E., and H. E. Wright, Jr. 1985. Simulation of the climate of 18,000 years B.P.: Results for the North American/North Atlantic/European sector and comparison with the geologic record of North America. Quat. Sci. Rev. 4:147–187.

Lebedev, A. N. (ed.). 1970. *The Climate of Africa.* Trans. by IPST staff. Jerusalem: Israel Program for Scientific Translations.

Livingstone, D. A. 1967. Postglacial vegetation of the Ruwenzori Mountains in equatorial Africa. Ecol. Monogr. 12:346–359.

———. 1971. A 22,000-year pollen record from the plateau of Zambia. Limnol. Oceanogr. 16:349–356.

———. 1975. Late Quaternary climatic change in Africa. Ann. Rev. Ecol. Syst. 6:249–280.

Livingstone, D. A., and W. D. Clayton. 1980. An altitudinal cline in tropical African grass floras and its paleoecological significance. Quat. Res. 13:392–402.

Livingstone, D. A., and T. van der Hammen. 1978. Palaeogeography and palaeoclimatology. Pp. 61–90 *in* A. Sasson and B. Hopkins (eds.), *Tropical Forest Ecosystems*. Natural Resources Research XIV. Paris: UNESCO.

Maley, J. 1982. Dust, clouds, rain types and climatic variations in tropical North Africa. Quat. Res. 18:1–16.

———. 1987. Fragmentation de la forêt dense humide africain et extension des biotopes montagnards au Quaternaire récent: Nouvelles donées polliniques et chronologiques. Implications paléoclimatiques et biogéographiques. Palaeoecol. Afr. 18:307–334.

Maley, J., and P. Brenac. 1987. Analyses polliniques préliminaires du Quaternaire récent de l'Ouest Cameroun: Mise en evidence de refuges forestiers et discussion des problèmes palaéoclimatiques. Mém. EPHE, Inst. de Montpellier, 17:129–142.

Maley, J., and D. A. Livingstone. 1983. Extension d'un élément montagnard dans le partie sud du Ghana (Afrique de l'Ouest) au Pléistocène supérieur et à l'Holocène inférieur: Premières données polliniques. C. R. Acad. Sci., Paris, T. 296, Ser. 2:1287–1292.

Maley, J., G. Caballe, and P. Sita. 1990. Etude d'un peuplement residuel à basse altitude de *Podocarpus latifolius* sur le flanc congolais du massif du Chaillu. Implications paléoclimatiques et biogéographiques. Etude de la pluie pollinique actuelle. Pp. 336–352 *in* R. Lanfranchi and D. Schwartz (eds.), *Paysages quaternaires de l'Afrique Centrale Atlantique*. Paris: ORSTOM.

Maley, J., D. A. Livingstone, P. Giresse, N. Thouveny, P. Brenac, K. Kelts, G. W. Kling, C. W. Stager, M. Haag, M. Fournier, Y. Bandet, D. Williamson, and A. Zogning. 1990. Lithostratigraphy, vulcanism, paleomagnetism and palynology of Quaternary lacustrine deposits from Barombi Mbo (West Cameroon): Preliminary results. J. Volcanol. Geother. Res. 42:319–335.

Manabe, S. 1967. General circulation of the atmosphere. Trans. Amer. Geophys. Union 48:427–431.

Manabe, S., and A. J. Broccoli. 1985. The influence of continental ice sheets on the climate of the Ice Age. J. Geophys. Res. 90:2167–2190.

———. 1990. Mountains and arid climates of middle latitudes. Science 247:192–195.

Manabe, S., and K. Bryan. 1969. Climate circulations with a combined ocean-atmosphere model. J. Atmos. Sci. 26:786–789.

Meadows, M. E., and K. F. Meadows. 1988. Late Quaternary vegetation history of the Winterberg Mountains, eastern Cape, South Africa. Suid-Afr. Tydskr. Wetensk. 84:253–259.

Moreau, R. E. 1963. Vicissitudes of the African biomes in the late Pleistocene. Proc. Zool. Soc. London 141:395–421.

Neftel, A., H. Oeschger, J. Schwander, B. Stauffer, and R. Zumbrunn. 1982. Ice core sample measurements give atmospheric CO_2 content during the past 40,000 yr. Nature 295:220–223.

Oeschger, H., J. Beer, U. Siegenthaler, B. Stauffer, W. Dansgaard, and C. C. Langway. 1984. Late glacial climate history from ice cores. Pp. 299–305 *in* J. E. Hansen and T. Takahashi (eds.), *Climate Processes and Climate Sensitivity.* Geophys. Monogr. 29, Maurice Ewing Vol. 5. Washington, D.C.: American Geophysical Union.

Olsen, P. E. 1986. A 40-million-year lake record of early Mesozoic orbital climatic forcing. Science 234:842–848.

Osmaston, H. A. 1975. Models for the estimation of firnlines of present and Pleistocene glaciers. Pp. 218–245 *in* R. Peel, M. Chisholm, and P. Haggett (eds.), *Processes in Physical and Human Geography.* London: Heinemann.

———. 1989. Glaciers, glaciations and equilibrium line altitudes on Kilimanjaro. Pp. 7–30 *in* W. C. Mahaney (ed.), *Quaternary and En-*

vironmental Research on East African Mountains. Rotterdam: Balkema.

Petit-Maire, N. 1988. Climatic change in Afroasian arid areas during the last 150,000 years. Pp. 182–190 *in* N. Petit-Maire (ed.), *Deserts: Past and Future Evolution*. Fuerteventura: International Geological Correlation Programme.

———. 1989. Interglacial environments in presently hyperarid Sahara: Paleoclimatic implications. Pp. 637–661 *in* M. Leinen and M. Sarnthein (eds.), *Paleoclimatology and Paleometeorology: Modern and Past Patterns of Global Atmospheric Transport*. Dordrecht: Kluwer.

Porter, S. C. 1981. Pleistocene glaciation in the southern lake district of Chile. Quat. Res. 16:263–292.

Preston, F. W. 1962. The canonical distribution of commonness and rarity. Ecology 43:185–215, 410–432.

Ramanathan, V., and W. Collins. 1991. Thermodynamic regulation of ocean warming by cirrus clouds deduced from observations of the 1987 El Nino. Nature 361:27–32.

Richardson, J. L., and R. A. Dussinger. 1986. Paleolimnology of mid-elevation lakes in the Kenya Rift Valley. Hydrobiol. 143:167–174.

Richardson, J. L., and A. E. Richardson. 1972. The history of an African Rift lake and its climatic implications. Ecol. Monogr. 42:499–534.

Richardson, J. L., T. J. Harvey, and S. A. Holdship. 1978. Diatoms in the history of shallow East African lakes. Pol. Arch. Hydrobiol. 25:341–353.

Ritchie, J. C. 1987. A Holocene pollen record from Bir Atrun, Northwest Sudan. Pollen & Spores 29:391–410.

Rosendahl, B. R. (ed.). 1988. Seismic Atlas of Lake Tanganyika, East Africa. Durham, N.C.: Duke University.

Rosendahl, B. R., and D. A. Livingstone. 1983. Rift lakes of East Africa: New seismic data and implications for future research. Episodes 1:14–19.

Sarnthein, M., G. Tetzlaff, B. Koopmann, K. Wolter, and U. Pflaumann. 1981. Glacial and interglacial wind regimes over the eastern subtropical Atlantic and north-west Africa. Nature 293:193–196.

Scholz, C. A., and B. R. Rosendahl. 1988. Low stands in Lakes Malawi and Tanganyika, East Africa, delineated with multifold seismic data. Science 240:1645–1648.

——— (eds.). 1989. *Seismic Atlas of Lake Malawi (Nyassa), East Africa*. Durham, N.C.: Duke University.

Scott, L. 1976. Preliminary palynological results from the Alexandersfontein Basin near Kimberley. Ann. S. Afr. Mus. 71:193–199.

Shackleton, N. 1967. Oxygen isotope analyses and Pleistocene temperatures re-assessed. Nature 215:15–17.

Stoffers, P., and S. A. Holdship. 1976. Diagenesis of sediments in an alkaline lake: Lake Manyara, Tanzania. IX Internat. Congr. Sedimentol., Nice, pp. 211–218.

Street-Perrott, F. A., and S. P. Harrison. 1984. Temporal variations in lake levels since 30,000 yr B.P.: An index of the global hydrologic cycle. Pp. 118–129 *in* J. E. Hansen and T. Takahashi (eds.), *Climate Processes and Climate Sensitivity*. Geophys. Monogr. 29, Maurice Ewing Vol. 5. Washington, D.C.: American Geophysical Union.

Street-Perrott, F. A., D. S. Marchand, N. Roberts, and S. P. Harrison. 1989. Global lake-level variations from 18,000 to 0 years ago: A palaeoclimatic analysis. Washington, D.C.: U.S. Department of Energy, Office of Energy Research, Office of Health and Environmental Research, Carbon Dioxide Research Program.

Sugden, J. M., and M. E. Meadows. 1989. The use of multiple discriminant analysis in reconstructing recent vegetation changes on the Nuweveldberg, South Africa. Rev. Palaeobot. Palynol. 60:131–147.

Talbot, M. R. 1981. Holocene changes in tropical wind intensity and rainfall: Evidence from southeast Ghana. Quat. Res. 16:201–220.

Talbot, M. R., and K. Kelts. 1986. Primary and methanic carbonates in the anoxic sediments of Lake Bosumtwi, Ghana. Geol. 14:912–916.

Talbot, M. R., and D. A. Livingstone. 1989. Hydrogen index and carbon isotopes of lacustrine organic matter as lake level indicators. Palaeogeogr. Palaeoclimatol. Palaeoecol. 70:121–137.

Talbot, M. R., D. A. Livingstone, P. G. Palmer, J. Maley, J. M. Melack, G. Delibrias, and S. Gulliksen. 1984. Preliminary results from sediment cores from Lake Bosumtwi, Ghana. Palaeoecol. Afr. 16:173–192.

Yemane, K., C. Robert, and R. Bonnefille. 1987. Pollen and clay mineral assemblages of a Late Miocene lacustrine sequence from the northwestern Ethiopian Highlands. Palaeogeogr. Palaeoclimatol. Palaeoecol. 60:123–141.

Chapter 16

Pleistocene Biogeography and Diversity in Tropical Forests of South America

Paul Colinvaux

I. Introduction

The outstanding property of lowland tropical forest is high species diversity. In equatorial Africa and South America, floras and faunas differ, but the vast numbers of species in each region present the same intellectual problem of prime cause, "Why are there so many species in the wet tropics?" This chapter advances the thesis that the answer must depend, at least in part, on the environmental histories of the wet tropics through Pleistocene time. Lyell's celebrated dictum should be reversed by saying that the "past is the key to the present."

Although many contemporary explanations of tropical diversity have a Pleistocene component, they have often been without the support of adequate paleoecological data, particularly for South America. Analysis of the few data that have recently become available suggests that the prime environmental forcing of tropical forests in ice-age America was cooling rather than aridity. The data also show that the Amazon rainforest was never broken into disparate patches as required by refugial theories, nor

Our South American research has been supported by the National Science Foundation, especially the Climate Dynamics and Ecology programs, and by the National Geographic Society. Both NSF and NGS have rallied at times with emergency funding to let us take advantage of critical windows of opportunity for field research. The research described here is largely the work of M. B. Bush, M. Steinitz-Kannan, K.-b. Liu, M. C. Miller, and D. Piperno, together with many contributions from members of my former laboratory at Ohio State University, including E. Asanza, I. Frost, P. De Oliveira, K. Olson, M. Riedinger, and M. Scheutzow. Responsibility for controversial points of view in this paper must remain mine.

 ISBN 0-300-05375-4

was the Isthmus of Panama ever penetrated by a savanna causeway in Pleistocene time. A new disturbance-vicariance hypothesis for tropical rainforest diversity is proposed.

A. Diversity Hypotheses

Several prominent diversity hypotheses postulate reduced extinction in the tropics. The earliest assumed equatorial environments so unchanging that species seldom suffered extinction from environmental accident (Wallace 1878; Fischer 1960; Sanders 1968). A more powerful differential extinction hypothesis then appeared with the concept of "intermediate disturbance" (Connell 1978; Hubbel 1979). This almost reversed the postulate of the equable tropics by relying on environmental instability to prevent extinction by competitive exclusion.

No ecologist can feel comfortable, however, with hypotheses that explain high diversity only as a result of reduced extinction rates, yet it has not been easy to find a mechanism for the obvious alternative, making speciation more rapid in the tropics. Ecologists have tended instead to concentrate on examining ways in which coexistence might be encouraged in equatorial environments. Closer niche-packing, though perhaps implied for high-diversity regions (Hutchinson 1959), could not be demonstrated satisfactorily in most circumstances (Schoener 1974), although plants in some Amazonian genera may well have saturated their habitats with species to the limit set by character displacements (Gentry 1982). Hypotheses of enhanced productivity (Connell and Orias 1964), complex ecosystem structure (MacArthur and MacArthur 1961), or large equatorial area (Terborgh 1973; Osman and Whitlach 1978) all invoke persuasive logic, though detailed objections can be made to each (Colinvaux 1973, 1986).

Only the biogeographically inspired Pleistocene refuge theory of Haffer (1969) departs from earlier hypotheses in offering enhanced species rates in the tropics. In this "refugial" model, centers of modern endemism are explained as reflecting patterns of tropical vicariance that change with each glacial cycle (Prance 1982; Whitmore and Prance 1987). For Haffer's mechanism to work, Neotropical environments must have alternated between episodes of aridity and moisture that fragmented and merged forest patches. The hypothesis is thus capable of being tested because it predicts the results of paleoclimatological studies in different parts of the Amazon basin.

II. On the Nonprevalence of Refugia

All Pleistocene refuge theories look on ice-age climates as phenomena to be avoided. They tacitly take the chronocentric view that species have long lived in communities closely comparable to those of modern times, with individual species required to live at densities, or in associations, similar to those we see. In particular, communities of our interglacial are deemed to represent a normality for which scope must be found in the alien climates of the glacials. The gross error to which this thinking can lead is nicely illustrated by the debate about the antiquity of the eastern forests of North America, now resolved by pollen analysis.

The deciduous forests of eastern North America are equalled in diversity only by those of eastern China, and they are composed of species many of which are known as fossils from the Tertiary. The northern habitats of the modern forest were close to the glacial front, posing the biogeographic and ecological question, "What happened to these diverse forests in an ice age?" Three hypotheses account for all possibilities:

1. The forest associations remained intact and in place over much of the modern range.
2. Forest associations persisted in refugia in the southern part of their range or in areas to which the whole community had migrated.
3. Modern forest associations did not exist in the ice age, but the component species were dispersed individually to different habitats throughout North America.

The first of these hypotheses was the prevailing view thirty years ago (Braun 1955). Those who accepted the early pollen evidence of Deevey (1949) as requiring tundra landscapes immediately south of the ice sheets, fell back on the second hypothesis of life in refugia, often requiring vegetation to have migrated southward (Martin 1958). And yet we now know that the third hypothesis is correct, for the history of American forest trees since the last glaciation has been reconstructed from an array of pollen sections throughout the region (Davis 1986; Webb 1986). No refugium ever existed for eastern American forests; rather the familiar forests we know were built by successive immigrations over most of the last 10,000 years.

These results are entirely in keeping with expectations of modern theoretical ecology in which communities are seen to represent only ephemeral accommodations between species able to share habitats. The species themselves are adapted to the range of environments experienced during the species span and can be expected to live in different associations as local climates change. The great ice ages were not passed in microcosms of the present, and all living plants are ice-age plants that have never needed refuges. Thus the concept of Pleistocene forest refuges in the tropics is as intellectually barren as its predecessor concept of forest refugia in North America.

III. Aridity or Cooling at the Ice-Age Equator?

Proxy data from ocean cores have been used to argue that glacial cooling was minimal at low latitudes (CLIMAP 1976). Instead, many investigators have found evidence for reduced precipitation, yielding the concept of tropical ice-age aridity. This concept has guided recent biogeographical thinking, particularly the forest refuge hypothesis for the ice-age Amazon and the hypothesis of savanna corridors across the Isthmus of Panama.

Ice-age aridity was described for Africa by Livingstone (1975), using the evidence of old water levels in ancient lakes, and elaborated by Street and Grove (1979). East African pollen diagrams are also consistent with reductions in ice-age precipitation that required large changes in the distribution and abundance of tropical forest species. But the pollen data do not require the existence of wet refugia for African forests as much as they suggest local persistence of forest species in low populations, perhaps widely dispersed in small favorable sites or along watercourses (see Chapter 15).

Ice-age aridity in tropical Africa is plausible because climate models of the ice-age earth predict that precipitation in their monsoonal climates should have been reduced by 10 to 20 percent during parts of glacial cycles (Kutzbach and Guetter 1984, 1986). In the Neotropics, however, monsoons are important only in the Caribbean and northern South America, the bulk of Amazonian rains coming from the South Atlantic. It is prudent, therefore, to base conclusions about Neotropical ice-age aridity only on data from America itself.

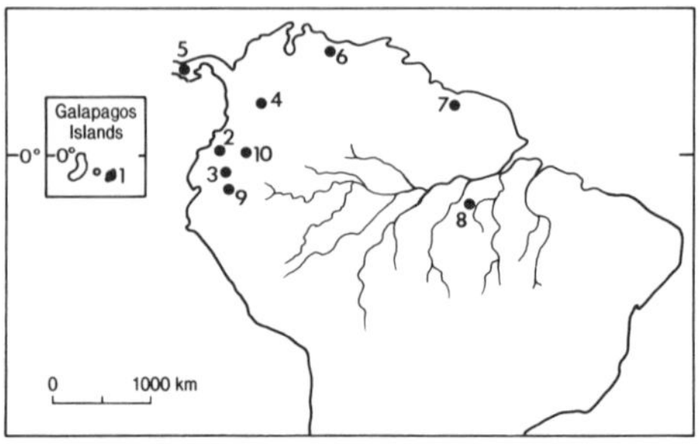

Figure 16.1. Location of glacial or late-glacial lake sediments. 1 = Lake El Junco, Isla San Cristobal, Galapagos; 2 = Yaguarcocha, Inter-Andean Plateau; 3 = Shell Mera and San Juan Bosco, Ecuadorian Amazonia. 4 = Sabana de Bogota; 5 = El Valle and Lake La Yeguada, Pacific lowlands of Panama; 6 = Lake Valencia; 7 = Guianan coastal deposits; 8 = Serra dos Carajas; 9 = Lake Kumpaka; 10 = Lakes on Napa-Aguarico system recording 800 B.P. flooding.

A. Galapagos and Pacific Coast Aridity

The first equatorial site to yield evidence of aridity at the American equator was on the Galapagos Islands, where El Junco lake dried completely sometime before 40,000 B.P., briefly ponded about 25,000 B.P., and did not become a permanent lake again until 10,000 B.P. (Colinvaux 1972). The cause of this increased Galapagos aridity is still not completely known. Colinvaux (1972) argued that the Intertropical Convergence Zone (ITCZ) must have failed to reach the islands at any part of the year, and Newell (1973) suggested that this could have happened if the ITCZ was displaced far to the south by increased Hadley cell circulation. Houvenaghel (1974) and Simpson (1975) invoked increased oceanic circulation that extended the present system of cold surface water and bottom-heavy, stable air masses.

Whatever the true cause, Galapagos aridity is clearly related to regional events in the eastern Pacific Ocean, rather than to a global condition of "tropical aridity." If regional, the effects might well be found on the western seaboard of equatorial South America, but not necessarily farther away. Data from the sediments of a lake, Yaguarcocha, close to the equator in the Inter-Andean Plateau of Ecuador are consistent with this expectation of late-glacial aridity (Colinvaux, Olson, and Liu 1988). Radiocarbon inversions give some doubt about the age of the lowest Yaguarcocha sediments, but the dating is consistent with a late-glacial age. Both pollen of Chenopodiaceae in the bottom sediments and diatom data strongly suggest a reduced lake in a semi-arid landscape (figures 16.1 and 16.2).

B. Pluvial Eastern Flank of the Andes

One other lake record from the tropical Andes below the paramo spans glacial times, the classic section from the Sabana de Bogota, at 5° N latitude and 2550-m elevation on the eastern (Amazonian) side of the cordillera (van der Hammen and Gonzalez

YAGUARCOCHA

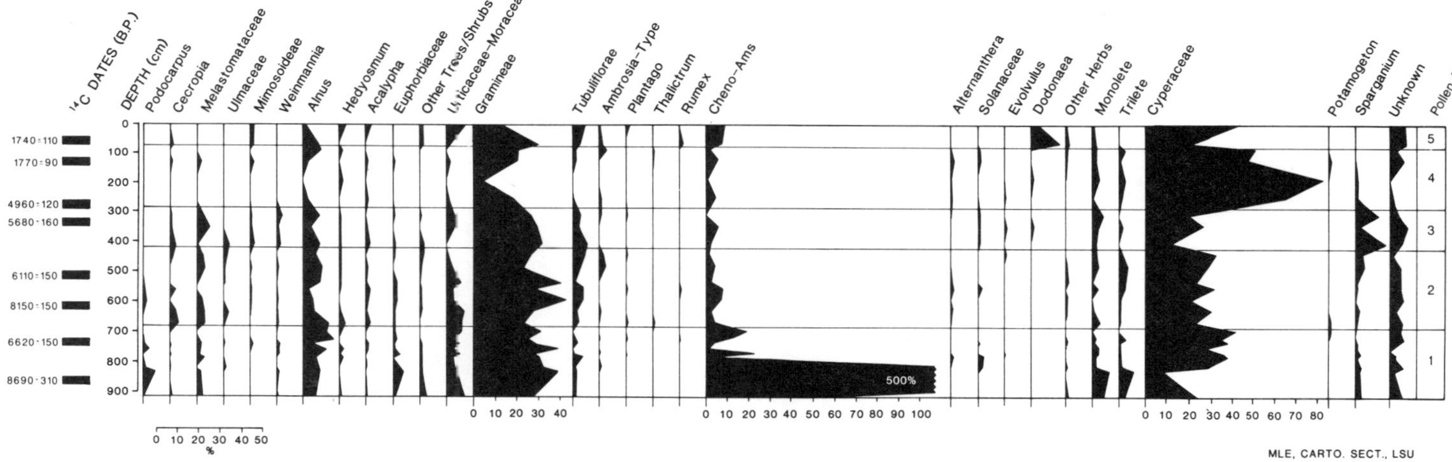

Figure 16.2. Pollen percentage diagram from Yaguarcocha. Pollen analysis by K. Olson, from Colinvaux, Olson, and Liu 1988.

1960; Hooghiemstra 1984, 1989). The lake is dry now (the result of stream downcutting), but it held water during the last glacial period. Pollen from *Isoetes* and other aquatic plants record a history of lake level rising and falling in synchrony with glaciations, but it is a "pluvial" story rather than a history of ice-age aridity, with the wet periods roughly coinciding with glaciations and the dry periods with interglacials like the present. This record suggests that it is unreasonable to extrapolate aridity cycles of the eastern Pacific across the Andes to the Amazonian side.

C. Ice-Age Aridity in the Monsoonal Caribbean

For Caribbean weather systems to the north of the South American landmass, patterns of reduced precipitation reminiscent of those in Africa are suggested by several records, including Lake Valencia in Venezuela, lakes of the Peten in Guatemala, and marsh deposits in the Guianas (Salgado-Labouriau 1980; Bradbury et al. 1981; Leyden 1984, 1985; Wijmstra and van der Hammen 1966). These data are consistent with climate models that predict 10 to 20 percent reductions in monsoonal rains in the Northern Hemisphere (Kutzbach and Guetter 1984, 1986). They suggest that dry seasons were prolonged over the northern part of South America as well as over the Caribbean.

D. Absence of Evidence for Amazonian Aridity

No unequivocal arid land features radiocarbon dated to glacial times have been identified from the Amazon basin, although surficial erosional features of uncertain age can be detected on side-scan radar images (Tricart 1977). Reported buried stone lines are not relics of desert ablation but concretions formed in deeply weathered tropical soils (Sternberg 1982; Irion 1982). The one provocative datum comes from the work of Damuth and Fairbridge (1970), who identified significant quantities of feldspar minerals in ice-age sediments from marine cores off the mouth of the Amazon. Feldspars do not survive humid weathering, and Damuth and Fairbridge suggested that the ice-age feldspars were delivered to the Amazon system by rivers draining the Guiana highlands to the north, when these were made even more arid than now by the reduction in monsoonal rains. To the extent that this explanation is correct, no arid demands are made on the Amazon lowlands. But a more cogent explanation for the feldspars is provided by Irion (1984) who suggests that feldspars were extracted from unweathered basements of deep channels cut by rivers when ice-age eustacy lowered sea levels in excess of 100 m.

Radiocarbon-dated pollen data to test the aridity hypothesis in the central Amazon lowlands are still lacking. A detailed section radiocarbon dated to glacial times has now been reported from the Brazilian Amazon, but it is on a 700-m plateau at Carajas covered with savanna-like vegetation in a strongly seasonal climate (Absy et al. 1989:figure 1). Mineral and pollen data both suggest episodes of decreased precipitation in the last glacial period. This does not imply widespread Amazonian aridity because the Carajas site is near the northern edge of the basin under the influence of northeast monsoons, the reduction of which also reduced precipitation in the Caribbean. Persistent tree pollen throughout the record, however, suggests tropical forest persisting in the region, presumably, as now, in bottom lands below the plateau.

Thus those who would believe, on biogeographic grounds, that the Amazon basin was semi-arid in glacial times should realize that we have not a single paleoecological datum that requires, or even suggests, aridity in what are now the forested bottom lands at times of northern glaciation (Salo 1987; Colinvaux 1987). Conversely, paleoecological data increasingly suggest that cooling was the primary climatic forcing of the ice-age Amazon.

E. The Andean Record of Cooling

By far the best paleoecological record of the Neotropics comes from the Sabana de Bogota cores. Not only have these sediments accumulated continuously throughout the Pleistocene but the proximity of the lake to tree line lets relative glacial state be inferred in the pollen diagrams. Van der Hammen and Gonzalez (1960) and Hooghiemstra (1984, 1989) calibrated the percentage of total arboreal pollen with altitude from surface samples, using the results to infer the elevation of tree line at all stages in a glacial cycle. That this procedure gave realistic results is confirmed by the data of individual Andean pollen taxa like *Wienmannia, Hedyosmum,* and *Podocarpus;* or by the *Quercus* pollen, the percentage of which falls in glacial maxima as the temperature-sensitive oaks are restricted to lower elevations. The overwhelming implication of this pollen history is that temperatures were depressed with each glacial cycle and raised to about those of the present day with each interglacial. Cool and relatively wet in glacial times, not warm and dry, is the message of the Sabana de Bogota cores.

Cool and moist glacial intervals for the Andes are fully in accord with the evidence for past firn lines and moraines that show glacial descents of about 1500 m, the same maximum descent shown by the tree line below them at Sabana de Bogota (Clapperton 1987; Hastenrath and Kutzbach 1985). Thus nearly all the well-established, radiocarbon-dated, proxy measures of the continental climate of the American equator in glacial times have long suggested cooling, and possibly moisture. Records suggesting moderate aridity are generally confined to the periphery of the basin, particularly on the Caribbean coast. Near the equator itself, only the Galapagos and Yaguarcocha (figure 16.1) records suggest aridity, and the parsimonious explanation of these is that the aridity was a local phenomenon driven by circulation patterns or air-mass movements in the eastern Pacific Ocean.

F. Cooling in the Ice-Age Amazon

While seeking ancient lakes in the Amazon rainforest of Ecuador in 1984, we found old lacustrine sediments rich in wood samples that were exposed by a road cut at Mera (figure 16.1). These yielded radiocarbon ages between 26,000 B.P. and 33,000 B.P. Both pollen and macrofossils suggested Andean taxa growing locally, requiring that temperature-sensitive trees had descended at least 700 m into what is now the upper part of the lowland tropical rainforest at 1100 m (Liu and Colinvaux 1985). Ecuadorians familiar with the forest told us of seeing other deposits with buried wood. In 1988 we undertook to search the rainforested foothills of the Ecuadorian Andes for additional sections. M. B. Bush and P. De Oliveira discovered a second site in a stream-cut near San Juan Bosco, about 100 km from Mera and at 950-m elevation (figure 16.1). Radiocarbon dates show that the section spanned part of the same time interval, 31,000 B.P. to 26,000 B.P. (Bush et al. 1990).

The main features of the Mera and San Juan Bosco pollen diagrams (figures 16.3 and 16.4) can be understood from surface pollen spectra arranged in order of altitude (figure 16.5). The Andean spectra from above 1000-m elevation are clearly separated from the Amazonian spectra below 1000 m. The ice-age spectra from the fossil sections, however, include elements of both lowland and upland floras, are unlike any modern surface spectra, but are closely comparable to each other. In both diagrams the high *Alnus* percentage is striking. Hooghiemstra (1984) and Grabandt (1980) regard such high alder percentages as indicating "*alnetum* meadows," which they find not to occur below 2500 m in Colombia. In surface samples from Ecuador we do not find high alder percentages below 3000 m (figure 16.5).

Wood samples from both sections, and phytoliths from Mera, record the local presence of the Andean taxa *Drimys,* Magnoliaceae, *Wienmannia,* and *Podocarpus,* as well as taxa appropriate to modern rainforest elevations like palms (Bush et al. 1990). The intrusion of wood and pollen of Andean taxa cannot have been by stream transport from high elevations because of the simultaneously high pollen concentrations of both Andean and rainforest taxa and because stream transport could not have moved pollen, phytoliths, and wood in unison to redeposit them together. The data thus require descent of Andean taxa, from their interglacial position of higher than 2500 m down to 1000-m elevation or lower, a descent consistent with the 1500-m descent of mountain glaciers and Colombian tree lines above them. Apparently the cooling long established for the high Andes was expressed at low elevations too (Bush et al. 1990).

The cooling can be measured by applying the logic of lapse rates. We use a rate of 5° C per thousand meters, which is the lapse rate shown by modern Ecuadorian meteorological records for all elevations from the Amazon lowlands to the high Andes, yielding a temperature depression of 7.5° C. This is a minimal estimate that allows highland taxa to descend only 1500 m to conform to the known tree line and glacial depressions.

The cooling cannot be simply a local effect caused by catabatic winds from ice at high elevation. The areal extent of glaciers on the highest Andes was still too small to make this hypothesis plausible, and we have shown comparable temperature depression in lowland Panama, where the highlands are yet too low to have held glacial ice (see "Glacial Cooling and the Exclusion of the Darien Refugium"). Paleoecological data from the Andes, therefore, now demonstrate ice-age cooling in excess of 7° C at all elevations, from snowline to the Amazon bottom lands.

G. Reconciliation of Proxy Data from Land and Sea

These data for significant cooling in equatorial America conflict directly with the CLIMAP (1976) reconstruction showing the

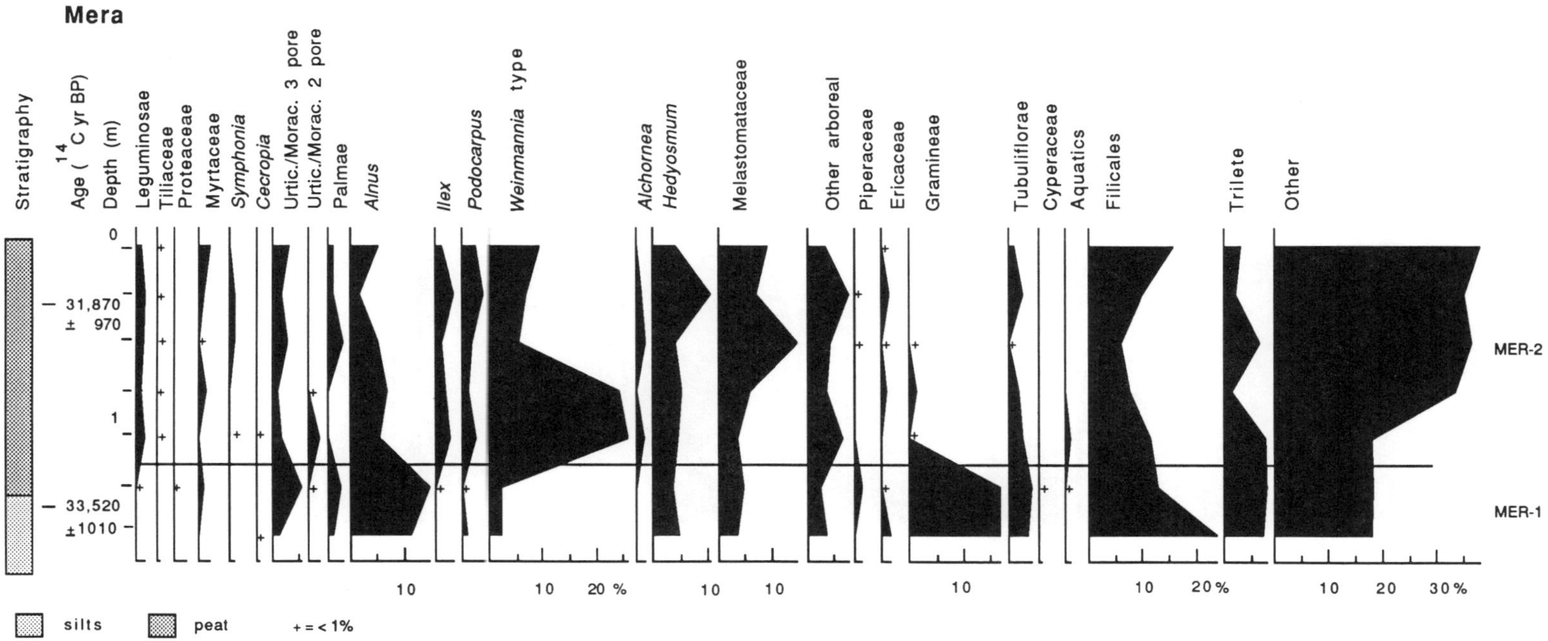

Figure 16.3. Pollen percentage diagram of glacial-age deposits at Shell Mera, Andean foothills, Ecuadorian Amazonia. Pollen analysis by M. B. Bush, from Bush et al. 1990.

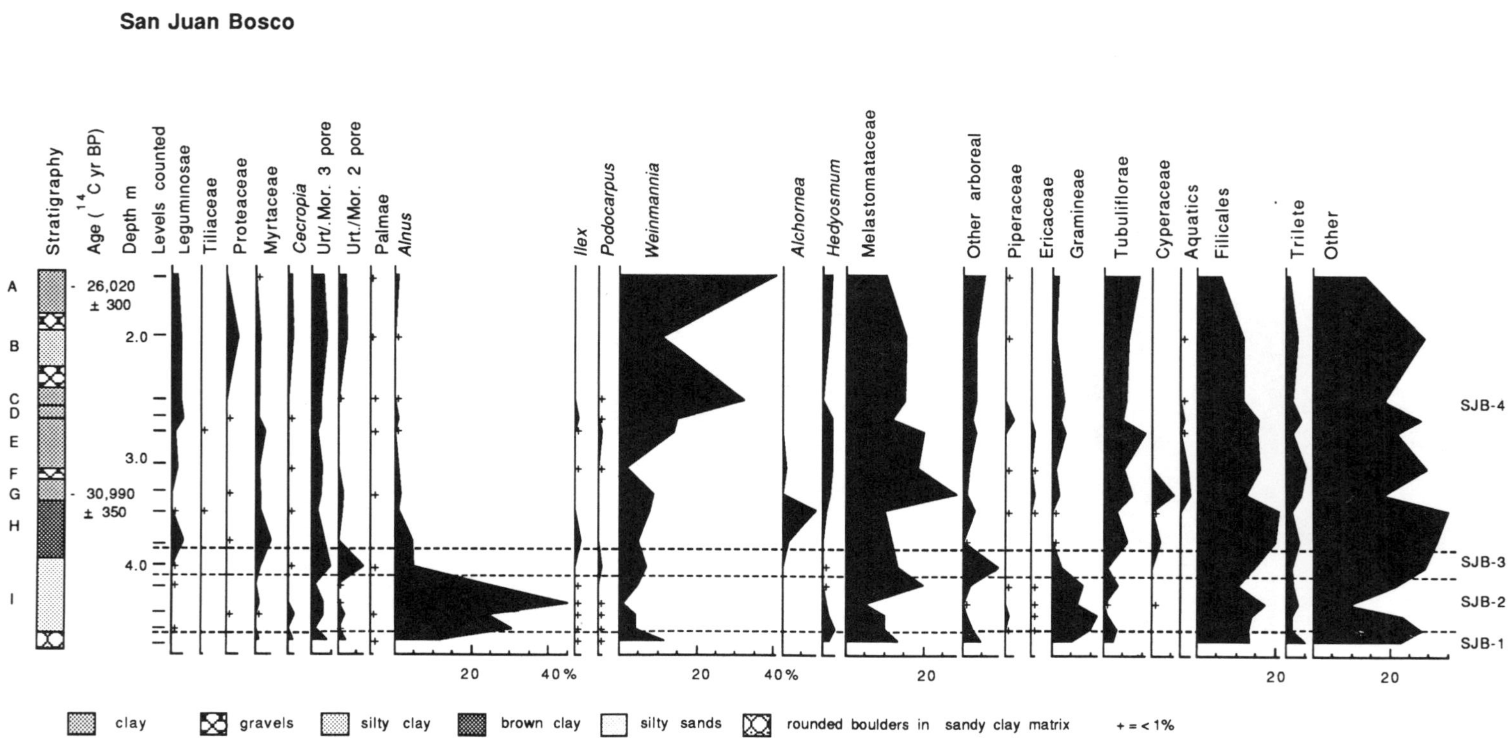

Figure 16.4. Pollen percentage diagram of glacial-age deposits at San Juan Bosco, Andean foothills, Ecuadorian Amazonia. Pollen analysis by M. B. Bush, from Bush et al. 1990.

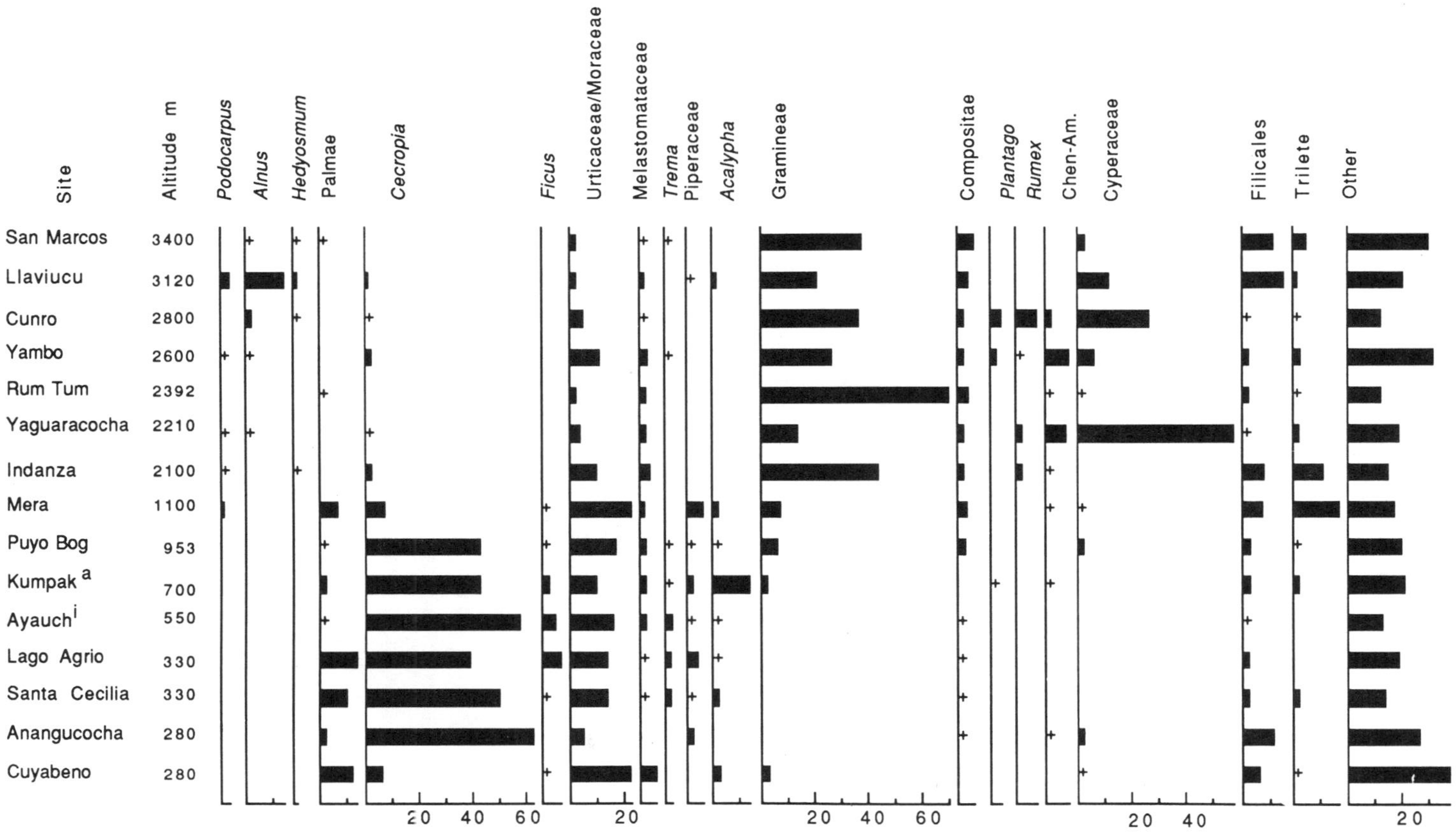

Figure 16.5. Surface pollen spectra from Ecuador. Pollen analyses by K-b. Liu and others, from Bush et al. 1990.

temperature of the equatorial sea surface as only modestly depressed. Our conclusions, however, are consistent with those of pollen analysts working in New Guinea (Flenley 1979; Walker and Flenley 1979) and Madagascar (see Chapter 15).

Before the recent results from Ecuadorian Amazonia were published, the conflict appeared to be merely between estimates of sea surface temperature and depressions of tree lines and snow lines at high elevations. Attempts were made to reconcile these two data sets by speculating that lapse rates were altitude-sensitive in glacial times, so that warm lowlands could persist under relatively colder highlands, a concept involving some difficult assumptions (Rind and Peteet 1985; Bonnefille et al. 1990). The new data showing temperature depression in the Amazon lowlands as well as in the high Andes appear to make these explanations untenable.

Proxy data for sea surface temperatures used by CLIMAP are assemblies of foraminiferans thought to be stenothermic. Temperature proxies from the land are forest trees, together with such geological benchmarks as old morraines and inferred firn lines. To the extent that the conflict pits Foraminifera and forest trees as alternative proxies for past temperature, the results could be reconciled by postulating that foraminiferal assemblies record mean annual temperature, whereas trees are sensitive to the coldest intervals in runs of years. This explanation requires that the amplitude of ice-age temperatures varied widely even as the mean remained close to the Holocene mean. In modern times, lowest recorded temperature at all elevations in the Andes tracks mean annual temperature quite closely (Bush et al. 1990). Furthermore, the explanation does nothing to address the issue of firn line and glacier depressions comparable to depressions of montane vegetation below them, since it requires that limits to mountain glaciers or firn lines should be set in a similar way by the coldest years or days. This seems unlikely.

An appealing alternative explanation is given by Livingstone in Chapter 15. This relies on the relative rarity of interglacial conditions like the present making unlikely the evolution of Foraminifera adapted to warmer water. As a result, modern Foraminifera of tropical surface oceans are postulated to be the same species that lived in colder oceans in glacial times, but which are now at the limits of their range, occupying ocean areas from which they would have been denied if more thermophilic species were available.

IV. Amazon Forests in Ice-Age Times

Cooling the Amazon by seven or more degrees celcius certainly had profound effects on the biota. Rainforest trees in bottom lands could not descend a further 1500 m, for this would put them below the surface of the sea. It follows that existing taxa of Amazonian rainforests are all adapted to climate regimes subject to air temperatures significantly colder than those of modern times, except for those few taxa that may have evolved in the last 10,000 years.

Community composition in Amazonian forests should have been different in ice-age times, as is illustrated by the Mera and San Juan Bosco records (figures 16.3 and 16.4). These show that communities replacing modern rainforests between the 900-m and 1100-m contours have no modern analogs. They are not merely high Andean forests displaced downslope but rather are assemblies built from selected species populations, many of which are now separated into lowland and highland floras. What appears to have happened is that glacial cooling allowed a number of Andean taxa to be able to compete at lower elevations, letting them arrive as immigrants into the upper reaches of the lowland forests. The relative abundance of lowland species then changed in response to competition from these immigrants, as well as directly to the cooling itself.

The predictable homes of the more thermophylic rainforest species throughout the colder intervals of glacial cycles are in the Amazonian bottom lands 100–300 m above modern sea level. Significant areas of Brazilian Amazonia lie close to 100-m elevation, thus providing a broad habitat some 800 m below our San Juan Bosco site. It is in this broad area that we should expect to find the densest ice-age populations of many Amazonian trees and their associated biota. A task of the next few years is to discover long pollen sections from lowland Amazonia in Brazil that can be used to test this hypothesis. Meanwhile we have pollen data only from our two sites in the Ecuadorian rainforest.

The Napo region of Ecuador probably holds the most species-rich forest in the world (Gentry 1988). Our records from Mera and San Juan Bosco show that this diverse rainforest was replaced in glacial times with moist forest of a kind for which there is no modern analog but in which both modern Andean elements and lowland taxa were prominent. Accordingly the Napo region was not a refugium in which modern rainforest persisted throughout a glacial cycle. Rather the Napo region of high species endemism was a place of invasion and counterinvasion, as the more temperature-sensitive plants extended ranges up and down slope with fluctuating temperatures of glacial cycles. The elevated Napo region, like all the other proposed "refugia," was thus a place of intense species interactions, apparently leading to the understandable outcome of being a center of speciation and endemism.

The Ecuador evidence for cooling further suggests that the more thermophilic of rainforest organisms had their ranges compressed to low elevations. Accordingly all the lower-lying parts of the Amazon basin that are forested now were probably forested throughout the glacial cycle, except for marginal regions with strongly seasonal climates. The forests of the central Amazon were probably not markedly fragmented. The Amazon basin as a whole was not arid, though savanna regions at the periphery were probably more extensive than now.

V. The Ice-Age Isthmus of Panama

Lake cores from La Yeguada and El Valle now provide a transglacial section for the Panamanian lowlands (Piperno et al. 1990; Bush and Colinvaux 1990; Bush et al. 1992). They are in a region of strongly seasonal climate where most of the 3–4 m of annual rain falls between January and June. Despite prolonged dry seasons at both sites, the local vegetation before human disturbance was tropical forest. The combined sections suggest that the overriding environmental events of glacial times were a lowering of mean annual temperature and the consequent compression of tropical forests into the lowlands, although some evidence of reduced precipitation is present, probably manifest as increased duration of dry seasons. A strong inference from the records is that the uplands of Darien were not a forest refugium and that moist tropical forest was present across the isthmus at all stages of a glacial cycle.

A. A Transglacial Pollen History from Lowland Panama

Laguna de La Yeguada lies at 650-m elevation on the Pacific side of the isthmus divide, which rises above the lake to 1500 m. Pollen and phytolith analyses are available for the complete sequence since the lake first formed with the emplacement of its dam at the start of the late-glacial period about 14,000 B.P. (Piperno et al. 1990; Bush et al. 1992). The pollen and phytolith data show that the local late-glacial forests were rich in what are now montane elements, including *Quercus, Magnolia, Ilex, Gunnera, Symplocus, Carex, Ranunculus,* and *Thalictrum*. Many of these taxa are now restricted to altitudes above 1500-m elevation, requiring downward displacements from present limits of at least 800 m in the late-glacial. Assuming the standard moist air lapse rate of 6° C per 1000 m, this yields a minimal temperature depression of 5° C. As with the Ecuadorian forest records, the montane taxa in the pollen diagram do not represent a total replacement of lowland vegetation by montane communities but rather an intermingling of montane and lowland elements into communities for which there are no modern analogs.

El Valle, site of an even older Panamanian environmental history, is a city built on the dried bed of a caldera lake, at 500-m elevation in a forested region of the Pacific slope of Panama. A 55-m sediment core of the underlying deposits spans from about 8000 B.P. at the top, through radiocarbon 30,000 B.P. by the tenth meter, to basal samples thought to represent the last interglacial of isotope substage 5e (Bush and Colinvaux 1990). The pollen diagram confirms the conclusion from the La Yeguada data that the late-glacial vegetation of the Panamanian lowlands included montane elements like oaks that were growing 800 m or more below their modern limits (figure 16.6). At El Valle, however, this record is extended into full glacial times. Temperature depressions ranging between 4° and 6° C are suggested for the whole glacial period.

B. Glacial Cooling and the Exclusion of the Darien Refugium

Temperature depression inferred from the La Yeguada and El Valle records strongly suggests that the cooling recorded at the Ecuadorian lowland sites was experienced throughout the entire Neotropical region. Certainly local explanations based on the descent of glacial ice cannot be applied to Panama, where no mountains were high enough to have been glaciated. The only discrepancy between Panamanian and Ecuadorian cooling is that temperature depression was slightly less in Panama, probably due to maritime effects from the nearby oceans.

In ice-age Panama, as in the Amazon, the more thermophilic plants of the tropical forests were restricted to bottom lands close to modern sea level. Higher elevations like the Darien were not refugia for tropical rainforests as has been argued (Brown 1987) but instead held moist forest communities resulting from cooler temperatures in which those

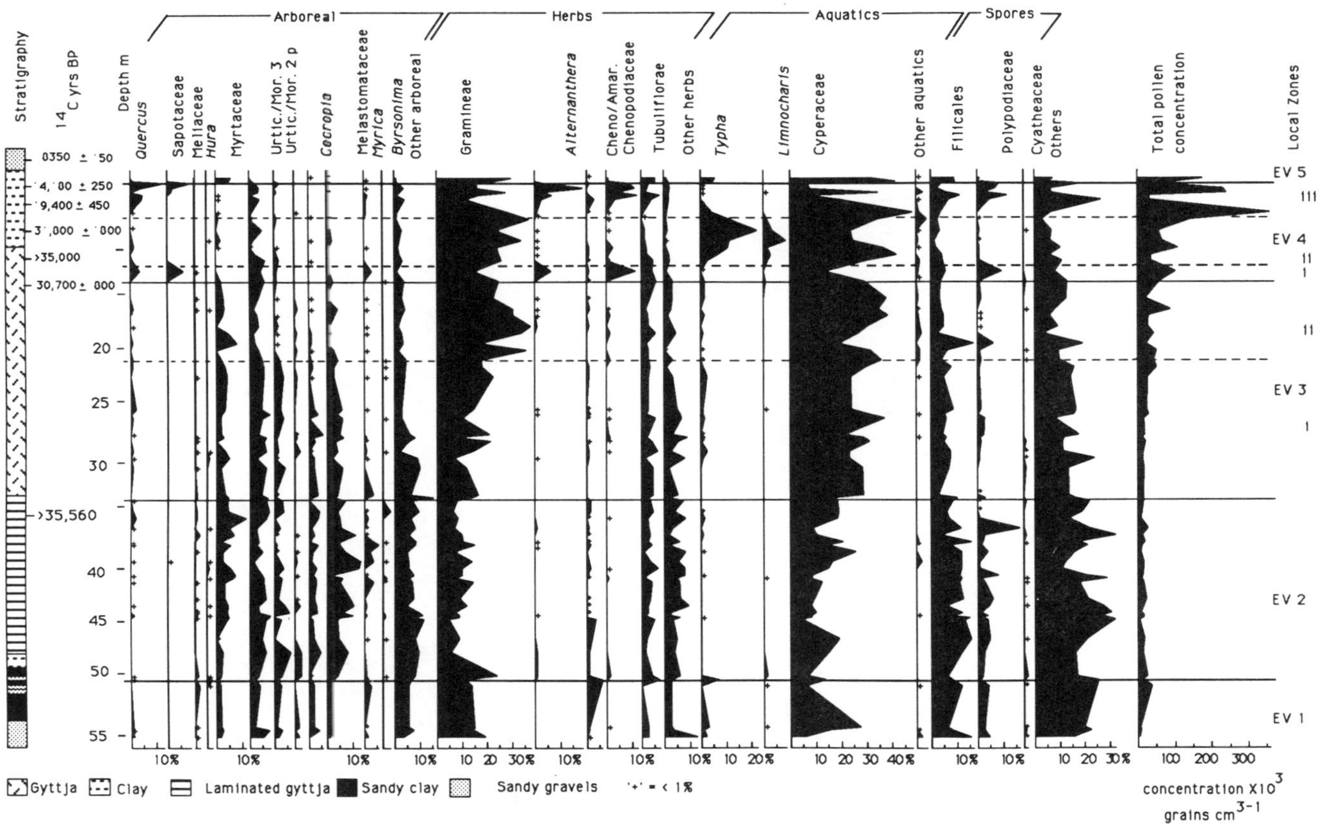

Figure 16.6. Pollen percentage diag-am of transglacial section of lake deposits from El Valle, Panama. Pollen analysis by M. B. Bush, from Bush and Colinvaux 1990.

tropical rainforest taxa with broadest temperature tolerances were associated with species now restricted to higher elevations.

The pronounced cooling was always insufficient, however, to allow modern montane taxa to be established in lowlands separating the elevated regions of Chiriqui and Darien. Taxa like *Quercus* that penetrated the isthmus, eventually reaching the Colombian Andes (Hooghiemstra 1984, 1989), yet had to disperse across gaps of lowland tropical forest. That tropical lowlands remained a severe barrier to dispersal explains why 11 oak species are known from Chiriqui but only 1 from Darien (Gentry 1989; Bush and Colinvaux 1990). That the gaps were crossed at all apparently was made possible because they were narrowed by the cool moist conditions of glacial times.

The data allow a strong inference that low-lying parts of the isthmus like the modern canal zone separating the elevated regions of Chiriqui and Darien, together with some coastal regions, were home to tropical forest through glacial and late-glacial times. Such is the variability of climate from one locale of Panama to another, however, that this lowland forest was doubtless a patchwork of forest communities, depending on position relative to rain shadows, though tropical rainforest should have been expected in the lowlands spanning the isthmus along the line of the modern canal.

C. Falsifying the Savanna Causeway Hypothesis

A consequence of having the Panamanian lowlands covered with tropical forest throughout entire glacial cycles is the exclusion of postulated savanna causeways for migrating fauna and humans. This exclusion is consistent with the paleontological evidence used by Webb (1976) to show that no savanna causeway can have existed through the isthmus for the last million years.

Localized, semi-arid habitats were probably available, however, on coastal strips of land exposed by eustacy but now submerged. Wide shelves of shallow sea now abut the southern (Pacific) shores of Panama and would have been exposed as dry land in glacial times. Because intermittent stretches of the modern south Panama coast are so strongly seasonal as to be semi-arid, it is not unreasonable to expect comparable aridity in the glacial period, thus providing extensive savanna habitats along the south coast. Where coastal aridity reflects rain shadows to northeast monsoons, the ice-age dry areas of the coast should have been even more extensive than those of modern times from the diminution of monsoon rains (Piperno et al. 1991a). Evidence of lowered water levels in the lake at El Valle concurs with the Lake Valencia and Peten records in suggesting this (Salgado-Labouriau 1980; Bradbury et al. 1981; Leyden 1984, 1985; Bush and Colinvaux 1990).

In modern times, semi-arid patches do not line the entire coast, there being rainforest to the sea's edge at the eastern (Colombian) end. Our pollen data suggest that this forest barrier was always present and may indeed have been even more extensive in glacial times (Piperno et al. 1991a, 1991b; Bush et al. 1992). It follows that all animals and plants using the Isthmus of Panama as a crossing between continents at any time in a glacial cycle necessarily penetrated stretches of lowland tropical forest.

VI. Holocene Dynamics and Diversity in the Neotropics

Many of the paleoecological data available to define the Holocene climates of Amazonia are reviewed extensively in Colinvaux (1987) and Salo (1987). The following is a condensation of those reviews, with the addition of more recent data.

Records of the last few millennia in Amazonia are consistent with the concept of "intermediate disturbance," with its implication that small-scale physical accidents prevent competitions running to the point of exclusion. This sense of recurrent disturbance is present in all dozen pollen diagrams so far available for parts of the Holocene in Amazonia, as is illustrated by pollen from Lake Kumpak[a], a crater lake (Liu and Colinvaux 1985; figure 16.7). In this diagram, nearly all pollen types, whether of disturbance taxa like *Cecropia* or taxa more suggestive of terra firme forests like *Ficus* or Moraceae, appear to oscillate around a mean for all 5000 years of the record. Pollen diagrams in which percentages of pollen types oscillate thus around a mean are typical of Holocene sections from Amazonia (Absy 1979, 1986; Bush and Colinvaux 1988; Frost 1988; Colinvaux, Frost, Frost, Liu, Steinitz-Kannan 1988; figure 16.8).

Disturbance through erosion by rivers, or by colluvial denudation of the soil surface, is also massive in the western Amazon. The most direct demonstration of the magnitude of this effect is for Peruvian rivers, where 27 percent of the total forest area occupies former floodplains no more than a few centuries old, and where 3.7 percent of the floodplain of one small river was shown to have been reworked in 13 years (Salo et al. 1986). River disturbance is necessarily concentrated at the lowest areas, suggesting that significant areas of terra firme forest escape (Dumont et al. 1990), but surficial denudation should penetrate where the rivers do not reach. Calculations by I. Frost based on estimates of sediment discharge from the Amazon system suggest that the western Amazon loses as much as 1.7 mm of land surface yearly and the rest of the basin an average of 0.06 mm a year (Colinvaux 1987). It is possible, therefore, that 5–10 cm of the soil surface may be removed from over the roots of the average tree of the western Amazon during its lifetime, with no compensating accumulation. This could be a significant factor in inducing tree fall.

Old riverine lakes of the Rio Napo and Aguarico systems in Ecuador hold a record of large channel movements, apparently associated with regional flooding in the interval 1300–800 B.P. that can be traced downstream to flooding episodes in two Brazilian Amazon sections of Absy's (1979) set, Costa da Terra Nova and Lake Caju (Colinvaux 1987). Partly because this flooding can be correlated with intervals of high lake level in the Ecuadorian Andes, the flooding has been attributed to increased precipitation on the flanks of the Andes and in the western Amazon (Colinvaux et al. 1985; Colinvaux, Frost, et al. 1988; Colinvaux, Olson, and Liu 1988; Frost and Miller 1987). The alternative hypothesis that the channel movements and associated flooding resulted from river damming and stream mechanics has been put forward by Räsänen and co-workers (1991). Whether the primary cause of the millennial flooding may turn out to be increased precipitation or stream mechanics,

LAKE KUMPAK[a]
Pollen Percentage Diagram

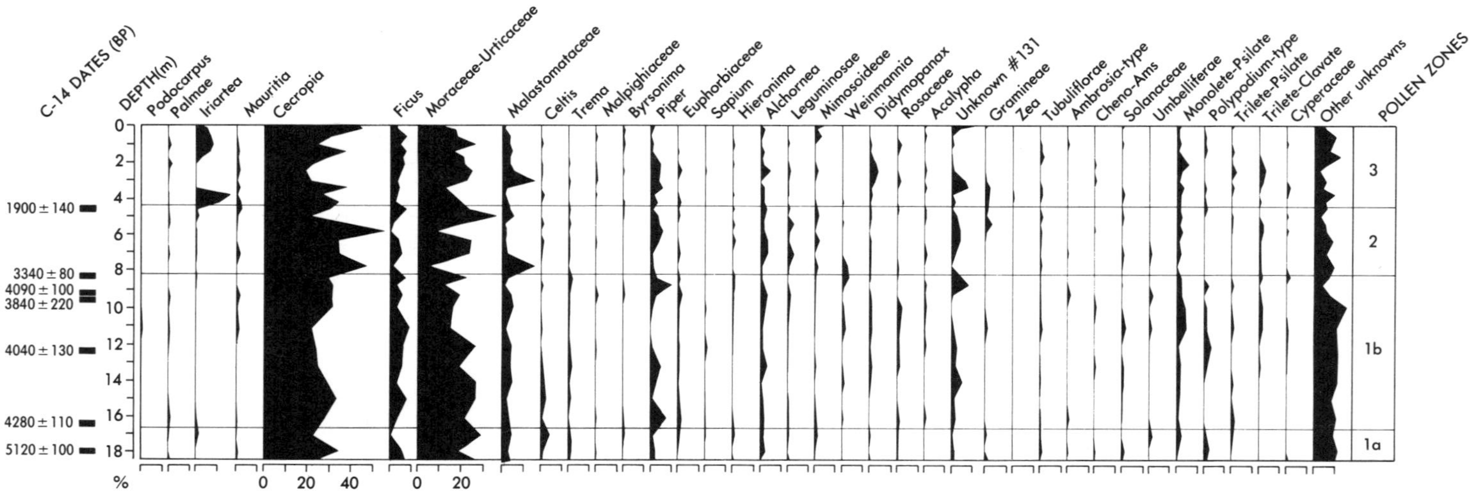

Figure 16.7. Pollen percentage diagram of sediments in Lake Kumpak[a], an explosion crater lake in Ecuadorian Amazonia. Pollen analysis by K-b. Liu, from Liu and Colinvaux 1985.

the occurrence itself illustrates an important way in which plant successions are repeatedly engendered in thc Amazon basin.

An alternative hypothesis of Amazonian disturbance has been put forward with the claim that in the early Holocene, or late in the last glacial cycle, the whole of the Amazonian lowlands were occupied by a giant freshwater lake, dubbed "Lake Amazonas" (Cambell and Frailey 1984; Frailey et al. 1988). This lake is postulated to have resulted when an ice-dam in the high Andes of Peru collapsed releasing the waters of a huge proglacial lake in a cascade onto the Amazon lowlands. The racing waters piled up sediment across the lower Amazon to make a dam, behind which the waters then remained for centuries or millennia. The evidence offered in support of this idea is dendritic patterns visible on landsat imagery that resemble dendritic deposits laid down under water in river estuaries. The hypothesis has the advantage of being testable, because it predicts the existence of both fossil shorelines and remnants of the dam. In addition, all surficial deposits within the Amazon system are predicted to have similar ages. Neither shorelines of the largest lake in history, nor the remnants of the necessary massive dam have been identified. Recent work by Räsänen and co-workers (1990) has shown that deposits exposed in Peruvian river cuts are of ages varying over orders of magnitude, thus directly falsifying the hypothesis.

Humans probably began disturbing Amazon forests early in the Holocene. The earliest paleoecological evidence for cultivation comes from the sediments of the Ecuadorian crater lake Ayauchi, where both pollen and phytolith data show that *Zea* was grown by 6000 B.P. (Bush et al. 1989). This discovery allows a reinterpretation of evidence for forest burning in the Amazon of Venezuela before 6000 B.P., which was described by Sanford and co-workers (1985) as probably denoting wildfire but which instead may well have been the result of fires set by human cultivators or hunters.

That humans burned tropical forests of the New World from an early date is shown by the charcoal record in the Lake La Yeguada cores from Panama. An abrupt rise in fine particulate carbon at 11,000 B.P. suggested widespread fires and was associated with phytoliths of weedy plants (Piperno et al. 1990).

An argument progresses in the anthropological literature over whether hunting and gathering peoples could have sustained themselves in tropical rainforests (Bailey et al. 1989; Colinvaux and Bush 1991). We take the evidence of charcoal at 11,000 B.P. in Panamanian forests, together with the evidence for limited *Zea* cultivation 5000 years later in the Amazon, to suggest that students of Amazonian diversity should be prepared for the possibility that the whole basin has been disturbed by hunting and burning for ten millennia. This would produce the type of people-induced gap dynamics postulated by Hubbel (1979). Such disturbance by rainforest peoples should have started much earlier in the Old World tropics.

The Holocene record from the Amazon suggests that all rainforests are disturbed on decadal, centennial, and millennial time scales. Most Amazonian ecosystems, most of the time, are in successional stages. This conclusion gives comfort to hypotheses like "intermediate disturbance" that rely on

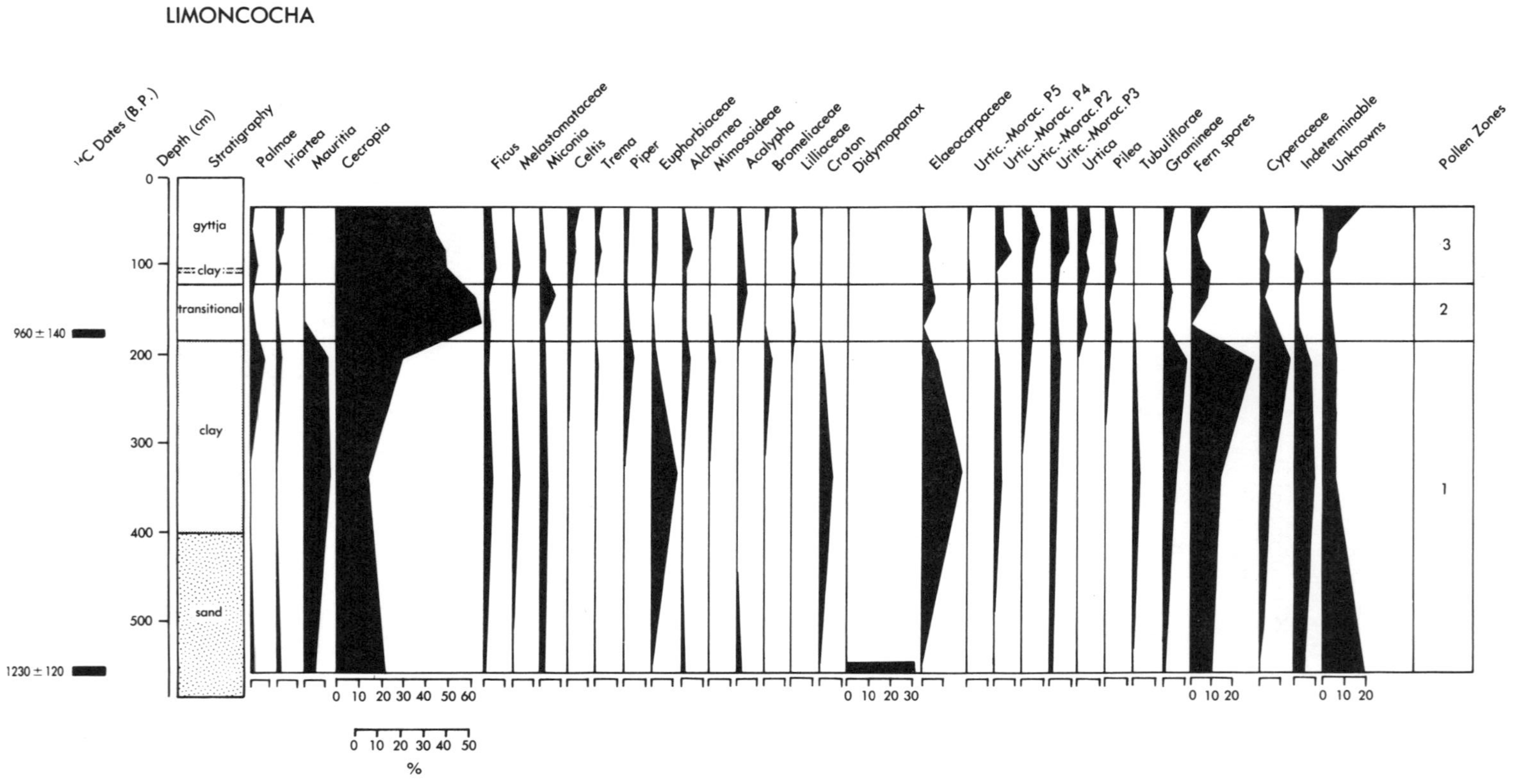

LIMONCOCHA
^{14}C Dates (B.P.)
Depth (cm)
Stratigraphy
Palmae
Iriartea
Mauritia
Cecropia
Ficus
Melastomataceae
Miconia
Celtis
Trema
Piper
Euphorbiaceae
Alchornea
Mimosoideae
Acalypha
Bromeliaceae
Lilliaceae
Croton
Didymopanax
Elaeocarpaceae
Urtic.-Morac. P5
Urtic.-Morac. P4
Urtic.-Morac. P2
Uritc.-Morac. P3
Urtica
Pilea
Tubuliflorae
Gramineae
Fern spores
Cyperaceae
Indeterminable
Unknowns
Pollen Zones
gyttja
clay
transitional
clay
sand
960 ± 140
1230 ± 120
0
100
200
300
400
500
3
2
1
0 10 20 30 40 50 60
0 10 20 30
0 10 20
0 10 20
0 10 20 30 40 50
%

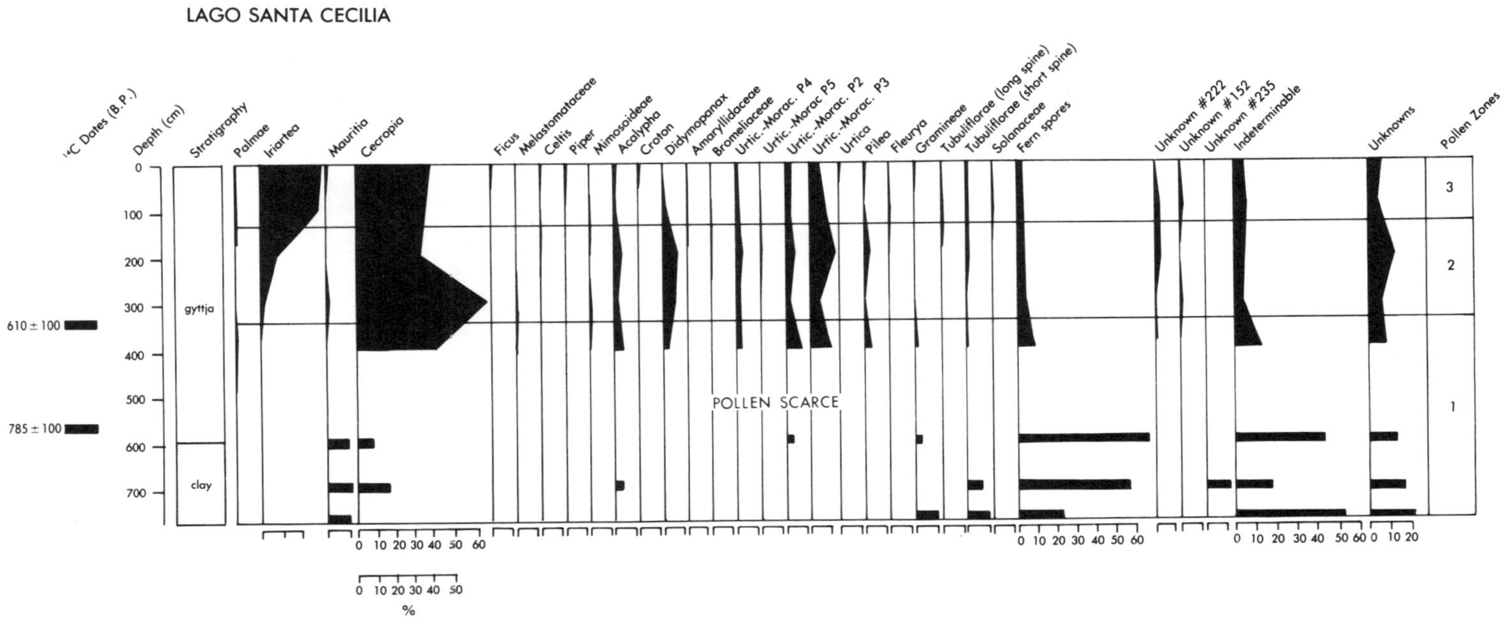

Figure 16.8. Pollen percentage diagrams from sediments of two lowland Amazonian lakes in Ecuador disturbed by the 800 B.P. flooding event, Lakes Limoncocha and Santa Cecilia. Pollen analyses by M. Frost and I. Frost, from Colinvaux, Frost, Frost, Liu, and Steinitz-Kannan 1988.

prevention of competitive exclusion, and less to hypotheses that rely on sharing resources at equilibrium.

VII. Reversing Lyell: The Past as the Key to the Present

To be used as an explanation for modern distributions, the past must be inferred independently of the evidence of modern distributions, or arguments become circular. This general hazard of circularity is compounded by the relative shortness of our interglacial compared with the long glacial period that preceded it. Glacials typically last far longer than brief interglacials like the present, requiring that a glacial regime should be regarded as the normal condition. The hazard of inferring the normal from data describing radical departure from normality is obvious, and yet that is what is done each time Pleistocene biogeography is reconstructed from speculations about the origins of modern communities.

Conversely, the biogeography of our present interglacial can be understood only by first knowing something of the distribution of habitats and populations in the long glacial period from which we have just emerged. This is what I mean by saying that we must reverse Lyell's celebrated dictum to declare that the past is the key to the present.

Taking the directly measured past as the key to the present in this way leads, as we have seen, to explanations of modern biogeography that are quite different from many current ideas. The biogeography of modern Panama, for instance, has come about not by emergence from refuges or travels over savanna, but primarily as a result of warming that allowed species of the lowland forests to extend their ranges upslope against the weakening competition of taxa like oaks that had been dominant at mid elevations in glacial times. Where oaks are present in modern Darien it is as remainders of the ice-age forests persisting in reduced populations at the edges of their range; my colleague Mark Bush calls them "die hards."

A. Loss of Species in Africa through Ice-Age Cooling

Our data suggest that cooling at the equator was so severe that the more thermophylic of Amazon rainforest taxa have their normal Pleistocene distributions, those of glacial times, confined to Amazon bottom lands lower than 300 m above modern sea level. This observation invites speculation about tropical tree distribution and diversity in Africa, where most of the rainforest of the Congo basin appears to lie above the modern 500-m contour, and where the west coast is without a wide continental shelf. Cold episodes comparable to those demonstrated for the western Amazon are thought to have been experienced in Africa also (Chapter 15). These should have cooled much of tropical Africa to below the tolerances of the more stenothermic of thermophilic species during the colder intervals of glaciations. Over the entire span of the Pleistocene, the shortage of low-elevation habitats might have led to random extinctions in the African rainforests, thus contributing to lower species richness than in Amazonia or Southeast Asia. Perhaps a measure of the validity of this speculation is given, however, by Gentry's suggestion (Chapter 17) that species diversity in African rainforests might not be very different from that in the Amazon after all.

VIII. The Disturbance-Vicariance Hypothesis

Denial of "refugial" status to regions of endemism in the Amazon allows a new hypothesis to account for their high species richness and hence high endemism. The regions are dynamic borderlands between the uplands and the low. Throughout most of the Pleistocene, both mean annual temperatures and low extremes of temperature are several degrees colder than at present. In the shorter intervals of extreme cold that come once or twice in each glacial cyle, means and extremes of temperature plunge further, to 7–9° C below modern levels. At those times, many rainforest taxa are excluded from the more elevated parts of their present range altogether. In the equally short, warm interglacials like the present, temperatures rise so that virtually all rainforest taxa can compete at elevations up to 1000 m. The result is an oscillation in species distribution and abundance in these apparent centers of endemism throughout the Pleistocene. Where the wet elevated regions abut onto true mountains, as in the Andean forelands, the dynamics are at their strongest as montane and lowland floras are squeezed together, repeatedly invading each other. It is thus understandable that the Andean foothills, as in the Napo region of Ecuador and Peru, appear to have the highest diversity of any part of the Amazon basin, or indeed of the world (Gentry 1988).

This is a model of endemism and speciation in the Amazon basin in which vicariance remains paramount; local regions of higher precipitation and relief are indeed separated as the purported "refugial" maps show. But they are separated only by lowland forest and rivers in which the dynamics of small-scale disturbances persist through all parts of a glacial cycle. Invasions, retreats, and reinvasions in the elevated isolates proceed with the rhythms of glacial advances as the air cools and warms, probably with accompanying local changes in precipitation or seasonality. Many of the interactions between the invaders are necessarily competitive, thus favoring species isolations.

The recent correlation by Nelson and co-workers (1990) of some refugialists' maps with plots of collecting intensity is consistent with this *disturbance-vicariance hypothesis*. Collectors might naturally have converged on the species-rich localities as the data collected by Nelson and co-workers (1990) suggest they have, but the underlying natural pattern might well remain that of wide distribution of genera and families in the lowlands, with isolates of species richness wherever particularly intense sequences of invasions and retreats over relief have been driven by temperature changes.

The concept of disturbance-vicariance is also applicable to all tropical rainforests on a smaller scale. Where gap dynamics are a prominent property, competitive invasions are the norm with climax or competitive exclusion being prevented. Because the forests occupy basins or plains of immense size, distant portions are always effectively isolated. In the Amazon, this general isolation is enhanced by the strong local isolations imposed by systems of rivers acting as barriers to many of the taxa in which local endemisms are best known, like forest birds or butterflies. It is unreasonable to expect panmixis across the large extents of rainforests, except in the most mobile species. The farther parts, therefore, are semi-isolates where *intermediate disturbance* forces endless invasions and hence opportunities for new species to evolve.

Gentry's (1988) demonstration that Amazon ecosystems are the richest on earth, even in forest trees, probably depends on the unique suitability of the Amazon basin to *disturbance-vicariance*. Not only is the area occupied by a larger and lower-lying tropical forest than that of other tropical ecosystems, but vicariance is enhanced through the extraordinary dissection of the Amazon basin by rivers. Perhaps more important still is the unique Amazon pattern of vast bottom lands surrounded, or punctuated, by immense hills. Climatic cycles force endless successions of invasion and dispersal on all sides of the basin. Vicariance ensures that the resulting swarms of local species shall persist as endemics to broad but separated areas of the Amazon basin.

Literature Cited

Absy, M. L. 1979. A palynological study of Holocene sediments in the Amazon basin. Ph.D. diss., University of Amsterdam.

———. 1986. Palynology of Amazonia: The history of the forests as revealed by the palynological record. Pp. 72–82 *in* G. T. Prance and T. Lovejoy (eds.), *Amazonia*. Oxford: Pergamon.

Absy, M. L., T. V. van der Hammen, F. Soubies, K. Suguio, L. Martin, M. Fournier, and B. Turcq. 1989. Data on the history of vegetation and climate in Carajas, Eastern Amazonia. Pp. 129–131 *in International Symposium on Global Changes in South America during the Quaternary: Past-Present-Future*. Saõ Paulo: ABEQUA/INQUA Special Publication.

Bailey, R. C., G. Head, M. Jenike, B. Owen, R. Rechtman, and E. Zechenter. 1989. Hunting and gathering in tropical rain forest: Is it possible? Amer. Anthropol. 91:59–82.

Bonnefille, R., J. C. Roeland, and J. Guiot. 1990. Temperature and rainfall estimates for the past 40,000 years in equatorial Africa. Nature 346:347–349.

Bradbury, J. P., B. Leyden, M. Salgado-Labouriau, W. Lewis, Jr., C. Schubert, M. Binford, D. Frey, D. Whitehead, and F. Weibezahn. 1981. Late Quaternary environmental history of Lake Valencia, Venezuela. Science 214:1299–1305.

Braun. E. L. 1955. The phytogeography of unglaciated Eastern United States and its interpretation. Bot. Rev. 21:297–375.

Brown, K. S., Jr. 1987. Soils and vegetation. Pp. 19–45 *in* T. C. Whitmore and G. T. Prance (eds.), *Biogeography and Quaternary History in Tropical America*. Oxford: Oxford Science Publications.

Bush, M. B., and P. A. Colinvaux. 1988. A 7000-year pollen record from the Amazon lowlands. Ecuador. Vegetatio 76:141–154.

———. 1990. A pollen record of a complete glacial cycle from lowland Panama. J. Veg. Sci. 1:105–118.

Bush, M. B., D. R. Piperno, and P. A. Colinvaux. 1989. A 6000-year history of Amazonian maize cultivation. Nature. 340:303–305.

Bush, M. B., P. A. Colinvaux, M. C. Wiemann, D. R. Piperno, and K-b. Liu. 1990. Late Pleistocene temperature depression and vegetation change in Ecuadorian Amazonia. Quat. Res. 34:330–345.

Bush, M. B., D. R. Piperno, P. A. Colinvaux, L. Krissek, P. E. De Oliveira, and M. C. Miller. 1992. A 14,300-year paleoecological profile of a lowland tropical lake in Panama. Ecol. Monog. 62:251–275.

Campbell, K. E., and D. Frailey. 1984. Holocene flooding and species diversity in southwestern Amazonia. Quat. Res. 21:369–375.

Clapperton, C. M. 1987. Maximal extent of the late Wisconsin glaciation in the Ecuadorian Andes. Pp. 165–180 *in* J. Rabassa (ed.), *Quaternary of South America and Antarctic Peninsula*, Vol. 5. Rotterdam: A. A. Balkema.

CLIMAP Project Members. 1976. The surface of the ice-age earth. Science 191:1131–1137.

Colinvaux, P. A. 1972. Climate and the Galapagos Islands. Nature 240:17–20.

———. 1973. *Introduction to Ecology.* New York: John Wiley & Sons.

———. 1986. *Ecology.* New York: John Wiley & Sons.

———. 1987. Amazon diversity in the light of the paleoecological record. Quat. Sci. Rev. 6:93–114.

Colinvaux, P. A., and M. B. Bush. 1991. The rain forest ecosystem as a resource for hunting and gathering. Amer. Anthropol. 93:153–160.

Colinvaux, P. A., M. C. Miller, K-b. Liu, M. Steinitz-Kannan, and I. Frost. 1985. Discovery of permanent Amazon lakes and hydraulic disturbance in the upper Amazon basin. Nature 313:42–45.

Colinvaux, P. A., M. Frost, I. Frost, K-b. Liu, and M. Steinitz-Kannan. 1988. Three pollen diagrams of forest disturbance in the western Amazon basin. Rev. Palaeobot. Palynol. 55:73–81.

Colinvaux, P. A., K. Olson, and K-b. Liu. 1988. Late-glacial and Holocene pollen diagrams from two endorheic lakes of the Inter-Andean plateau of Ecuador. Rev. Palaeobot. Palynol. 55:83–100.

Connell, J. H. 1978. Diversity in tropical rainforests and coral reefs. Science 199:1302–1310.

Connell, J. H., and E. Orias. 1964. The ecological regulation of species diversity. Amer. Natural. 98:387–414.

Damuth, J. E., and R. W. Fairbridge. 1970. Equatorial Atlantic deep-sea arkosic sands and ice-age aridity in tropical South America. Bull. Geol. Soc. Amer. 81:189–206.

Davis, M. B. (ed.). 1986. Vegetation—climate equilibrium. Vegetatio 67, no. 2 (Spec. Iss.):141.

Deevey, E. S. 1949. Biogeography of the Pleistocene. Bull. Geol. Soc. Amer. 60, no. 9:1315–1416.

Dumont, J. F., S. Lamotte, and F. Kahn. 1990. Wetland and upland forest ecosystems in Peruvian Amazonia: Plant species diversity in the light of some geological and botanical evidence. Forest Ecol. Manag. 33, 34:125–139.

Fischer, A. 1960. Latitudinal variations in organic diversity. Evolution 14:64–81.

Flenley, J. R. 1979. *A Geological History of Tropical Rainforest.* London: Butterworths.

Frailey, C. D., E. L. Lavina, A. Rancy, and J. P. de Souza Filho. 1988. A proposed Pleistocene/Holocene lake in the Amazon Basin and its significance to Amazonian Geology and Biogeography. Acta Amaz. 3–4, no. 18:119–143.

Frost, I. 1988. A Holocene sedimentary record from Anangucocha in the Ecuadorian Amazon. Ecology 69:66–73.

Frost, I., and M. C. Miller. 1987. Late Holocene flooding in the Ecuadorian rain forest. Freshwater Biology 18:443–453.

Gentry, A. H. 1982. Neotropical floristic diversity: Phytogeographical connections between Central and South America, Pleistocene climatic fluctuations, or an accident of the Andean orogeny? Ann. Missouri Bot. Gard. 69:557–593.

———. 1988. Tree species richness of upper Amazonian forests. Proc. Natl. Acad. Sci. U.S.A. 85:156–159.

———. 1989. Speciation in tropical forests. Pp. 113–134 *in* L. B. Holm-Nielsen, I. C. Nielsen, and H. Balslev (eds.), *Tropical Forests.* London: Academic Press.

Grabandt, R. A. J. 1980. Pollen rain in relation to arboreal vegetation in the Colombian Cordillera Oriental. Rev. Palaeobot. Palynol. 29:65–147.

Haffer, J. 1969. Speciation in Amazonian forest birds. Science 165:131–137.

Hastenrath, S., and J. Kutzbach. 1985. Late Pleistocene climate and water budget of the South American Altiplano. Quat. Res. 24:249–256.

Hooghiemstra, H. 1984. *Vegetational and Climatic History of the High Plain of Bogota, Colombia: A Continuous Record of the Last 3.5 Million Years.* Vaduz: Verlag.

———. 1989. Quaternary and upper Pliocene glaciations and forest development in the tropical Andes: Evidence from a long high-resolution pollen record from the sedimentary basin of Bogotá, Colombia. Palaeogeogr. Palaeoclimatol. Palaeoecol. 72:11–26.

Houvenaghel, G. T. 1974. Equatorial undercurrent and climate in the Galapagos Islands. Nature 250:565–566.

Hubbel, S. P. 1979. Tree dispersion, abundance, and diversity in a tropical dry forest. Science 203:1299–1309.

Hutchinson, G. E. 1959. Homage to Santa Rosalia or why are there so many kinds of animals? Amer. Natural. 93:145–159.

Irion, G. 1982. Mineralogical and geochemical contribution to climatic history in central Amazonia during Quaternary time. Trop. Ecol. 23:76–85.

———. 1984. Sedimentation and sediments of Amazonian rivers and evolution of the Amazonian landscape since Pliocene times. *In* H. Sioli (ed.), *The Amazon Limnology and Landscape of a Mighty Tropical River and Its Basin*. Dordrecht: Dr. W. Junk.

Kutzbach, J. E., and P. J. Guetter. 1984. The sensitivity of monsoon climates to orbital parameter changes for 9,000 years B.P.: Experiments with the NCAR General Circulation Model. Pp. 801–820 *in* A. L. Berger et al. (eds.), *Milankovitch and Climate*, Part 2. Boston: Reidel.

———. 1986. The influence of changing orbital parameters and surface boundary conditions on climate simulations for the past 18,000 yrs. J. Atmos. Sci. 43:1726–1759.

Leyden, B. 1984. Guatemalan forest synthesis after Pleistocene aridity. Proc. Nat. Acad. Sci. U.S.A. 81:4856–4859.

———. 1985. Late Quaternary aridity and holocene moisture fluctuations in the Lake Valencia basin, Venezuela. Ecology 66:1279–1295.

Liu, K.-b., and P. A. Colinvaux. 1985. Forest changes in the Amazon basin during the last glacial maximum. Nature 318:556–557.

Livingstone, D. A. 1975. Late Quaternary climatic change in Africa. Ann. Rev. Ecol. Syst. 6:249–280.

MacArthur, R. H., and J. MacArthur. 1961. On bird species diversity. Ecology 42:594–598.

Martin, P. S. 1958. Pleistocene ecology and biogeography of North America. Pp. 375–420 *in* C. L. Hubbs (ed.), *Zoogeography*, Publ. 51. Washington, D.C.: American Association for the Advancement of Science.

Nelson, B. W., C. A. C. Ferreira, M. F. da Silva, and M. L. Kawasaki. 1990. Endemism centres: Refugia and botanical collection density in Brazilian Amazonia. Nature 345:714–716.

Newell, R. E. 1973. Climate and the Galapagos Islands. Nature 245:91–92.

Osman, R. W., and R. B. Whitlach. 1978. Patterns of species diversity: Facto or artifact? Paleobiology 4:41–54.

Piperno, D. R., M. B. Bush, and P. A. Colinvaux. 1990. Paleoenvironments and human settlement in late-glacial Panama. Quat. Res. 33:108–116.

———. 1991a. Paleoecological perspectives on human adaptation in Panama: I. The Pleistocene. Geoarch. 6:201–226.

———. 1991b. Paleoecological perspectives on human adaptation in Panama: II. The Holocene. Geoarch. 6:227–250.

Prance, G. T. 1982. Forest refuges evidence from woody angiosperms. Pp. 137–156 *in* G. T. Prance (ed.), *Biological Diversification in the Tropics*. New York: Columbia University Press.

Räsänen, M. E., J. S. Salo, H. Jungner, and L. Romero Pittman. 1990. Evolution of the Western Amazon lowland relief: Impact of Andean foreland dynamics. Terra Nova 2: 320–332.

Räsänen, M. E., J. S. Salo, and H. Jungner. 1991. Holocene floodplain lake sediments in the Amazon: 14C dating and palaeolecological use. Quat. Sci. Rev. 10:363–372.

Rind, D., and D. Peteet. 1985. Terrestrial conditions at the last glacial maximum and CLIMAP sea-surface temperature estimates: Are they consistent? Quat. Res. 24:1–22.

Salgado-Labouriau, M. L. 1980. A pollen diagram of the Pleistocene-Holocene boundary of Lake Valencia, Venezuela. Rev. Palaeobot. Palynol. 30:297–312.

Salo, J. 1987. Pleistocene forest refuges in the Amazon: Evaluation of the biostratigraphical, lithostratigraphical and geomorphological data. Ann. Zool. Fennici 24:203–211.

Salo, J., R. Kalliola, L. Hakkinen, Y. Kakinen, P. Niemela, M. Puhakka, and P. D. Coley. 1986. River dynamics and the diversity of Amazon lowland forest. Nature 322:254–258.

Sanders, H. L. 1968. Marine benthic diversity: A Comparative Study. Amer. Natural. 102:243–282.

Sanford, R. L., J. Saldarriaga, K. E. Clark, C. Uhl, and R. Herrera. 1985. Amazon rain-forest fires. Science 227:53–55.

Schoener, T. W. 1974. Resource partitioning in ecological communities. Science 185:27–39.

Simpson, B. B. 1975. Glacial climates in the eastern tropical South Pacific. Nature 253:34–36.

Sternberg, H. 1982. Refugial theory and Amazonian environment. Pp. 997–998 *in* V. A. Morak and J. M. Likovsky (eds.), *Evolution and Environment*. Prague: ČSAV.

Street, F. A., and A. T. Grove. 1979. Global maps of lake level fluctuations since 30,000 yr B.P. Quat. Res. 12:83–118.

Terborgh, J. 1973. Chance, habitat, and dispersal in the distribution of birds in the West Indies. Evolution 27:338–349.

Tricart, J. 1977. Apercus sur le quaternaire Amazonien: Recherches françaises sur le quaternaire. INQUA, Supplement au Bulletin AFEQ 265–271.

van der Hammen, T., and E. Gonzalez. 1960. Upper Pleistocene and Holocene climate and vegetation of the Sabana de Bogota (Colombia, South America). Leidse. Geol. Med. 25:126–315.

Walker, D., and J. R. Flenley. 1979. Late Quaternary vegetational history of the Enga Province of upland Papua New Guinea. Philos. Trans. Roy. Soc. London, B 286:265–344.

Wallace, A. R. 1878. *Tropical Nature and Other Essays*. London: Macmillan.

Webb, S. D. 1976. Mammalian faunal dynamics of the great American interchange. Paleobiology 2:220–234.

Webb, T., III. 1986. Is vegetation in equilibrium with climate? How to interpret late Quaternary pollen data. Vegetatio 67:75–91.

Whitmore, T. C., and G. T. Prance. 1987. *Biogeography and Quaternary History in Tropical America.* Oxford: Oxford University Press.

Wijmstra, T. A., and T. van der Hammen. 1966. Palynological data on the history of tropical savannas in northern South America. Leidse. Geol. Med. 38:71–90.

Chapter 17

Diversity and Floristic Composition of Lowland Tropical Forest in Africa and South America

Alwyn H. Gentry

I. Introduction

The last volume that focused on relationships between Africa and South America included floristic overviews by Thorne (1973), Smith (1973), and Richards (1973), plus reviews of resin-producing legumes (Langenheim 1973), Velloziaceae (Ayensu 1973), and palms (Moore 1973). With the exception of Moore's (1973) palm paper, which came out strongly for a shared West Gondwanaland (= Africa + South America) origin of the palms, the other authors emphasized differences between the continents and focused on long-distance dispersal to explain the present disjunct distributions. The consensus, as stated by Thorne and Smith, was that South America and Africa have little in common floristically; what similarities there are derive from immigration of most of the same families to the two continents, supplemented by a few subsequent transatlantic long-distance dispersal events.

I thank the National Geographic Society for a series of grants that supported much of the research summarized here. Additional data were gathered with support from the Mellon Foundation, the National Science Foundation (BSR 86-07113), and the MacArthur Foundation. I thank R. Clinebell for assistance with data analysis, B. Boyle for preparation of figure 17.4, J. Lovett, J. M. Fay, and G. Schatz for making available unpublished African data, and the Pew Foundation for support from the Pew Scholars Program during the writing of this chapter. Among the colleagues who collaborated in gathering the African data reported here were D. Thomas and F. Namata (Cameroon), J. M. Fay and D. Harris (Central African Republic), L. Mwasumbe (Tanzania), and G. Schatz (Madagascar). Assistance with collecting Neotropical data included in this review has been previously acknowledged.

 ISBN 0-300-05375-4

Although Thorne produced an impressive list of 111 seed plant genera restricted to tropical and subtropical Africa and Madagascar and the Neotropics and 12 families similarly restricted, he emphasized that most of these families are only tenuously represented in Africa and that both the lists of shared families and shared genera represent very small fractions of the total flora of either continent. The uniquely shared genera are fewer than 2.5 percent of the 4500 genera found on only one or the other continent; even if more wide-ranging genera are considered, Thorne estimated that the genera shared between Africa and South America constitute less than 10 percent of the total generic flora. Thorne also pointed out that many of the large speciose genera especially characteristic of either continental region are completely unrepresented in the other and that even the shared genera tend to be very asymmetrically distributed, typically with one or a very few species on one of the continents, hinting at recent arrival, 15 mostly American genera having very few species in Africa. Of the 7 African genera with minimal South American representation, *Gambeya* is not in South America at all to my knowledge and *Jaumea* is largely subtropical and about equally represented in both regions, leaving only *Acioa* (now generically split between the two continents), *Loudetia* (whose Arundinelloid relatives *Loudetiopsis* and *Tristachya* are also mostly African with a single Neotropical species that could be added to the list [Zuloaga, personal communication, 1990]), *Mostuea, Raphia* (sometimes thought to represent a human introduction), and *Symphonia* with this pattern.

In several of the unequally represented shared genera the sole African species is conspecific with one of the American ones. Most of these almost certainly represent recent introductions as do most of the 108 species native both to West Africa and tropical America, half of which are aquatics and another 20 weeds. With a few exceptions, like *Ceiba pentandra,* which is surely native in both continents (personal observation), these plants are largely incidental to a comparison of the origins of the African and South American floras.

Based on this kind of taxonomic "head count" the essentially independent derivation of the two floras postulated by Thorne (1973) seems eminently logical. Indeed Thorne (1973:45) suggested that South America's floristic links with Asia are stronger than with Africa and concluded: "The botanical evidence just presented argues against continental drift as a significant factor in explaining the distribution of seed plants between Africa and South America."

A second kind of floristic difference between the two continents was also noted by several contributors to the 1973 volume *Tropical Forest Ecosystems*. This is a marked tendency for Africa to have fewer, more wide-ranging species overall and a strikingly sparse representation of many characteristic pantropical plant taxa such as palms orchids, Lauraceae, Myrtaceae, and Myristicaceae. This apparent floristic imbalance led Richards (1973) to term Africa "the odd man out." Richards suggested that the peculiarities of Africa's flora might have been caused either by a more extensive history of human occupation, much of today's forest, especially in West Africa, being essentially old second growth, or by greater climatic vicissitudes associated with Africa's generally drier climate, or both. The latter interpretation gained prominence with publication of the now-classic Raven and Axelrod (1974) synthesis of angiosperm biogeography wherein much evidence was assembled that clearly supported a West Gondwanaland angiosperm origin and early radiation prior to the final separation of South America from Africa. Raven and Axelrod (1974) suggested that today's relatively depauperate tropical African flora was largely a result of extinction during the Neogene climatic deterioration culminating in the dry periods associated with Pleistocene glacial advances at higher latitudes. In support of this view, they note the relatively rich representation of many of the conspicuously underrepresented African forest taxa on Madagascar and even to some extent in the Canary Islands. The paleobotanical evidence marshalled by Raven and Axelrod (1974) and Axelrod and Raven (1978) and subsequently reinforced by additional discoveries of African fossils of taxa now absent on Africa but still present on Madagascar, like Winteraceae and the Madagascar endemic family Sarcolaenaceae (Coetzee and Muller 1984), strongly supports this interpretation.

To recapitulate the salient points of the now widely accepted consensus, the angiosperms arose in then centrally located West Gondwanaland during the latter part of the Lower Cretaceous (probably in Barremian times) and underwent much early ordinal and familial (and some generic) differentiation prior to the definitive separation of South America from Africa, perhaps as late as the Paleocene, as the two continents initially slid past each other. Madagascar separated from the other side of Africa earlier, perhaps in the Jurassic (Rabinowitz et al. 1983), but apparently slid south along the east African coast for most of the Cretaceous rather freely interchanging plant and animal taxa with the mainland at least until 95 Ma and perhaps intermittently thereafter. During the Paleogene the African and South American floras underwent some differentiation, perhaps in part in response to the different evolutionary milieus induced by the presence of placental mammals on Africa but only of marsupials in South America prior to mid-Tertiary. However, the earth's climatic deterioration during the latter half of the Cenozoic, coupled with the geologic uplift of Africa, led to a significantly drier climate and major impoverishment of the African forest flora, culminating in massive extinction during the Pleistocene dry periods and today's very different floras. Countering this loss of species, numerous taxa that had evolved in the Neotropics subsequent to Gondwanan breakup reached Africa via increasingly long-distance dispersal throughout the Cenozoic, enriching the familial composition of Africa's flora significantly but its species-level diversity only minimally.

This chapter addresses a different, and much less widely known, aspect of the Afro-Madagascar and South American floras—their composition at the local community level—which suggests dramatically different conclusions. The idea for this analysis began to take form in 1981 on my first visit to tropical Africa. Given the literature's emphasis on Africa's status as "odd man out" and Thorne's (1973) statistical analysis of the floristic dissimilarity of the two continents, I was pleasantly surprised to discover how many African taxa were familiar from my previous Neotropical experience. Not only did virtually all of the flora belong to familiar Neotropical families, but I even recognized many of the genera. Moreover, most of the families that were supposed to be conspicuously lacking in Africa—Lauraceae, Myristicaceae, palms—were very much in evidence. Then came another surprise: many of the genera I thought I recognized turned out to go by different names in Africa. Clearly the African and Neotropical floras as observed in the field were far more similar than indicated by the extant literature.

During the past decade I have not only visited many forests in Africa and the Neotropics (as well as many other parts of the world) but have also attempted to document the floristic composition of many of these forests by a replicated program of ecological sampling. The data available for this analysis of the floristic composition of African and South American forests include a series of 92 0.1-ha transect samples, 11 in Africa and 3 in Madagascar (table 17.1), as well as 78 in lowland Latin America in which all plants with a diameter greater than 2.5 cm have been identified and ennumerated (Gentry 1982b, 1988a). (For complete site data, see Gentry 1988a, 1991a:appendix.) These data are supplemented by 16 1-ha tree plots (table 17.2) in which all trees and large lianas with a diameter at breast height (dbh) greater than 10 cm have been identified (Gentry 1988a, 1988c, 1990b) and by data compiled from a series of 11 more or less completely censused local florulas. The primary data set, the 0.1-ha samples, now includes data from 160 sites from 37 countries and all continents except Antarctica (figure 17.1).

In addition to my own tree plot data (Gentry 1988a, 1988c; Faber-Langendoen and Gentry 1991), there are comparable published data for several 1-ha tree plots in Gabon (Reitsma 1988), near Manaus (Prance 1990), and elsewhere in Amazonia (Campbell et al. 1986; Boom 1987); similar unpublished data for three 1-ha tree plots at La Selva, Costa Rica, for appropriate subsets of the 50-ha plot on Barro Colorado Island, Panama, and for four 1-ha tree plots at Ndakan, Central African Republic, have kindly been made available by D. Lieberman, S. Hubbell, and M. Fay, respectively (see Gentry 1990b). Unfortunately, until recently there has been little stabilization of sampling technique so that many of the published data sets (e.g., Pierlot 1966; Hall and Swaine 1981; Puig 1979; Mori and Boom 1987; Doumenge 1990) are difficult to compare with the tropical forest phytosociological "industrial standard" (cf., Whitmore 1985; see also Prance 1984; Campbell 1989) of 1-ha plots sampling trees *and lianas* used here.

In this chapter I analyze only the lowland (less than 1000 m) subset of the data. I will compare the structure, floristic composition, and diversity of lowland African, Madagascan, and South American forests. I will also draw some tentative conclusions about ecological similarities and differences between these forests. Finally, I will reconcile the observed local pattern of remarkably similar plant communities with the strikingly different continent-level patterns.

II. Structure

Tropical forests in different parts of the world have been generally assumed to be structurally similar. They share such features as three reputed strata of trees, similar canopy height and tree densities, high biomass and

Table 17.1. Data from 0.1-ha samples of African and Madagascan forests (plants ≥ 2.5 cm diameter).

	Number of species					Number of individuals				
	Families (+ indet)	Total spp.	Liana spp.	Trees ≥2.5 cm dbh	Trees ≥10 cm dbh	Total indiv.	Lianas (+ strang.)	Trees ≥2.5, ≤10 cm	Trees ≥2.5 cm	Trees ≥10 cm dbh (+lianas)
AFRICA										
Makokou, Gabon 1	39(+3)	135	47	88	28(+9)	339	103	185	236	51(+14)
Makokou, Gabon 2	33(+2)	107	38	69	23(+2)	326	82	203	244	41(+4)
Belinga, Gabon[a] (500 m²)		(88)	(15)		(21+1)	400	46	278	344	66(+2)
Ndakan upland, CAR	37(+3)	128(−)	35(−)	ca.93	34(+3)	400	97(+1)	248	302	54(+3)
Ndakan inundated, CAR	32	96	36(+2)	60	24(+2)	369	154(+2)	161	215	54(+2)
Korup, Cameroon	43	139	37(+2)	100	35(+2)	424	72(+2)	288	350	62(+2)
Mt. Cameroun, Cameroon	38(+2)	128	45(+1)	82	38(+4)	397	129(+1)	205	267	62(+4)
Banyong, Cameroon[a] (800 m²)	40+	ca.180	42+	122+	49(+3)	468	85	303	383	80(+4)
Moist forest Average[b]	37	130	40	88	33(+4)	390	96	234	293	59(+4)
Omo, Nigeria	30	73	28	45	21	331	76	170	255	85
Pugu, Tanzania	28(+2)	75–85	ca.33	ca.42	21(+8)	454	168	228	286	58(+9)
Pande, Tanzania[a] (500 m²)	(15+)	(ca.45)	(21+1)	(ca.22)	(10+1)	298	102(+2)	128	194	66(+2)
Dry forest average[b]	29	77	31	44	21(+4)	361	115	175	245	70(+4)
Overall average[b]	36	118	38	78	30(+4)	382	101	218	280	62(+4)
MADAGASCAR										
Perinet	53(+5)	ca.197	37(+1)	159	ca.76	750	115(+1)	512	634	122
Nosy Mangabe	50(+5)	ca.211	ca.38(+2)	ca.171	ca.80(+1)	716	133(+3)	420	580	160(+1)
Moist forest average	52	204	38	165	78(+1)	733	124	466	607	141(+1)
Ankarafantsika (dry forest)	40(+1)	94	30	64	28(+1)	459	122	276	337	61(+2)

Source: For site descriptions, see Gentry 1991a.

[a]Incomplete sample; number of individuals extrapolated to 0.1 ha; number of species extrapolated to 0.1 ha or actual number of sampled species indicated in parentheses.

[b]Excluding parenthetical values; anomalous Nigerian Omo Forest site, perhaps successional, included with dry forests (see discussion in Gentry 1988a).

Table 17.2. Tree plot summary data for African and South American 1-ha plots.

SOUTH AMERICA	Trees ≥ 10 cm dbh	Lianas ≥ 10 cm dbh	Trees ≥ 20 cm dbh	Trees ≥ 30 cm dbh	Trees ≥ 50 cm dbh	Trees ≥ 100 cm dbh	Basal area (m^2/ha)	Tree spp. ≥ 10 cm	Liana spp. ≥ 10 cm	Total spp. ≥ 10 cm	Families
Yanamono (Explorama), Peru	580	26	236	109	36	1	32.6	283	17	300	58
Mishana, Peru	842	16	260	83	12	1	—	275	14	289	50
Cabeza de Mono, Peru	520	24	180	81	31	4	—	169	16	185	40
Cocha Cashu, Peru	650	23	178	110	35	5	37.1	189	15	204	48
Tambopata alluvial, Peru	526	16	245	100	21	3	—	155	10	165	41
Tambopata lateritic 2, Peru	545	19	254	84	23	1	—	175	10	185	45
Tambopata lateritic 1, Peru	ca.547	17	(238+)	(71+)	(17+)	2	—	162	12+	174+	41+
Tambopata swamp, Peru	526	6	244	136	45	2	43.2	152	6	158	37
Tambopata upland sandy 1, Peru	585	17	181	80	24	2	—	167	13	179	42
Tambopata upland sandy 2, Peru	549	18	199	80	24	2	—	152	10	162	45
Cuzco Amazonico 1-E, Peru	495	44(+1)	183	86	24	4	—	—	—	—	—
Cuzco Amazonico 1-U, Peru	513	30(+7)	187	74	25	3	—	—	—	—	—
Cerro Neblina, Venezuela	493	20	211	84	33	1	—	89	13	102	32
Bajo Calima, Colombia[a]	664	11	197	83	12	0	25.0	252	6(+)	258	45(+)
Average	574	21	212	92	27	2	34.5	185	12	197	44
St. Elie, French Guiana[b]	—	—	255	—	—	—	—	—	—	—	—
Arataye, French Guiana[c]	—	—	199	—	—	—	—	—	—	—	—
Arataye, French Guiana[c]	—	—	172	—	—	—	—	—	—	—	—
Arataye, French Guiana[c]	—	—	212	—	—	—	—	—	—	—	—

(continued)

Table 17.2. *Continued.*

SOUTH AMERICA	Trees ≥ 10 cm dbh	Lianas ≥ 10 cm dbh	Trees ≥ 20 cm dbh	Trees ≥ 30 cm dbh	Trees ≥ 50 cm dbh	Trees ≥ 100 cm dbh	Basal area (m^2/ha)	Tree spp. ≥ 10 cm	Liana spp. ≥ 10 cm	Total spp. ≥ 10 cm	Families
Data from Point-Quarter method[d]											
Alto Ibon, Bolivia	649	—	197	54	13	0	21.5	—	—	—	—
Añangu, Ecuador	728	—	266	89	18	0	31.6	—	—	—	—
Bahia, Brazil	891	—	401	187	40	0	46.3	—	—	—	—
Camaipi, Amapa, Brazil	540	—	254	132	38	3	34.8	—	—	—	—
Rio Falsino, Amapa, Brazil	492	—	209	110	48	7	39.3	—	—	—	—
Saul, French Guiana	619	—	306	180	64	13	53.0				
Average	653	—	272	125	37	4	37.8				
South American average	598	21	227	102	30	3	36.4	185	12	197	44
AFRICA											
Oveng, Gabon[e]	485	12	210	111	43	5	36.4	123	8	131	30(+1)
Doussala, Gabon[e]	413	12	181	112	37	8	35.7	100	9	109	32(+1)
Lopé, Gabon[e]	392	4	179	109	42	9	35.8	65	4	69	26
Ekobakoba, Gabon[e]	429	9	217	121	46	5	42.9	77	8	85	28
Korup, Cameroon[f]	—	—	—	—	—	—	—	79(+)	—	—	—
Korup, Cameroon[g]	471	12	—	—	—	—	27.5	75	—	—	—
Korup, Cameroon[h]	—	—	190	—	—	—	—	—	—	—	—
Makokou, Gabon[h]	—	—	202	—	—	—	35	—	—	—	—

Makokou, Gabon[h]	—	—	208	—	—	—	—	—	—	—	—
Odzala, Congo[h]	—	—	140	—	—	—	—	—	—	—	—
Odzala, Congo[h]	—	—	167	—	—	—	—	—	—	—	—
Odzala, Congo[h]	—	—	161	—	—	—	—	—	—	—	—
Ghana (overall average)[i]	487	—	—	—	—	—	—	—	—	—	—
Ghana (wet evergreen, 12 sites)[i]	445	—	—	—	—	—	25.5	—	—	—	—
Kibungwe, Zaire[j]	—	—	287	—	—	—	—	—	—	—	—
Kogogo, Zaire[j]	—	—	229	—	—	—	—	—	—	—	—
Muhunzi, Zaire[j]	—	—	114	—	—	—	—	—	—	—	—
Obiloto, Zaire[j]	—	—	118	—	—	—	—	—	—	—	—
Ndakan, CAR[k]	371	—	138	65	18	1	20.0	103	—	—	—
Ndakan, CAR[k]	451	—	130	67	26	2	24.6	106	—	—	—
Ndakan, CAR[k]	417	—	174	93	30	2	27.1	106	—	—	—
Ndakan, CAR[k]	403	—	144	84	23	2	25.2	104	—	—	—
African average	433	10	177	95	33	4	30.5	94	7	99	29

Source: South American sites mostly from Gentry 1988a, 1990b.
[a]Faber-Langendoen and Gentry 1991
[b]Puig 1979
[c]Maury-Lechon and Ponce 1986
[d]Mori and Boom 1987
[e]Reitsma 1988
[f]D. Thomas, personal communication, 1987.
[g]Gartlan et al. 1986, mean in 0.64 ha.
[h]Doumenge 1990
[i]Hall and Swaine 1981
[j]Pierlot 1966
[k]M. Fay, personal communication, 1990.

Figure 17.1. Map of 0.1-ha study sites and local florula sites. Dots = 0.1-ha samples; arrows = local florulas.

productivity, leaf size and shape, and the prevalence of unusual life forms such as lianas and epiphytes, and such distinctive features as buttresses, stilt roots, and cauliflory (Dawkins 1959; Richards 1952; Leigh 1975; Gentry 1990d). On the other hand, several recent studies have suggested that there may be intercontinental differences in the structure of tropical forests (Emmons and Gentry 1983; Richards 1973; Gentry 1990a).

Lianas are often singled out as one of the most characteristic features of a tropical rainforest (Gentry 1991a; Croat 1978). Richards (1973) suggested that one of the "odd man out" features of African forests is that they have relatively few specialized life forms like epiphytes and lianas, although no actual data were presented. On the other hand, a limited data set for a total of only 0.37 ha of African forest led to the opposite suggestion that African forests have *more* lianas than forests on other continents (Emmons and Gentry 1983).

Since there is such a fundamental divergence of opinion, I have reexamined the data, which clearly confirm the status of African forests as exceptionally liana rich: these forests have an average of 106 lianas with a diameter greater than 2.5 cm per 0.1 ha versus 70 in the Neotropics ($p < .001$ versus both cis-Andean and trans-Andean data sets) (table 17.3). Madagascar also has unusually high liana densities of 122 stems ≥2.5 cm diameter per 0.1 ha. Lianas often constitute approximately 19 percent of the total flora at a Neotropical site; again these values seem to be higher in Africa, where 24 percent of the Makokou Gabon flora is lianas (Gentry 1991a), and where Hall and Swaine (1981) report that lianas constitute 40 percent of the total flora on small sample plots in Ghana. This difference probably applies only to smaller lianas since Rollet (1974) and Reitsma (1988), respectively, report averages of only 9 and 9.25 lianas ≥10 cm diameter per ha for a series of African sites.

Another structural difference that has been suggested involves the densities and

Table 17.3. Comparison of some structural characteristics of Neotropical, African, and Madagascan forests.

	Neotropics	Africa	Madagascar
Lianas ≥ 2.5 cm diameter/0.1 ha	70 (average)[a] 68 (moist + wet forest)[b] 68 (pluvial forest)[b] 82 (dry forest)[c]	106[a]	122[a]
Climbers as percentage of local flora	19%[d]	40%[e] 23.5%[f]	
Lianas ≥ 10 cm diameter/ha	21[g] 33[a] (dry forest)	9[h] 9.25[i] 40 (moist forest)[j] 40 (dry forest)[j]	10 (moist forest)[j] 20 (dry forest)[j]
Trees ≥ 10 cm dbh/ha	574[g] 640 (moist + wet forest)[b] 880 (pluvial forest)[b] 623 (dry forest)[c]	427[k] 590 (moist forest)[j]	1410 (moist forest)[j]
Trees ≥ 2.5 cm dbh/0.1 ha	304 (moist + wet forest)[b] 400 (pluvial forest)[b] 292 (trop. S. Amer. dry forest)[c]	293 (moist forest)[j] 245 (dry forest)[j]	607 (moist forest)[j]
Total plants ≥ 2.5 cm dbh/0.1 ha	373 (moist + wet forest)[b] 510 (pluvial forest)[b] 374 (trop. S. Amer. dry forest)[c]	390 (moist forest)[j] 361 (dry forest)[j]	
Trees ≥ 50 cm dbh/ha	27[l] 37[m]	42[i]	
Trees ≥ 100 cm dbh/ha	2[l] 4[m]	7[i] 4[g]	
Basal area (m^2/ha)	36[g] 37.8[m]	33.8[n] 30.5[g]	

Source: Data mostly summarized from table 17.1 (African 0.1-ha plots), table 17.2 (1-ha tree plots), Gentry 1986 (Neotropical moist and pluvial forest 0.1-ha plots), and Gentry 1993a (Neotropical dry forest 0.1-ha plots).

[a]Gentry 1991a
[b]Gentry 1986
[c]Gentry 1993a
[d]Gentry 1991a; average for 7 sites
[e]Hall and Swaine 1981
[f]Gentry 1991a, Makokou, Gabon
[g]Table 17.2.
[h]Rollet 1974
[i]Reitsma 1988
[j]Table 17.1
[k]Table 17.2; 10 sites, excluding all-Ghana average
[l]Table 17.2; Gentry plots
[m]Table 17.2, Mori and Boom data
[n]Table 17.2, excluding Fay data from "open" forest type.

relative frequencies of different size classes of trees in African and South American forests. In Africa, the density of trees ≥10 cm dbh tends to be relatively low, but the biomass tends to be relatively high (Reitsma 1988; Gentry 1988a); the difference is presumably due to the greater prevalence of very large trees in African forests (e.g., 42 trees ≥50 cm dbh per ha in Reitsma's (1988) data vs. 27 in my series of Neotropical tree plots; see table 17.2. For example, Reitsma (1988) found only 392–485 ($\bar{X}$ = 430) trees ≥10 cm dbh in his four study hectares in Gabon, (296–497 stems ≥ 10 cm dbh [$\bar{X}$ = 439] including lianas). Density estimates for trees ≥10 cm dbh per ha at other Gabonese sites include 427 at Makokou (Hladik 1978) and 495 at Belinga (Aubreville 1967). Hall and Swaine (1981) found slightly higher densities averaging 487 trees ≥10 cm per ha in a wide variety of Ghanan forest types. Perhaps more to the point, Hall and Swaine (1981) found wet evergreen forest, the Ghana forest type most closely approximating the Gabon sites, to have the lowest density, averaging 445 trees ≥10 cm dbh per ha, almost exactly the average for the Gabon sites.

These values contrast with 493–842 ($\bar{X}$ = 574) trees ≥10 cm dbh (513–858 total stems ≥10 cm; $\bar{X}$ = 594) in my series of 14 South American hectares (table 17.2) and an average tree density of trees ≥10 cm dbh of 517 per ha calculated by Campbell and coworkers (1986) for a different series of Neotropical samples. In Amazonian Ecuador Balslev and co-workers (1987) reported densities of 417 and 728 trees ≥10 cm dbh per ha for two study areas. The difference between high densities of smaller trees (≥10 cm dbh or ≥20 cm dbh) in the Neotropics and low densities in Africa is highly significant (Mann-Whitney U, $p > .001$).

There is little correlation of basal area with tree density. Thus Hall and Swaine's (1981) 12 low density wet evergreen Ghanian sites had a relatively high basal area compared to other Ghana forests, averaging 25.5 m^2 per ha and slightly exceeded in biomass only by upland evergreen forest. The Gabon sites also have high basal areas despite their low tree densities (e.g., 36.4 m^2 per ha at Oveng, 35.7 at Doussala, 35.8 at Lope, 42.9 at Ekobakaba). Hladik (1982) reports a basal area value at Makokou of 35 m^2 per ha and Pierlot (1966) reports values of 30–35 m^2 per ha in Zaire. Fay (personal communication, 1990) reports lower values of 20.0–27.1 m^2 per ha at his Central African Republic (CAR) study sites. African basal area values average about 34 m^2 per ha, excluding Fay's anomalously open forest (table 17.2).

Basal areas reported for Neotropical forests range from 21.5–53 m^2 per ha (Mori et al. 1983; Mori and Boom 1987; Faber-Langendoen and Gentry 1991), averaging about 35–36 m^2 per ha, the same values found by Balslev and co-workers (1987). Thus basal areas on the two continents are virtually identical.

Tree density is significantly different between Africa and the Neotropics, but this is largely due to more smaller trees in the Neotropics. I conclude that basal areas (and presumably biomass) of African and Neotropical lowland forests are remarkably similar but that a greater portion of the African forest basal area comes from large diameter trees.

Although epiphytes are not quantified in any African study, Johansson (1974) found many epiphytes in his wet Liberian study sites, up to 22 species in a single tree. Reitsma (1988) notes that in his Gabon 0.02-ha samples, epiphytes, although not quantified, constituted about 5 percent of the species, a value similar to that found by Gentry and Dodson (1987) for similarly dry forests in western Ecuador.

Thus the overall picture is of similarly constructed forests on the two continents. Moreover, apparent structural differences that do exist between South American and African forests tend to be opposite to those suggested by Richards (1973). More telling, all are in the direction that would be predicted by modern ecology. African forests have similar epiphyte loads to Neotropical forests with similar precipitations. Although African forests overall have more lianas, they also tend to cluster at what would be considered in the Neotropics to be the transition from moist to dry forest, exactly where the most liana-rich Neotropical forests occur (Gentry 1991a). Typically the fewer small-diameter trees in African forests are compensated for by more large-diameter ones, so forests on both continents have remarkably similar basal areas.

III. Floristic Composition

A. Taxonomic Artifacts

African and South American forests are also more alike floristically than often supposed. As indicated above, when I first visited tropical Africa, I found that I could recognize to family and even to genus most of the plants I encountered, using Neotropical taxonomic concepts, although many of the plants subsequently turned out to masquerade under different generic names in Africa. Although there are often technical characters which purportedly distinguish such transatlantic pairs of genera, it seems clear that many of them would be universally regarded as constituting a single genus did they not occur on opposite sides of the Atlantic. Obviously, the necessary taxonomic revisions of all these families are beyond the scope of this chapter. Whatever the ultimate taxonomic resolution of such cases, even a superficial consideration is adequate to indicate that many genera on opposite sides of the Atlantic are closer to each other than either is to other genera on the same continent. Some of these pairs are listed in table 17.4. The implications for a comparison of Africa–South American floristics are obvious: to the degree that generic taxonomy is a biogeographical artifact based on provincial taxonomy, the floristic relationships between the two continents may be much closer than reflected by comparisons of lists of genera.

B. Community Patterns

One way to get around this kind of problem is with direct comparisons of individual communities.

One of the most striking aspects of a comparison of African and Neotropical forests is their floristic similarity at the community level. One indicator of this similarity is the shared familial and generic composition of forests on the two continents. At the level of 0.1-ha samples, the familial composition is virtually identical. The only families represented in my African samples that are not also represented in the Neotropics are Ancistrocladaceae, Medusandraceae, Scytopetalaceae, Pandaceae, Irvingiaceae, and Ixonanthaceae, and of these only Irvingiaceae (a sometime segregate of Simaroubaceae), Pandaceae (sometimes included in

Table 17.4. Some genera of forest trees and lianas shared between Africa and South America and closely related pairs of woody genera that may be closer to each other than to other genera on same continent.

Family	Africa	South America
Acanthaceae	*Mendoncia*	*Mendoncia*
Anacardiaceae	*Pseudospondias*	*Spondias*
Annonaceae	*Annonidium* (*Annona*)	*Annona*
	Pachypodanthium	*Duguetia*
	Xylopia	*Xylopia*
	Dictyophlebia/*Landolphia*	*Pacouria*
	Tabernaemontana	*Tabernaemontana*
Araliaceae	(*Schefflera*)	(*Schefflera*)
Bombacaceae	*Ceiba*	*Ceiba*
	Bombax and Pochota	*Pochota* (incl. *Bombacopsis*)
Boraginaceae	*Cordia*	*Cordia*
Burseraceae	*Dacryodes*	*Dacryodes*
Celastraceae	(*Maytenus*)	(*Maytenus*)
Chrysobalanaceae	*Maranthes*	*Maranthes* (*Central America*)
	Parinari	*Parinari*
Combretaceae	*Combretum*	*Combretum*
	Terminalia	*Terminalia*
Connaraceae	*Connarus*	*Connarus*
	Byrsocarpus	*Rourea*
Convolvulaceae	*Calycobolus*	*Prevostoea*
Dichapetalaceae	*Dichapetalum*	*Dichapetalum*
Dilleniaceae	*Tetracera*	*Tetracera*
Dioscoreaceae	(*Dioscorea*)	(*Dioscorea*)
Ebenaceae	*Diospyros*	*Diospyros*
Euphorbiaceae	*Acalypha*	*Acalypha*
	Alchornea	*Alchornea*
	Croton	*Croton*
	Drypetes	*Drypetes*
	Conceveiba	*Conceveiba*
	Leeuwenbergia	*Johannesia*
	Maprounea	*Maprounea*
	(*Phyllanthus*)	(*Phyllanthus*)
	Sapium	*Sapium*
	Tetrorchidium	*Tetrorchidium*
Flacourtiaceae	*Casearia*	*Casearia*

Family	Africa	South America
	Homalium	*Homalium*
	Lindackeria	*Lindackeria*
Gnetaceae	*Gnetum*	*Gnetum*
Guttiferae	*Garcinia*	*Garcinia* (ex *Rheedia*)
	Mammea	*Mammea* (Antilles only)
	Symphonia	*Symphonia*
Hippocrateaceae	*Hippocratea*	*Hippocratea*
	Salacia	*Salacia*
	Vismia (and *Harunga*)	*Vismia*
Lauraceae	*Beilschmiedia*	*Beilschmiedia*
	Ocotea	*Ocotea*
Leguminosae	*Anthonotha/Gilbertiodendron/ Neochevalieradendron* etc.	*Macrolobium*
	Copaifera	*Copaifera*
	Crudia	*Crudia*
	Cynometra	*Cynometra*
	Dialium	*Dialium*
	Swartzia	*Swartzia*
	Hymenaea (ex *Trachylobium*)	*Hymenaea*
	Acacia	*Acacia*
	Albizia	*Albizia*
	Entada	*Entada*
	Parkia	*Parkia*
	Pentaclethra	*Pentaclethra*
	Newtonia	*Newtonia*
	Dalbergia	*Dalbergia*
	Dioclea	*Dioclea*
	Lonchocarpus	*Lonchocarpus*
	Milletia	*Milletia* (cf. Lonchocarpus)
	Mucuna	*Mucuna*
	Pterocarpus	*Pterocarpus*
Lepidobotryaceae	*Lepidobotrys*	*Ruptiliocarpum*
Liliaceae	(*Dracaena*)	(*Dracaena*) (only Central America)
Loganiaceae	*Anthocleista*	*Potalia*
	(*Mostuea*)	(*Mostuea*)
	Strychnos	*Strychnos*
Melastomataceae	*Memecylon*	*Mouriri*

(*continued*)

Table 17.4. *Continued.*

Family	Africa	South America
Meliaceae	*Entandophragma*	*Swietenia*
	Guarea	*Guarea*
	Trichilia	*Trichilia*
Moraceae	*Ficus*	*Ficus*
	Musanga	*Cecropia*
	(*Dorstenia*)	(*Dorstenia*)
Myristicaceae	*Coelocaryon*	*Osteophloem*
	Pycnanthus/Scyphocephalium	*Virola*
	Staudtia	*Iryanthera*
Myrsinaceae	(*Ardisia*)	(*Ardisia*)
Myrtaceae	*Eugenia*	*Eugenia*
Ochnaceae	*Campylospermum*	*Ouratea*
	Lophira	*Cespedezia*
Olacaceae	*Aptandra*	*Aptandra*
	Heisteria	*Heisteria*
	Strombosia/Strombosiopsis	*Tetrastylidium*
Opiliaceae	*Opilia*	*Agonandra*
Palmae	(*Elaes*)	(*Elaes*)
	(*Raphia*)	(*Raphia*)
Piperaceae	*Piper*	*Piper*
Rhamnaceae	*Gouania*	*Gouania*
Rhizophoraceae	*Cassipourea*	*Cassipourea*
Rubiaceae	*Gardenia*	*Randia*
	Bertiera	*Bertiera*
	(*Morinda*)	(*Morinda*)
	Psychotria	*Psychotria*
	Sabicea	*Sabicea*
	Uncaria	*Uncaria*
Rutaceae	*Zanthoxylon* (incl. *Fagara*)	*Zanthoxylon*
Sapindaceae	*Allophylus*	*Allophylus*
	Deinbollia	*Talisia*
	Laccodiscus	*Cupania*
Sapotaceae	*Chrysophyllum*	*Chrysophyllum*
	Donella	*Micropholis*
	Manilkara	*Manilkara*
	Omphalocarpum	*Pouteria*
	Sideroxylon	*Sideroxylon*

Family	Africa	South America
Simaroubaceae	*Quassia*	*Quassia*
Smilacaceae	*Smilax*	*Smilax*
Sterculiaceae	*Byttneria*	*Byttneria*
	Cola	*Theobroma/Herrania*
	Pterygota	*Pterygota*
	Sterculia	*Sterculia*
Ulmaceae	*Celtis*	*Celtis*
	Trema	*Trema*
Urticaceae	*Urera*	*Urera*
Verbenaceae	*Clerodendron*	*Clerodendron/Aegiphila*
	Vitex	*Vitex*
Violaceae	*Rinorea*	*Rinorea*
Vitaceae	*Cissus*	*Cissus*
Ferns	*Cyathea*	*Cyathea*

Note: Only woody forest plants included, with a few taxa of open areas or swamps and a few subwoody primary forest taxa indicated parenthetically. Very many other genera occur outside the forest on both continents, mostly as weeds, or as epiphytes or herbs.

Euphorbiaceae), and Scytopetalaceae are represented by more than one or two individuals in all of the African samples taken together.

If the complete florula of Makokou Gabon (Florence and Hladik 1980 and included references; see also Gentry and Dodson 1987) is similarly analyzed, this near-identity is equally obvious (tables 17.5a and 17.5b). Of the 104 families represented at Makokou, only 7, accounting for a grand total of 15 of the 1131 local species, are not also Neotropical, and all but 2 of these—Scytopetalaceae with 3 species, and Tecophilaeaceae with 1—are segregates that are often not accepted: Irvingiaceae from Simaroubaceae, Ixonanthaceae from Linaceae, Leeaceae from Vitaceae, Octoknemaceae from Olacaceae, Pandaceae from Euphorbiaceae. To these might be added Balsaminaceae (5 spp. at Makokou), which has a few Central American species but is represented in South America by only 2 escaped species.

Madagascar is equally similar floristically to the Neotropics (tables 17.5a, 17.5b, 17.6a, 17.6b). Of the 92 families represented at Nosy Mangabe (Schatz, personal communication, 1990), only 6, accounting for a total of 8 of the 490 local species, are not also Neotropical: Flagellariaceae, Leeaceae, Pandanaceae, Physenaceae, Pittosporaceae, Sphaerosepalaceae. To these might be added Balsaminaceae (2 spp. at Nosy Mangabe), Goodeniaceae (1 sp.) and

Table 17.5a. Largest families, including weeds, in Neotropical local florulas.

La Selva, Costa Rica		BCI, Panama		Cocha Cashu, Peru		Ducke, Brazil	
Family	Spp.	Family	Spp.	Family	Spp.	Family	Spp.
(Ferns)	169	Leguminosae	112	Leguminosae	102	Leguminosae	104
Orchidaceae	114	(Mimosaceae)	37	(Mimosaceae)	48	(Mimosaceae)	33
Araceae	99	(Caesalpiniaceae)	16	(Caesalpiniaceae)	16+	(Caesalpiniaceae)	28
Rubiaceae	99	(Fabaceae)	59	(Fabaceae)	28+	(Fabaceae)	43
Leguminosae	79	(Ferns)	102	Moraceae	66	Sapotaceae	48
(Mimosaceae)	31	Orchidaceae	90	Rubiaceae	65	Rubiaceae	42
(Caesalpiniaceae)	10	Gramineae	79	(Ferns)	65	Chrysobalanaceae	40
(Fabaceae)	38	Rubiaceae	66	Orchidaceae	45	Lauraceae	38
Piperaceae	79	Araceae	46	Acanthaceae	44	(Ferns)	37
Melastomataceae	71	Compositae	42	Sapindaceae	40	Annonaceae	36
Gramineae	63	Moraceae	36	Bignoniaceae	38	Moraceae	35
Compositae	48	Melastomataceae	35	Araceae	38	Lecythidaceae	30
Euphorbiaceae	42	Piperaceae	32	Solanaceae	38	Melastomataceae	26
Cyperaceae	32	Bignoniaceae	29	Myrtaceae	37	Palmae	25
Moraceae	31	Cyperaceae	28	Piperaceae	36	Orchidaceae	24
Palmae	30	Euphorbiaceae	28	Sapotaceae	34	Apocynaceae	22
Gesneriaceae	27	Sapindaceae	26	Annonaceae	30	Euphorbiaceae	21
Acanthaceae	27	Solanaceae	23	Euphorbiaceae	29	Myristicaceae	20
Lauraceae	27	Apocynaceae	20	Compositae	27	Burseraceae	18
Solanaceae	26	Malpighiaceae	20	Lauraceae	24	Bignoniaceae	18
Bignoniaceae	24	Bromeliaceae	18	Gramineae	24	Flacourtiaceae	15
Bromeliaceae	23	Flacourtiaceae	16	Malpighiaceae	21	Guttiferae	12
Guttiferae	22	Palmae	16	Palmae	20	Bombacaceae	10
Marantaceae	21			Melastomataceae	20		
				Meliaceae	20		
121 families 1,668 species		118 families 1,320 species		119 families 1,370 species		88 families 688 species	

Source: Gentry 1990b.

Table 17.5b. Twenty largest families in African and Madagascan local florulas.

Makokou, Gabon		Nosy Mangabe, Madagascar	
Family	Spp.	Family	Spp.
Leguminosae	106	Rubiaceae	42
(Mimosoideae)	19	Euphorbiaceae	22
(Caesalpinioideae)	52	(Fern)	21
(Faboideae)	35	Ebenaceae	21+
Rubiaceae	102	Guttiferae	21
(Fern)	69	Lauraceae	18+
Annonaceae	54	Melastomataceae	17
Euphorbiaceae	52	Flacourtiaceae	16
Cyperaceae	45	Apocynaceae	16
Apocynaceae	37	Leguminosae	13+
Moraceae	34	Myrsinaceae	12
Acanthaceae	32	Gramineae	12
Orchidaceae	29	Orchidaceae	10
Dichapetalaceae	23	Sapindaceae	10
Hippocrateaceae	23	Anacardiaceae	9
Gramineae	23	Annonaceae	9
Sterculiaceae	20	Myrtaceae	9+
Sapindaceae	17	Sapotaceae	9
Connaraceae	16	Verbenaceae	9
Sapotaceae	16	Bignoniaceae	8
Loganiaceae	15	Moraceae	8

Sources: Makokou, Gabon, compiled from Florence and Hladik 1980 et prec. (see Gentry and Dodson 1987); Nosy Mangabe, Madagascar, compiled from Schatz, personal communication, 1990.

Hamamelidaceae (2 spp.), which are represented only marginally in South America.

Of the Nosy Mangabe genera 43 percent are also Neotropical. However, different genera have radiated on the different continents. Twelve of the 18 largest genera at Nosy Mangabe are not known at all from the Neotropics (tables 17.6a and 17.6b). Even in the third of the large Nosy Mangabe genera which are also found in the Neotropics, only 3 are ever large in Madagascar forests—*Ficus, Psychotria,* and *Eugenia/Syzygium.* The reverse pattern also occurs with *Diospyros, Drypetes,* and *Symphonia* never having more than a single species (rarely 2 in *Diospyros*) at any given Neotropical site.

At Makokou, while only 30 percent of the genera are also in the Neotropics, 85 percent of the largest genera (17 of 20) are also Neotropical. If the 26 largest Makokou

Table 17.6a. Largest genera of vascular plants, including weeds, in Neotropical local florulas.

Rio Palenque	*N*	La Selva	*N*	BCI	*N*	Cocha Cashu	*N*	Manaus (Ducke)	*N*
Piper	22	Piper	44	Piper	21	Ficus	34	Licania	21
Ficus	18	Psychotria	38	Psychotria	20	Inga	26	Inga	17
Solanum	18	Philodendron	31	Inga	18	Piper	25	Protium	14
Peperomia	15	Anthurium	25	Ficus	16	Pouteria	21	Eschweilera	13
Philodendron	15	Miconia	25	Miconia	14	Paullinia	19	Swartzia	13
Pleurothallis	15	Thelypteris	17	Polypodium	13	Psychotria	14	Aniba	12
Anthurium	14	Inga	16	Philodendron	13	Miconia	13	Miconia	12
Epidendrum	14	Peperomia	16	Epidendrum	13	Eugenia	13	Ocotea	11
Maxillaria	13	Calathea	15	Anthurium	12	Philodendron	12	Casearia	10
Thelypteris	11	Ficus	15	Cyperus	11	Trichilia	12	Couepia	10
Heliconia	10	Pleurothallis	15	Passiflora	11	Calathea	11	Palicourea	10
Columnea	9	Trichomanes	14	Solanum	11	Passiflora	11	Pouteria	10
Drymonia	9	Passiflora	13	Panicum	10	Thelypteris	11	Sloanea	10
Passiflora	8	Ocotea	13	Peperomia	10	Peperomia	10	Virola	10
Adiantum	7	Clidemia	13	Trichomanes	9	Neea	10	Iryanthera	8
Cyperus	7	Cyperus	12	Adiantum	9	Anthurium	10	Brosimum	7
Calathea	7	Panicum	12	Paspalum	9	Solanum	10	Hirtella	7
Cordia	7	Heliconia	12	Desmodium	9	Epidendrum	9	Micropholis	7
Hyptis	7	Epidendrum	12	Paullinia	9	Adiantum	8	Guatteria	7
Ocotea	7	Solanum	12	Serjania	9	Justicia	8	Duguetia	6
Psychotria	7	Asplenium	11	Thelypteris	8	Alibertia	8	Licaria	6
Pilea	7	Diplazium	11	Heliconia	8	Calyptranthes	8	Mouriri	6
Inga	6	Tillandsia	11	Calathea	8			Lecythis	6
Tillandsia	6	Maxillaria	10	Maxillaria	8			Trichomanes	6
		Drymonia	10	Dioscorea	7				
		Selaginella	9	Machaerium	7				
		Polypodium	9	Casearia	7				
		Monstera	9						
		Costus	9						
		Clusia	9						
		Conostegia	8						
		Pilea	8						

Source: Gentry 1990b.

Table 17.6b. Largest genera of vascular plants, including weeds, in African and Madagascan local florulas.

Makokou, Gabon		Nosy Mangabe, Madagascar	
Ficus	29	Diospyros	21+
Dichapetalum	23	Oncostemum	12
Begonia	14(−)	Eugenia/Syzygium	9+
Asplenium	12	Landolphia	7
Cyperus	10	Tambourissa	6
Strychnos	10	Memecylon	6
Bertiera	10	Dracaena	5
Uvaria	9	Cyperus	5
Diospyros	9	Ficus	5
Salacia	9	Psychotria	5
Artabotrys	8	Protorhus	4
Xylopia	8	Polyscias	4
Culcasia	8	Dypsis	4
Psychotria	8	Asplenium	4
Sabicea	8	Symphonia	4
Scleria	7	Malleastrum	4
Hypolytrum	7	Garcinia	4
Drypetes	7	Potameia	4+
Beilschmiedia	7	Drypetes	3–4
Bulbophyllum	7		
Monanthotaxis	6		
Phyllanthus	6		
Garcinia	6		
Baphia	6		
Dracaena	6		
Geophila	6(−)		
Tricalysia	6		

Sources: Makokou, Gabon, compiled from Florence and Hladik 1980, et prec. (see Gentry and Dodson 1987); Nosy Mangabe, Madagascar, compiled from Schatz, personal communication, 1990.

genera are considered, 78 percent (21 of 27) are also Neotropical. However, even though the same genera are *present,* very few of the large Makokou genera are genera that are ever large at a Neotropical site, the three exceptions being ubiquitous *Ficus* and *Psychotria,* as well as *Cyperus* (tables 17.6a and 17.6b). Several of these genera—*Begonia, Asplenium, Strychnos, Salacia, Xylopia, Sabicea, Scleria, Phyllanthus, Garcinia*—sometimes have several species in a given Neotropical forest, unlike their Madagascan equivalents. Even though a lower percentage of its genera are shared, there is a much greater tendency for those continental African genera that are shared to have similar patterns of diversification in Africa and South America. Three of the nonshared large genera are the liana Annonacs *Uvaria, Artabotrys,* and *Monanthotaxus,* part of a major paleotropical radiation that has no Neotropical equivalent (Gentry 1991a). Another is *Culcasia,* which plays a role similar to Neotropical *Philodendron.* Among the shared genera with conspicuously asymmetrical radiations, *Dichapetalum* is most striking with 23 Makokou species and only 1 at any Neotropical florula site. *Bertiera* (10 spp.), *Diospyros* (9 spp.), *Drypetes* (7 spp.), *Beilschmiedia* (7 spp.), and *Bulbophyllum* (7 spp.) are the most striking of the genera disproportionately better represented at the local florula level in Africa.

A similar analysis of floristic affinities can be carried out at the level of the 0.1-ha samples (tables 17.7a and 17.7b; figure 17.2). Exactly comparable data are available for 11 African sites, 3 in Madagascar, and 78 in lowland Latin America. In addition data are available from both continents for a number of 1-ha plots in which plants ≥10 cm dbh have been censused. Data from 16 such plots are available from the Neotropics (in part supplied by Hubbell (personal communication, 1988) and D. Lieberman (personal communication, 1988) and from 5 African plots (4 of which were surveyed by Reitsma 1988; see figure 17.3). Again the picture is one of similarity.

If only the 0.1-ha samples are compared, the continental sites average 31 percent shared Neotropical genera (28–34%). Madagascar is even more similar at the generic level. The three sites average 40 percent shared genera, and the *lowest* Madagascar value for shared genera (34% at Perinet) is the same as the highest for continental Africa. The only families represented in the Madagascar samples but not the Neotropics are Physenaceae, Sphaerosepalaceae, Pandanaceae, Montiniaceae, and Pittosporaceae.

One of the most striking South America–Africa continental similarities is the dominance of legumes. Legumes dominate virtually all Neotropical lowland 0.1-ha samples. Of the 57 continental lowland Neotropical sites between 23.5° N and S latitudes (tables 17.7a and 17.7b; see also Gentry 1993a: table 5 for dry forest sites) 44, or 77 percent, had legumes as the most speciose family, and legumes were second most speciose in 9 other samples. Leguminosae were edged out as the most diverse family by Bignoniaceae in 7 especially liana-rich sites, by Moraceae in 4 sites with unusually rich soil, by Lauraceae in 1 Andean foothill cloud-forest, by Myrtaceae in 1 coastal Brazil sample, and by Rubiaceae in 1 of the 2 Choco pluvial forests. The dominance of Leguminosae in

Amazonian forests is even more dramatic with 21 of the 24 samples having more legumes than any other family and the other 3 having legumes second.

Other families also show remarkably consistent representation in different Neotropical forests. The same 11 families contribute about half the species to 0.1-ha lowland Neotropical samples: Leguminosae, Lauraceae, Annonaceae, Rubiaceae, Moraceae, Myristicaceae, Sapotaceae, Meliaceae, Palmae, Euphorbiaceae, and Bignoniaceae (Gentry 1988a). At least 8 of these families are always among the 10 most diverse families in any lowland Neotropical moist or wet forest.

There are also some significant differences in floristic composition between different Neotropical lowland forests, mostly between dry and moist forest and between sandy poor soil sites and richer lateritic and alluvial sites. Most families are poorer in species and individuals in dry than in wet forest; the only families actually richer in species in dry than in moist or wet forest are Cactaceae, Zygophyllaceae, Capparidaceae, Erythroxylaceae, and Julianaceae (Gentry 1993a). There are also some significant differences between the floristic compositions of poor-soil and richer-soil Neotropical forests. For my data set, three families, Moraceae and Combretaceae (both $p < .01$, Mann-Whitney U) and Flacourtiaceae ($p < .05$) have significantly more individuals on intermediate to rich soils and one, Euphorbiaceae ($p < .05$) on poor soils. In addition, if a one-tailed test is used, Rubiaceae, Myristicaceae, Melastomataceae, and Tiliaceae all have significantly more individuals on rich soils.

If average numbers of species are compared instead, 16 of the most prevalent 52 families have significantly different representations on different soil types in Amazonia. Chrysobalanaceae, Sapotaceae, Lauraceae, and marginally Apocynaceae (one-tailed test only) have significantly more species in sandy poor-soil forests, while Moraceae, Combretaceae, Flacourtiaceae, Monimiaceae, Nyctaginaceae, Piperaceae, Polygonaceae, Violaceae, Hippocrateaceae, and Guttiferae (and with a one-tailed test also Palmae and Tiliaceae) have more species in rich to intermediate clay or alluvial soils.

In contrast, biogeography contributes relatively little to floristic differences between most different lowland continental Neotropical forests. For example, the only significant difference in average number of individuals per family between trans-Andean and Amazonian sites is that Chrysobalanaceae and Menispermaceae are better represented in Amazonia. In addition, in coastal Brazil, Myrtaceae are better represented (Peixoto and Gentry 1990; Mori et al. 1983; table 17.7b).

Discriminating by number of species, rather than individuals, per family, 16 families are differently represented in trans-Andean than in Amazonian forests. Rubiaceae, Myrsinaceae, Verbenaceae, and Palmae average more species in the trans-Andean samples, whereas Burseraceae, Sapotaceae, Dichapetalaceae, Myristicaceae, Menispermaceae, Violaceae, Hippocrateaceae, and Connaraceae (and with a one-tailed test also Leguminosae, Sterculiaceae, Nyctaginaceae, and Flacourtiaceae) are more speciose in Amazonian forests. We may

Table 17.7a. Major families in 0.1-ha samples in Neotropics: Amazonia.

	Jatun Sacha, Ecu.	Indiana, Peru	Allpahuayo, Peru	Yanamono 1, Peru	Yanamono 2, Peru	Yanamono Tahuampa, Peru	Cocha Cashu, Peru	Sucusari, Peru	Shiringamazu, Peru	Jenaro Herrera, Peru	Res. Ducke, Brazil
Leguminosae	21	22	ca. 37	26	29	15	14	23	30	20	25
Annonaceae	8	10	ca. 12	12	13	12	13	22	5	10	14
Moraceae	22	18	ca. 17	13	14	13	10	7	6	14	14–15
Bignoniaceae	7	8	9	9	8	7	11	5	6	6	5
Lauraceae	15	5	9	5	6	2	7	12	11	16	ca. 15
Rubiaceae	17	8	7–8	13	16	8	4	8	7	4	5
Sapotaceae	11	8	11	5	4	7	8	8	12	15	ca. 15
Palmae	6	6	11	8	5	7	4	11	5	9	9
Euphorbiaceae	7	10	10	7	7	7	5	7	4	5	2
Myristicaceae	10	10	8	10	12	3	4	12	8	8	5
Meliaceae	6	8	7	10	10	1	9	5	5	6	5–6
Sapindaceae	3	9	ca. 11	5	6–7	4	5	7	4	1	5
Apocynaceae	4	1	2	2	5	5	4	5	2	7	7
Burseraceae	9	6	ca. 8	2	4	1	1	10	9	15	20
Chrysobalanaceae	1	1	2	1	1	3	2	9	6	13	4
Flacourtiaceae	9	5	9	8	8	1	4	5	1	3	4
Myrtaceae	4	11	6	6	5	7	1	5	3	1	4
Lecythidaceae	2	2	3–4	4	6–7	5–6	—	5–6	2	8–9	ca. 13
Guttiferae	4	—	ca. 6	2	5	2	2	2	8	4	1
Melastomataceae	8	2	1	3	1	9	3	2	4	10	6
Malpighiaceae	2	6	3	4	2	1	4	3	1	1	1
Hippocrateaceae	1	4	4	3	3	5	2	3	4	2	1
Violaceae	3	5	4	3	3	2	2	5	—	5	3
Menispermaceae	3	2	5	1	1	1	4	5	4	4	3
Sterculiaceae	6	3	5	3	2	4	4	3	2	1	2
Bombacaceae	4	1	2	2	2	3	3	7	1	2	3
Dilleniaceae	4	2	1	1	1	1	1	2	2	1	1
Nyctaginaceae	3	3	3–4	4	5	2	3	5	1	1	1
Olacaceae	1	2	2	—	—	1	—	6	2	1	1

Von Humboldt, Peru	Tambopata lateritic, Peru	Tambopata upland, Peru	Saul, French Guiana	Mocambo, Brazil	Madidi, Bolivia	Madidi Ridge, Bolivia	Cabeza de Mono, Peru	Mishana lowland, Peru	Mishana tahuampa, Peru	Mishana white sand, Peru	Cerro Neblina 1, Venezuela	Cerro Neblina 2, Venezuela
16	13	10	18	20	ca. 18	ca. 19	19	19	21	24	23	17
7	8	5	5	4	4	3	5	16	8	16	6	5
9	9	8	8	4	18	4	4–5	11	8	3	5	8
12	14	13	7	7	16	7	6	12	4	4	5	3
8–9	8	8	11	7	ca. 9	7	9	15	11	15	6	2
3	6	5	1	1	3	18	3	7	ca. 8	4	2	2
5	2	4	5	6	ca. 8	ca. 7	13	8	8	ca. 17	4	4
6	7	6	2	5	4	3	3	9	3	3	3	2
3	3	1	3	2	4	7	7	12	6	ca. 15	3	4
4–5	6	5–6	2	3	6	3	7	10	3	ca. 10	1	2
6	4	1	6	3	7	9	6	9	2	4	1	—
4	2	2	9	2	4	6–7	1	4	6	5	5	4
6	1	1	7	3	3	—	6	5	3	ca. 6	2	4
4	4	3	10	10	1	4	5	6	2	3	5	4
7–8	3	4	2	7	5	2	2	8	7	3	2	2
4	3	2	—	—	3	4	1	3	1	—	2	2
1	3	—	6	4	6	5	1	5	9	3	1	1
—	2	2	9	3	1	—	5–6	5	7	2	—	1
—	1	3	4	4	1	2–3	5	7	3	5	1	1
6	1	7	1	2	8	3	4	4	2	2	—	—
1	2	3	1	—	2	3	—	2	2	6	1	2
3	4	3	3	1	—	3	3	6	6	3	—	1
2	3	1	3	1	4	1	—	2	1	—	—	—
1	3	1	1	—	3	4	1	7	1	2	3	—
2	1	1	2	3	4	—	1	4	1	—	1	—
1	2	1	3	—	2	6	2	—	3	2	2	1
4	1	—	—	1	3	—	2	3	2	1	1	1
—	3	2	—	—	1	4	1	2	—	4	—	—
—	—	2	1	2	1	2	—	2	3	2	—	—

(continued)

Table 17.7a. *Continued.*

	Jatun Sacha, Ecu.	Indiana, Peru	Allpahuayo, Peru	Yanamono 1, Peru	Yanamono 2, Peru	Yanamono Tahuampa, Peru	Cocha Cashu, Peru	Sucusari, Peru	Shiringamazu, Peru	Jenaro Herrera, Peru	Res. Ducke, Brazil
Loganiaceae	2	1	1	2	3	3	2	3–4	1	2	1
Icacinaceae	—	1	4	2	2	—	—	1	3	2	—
Elaeocarpaceae	1	1	2	—	1	—	—	1	6	4	1
Connaraceae	1	1	3	1	3	4	—	4	2	2	3–4
Myrsinaceae	—	—	—	2	—	1	—	—	—	1	—
Combretaceae	1	1	4	3	2	2	3	—	2	1	1
Boraginaceae	1	2	3	1	—	1	2	2	1	1	3
Total species	240	225	256	212	225	163	165	ca. 240	182	240	233
Number of families[a]	61	61 (+1)	52	50 (+4)	50 (+2)	50	49	47	50	58	53

Source: Original data for plants ≥ 2.5 cm dbh (greatest diameter for lianas); sites listed in Gentry 1991a.
Notes: ≡ = Numbers for most speciose family
= = Numbers for second most speciose family
— = Numbers for third to tenth most speciose families
[a]Familial undetermined species in parentheses.

conclude that although the great majority of Neotropical forests are composed of a shared group of taxa, there are also significant floristic differences determined by intracontinental geography and ecology.

Although insufficient African sites are available for a similar statistical analysis, there are equally strong between-site similarities. The general between-site floristic similarity of both South America and Africa is paralleled in an equally striking between-continent floristic similarity. The families represented in the 0.1-ha samples from the Neotropics and Continental Africa are virtually identical (figure 17.2; table 17.8), as already noted by Gentry (1988a). Each African sample includes 1 to 4 families not represented in the Neotropics, but these represent an average of only two species and four individuals per sample contributed by families not represented in the equivalent Neotropical data. Ancistrocladaceae and Medusandraceae were represented by one individual each at two sites, Ixonanthaceae by one indi-

Von Humboldt, Peru	Tambopata lateritic, Peru	Tambopata upland, Peru	Saul, French Guiana	Mocambo, Brazil	Madidi, Bolivia	Madidi Ridge, Bolivia	Cabeza de Mono, Peru	Mishana lowland, Peru	Mishana tahuampa, Peru	Mishana white sand, Peru	Cerro Neblina 1, Venezuela	Cerro Neblina 2, Venezuela
—	1	—	1	2	2	1	—	1	3	1	1	—
—	—	1	1	3	1	—	1	2	—	—	2	—
1	1	1	—	—	2–3	—	2	2	—	1	1	—
1	2	2	1	3	2	1	2	4	3	1	—	—
—	1	1	2	—	1	1	2	1	1	1	1	1
—	2	2	2	—	2–3	3	1	—	2	—	1	—
2	1	1	1	—	2	—	—	1	—	1	—	—
154	149	132	160	130	204	175	148	269 (+)	ca. 168	140	98	85
43	48 (+2)	43 (+1)	46 (+2)	40 (+1)	61	49 (+2)	42	58 (+3)	40 (+2)	46 (+3)	31 (+1)	28 (+1)

vidual at one site, and Scytopetalaceae at three sites. Only in the latter case did an endemic family contribute significantly to a site's diversity with *Ouabangia alata* the fifth most common species at Korup and *Rhaptopetalum* cf. *coriaceum* represented by three individuals at the same site. Moreover, several of the endemic African families included in these analyses are somewhat questionable segregates: Pandanaceae from Euphorbiaceae, Irvingiaceae from Simaroubaceae, and Ixonanthaceae from Linaceae.

In Africa, as in the Neotropics, legumes are the most speciose family at most sites. Legumes are most speciose in 7 of the 11 African samples, second in another, and third in another. The only exceptions to legume prevalence are a putatively disturbed forest at Omo and the richest soil forest on Mt. Cameroon. Hladik (1986) also emphasized the similarly important role of Leguminosae in both continents, based in large part on the Gentry (1982b, 1988a) data. The preeminent role of legumes on both continents is even more dramatic when seen in the context of Australasia, where 11 similar 0.1-ha sites from lowland moist and wet forests

Table 17.7b. Major families in 0.1-ha samples in Neotropics: Chocó, Coastal Brazil, and Trans-Andean moist and wet forests.

	Chocó pluvial		Coastal Brazil			Trans-Andean moist and wet forest											
	Bajo Calima, Colombia	Tutunendo, Colombia	Linhares, ES, Brazil	Jacarepagua, RJ, Brazil	Boraceia, SP, Brazil	Osa, Corcovado, Costa Rica	Pipeline Rd., Panama	Centinela, Ecuador	Rio Palenque 1, Ecuador	Rio Palenque 2, Ecuador	Madden Forest, Panama	Jauneche, Ecuador	Esmeraldas, Ecuador	La Raya, Colombia	Bosque de la Cueva, Colombia	Mariquita, Colombia
Leguminosae	17	25	32	18	7–8	13	10	7	8	8	16	9	7–8	22	8	18–19
Annonaceae	13	15	6	3	1	6	7	5	2	2	3	2	2–3	6	1	3
Moraceae	7	11	4	8	1	16	9	9	11	12	2	9	9	14	8	6
Bignoniaceae	1	2	11	6	3	1	6	2	4	5	7	11	10	6	9	6
Lauraceae	11	9	8–9	8	12	3	5	ca. 10	3	7	3	2	3	4	8	2
Rubiaceae	19	18	5	13	8	3	6	8	7	4	14	3	6	9	5	6
Sapotaceae	17	13	13	11	3	5	3	1	1	2	1	1	3	3	—	1
Palmae	17	17	3	3	2	6	10	8–9	6	6	4	3	2	8	6	2
Euphorbiaceae	6	7	8	6	2	6	1	2	4	3	3	6	3	3	4	2
Myristicaceae	11	7	1	2	—	5	3	—	4	5	1	1	2	4	1	1
Meliaceae	2	4	3	4	2	4–5	2	6	4	6	4	2	2	2	1	—
Sapindaceae	1	6	5	5	5	8	7	4	1	4	5	4	4	5	1	6
Apocynaceae	6	4	6	7	1	2	6	2	—	1	2	3	2	8	3	3
Burseraceae	4	4	4	2	1	2	4	1	—	—	2	—	—	4	1	1–2
Chrysobalanaceae	6	7	1	2	2	1	2	2	1	1	1	1	—	1	—	2

Flacourtiaceae	3	2	4	1	—	4	1	1	1	2	8	2	ca. 3	3	—	3
Myrtaceae	10	6	ca. 31	12–14	ca. 30	—	2	3	—	—	3	1	ca. 4	1	—	4
Lecythidaceae	7	8	1	1	—	1	3	5	—	2	1	1	1	3	—	—
Guttiferae	16	12	2	2	2	4	5	4	2	3	—	—	—	5	—	2
Melastomataceae	16	14	—	4	1	1	7	6	2	4	2	—	—	5	2	1
Malpighiaceae	1	1	2	3	2	1	8	1	—	1	4	4	3	1	1	—
Hippocrateaceae	4	1	8	8	2	—	2	—	—	—	—	2	1	2	—	—
Violaceae	1	2	3	1	—	1	3	1	—	—	2	—	—	2	—	—
Menispermaceae	2	2	2	3	2	2	1	2	3	—	—	—	1	1	—	1
Sterculiaceae	1	1	2	—	—	—	—	1	—	1	1	—	1	1	2	1
Bombacaceae	6	5	3	2	1	2	—	5	3	3	2	1	1	1	—	—
Dilleniaceae	5	2	2	1	—	3	2	—	—	—	4	1	—	4	1	1
Nyctaginaceae	1	1	ca. 5	4	2	—	1	—	1	1	—	2	1	1	2	1
Olacaceae	1	1	2	1	—	2	1	1	1	1	—	—	—	1	—	—
Loganiaceae	4	3	—	—	—	1	1	1	—	—	—	—	—	1	—	—
Icacinaceae	1	—	2	—	1	—	—	1	1	—	—	—	—	1	—	—
Elaeocarpaceae	1	3	2	1	1	1	1	1	—	1	1	—	—	2	—	—
Connaraceae	—	—	—	—	—	2	2	1	—	—	1	1	—	1	—	—
Myrsinaceae	3	5	—	2	5	—	3	2	2	1	—	—	1	1	3	1
Combretaceae	—	1	2	1	—	—	2	1	—	—	1	—	1	—	3	—
Boraginaceae	1	1	1	—	2	1	2	1	1	—	1	1	1	4	1	1
Total species	265	258	212	ca. 159	125	130	168	140	119	121	126	96	ca. 96	160	93	ca. 94
Number of families[a]	58	53	53 (+3)	45	43	46	59	55	50	43	45 (+2)	31 (+1)	28 (+1)	51	35	39

Source: Original data for plants ≥ 2.5 cm dbh (greatest diameter for lianas); sites listed in Gentry 1991a.

Notes: ≡ = Numbers for most speciose family

═ = Numbers for second most speciose family

— = Numbers for third to tenth most speciose families

[a]Species unidentified to family in parentheses.

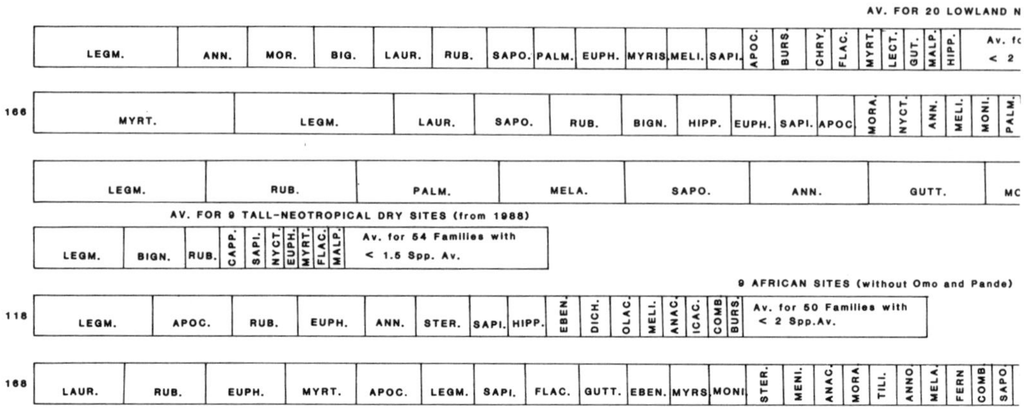

Figure 17.2. Comparison of average number of species per family for 0.1-ha samples from subsets of different lowland Neotropical forest types with equivalent data for Africa and Madagascar. Shortest column segments are two species tall. Family codes are first four letters of family names.

include an average of 5.7 legumes. For sites south of Wallace's Line, legumes are even sparser, averaging only 2 species per 0.1 ha. Only in seasonally dry forests, as in central and northern Thailand, do legumes become a dominant element of Asian forests.

All of the families most prevalent in South America (table 17.7a and 17.7b) are represented in African transects except Malpighiaceae, Nyctaginaceae, Elaeocarpaceae, and Myrsinaceae. Moreover, essentially the same set of families dominates on both continents. Of the 20 most-prevalent Neotropical families, all but Bignoniaceae, Chrysobalanaceae, Lauraceae, Myrtaceae, Lecythidaceae, and Melastomataceae were sometimes among the 10 locally most speciose families in African forests; even these are all represented in some of the African samples.

We may examine the significance of the floristic differences by comparing average percentage representation of the distinct families on the two continents with Mann-Whitney U tests. If number of individuals per sample is used as a criterion, 29 of the relatively well represented families show no significant difference, whereas 8—Anacardiaceae, Apocynaceae, Dichapetalaceae, Ebenaceae, Euphorbiaceae, Hippocrateaceae, Olacaceae, and Sterculiaceae—are significantly better represented in Africa and 5 in South America—Bignoniaceae, Lauraceae, Melastomataceae, Moraceae, and Myristicaceae. Chrysobalanaceae and Lecythidaceae are also marginally better represented in South America and Flacourtiaceae in Africa as indicated by a one-tailed test.

If number of species per family at each site is compared, the results are almost exactly the same. For example, comparing number of species per site for Africa with number of species per site for the most ecologically equivalent subset of lateritic-soiled South American sites generates exactly the same list of families significantly differently

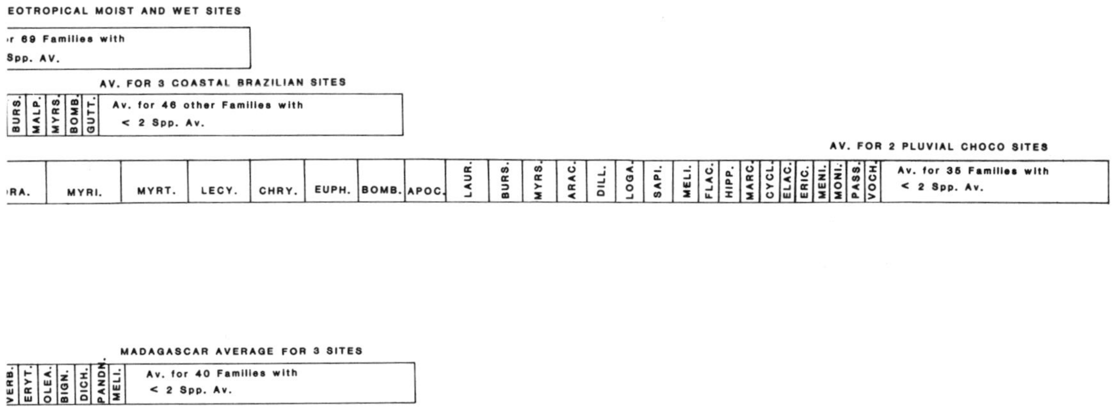

represented in the forests of the two continents as does the comparison based on number of individuals, except that Chrysobalanaceae is added to the families significantly better represented in South America.

If the 8 continental African samples are compared with all 13 of the appropriate South American lowland moist forest samples, the intercontinental differences become more pronounced. In this analysis 22 families differ significantly between the two continents and 21 are the same. The families better represented in Africa include Anacardiaceae, Apocynaceae, Dichapetalaceae, Ebenaceae, Euphorbiaceae, Hippocrateaceae, Olacaceae, Rubiaceae, Sterculiaceae, and Styracaceae (all $p < .01$ except Rubiaceae where $p < .05$). Better represented in South America are Bignoniaceae, Bombacaceae, Chrysobalanaceae, Lauraceae, Melastomataceae, Monimiaceae, Moraceae, Myrtaceae, Myristicaceae, Palmae, Piperaceae ($p < .01$), and Nyctaginaceae (the latter only at $p < .05$). In addition, with a one-tailed Mann-Whitney U test, Loganiaceae, Sapindaceae, and Vitaceae are better represented in Africa and Meliaceae in South America.

From analysis of the 0.1-ha data set, we may conclude that while almost exactly the same set of families is represented in equivalent forests on the two continents, there are nevertheless some intrinsic differences in how African and Neotropical lowland forest communities are put together.

Elsewhere I have suggested (Gentry 1988a) that Madagascar is floristically more like the Neotropics than Africa is. There are too few Madagascar samples, and they are too variable among themselves (one from dry forest, one from lowland wet forest, one from premontane forest), to make statistical comparisons very instructive. If the three Madagascar samples (table 17.9) are lumped, only Lauraceae and Monimiaceae are significantly better represented

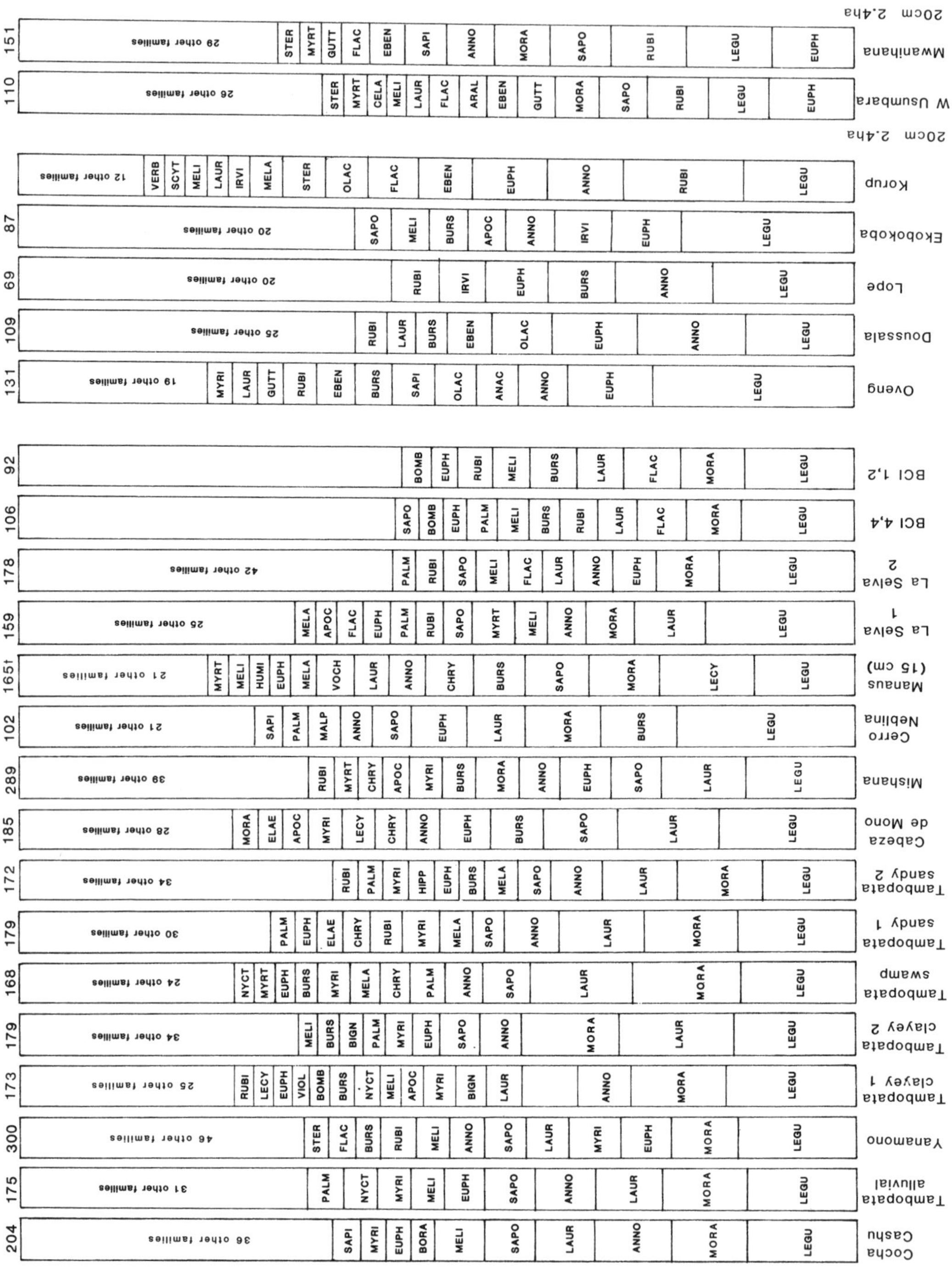

Figure 17.3. Familial compositions of some Neotropical and African 1-ha tree plots (plants ≥10 cm dbh). Amazonia (bars 1–12), southern Central America (bars 13–16), and Central Africa (17–21). Two right-hand bars are data for discontinuous larger samples of mid-elevation East African montane forest with 20 cm dbh cut off (Lovett, personal communication, 1990). Amazonian sites from original data, except the Manaus one (Prance et al. 1976). Central American sites are from data of D. Lieberman and S. Hubbell as reported in Gentry (1990b). First four African sites are from Reitsma (1988); Korup data are from D. Thomas (personal communication, 1987). Family codes are first four letters of family names.

Table 17.8. Major families of plants ≥ 2.5 cm dbh (2.5 cm greatest diameter for lianas) in 0.1-ha samples in Africa.

	Mt. Cameroun	Korup	Banyong (800m²)	Makokou 1	Makokou 2	Belinga (500m²)	Ndakan upland	Ndakan inundated	Omo	Pugu	Pande (500 m²)
Leguminosae	5	22	12	23	22	7	15	15	5	10	7
Apocynaceae	11	8–10	15	12	8	7	8–9	8	6	6	6
Euphorbiaceae	8	12	11–12	7	6	10	12	7	7	2	7
Rubiaceae	14	10–11	6	4	3	7	5–6	3	8	ca. 9	1
Annonaceae	7	6	8	8	5	3	6–7	9	3	3	3
Sterculiaceae	7	9	15	7	3	6	4	2	3	ca. 6	2
Sapindaceae	3–4	4	8	7	7	5	2	2	—	2–3	3
Hippocrateaceae	6	—	6	7	6	2	3	6	5	ca. 4	2
Dichapetalaceae	5	1	6	5	4	4	—	2	1	4	—
Olacaceae	6	4	4	2	4	4	7	1	1	—	—
Ebenaceae	6	5	3	2	1	4	5	4	7	2	—
Burseraceae	2	1	2	6	4	1	2	—	7	—	—
Connaraceae	1	3	3	2	3	1	2	2	3	—	—
Anacardiaceae	1	2	7	2	3	3	2	1	—	2	—
Icacinaceae	7	2	5	1	3	—	2	3	1	—	—
Guttiferae	2	1	5	2	2	5	—	—	1	—	—
Myristicaceae	3	5	1	1	2	—	—	1	—	—	—
Flacourtiaceae	3	2	2	2	2	—	1	—	—	2	—
Meliaceae	3	2	4	2	1	4	9–10	2	—	—	—
Menispermaceae	2	1	2	1	1	—	2–3	4	1	1	—
Violaceae	4	3	3	2	—	—	2	—	2	3	—
Sapotaceae	—	1	4	4	2	2	2	—	—	1	2
Moraceae	1	2	2	1	—	—	3	4	2	2	1
Combretaceae	—	1	1	4	3	—	4	3	1	4–5	ca. 5
Lauraceae	1	2	1	1	3	1	2	1	—	—	—
Irvingiaceae	1	2	3	2	1	1	1	1	—	—	—
Tiliaceae	1	—	2	—	2	—	2	3	1	1	—
Loganiaceae	2	1	3	—	3	—	1	—	2	1–2	2
Pandaceae	1	2	2	2	1	—	1	2	—	—	—

(continued)

Table 17.8. *Continued.*

	Mt. Cameroun	Korup	Banyong ($800 m^2$)	Makokou 1	Makokou 2	Belinga ($500 m^2$)	Ndakan upland	Ndakan inundated	Omo	Pugu	Pande ($500 m^2$)
Linaceae	1	1	2	—	—	—	—	1	1	1	1
Lecythidaceae	1	1	1	2	—	—	1	—	—	1	—
Total species	128	137	ca. 180	135	114	88 (+)	108	96	73	75–80	46 (+)
Number of families[a,b]	38 (+2)	43	40 (+)	39 (+3)	34 (+2)	25 (+4)	38 (+3)	32	30	28 (+2)	15 (+)

Source: Original data; sites listed in Gentry 1991a.
Notes: ≡ = Numbers for most speciose family
= = Numbers for second most speciose family
— = Numbers for third to tenth most speciose families
[a]Species unidentified to family in parentheses.
[b]71 families in total sample.

in Madagascar than in Africa, and only Leguminosae is better represented in Africa than in Madagascar. Other familial differences, such as the much greater representation of Myrsinaceae and Myrtaceae in moist Madagascan than in African forests, are not statistically apparent from these data. Perhaps the most interesting aspect of the Madagascan 0.1-ha data set is that at the generic level it is closer to the Neotropics, with an average of 40 percent shared genera per site, than to tropical Africa, with an average of 31 percent shared genera per site.

In addition to the between-site differences and similarities of 0.1-ha samples, data from 1-ha tree plots can also be used to characterize both African and Neotropical forests. Comparable floristic data are available for trees (and lianas) ≥10 cm dbh in 11 Amazonian plots, 4 Central American ones, and 5 central African ones (tables 17.10a and 17.10b; figure 17.3). In addition, data on trees ≥20 cm dbh in noncontiguous variable-area plots totaling approximately 2.4 ha each are available for two forests just above 1000 m elevation in Tanzania (Lovett, personal communication, 1990; figure 17.3). Again the overwhelming pattern is of similarity, both within and between continents. Leguminosae are the most speciose family in all the Neotropical tree plots across a wide variety of soil types and in both Central America and South America. Moraceae are the second most speciose tree

Table 17.9. Major families of plants ≥ 2.5 cm dbh (2.5 cm greatest diameter for lianas) in 0.1-ha samples in Madagascar.

	Nosy Mangabe	Perinet	Ankarafantsika (dry forest)
Lauraceae	16	17–18	—
Rubiaceae	11–12	15	4
Euphorbiaceae	11	15	3
Myrtaceae	13	10	2
Apocynaceae	10	ca. 10	2
Leguminosae	4	4	11
Sapindaceae	9	6	4
Flacourtiaceae	13	2	3
Guttiferae	9	7	1
Ebenaceae	13	1	1
Monimiaceae	4–6	7	1
Myrsinaceae	7	7	—
Sterculiaceae	2	ca. 6	5
Moraceae	3	6	1
Menispermaceae	8	—	3
Anacardiaceae	ca. 6	1	3
Annonaceae	2	5	2
Tiliaceae	2	3	5
Melastomataceae	5	3	1
Combretaceae	3	1	4
Fern	1	8–9	—
Sapotaceae	4	3	—
Verbenaceae	1	3	3
Hippocrateaceae	1	1	3
Bignoniaceae	3	1	2
Connaraceae	5	1	—
Dichapetalaceae	3	1	2
Pandanaceae	2–3	3	—
Total species	ca. 211	ca. 198	94
Number of families[a,b]	51 (+6)	52 (+5)	40 (+1)

Source: Original data; sites listed in Gentry 1991a.

Notes: ≡ = Numbers for most speciose family
═ = Numbers for second most speciose family
— = Numbers for third to tenth most speciose families
[a]Species unidentified to family in parentheses.
[b]70 families in total sample.

family in all the South American samples on relatively good (= rich lateritic and alluvial) soil and in all but one of the Central American samples, which are also from forests on relatively rich soil. The only rich-soil exception to Moraceae as the second most diverse family is one La Selva, Costa Rica, plot where they are edged out by Lauraceae. Lauraceae, which is also the second most speciose family in all the samples from the poorest sandy soils, is always among the 10 most speciose families and is usually third or fourth in diversity. Annonaceae and Sapotaceae are also repeatedly important in all South American samples, usually rounding out the top 5 and always among the 10 most diverse families; these 2 families are notably less species-rich in the Central American samples, and on Barro Colorado Island are usually not even among the 10 most speciose families (Gentry 1990b).

African tree plots are generally similar in floristic composition to Neotropical ones with a few striking exceptions. Leguminosae are always the most speciose family, just as in the Neotropics, although Caesalpinioids are more important (e.g., averaging ca. 72% of the legumes in Reitsma's [1988] Gabon plots). Annonaceae, always among the top 4

Table 17.10a. Major families in 1-ha tree plots in Amazonia.

	Cocha Cashu, Peru	Ducke Reserve, Manaus, Brazil[a]	Tambopata alluvial, Peru	Tambopata swamp, Peru	Tambopata sandy 1, Peru	Tambopata sandy 2, Peru	Tambopata clayey 2, Peru	Tambopata clayey 1, Peru	Yanamono, Peru	Mishana, Peru	Cerro Neblina, Venezuela	Cabeza de Mono, Peru
Leguminosae	27	19	23	24	26	19	26	ca. 28	42–44	37–38	23	30–31
Moraceae	18	13	16	22	19	18	21	18	24	15	9	5–6
Annonaceae	18	7	13	8	11	10	11	11–12	12	15	4	7
Lauraceae	15–16	7	14	23	16	14	24	6–7	15	30	7	22
Sapotaceae	14	11	12	9	8	7	8	12	16	16	3	16
Meliaceae	11	4	7	1	0	1	4	5	11	7	2	4
Myristicaceae	7	3	7	6	6	7	6	6	ca. 18	13–14	2	7–8
Bombacaceae	7	2	3	1	3	1	1	4	6	2	2	1
Sapindaceae	7	2	1	1	3	2	1	1	4	4	3	4
Euphorbiaceae	6	4	6	4	6	5	6	4	18	15	6	11
Palmae	4	0	6	7	5	6	5	5	6	6	3	1
Bignoniaceae	4	0	4	2	3	4	5	7	1	1	0	4
Nyctaginaceae	3–4	2	6	4	3	2	3–4	ca. 5	5	3	0	3
Chrysobalanaceae	3	10	2	6	5–6	5	2	1	3	9	2	7
Rubiaceae	2	3	3–4	2	6	5	1	4	13–14	9	2	2
Melastomataceae	1	5	0	6	7–8	7	0	2	0	2	0	0
Burseraceae	1	10	3	5	3	3	4	4–5	9	13	9	12–13
Lecythidaceae	1	18	3	2	3	3	3	4	7–9	5	0	8–9
Vochysiaceae	0	8	0	0	0	0	0	0	0	0	0	0
Flacourtiaceae	3	3	3	3	3	2	3	5	8	1	1	1
Olacaceae	2	2	1	0	2	2	1	2	2	2	0	1
Irvingiaceae	0	0	0	0	0	0	0	0	0	0	0	0
Ebenaceae	1	0	0	2	1	2	1	0	1	0	0	0
Anacardiaceae	1	0	3	0	1	3	1	2–3	2	1	1	0
Guttiferae	2	2	4	2	4	3	3	2	4	7	0	4
Apocynaceae	3	3	6	3	4	3	3	5	4	10	2	6

Source: Gentry 1990b.

Notes: ≡ = Numbers for most speciose families.

= = Numbers for second most speciose families.

— = Numbers for third to tenth most speciose families.

[a]Prance et al. 1976.

Table 17.10b. Major families in 1-ha tree plots in Africa.

	Oveng, Gabon[a]	Douassola, Gabon[a]	Lope, Gabon[a]	Ekobakoba, Gabon[a]
Leguminosae	32	14	12	18
Moraceae	—	—	—	—
Annonaceae	8	14	8	5
Lauraceae	4	4	1	2
Sapotaceae	3	—	—	4
Meliaceae	2	2	1	4
Myristicaceae	4	1	3	3
Bombacaceae	—	—	—	—
Sapindaceae	7	3	1	2
Euphorbiaceae	13	11	5	7
Palmae	—	—	—	—
Bignoniaceae	—	—	—	—
Nyctaginaceae	—	—	—	—
Chrysobalanaceae	2	2	1	1
Rubiaceae	5	4	4	3
Melastomataceae	1	2	1	—
Burseraceae	6	4	6	4
Lecythidaceae	1	—	—	1
Vochysiaceae	1	1	1	—
Flacourtiaceae	2	2	2	1
Olacaceae	7	8	3	3
Irvingiaceae	3	3	4	6
Ebenaceae	6	6	—	1
Anacardiaceae	7	—	2	3
Guttiferae	4	3	2	1
Apocynaceae	1	3	—	4

Source: Reitsma 1988.

Notes: ≡ = Numbers for most speciose families.

= = Numbers for second most speciose families.

— = Numbers for third to tenth most speciose families.

families, are also about as well represented as in Neotropical forests. Euphorbiaceae, although not differing significantly in absolute numbers from Neotropical forests, are proportionally more important, typically being the second most diverse tree family (and as diverse or more diverse than legumes in Lovett's [personal communication, 1990] midelevation African sites). In contrast, Moraceae are dramatically more poorly represented in the African samples and, in fact, are hardly represented at all. Another conspicuously underrepresented family in African tree plots is Palmae, a well-known phenomenon (Richards 1973; Emmons and Gentry 1983), although rattans are often conspicuous and tree palms like *Elaeis* and *Raphia* are dominant components of certain vegetation types. Although Lauraceae and Sapotaceae are not as prevalent as in the Neotropics, the former is often among the 10 most speciose families and the latter sometimes is. Myristicaceae, like Lauraceae sometimes cited as a family poorly represented in Africa, is usually among the 10 most species-rich families and is actually about as prevalent as in the Neotropics.

Most of the families that are most conspicuously underrepresented in African tree plots are not the families that have been cited on the basis of continent-level or regional species counts as being depauperate in Africa. Besides Palmae and Moraceae, they are Bombacaceae, Bignoniaceae, Nyctaginaceae, which occur in none of the African tree plots, and perhaps Lecythidaceae, which occurs only twice in the available tree plot data set. On the other hand, several families are noticeably better represented in African than in South American tree plots, the most striking being Olacaceae, which averages more than 5 species per plot in Africa, and the endemic family Irvingiaceae, which averages almost 4. Ebenaceae and Anacardiaceae, averaging about 3 species per plot, and Apocynaceae and Guttiferae, averaging 2–2.5 are marginally more prevalent in Africa.

Hladik (1986) cited data from CTFT surveys of trees > 20 cm dbh in even larger areas (ca. 41 ha) which show similarly striking between-continent familial similarity of trees in samples from Gabon and French Guiana.

Lianas are less similar taxonomically between continents than trees and account for much of the greater intercontinental floristic difference that exists between 0.1-ha and 1-ha plots. African liana species richness is almost the same as in Amazonia (41 vs. 39 spp./0.1 ha), but different families tend to be important (Gentry 1991a). The biggest difference is the near absence of Bignoniaceae lianas in African forests, whereas this is the predominant Neotropical liana family. The other striking floristic anomaly is the prevalence of Apocynaceae in African climbing-plant communities. Apocynaceae average 7 species per 0.1-ha sample and are the most speciose liana family in five of the eight African samples and second in the other three. Apocynaceae essentially replace Bignoniaceae as the dominant lianas in Africa.

Most of the other climbers in African forests are the same families, and even many of the same genera, as in the Neotropics. Leguminosae are runners-up in dominance with an average of 5 species per 0.1-ha sample, exactly as in the Neotropics. Hippocrateaceae (5 spp./0.1 ha), Combretaceae (2 spp./0.1 ha), Annonaceae (1.5 spp./0.1 ha), and perhaps Connaraceae (2 spp./0.1 ha) are more speciose in African forests. The greater role of Hippocrateaceae in African forests is most noteworthy. Hippocrateaceae are the most speciose climbing family at one site, runners-up at three others, and in the top 3(4) families at all sites except the one with the poorest soils (Korup, Cameroon) (Gentry 1991a).

Madagascar is similar to both Africa and Amazonia in its overall liana diversity but more similar to Africa in the dominance of Apocynaceae climbers and the prevalence of Dichapetalaceae and Hippocrateaceae. Sterculiaceae (*Byttneria*) and Menispermaceae may be slightly more important among Madagascar climbers than on the African continent.

IV. Floristic Summary

In general tropical African and Neotropical forest plant communities are put together in strikingly similar ways. Not only their diversity, but also their taxonomic composition, tends to be similar given similar environmental constraints. Although Africa may be depauperate at the continental level and may be especially poor in many of the characteristic woody taxa of lowland tropical forests elsewhere, at the level of the community this is rarely apparent. Essentially the same families and about a third or more of the same genera are represented in Florulas or various kinds of samples from the two continents, as well as from Madagascar.

The most speciose woody families in lowland tropical American forests are, in declining order, Leguminosae, Moraceae, Lauraceae, Annonaceae, and Sapotaceae. Drier Neotropical forests essentially represent a depauperate sample of the families and genera of wetter Neotropical forests. Coastal Brazilian forests differ especially in a greater representation of Myrtaceae, Central American forests in a lesser representation of Chrysobalanaceae and Menispermaceae. Forests on the poorest Neotropical soils have more Chrysobalanaceae, Sapotaceae, and Lauraceae, those on richer soils more Moraceae, Nyctaginaceae, Hippocrateaceae, and Guttiferae.

African tree communities also consist of the same families, differing most notably in greater representation of Euphorbiaceae, Sterculiaceae, and Ebenaceae and lesser representation of Moraceae, Lauracae, Sapotaceae, and Palmae; African liana communities are dominated by Apocynaceae, unlike Bignoniaceae-dominated Neotropical ones. There are also interesting intercontinental differences within the same family. For example, Annonaceae are mostly lianas in Africa, nearly all arborescent in the Neotropics, whereas Sapindaceae are predominantly scandent in the Neotropics but nearly all erect in Africa. Lowland moist and wet forests in Madagascar are floristically more similar to mid-elevation Andean forests than to lowland Amazonian forests, especially in the prevalence of Lauraceae and the relative unimportance of Leguminosae. In contrast, legumes dominate in Madagascan dry forests, just as in the Neotropics, but Madagascan forests have more species of Sterculiaceae and Tiliaceae and fewer of Capparidaceae and Bignoniaceae. Nevertheless the intracontinental floristic differences are at least as pronounced overall as the intercontinental ones.

V. Diversity Patterns

A number of community-level Neotropical diversity patterns have been suggested by Gentry (1982b, 1988a). Wetter tropical forests tend to have more species than drier ones (Gentry 1982b, 1988a, 1993a). Forests on poorer soils may generally have fewer species than forests on richer soils, but the relationship is tenuous and diversity is largely independent of soil fertility (Gentry 1988a; Stark et al. In prep., contrary to Huston 1979, 1980 and Ashton 1992). Low elevation forests have more species than high-elevation forests, although there is little decrease in diversity up to approximately 1500 m elevation (Gentry 1988a).

In figure 17.4 the African and Madagascan data for species richness in 0.1-ha samples versus precipitation are plotted against the Neotropical data (from Gentry 1988a). These samples show no apparent difference in diversity from the Neotropical data, except that two samples from the high-rainfall area near the base of Mt. Cameroon are less species-rich than their Neotropical equivalents from the Colombian Choco. However, both of these sites have independent ecological constraints that might be expected to correlate with anomalously low diversity. One from Korup National Park is on very nutrient-poor ultra-leached skeletal soil; the other is on the slopes of Mt. Cameroon, an active volcano, and old lava flows had to be dodged in laying out the transect (which could not be located elsewhere due to human perturbations). The high-rainfall lateritic soil site at Banyong, north of Mt. Cameroon, is as species-rich (ca. 180 spp. $\geq$2.5 cm diameter) as many of the richer Neotropical sites, just as are the Madagascan wet forests.

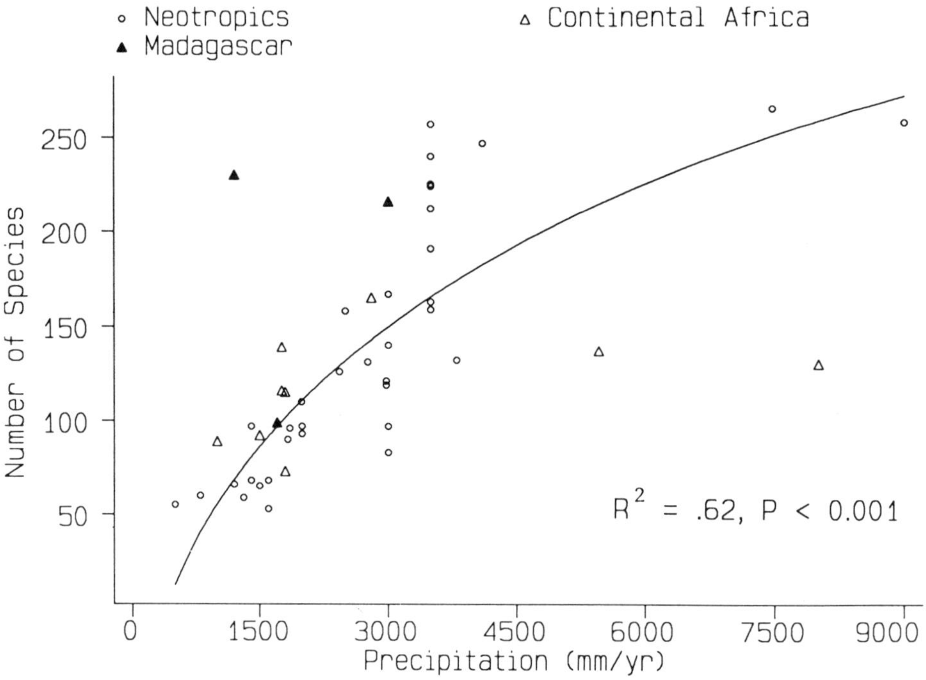

Figure 17.4. Change in species richness of 0.1-ha samples with precipitation. African sites (△) and Madagascan sites (▲) plotted against square-root-fitted Neotropical regression (No. spp. = −0.016 (ppt.) + 5.47 (ppt.)$^{\frac{1}{2}}$ − 101.56). A few recent samples that have not yet been entered into the computerized data base are not included. Note that the highest diversity African site at Banyong, Cameroon, records data for only 8 transects; the full 0.1-ha sample is projected to have 180 species.

At the community level Africa is not depauperate at all. Contrary to the concept of Africa as "odd man out," community-level diversity and patterns of diversity in Africa are quite similar to those of the Neotropics.

Another pattern that is also shared by the two continents (and by tropical forests in general) is the striking preponderance of rare or locally distributed species in these kinds of samples. This phenomenon seems to be at a similar level on the two continents. For example in Reitsma's four Gabon tree plots, 70 percent of the 272 species (including 26 lianas) occur at a single site, and 40 percent are represented by single individuals; only 8 species (3%) occur in all 4 plots and only 14 species have 16 or more individuals on any one plot. For a similar set of seven 1-ha tree plots on different substrates at Tambopata, Peru (Gentry 1988a, 1993b), there are approximately 600 species ≥10 cm diameter (including 50 lianas). However, only 32 species have more than 10 individuals in any one of the six primary forest plots and only 18 species have 16 or more individuals on any one plot. Each plot averages only 5 species with 10 or more individuals. Moreover, there are only 67 species represented by 10

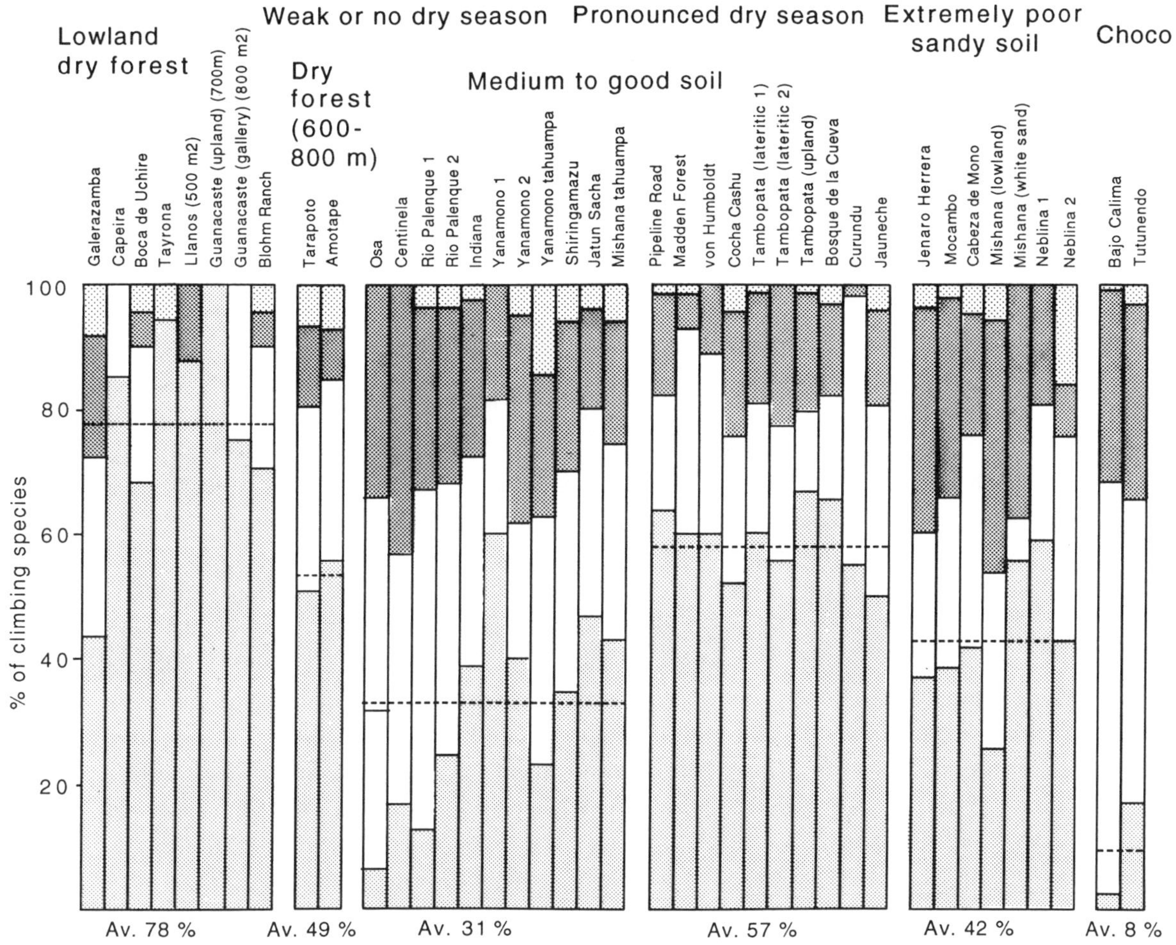

Figure 17.5. Dispersal of Neotropical lianas. Shading from base to top: wind dispersal; bird dispersal; mammal dispersal; autochorous and water dispersal. From Gentry 1991b:figure 15.1.

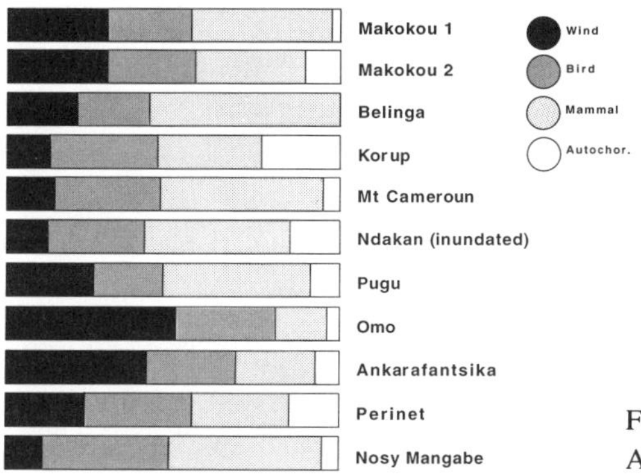

Figure 17.6. Dispersal of paleotropical lianas. Autochorous category includes water dispersal.

or more individuals in the seven plots taken together. All 21 of the commonest species are represented by at least 10 individuals on at least one plot, however.

VI. Patterns of Dispersal and Pollination

The intercontinental differences in some other aspects of tropical plant community composition are much greater than the differences in diversity and floristic composition. For example, dispersal ecology shows significant differences between Africa and South America. The two most striking differences are that Africa has much less wind dispersal and has many species with large elephant-dispersed fruits.

Lianas show these differences dramatically. In the Neotropics (figure 17.5), wind-dispersed lianas average 78 percent in dry forests, 57 percent in moist forests on good to medium-poor soil, 42 percent in moist forests on very poor soil, 31 percent in wet forests, and 8 percent in Choco pluvial forests. African lowland dry and moist forest lianas (figure 17.6) have only 24 to 30 percent wind-dispersed species. Madagascan lianas may be closer to the Neotropical dispersal pattern with 42 percent wind-dispersal in the only dry forest sample, but there are too few data to support broad generalization.

Erect plants (figure 17.7) show similar, but less dramatic, patterns. The most striking dispersal difference between African and Neotropical trees is the large number of elephant-dispersed species, with genera like *Irvingia, Kleinedoxa, Omphalocarpum, Massularia,* and *Picralima* conspicuous elements of the forest (figure 17.8). In Madagascar, lemur dispersal is a major theme, with lemur-dispersed taxa like tribe Coleeae of the Bignoniaceae having undergone strong evolutionary radiations (Gentry 1988b).

Although the data are few and speculative, pollination systems also appear to differ to some extent between Africa and South America. Pollination by settling moths is likely to be more important in Africa and Madagascar because of the greater prevalence of Apocynaceae and Rubiaceae. Beetle pollination is probably relatively constant with beetle-pollinated Annonaceae lianas, replacing beetle-pollinated Cyclanthaceae and aroids (Schatz 1990; Chapter 5; Gentry 1991b).

Bird pollination occurs on both continents but involves only perching birds, especially sunbirds, in Africa and both perching birds and, more importantly, hummingbirds in the Neotropics (Stiles 1981); hummingbirds are associated with the much greater number of bird-pollinated plant species in the Neotropics. There are at least 450 nectarivorous Neotropical birds, 315 of them hummingbirds; in contrast Africa has only 150 nectarivorous passerines, all perching (Stiles 1981).

Although bat pollination occurs on both continents, like bird pollination it is intrinsically different in Africa than in the Neotropics. The flower-visiting paleotropical bats are relatively large clumsy Megachiroptera, whereas in the New World, where only Microchiropteran bats occur, nectarivorous bats are smaller and more mobile. Perhaps such differences are correlated with the different evolutionary patterns of bat-pollinated plants on the two continents. I know of no African example of a bat-pollinated genus

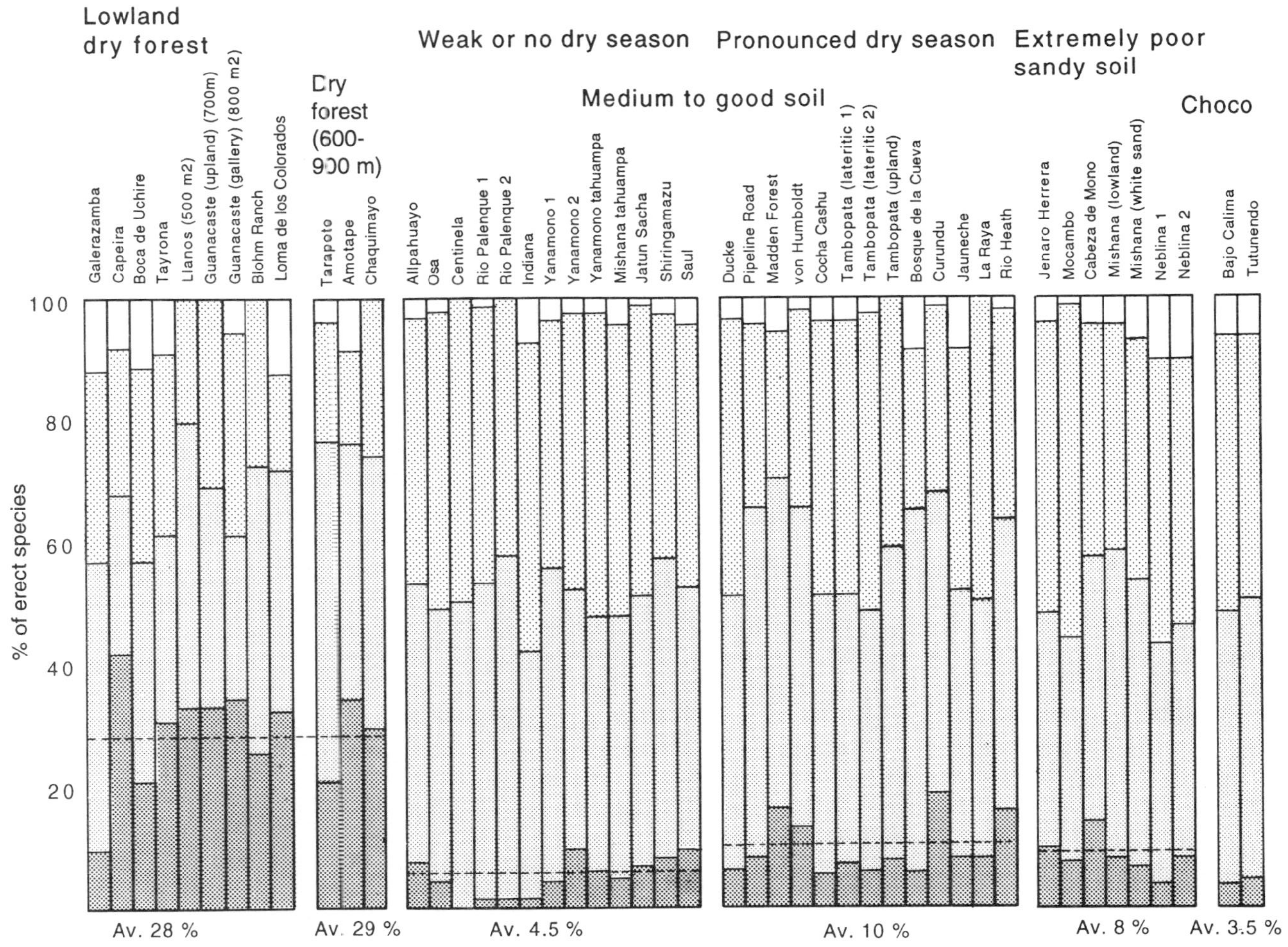

Figure 17.7. Dispersal of Neotropical trees. Dark shading = wind dispersal; medium shading = bird dispersal; light shading = mammal dispersal; white = autochorous and water dispersal.

Figure 17.8. Some elephant-dispersed fruits from Makokou, Gabon. Left to right, top row: *Picralima* (Apocynaceae), *Massularia* (Rubiaceae), *Omphalocarpum* (Sapotaceae); bottom row: *Strychnos* (Loganiaceae), *Kleinedoxa* (Irvingiaceae). Plant shears are 21 cm long.

with extensive speciation, whereas a number of bat-pollinated Neotropical genera (e.g., *Amphitecna, Parmentiera, Marcgravia*) have radiated profusely.

Pollination by large- to middle-sized bees is apparently much lower in Africa to judge from the much lesser extent of the standard zygomorphic Tubiflorae-type flowers that are associated with this syndrome (e.g., Bignoniaceae, Gesneriaceae), as might be expected from the relatively impoverished African apifauna (Michener 1979). The same phenomenon might account for the relatively few orchid species in Africa, since the Euglossine-pollination syndrome that accounts for so many species in the Neotropics is absent.

VII. Conclusion

Africa and South America are clearly more alike at the community level than at the continental level. Moreover, the strong floristic similarities seen when individual sites or communities are compared strongly suggest that today's forests result to a much greater extent from common origin than from chance immigration. Indeed, the similarities are so striking that they seem to imply that some degree of stabilizing selection and perhaps even similar niche specificities have been involved in the evolutionary histories of tropical lowland forest species on the two continents.

This pattern contrasts strikingly with the continental one of a relatively depauperate tropical African flora. However, there are also some noteworthy community-level differences, for example, in prevalence of different dispersal and pollination systems. The now-traditional explanation of the continental floristic differences has been that the African flora suffered massive extinction during the Cenozoic and Pleistocene due to the drier climate in most of Africa.

From the evidence summarized above, I would propose two alternate, although not necessarily exclusive, hypotheses to explain the intercontinental floristic differences that we see today. One is simply that the observed patterns merely reflect current ecological conditions. Most African forests are drier than most Neotropical ones and should hardly be expected to harbor wet-forest taxa like Lauraceae, Moraceae, or Palmae, which are also sparse or absent from similarly dry forests in the Neotropics. Similarly the prevalence of lianas in Africa may merely reflect the fact that most African forests are close to the dry forest–moist forest transition where lianas are equally prevalent in the Neotropics. Moreover, there is a marked tendency for dry-forest taxa to be more widespread than wet-forest ones, perhaps related to the prevalence of wind dispersal as opposed to mammal dispersal (Gentry 1982c). If African species are generally wider ranging as a correlate of their drier-forest habitat, then we might expect exactly the observed pattern, with fewer species at the continental level merely because the same sets of taxa tend to make up the plant communities in different parts of the continent.

There is also a second line of reasoning that might reconcile the community-level floristic similarities with the continental dissimilarity in diversity. Perhaps what we see today is more the result of different patterns of evolutionary radiation than differential extinction. At least in the taxa I have worked with, the morphological diversity within families and genera in Africa is not less than in South America. Rather they are differently apportioned into species (Gentry 1990c). An apparent failure to speciate seems to be the prevalent theme in African Bignoniaceae, for example.

Bird pollination is a good example. Hummingbird pollination is related to greater pollinator specificity than pollination by perching birds (e.g., Neill 1987; Snow and Snow 1980; Stein 1987), again suggesting differential evolution. Similarly the extensive radiations of microchiroptera-pollinated Neotropical genera like *Marcgravia* and *Amphitecna* has no parallel among Megachiroptera-pollinated African plants. The relatively depauperate African orchid flora may reflect the absence of coevolutionary radiations with Euglossine bees more than it does massive extinction.

Similarly, differences in dispersal systems on the two continents may be correlated with differences in speciation. For example, elephants are important seed dispersers in Africa, but reached South America only fleetingly subsequent to closure of the Panamanian isthmus, and can have had little influence on the evolution of South American plants. Elephants are individually wide-ranging, and dispersal-induced gene flow might be greater in elephant-dispersed African species than in their agouti-dispersed or primate-dispersed Neotropical equivalents.

Even though the higher plant taxa, and to a large extent the communities they compose, were already in existence prior to Gondwanan breakup, many of their current pollinators and dispersers were not. Thus part of the intercontinental difference in floristic richness may be a geographic accident from the botanical viewpoint, reflecting coevolutionary interactions with different faunistic elements on the two continents.

One could even suggest that the different continental milieu is causally correlated with the intercontinental floristic differences. For example, wind dispersal is clearly more prevalent in South America than in Africa, especially among lianas (Gentry 1991b). Many Neotropical forests, especially in Central America and near the base of the Andes, tend to be on geologically younger, richer soils than are most African (or central Amazonian) forests. Perhaps higher turnover of rich-soil forests correlates with more frequent gaps and a more broken forest structure in the Neotropics, a situation that in turn favors wind-dispersed plants like the dominant Bignoniaceae lianas.

The Andes, far more massive and extensive than any other tropical mountains, may have played a major role in giving rise to today's intercontinental floristic differences.

For example, both hummingbird pollination and orchids are concentrated in Andean cloud forests where evolutionary divergence may have been speeded up (Gentry and Dodson 1987), and there is little equivalent habitat in Africa. Most of the "excess" species of the Neotropics are concentrated in the mountainous areas of northwestern South America and adjacent Central America (Gentry 1982a). Moreover, most of these species do not represent uniquely South American families at all, but rather distinctive radiations of taxa shared with Africa. Greater speciation in South America rather than more extinction in Africa may be the operant factor (Gentry 1989).

The continental pattern reflects the macroevolutionary one; the community-level pattern does not. If today's African rainforest communities are reconstituted from the few taxa that managed to survive massive extinction, then they might be expected to show some of the patterns of community impoverishment and floristic imbalance that characterize islands. This may be true in West Africa, to judge from my Omo forest sample, but there is independent evidence that this region was devastated more recently by large human populations. In contrast, the individual central African plant communities seem put together in predictable and consistent ways and are remarkably similar to their Neotropical equivalents. The difference between the two continents is that different African plant communities are composed of the same species to a much greater extent than are Neotropical communities.

The importance of extinction due to climatic vicissitudes as a causative factor in producing today's relatively impoverished African flora remains to be worked out. Whatever the relative roles of differential extinction, speciation, and modern ecology in giving rise to the intercontinental differences, at the community level the floras of Africa and South America are very much alike indeed. Any theory purporting to explain the large-scale differences must also take into account the local similarities in diversity and floristic composition of lowland tropical forest on the two continents.

Literature Cited

Ashton, P. 1992. Ecological studies in the mixed dipterocarp forest of northwest Borneo III. Patterns of species richness. J. Ecol. In press.

Aubréville, A. 1967. La forêt primaire des montagnes de Bélinga. Biol. Gabonica 3(2):95–112.

Axelrod, D. I., and P. Raven. 1978. Late Cretaceous and Tertiary vegetation history of Africa. Pp. 77–130 *in* M. J. A. Werger (ed.), *Biogeography and Ecology of Southern Africa.* The Hague: W. Junk.

Ayensu, E. S. 1973. Phytogeography and evolution of the Velloziaceae. Pp. 105–119 *in* B. Meggers, E. Ayensu, and W. Duckworth (eds.), *Tropical Forest Ecosystems in Africa and South America: A Comparative Review.* Washington, D.C.: Smithsonian Institution Press.

Balslev, H., J. Luteyn, B. Øllgaard, and L. B. Holm-Nielsen. 1987. Composition and structure of adjacent unflooded and floodplain forest in Amazonian Ecuador. Opera Bot. 92:37–57.

Boom, B. 1987. Ethnobotany of the Chacobo Indians, Beni, Bolivia. Adv. Econ. Bot. 8:57–76.

Brenan, J. P. M. 1978. Some aspects of the phytogeography of tropical Africa. Ann. Missouri Bot. Gard. 65:437–478.

Campbell, D. G. 1989. Quantitative inventory of tropical forests. Pp. 523–533 *in* D. Campbell and H. D. Hammond (eds.), *Floristic Inventory of Tropical Countries.* New York: New York Botanical Garden.

Campbell, D. G., D. C. Daly, G. Prance, and U. N. Maciel. 1986. Quantitative ecological inventory of terra firme and varzea tropical forest on the Rio Xingu, Brazilian Amazon. Brittonia 38:369–393.

Coetzee, J. A., and J. Muller. 1984. The phytogeographic significance of some extinct Gondwana pollen types from the Tertiary of the southwestern Cape (South Africa). Ann. Missouri Bot. Gard. 71:1088–1099.

Croat, T. B. 1978. *Flora of Barro Colorado Island*. Stanford, Calif.: Stanford University Press.

Dawkins, H. C. 1959. The volume increment of natural tropical high-forest and limitations on its improvements. Empire Forest. Rev. 38:175–180.

Doumenge, C. 1990. Contribution à l'étude des structures de populations d'arbres des forêts d'Afrique Centrale (exemples du Gabon, Cameroun et Congo). Ph.D. diss., Université Montpellier, France.

Emmons, L. H., and A. H. Gentry. 1983. Tropical forest structure and the distribution of gliding and prehensile-tailed vertebrates. Amer. Natural. 121:513–524.

Faber-Langendoen, D., and A. Gentry. 1991. The structure and diversity of rainforests at Bajo Calima, Chocó Region, western Colombia. Biotropica 23:2–11.

Florence, J., and A. Hladik. 1980. Catalogue de phanerogames et des pteridophytes du Nord-est du Gabon (Sixieme liste). Adansonia, ser. 2, 20:235–253.

Gartlan, J. S., D. M. Newbery, D. W. Thomas, and P. G. Waterman. 1986. The influence of topography and soil phosphorus on the vegetation of Korup Forest Reserve, Cameroun. Vegetatio 65:131–148.

Gautier-Hion, A., J.-M. Duplantier, R. Quris, F. Feer, C. Sourd, J.-P. Decoux, G. Dubost, L. Emmons, C. Erard, P. Hecketsweiler, A. Moungazi, C. Roussilhon, and J.-M. Thiollay. 1985. Fruit characters as a basis of fruit choice and seed dispersal in a tropical forest vertebrate community. Oecologia 65:324–337.

Gentry, A. H. 1982a. Neotropical floristic diversity: Phytogeographical connections between Central and South America: Pleistocene climatic fluctuations, or an accident of the Andean orogeny? Ann. Missouri Bot. Gard. 69:557–593.

———. 1982b. Patterns of neotropical plant species diversity. Evol. Biol. 15:1–84.

———. 1982c. Phytogeographic patterns as evidence for a Choco Refuge. Pp. 112–136 *in* G. Prance (ed.), *Biological Diversification in the Tropics*. New York: Columbia University Press.

———. 1986. Species richness and floristic composition of Chocó region plant communities. Caldasia 15:71–91.

———. 1988a. Changes in plant community diversity and floristic composition on environmental and geographical gradients. Ann. Missouri Bot. Gard. 75:1–34.

———. 1988b. Distribution and evolution of the Madagascar Bignoniaceae. Monogr. Syst. Bot. Missouri Bot. Gard. 25:175–185.

———. 1988c. Tree species richness of upper Amazonian forests. Proc. Natl. Acad. Sci. U.S.A. 85:156–159.

———. 1989. Speciation in tropical forests. Pp. 113–134 *in* L. Holm-Nielsen, I. C. Nielsen, and H. Balslev (eds.), *Tropical Forests: Botanical Dynamics, Speciation, and Diversity*. London: Academic Press.

———. 1990a. Evolutionary patterns in neotropical Bignoniaceae. Mem. N.Y. Bot. Gard. 55:118–129.

——— (ed.). 1990b. *Four Neotropical Rainforests*. New Haven: Yale University Press.

———. 1990c. Herbarium taxonomy versus field knowledge: Is there an attainable solution? Flora Males. Bull., sp. vol. 1:31–36.

———. 1990d. Tropical forests. Pp. 35–43 *in* A. Keast (ed.), *Biogeography and Ecology of Forest Bird Communities*. The Hague: SPB Academic.

———. 1991a. The distribution and evolution of climbing plants. Pp. 3–41 *in* F. Putz and H. Mooney (eds.), *The Biology of Vines*. Cambridge: Cambridge University Press.

———. 1991b. Reproductive strategies of lianas. Pp. 385–415 *in* F. Putz and H. Mooney (eds.),

The Biology of Vines. Cambridge: Cambridge University Press.

———. 1993a. Diversity and floristic composition of neotropical dry forests. In press *in* S. Bullock, H. Mooney, and E. Medina (eds.), *Tropical Dry Forest Ecosystems*. Cambridge: Cambridge University Press.

———. 1993b. Biodiversity of Peruvian Amazonia: Problems and opportunities. In press *in* K. Mejia (ed.), *IIAP Symposium, Folia Amazonica,* Iquitos.

Gentry, A. H., and C. Dodson. 1987. Diversity and biogeography of neotropical vascular epiphytes. Ann. Missouri Bot. Gard. 74:205–233.

Hall, J. B., and M. D. Swaine. 1981. Distribution and ecology of vascular plants in a tropical rain forest: Forest vegetation in Ghana. The Hague: W. Junk.

Hladik, A. 1978. Phenology of leaf production in rain forest of Gabon: Distribution and composition of food for folivores. Pp. 51–71 *in* G. Montgomery (ed.), *The Ecology of Arboreal Folivores*. Washington, D.C.: Smithsonian Institution Press.

———. 1982. Dynamique d'une forêt équatoriale africaine: Mesures en temps réel et comparaison du potenciel de croissance des différentes espèces. Acta Oecologia, Oecologia Generalis 3:373–392.

———. 1986. Données comparatives sur la richesse specifique et les structures des peuplements des forêts tropicales d'Afrique et d'Amérique. Vertébrés et forêts tropicales humides d'Afrique et d'Amérique. Mém. Mus. Natn. Hist. Nat., n.s. 132:9–17.

Huston, M. 1979. A general hypothesis of species diversity. Amer. Natural. 113:81–101.

———. 1980. Soil nutrients and tree species richness in Costa Rican forests. J. Biogeogr. 7:147–157.

Johansson, D. 1974. Ecology of vascular epiphytes in West African rain forest. Acta Phytogeogr. Suec. 59:1–127.

Langenheim, J. H. 1973. Leguminous resin-producing trees in Africa and South America. Pp. 89–104 *in* B. Meggers, E. Ayensu, and W. Duckworth (eds.), *Tropical Forest Ecosystems in Africa and South America: A Comparative Review.* Washington, D.C.: Smithsonian Institution Press.

Leigh, E. 1975. Structure and climate in tropical rain forest. Ann. Rev. Ecol. Syst. 6:67–86.

Lovett, J. C. 1988. Endemism and affinities of the Tanzanian montane forest flora. Monogr. Syst. Bot. Missouri Bot. Gard. 25:591–598.

Maury-Lechon, G., and O. Ponce. 1986. Dynamique forestière sur 6 hectares de forêt dense humide de Guyane française, à partir de quelques espèces de forêt primaire et de cicatrisation. C. R. Vertébrés et forêts tropicales d'Afrique et d'Amérique, Mem. Mus. Natn. Hist. Nat., Paris, n.s. (A) Zool. 132:211–242.

Michener, C. D. 1979. Biogeography of the bees. Ann. Missouri Bot. Gard. 66:277–347.

Moore, H. E., Jr. 1973. Palms in the tropical forest ecosystems of Africa and South America. Pp. 63–88 *in* B. Meggers, E. Ayensu, and W. Duckworth (eds.), *Tropical Forest Ecosystems in Africa and South America: A Comparative Review.* Washington, D.C.: Smithsonian Institution Press.

Mori, S. A., and B. Boom. 1987. The forest. Pp. 9–29 *in* S. Mori, *The Lecythidaceae of a Lowland Neotropical Forest: La Fumee Mountain, French Guiana.* Mem. N.Y. Bot. Gard. 44.

Mori, S. A., B. Boom, A. M. de Carvalho, and T. S. dos Santos. 1983. Southern Bahian moist forest. Bot. Rev. 49:155–232.

Neill, D. A. 1987. Trapliners in the trees: Hummingbird pollination of *Erythrina* sect. *Erythrina* (Leguminosae: Papillionoideae). Ann. Missouri Bot. Gard. 74:27–41.

Peixoto, A., and A. Gentry. 1990. Diversidade e composição floristica da mata de tabuleiro da Reserva Florestal de Linhares (Espirito Santo, Brasil). Rev. Bras. Bot. 13:19–25.

Pierlot, R. 1966. Structure et composition des forêts denses d'Afrique centrale, specialement celles du Kivu. Mém. Acad. Roy. Sci. d'Outre-Mer, n.s. 16, 4. Brussels: ARSOM.

Prance, G. P. 1984. Completing the inventory. Pp. 365–397 *in* V. Heywood and D. Moore (eds.), *Current Concepts in Plant Taxonomy.* London: Academic Press.

———. 1990. The floristic composition of the

forests of Central Amazonian Brazil. Pp. 112–140 *in* A. Gentry (ed.), *Four Neotropical Rainforests*. New Haven: Yale University Press.

Prance, G. P., W. A. Rodriguez, and M. F. da Silva. 1976. Inventario floristico florestal de um hectare de mata de terra firme km. 30 de estrada Manaus-Itacoatiara. Acta Amaz. 6:9–35.

Puig, H. 1979. Production de litière en forêt guyanaise: Résultats préliminaires. Bull. Soc. Hist. Nat. Toulouse 115:338–346.

Rabinowitz, P. D., M. F. Coffin, and D. Falvey. 1983. The separation of Madagascar and Africa. Science 220:67–69.

Raven, P. H., and D. I. Axelrod. 1974. Angiosperm biogeography and past continental movements. Ann. Missouri Bot. Gard. 61:539–673.

Reitsma, J. M. 1988. Vegetation forestière du Gabon. Tropenbos Tech. ser. 1:1–142.

Richards, P. W. 1952. *The Tropical Rain Forest*. Cambridge: Cambridge University Press.

———. 1973. Africa, the "odd man out." Pp. 21–26 *in* B. Meggers, E. Ayensu, and W. Duckworth (eds.), *Tropical Forest Eco-Systems in Africa and South America: A Comparative Review*. Washington, D.C.: Smithsonian Institution Press.

Rollet, B. 1974. *L'architecture des forêts denses humides sempervirentes de plaine*. Nogent-sur-Marne, France: CTFT.

Schatz, G. 1990. Some aspects of pollination biology in Central American forests. Pp. 69–84 *in* K. Bawa and M. Hadley (eds.), *Reproductive Ecology of Forest Plants*, MAB series. Paris: UNESCO, and Carnforth: Parthenon.

Smith, A. C. 1973. Angiosperm evolution and the relationship of the floras of Africa and America. Pp. 49–62 *in* B. Meggers, E. Ayensu, and W. Duckworth (eds.), *Tropical Forest Ecosystems in Africa and South America: A Comparative Review*. Washington, D.C.: Smithsonian Institution Press.

Snow, D. W., and B. K. Snow. 1980. Relationships between hummingbirds and flowers in the Andes of Colombia. Bull. Brit. Mus. (Nat. Hist.) Zool. ser. 38:105–139.

Stein, B. 1987. Systematics and Evolution of *Centropogon* subgenus *Centropogon* (Campanulaceae, Lobelioideae). Ph.D. diss., Washington University, St. Louis.

Stiles, F. G. 1981. Geographical aspects of bird-flower coevolution, with particular reference to Central America. Ann. Missouri Bot. Gard. 68:323–351.

Thorne, R. F. 1973. Floristic relationships between tropical Africa and tropical America. Pp. 27–48 *in* B. Meggers, E. Ayensu, and W. Duckworth (eds.), *Tropical Forest Ecosystems in Africa and South America: A Comparative Review*. Washington, D.C.: Smithsonian Institution Press.

Whitmore, T. C. 1985. Particular plant aspects of species diversity. Paper presented to the Third International Congress of Systematic and Evolutionary Biology, University of Sussex. (Not seen; noted by Campbell 1989).

Chapter 18

Comparative Ecology of African and South American Arid to Subhumid Ecosystems

Samuel J. McNaughton, Osvaldo E. Sala, and Martin Oesterheld

I. Introduction

The arid to subhumid ecosystems of Africa and South America have identical origins in vast tracts of West Gondwana before the two continents were separated by continental drift and seafloor spreading (Raven and Axelrod 1974). Monocotyledons, including the Poaceae, appear to have originated during this period, although the grasses may have spread into South America later, in the Paleocene or Upper Cretaceous (Stebbins 1981). The earliest grass fossils, customarily taken as diagnostic of arid to subhumid ecosystems, appear in Patagonian deposits of Eocene age (Frenguelli 1930; Teruggi 1955), but they did not become common until the Miocene, by which time the Poaceae was almost fully differentiated, suggesting previous radiation in rare habitats or locations not lending themselves to fossil formation (Stebbins 1981). On both continents, the expansion and contraction of wet tropical forest associated with polar shifts and glaciation at higher elevations resulted in periodic fragmentation and subsequent reemergence of contiguous dryland ecosystems (Aubréville 1962). There were also substantial radiations of large, mammalian browser and grazer faunas in both South America and Africa contemporaneous with these climatic and vegetation oscillations. Thus, the arid to subhumid ecosystems of the two continents apparently had identi-

The research described here was supported by the U.S. National Science Foundation Ecosystem Studies Program and Consejo Nacional de Investigaciones Cientificas y Técnicas, Republic of Argentina, and the University of Buenos Aires. We thank Raul Lavado and Jorge Adamoli for assistance with literature survey.

 ISBN 0-300-05375-4

cal origins and very similar developments through geological time.

II. Ecosystems Considered

We concentrate here on ecosystems where mean annual precipitation ranges from an approximate lower limit of 20 cm to an upper limit near 150 cm, although, like all such boundaries, these are subject to modification by local edaphic conditions, evaporative demand, altitude, and topography. In general, we consider ecosystems where members of the Poaceae are important in the herbaceous layer and woody vegetation can range from scattered dwarf shrubs to an almost continuous canopy of small-statured trees.

South American ecosystems considered include the Patagonian steppe, the pampas, open grassland called "campo," grass and savanna woodland known as "llano" or "cerrado," the open dry thorny woodlands called "chaco," and the dry thorn scrub named "caatinga." African ecosystems of these types (Huntley 1982; Werger 1983) would simply be known as grassland, bushland, or by the indefinite Spanish term derived from Amerindian languages (Sarmiento 1984), "savanna," and are commonly distinguished from woodland vegetation called "miombo."

Approximately 40 percent of earth's total area covered by arid to subhumid ecosystems occurs on the two continents considered here (FAO 1987). Therefore, these two continents are major sites of arid to subhumid ecosystems, although the proportion of continental area covered by them is less than in Australia. In Africa, they fringe the Mediterranean shore before being interrupted by the Sahara Desert, then sweep across the continent south of there to be bifurcated by rainforest except along the east, and continue southward more or less uninterrupted. In South America, they occur as disjunct areas north of the Amazon basin, scattered areas throughout that basin, and then sweep southward to near the continent's southern tip (figure 18.1).

The contemporary African and South American ecosystems are remarkably similar. Although the species of major herbaceous and woody plants are largely different, many grass and dicotyledonous genera and families appear on both continents, and the physiognomy of the vegetation on both continents is strikingly similar and varies in a similar fashion with precipitation and thermal climate. There is, however, one notable exception to the general similarity between African and South American ecosystems: South America's diverse and abundant browser and grazer fauna was comparable to that of Africa through the Tertiary, but it almost totally failed to survive the Pleistocene extinctions of large herbivorous mammals, whereas Africa preserves earth's greatest remnant of that fauna (Keast et al. 1972). The reasons for this notable difference are by no means clear (see Chapter 13), but it has significant implications for ecosystem structure and function on the two continents.

The ecosystems considered here are associated with climates characterized by pronounced dry seasons, high fire frequencies, and highly variable, but often anomalous, soil properties (Werger 1979; Walter 1979; Huntley and Walker 1982; Sarmiento 1984; Cole 1986). It is far beyond the scope of this chapter to consider such patterns in detail; instead we concentrate on variation

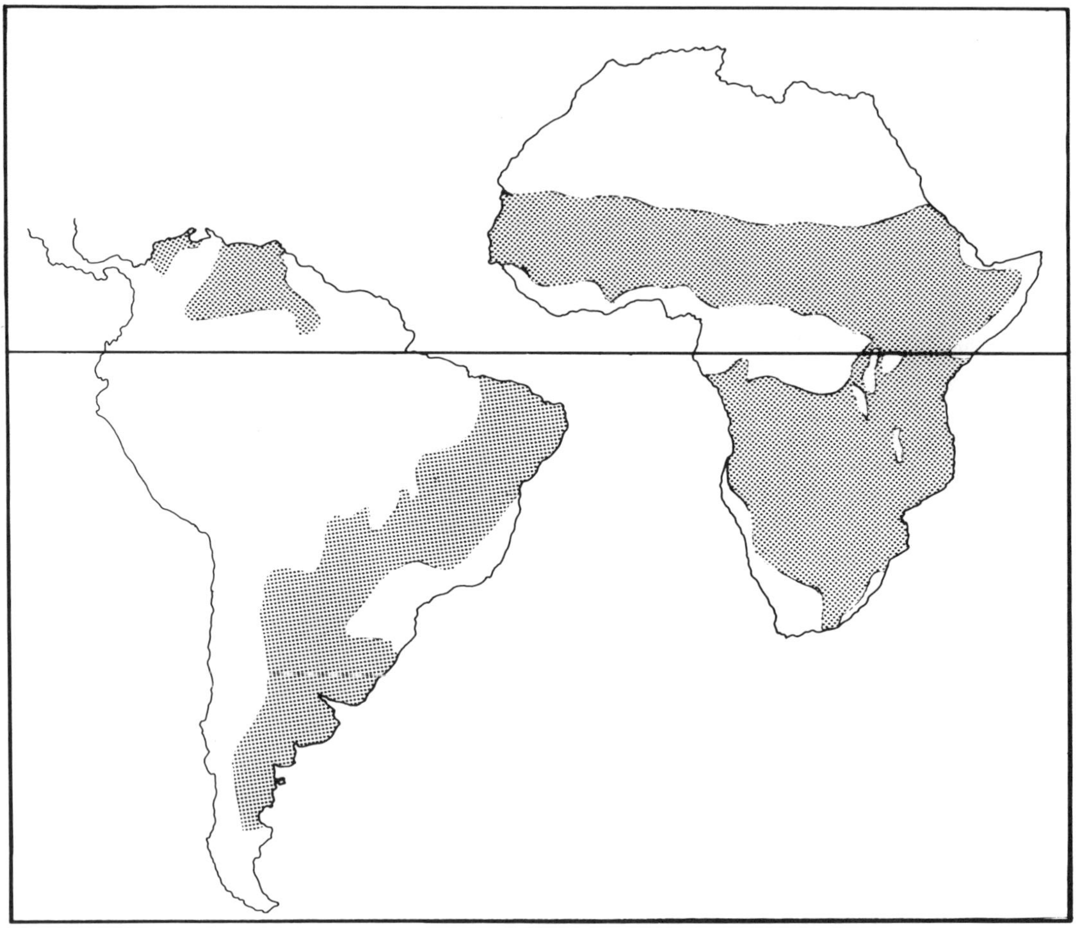

Figure 18.1. Map of South America and Africa with regions occupied by extensive arid to subhumid ecosystems indicated with dark shading.

in major structural and functional properties of semi-arid to subhumid ecosystems, comparing and contrasting the two continents ecologically.

III. Vegetation Structure

On both continents, the herbaceous layer is commonly dominated by grasses, species with the C_4 photosynthetic pathway most common in equatorial locations and species with the C_3 pathway becoming more important away from the tropics. In subtropical to temperate localities, such as the Argentinean pampas, there often is seasonal alternation of dominance by C_3 and C_4 grasses (Sala et al. 1982, Oesterheld and Léon 1987). Peak productivity of C_3 species occurs in early spring, whereas C_4 species have maximum productivity in late spring or early summer, a pattern also observed in mixed grass prairies of the northern Great Plains in North America (Ode et al. 1980).

Walter (1971) hypothesized that there is a vertical partitioning of soil resources, comparable to the two-phase aboveground structure, which reduces niche overlap between the herbaceous layer and the shrubs and trees. This could be a major attribute of these ecosystems tending to augment and stabilize productivity compared to a uniphasic ecosystem containing only herbs or only woody plants (Walker and Noy-Meier 1982). This hypothesis states that grasses absorb water from the upper soil layers, whereas trees absorb water from both upper and lower layers. It is based upon the morphological differences in the roots of grasses and woody plants. Grasses have finely branched roots that densely permeate the surface soil, whereas woody species have an extensive root system with coarse roots extending much farther into the soil profile.

Experimental data from southern African savannas provide support for the hypothesis (Knopp and Walker 1985). Removal of the herbaceous layer had a negligible effect on tree growth in a broad-leaved savanna with a low ratio of herbaceous to woody cover. In an *Acacia* savanna with sparser tree cover and a denser herbaceous layer, however, removal of the herbaceous layer resulted in significant increases in tree growth. Soil water and nutrient data suggested that this effect was due to the grasses intercepting percolating water in shallow soil layers, thus preventing water from light showers from reaching tree roots lower in the profile.

The two-layer hypothesis has also been extended to the shrub-grass balance in temperate steppes and tested with a removal experiment in Patagonia (Sala et al. 1989). Results support the overall hypothesis since shrubs and grasses used mainly different water resources. The data suggest a different distribution of resources between the two life forms. Patagonian shrubs absorb water almost exclusively from lower soil layers, whereas grasses take up water mostly from the upper layers. The major effect of grasses, then, is indirect: they intercept water that, in their absence, would percolate to lower layers where shrub roots have access to them. Consequently, grasses are not able to use the resources freed by shrub removal but shrubs can use the water freed by grass removal, but with a very low efficiency, about 25 percent.

These patterns of soil water partitioning pose similar questions regarding the mineral nutrition of grasses and woody plants in arid to subhumid ecosystem: do both life forms absorb nutrients from the same location in the soil at which water absorption is concentrated? If so, the woody plants would be at a clear disadvantage in competition for mineral nutrients since mineral cycles in these ecosystems are concentrated very near the soil surface (Clark 1977; Cole et al. 1977; Woodmansee et al. 1981). Data from tracer experiments in Patagonia (Sala et al. 1991) confirm the idea that shrubs absorb nutrients from the lower soil layers whereas grasses absorb principally from upper, nutrient-rich layers. These results suggest that shrubs must have effective strategies for coping with the differential mineral nutrient supply: (a) conservation and retranslocation of acquired nutrients at higher efficiency than grasses, (b) exploitation of nutrients from sporadic leaching events associated with exceptionally high rainfall, or (c) by percolation during periods when grasses are not actively growing.

The replacement of grasslands by shrublands is a serious worldwide problem in arid to subhumid ecosystems, a phenomenon commonly referred to as bush encroachment. The biphasic models just discussed suggest that intense and selective removal of grasses by livestock can lead to a redistribution of resources making conditions more favorable to the woody plants. This would probably lead to an associated decline of net primary productivity due to the low efficiency with which woody plants use the resources freed by grass decline. The existence of underexploited resources under the shrub-dominated condition suggests that this phase could be reversed by reducing herbivore pressure. However, the shrubs' advantage in aboveground competition would place them at a decided advantage once encroachment had become pronounced, perhaps leading to irreversibility in the absence of fire to kill back aboveground portions of woody plants.

At moderate to low precipitation levels, tree spacing in savannas appears to result from competition between trees for soil resources (Smith and Goodman 1986). In studies at a location with mean annual precipitation of 610 mm, a variety of lines of evidence implicated competition in tree spacing. First, there was a positive correlation between nearest neighbor distance of canopy trees and the combined canopy volume of those neighbors. Second, the slope of the regression between combined canopy cover of nearest neighbors and distance between them was positively related to both water available from the soil and total exchangeable bases. And, third, removal of surrounding trees resulted in significantly enhanced growth of the remaining experimental trees, in spite of the fact that canopies were widely spaced and nonoverlapping. These results indicate that the spacing of trees in some savannas is due to competition for soil-borne resources.

It is also important to recognize the potential importance of fire in maintaining the balance between the herb layer and the arborescent canopy (Vesey-Fitzgerald 1971), particularly in higher rainfall locations. Experimental fire plots established in Zambia in 1933 at a location with a mean rainfall near 1200 mm included annual treatments of (a) a late dry season, hot burn, (b) an early dry season, cool burn, and (c) no burning. The vegetation on these plots came to be decidedly different, documenting the role of fire in the balance of arborescent and herbaceous vegetation in subhumid ecosystems (Trapnell 1959; Werger 1983). Hot fires almost completely destroyed the woody overstory, resulting in a very open savanna with an extremely sparse overstory of large trees. Cool fires produced the typical open miombo woodland that is characteristic of that region of Africa. The unburned vegetation developed into a closed woodland of dry evergreen forest. These long-term studies provide convincing evidence that fire plays a central role in maintaining the open character of savannas toward the higher end of the rainfall gradient. Early burning, at the wet-dry season transition when grass biomass still has a good deal of moisture and fires are, therefore, "cool" is now a common management practice throughout Africa to maintain the tree canopy (Stronach and McNaughton 1989).

Regular flooding also can influence the balance between grasses and woody vegetation in arid to subhumid ecosystems by leading to open grassland, whereas better-drained areas support savanna or woodland. These hydrologic grasslands are common in the Brazilian Mato Grosso and Venezuelan-Colombian llanos (Sarmiento 1984). Conspicuous African examples are the floodplain of the Kafue flats in Zambia (Werger and Ellenbrook 1980), the Okavango Delta in Botswana (Cole 1986), and the Ndabaka Plains in the western Serengeti ecosystem (McNaughton 1983). Both the South American and the African ecosystems of this type are characterized by inundation during the wet season and severe drought during the dry season. Anoxia for significant periods appears to deter the colonization of and suppress the growth of trees much more severely than it affects grasses. Although it has been argued that the pampas were anthropogenic grasslands maintained through burning by the aboriginal Amerindian population (Schmieder 1927), they are more plausibly recognized as edapho-climatic grasslands (Walter 1979).

Even gentle topographic relief in hydrologic grasslands leads to alternating patches flooded for extended periods and unflooded areas (Leon et al. 1984; Poiani and Johnson 1989; Oliff et al. 1988). In Venezuela, flood-intolerant C_4 species are most common on elevated areas, whereas commonly inundated swales are dominated by flood-tolerant C_3 plants (Medina and Motta 1990). This pattern is reversed in the more temperate flooding pampa where elevated communities are dominated by C_3 species and neighboring lower patches are dominated by C_4 species (Sala et al. 1986). These regional differences may be explained by ecophysiological and climatic differences. Pampean C_4 species have lower reductions in leaf water potential than C_3 species in response to flooding (Sala et al. 1982).

Grasses from the flooding pampa and from flooded regions of Africa's Serengeti ecosystem exhibit a variety of common responses to flooding (Oesterheld 1990). Greater flooding tolerance in both floras is associated with greater plant height, lack of roots above the soil surface, higher root porosity, and the ability to maintain higher stomatal conductance and photosynthesis when flooded.

Large grazing mammals, whether livestock in South America (Sala et al. 1986) or native grazers in Africa (McNaughton 1984), have a pronounced effect upon vertical structure of savanna grasslands. Grazing compresses the canopy into a shallow layer within about 5 cm of the soil surface that is inaccessible to large grazers. In ungrazed grasslands, the major green canopy layer is substantially above the ground surface, and many short-statured species or ecotypes are eliminated from the ground layer by shading in the absence of grazing (McNaughton 1983; Sala et al. 1986).

When grazing pressure is not intense, grass swards develop a lateral structure of bimodal height classes partitioned between tall, ungrazed patches and short, grazed patches (Vesey-Fitzgerald 1974). Grazed patches are returned to repeatedly by grazers, but the tall patches are ignored. The high quality of the grazed patches makes them preferred foraging areas, whereas the patches that escaped grazing decline in quality and in yield per bite as growth proceeds (McNaughton 1984).

Africa's abundant browsers also have significant influences upon the tree-grass balance (Laws 1970; McNaughton and Sabuni 1988). Elephants (*Loxodonta africana*) can severely deplete the tree layer by direct destruction and bark-stripping, browsers suppress the height growth of seedings and saplings, keeping them in the fire-susceptible height range, and fire burns potential recruitment classes back to ground level (Norton-Griffiths 1979). These interacting factors often lead to tree size-frequency distributions that are decidedly bimodal (McNaughton and Sabuni 1988), with many small trees in the browser- and fire-susceptible class and a very few quite large trees that managed to surpass this bottleneck at some time in their life.

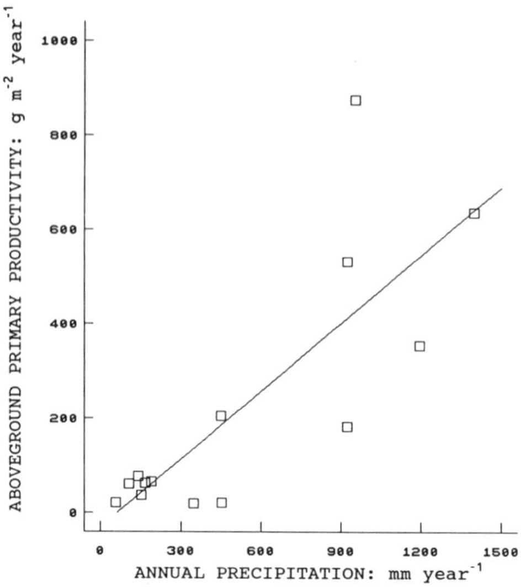

Figure 18.2. Relationship between annual precipitation and net aboveground primary productivity for South American arid to subhumid ecosystems.

IV. Primary Production

Primary productivity, the rate at which plants accumulate biomass, is a fundamental ecosystem variable that acts as a major indicative integrator of many other ecosystem properties (McNaughton et al. 1989). A comparison of correlations between primary productivity and precipitation for African and South American data sets reveals essentially identical relationships between the two variables. For 14 South American ecosystems collated for this chapter (San José and Medina 1975; Fonseca et al. 1976; Bulla et al. 1981; Sala et al. 1982; Bertiller 1984; Distel and Fernandez 1986; Oesterheld and Léon 1987; Defosse et al. 1990; Fernandez et al. 1991), the relationship was (see figure 18.2)

$$\text{ANPP} = 0.48 \text{ mm} - 30,$$

for 20 locations in the Serengeti ecosystem in East Africa (McNaughton 1985),

$$\text{ANPP} = 0.69 \text{ mm} - 102,$$

for 33 combinations of locations and years in East and southern Africa (Deshmukh 1984),

$$\text{ANPP} = 0.85 \text{ mm} - 20,$$

and for 45 North African ecosystems (Le Houérou and Hoste 1977),

$$\text{ANPP} = 0.26 \text{ mm} - 11,$$

where ANPP is aboveground net primary productivity in g m^{-2} $year^{-1}$ and mm is annual precipitation (figure 18.3). Although a wide variety of investigators and methods were involved in data collection in all data sets except the Serengeti ecosystem, these relationships suggest a remarkable uniformity of production-precipitation relationships. For 100 locations across the North American Great Plains (Sala, Parton, et al. 1988),

$$ANPP = 0.6 \text{ mm} - 34.$$

Therefore, the relationship between primary productivity and precipitation in arid to subhumid ecosystems is widely similar across many different geographic regions, with an increment of between one-half and three-fourths of a gram of production per square meter annually for each millimeter of precipitation (figure 18.3).

Using the X-intercept as an index of the lower limit of these ecosystems (where potential productivity equals zero), values of rainfall limit are 63 mm for South America, 148 mm for the Serengeti ecosystem, 23 mm for Africa, 42 mm in the Sahel, and 57 mm for North America. Of course, the ecosystems considered here merge into desert past their lower limits, so these data suggest that semi-arid ecosystems become desert at rainfalls of less than approximately 100 mm annually, near our initially defined limit of around 200 mm annually.

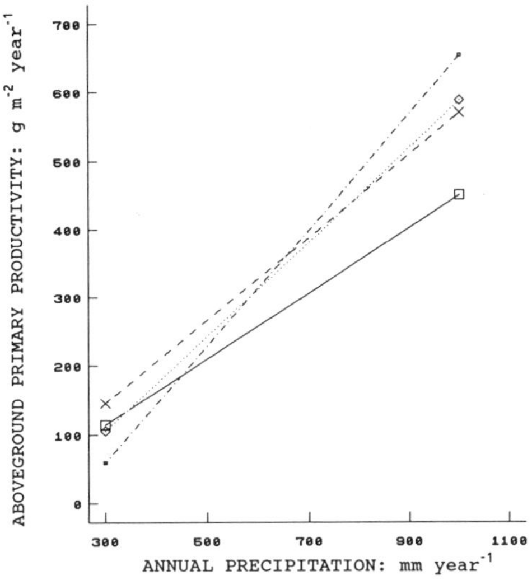

Figure 18.3. Best fit lines relating annual precipitation to net aboveground primary productivity from several geographic compilations for arid to subhumid ecosystems in Africa and South America.

Although the similarity of these general trends is of major ecological significance, substantial deviation does occur due to finer-scale ecosystem differentiation. In the Serengeti ecosystem, for example, productivity of vegetation growing in locations with high rainfall on old, leached sands derived from Precambrian deposits that contain little organic matter and nitrogen falls substantially below the regression line (McNaughton 1985). ANPP at five locations in the Orinoco Llanos in Venezuela with a rainfall of 1257 mm ranged from 123 to 482 g m^{-2} yr^{-1} (San José et al. 1985), whereas values predicted from the above equations range from 559 to 1022 g m^{-2} yr^{-1}. The exploitable soil volume and soil nutritional factors seemed important across these localities. Similarly, Sarmiento's (1984) tabulation of 23 savanna ANPP values in Venezuela over a rainfall range from 1017 to 1350 mm revealed a productivity range from 229 to 731 g m^{-2} yr^{-1}, and no evident association with precipitation.

Sala, Parton, and co-workers (1988) demonstrated for the North American Great Plains that the factors controlling primary productivity changed with spatial scale considered. At a coarse scale, broad regional patterns of primary productivity were explained principally by annual precipitation. At finer spatial scales, however, additional environmental properties had to be incorporated, particularly soil properties. At fine scales they confirmed an inverse-texture hypothesis: below an annual rainfall threshold of 370 mm, production is greater on coarse-textured soils (sands with low water-holding capacities), but above this threshold, production is greater on finer-textured soils (clays with greater water-holding capacities).

Therefore, although there are broad regional patterns of association between primary productivity and precipitation, fine-scale differentiation within narrow precipitation ranges reveals considerable heterogeneity in the productive process. Localized edaphic factors are particularly important causes of such deviations in both South America and Africa.

Recent comparative studies of tropical grasslands at four locations, one each in Central America, South America, Africa, and Asia, indicate that many estimates of ANPP, including most of those discussed above, undercalculate ANPP by 5 to 39 percent compared to estimates that include tissue turnover (Long et al. 1989). Total productivity estimates, including the productivity of belowground organs, were 37 to 80 percent below values including estimated tissue turnover. In contrast, recent analyses of the methods commonly used to calculate above- and belowground production indicate that most estimates represent overestimates of the true production value (Singh et al. 1984; Sala, Biondini, et al. 1988). Random errors associated with the estimates of plant biomass, which are then used to calculate net production, always result in a positive bias in estimating primary production. Although methods that include tissue turnover may be closer to actual values, they are also likely to be positively biased by small random errors. Therefore, more information on productivity that incorporates tissue turnover is desirable, but we think this information is unlikely to modify the fundamental trends presented here.

V. Herbivory

The most conspicuous difference between native South American and African semiarid to subhumid ecosystems is the prevalence of leaf-harvesting ants as principal herbivores in the former (Bucher 1982) and large mammals as major herbivores in the latter (McNaughton and Georgiadis 1986). Large mammals, of course, have been reintroduced as livestock throughout the arid to subhumid ecosystems of South America, where they currently constitute the major herbivore class.

The forage-harvesting attine ants are now categorized as major "pests" of Neotropical agriculture, and most estimates of their importance are related to this status (Fowler et al. 1986). This status, therefore, may lead to overestimates of their importance compared to natural ecosystems, since there may be a tendency to concentrate measurements where the influence is most pronounced. Nevertheless, studies in pastures indicate that they can be formidable competitors with livestock (Robinson and Fowler 1982). Ants have a direct inhibitory effect on cattle grazing. Cattle avoid areas where ants are foraging and, when confined to those areas by staking, graze less and walk more than when staked in control areas (Fowler and Saes 1986). In addition, confined cattle fed forage with ants on it consume significantly less than if they are absent, suggesting that thoracic spines or other properties of the ants act as direct feeding deterrents to large mammals.

Estimates of the level of herbivory by ants are indirect and conspicuously variable but, nonetheless, indicate that they are major herbivores in Neotropical arid to subhumid ecosystems (Fowler et al. 1986). Methods of estimating consumption include (a) multiplying counts of foraging workers by weights of loads carried, (b) calculating the weight ratio of fresh to spent fungal substrate, and (c) comparing exclusion plots, where ants are poisoned, with untreated plots. In pastures in Paraguay, exclusion indicated that ants consumed from 2 to 37 percent of the standing crop, compared to cattle consumption of 20 to 65 percent; in all cases, consumption was conspicuously lower in plots where both ants and cattle had access, suggesting mutual interference (Robinson and Fowler 1982). A compilation of consumption values from subtropical South American grasslands indicates that ants consume up to 900 g m^{-2} yr^{-1} (Fowler et al. 1986). Therefore, leaf-harvesting ants appear to be the natural herbivores in South America occupying the grazing niche that is occupied by large mammals in Africa. In fact, the fragmentary data on herbivory indicate that they may sometimes attain levels of consumption comparable to the large mammals. We believe that one of the great gaps in a comparative understanding of South American and African arid to subhumid ecosystems is the paucity of data on consumption by leaf-harvesting ants in South America.

Excluding primates, more than 90 species of large (weighing more than 5 kg) mammalian herbivore species occur on the African continent (Maglio and Cooke 1978). More than 10 species coexist in most game areas, and large reserves commonly have more than 20 species of large mammalian herbivores (Cumming 1982). The highest levels of sustained herbivory in natural ecosystems have been consistently recorded in Africa, with estimates commonly in the range of 33 to 66 percent of aboveground productivity consumed, although some values range up to 95 percent (McNaughton and Georgiadis 1986). For 28 study sites spread across the Serengeti ecosystem (McNaughton 1985), the proportion of ANPP consumed annually ranged from a minimum of 17 percent to a maximum of 94 percent, with a mean value of 66 percent and a median of 71 percent (figure 18.4). These data were from studies employing temporary exclosures and unfenced plots to estimate both

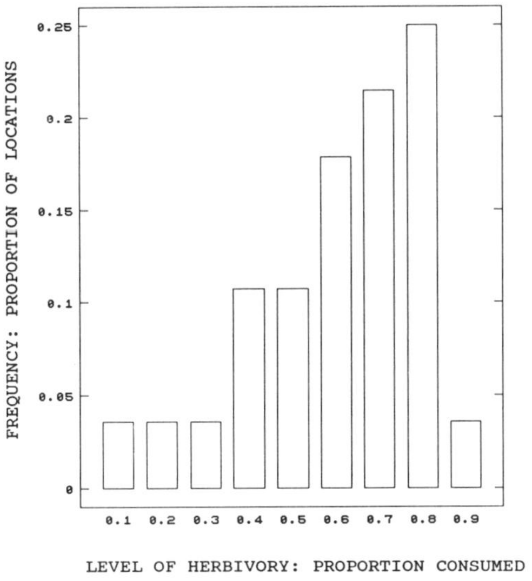

Figure 18.4. Frequency of proportion of net aboveground primary productivity consumed at different locations in the Serengeti ecosystem (after McNaughton 1985).

ANPP and herbivory. Using the common comparison of peak standing crop with minimum standing crop indicated that 92 percent of aboveground production in the Serengeti ecosystem disappeared by the end of the growing season due to herbivory (including both mammals and termites), fire, and trampling into the soil. Therefore, the residual standing crop in African ecosystems is extremely low by comparison with most other systems, where the bulk of primary production enters detritus food webs. In African ecosystems where herbivores are abundant, there is virtually no sustained accumulation of residual standing dead plant material, and turnover time of standing crop is very short compared to plots that are fenced and unburned (McNaughton et al. 1988). The vast majority of primary productivity each year is consumed by herbivores or termites, burned, or trampled into the soil where it is rapidly decomposed.

VI. Primary Productivity and Large Herbivore Biomass

All significant ecosystem properties of the herbivore tropic level, biomass, consumption, and productivity are significantly correlated with primary productivity in terrestrial ecosystems ranging from desert and tundra to tropical forest and salt marsh (McNaughton et al. 1989). However, tropical grasslands depart from the general trend by having higher levels of consumption and lower levels of biomass per unit of primary productivity than comparable ecosystems with invertebrates as the major consumer class (McNaughton et al. 1991).

A compilation of livestock biomass (*B*) across Argentina (Mininsterio de Economia

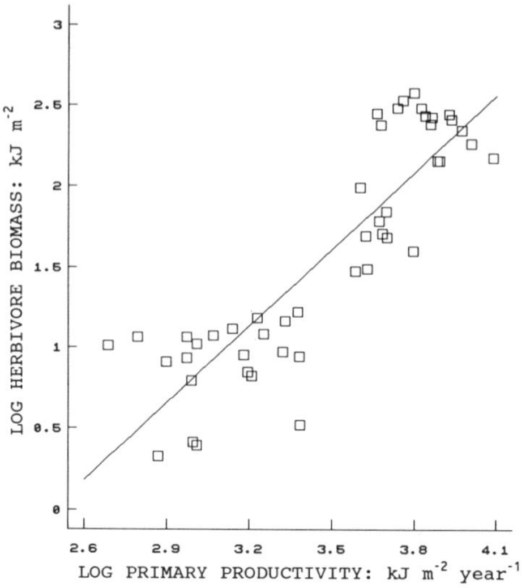

Figure 18.5. Relationship between large herbivore (livestock) biomass and net aboveground primary productivity of South American ecosystems.

1974), with primary productivity estimated from the ANPP/precipitation relationship documented above, reveals a similar association (figure 18.5):

$$\log B = 1.64 \log \text{ANPP} - 3.75,$$

where B is expressed in kJ m^{-2} and ANPP in kJ m^{-2} yr^{-1}. However, a comparison of this pattern with the best fit line from natural ecosystems reveals a major difference (figure 18.6). The biomass of livestock supported per unit of primary production is about an order of magnitude above the level of natural herbivores. This indicates that agricultural management practices, including dietary supplementation, veterinary practices, and elimination of predators, significantly increases the carrying capacity of ecosystems for herbivores. To our knowledge, this is the first quantitative documentation of

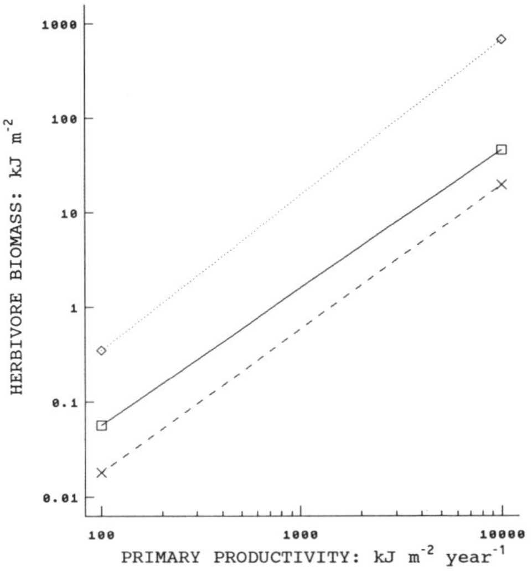

Figure 18.6. Best fit lines relating herbivore biomass to net aboveground productivity. Diamond: South American livestock; box: African ungulates; x: all native herbivores.

this phenomenon over a broad range of primary productivities (Oesterheld et al. 1992).

In addition to the positive association between total mammal herbivore biomass and vegetation productivity, data from African game reserves indicate that the biomass of 19 of 23 individual species is also positively associated with rainfall (East 1984). At the highest rainfall levels, however, there were tendencies both for the biomass of many species to decline and for some species to disappear altogether. Moreover, the trends were accompanied by a pattern of change in average body size of major herbivores. At progressively higher rainfall levels and, therefore, higher levels of net primary production, larger species (e.g., elephant and buffalo, *Sycerus caffer*) become increasingly important elements of the herbivore trophic level.

A similar pattern of variation in livestock size with net primary productivity was evident in South American livestock (figure 18.7). The proportion of sheep, compared to cattle, present in livestock herds declined with increasing ANPP from near 100 percent sheep at the lowest productivity levels in Patagonian areas receiving less than 200 mm of precipitation annually to near zero at the highest productivities in subtropical regions with annual precipitation levels near 1500 mm. Small-bodied livestock predominate at low levels of rainfall and, therefore, primary productivity, whereas large-bodied livestock predominate at higher rainfalls, just as in African wildlife communities.

Gut volume of mammals is a constant proportion of body weight, but maintenance metabolism is a fractional power of that weight (Demment and Van Soest 1985). Thus, the metabolic requirement per unit of gut capacity decreases as body weight in-

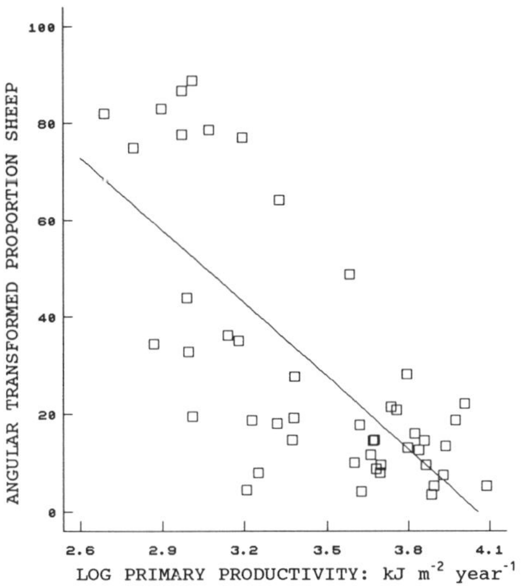

Figure 18.7. Proportion of sheep in South American livestock biomass in relation to net aboveground primary productivity.

creases and the fiber content of an herbivore diet increases with body size. Since a positive relationship probably exists between fiber content of the vegetation and primary production (Demment and Van Soest 1985), herbivores of larger body size, whether wild or domesticated, are more capable of using the more abundant but lower-quality forage produced at higher levels of rainfall.

VII. Plant Nutrition

It has long been known that the savanna vegetation of the Neotropics is often associated with extraordinarily infertile soils (Beard 1953; Lopes and Cox 1977). However, this is not true of all arid to subhumid ecosystems there, and the Argentinean pampas occur on soils that are among earth's most fertile (Scobie 1964; Hall et al. 1991). African arid to subhumid ecosystems are commonly characterized by a gradient from arid-eutrophic to moist-dystrophic systems (MacVicar 1977; Huntley 1982). This seems most applicable to eastern and southern Africa, less so to northern and western Africa.

The Malian-Dutch Sahel Primary Production project, spanning rainfall levels near and at the lower limits considered here, concluded that low nutrient availability, particularly of phosphorus and nitrogen, was a more serious limitation on productivity than low rainfall (Bremen and de Wit 1983). Experimental fertilization in the Sahel produced increases in ANPP of three- to fivefold (De Vries et al. 1980). These increases were attained on a wide range of soil types and with vegetation varying from dominance of the grass layer by annuals to dominance by perennials. In some instances, the vegetation response to nitrogen appeared linear to application levels nearing 250 kg N ha^{-1}. At nearer the humid-dystrophic end of the African nutritional continuum, fertilization of *Hyparrhenia* grasslands in Kenya resulted in linear increases in production up to application levels of 300 kg N ha^{-1}, and additional increments of production due to the addition of phosphorus at higher levels of nitrogen fertilization (Keya 1973).

Fertilization of Venezuelan llanos dominated by *Trachypogon* with 67 kg N ha^{-1} or 79 kg P ha^{-1} produced aboveground yield increases of 30 percent and 18 percent, respectively (Medina et al. 1977). An array of grassland fertilization experiments distributed throughout Argentina serve to characterize the fertility features of the region at a coarse scale (Berardo and Darwich 1969; Bruno et al. 1982; Marchegiani et al. 1982; Mendoza et al. 1983; Ginzo et al. 1986). The sites encompassed a variety of climates from subtropical to temperate. Nitrogen fertilization at levels between 80 and 360 kg N ha^{-1} resulted in primary productivity increases up to 87 percent. The response to phosphorus fertilization was larger than the response to nitrogen. Additions of phosphorus ranging between 20 and 140 kg P ha^{-1} resulted in production increases up to 135 percent.

For each site, there was a linear increase in yield that corresponded to the level of nitrogen or phosphorus fertilization. However, when all sites are pooled together, the large variability in soils and climates masks a clear relationship between yield and fertilization dose. Again, as in the case of the pattern and controls of primary production associated with climate, soil features vary at a scale finer than the regional patterns.

In general, the larger response to phosphorus than to nitrogen fertilization can be considered a regional characteristic of South American arid to subhumid ecosystems. Phosphorus is primarily stored in the soil and its availability is mainly controlled by properties of the parent material. In contrast, the largest nitrogen pool is located in the atmosphere and the parent material controls its availability primarily through the effects of soil particle size (Parton et al. 1988).

Much of the Kalahari ecosystem in Botswana and southwest Africa occurs on sandy soils of extremely low fertility, perhaps an extreme of the dystrophic status, and also at lower levels of rainfall (Cole 1986), that is, it is an arid-dystrophic ecosystem. However, the sands are often covered with cyanobacterial crusts and acetylene-reduction assays indicate that these crusts have a substantial nitrogen-fixing capacity shortly after being moistened (Skarpe and Henriksson 1987). In addition, important grasses in the Kalahari have rhizosphere sheaths containing *Azospirillum* with acetylene-reducing capacity. Thus, much of the nitrogen flow in this arid-dystrophic ecosystem on deep, porous (95 percent sand) soils may be derived from direct fixation fluxes rather than from recycling of residual soil pools.

VIII. Animal Nutrition

Forages of South American ecosystems are commonly so nutrient-deficient for livestock that complete mineral supplementation is advocated as a standard husbandry practice (McDowell et al. 1983). In fact, it is unlikely that there would be a substantial livestock industry in South America without such supplementation. For instance, 80 to 90 percent of the ranches in the Brazilian cerrado supply common salt to livestock and 5 to 30 percent also supplement ground bone (Saturino et al. 1977). Since South America did have a diverse large mammal fauna until the Pleistocene, however, it seems implausible to argue that their historical absence was due to soil-related nutritional deficiency (Sanchez and Buol 1975). Moreover, many areas of Africa require mineral supplementation for livestock, particularly higher rainfall areas supporting moist-dystrophic ecosystems (McDowell et al. 1983), where native animal biomass is lower than in arid-eutrophic systems and consists principally of large-bodied species (Bell 1982).

Bell's (1982) tabulation of animal biomass, rainfall, and geological substrate in African game parks and wildlife areas provides a means of examining the role of nutritional factors, as indexed by soil parent material, in ecosystem carrying capacity for large mammals. Data in his table 3 characterize parent material as (1) volcanic, (2) rift sediments washed from highlands associated with faulting, (3) marine sediments, and (4) basement rocks, largely granite of Precambrian age (he also included single examples of Kalahari sands and floodplains, which we excluded). The geographic range was from Sudan to Zaire to South Africa, and we utilized the rainfall span from a minimum mean annual value of 165 mm to a maximum of 1550 mm.

Animal biomass carrying capacity per unit of rainfall, expressed as kg km^{-2} mm^{-1}, varies substantially in African ecosystems ($F_{3,39} = 5.40$, $p < 0.01$) with geometric means of: basement = 2.74^{a}, rift sediments = $4.14^{a,b}$, marine sediments = $7.50^{a,b}$, and volcanic = 9.41^{b} (values with different superscripts differ significantly from one another by Sheffé's test, at least at $p = .05$). Therefore, the large mammal carrying capacities per unit of rainfall in African arid to subhumid ecosystems vary more than threefold from minima in regions where soils are derived from old, heavily weathered and leached basement rocks to maxima in regions where soils are of volcanic derivation associated with tectonic activity and the continental rifting system.

African ecosystems contain mammalian herbivore populations that exploit the environment in two different ways: resident herds occupy spatially fixed home ranges and migratory herds move between distinct seasonal ranges (McNaughton and Georgiadis 1986). Resident herds are heterogeneously distributed: some areas contain concentrations of several species, whereas nearby areas are only sparsely occupied (McNaughton 1988). Some migrants congregate around permanent surface water in the dry season and disperse to more arid locations in the wet season; others move along rainfall gradients, concentrating in more arid areas in the wet season and moving to wetter areas in the dry season (McNaughton 1990). Migratory animals typically dominate community biomass wherever they occur (Fryxell et al. 1988).

Both the distribution of resident animals (McNaughton 1988) and the seasonal movements of migratory herds (McNaughton 1990) are related to forage mineral nutrient concentrations in the Serengeti ecosystem. The nutritional requirements of pregnant and lactating females and of young, growing animals appear particularly important to resident herd distribution and the seasonal utilization patterns of migrants. Resident herds are concentrated in areas supporting forages of high magnesium, phosphorus, and sodium content (McNaughton 1988). Migratory herds spend the period of major lactation demand for females and growth demand for young animals in areas supporting forages of higher calcium, copper, magnesium, nitrogen, phosphorus, sodium, and zinc levels (McNaughton 1990).

These results indicate that areas of Africa supporting dense herds of native herbivores contain critical habitats supporting forages with nutritional properties capable of meeting animal demands during critical life stages. They also suggest that the sparser populations of large-bodied species in humid-dystrophic ecosystems may be due to a paucity of such habitats, which forces the animals to forage over larger areas and to be more selective in balancing their diets.

Although the movements of migrants in the Serengeti are related to regional soil fertility gradients (McNaughton 1985), there was no evidence that the soils supporting resident herds are intrinsically more fertile than soils in adjacent areas lacking resident concentrations (McNaughton 1988). However, since the activity of animals can facilitate nutrient recycling (McNaughton et al. 1988) and create localized areas of high nutritional status (Georgiadis and McNaughton 1990), it is possible that certain areas predisposed to animal occupance by the presence of permanent water or mineral licks could, through the activities of the animals themselves, develop nutritional sufficiency.

An explanation for the scarcity of large mammalian herbivores in South American compared to African ecosystems still remains enigmatic, although lack of suitable recolonization routes following the more pronounced Pleistocene extinctions in South

America seems a plausible explanation (see Chapter 13). Nevertheless, a diverse and locally abundant fauna of Camelidae does occur in South America. Given the apparent importance of isolated high nutritional areas in African wildlife distribution and abundance, it would be interesting to determine whether this camelid fauna is restricted to similar areas of nutritional sufficiency and whether such areas are lacking throughout much of the South American arid to subhumid zone.

IX. Concluding Comparisons

South American and African arid to subhumid ecosystems share many structural and functional similarities at broad regional levels, but differentiation in properties at finer scales is also evident on both continents. Vegetation structural properties are similar and there is strong evidence for belowground partitioning of soil resources due to different rooting depths of the herbaceous and arborescent layers. Grasses can take advantage of rain due to their shallow rooting, but woody plants have a refuge from competition when heavy rainfall results in soil water percolation into deeper soil layers. Relationships between primary productivity and precipitation are almost identical over broad precipitation ranges, but on both continents local edaphic constraints result in departures from the regional patterns, particularly due to low fertility or flooding. Native large mammals are much more important herbivores in Africa than in South America, where attine ants have evolved to fill the primary grazer niche. Mammalian herbivore biomass is closely correlated with primary productivity on both continents, and there is a transition from dominance by smaller-bodied species at low precipitation levels to larger-bodied species at higher levels, with a decline in biomass at the highest levels of rainfall on both continents. However, livestock biomass in South American ecosystems is about an order of magnitude above the level of biomass in natural ecosystems due to husbandry practices. Localized areas containing forages of high nutritional content are important determinants of animal distribution and abundance in Africa, but nothing is known of these in South America. Mineral supplementation of livestock suggests, however, that such habitats are rare in South America. The post-Pleistocene differences between the abundances of large, mammalian herbivores on the two continents may have been related to different frequencies of nutritionally sufficient habitats.

Literature Cited

Archer, S., C. J. Scifres, C. R. Bassham, and R. Maggio. 1988. Autogenic succession in a subtropical savanna: Rates, dynamics and processes in the conversion of grassland to a thorn woodland. Ecol. Monogr. 58:111–127.

Aubréville, A. 1962. Savanisation tropicale et glaciations quarternaires. Adansonia 2:16–84.

Beard, J. S. 1953. The savanna vegetation of northern tropical America. Ecol. Monogr. 23:149–215.

Bell, R. H. V. 1982. The effect of soil nutrient availability on community structure in African ecosystems. Pp. 193–216 *in* B. J. Huntley and B. H. Walker (eds.), *Ecology of Tropical Savannas*. New York: Springer-Verlag.

Berardo, A., and N. Darwich. 1969. Fertilizactio de pasturas en el sudeste bonaerense. Actas V Reunion Argentina Ciencia el Suelo, pp. 642–652.

Bertiller, M. 1984. Specific primary productivity

dynamics in arid ecosystems: A case study in Patagonia, Argentina. Acta Oecol. 5:365–381.

Bremen, H., and C. T. de Wit. 1983. Rangeland productivity and exploitation in the Sahel. Science 221:1341–1347.

Bruno, O. A., J. L. Fossati, J. L. Panigatti, S. Gambaudo, H. F. Fenoglio, and O. Quaino. 1982. Fertilizactio de grama rhodes (*Chloris gayana* Kunth.) en los submeridionales. Publicación Técnica No. 19. Rafaela, Argentina: INTA EERA.

Bucher, E. H. 1982. Chaco and caatinga: South American arid savannas, woodlands and thickets. Pp. 48–79 *in* B. J. Huntley and B. H. Walker (eds.), *Ecology of Tropical Savannas*. New York: Springer-Verlag.

Bulla, L., J. Pacheco, and R. Miranda. 1981. A simple model for the measurement of primary production in grasslands. Bol. Soc. Venez. Ci. Nat. 35:281–304.

Clark, F. E. 1977. Internal recycling of ^{15}N in shortgrass prairie. Ecology 58:1322–1333.

Cole, C. V., J. S. Innis, and J. W. B. Stewart. 1977. Simulation of phosphorus cycling in semiarid grasslands. Ecology 58:1–5.

Cole, M. M. 1986. *The Savannas: Biogeography and Geobotany.* New York: Academic Press.

Cumming, D. H. M. 1982. The influence of large herbivores on savanna structure in Africa. Pp. 217–245 *in* B. J. Huntley and B. H. Walker (eds.), *Ecology of Tropical Savannas*. New York: Springer-Verlag.

Defosse, G. E., M. B. Bertiller, and J. O. Ares. 1990. Aboveground phytomass dynamics in a grassland steppe of Patagonia, Argentina. J. Range Manage. 43:157–160.

Demment, M. W., and P. J. Van Soest. 1985. A nutritional explanation for body-size patterns of ruminant and nonruminant herbivores. Amer. Natural. 125:641–672.

Deshmukh, I. K. 1984. A common relationship between precipitation and grassland peak biomass for east and southern Africa. Afr. J. Ecol. 22:181–186.

de Vries, F. W. T. P., J. M. Krul, and H. van Keulen. 1980. Pp. 95–113 *in* T. Rosswall (ed.), *Nitrogen Cycling in West African Ecosystems*. Stockholm: SCOPE/UNEP International Nitrogen Unit, Royal Swedish Academy of Sciences.

Distel, R. A., and O. A. Fernandez. 1986. Productivity of *Stipa tenuis* Phil. and *Piptochetium napostaense* (Speg.) Hack in semiarid Argentina. J. Arid Environ. 11:93–96.

East, R. 1984. Rainfall, soil nutrient status and biomass of large African savanna mammals. Afr. J. Ecol. 22:245–270.

FAO. 1987. *Production Yearbook 1986*. Rome: United Nations Food and Agriculture Organization.

Fernadez, A., R. J. Leon, O. E. Sala, and R. A. Golluscio. 1991. Woody and herbaceous aboveground production in a Patagonian steppe. J. Range Manage. 44:434–437.

Fonseca, E. A., E. E. Gobbee, and O. E. Sala. 1976. Estimación de la biomasa aerea de un pastizal natural de la Depresión del Salado. Monografias CIC 6:11–29.

Fowler, H. G., and N. B. Saes. 1986. Dependence of the activity of grazing cattle on foraging grass-cutting ants (*Atta* spp.) in the southern Neotropics. J. Appl. Ent. 101:154–158.

Fowler, H. G., L. C. Forti, V. Pereira-da-Silva, and N. B. Saes. 1986. Economics of grass-cutting ants. Pp. 18–35 *in* C. S. Lofgren and R. K. Vander Meer (eds.), *Fire Ants and Leaf-cutting Ants*. Boulder, Colo.: Westview Press.

Frenguelli, J. 1930. Particulas de silice organizada en el loess y en los limos pampeanos. Ann. Soc. Ci. S. Fé 2:1–47.

Fryxell, J. M., J. Greever, and A. R. E. Sinclair. 1988. Why are migratory ungulates so abundant? Amer. Natural. 131:781–798.

Fuentes, E., A. J. Hoffmann, A. Poiani, and M. C. Alliende. 1986. Vegetation change in large clearings: Patterns in the Chilean matorral. Oecologia 68:356–366.

Georgiadis, N. G., and S. J. McNaughton. 1990. Elemental and fibre contents of savanna grasses: Variation with grazing, soil type, season and species. J. Appl. Ecol. 27:623–634.

Ginzo, H. D., M. B. Collantes, and O. H. Caso. 1986. Fertilization of a halophytic grassland in Argentina: Herbage dry matter, botanical composition, and mineral content. Turrialba 36:453–459.

Hall, A. J., C. M. Rebella, C. M. Ghersa, and J. P. Culot. 1991. Field crop systems of the pampas. Pp. 413–450 *in* C. J. Pearson (ed.), *Field Crop Ecosystems: Ecosystems of the World.* Amsterdam: Elsevier.

Huntley, B. J. 1982. Southern African savannas. Pp. 101–119 *in* B. J. Huntley and B. H. Walker (eds.), *Ecology of Tropical Savannas.* New York: Springer-Verlag.

Huntley, B. J., and B. H. Walker (eds.). 1982. *Ecology of Tropical Savannas.* New York: Springer-Verlag.

Keast, A., F. C. Erk, and B. Glass (eds.). 1972. *Evolution, Mammals, and Southern Continents.* Albany: SUNY Press.

Keya, N. C. O. 1973. The effect of N.P. fertilizers on the productivity of *Hyparrhenia* grasslands. E. Afr. Agric. For. J. 39:195–200.

Knopp, W. T., and B. H. Walker. 1985. Interactions of woody and herbaceous vegetation in a southern African savanna. J. Ecol. 73:235–253.

Laws, R. M. 1970. Elephants as agents of habitat and landscape change in East Africa. Oikos 21:1–15.

Le Houérou, H. N., and C. H. Hoste. 1977. Rangeland production and annual rainfall relations in the Mediterranean basin and in the African Sahelo-Sudanian zone. J. Range. Manage. 30:181–189.

Leon, R. J. C., G. M. Rusch, and M. Oesterheld. 1984. Pastizales pampeanos: Impacto agropecuario. Phytocoenologia 12:201–218.

Long, S. P., E. G. Moya, S. K. Imbamba, A. Kamnalrut, M. T. F. Piedade, J. M. O. Scurlock, Y. K. Shen, and D. O. Hall. 1989. Primary productivity of natural grass ecosystems of the tropics: A reappraisal. Plant and Soil 115:155–166.

Lopes, A. S., and F. R. Cox. 1977. A survey of the fertility status of surface soils under "cerrado" vegetation in Brazil. J. Soil Sci. Soc. Amer. 41:742–747.

McDowell, L. R., J. H. Conrad, G. L. Ellis, and J. K. Loosli. 1983. Minerals for grazing ruminants in tropical regions. Gainesville: Department of Animal Science, Center of Tropical Agriculture, University of Florida.

McNaughton, S. J. 1983. Serengeti grassland ecology: The role of composite environmental factors and contingency in community organization. Ecol. Monogr. 53:291–320.

———. 1984. Grazing lawns: Animals in herds, plant form, and coevolution. Amer. Natural. 124:863–886.

———. 1985. Ecology of a grazing ecosystem: The Serengeti. Ecol. Monogr. 55:259–294.

———. 1988. Mineral nutrition and spatial concentrations of African ungulates. Nature 334:343–345.

———. 1990. Mineral nutrition and seasonal movements of African migratory ungulates. Nature 345:613–615.

McNaughton, S. J., and N. J. Georgiadis. 1986. Ecology of African grazing and browsing mammals. Ann. Rev. Ecol. Syst. 17:39–65.

McNaughton, S. J., and G. A. Sabuni. 1988. Large African mammals as regulators of vegetation structure. Pp. 339–354 *in* M. J. A. Werger, P. J. M. van der Aart, H. J. During, and J. T. A. Verhoeven (eds.), *Plant Form and Vegetation Structure.* The Hague: SPB Academic.

McNaughton, S. J., R. W. Ruess, and S. W. Seagle. 1988. Large mammals and process dynamics in African ecosystems. BioScience 38:794–800.

McNaughton, S. J., M. Oesterheld, D. A. Frank, and K. J. Williams. 1989. Ecosystem-level patterns of primary productivity and herbivory in terrestrial habitats. Nature 341:142–144.

———. 1991. Primary and secondary production in terrestrial ecosystems. Pp. 120–139 *in* J. J. Cole, G. M. Lovett, and S. E. G. Findlay (eds.), *Comparative Analyses of Ecosystems: Patterns, Mechanisms, and Theories.* New York: Springer-Verlag.

MacVicar, C. N. 1977. *Soil Classification: A Binomial System for South Africa.* Pretoria: Department of Agricultural and Technical Services.

Maglio, V. J., and H. B. S. Cooke (eds.). 1978. *Evolution of African Mammals*. Cambridge: Harvard University Press.

Marchegiani, G. J., O. V. Satorre, A. T. A. Torales. 1982. Fertilización fosforica e introducción de leguminosas en un pastizal natural. Prod. Ani. 9:135–146.

Medina, E., and N. Motta. 1990. Metabolism and distribution of grasses in tropical flooded grasslands in Venezuela. J. Trop. Ecol. 6:77–89.

Medina, E., A. Mendoza, and R. Montes. 1977. Balance nutricional y producción de materia orgánica en las sabanas de *Trachypogon* de Calabozo, Venezuela. Bol. Soc. Venez. Ci. Nat. 134:101–120.

Mendoza, R. E., D. H. Cogliatti, M. B. Collantes, and M. Kade. 1983. Efecto de la fertilización nitrogeno-fosfatada sobre el crecimiento otono-invernal y la absorción de fosforo en tepes de un pastizal natural. Turrialba 33:311–320.

Ministerio de Economia. 1974. Empadronamiento Nacional Agropecuario y Censo Ganadero. Buenos Aires: Ministerio de Economia.

Norton-Griffiths, M. 1979. The influence of grazing, browsing, and fire on the vegetation dynamics of the Serengeti. Pp. 310–353 *in* A. R. E. Sinclair and M. Norton-Griffiths (eds.), *Serengeti: Dynamics of an Ecosystem*. Chicago: University of Chicago Press.

Ode, D. J., L. L. Tieszen, and J. C. Lerman. 1980. The seasonal contribution of C_3 and C_4 plant species to primary production in a mixed prairie. Ecology 61:1304–1311.

Oesterheld, M. 1990. The effect of grazing and flooding on plant growth and allocation. Ph.D. diss., Syracuse University.

Oesterheld, M., and R. J. C. Léon. 1987. El envejecimiento de las pasturas implantadas: Su efecto sobre la productividad primaria. Turrialba 37:29–35.

Oesterheld, M., O. E. Sala, and S. J. McNaughton. 1992. Effect of animal husbandry on herbivore-carrying capacity at a regional scale. Nature 356:234–236.

Oliff, H., J. P. Bakker, and L. F. M. Fresco. 1988. The effect of fluctuations in tidal inundation frequency on a salt-marsh vegetation. Vegetatio 78:21–26.

Parton, W. J., J. W. B. Stewart, and C. V. Cole. 1988. Dynamics of C, N, P, and S in grassland soils: A model. Biogeochem. 5:109–131.

Poiani, K. A., and W. C. Johnson. 1989. Effect of hydroperiod on seed-bank composition in semi-permanent prairie wetlands. Canad. J. Bot. 67:856–864.

Raven, P. H., and D. I. Axelrod. 1974. Angiosperm biogeography and past continental movements. Ann. Missouri Bot. Gard. 61: 539–673.

Robinson, S. W., and H. G. Fowler. 1982. Foraging and pest potential of Paraguayan grass-cutting ants (*Atta* and *Acromyrmex*) to the cattle industry. Z. Ang. Ent. 93:42–54.

Sala, O. E., A. Soriano, and S. Perelman. 1982. Relaciones hídricas de algunos componentes de un pastizal de la Depresión del Salado. Rev. Fac. Agron. 2:1–10.

Sala, O. E., M. Oesterheld, R. J. Leon, and A. Soriano. 1986. Grazing effects upon plant community structure in subhumid grasslands of Argentina. Vegetatio 67:27–32.

Sala, O. E., M. E. Biondini, and W. K. Lauenroth. 1988. Bias in estimates of primary production: An analytical solution. Ecol. Model. 44:43–55.

Sala, O. E., W. J. Parton, L. A. Joyce, and W. K. Lauenroth. 1988. Primary production of the central grassland region of the United States. Ecology 69:40–45.

Sala, O. E., R. A. Golluscio, W. K. Lauenroth, and A. Soriano. 1989. Resource partitioning between shrubs and grasses in the Patagonian steppe. Oecologia 81:501–505.

Sala, O. E., R. A. Golluscio, W. K. Lauenroth, D. G. Milchunas, and I. C. Burke. 1991. Coupling of water and nutrient capture strategies in the Patagonian steppe. Bull. Ecol. Soc. Amer. 72, no. 2 (Suppl.):238.

Sanchez, P. A., and S. W. Buol. 1975. Soils of the tropics and the world food crisis. Science 188:598–603.

San José, J. J., and E. Medina. 1975. Effects of fire on organic matter production and water balance in a tropical savanna. Pp. 251–264 *in*

F. B. Golley and E. Medina (eds.), *Tropical Ecological Systems: Trends in Terrestrial and Aquatic Research*. New York: Springer-Verlag.

San José, J. J., R. Montes, J. García-Miragaya, and B. E. Orihuela. 1985. Bio-production of trachypogon savannas in a latitudinal cross-section of the Orinoco Llanos, Venezuela. Acta Oecol., Oecol. Gener. 6:25–43.

Sarmiento, G. 1984. *The Ecology of Neotropical Savannas*. Cambridge: Harvard University Press.

Saturnino, H. M., J. Mottoso, and A. S. Correa. 1977. Sistema de produção pecuaria em uso nos Cerrados. Pp. 59–84 *in* M. G. Ferri (ed.), *Simposio Sobre o Cerrado*. San Pablo: Editora Universidad San Pablo.

Schmieder, O. 1927. The Pampa: A natural or culturally induced grassland? Univ. Calif. Publ. Geog. 2:303–321.

Scobie, J. R. 1964. *Revolution on the Pampas*. Austin: Institute for Latin-American Studies, University of Texas.

Singh, J. S., W. K. Lauenroth, H. W. Hunt, and D. M. Swift. 1984. Bias and random errors in estimation of net root production: A simulation approach. Ecology 67:577–579.

Skarpe, C., and E. Henriksson. 1987. Research note: Nitrogen fixation by cyanobacterial crusts and by associative-symbiotic bacteria in Western Kalahair, Botswana. Arid Soil Res. Rehab. 1:55–59.

Smith, T. M., and P. S. Goodman. 1986. The effect of competition on the structure and dynamics of *Acacia* savannas in southern Africa. J. Ecol. 74:1031–1044.

Stebbins, G. L. 1981. Coevolution of grasses and herbivores. Ann. Missouri Bot. Gard. 68:75–86.

Stronach, N. R. H., and S. J. McNaughton. 1989. Grassland fire dynamics in the Serengeti ecosystem, and a potential method of retrospectively estimating fire energy. J. Appl. Ecol. 26:1025–1033.

Teruggi, M. E. 1955. Algunas observaciones microscópicas sobre vidrio volcánico y ópalo organogeno en sedimentos pampianos. Notas Mus. La Plata (Geol.) 18:17–26.

Trapnell, C. G. 1959. Ecological results of woodland burning experiments in northern Rhodesia. J. Ecol. 47:129–168.

Vesey-Fitzgerald, D. 1971. Fire and animal impact on vegetation in Tanzanian national parks. Proc. Ann. Tall Timbers Fire Ecol. Conf., pp. 297–317.

———. 1974. Utilization of the grazing resources by buffaloes in the Arusha National Park, Tanzania. E. Afr. Wildl. J. 12:107–134.

Walker, B. H., and I. Noy-Meier. 1982. Pp. 577–590 *in* B. J. Huntley and B. H. Walker (eds.), *Ecology of Tropical Savannas*. New York: Springer-Verlag.

Walter, H. 1971. *Natural Savannas: Ecology of Tropical and Subtropical Vegetation*. Edinburgh: Oliver & Boyd.

———. 1979. *Vegetation of the Earth*. New York: Springer-Verlag.

Werger, M. J. A. 1979. Vegetation-soil relationships in the southern Kalahari. Docum. Phytosociol. n.s. 4:967–981.

———. 1983. Tropical grasslands, savannas, woodlands: Natural and man-made. Pp. 107–137 *in* W. Holzner, M. J. A. Werger, and I. Ikusima (eds.), *Man's Impact on Vegetation*. The Hague: W. Junk.

Werger, M. J. A., and G. A. Ellenbroek. 1980. Water resource management and floodplain ecology: An example from Zambia. Pp. 693–702 *in* J. I. Furtado (ed.). *Tropical Ecology and Development*. Kuala Lumpur: International Society for Tropical Ecology.

Woodmansee, R. G., I. Vallis, and J. J. Mott. 1981. Grassland nitrogen. Pp. 443–462 *in* F. E. Clark and T. Rosswall (eds.), *Terrestrial Nitrogen Cycles*. Stockholm: Ecological Bulletin.

Chapter 19

Ant-Plant Symbioses in Africa and the Neotropics: History, Biogeography and Diversity

Doyle McKey and Diane W. Davidson

I. Introduction

Symbiotic ant-plant relationships afford an excellent opportunity to analyze the effects of both historical and ecological factors on the evolution of mutualisms. Occurring in tropical forests throughout the world, all myrmecophytic plants provide food and permanent housing to ants; the ants, in turn, are known or presumed to protect their hosts from herbivores or competition or to provision them with nutrients. Despite this underlying similarity, ant-plant symbioses differ in diversity and nature on different landmasses. These disparities could have arisen due to either unique features of the separate biotas or to distinctive climatic and other environmental variables that have influenced the net outcome or stability of the inter-

We are grateful to R. R. Snelling for numerous ant identifications, graciously fit into an overburdened schedule. We also thank J. Bronstein, J. Longino, and P. Ward for critical comments on the manuscript, and M. Hossaert for help in conducting our extensive e-mail correspondence, sorting out computer problems, and otherwise facilitating our interface with technology. Our manuscript benefitted immeasurably from consultations with myrmecologists and botanists specializing in particular taxa or floras. These included: D. Agosti, C. C. Berg, B. Bolton, W. L. Brown, Jr., W. Burger, J. Dransfield, A. H. Gentry, W. Judd, J. Longino, J. Miller, D. Olson, S. Renner, J. C. Solomon, C. M. Taylor, H. van der Werff, P. S. Ward, and J. L. Zarucchi. This work was supported by the Swiss Natural History Society, the University of Miami, the National Geographic Society, and the Mission "Radeau des Cimes—Fondation Elf, Afrique 91" (to McKey), and National Science Foundation grants RII-8310359 and BSR-9003079, the Guggenheim Foundation, and the Christensen Research Institute (to Davidson).

 ISBN 0-300-05375-4

actions. In contrast, similar selection pressures could have produced strongly parallel or convergent evolution in independent and geographically isolated ant-plant relationships. Cross-continental comparisons of ant-plant symbioses might then reveal the types of selection pressures driving plants and ants into intimate and mutually beneficial associations, as well as the kinds of plant and ant traits that could have facilitated such transitions.

Here we compare radiations of myrmecophytes and specialized plant-ants (ants that regularly live in symbiosis with a comparatively restricted set of host plants) in Africa and the Neotropics. We begin by assessing the extent to which patterns in the representation of ant-plants and plant-ants on different tropical land masses are correlated with trends in the diversity of floras or ant faunas as a whole. For both plants and ants, we then examine existing phylogenetic evidence bearing on the relationships between both ant-plants and plant-ants in Africa and America. Has the evolution of ant-plant associations occurred independently in the two regions, or can associations in Africa and America be traced to a common evolutionary past? Elsewhere (Davidson and McKey 1993), we argue that similar selection pressures may have driven strongly parallel and convergent evolution in ant-plant symbioses on different tropical landmasses and that selection operated on similarly preadapted plants and ants of Africa and the Neotropics. This evidence, weighed in the context of the historical data presented here, allows us to infer how major differences in the climatic and geologic histories of Africa and America may have profoundly influenced contemporary patterns in the number and distribution of ant-plant associations on the two continents.

Focusing on the more highly specialized participants in *symbiotic* ant-plant relationships of Africa and the Neotropics, we restrict our analyses to trees, shrubs, and hemiepiphytes with ant domatia. Domatia are regularly occurring cavities, either in stems or stem-like structures, or enclosures formed by invagination, by folding, for example, of leaves, stipules, and similar organs. Our discussion frequently includes the Asian and Australian ant-plant symbioses, as these often serve as useful reference points. Throughout our contribution, we use the terms *myrmecophyte, myrmecophytic,* and *ant-plant* to describe plants in which ants regularly nest, without implying that plants either benefit from the ants or have traits evolved principally as ant attractants. Existing evidence is often too meager for a convincing assessment of the benefits of particular associations to the plant (Davidson and McKey 1993).

A. The Ecological Diversity of Ant-Plant Symbioses

Janzen's pioneering study of ant-acacias gave us the first clear evidence for the mutualistic nature of an ant-plant symbiosis, by demonstrating conclusively the advantages to the plant of occupancy by specialist *Pseudomyrmex* acacia-ants. These ants protect the host from insect and other herbivores and prune neighboring plants, preventing engulfment of their host by vines (Janzen 1966). Over the past 25 years, studies have generalized these basic findings to a number of other ant-plant systems, while at the same time showing that considerable diversity lurks beneath the surface of these fundamentally similar interactions (see Beattie 1985; McKey 1988; Hölldobler and Wilson 1990). For example, in addition to protection, ant associates of myrmecophytic epiphytes and ant-garden plants can make important contributions to the mineral nutrition of their hosts (Janzen 1974b; Huxley 1980; Rickson 1979; Davidson and Epstein 1989). Even in systems in which the principal benefits to the plant are protection against herbivores and vines, there is substantial diversity in functionally important aspects of the symbiosis. This includes variation in ant traits such as worker size and number, aggressiveness toward different intruders, and distribution of patrolling activity over the plant (McKey 1988). Variation in plant traits occurs in the type of food rewards produced for ants and the rate at which plants supply resources to ants (Davidson et al. 1991; Davidson and Fisher 1991). Plant-ants also vary in the degree to which they are host-specific. Some plant-ants occupy a number of unrelated plants that are morphologically and ecologically similar, though phylogenetically distant. Other plant-ants (e.g., acacia-ants) occupy several related host species. Cases of absolute species-specificity in host relationships appear to be exceptional and usually involve taxonomically depauperate systems (McKey 1989). Patterns of specificity and matches between plant and ant associates appear to be due not simply to pairwise interaction between plant and ant, but to interspecific competition among ants for plant resources (Davidson et al. 1988, 1989; Davidson and McKey 1993).

B. Similarities in Form of Ant-Plant Symbioses between Africa and the Neotropics

Emerging patterns are beginning to define the ecological and functional diversity of ant-plant symbioses. Recognizable guilds of ant-plants and plant-ants occur on more than one continent. The term *guild* is used here to designate subsets of ant-plants and plant-ants differing in the traits that affect functioning of ant-plant interactions. Such characteristics include competitive abilities, behaviors, colony sizes, and energy requirements of ants, and the habitats, life forms, growth rates, and myrmecophytic attributes of plants. Between Africa and the Neotropics, in particular, remarkable similarities exist in the forms of symbiotic ant-plant relationships (Davidson and McKey 1993). Thus, on both continents, large colonies of active and aggressive ants occupy fast-growing and light-demanding pioneer trees, such as the Neotropical *Cecropia* and the African *Macaranga.* (Hollow stemmed Asian *Macaranga* are even closer analogues of *Cecropia.*) In contrast, timid ants inhabit small, slow-growing understory shrubs or treelets. The domatia of these ant-plants are often leaf-pouches covered with long, erect, and dense trichomes (as in African *Magnistipula, Delpydora, Cola,* and *Scaphopetalum* and Neotropical *Hirtella, Duroia,* and many melastomes). Finally, myrmecophytic trees of secondary forests and forest light gaps (African *Barteria* and Neotropical *Triplaris*) grow in circular clearings made by pseudomyrmecine ants, which attack vegetation in the neighborhood of their hosts.

Elsewhere (Davidson and McKey 1993) we argue that these remarkable similarities are the product of convergence and parallelism on a grand scale in different tropical regions. Both convergent and parallel evolution have been driven by similarities in the organization of arboreal ant communities and have led to similar selective forces on the participants in ant-plant symbioses. On both continents, competitively dominant ants are associated with plants that provide resources at the high rates required to fuel the energy-intensive life-style that competitive dominance entails. In contrast, competitively subordinate ants of different continents are restricted to plants that supply resources at low rates or to those from which the dominants are excluded by traits of ants (e.g., pruning of neighboring vegetation: Davidson et al. 1988), plants (e.g., long, dense trichomes: Davidson et al. 1989), or both. Leading to further parallels between relationships on different continents, the evolution of these symbioses has also been facilitated by the same set of ant and plant preadaptations in both regions.

However, a simpler hypothesis to account for similarities between ant-plant symbioses in the two regions would be that they (or their participants) share a common origin. Our contribution addresses this alternative hypothesis. We begin by describing the myrmecophyte and plant-ant biotas of the two continents, comparing first their diversity and then the taxonomic composition of their participants in an effort to distinguish between hypotheses of shared and independent origins of African and Neotropical ant-plant symbioses.

II. Diversity of Myrmecophytes and Plant-Ants in Africa and America

A. Plants

1. Myrmecophytes in Relation to Floristic Diversity Patterns in the diversity of myrmecophytes across tropical landmasses mirror the trends for floras as a whole (table 19.1 and the appendix): plants are less diverse in tropical Africa than in tropical America. The flora of tropical Africa is estimated to include about 2500 genera and 30,000 species, whereas that of tropical America is much richer, with 40,000 species estimated for Brazil alone (Brenan 1978). A total of 4200 genera and perhaps as many as 90,000 species may exist in the Neotropical region (Gentry 1982). As table 19.1 shows, tropical Africa is also poorer in the absolute number of genera of myrmecophytes than is tropical America. However, the *proportion* of plant genera including myrmecophytes is similar on the two continents. This similarity holds whether we compare entire biogeographic regions or mesic tropical forest regions. With the exception of *Acacia* spp., myrmecophytes are virtually restricted to mesic tropical forest habitats in both regions.

On both continents, among genera that have produced ant-plants, there is great variation in the number of myrmecophytes per genus (see the appendix). Both regions offer examples of species-rich genera that contain only one or very few myrmecophytic species. *Cola* (Sterculiaceae), with ca. 140 species (Nkongmeneck 1985), possibly the largest plant genus endemic to Africa, has a single myrmecophytic species (Schnell and Beaufort 1966). The same is true for *Diospyros* (Ebenaceae), with 49 mesic-forest species in Africa (White 1979). In the Neotropics, *Cordia* (Boraginaceae) contains 245–250 species but is represented by only 2 ant-plants (J. Miller, personal communication, 1991). *Besleria* (Gesneriaceae) and *Hoffmannia* (Rubiaceae), both with 100 or more species, each includes only a single myrmecophyte (Benson 1985). Among plant genera with ant-plants, myrmecophytism appears to be restricted to a single species in

Table 19.1. Generic diversity and endemicity by biogeographic region for plants as a whole and myrmecophytic plants.

Measure	Region		
	Ethiopian	Neotropical	Oriental and Australian
All Plant Genera			
Total/mesic tropical region	2497[a]	?	>2178[b]
Total/biogeographic region	?	4200[c]	?
Endemic to a biogeographic region	1500[d]	?	?
Percentage endemic	?	?	?
Ant-Plant Genera			
Total/biogeographic region	31	39	52
Percentage of mesic tropical genera	1.2	?	<2.1
Percentage of regional genera	?	0.9	?
Endemic to a biogeographic region	17	30	32
Percentage endemic	54.8	78.9	61.5

[a]Tropical Africa: Brenan 1978
[b]Figure given is the number of accepted generic names in the area covered by Flora Malesiana: van Steenis 1987
[c]Gentry 1982
[d]Tropical Africa, Madagascar, and southern Africa: Thorne 1973

a greater fraction of African genera (18 of 31, or 58%) than of Neotropical genera (12 of 39, or 31%).

Ant-plants from the Oriental and Australian regions are included in our tabulations for comparison (table 19.1 and the appendix). Species of 51 genera have been reported to be frequently or regularly occupied by ants, indicating a rich potential source of material for testing many of the generalizations we will offer from analysis of African and Neotropical cases. Many of the Oriental and Australian examples, however, are very poorly known at present, though work on these systems is accelerating (Fiala and Maschwitz 1990; Fiala et al. 1989; Maschwitz et al. 1989, 1991).

Sizable radiations of ant-plants within genera are more frequent in the Neotropics than in Africa. When Neotropical genera are compared with closely related or ecologically analogous African genera, the number of congeneric myrmecophytic species is consistently lower in Africa. In part, this simply reflects the smaller size of many African genera. In the Chrysobalanaceae, for example, the African genus *Magnistipula* contains only a single ant-plant, whereas closely related *Hirtella* has 6–7 Neotropical myrmecophytes (see the appendix), which Prance (1972, 1989) treats as a separate section of close relatives. The proportions of myrmecophytic species are similar, however, since *Magnistipula* contains only 6 species in all (Letouzey and White 1978), whereas *Hirtella* has 88 species, all but 2 of them Neotropical (Prance 1972). Similarly, the African genus *Musanga* (Cecropiaceae) contains only 2 species (de Ruiter 1976). One of these, *M. cecropioides* (frequently associated with ants [Duviard and Segeren 1974], though probably not a true myrmecophyte), is widely distributed, whereas the other is restricted to montane forest on the eastern fringe of the central African forest zone. In contrast, the closely related and ecologically analogous American genus *Cecropia* has about 75 currently recognized species (Mabberley 1987), though this number is likely to decrease to 60–70 with further revision (C. C. Berg, personal communication, 1992). Probably four-fifths of *Cecropia* spp. are myrmecophytes (Benson 1985; C. C. Berg, personal communication, 1992).

In addition to *Cecropia* and *Hirtella,* several Neotropical genera without obvious taxonomic counterparts in Africa have produced substantial numbers of myrmecophytic species. These radiations include *Tachigali* (Fabaceae), *Triplaris* (Polygonaceae), and *Ocotea* (Lauraceae), as well as *Clidemia, Maieta,* and *Tococa* (all Melastomataceae). For some of these we can identify unrelated African plants as ecological analogues. Here too, comparisons show that African genera are smaller and contain fewer myrmecophytic species than do their Neotropical counterparts. Neotropical *Triplaris* spp. (Polygonaceae) and African *Barteria fistulosa* (Passifloraceae) are both trees of secondary forest and forest light gaps and have evolved associations with vegetation-pruning pseudomyrmecine ants. *Barteria* includes only one other species, *B. nigritana,* a small tree associated with *Crematogaster* ants. With 18 species (Brandbyge and Ollgaard 1984; Brandbyge 1990), *Triplaris* has undergone a much more extensive radiation. All but 1 of these species are myrmecophytes associated with pseudomyrmecines or (1 or 2) with a timid species of *Azteca* (Davidson et al. 1989; Ward 1991). Similarly, ant-plants with hairy domatia (most often in leaf pouches) occur in both African and Neotropical forests. In Africa, this group includes only 8 species in 7 genera of five families, whereas in the Neotropics, more than 80 species from 19 genera and eight families possess such domatia (Davidson and McKey 1993). Together, 3 of these genera in the Melastomataceae, *Clidemia* (*sensu stricto*), *Maieta,* and *Tococa,* contain more than 75 myrmecophytic species (W. Judd, personal communication, 1992).

Comparisons of the number of myrmecophytes in African and Neotropical genera allow two conclusions: (1) On both continents, myrmecophytism usually characterizes only a small proportion of the species of a genus. When radiations of myrmecophytic lineages occur, these are usually modest at best. (2) Exceptions to this general rule are more frequent in the Neotropics than in Africa. The only African genera with sizable numbers (5 or more) of myrmecophytic species are *Canthium, Cuviera* (both Rubiaceae), and possibly *Clerodendrum* (Verbenaceae), all poorly studied.

2. Distributional Ranges in Africa and America Comparisons of plant diversity in Africa and the Neotropics often emphasize what seems to be a major disparity in ranges: not only is the tropical African flora less species-rich, but many of its species are also more widely distributed. In African forests, many tree species have very large geographic ranges, sometimes occupying the whole of the forest zone (Brenan 1978; White 1979). Some African myrmecophytes also exhibit this pattern. *Barteria fistulosa,* for example, occurs throughout the Central African forest zone, as do *Musanga cecropioides* and, among ant-plants with leaf-pouch domatia, *Diospyros conocarpa.* In contrast, in the Neotropics, Gentry (1990) has noted that often few plant species are shared between disjunct forests despite much overlap in the dominant genera. Correspondingly, the Neotropical analogues of widespread African myrmecophytes are genera rich in species; many of these species have restricted geographic ranges. For example, most species of *Triplaris* occupy only a small portion of the collective range of the genus, with fully 9 of the 17 myrmecophytic *Triplaris* species confined to western Amazonia (Brandbyge 1986, 1990). Three of 6–7 myrmecophytic *Hirtella* (*dorvalii, revillae,* and *vesiculosa*) are restricted to very narrow ranges in central Amazonia (Prance 1972, 1989). Although most *Cecropia* have medium-sized ranges, a number also appear to have small to very narrow endemic ranges, especially in western Amazonia (C. C. Berg, personal communication, 1992).

African myrmecophytes reflect two additional patterns within the African tropical forest zone: the poverty of species in the West African forest block compared to the Central African block (Gentry 1988), and the concentration of diversity in Lower Guinea, the coastal forest from Nigeria through Cameroon to Gabon in the western part of the Central African block (White 1979). Only a few myrmecophytes are known from African forests west of the Dahomey gap. Those for which information is available (e.g., *Canthium glabriflorum* and other species of *Canthium, Cuviera* spp., and *Gardenia imperialis*) are inhabited by *Crematogaster* spp. that are probably not host-specific, suggesting that these may be recently evolved relationships. *Gardenia imperialis* is the only myrmecophyte with leaf-pouch domatia known from the West African forest block. Its pouches are little more than enlarged nectary-bearing pits over which carton-nesting *Crematogaster* build small carton shelters (Schnell and Beaufort 1966). The other possible myrmecophytes from this region are *Stereospermum kunthianum* and *Uncaria africana.* Both are wide-ranging and morphologically highly variable species that have been noted to be associated with ants (probably not host specific) in the Congo basin or further east, but whose biology in West Africa is unknown. With the possible exception of some *Canthium* and *Cuviera* species, none of the West African myrmecophytes is endemic to this forest block.

All other African myrmecophytes (except those found in drier vegetation types: *Acacia* and *Stereospermum*) are restricted to the central African rainforest block extending from Nigeria to the eastern rim of the Congo Basin. A number of these are distributed rather generally throughout this region: *Diospyros conocarpa, Macaranga saccifera, Buchnerodendron speciosum, Barteria fistulosa, Bertiera simplicicaulis, Nauclea vanderguchtii, Rothmannia macro-*

carpa, Cola marsupium, Scaphopetalum thonneri, and myrmecophytic *Vitex* lianes. In addition to these, numerous species are found only in the western part of this forest block, the Lower Guinea coastal forest. These include *Magnistipula bimarsupiata, Calpocalyx cauliflorus, Leonardoxa africana* and other congeneric species soon to be described, *Barteria nigritana, Heinsia myrmoecia, Ixora hippoporifera, Rothmannia lujae,* and *Delpydora macrophylla.* Thus myrmecophytes belonging to virtually all of the 31 African genera are found in Lower Guinea, and species of at least 8 genera are found only there. This relatively small area harbors a disproportionately large share of African myrmecophytes.

In the Neotropics (Gentry 1986a, 1990), regional endemism is comparatively high, and geographic trends in plant species diversity are closely mimicked by patterns in myrmecophyte diversity. Tree forms in the genera *Cecropia, Triplaris,* and *Tachigali* have their centers of diversity in Amazonia, whereas myrmecophytic shrubs and climbers, for example, *Ocotea, Piper, Acacia, Besleria, Hoffmannia, Clidemia* and *Conostegia,* are best represented in Central America or Andean foothills and ridges, or both (Gentry 1982). Like many endemic plants in tropical communities (Gentry 1986a), narrowly endemic ant-plants of the Neotropics appear to be largely neoendemics, products of perhaps recent and rapid speciation, rather than paleoendemic relicts of formerly widespread taxa. Consistent with Gentry's (1986a) view that habitat specialization has been the "prevalent evolutionary theme in giving rise to local endemics in some Amazonian families," a large component of the species diversity in at least some myrmecophyte genera accumulates across habitat boundaries. For example, in southeastern Peru, one or more *Cecropia* species are typical of recently building sandy river beaches (*C. membranacea*), oxbow lakes (*C. latiloba*), banks of quebradas or small streams (*C. engleriana* and *C. polystachya*), land-slips along steep riverbanks and ridges (*C. polystachya* and an unidentified species), and aguajals and *Ficus trigona* swamps (provisionally *C. "tessmannii"*; Davidson et al. 1991). Both the narrowly endemic *Hirtella* ant-plants and myrmecophytic *Duroia* species are habitat specialists confined to white sand substrates or other soils of low fertility (Prance 1972, 1989; A. Gentry, personal communication, 1991).

In summary then, not only the diversity but also the distributions of myrmecophytes mirror general differences between African and Neotropical floras. Comparing tropical floras as a whole, African genera usually *lack* the proliferation of geographically restricted species that characterize radiations of many American genera in both western Amazonia and Central America and northwestern South America (Gentry 1986a). Although wide-ranging myrmecophytes (e.g., *Triplaris americana, Cecropia peltata,* and *Cordia alliodora;* Brandbyge 1986; Berg 1978; Gibbs and Taroda 1983) are not lacking in the Neotropics, species-rich Neotropical genera also include many ant-plants that are geographically restricted and often very specialized in their habitat associations.

3. Local Diversity Differences in ant-plant diversity on a regional scale are reflected at the level of local diversity (alpha-diversity), which can be two-fold higher in western Amazonia than in Africa. The coastal forests of Cameroon are richer in ant-plants than any other part of the African forest zone, but even here no more than about 6 or 7 species of ant-plants are usually found in a single site. In the richest site known to us, near Ebodié, about 30 km north of the Cameroon–Equatorial Guinea border in the Campo Forest Reserve, at least 9 myrmecophytes occur sympatrically (i.e., can be found within a radius of ca. 7 km): 2 species of *Barteria,* 1 species each of *Leonardoxa, Delpydora, Diospyros, Ixora, Cola,* and *Scaphopetalum,* and 1 or more species of *Cuviera.* Further north, in Korup National Park, where with the exception of *Diospyros conocarpa,* leaf-pouch myrmecophytes appear to be absent, the list would include 1 species each of *Barteria, Leonardoxa, Diospyros, Calpocalyx, Vitex, Canthium,* and *Cuviera.* In southeastern Cameroon, away from the Lower Guinea coastal forest, frequently encountered myrmecophytes are 1 species each of *Barteria, Vitex,* and *Cuviera.* Thus, at a local community level, the number of myrmecophytes at least potentially involved in ecological interactions with arboreal ants, and indirectly with each other, often does not exceed 3 to 6 species in African forests.

In contrast, in western Amazonia the number of co-occurring ant-plants can range to at least 16 species. At or near both Cocha Cashu and Reserva Tambopata, two sites in Madre de Dios, Peru, local myrmecophytes include at least 6 *Cecropia* species (8 at Cocha Cashu), 2 species each of *Cordia* and *Triplaris,* 1 *Pleurothyrium,* 3 melastomes with leaf-pouch domatia (1 *Tococa,* 1 *Maieta,* and 1 *Clidemia*), and at least 11 species of ant-garden epiphytes (Davidson 1988; Davidson et al. 1989, 1991). Nevertheless, as indicated above, many of these species exhibit strict habitat associations. Only *Cordia nodosa* and *Triplaris americana* are typical of both low-lying floodplain forests and terra firme forests. Thus, it seems that the high local myrmecophyte diversity may be largely a function of the complex mosaic of habitats typical of western Amazonia (Räsänen et al. 1987). Although many of the associated plant-ants are restricted to particular host genera or even species, some are not (Davidson et al. 1989, 1991). Where distinctive habitat types are closely adjacent, as in western Amazonia, there should be considerable potential for myrmecophytes of one habitat to be colonized (at least occasionally) by ants characteristic of other ant-plants in contiguous habitats.

B. Ants

Among specialized plant-ants, generic richness across tropical biogeographic regions does not seem to parallel that for ant faunas as a whole (table 19.2). The greatest contrast is between ants of the Oriental and Australian regions, on the one hand, and ants of the Ethiopian and Neotropical regions, on the other. Although generic richness of the Oriental and Australian ant faunas is greater than that of Africa and comparable to that of the Neotropics, fewer Oriental and Australian genera (apparently only 5–7%) include specialized plant-ants. (This may change with further study of many poorly investigated Oriental and Australian ant-plants and revisionary work on their ant occupants.) In the Ethiopian and Neotropical biogeographic regions, 10 to 12 percent of genera include plant-ants, and the fraction increases to 13 to 15 percent of genera in

Table 19.2. Generic diversity and endemicity by biogeographic region (Hölldobler and Wilson 1990) and mesic tropical regions (Brown 1973), for ants as a whole and obligate plant-ants.

Measure	Ethiopian	Neotropical	Oriental	Australian
All Ant Genera				
Total/mesic tropical region	75	94	96	88
Total/biogeographic region	92	120	124	109
Endemic to a biogeographic region	31	57	31	29
Percentage endemic	33.3	47.5	25.0	26.6
Plant-Ant Genera				
Total/biogeographic region	11	12	7	8
Percentage of mesic tropical genera	14.5	12.8	7.3	9.1
Percentage of regional genera	12.0	10.0	5.6	7.3
Endemic to a biogeographic region	5	5	1	2
Percentage endemic	45.5	41.7	14.3	25.0

Note: Australian includes Oceania. Classification of these regions follows Brown (1973) and departs slightly from Hölldobler and Wilson (1990). Endemism is calculated in relation to the seven zoogeographic regions of Darlington (1957) and includes Palearctic, Nearctic, and Madagascan regions, as well as those listed here. Calculations based on genera and distributions given by Brown (1973) and Hölldobler and Wilson (1990) were updated to account for revisionary work by Ward (1990) and Shattuck (1991, 1992a, 1992b). Fractions of regional genera with specialized plant-ants include ants whose specialization remains questionable (see table 19.3).

mesic tropical areas, to which plant-ants are mostly restricted.

Disparities between Africa and the Neotropics in the frequency of plant-ants are less marked. For all ant genera and for endemic genera, richness is greater in the Neotropics than in the Ethiopian region. But although mesic Neotropical forests have 1.2 times the number of ant genera as inhabit mesic tropical forests of Africa, the number of genera known to include specialized plant-ants is almost the same for the two areas. Thus the proportion of ant genera with specialized plant-ants is slightly higher in Africa.

As for myrmecophytic plants, the most striking differences in the diversity of African and American plant-ants occur at the species level (table 19.3). Whereas 5 Neotropical genera, representing 4 different subfamilies, are rich in plant-ants (*Pseudomyrmex, Allomerus, Pheidole, Azteca,* and *Myrmelachista*), this is true in Africa only for the pseudomyrmecine genus *Tetraponera* and probably the dolichoderine genus *Technomyrmex*. Even in these cases, the number of plant-ant species is markedly lower than in the Neotropical members of the respective subfamilies. For example, in the Pseudo-

Table 19.3. Present-day distributions of (wholly tropical) obligate plant-ants by subfamily, tribe, and genus.

Taxon	Ethiopian	Neotropical	Oriental	Australian	References
Ponerinae					
Ponerini					
Pachycondyla	×	×(3–4+)	×	×	Davidson & Fisher 1991; J. Longino PC
Pseudomyrmecinae					
Myrcidris	—	×(1)	—	—	Ward 1990
Pseudomyrmex	—	×(32)	—	—	Ward 1991; P. Ward PC
Tetraponera	×(5+)	—	×(2)	×(1)	Ward 1991; P. Ward PC
Myrmicinae					
Cephalotini					
Zacryptocerus	—	×(1+)	—	—	Longino 1989a
Crematogastrini					
Crematogaster	×(3+)	×(2+)	×(8+)	×(1+)	Eth = DM; Neo = Davidson et al. 1989; Davidson & Fisher 1991; Vasconcelos 1990; Ori = Fiala & Maschwitz 1990; J. Longino PC; DD; Aus = Hölldobler & Wilson 1990
Leptothoracini					
Leptothorax	×(1)	×	×	×	R. Snelling PC; DM
Pheidolini					
Pheidole	×	×(6+)	×	×(1+)	Neo = Benson 1985; Davidson et al. 1989; Letourneau 1983; Aus = Hölldobler & Wilson 1990
Solenopsidinae					
Allomerus	—	×(8+)	—	—	Wheeler 1942; Benson 1985
Solenopsis	×	×(1+)	×	×	Benson 1985
Tetramoriini					
Tetramorium	×(1+)	×	×	×	R. Snelling PC; DM
Dacetini					
(Strumigenys)	×	×(1+)	×	×	Brown 1962; W. L. Brown, Jr. PC
Tribes unclassified					
Cataulacus	×(1+??)	—	—	—	R. Snelling PC; DM

Podomyrma	—	—	—	×(2+)	G. Monteith & P. Flower PC; DD
Atopomyrmex	×(1)	—	—	—	R. Snelling PC; DM
Dolichoderinae					
Tapinomini					
Anonychomyrma[a]	—	—	—	×(3+)	DD
Axinidris	×(2+??)	—	—	—	R. Snelling PC; DM
Azteca	—	×(20++)	—	—	Wheeler 1942; Benson 1985; Longino 1989a, 1989b, 1991
Philidris[b]	—	—	×(1+)	×(1+)	Ori = Ridley 1910; R. Snelling PC; DD; Aus = DD
(Tapinoma)	×(1?)	×	×(1?)	×	Eth = R. Snelling PC; DM; Ori = T. Musthak Ali PC; DM
Technomyrmex[c]	×(6+??)	×	×(1+)	×(1+)	Eth = R. Snelling PC; DM; Ori = D. Samson PC; R. Snelling PC; Aus = DD
Formicinae					
Plagiolepidini					
(Plagiolepis)	×(1+)	—	×	×	Bequaert 1922
Myrmelachistini					
Aphomomyrmex	×(1)	—	—	—	R. Snelling PC; DM
Cladomyrma	—	—	×(4+)	—	Agosti 1991; Maschwitz et al. 1991; DD
Myrmelachista	—	×(6++)	—	—	J. Longino PC; DD
Petalomyrmex	×(1)	—	—	—	Snelling 1979; McKey 1984
Camponotini					
Camponotus	×	×(3+)	×(7+)	×(2+)	Neo = Davidson & Fisher 1991; Ori = Hölldobler & Wilson 1990; DD; Aus = Monteith 1987; Hölldobler & Wilson 1990

Note: Presence (×) or absence (—) by zoogeographic region. Taxonomy and distribution from Hölldobler and Wilson (1990). Obligacy of symbiotic association with plants is uncertain for genera in parentheses. Rough estimates of numbers of plant-ant species are indicated in parentheses; these are likely to increase moderately (+) or substantially (++) with additional collection and taxonomic revision. Abbreviations used in references: Aus = Australian; DD and DM = respective author's observations; Eth = Ethiopian; Neo = Neotropical; Ori = Oriental; PC = personal communication.

[a]Recently segregated from *Iridomyrmex* (Shattuck 1992b).

[b]Recently segregated from *Iridomyrmex* (Shattuck 1992b). Most or all of these plant-ants occupy ant-epiphytes.

[c]Includes *Engramma* (S. Shattuck 1992a).

myrmecinae, only 3 of 12 hypothesized independent origins of plant-ants in the Pseudomyrmecinae have taken place in Africa, and only 4 of 37 confirmed pseudomyrmecine plant-ants occur there (Ward 1991). All but 1 of the remaining species and origins are Neotropical. (To these published records must be added an undescribed plant-ant species of *Tetraponera* recorded from *Mimosa* sp. in Madagascar; this apparently represents an additional independent origin of the habit; see P. Ward, personal communication, 1991). *Technomyrmex* appears to have produced a modest number of plant-ants in Africa (table 19.3), and their hosts include 8 genera of 5 plant families (Davidson and McKey 1993). The number of plant-ant species, and the number of independent origins of the plant-ant habit, are unclear in this case. Among Africa's 5 endemic plant-ant genera, both *Aphomomyrmex* and *Petalomyrmex* are monotypic, and the other 3 contain only 1 or 2 confirmed plant-ant species. All 5 plant-ant genera endemic to Africa are associated with 1 or very few host genera. In contrast, all 4 genera of Neotropical endemic plant-ants inhabit hosts in multiple plant families (see below) and probably include multiple radiations of plant-ants. Especially in *Pseudomyrmex,* discussed above, but also in *Azteca,* radiations of plant-ants have been explosive, giving rise to specialists on plants in a broad range of wet and even dry forest habitats (Benson 1985; Ward 1991). *Azteca* includes at least 20 plant-ant species (table 19.3), occupying at least 13 plant genera in 8 families (Davidson and McKey 1993). *Allomerus, Pheidole,* and *Myrmelachista* each have at least 6 to 8 plant-ant species in the Neotropics (table 19.3) and occupy a diversity of hosts (in each case at least 6 genera in at least 4 families; Davidson and McKey 1993).

The apparent lower species richness of African plant-ant genera may be due in part to less thorough collecting. Despite some work in the former Belgian Congo (Bequaert 1922), collections of African plant-ants have been much less thorough and more narrow in geographic scope than collections in the Neotropics (Wheeler 1942), and this difference continues today. Further field work and taxonomic revision will probably tend to reduce the gap between the numbers of African and Neotropical plant-ants. An increase in the number of African plant-ant species is likely to be concentrated in the dolichoderine genera *Axinidris* and *Technomyrmex* and to result from both description of new species (especially *Technomyrmex*) and work on the natural history of known species. The dolichoderine genus *Axinidris* is a case in point. Before Shattuck's (1991) revision, this genus was considered monotypic, but it now contains 13 named species. Prior to the collection (in 1989) of 2 species from swollen internodes of *Leonardoxa* in Cameroon, *Axinidris* was not known to include plant-ants, since it had been collected only as stray workers and its habits were undescribed. Despite such indications that the African fauna is poorly studied, we believe that the intercontinental difference in species richness of plant-ant genera is real. New Neotropical plant-ants have also recently been found in the genera *Pachycondyla, Crematogaster, Myrcidris,* and *Azteca* (Davidson and Fisher 1991; Ward 1990; Longino 1989b; J. Longino, personal communication, 1991). Furthermore, if myrmecophyte species richness gives any indication of plant-ant species richness, the

Neotropical region would be expected to contain more plant-ants. The two continents differ dramatically in myrmecophyte species richness, and this difference is unlikely to be greatly altered by further work.

As in ant-plants, much of the diversity of plant-ants appears to be attributable to habitat specificity. Even a single plant host species often harbors different ant associates in different habitats (Benson 1985; Harada and Benson 1988; Longino 1989b; Davidson et al. 1989, 1991). Similarly, related ant-plants that are specialized on different habitats, like the *Cecropia* spp. discussed above, are occupied by different suites of ants. The ant associates of members of a plant lineage in different habitats are often not closely related, belonging to different species groups, genera, or even subfamilies (Benson 1985; Davidson et al. 1991; Davidson and Fisher 1991; Davidson and McKey 1993). This suggests that habitat-dependent host shifts, and habitat-dependent secondary colonizations of ant-plants by multiple ant lineages, have been important processes. Parallel diversification of associated ant and plant lineages, possibly involving parallel patterns of habitat specialization in ants and plants, seems to have occurred less frequently. Some examples, however, are suggestive. In the genus *Pseudomyrmex,* closely related (sibling) species often have broadly overlapping distributions, and allopatric pairs of sister species are uncommon (Ward 1989). As many as 4 species of *Pseudomyrmex* co-occur locally on both Peruvian *Triplaris* (*P. viduus* group) and Central American *Acacia* (*P. ferrugineus* group). Ward implies that speciation in *Pseudomyrmex* has been largely allopatric, with rapid secondary overlap. Across the three *Azteca* lineages associated with *Cecropia,* three different distributional patterns are apparent (Longino 1989a, 1989b; J. Longino, personal communication, 1991). The newly separated *A. alfari* and *A. ovaticeps* have broadly overlapping geographic ranges extending throughout most of the Neotropics (Longino 1989b). Also, two species of the *A. constructor* complex are sympatric in the Choco region of Colombia (J. Longino and T. M. Aide, personal communication, 1991). In contrast, closely related *A. xanthochroa* and *A. australis* are allopatric, the former occurring in Central America and the latter in South America. Without additional revisionary work in many ant and plant groups, it is impossible to judge the degree to which the distributional patterns and speciational processes are alike in associated ant and plant taxa.

C. Intercontinental Differences in Ant-Plant and Plant-Ant Diversity

As summarized in table 19.1 and the appendix, the major tropical regions differ in diversity of ant-plant symbioses. The Oriental and Australian regions combined have the largest number of reported ant-plant genera, but only a few have been well studied (e.g., Fiala and Maschwitz 1990; Fiala et al. 1989; Maschwitz et al. 1989, 1991). Among the 52 Oriental and Australian genera, only 25 (48%) have been sufficiently well-documented to establish the presence or absence of specialized structures to house ants (Davidson and McKey 1993). On the whole, a greater proportion of African and Neotropical genera appear to be at least minimally understood. In only 2 African genera (6%) and 10 Neotropical genera (26%) do we judge that uncertainty remains about the presence or absence of specialized domatia. Future studies may yet reveal that the greater uncertainty about Australian and Oriental ant-plants is merely a consequence of the difficulty of ruling out domatia, where obvious domatia are lacking. Here, however, we restrict our comparisons to Africa and the Neotropics.

What factors account for myrmecophytes and specialized plant-ants radiating more extensively in the Neotropics than in Africa? Are discrepancies in diversity the consequence of regional differences affecting the biology of ants or plants in particular, or are they one manifestation of differences that have had general consequences for rates of species extinction or origination in many taxa on the two continents? We first consider the former possibility.

1. Explanations Specific to Ants and Ant-Plant Interactions Carroll (1979) documented intercontinental differences in diversity of the stem-nesting guild of arboreal ants and proposed a hypothesis to explain his findings. Because most specialized plant-ants are derived from ants in this guild, patterns of diversity in plant-ants may be a reflection of underlying differences in the stem-nesting guild.

Compared to African wet forests (Carroll 1979), Central American forests are richer in both biomass and species of arboreal stem-nesters. Carroll (1979) attributes the lower abundance and diversity of African stem-nesters to greater interference from aggressive dominants (genera *Oecophylla, Tetramorium,* and *Crematogaster*), whose construction of nests from leaves or carton frees them from nest-site limitation. Tending of Homoptera by these same ants provides an abundant resource and funds the aggression that prevents behaviorally subordinate species, including most stem-nesters, from coexisting with dominants. Because aggressive and dominant *Camponotus, Azteca, Hypoclinea* (= *Dolichoderus;* Shattuck 1992a) and *Crematogaster* of South America are ecologically equivalent to the African dominants, Carroll's hypothesis would prove inadequate if South American stem-nesters exhibited biomass and diversity equivalent to that in Central America. Although exactly comparable data are not available, the remarkable abundance and diversity of arboreal ants in Peruvian rainforests appear to contradict the hypothesis. From two censuses of a single leguminous tree in secondary forest at Reserva Tambopata in Madre de Dios, Peru, Wilson (1987) identified 26 genera and 43 species of arboreal ants. Although not all of the species are stem-nesters, we estimate that at least 26 of the species in this tree nest regularly in twigs and branches. The results of Tobin's (1991) studies in a nearby forest within the Manu National Park suggest that Wilson's figures are not anomalously high. In canopy fogging samples from two isolated rainforest trees, Tobin found 52 ant species (not distinguished by nesting sites), accounting for 70 percent of the arthropod numbers and probably more than 50 percent of arthropod biomass. Four dolichoderine species, together comprising more than 90 percent of ant biomass, might qualify as dominants comparable to those identified from African forests. Considering the abundance of ecologically analogous aggressive dominant ants in both continents, we find little evidence to support the view that more intense interaction with dominant ants has led to lower abundance and diversity of stem-nesting arboreal ants in Africa. Nevertheless, there may be differences between the two continents in abundance of some groups of small-bodied stem-nesters, most notably those belonging to *Crematogaster*. This genus appears to be an especially important component of arboreal ant faunas in the Old World tropics, and its members are often competitive dominants (Davidson and McKey 1993). If twig-nesting *Crematogaster* are more abundant in Africa, this might have important consequences for other twig-nesters. Canopy fogging samples from African forests might shed further light on this issue.

2. General Explanations of Diversity Differences Intercontinental disparities in the diversity of ant-plants closely mirror those for plants as a whole, as well as for other kinds of organisms (Amadon 1973; Laurent 1973; Roberts 1973; Chapter 9). For ants, the same is true in somewhat lesser degree; the correspondence is more evident at the species level than at the generic level, and is more evident for the Neotropical/Ethiopian pair than for other intercontinental comparisons. We find it difficult to avoid the conclusion that differences in the diversity of these symbiotic associates are just one manifestation of a general trend toward greater species richness in the Neotropics. Such a broad pattern likely has general causal mechanisms, and several have been proposed. As Duellman (Chapter 9) points out, the area of mesic tropical forest habitat in America is approximately 2.8-fold greater than that in Africa. Extinction rates on the two continents might have differed dramatically during past periods of climatic deterioration, due to differential effects of Pleistocene cooling and drying cycles on habitat reduction in Africa and South America (Raven and Axelrod 1974; Axelrod and Raven 1978; Gentry 1986a, 1988). Unlike Amazonia, which is shielded by the Andes, tropical Africa lacks protection from the drying effect of a cold western ocean current. The climate of tropical Africa is also driven largely by the Asian and African monsoons, which weakened greatly during northern glaciations. Consequently, rainfall in Africa at 18,000 B.P. may have been reduced by half, while Amazonian rainfall, dependent mainly on tropical sea surface temperature, might have been reduced only by 10 to 20 percent (Chapter 15). African forests may then have experienced greater reduction, and persisted in a smaller number of refugia, during the Pleistocene (Maley 1987, 1989; Maley et al. 1990; Chapters 15 and 16). Forest reduction was most extreme in West Africa, where rainfall was highly dependent on the African monsoon, and least in the Lower Guinea coastal forests, where the equatorial countercurrent maintained relatively high sea surface temperatures and hence relatively high rainfall even during glacial maxima (Maley 1989). This difference may explain the generally low plant diversity of the West African forest block and the concentration of diversity in Lower Guinea, two patterns reflected in the diversity of ant-plants.

Frequent and severe African droughts might also have taken their toll on ant diversity and on symbiotic species associations. The majority of obligate plant-ants appear to have evolved from ancestors that tended Homoptera inside cavities of live plants, and droughts could have limited the diversification of such ants in Africa. Compatible with this hypothesis are the findings of Terron

(1970), who documented a strong impact of seasonal drought on rates of egg-laying, älate production, and larval death in African *Tetraponera anthracina,* a homopteran-tending stem-nester confined to comparatively mesic rainforests and gallery forests. In addition, since the net effects of species interactions on individual participants are usually context-dependent (Davidson and McKey 1993), the greater range of environmental change in Africa might have been more likely to convert mutualisms to parasitisms and to lead to the dissolution of symbiotic mutualisms, even in the absence of species extinctions.

Most attempts to explain differences in diversity between African and South American tropical forest ecosystems have focused on contemporary and historical differences in climate. Several observations, however, make it seem unlikely that these differences fully account for the intercontinental disparities in floral and faunal richness. First, the case of the Cape Province of southern Africa, with its striking plant diversity and endemism, illustrates that even a relatively small area can accumulate high species richness over comparatively brief intervals of geologic time. Second, climate change here seems to have enhanced rather than reduced diversity. Swarms of endemic plant species in the fynbos and other distinctive communities of the Cape Province have evolved in the last 2 to 5 million years, since the establishment of the cold Benguela current that brought mediterranean climates to the Cape (decimating the existing subtropical flora), and hyperaridity to the Namib (Chapter 3). Although comparisons of African and Neotropical diversity have tended to emphasize the negative effects of climate change, such fluctuations may have helped to magnify diversity here as well (Chapter 16), as geographic ranges fractionated and organisms encountered new habitats during the range re-expansions that would have occurred in periods of climatic amelioration.

Third, major differences in geologic activity on the two continents, and in correlated rates of habitat disturbance, are likely to have affected both the generation of diversity over evolutionary time (e.g., Vanzolini 1973) and its maintenance in ecological time. We believe that differences between the two continents in the structure of historical and contemporary landscapes, resulting from their contrasting geological histories, are an overlooked and crucial factor in explaining differences in their biological diversity. Africa's less eventful geological history has produced a less diverse mosaic of landforms and habitats than in tropical South America and has presented fewer opportunities for habitat specialization.

As the keystone of Gondwana, Africa was at the interior of a supercontinent that had been exposed to weathering for a very long time and was probably relatively flat, as it is now. According to Wright and coworkers (1985:1), "almost the whole of Africa has been unaffected by major mountain-building (orogenic) earth movements since the end of the Precambrian. . . . Africa is predominantly a continent of plains and plateaux and intervening escarpments, the result of erosion and planation lasting many hundreds of millions of years. Topographic relief is low over vast areas." While most of Africa remained flat during drift, the Andes pushed up along the leading edge of South America, with three important consequences for the subsequent evolution of the biota of this continent. First, the Andes blocked the drying effects of a cold ocean on tropical South America. Second, in western Amazonia the Andes provided habitat diversity in the form of topographic relief, with its consequences for temperature, rainfall, and cloudiness regimes. Third, by lifting vast amounts of rock of varying composition and producing the orographic rainfall that erodes it, "the rising Andes became an inexhaustible source of vast masses of sediment" (Klammer 1984:77). The tectonic activity associated with the Andes generated a fluvial landscape and an edaphic mosaic whose complexity (Salo et al. 1986, Räsänen et al. 1987; Brown 1987c) is unmatched in geologically relatively stable Africa.

With the conspicuous exception of East African mountains and rift valleys, lying mostly outside the mesic forest zone, most of tropical Africa is a relatively featureless plateau. Elevational relief contributes little to habitat heterogeneity in this region of long-term geologic stability. Africa has little cloud forest habitat, in marked contrast to the expanses of mid-elevation, submontane, and montane forest present in the coastal moist forest zone of the northern Andes, in southern Central America, and in western Amazonia near the base of the Andes. These areas harbor a tremendous diversity of plant species, many of them endemic and apparently produced by relatively recent and explosive speciation (Gentry 1986a). Gentry (1986a:168–169) considers this very active speciation, typical of montane areas throughout the tropics, to be "somehow related to the peculiarities of the broken terrain and/or the complex juxtaposition of different vegetation types." Adding to the effects of elevational habitat heterogeneity per se is Janzen's (1967) observation that elevational relief creates especially effective barriers and isolating mechanisms for tropical organisms, which tend to have narrow thermal tolerances.

Mountain-building activity also creates habitat diversity far beyond the montane and submontane areas immediately affected by orogeny (Salo et al. 1986). Lack of such activity in Africa has resulted in an edaphic landscape that, at least in the mesic forest zone, is much less complex than that of tropical South America. Missing in Africa are areas of both elevated continental shield (e.g., South America's Guyanan shield), and the expanses of acid white-sand soils created by the erosion of these ancient rocks (as in the Rio Negro basin) (Sombroek 1984), or resulting from former beaches of oceans, inland seas, and rivers (Prance 1987). Lacking as well is the edaphic mosaic produced by Andean orogeny: volcanic soils, soils with peculiar mineral imbalances (Chapter 17), and the great sheets of alluvial soils produced as young Andean rocks have been eroded by prodigious orographic rainfall. Particularly in western Amazonia, long-term fluvial perturbance over the Tertiary and Quaternary has produced a temporally structured and highly complex mosaic of fossil and present floodplains (Räsänen et al. 1987). The scale of riverine disturbance in western Amazonia is illustrated by the finding of Salo and co-workers (1986) that 26.6 percent of the modern lowland forests show evidence of recent erosional and depositional activity and that 12 percent is currently in succession. Like elevational diversity in the Neotropics, this edaphic mosaic contributes to great habitat diversity, which has probably magnified species diversity, especially in northwestern Amazonia (Gentry 1986a, 1986b, 1988; Salo et al. 1986).

Compared to the Neotropics then, the African forest zone seems to offer relative monotony along many environmental dimensions. Greater habitat diversity may explain why regional species diversity and total species richness is greater in the Neotropical forest flora, even though within-site plant species diversity is fairly similar in African and Neotropical forests with comparable rainfall (Gentry 1988; Chapter 17).

Mounting evidence points to an important role of habitat specialization in the evolution of tropical plant species diversity (Gentry 1986a, 1988). We postulate that the African forest zone offers a lower diversity of habitats and hence fewer opportunities for habitat specialization. If Colinvaux (Chapter 16) is correct in hypothesizing a "speciation pump" primed by range re-expansion into new habitats, rates of species origination during periods of past climatic change may have been higher in the Neotropics. First, the number of refugia from which forest species could re-expand, and the size of refugia, are both postulated to have been greater in the Neotropics than in Africa (compare maps in Brown 1987a and Maley et al. 1990). Thus the number of species "available" for speciation may have been greatly reduced in Africa. Second, as climates warmed and became wetter, and ranges of mesic-forest organisms expanded, plants in the Neotropics were more likely than those in Africa to encounter elevational, climatic, and edaphic conditions sufficiently new to stimulate evolutionary change. They may also have encountered more environmental boundaries sharp enough to allow evolutionary divergence from parent populations. Gentry's (1986a) argument that many endemic Amazonian plants appear to be neoendemics, rather than paleoendemics of old refugia, lends credence to the view that climatic change and habitat heterogeneity could have been major contributors to evolutionary diversification in the Neotropics. Moreover, contemporary ecological differences in habitat diversity on the two continents may be as important as differences in the amount and seasonal distribution of rainfall (Gentry 1988) in maintaining disparities in species richness.

What is the particular role of habitat specialization in the evolutionary diversification of myrmecophytes and specialized plant-ants? As noted previously, outcomes of interspecies interaction may vary markedly among environments. Consequently, the match between ant and plant associates, largely determined by interspecific competition among ants for plant resources, might be particularly sensitive to habitat differences that affect the rates at which plants supply resources to ants (Davidson et al. 1991; Davidson and Fisher 1991). This topic will be treated in detail elsewhere (Davidson and McKey 1993).

III. Relationships between Plant-Ants and Ant-Plants on Different Landmasses

In a subsequent paper (Davidson and McKey 1993), we argue that intercontinental similarities in ant community organization, and in ant and plant preadaptations, have driven strong parallel and convergent evolution in the ant-plant symbioses of Africa and the Neotropics. The strength of such inferences about evolutionary processes based on ecological similarities depends on a confident answer to the following question: Have ant-plant symbioses evolved independently on the two continents, or do similarities in their form simply indicate a single common origin?

The first step in understanding these striking commonalities is to determine the extent to which they reflect the descent of plants, ants, or even ant-plant associations, from common ancestors. Two kinds of evidence bear on this issue: the extent to which plant groups, ant groups, and symbiotic relationships are shared between the two continents, and fossil evidence dating the first appearance of genera involved in ant-plant symbioses.

A. Intercontinental Relationships of Ants, Plants, and Symbiotic Associations

In a seminal volume comparing tropical forest ecosystems of Africa and South America, Thorne (1973) and Brown (1973) examined the biological relationships in plants and ants, respectively. They reached similar conclusions: relationships between the biotas of Africa and the Oriental-Australian regions are closer than those between Africa and South America. Although their analyses were not explicitly cladistic-biogeographic (lack of data would have precluded such an analysis), the patterns at lower taxonomic levels such as genera and species-groups are clear enough to deserve our general confidence. The pattern of closer relationships between the biotas of tropical Africa and tropical Asia than between tropical Africa and tropical America is widespread among taxa and accords well with paleogeography as summarized by Raven and Axelrod (1974), Rosen (1978), and Barron and coworkers (1981). The closer ties between the African and Asian tropics may have two explanations: the separation of Afroindia from the rest of Gondwanaland, and the possibility of biotic exchange between tropical Africa and tropical Laurasia (with India now accreted) until the global cooling and drying of about 50 Ma (Goldblatt 1978; McGowran 1990), and much more recently as well, following reestablishment of direct connection between Africa and Eurasia some 17 Ma (Raven and Axelrod 1974).

Lists of plant genera with myrmecophytes (appendix) and ant genera including obligate plant-ants (table 19.3) follow the general pattern. Thorne (1973) estimated that only 111 plant genera are common to Africa and tropical America and found only there, compared to 4500 genera limited to one of the two continents. Myrmecophytes are represented in only three plant genera shared by Africa and South America (*Bertiera* [Rubiaceae], *Hirtella* [Chrysobalanaceae], and *Ocotea* [Lauraceae]), and none of these has myrmecophytic species in both continents. Each genus has one or more myrmecophytic species only in the continent in which it is more diverse. The only genus with myrmecophytic species in both Africa and the Neotropics is the widespread genus *Acacia,* and here evolutionary radiations of myrmecophytes are clearly independent, though somewhat parallel, in African and Neotropical species (McKey 1989). In contrast, links between Africa and Asia are stronger, for both seed plants in general and myrmecophytes in particular. About 500 genera of seed plants have a distribution limited to Africa, Asia, and associated islands (Thorne 1973). Several of these genera include myrmecophytes, and some of them

(*Macaranga, Stereospermum, Canthium, Clerodendrum,* and *Psychotria*) have myrmecophytic species in both tropical Africa and tropical Asia. In addition, tropical Asia harbors at least one myrmecophytic *Acacia,* a south Indian species apparently closely related to East African ant-acacias (D. McKey, unpublished field notes, 1991). Analysis at the generic level thus yields not a single case in which we might suspect that African and American myrmecophytes share a common myrmecophytic ancestor. In contrast, the existence of several African-Asian genera suggests that some ant-plants in these two continents could either have a common origin or have evolved in parallel from congeneric non-myrmecophytic ancestors.

Analysis only at the generic level may obscure intercontinental relationships. Gentry (Chapter 17) believes that in many cases, an African genus is very closely related to one in America. Is our conclusion that myrmecophytes have evolved independently in Africa and America likely to be altered by progress in understanding phylogenetic relationships of plants in the two continents? We think not, for two reasons. First, only two cases involve African and American ant-plants so closely related that they might conceivably be lumped into a single genus. These are *Hirtella* and *Magnistipula* (Chrysobalanaceae), which Prance (1972) holds to be "good" genera (i.e., monophyletic), and two genera of Cecropiaceae, *Cecropia* and *Musanga* (the latter of which is less consistently associated with ants). Second, as pointed out previously, myrmecophytism often characterizes only one or a few species of a genus. Although cladistic treatments are lacking, our strong impression is that myrmecophytic species usually represent modest, and relatively recent, radiations. Even if a few closely related African-American pairs of genera were to be lumped, we believe that analysis of infrageneric relationships would show that myrmecophytes evolved independently in the two continents. In contrast, taxonomic progress might strengthen the apparent relationships between African and Asian ant-plant symbioses. In addition to the congeneric African and Asian ant-plants listed above, African myrmecophytes in *Nauclea* are closely related to the Asian myrmecophytes placed in *Neonauclea* and *Myrmeconauclea* by Ridsdale (1978). We surmise that many of these are cases of parallel evolution of myrmecophytes in Asia and Africa, but without further taxonomic work the alternative of descent from a myrmecophytic common ancestor cannot be ruled out.

For ants, the patterns of intercontinental relationships are similar (Brown 1973, updated by Hölldobler and Wilson 1990 and recent revisionary work; see tables 19.2 and 19.3). No ant genera are restricted to the Ethiopian plus Neotropical regions, but 27 genera are limited to the Ethiopian plus the Oriental-Australian regions. These include 2 genera that have produced specialized plant-ants (*Tetraponera* and *Plagiolepis*), though only *Tetraponera* is represented by such specialists in more than one region (table 19.3). About the same number of ant genera (31) are "tropicopolitan," including 10 genera that have produced specialized plant-ants. Six of these groups (*Pachycondyla, Leptothorax, Pheidole, Solenopsis, Tetramorium,* and *Strumigenys*) contain obligate plant-ants in only one area, but 4 genera (*Crematogaster, Tapinoma, Technomyrmex,* and *Camponotus*) include such specialists in two or more zoogeographic regions. Of these, only *Crematogaster* is represented by specialist plant-ants in both Africa and the Neotropics, as well as in the Oriental and Australian regions. Because obligate plant-ants account for such a small fraction of the *Crematogaster* fauna in each tropical region, it seems very likely that specialists have evolved independently in Africa and the Neotropics. There are only two known specialist plant-ant *Crematogaster* in the Neotropics, and taxonomic position of at least one of these appears to confirm its independent evolution of the plant-ant habit. This species, a so-far unnamed specialist on *Cecropia* in northeastern Peru, is a derivative of the widespread Neotropical *C. curvispinosa,* a generalized stem-nester (J. Longino, personal communication, 1991). Further systematic studies may eventually reveal whether or not African and Asian plant-ants in this genus have had independent origins. *Crematogaster* species occur on *Macaranga* in both continents, as well as on closely related plant genera in the two regions. These taxa include African *Nauclea,* compared with Asian *Myrmeconauclea* and *Neonauclea,* and perhaps African *Leonardoxa,* compared with Asian *Humboldtia.*

Based strictly on comparisons at the generic level, plant-ants in Africa and America appear to have arisen wholly independently. However, this conclusion must be qualified somewhat in view of the marked concentration of African and Neotropical plant-ants into three ant tribes, each represented on each continent by (usually) endemic genera rich in plant-ants or by monotypic plant-ant genera. With plant-ants representing 16 percent of 230 pseudomyrmecines known to Ward (P. Ward, personal communication, 1991), for example, this subfamily consists of New World *Pseudomyrmex* and *Myrcidris* (a monotypic plant-ant), and Old World *Tetraponera* (Ward 1990). *Tetraponera* and *Pseudomyrmex* are a possible example of a sister-group relationship that may reflect an old relationship between tropical Africa and South America (Ward 1990).

The most important tapinomine (Dolichoderinae) plant-ants are either *Technomyrmex* (widespread, with most plant-ant species in Africa) or endemic Neotropical *Azteca.* Although plant-ants contribute substantially to each of these genera, exact percentages of plant-ants are not yet known. The genus *Tapinoma* presents a case that may be suggestive of an old origin of the plant-ant habit. Species of *Tapinoma* have been recorded both from *Leonardoxa* (Detarieae) in Africa and *Humboldtia* (in the closely related caesalpinioid tribe Amherstieae) in Asia. While the record of *Tapinoma* in *H. laurifolia* is suspect because it came from a botanical garden (Ridley 1910), two species of *Tapinoma* have been collected from swollen internodes of *H. brunonis* in evergreen forest of the Western Ghats in south India (McKey, unpublished field notes, 1990, and T. Musthak Ali, personal communication, 1990).

The formicine genera currently recognized as the tribe Myrmelachistini (Hölldobler and Wilson 1990) offer a third example suggestive of a very early origin of specialization on live plant hosts. This tribe includes endemic African *Aphomomyrmex* and *Petalomyrmex,* and endemic Neotropical *Myrmelachista.* Both African genera appear to be monotypic plant-ants, and *Myrmelachista* appears to be rich in plant-ant species. A recent phylogenetic study confirms the close relationships among these genera, placing them all (along with nine other genera) in the *Pseudolasius* genus-group (Agosti 1991). Asian plant-ants in the genus *Cladomyrma* have also been placed in this tribe (Hölldobler and Wilson 1990). Of five described species of *Cladomyrma* (Agosti 1991), at least four, and possibly all five, are plant-ants (Maschwitz et al. 1989, 1991; see table 19.3 and Davidson and McKey 1993). As in pseudomyrmecines and tapinomines, then, both the general importance and the widespread generic distributions of plant-ants among myrmelachistines suggest that specialization on live plant hosts may have evolved early, and not necessarily independently on different continents. Recent work by Agosti (1991), however, suggests that *Cladomyrma* may in fact not be closely related to the African and Neotropical genera placed in the Myrmelachistini. His classification, placing *Cladomyrma* in a different lineage (the *Lasius* genus-group, with six other genera, none including plant-ants), implies that the plant-ant habit evolved at least twice in "myrmelachistines": at least once in *Cladomyrma* and at least once in the lineage including *Aphomomyrmex, Petalomyrmex,* and *Myrmelachista.* If Agosti's (1991) conclusion is correct, we are confronted with yet another striking case of parallel evolution of ant-plant symbioses. The hosts of African *Petalomyrmex* and *Aphomomyrmex* and of two Asian *Cladomyrma* species include legumes of the caesalpinioid tribe Detarieae, and these ant-plants (*Leonardoxa* in Africa and *Saraca* in Asia) present many morphological and ecological similarities. Distinguishing among common origin (homology), parallel evolution from similar starting material (homoiology), and convergence as explanations for the similarities in this set of symbioses in which both ants and plants are somewhat closely related will be a challenging task.

In all these cases, no firm conclusion can be reached without further phylogenetic study. Nevertheless, outside of the subfamily Pseudomyrmecinae and the tribes Tapinomini and Myrmelachistini, and possibly within these groups as well (Ward 1991), plant-ants have clearly had multiple and independent origins in Africa and the Neotropics. Thus, for both myrmecophytes and their associated plant-ants, evidence from extant taxa strongly supports the hypothesis that symbiotic ant-plant associations arose independently on the two continents.

B. The Fossil Record

The independent evolution of myrmecophytes and plant-ants in Africa and the Neotropics is also supported by the fossil record. This evidence is most clearly interpretable for ants (Davidson and McKey 1993). Ants first appeared in the fossil record in the lower Cretaceous (Brandão et al. 1989), and a major adaptive radiation had taken place by the early Tertiary period (Hölldobler and Wilson 1990), approximately 65 Ma. Although this radiation was eventually explosive, faunas taxonomically and ecologically similar to those of contemporary times are not represented by fossils until the middle Tertiary (e.g., Wilson 1985). Elsewhere (Davidson and McKey 1993) we discuss the earliest fossil records of all ant genera containing one or more specialized plant-ants; these findings will be summarized here. Both widely distributed genera and the few geographically restricted groups with paleontological data typically first appeared in fossil deposits between the early Oligocene and the late Miocene (35–12 Ma) (Hölldobler and Wilson 1990), long after Africa and South America split apart (130–85 Ma: Chapter 2). For the three ant groups rich in plant-ants, one (Myrmelachistini) has no fossil record, and the others first appeared in fossil deposits of the early Oligocene (*Tetraponera* of the Pseudomyrmecinae) or the Miocene (*Tapinoma* and *Technomyrmex* of the Tapinomini). Tending of Homoptera and of extrafloral nectaries may be similarly recent. The earliest fossil record of homopteran tending is that of *Iridomyrmex* and aphids in the early Oligocene Baltic amber (Wheeler 1914). The first paleontological evidence for extrafloral nectaries comes from leaves of Oligocene *Populus* specimens in Colorado's Florissant shales (Pemberton 1990).

For plants, we know of no direct fossil evidence of structures that regularly housed symbiotic ants. Our conclusions must thus be based on inferences from pollen and other fossils of groups that include extant myrmecophytes. (It is interesting but currently idle to speculate about how many of these symbioses have evolved and disappeared without a trace.) The first records of pollen ascribed to extant angiosperm families are from the period 113–88 Ma (Chapter 4). Thus, flowering plant families were becoming well differentiated during the period when Africa and South America were splitting apart. Considering that radiations of myrmecophytes usually characterize taxa at levels no higher than genera or, more commonly, species groups within genera, it seems a priori unlikely that any extant myrmecophyte lineages had appeared by this time. In some cases, fossil pollen of great age has been attributed to a genus that includes myrmecophytes. Pollen attributed to *Cupaniopsis* (Sapindaceae), for example, is known from the Upper Cretaceous of South America, Africa, and Asia (Chapter 3). But since this genus includes only one extant myrmecophytic species, which is of limited distribution (appendix), there is little reason to suspect that the age of the myrmecophyte habit even approaches the age of this genus.

Patterns in the distribution of symbiotic plants and ants across continents, taxonomic composition of ant-plants and plant-ants, and the fossil record all have important implications for our analysis. Although some comparisons between Africa and Asia suggest common descent of ant-plants, plant-ants, or both, in Africa and the Neotropics both myrmecophytes and specialist plant-ants evolved largely independently. With the possible exception of pseudomyrmecines, tapinomines, and myrmelachistines, in which the plant-ant habit may be ancient, similarities in symbiotic associations between Africa and the Neotropics are not due to common descent of one or both partners from an association that predated continental separation or other vicariance events or that migrated intact from one continent to the other. Similarities must be due to some combination of (1) parallel evolution of ants, plants, or both from similar starting material, (2) evolutionary convergence, and (3) the matching of symbiotic partners according to some set of rules shared between the two continents. Elsewhere (Davidson and McKey 1993), we summarize strong evidence supporting each of these hypotheses.

IV. Why Are Ant-Plant Systems More Diverse in the Neotropics than in Africa?

At the generic level, both myrmecophytes and specialized plant-ants exhibit comparable diversity in tropical Africa and America. However, within genera of each of these groups, radiations of species living regularly in symbiotic partnerships have been more numerous or more expansive in the Neotropics. To a great extent, these regional discrepancies in diversity mirror general trends in floras and faunas of the two continents, and we suggest that they reflect underlying general causes. According to the view developed here, biological diversity in tropical forest ecosystems is the outcome of complex plant and animal responses to both geological and climatic history, as well as to the structure of the landscape generated by this history. Likely explanations for lower African diversity are greater extinction of tropical rainforest species during periods of past climatic deterioration and lower rates of species origination as species re-expanded into novel habitats. A combination of orogenic activity, riverine disturbance, large expanses of ancient, impoverished shield substrate, and white sand products of erosion have created much more complex habitat mosaics in South America than in Africa. Diversity of both myrmecophytes and their attendant ants appears to accumulate mainly across habitats, rather than across biogeographic regions, and narrow endemics of unique or isolated habitats are more common in America than in Africa.

Because intercontinental differences in geomorphology and habitat diversity are ancient, the forest biota of Africa was probably already poorer than that of Amazonia long before periods of increased aridity in the Pleistocene. Thus, since drought-associated contraction of mesic forest habitats was greater in Africa than in South America, the Pleistocene probably brought to Africa greater extinction in an already less diverse forest biota. Specialist plant-ants and ant-plants were probably more susceptible to extinction during periods of climatic deterioration than were generalist arboreal ants and plants not dependent on symbionts. If ants and plants of moist, shady forest understory were decimated disproportionately, this might explain why leaf-pouch ant-plants and their typical *Technomyrmex* occupants are absent from the West African forest block (Bequaert 1922). These same historical circumstances may account for the predominance of *Crematogaster* as associates of *Canthium, Cuviera,* and some other African ant-plants of forest understory (Davidson and McKey 1993), since the ancestors of these ants probably inhabited more open habitats. Other plant-ants and their hosts may have been restricted to refugia such as the coastal forest of Lower Guinea, where they persist today as endemic monotypic genera (e.g., *Aphomomyrmex* and its probable derivative *Petalomyrmex*).

Drought may not have been the only nor the most widespread problem facing African ant-plant symbioses during Pleistocene climatic fluctuations. Current evidence (e.g., Bengo and Maley 1991) indicates that montane forest extended into the central African lowlands during several periods over the last 135,000 years as temperature cooled but conditions remained sufficiently moist to support forest. The resulting cool, moist conditions may not have been conducive to poikilothermic surface foragers such as ants (e.g., Janzen 1973). The African mesic forest zone occupies a plateau whose mean elevation is probably on the order of 500 m higher than the South American tropical forest zone. African ant-plant symbioses may have been harder hit by this type of climatic shift than were their Neotropical counterparts.

Origination of new plant-ant and ant-plant species may have occurred primarily as ranges re-expanded during periods of climatic amelioration. Speciation rates should have been higher in Amazonia, since range expansion would be more likely to place ants, plants, or both into habitats with soil, rainfall, elevation, or other characteristics sufficiently novel to shift selective regimes. Regardless of whether speciation occurred in refugia or during range re-expansion, the potential for the evolution of new associations via host shifts and secondary colonization depended in part on the size of ant-plant and plant-ant guilds that could interact locally. In Amazonia, this potential was higher for two reasons. First, guilds were less likely to have been decimated by past climatic deterioration. Second, the closely packed, small-scale habitat mosaic of the dynamic fluvial landscape (Salo et al. 1986) meant that much of the higher regional diversity of Amazonia was represented at a local level. Close juxtaposition of diverse habitats has allowed exchange of ants and plants across habitat boundaries (Davidson et al. 1991; Davidson and Fisher 1991; Davidson and McKey 1993). Major differences in landscape structure have been associated with a greater effect of beta-diversity on ecological and evolutionary interactions within each habitat in Amazonia than in Africa. To a greater extent in Amazonia than in Africa, (beta-)diversity has almost certainly been a major factor in enhancing (alpha-)diversity.

V. Conclusions

The symbiotic ant-plant relationships have evolved independently in Africa and the Neotropics. Specialized plant-ants may have evolved comparatively early in some lineages, and not necessarily independently in the major tropical regions. Nevertheless, such specialists form a minor component of numerous distinct taxa and have certainly evolved independently many times. Extant myrmecophytes in Africa and the Neotropics appear to have had separate origins in every case examined. In view of their independent histories, the similarities in ant-plant symbioses of these two regions are remarkable, and they appear to be due to similar selection regimes acting on similarly preadapted plants and ants to produce parallel and convergent evolution in symbiotic relationships of the two continents (Davidson and McKey 1993).

Our analysis of the evolutionary dynamics of ant-plant symbioses leads us to propose a new hypothesis to explain the much greater diversity of ant-plants and plant-ants in the Neotropics than in Africa, a difference that is also seen in many other groups of organisms. Previous explanations for differences in diversity of continental floras and faunas, and even ant-plants (Benson 1985), have emphasized the contrasting climatic histories of these two regions. Focusing on range contractions during periods of unfavorable climate, these explanations postulate that Africa's lower diversity is due to greater extinction during the Pleistocene, as Africa's climate became drier and refugia were fewer, than in Amazonia. We propose that differences between the two regions in the rates of species origination may be at least as important as extinction rates. The relatively stable geological history of most of Africa, including the rainforest zone, has created a landscape with relatively little elevational relief (hence few sharp spatial contrasts in temperature and rainfall), relatively little edaphic variation, and relatively infrequent and spatially limited fluvial disturbance. In contrast, the Andean orogeny, and sub-Andean tectonic activity, have helped to create a landscape of great elevational, climatic, and edaphic complexity, especially in western Amazonia. This has resulted in a complex and dynamic mosaic of habitats. Colinvaux (Chapter 16) suggests that species origination usually takes place when ranges are re-expanding during periods of climatic amelioration. If this is so, then in the Neotropics, especially in western Amazonia, range expansion would be much more likely than in the African forest zone to place ants and plants into novel habitats, leading to speciation, the formation of new associations, or both.

This hypothesis will, we hope, contribute to a fruitful resolution of the current debate about the roles of history and contemporary ecology in explaining patterns of diversity. As Brown (1987b) argues, progress is likely to come when these two determinants of diversity are viewed not as simple antithetical alternatives, but as interacting components of a synthetic model. The attention we draw to contrasts in landscape structure between these two tropical regions may have broader implications for understanding intercontinental differences in biological diversity within the tropics. Plate tectonics "is probably an overriding factor in the geomorphological diversity of the humid tropics," and "tectonic contrasts have produced morphological contrasts" (Douglas 1978:174). The pedological and geomorphological contrasts Douglas (1978) presents between the very young, rapidly evolving landscapes of areas included within the great Tertiary mountain arcs and the old, static landscapes of Gondwanaland remnants may be of great importance to understanding patterns of tropical biological diversity.

Appendix

Present-day distribution of plant genera containing species that are regularly symbiotic with plant-ants (taxonomy, distribution, and order of families from Mabberley 1987). The table lists all plants known to have been reported as regularly occupied by plant-ants and includes plants that possess no obvious adaptations for housing ants, as well as plants with myrmecophytic traits. Elsewhere (Davidson and McKey 1993) we present detailed information on these traits. Presence (×) or absence (−) by biogeographic region. Approximate estimates of the number of ant-plant species are indicated in parentheses. Included in ant-plants of the Ethiopian region is *Mimosa* sp. from Madagascar, the only myrmecophyte so far known from the Madagascan region. Supplemental information on distribution within the Asian tropics was taken from van Steenis and de Wilde (1950–1989). If ant-plants within a genus are restricted to either the Oriental or the Australian region, an "O" or an "A" follows the number in parentheses. If ant-plants are known to be found in both regions, the region with the larger number of ant-plant species is indicated first. If distribution of ant-plant species within the Oriental and Australian tropics is uncertain, no information is given following the number of ant-plant species. Abbreviations used: Aus = Australian; Eth = Ethiopian; Neo = Neotropical; Ori = Oriental. Numbered references refer to the list given at the end of the appendix.

Taxon	Ethiopian	Neotropical	Oriental and Australian	References
Myristicaceae				
Myristica	—	—	× (1–3, A)	3, 26
Monimiaceae				
Kibara	—	—	× (4, A)	45
Steganthera[a]	—	—	× (4, A)	44
Lauraceae				
Cryptocarya	—	—	× (?)	25
Licaria	—	× (> 1)	—	10, 73
Ocotea	×	× (6–10)	—	10, 81
Pleurothyrium	—	× (3–4)	—	81
Piperaceae				
Piper	×	× (3+)	× (1, O)	Neo = 9, 67; Ori = 12
Nyctaginaceae				
Pisonia	×	×	× (1)	25
Polygonaceae				
Ruprechtia	—	× (1–3)	—	4, 8
Triplaris	—	× (17)	—	7
Actinidiaceae				
Saurauia	—	×	× (1?)	28
Elaeocarpaceae				
Elaeocarpus	—	×	× (1?)	11, 28
Sterculiaceae				
Cola	× (1)	—	—	52
Scaphopetalum	× (2–3)	—	—	52
Cecropiaceae				
Cecropia	—	× (45–55)	—	4, 66
Coussapoa	—	× (1–2)	—	4
Musanga	× (1)	—	—	19
Poikilospermum	—	—	× (1, O)	80
Pourouma	—	× (2–4)	—	4, 6
Flacourtiaceae				
Buchnerodendron	× (1?)	—	—	5
Ryparosa[b]	—	—	× (1+, O)	25
Tetrathylacium[c]	—	× (1)	—	55
Passifloraceae				
Barteria	× (2)	—	—	5, 84
Sapotaceae				
Delpydora	× (1)	—	—	84
Ebenaceae				
Diospyros	× (1)	×	×	32, 84
Symplocaceae				
Symplocos	—	×	× (1?, A)	71
Myrsinaceae				
Tapeinosperma	—	—	× (1, A)	25, 54

(*continued*)

Appendix. *Continued.*

Taxon	Ethiopian	Neotropical	Oriental and Australian	References
Chrysobalanaceae				
Hirtella	×	× (6)	—	46, 47
Magnistipula	× (1)	—	—	33, 34
Fabaceae (Leguminosae)				
Acacia	× (ca. 15)	× (12)	× (1, O)	Eth = 24; Neo = 27; Ori = 78, 84
Archidendron	—	—	× (1, A)	13, 58, 83
Calpocalyx	× (1)	—	—	59, 84
Humboldtia	—	—	× (2+, O)	38, 84
Leonardoxa	× (2)	—	—	37, 84
Millettia	×	—	× (1, O)	1, 40
Mimosa[d]	× (1)	×	×	82
Ormosia	—	× (1)	×	4
Pithecellobium	—	× (1)	—	60, 62, 82
Platymiscium	—	× (2)	—	63
Pterocarpus	×	× (1)	×	51
Saraca	—	—	× (1, O)	1, 41
Tachigali[e]	—	× (ca. 20)	—	4
Crypteroniaceae				
Crypteronia	—	—	× (1?, O)	1, 41
Thymelaeaceae				
Wikstroemia	—	—	× (1, O)	28
Myrtaceae				
Myrcia	—	× (1)	—	61, 62
Syzygium	×	—	× (2, A)	26, 42, 76
Melastomataceae				
Allomaieta[f]	—	× (1)	—	70, 79
Blakea	—	× (1)	—	4
Clidemia s.s.[g]	—	× (15–20)	—	29, 72
Conostegia	—	× (1–2)	—	72, 79
Henriettea[h]	—	× (1?)	—	72, 79
Maieta	—	× (ca. 15)	—	29, 72
Medinilla	×	—	× (1–2, O)	53
Miconia[i]	—	× (4)	—	4
Sagraea[j]	—	× (2 + ?)	—	72, 79
Tococa[k]	—	× (ca. 40–45)	—	72, 79
Topobea	—	× (1)	—	79
Euphorbiaceae				
Drypetes	×	×	× (1?, A)	71
Endospermum	—	—	× (2+, A)	2, 5
Glochidion	—	×	× (1?, A)	71
Homalanthus	—	—	× (1?, A)	71
Mabea	—	× (1–2?)	—	4

Taxon	Ethiopian	Neotropical	Oriental and Australian	References
Macaranga	× (2)	—	× (ca. 20, O)	Eth = 5; Ori = 20, 21, 64, 65
Pimelodendron	—	—	× (1?, A)	71
Sapium	×	× (1–2?)	×	4
Sapindaceae				
Cupaniopsis	—	—	× (1, A)	62
Harpullia	—	—	× (3, AO)	31
Mischocarpus	—	—	× (1)	71
Anacardiaceae				
Euroschinus	—	—	× (1, A)	14
Lannea	×	—	× (1, O)	28
Semecarpus	—	—	× (4, A)	14, 25
Simaroubaceae				
Picrolemma	—	× (2–3)	—	4
Meliaceae				
Aglaia	—	—	× (?)	25
Aphanamixis	—	—	× (1, A)	36
Chisocheton	—	—	× (6, OA)	35
Guarea	×	× (3+)	—	43, 73
Rutaceae				
Zanthoxylum	×	×	× (5+, AO)	23, 69, 74
Loganiaceae				
Anthocleista	× (1 + ?)	—	—	84
Fagraea	—	—	× (1+, O)	28, 30
Gentianaceae				
Tachia	—	× (5)	—	4, 39, 84
Boraginaceae				
Cordia	×	× (2)	×	63, 75
Verbenaceae				
Callicarpa	×	×	× (1)	57
Clerodendrum	× (3–5)	—	× (3, O)	Eth = 52; Ori = 3, 26
Vitex	× (4)	×	×	52
Scrophulariaceae				
Wightia	—	—	× (1?, O)	28
Gesneriaceae				
Besleria	—	× (1)	—	4
Bignoniaceae				
Stereospermum	× (1)	—	× (1, O)	5, 28, 84
Rubiaceae				
Bertiera	× (1?)	×	—	52
Canthium	× (3–6+)	—	× (1, A)	Eth = 52; Aus = 77

(*continued*)

Appendix. ***Continued.***

Taxon	Ethiopian	Neotropical	Oriental and Australian	References
Cuviera	× (8+)	—	—	52
Duroia	—	× (2–3)	—	4, 26
Gardenia	× (1?)	—	×	52
Heinsia	× (1)	—	—	52
Hoffmannia	—	× (1)	—	4
Ixora	× (1)	×	×	83
Myrmeconauclea	—	—	× (2–3, O)	40, 49
Nauclea[l]	× (1)	—	×	40, 49
Neonauclea	—	—	× (4+, O)	40, 49, 83
Patima	—	× (1)	—	63
Psychotria[m]	× (1)	×	× (1, A)	Eth, Aus = 5
Remijia	—	× (1–3)	—	4, 26
Rothmannia	× (2)	—	—	52
Tricalysia	× (2)	—	—	22, 50
Uncaria	× (1)	×	×	5, 52
Vangueriopsis	× (2+)	—	—	25, 52
Palmae				
Calamus	×	—	× (2+, O)	5, 16, 17, 56
Daemonorops	—	—	× (4 +, O)	5, 48, 56
Eremospatha	× (1+)	—	—	68
Korthalsia	—	—	× (7+, O)	17, 18, 56
Laccosperma	× (4+)	—	—	56, 68
Pogonotium[n]	—	—	× (1+, O)	16, 68
Gramineae				
Gigantochloa[o]	—	—	× (1+, O)	15
Schizostachyum[o]	×	—	× (1+, O)	15

Sources:

A. Publications

1. Agosti 1991
2. Airy Shaw 1980
3. Beccari 1884
4. Benson 1985
5. Bequaert 1922
6. Berg et al. 1990
7. Brandbyge 1986
8. Brandbyge 1990
9. Burger 1971
10. Burger and van der Werff 1990
11. Coode 1981
12. de Candolle 1916
13. de Wit 1942
14. Ding Hou 1978
15. Dorow and Maschwitz 1990
16. Dransfield 1978
17. Dransfield 1981
18. Dransfield 1984
19. Duviard and Segeren 1974
20. Fiala and Maschwitz 1990
21. Fiala and Maschwitz 1991
22. Hallé 1970
23. Hartley 1966
24. Hocking 1970
25. Hölldobler and Wilson 1990
26. Huxley 1986
27. Janzen 1974b
28. Jolivet 1986
29. Judd 1989
30. Leenhouts 1972
31. Leenhouts and Vente 1982
32. Letouzey and White 1970
33. Letouzey and White 1976
34. Letouzey and White 1978

35. Mabberley 1979
36. Mabberley 1985
37. McKey 1984
38. McKey 1989
39. Maguire and Weaver 1975
40. Maschwitz et al. 1989
41. Maschwitz et al. 1991
42. Monteith 1986
43. Pennington 1981
44. Philipson 1984
45. Philipson 1985
46. Prance 1972
47. Prance 1989
48. Rickson and Rickson 1986
49. Ridsdale 1978
50. Robbrecht 1979
51. Rojo 1972
52. Schnell and Beaufort 1966
53. Solereder 1920
54. Stove and Whitmore 1970
55. Tennant 1989
56. Uhl and Dransfield 1987
57. van Steenis 1967
58. Verdcourt 1979
59. Villiers 1984
60. Ward 1989
61. Ward 1990
62. Ward 1991
63. Wheeler 1942
64. Whitmore 1973
65. Whitmore 1975

B. Personal Communications

66. C. C. Berg 1992
67. W. Burger 1991
68. J. Dransfield 1992
69. B. Fiala 1991
70. A. Gentry 1991
71. M. Jebb 1989
72. W. Judd 1992
73. J. Longino 1991
74. U. Maschwitz 1991
75. J. Miller 1990
76. G. Monteith 1990
77. G. Monteith and P. Flower 1990
78. T. Musthak Ali 1991
79. S. Renner 1991
80. D. Samson 1991
81. H. van der Werff 1990
82. P. Ward 1991

C. Personal Observations of the Authors

83. D. W. Davidson
84. D. McKey

[a]*Anthobembix hospitans* (Becc.) Perkins, cited by Bequaert (1922) is a synonym of *Steganthera hospitans* (Becc.) Kan. & Hat., cited by Hölldobler and Wilson (1990) (Philipson 1984).

[b]*Gertrudia—G. amplifolia* of Bequaert (1922) and Hölldobler and Wilson (1990)—is a synonym of *Ryparosa* (Mabberley 1987).

[c]Twigs of herbarium specimens of *T. macrophyllum* Poeppig are consistently hollow (DM); all ants so far recorded for the plant are known not to be specialist plant-ants (J. Longino, personal communication, 1991).

[d]Madagascar.

[e]Includes *Sclerolobium* (J. Zarucchi, personal communication, 1986).

[f]A cloud-forest shrub of Colombia, *Allomaieta* looks like *Maieta* but has poorly developed or no domatia (A. Gentry, personal communication, 1991).

[g]Includes *Myrmidone* (Judd and Skean 1991).

[h]S. Renner (see also Benson 1985; Huxley 1986) reports myrmecophytism in *Henriettella*, which Judd (1986) has now placed in *Henriettea*.

[i]Includes *Pterocladon sprucei* Hook. f. ex Cogn. (S. Renner, personal communcation, 1991), also reported to be myrmecophytic (Forel 1904).

[j]Includes *Ossaea* p.p. (W. Judd, personal communication, 1991).

[k]Includes *Microphysca* (W. Judd, personal communication, 1991).

[l]All Asian myrmecophytic species of *Nauclea* listed by Bequaert (1922) and Hölldobler and Wilson (1990) are treated by Ridsdale (1978) as *Myrmeconauclea* or *Neonauclea.*

[m]*Grumilea* (Hölldobler and Wilson 1990) is included in *Psychotria* (Mabberley 1987).

[n]*Pogonotium ursinum* (Becc.) J. Dransfield is the current name of *Daemonorops ursina* Becc., for which relationships with ants were described by Dransfield (1978).

[o]These two bamboos apparently lack domatia but are inhabited by host-specific plant-ants that build pavilions on leaf surfaces.

Literature Cited

Agosti, D. 1991. Revision of the oriental ant genus *Cladomyrma*, with an outline of the higher classification of the Formicinae (Hymenoptera: Formicidae). Syst. Ent. 16:293–310.

Airy Shaw, H. K. 1980. The Euphorbiaceae of New Guinea. Kew Bull. Addit. Series VIII.

Amadon, D. 1973. Birds of the Congo and Amazon forests: A comparison. Pp. 267–277 *in* B. J. Meggers, E. S. Ayensu, and W. D. Duckworth (eds.), *Tropical Forest Ecosystems in Africa and South America: A Comparative Review.* Washington, D.C.: Smithsonian Institution Press.

Axelrod, D. I., and P. H. Raven. 1978. Late Cretaceous and Tertiary vegetation history of Africa. Pp. 77–130 *in* M. J. A. Werger (ed.), *Biogeography and Ecology of Southern Africa,* Vol. 1. The Hague: W. Junk.

Barron, E. J., C. G. A. Harrison, J. L. Sloan II, and W. W. Hay. 1981. Paleogeography, 180 million years ago to the present. Eclogae Geol. Helveticae 74:443–470.

Beattie, A. J. 1985. *The Evolutionary Ecology of Ant-Plant Mutualisms.* Cambridge: Cambridge University Press.

Beccari, O. 1884. *Malesia,* Vol. 2: *Piante Ospitatrici.* Genoa.

Bengo, M. D., and J. Maley. 1991. Analyse des flux polliniques sur la marge sud du Golfe de Guinée depuis 135,000 ans. Compt. Rend. Acad. Sci. Paris, 2d ser., 313:843–849.

Benson, W. W. 1985. Amazon ant-plants. Pp. 239–266 *in* G. T. Prance and T. E. Lovejoy (eds.), *Amazonia.* New York: Pergamon Press.

Bequaert, J. 1922. Ants of the American Museum Congo Expedition: A contribution to the myrmecology of Africa, IV. Ants in their diverse relations to the plant world. Bull. Amer. Mus. Nat. Hist. 45:333–583.

Berg, C. C. 1978. Espécies de *Cecropia* da Amazônia Brasileira. Acta Amaz. 8:149–182.

Berg, C. C., R. W. A. P. Akkermans, and E. C. H. van Heusden. 1990. Cecropiaceae: *Coussapoa* and *Pourouma* with an introduction to the family. Flora Neotropica, monograph 51.

Brandão, C. R. F., R. G. Martins-Neto, and M. A. Vulcano. 1989. The earliest known fossil ant (first southern hemisphere Mesozoic record) (Hymenoptera; Formicidae: Myrmeciinae. Psyche 96:195–208.

Brandbyge, J. 1986. A revision of the genus *Triplaris* (Polygonaceae). Nord. J. Bot. 6:545–570.

———. 1990. Woody Polygonaceae from Brazil: New species and a new interpretation. Nord. J. Bot. 10:155–160.

Brandbyge, J., and B. Ollgaard. 1984. Inflorescence structure and generic delimitation of *Triplaris* and *Ruprechtia* (Polygonaceae). Nord. J. Bot. 4:765–769.

Brenan, J. P. M. 1978. Some aspects of the phytogeography of tropical Africa. Ann. Missouri Bot. Gard. 65:437–478.

Brown, K. S., Jr. 1987a. Areas where humid tropical forest probably persisted. Pp. 44–45 *in* T. C. Whitmore and G. T. Prance (eds.), *Biogeography and Quaternary History in Tropical America.* Oxford: Clarendon Press.

———. 1987b. Conclusions, synthesis, and alternative hypotheses. Pp. 175–196 *in* T. C. Whitmore and G. T. Prance (eds.), *Biogeography and Quaternary History in Tropical America.* Oxford: Clarendon Press.

———. 1987c. Soils. Pp. 24–28 *in* T. C. Whitmore and G. T. Prance (eds.), *Biogeography and Quaternary History in Tropical America.* Oxford: Clarendon Press.

Brown, W. L., Jr. 1962. The neotropical species of the ant genus *Strumigenys* Fr. Smith: Synopsis and keys to the species. Psyche 69:238–267.

———. 1973. A comparison of Hylean and Congo-West African rain forest ant faunas. Pp. 161–185 *in* B. J. Meggers, E. S. Ayensu, and W. D. Duckworth (eds.), *Tropical Forest Ecosystems in Africa and South America.* Washington, D.C.: Smithsonian Institution Press.

Burger, W. C. 1971. Piperaceae. Pp. 5–218 *in Flora Costaricensis,* Vol. 35. Chicago: Field Museum of Natural History.

Burger, W. C., and H. van der Werff. 1990. Flora

Costaricensis, Family Number 80, Lauraceae. Fieldiana: Botany 23:1–138.

Carroll, C. R. 1979. A comparative study of two ant faunas: The stem-nesting ant communities of Liberia, West Africa and Costa Rica, Central America. Amer. Natural. 113:551–561.

Coetzee, J. A. (This volume). African flora since the terminal Jurassic.

Colinvaux, P. (This volume). Pleistocene biogeography and diversity in tropical forests of South America.

Coode, M. J. E. 1981. Elaeocarpaceae. Pp. 38–185 *in* E. E. Henty (ed.), *Handbooks of the Flora of Papua New Guinea,* Vol. 2. Carleton, Victoria: Melbourne University Press.

Darlington, P. J., Jr. 1957. *Zoogeography: The Geographical Distribution of Animals.* New York: John Wiley & Sons.

Davidson, D. W. 1988. Ecological studies of neotropical ant-gardens. Ecology 69:1138–1152.

Davidson, D. W., and W. W. Epstein. 1989. Epiphytic associations with ants. Pp. 200–233 *in* U. Lüttge (ed.), *Vascular Plants as Epiphytes.* Berlin: Springer-Verlag.

Davidson, D. W., and B. L. Fisher. 1991. Symbiosis of ants with *Cecropia* as a function of the light regime. Pp. 289–309 *in* C. R. Huxley and D. K. Cutler (eds.), *Ant-plant Interactions.* Oxford: Oxford University Press.

Davidson, D. W., and D. McKey. 1993. The evolutionary ecology of symbiotic ant-plant relationships. J. Hymenoptera Res. Vol. 2

Davidson, D. W., J. T. Longino, and R. R. Snelling. 1988. Pruning of host plant neighbors by ants: An experimental approach. Ecology 69:801–808.

Davidson, D. W., R. R. Snelling, and J. T. Longino. 1989. Competition among ants for myrmecophytes and the significance of plant trichomes. Biotropica 21:64–73.

Davidson, D. W., R. B. Foster, R. R. Snelling, and P. W. Lozada. 1991. Variable composition of some tropical ant-plant symbioses. Pp. 145–162 *in* P. W. Price, T. M. Lewinsohn, G. W. Fernandes, and W. W. Benson (eds.), *Plant-Animal Interactions: Evolutionary Ecology in Tropical and Temperate Regions.* New York: John Wiley & Sons.

de Candolle, C. 1916. Piperaceae Philippinenses novae vel nuper repertae. Philip. J. Sci. (Bot.) 11:207–225.

de Ruiter, G. 1976. Revision of the genera *Myrianthus* and *Musanga* (Moraceae). Bull. Jard. Bot. Nat. Belg. 46:471–510.

de Wit, H. C. D. 1942. Conspectus of the genus *Archidendron* F. von Mueller (Legum.). Bull. Bot. Gard. Buitenzorg 17:256–272.

Ding Hou. 1978. Anacardiaceae. Pp. 395–548 *in* C. G. G. J. van Steenis and W. J. J. O. de Wilde (eds.), *Flora Malesiana Series I,* Vol. 8. Dordrecht: Kluwer.

Dorow, W. H. O., and U. Maschwitz. 1990. The arachne-group of *Polyrachis* (Formicidae, Formicinae): Weaver ants cultivating Homoptera on bamboo. Insect. Soc. 37:73–89.

Douglas, I. 1978. Tropical geomorphology: Present problems and future prospects. Pp. 162–184 *in* C. Embleton, D. Brunsden, and D. K. C. Jones (eds.), *Geomorphology: Present Problems and Future Prospects.* Oxford: Oxford University Press.

Dransfield, J. 1978. New records of rattan (Palmae: Lepidocaryoideae) for peninsular Malaysia. Malay. Nat. J. 30:523–531.

——— 1981. A synopsis of the genus *Korthalsia* (Palmae: Lepidocaryoideae). Kew Bull. 36: 163–194.

———. 1984. *The Rattans of Sabah,* Sabah Forest Record No. 13. Sabah: Forest Department.

Duellman, W. E. (This volume). Amphibians in Africa and South America: Evolutionary history and ecological comparisons.

Duviard, D., and P. Segeren. 1974. La colonisation d'un myrmécophyte, le parasolier, par *Crematogaster* spp. (Myrmicinae) en Côte-d'Ivoire forestière. Insect. Soc. 21:191–212.

Fiala, B., and U. Maschwitz. 1990. Studies on the South East Asian ant-plant association *Crematogaster borneensis/Macaranga:* Adaptations of the ant partner. Insect. Soc. 37:212–231.

———. 1991. Extrafloral nectaries in the genus *Macaranga* (Euphorbiaceae) in Malaysia: Comparative studies of their possible significance as predispositions for myrmecophytism. Biol. J. Linn. Soc. 44:287–305.

Fiala, B., U. Maschwitz, Y. P. Tho, and A. J. Helbig. 1989. Studies of a South East Asian ant-plant association: Protection of *Macaranga* trees by *Crematogaster borneensis*. Oecologia 79:463–470.

Forel, A. 1904. In und mit Pflanzen lebende Ameisen aus dem Amazonas-Gebiet und aus Peru. Zool. Jahrb. (Abtheil. f. Syst.) 20:677–707.

Gentry, A. H. 1982. Neotropical floristic diversity: Phytogeographical connections between Central and South America, Pleistocene climatic fluctuations, or an accident of the Andean orogeny? Ann. Missouri Bot. Gard. 69:557–593.

———. 1986a. Endemism in tropical vs. temperate plant communities. Pp. 153–181 *in* M. E. Soulé (ed.), *Conservation Biology: The Science of Scarcity and Diversity.* Sunderland, Mass.: Sinauer.

———. 1986b. An overview of neotropical phytogeographic patterns with an emphasis on Amazonia. An. 10 Simposio do Tropico Umido 2:19–35.

———. 1988. Changes in plant community diversity and floristic composition on geographical and environmental gradients. Ann. Missouri Bot. Gard. 75:1–34.

———. 1990. Floristic similarities and differences between southern central America and upper and central Amazonia. Pp. 141–157 *in* A. H. Gentry (ed.), *Four Neotropical Rainforests.* New Haven: Yale University Press.

———. (This volume). Diversity and floristic composition of lowland tropical forest in Africa and South America.

Gibbs, P. E., and N. Taroda. 1983. Heterostyly in the *Cordia alliodora—C. trichotoma* complex in Brazil. 1983. Rev. Brasil. Bot. 6:1–10.

Goldblatt, P. 1978. An analysis of the flora of southern Africa: Its characteristics, relationships, and origins. Ann. Missouri Bot. Gard. 65:369–436.

Hallé, N. 1970. Rubiacées (2e partie). Pp. 1–335 *in* A. Aubréville and J.-F. Leroy (eds.), *Flore du Gabon,* Vol. 17. Paris: Muséum National d'Histoire Naturelle.

Harada, A. Y., and W. W. Benson. 1988. Espécies de *Azteca* (Hymenoptera, Formicidae) especializadas em *Cecropia* (Moraceae): Distribuição geográfic e considerações ecológicas. Rev. Bras. Ent. 32:423–435.

Hartley, T. G. 1966. A revision of the Malesian species of *Zanthoxylum* (Rutaceae). J. Arnold Arbor. 47:171–221.

Hocking, B. 1970. Insect associations with the swollen thorn acacias. Trans. R. Ent. Soc. London. 122:211–255.

Hölldobler, B., and E. O. Wilson. 1990. The Ants. Cambridge, Mass.: Harvard University Press, Belknap Press.

Huxley, C. R. 1980. Symbiosis between ants and epiphytes. Biol. Rev. 55:321–340.

———. 1986. Evolution of benevolent ant-plant relationships. Pp. 257–282 *in* B. Juniper and Sir R. Southwood (eds.), *Insects and the Plant Surface.* London: Edward Arnold.

Janzen, D. H. 1966. Coevolution of mutualism between ants and acacias in central America. Evolution 20:249–275.

———. 1967. Why mountain passes are higher in the tropics. Amer. Natural. 101:233–249.

———. 1973. Sweep samples of tropical foliage insects: Effects of seasons, vegetation types, elevation, time of day, and insularity. Ecology 54:687–708.

———. 1974a. Epiphytic myrmecophytes in Sarawak: Mutualism through the feeding of plants by ants. Biotropica 6:237–259.

———. 1974b. Swollen-thorn acacias of Central America. Smithsonian Contr. Bot. 13:1–131.

Jolivet, P. 1986. Les Fourmis et les plantes: Un Exemple de coévolution. Paris: Boubée.

Judd, W. S. 1986. Taxonomic placement of *Calycogonium squamulosum* (Melastomataceae: Miconieae). Brittonia 38:238–242.

———. 1989. Taxonomic studies in the Miconieae (Melastomataceae), III: Cladistic analysis of axillary-flowered taxa. Ann. Missouri Bot. Gard. 76:476–495.

Judd, W. S., and J. D. Skean, Jr. 1991. Taxonomic studies in the Miconieae (Melastomataceae), IV: Generic realignments among terminal-flowered taxa. Bull. Florida Mus. Nat. Hist. Biol. Sci. 36:25–84.

Klammer, G. 1984. The relief of the extra-Andean Amazon basin. Pp. 47–83 *in* H. Sioli (ed.), *The Amazon: Limnology and Landscape Ecology of a Mighty Tropical River and Its Basin*. Dordrecht: Dr. W. Junk.

Laurent, R. F. 1973. A parallel survey of equatorial amphibians and reptiles in Africa and South America. Pp. 259–266 *in* B. J. Meggers, E. S. Ayensu, and W. D. Duckworth (eds.), *Tropical Forest Ecosystems in Africa and South America: A Comparative Review*. Washington, D.C.: Smithsonian Institution Press.

Leenhouts, P. W. 1972. Loganiaceae. Pp. 293–387 *in* C. G. G. J. van Steenis and W. J. J. O. de Wilde (eds.), *Flora Malesiana Series I*, Vol. 6. Dordrecht: Kluwer.

Leenhouts, P. W., and M. Vente. 1982. A taxonomic revision of *Harpullia* (Sapindaceae). Blumea 28:1–51.

Letourneau, D. K. 1983. Passive aggression: An alternative hypothesis for the *Piper-Pheidole* association. Oecologia 60:122–126.

Letouzey, R., and F. White. 1970. Ébénacées. Pp. 3–184 *in* A. Aubréville and J.-F. Leroy (eds.), *Flore du Cameroun*, Vol. 11. Paris: Muséum National d'Histoire Naturelle.

———. 1976. Chrysobalanacées nouvelles du Cameroun et du Gabon. Adansonia, ser. 2, 16:229–243.

———. 1978. Chrysobalanacées. Pp. 3–138 *in* A. Aubréville and J.-F. Leroy (eds.), *Flore du Cameroun*, Vol. 20. Paris: Muséum National d'Histoire Naturelle.

Livingstone, D. A. (This volume). Evolution of African climate.

Longino, J. T. 1989a. Geographic variation and community structure in an ant-plant mutualism: *Azteca* and *Cecropia* in Costa Rica. Biotropica 21:126–132.

———. 1989b. Taxonomy of the *Cecropia*-inhabiting ants in the *Azteca alfari* species group (Hymenoptera: Formicidae): Evidence for two broadly sympatric species. Nat. Hist. Mus. L.A. Co., Contrib. Sci. 19: 412:1–16.

———. 1991. *Azteca* ants in *Cecropia* trees: Taxonomy, colony structure and behavior. *In* C. R. Huxley and D. K. Cutler (eds.), *Ant-Plant Interactions*. Oxford: Oxford University Press.

Mabberley, D. J. 1979. The species of *Chisocheton* (Meliaceae). Bull. Brit. Mus. Nat. Hist. (Bot.) 6:301–386.

———. 1985. Florae malesianae praecursores LXVII: Meliaceae (divers genera). Blumea 31:129–152.

———. 1987. *The Plant-Book*. Cambridge: Cambridge University Press.

McGowran, B. 1990. Fifty million years ago. Amer. Sci. 78(1):30–39.

McKey, D. 1984. Interaction of the ant-plant *Leonardoxa africana* (Caesalpiniaceae) with its obligate inhabitants in a rainforest in Cameroon. Biotropica 16:81–99.

———. 1988. Promising new directions in the study of ant-plant mutualisms. Pp. 335–355 *in* W. Greuter and B. Zimmer (eds.), *Proceedings of the XIV International Botanical Congress*. Königstein/Taunus: Koeltz.

———. 1989. Interactions between ants and leguminous plants. Pp. 673–718 *in* C. H. Stirton and J. L. Zarucchi (eds.), *Advances in Legume Biology*. Monogr. Syst. Bot. Missouri Bot. Gard. 29.

Maguire, B., and R. E. Weaver, Jr. 1975. The Neotropical genus *Tachia* (Gentianaceae). J. Arnold Arbor. 56:103–125.

Maley, J. 1987. Fragmentation de la forêt dense humide africaine et extension des biotopes montagnards au Quaternaire récent: Nouvelles données polliniques et chronologiques, implications paléoclimatiques et biogéographiques. Paleoecol. Afr. 18:307–334.

———. 1989. Late Quaternary climatic changes in the African rain forest: Forest refugia and the major role of sea surface temperature variations. Pp. 585–616 *in* M. Leinen and M. Sarnthein (eds.), *Paleoclimatology and Paleometeorology: Modern and Past Patterns of Global Atmospheric Transport*. Dordrecht: Kluwer.

Maley, J., D. A. Livingstone, P. Giresse, N. Thouveny, P. Brenac, K. Kelts, G. Kling, C. Stager, M. Haag, M. Fournier, Y. Bandet, D. Williamson, and A. Zogning. 1990. Lithostratigraphy, volcanism, paleomagnetism and palynology of Quaternary lacustrine deposits from Barombi Mbo (West Cameroon): Preliminary results. J. Volcanol. Geotherm. Res. 42:319–335.

Maschwitz, U., B. Fiala, Y. F. Lee, V. K. Chey, and F. L. Tan. 1989. New and little known myrmecophytic associations from Bornean rainforests. Malay. Nat. J. 43:106–115.

Maschwitz, U., B. Fiala, J. Moog, and L. G. Saw. 1991. Two new myrmecophytic associations from the Malay Peninsula: Ants of the genus *Cladomyrma* (Formicidae, Camponotinae) as partners of *Saraca thaipingensis* (Caesalpiniaceae) and *Crypteronia griffithii* (Crypteroniaceae), 1: Colony foundation and acquisition of trophobionts. Insect. Soc. 38:27–35.

Monteith, G. B. 1986. Presidential Address: Some curious insect-plant associations in Queensland. Queensland Naturalist 26:105–114.

———. 1987. *Endospermum,* ants and uraniid moths in Australia. Qd. Nat. 28:35–41.

Nkongmeneck, B.-A. 1985. Le genre *Cola* au Cameroun. Ann. Fac. Sci. Biol.-Biochim. 3(3):5–27.

Pemberton, R. W. 1990. Fossil extrafloral nectaries, evidence for ant-guard antiherbivore defense in a fifty-million-year-old plant. Poster session, V International Congress of the International Association for Ecology, Yokohama, Japan, 1990.

Pennington, T. D. 1981. Meliaceae. Pp. 1–470 *in Flora Neotropica,* Monograph No. 28. New York: New York Botanical Garden.

Philipson, W. R. 1984. A synopsis of the Malesian species of *Steganthera* (Monimiaceae). Blumea 29:481–497.

———. 1985. A synopsis of the Malesian species of *Kibara* (Monimiaceae). Blumea 30:389–415.

Pitman, W. C., III, S. Cande, J. LaBrecque, and J. Pindell. (This volume). Fragmentation of Gondwana: The separation of Africa from South America.

Prance, G. T. 1972. The Chrysobalanaceae. Pp. 1–410 *in Flora Neotropica,* Monograph No. 9. New York: New York Botanical Garden.

———. 1987. Vegetation. Pp. 28–45 *in* T. C. Whitmore and G. T. Prance (eds.), *Biogeography and Quaternary History in Tropical America.* Oxford: Clarendon Press.

———. 1989. Chrysobalanaceae. Pp. 1–267 *in Flora Neotropica,* Monograph No. 9S. New York: New York Botanical Garden.

Räsänen, M. E., J. S. Salo, and R. Kalliola. 1987. Fluvial perturbance in the western Amazon basin: Regulation by long-term sub-Andean tectonics. Science 238:1398–1401.

Raven, P., and D. I. Axelrod. 1974. Angiosperm biogeography and past continental movements. Ann. Missouri Bot. Gard. 61:539–673.

Rickson, F. R. 1979. Absorption of animal tissue breakdown products into a plant stem: The feeding of a plant by ants. Amer. J. Bot. 66: 87–90.

Rickson, F. R., and M. M. Rickson. 1986. Nutrient acquisition facilitated by litter collection and ant colonies on two Malaysian palms. Biotropica 18:337–343.

Ridley, H. N. 1910. Symbiosis of ants and plants. Ann. Bot. 24:457–483.

Ridsdale, C. E. 1978. A revision of the tribe Naucleeae s.s. (Rubiaceae). Blumea 24: 307–366.

Robbrecht, E. 1979. The African genus *Tricalysia* A. Rich. (Rubiaceae-Coffeeae), 1: A revision of the species of subgenus *Empogona.* Bull. Jard. Bot. Nat. Belg. 49:239–360.

Roberts, T. R. 1973. Ecology of fishes in the Amazon and Congo basins. Pp. 239–254 *in* B. J. Meggers, E. S. Ayensu, and W. D. Duckworth (eds.), *Tropical Forest Ecosystems in Africa and South America: A Comparative Review.* Washington, D.C.: Smithsonian Institution Press.

Rojo, J. P. 1972. *Pterocarpus* (Leguminosae-Papilionaceae) Revised for the World, Phanerogamarum Monographiae 5. Leutershausen: Verlag von J. Cramer.

Romero, E. J. (This volume). South American paleofloras.

Rosen, D. 1978. Vicariant patterns and historical explanations in biogeography. Syst. Zool. 27:159–188.

Salo, J., R. Kalliola, I. Häkkinen, Y. Mäkinen, P.

Niemelä, M. Puhakka, and P. D. Coley. 1986. River dynamics and the diversity of Amazon lowland forest. Nature 322:254–258.

Schnell, R., and F. Grout de Beaufort. 1966. Contribution à l'étude des plantes à myrmécodomaties de l'Afrique intertropicale. Mémoires de l'Institut Fondamental d'Afrique Noire, No. 75. Mélanges Botaniques. Dakar: Institut Fondamental d'Afrique Noire.

Shattuck, S. O. 1991. Revision of the dolichoderine ant genus *Axinidris* (Hymenoptera: Formicidae). Syst. Ent. 16:105–120.

———. 1992a. Generic revision of the ant subfamily Dolichoderinae (Hymenoptera: Formicidae). Sociobiology 21:1–181.

———. 1992b. Review of the dolichoderine ant genus *Iridomyrmex* Mayr with descriptions of three new genera (Hymenoptera: Formicidae). J. Aust. Ent. Soc. 31:13–18.

Snelling, R. R. 1979. *Aphomomyrmex* and a related new genus of arboreal African ants (Hymenoptera: Formicidae). Contrib. Sci. Los Angeles County Mus. Nat. Hist. 316:1–8.

Solereder, H. 1920. Über eine heterophylle philippinische Ameisenpflanze aus der Familie der Melastomataceae, nebst Bemerkungen über das Auftreten von Amylodextrin-Körnern in den sog. Perldrüsen. Naturwiss. Wochenschr. 35:689–691.

Sombroek, W. G. 1984. Soils of the Amazon region. Pp. 521–535 *in* H. Sioli (ed.), *The Amazon: Limnology and Landscape Ecology of a Mighty Tropical River and Its Basin*. Dordrecht: Dr. W. Junk.

Stone, B. C., and T. C. Whitmore. 1970. Notes on the systematy of Solomon Islands' plants and some of their New Guinea relatives, XI: *Tapeinosperma* (Myrsinaceae). Reinwardtia 8:3–11.

Tennant, L. E. 1989. A new ant-plant, *Tetrathylacium costaricense*. Abstracts, Symposium on Interactions between Ants and Plants, Oxford, England, July 1989.

Terron, G. 1970. Recherches morphologiques et biologiques sur *Tetraponera anthracina* Santschi et sur son parasite social temporaire *Tetraponera ledouxi* Terron. Ph.D. diss., Université de Toulouse.

Thorne, R. F. 1973. Floristic relationships between tropical Africa and tropical America. Pp. 27–47 *in* B. J. Meggers, E. S. Ayensu, and W. D. Duckworth (eds.), *Tropical Forest Ecosystems in Africa and South America: A Comparative Review.* Washington, D.C.: Smithsonian Institution Press.

Tobin, J. E. 1991. A neotropical rainforest canopy ant community: Some ecological consideration. Pp. 536–538 *in* C. R. Huxley and D. K. Cutler (eds.), *Ant-plant Interactions.* Oxford: Oxford University Press.

Uhl, N. W., and J. Dransfield. 1987. *Genera Palmarum.* Lawrence: Allen Press.

van Steenis, C. G. G. J. 1967. Miscellaneous botanical notes XVIII. Blumea 15:145–155.

———. 1987. *Checklist of Generic Names in Malesian Botany Flora.* Leiden: Malesiana Foundation, Rijksmuseum.

van Steenis, C. G. G. J., and W. J. J. O. de Wilde (eds.). 1950–1989. *Flora Malesiana.* Dordrecht: Kluwer.

Vanzolini, P. E. 1973. Paleoclimates, relief, and species multiplication in equatorial forests. Pp. 255–258 *in* B. J. Meggers, E. S. Ayensu, and W. D. Duckworth (eds.), *Tropical Forest Ecosystems in Africa and South America: A Comparative Review.* Washington, D.C.: Smithsonian Institution Press.

Vasconcelos, H. L. 1990. The relationship between plant size and ant associates in the ant-plant *Maieta guianensis*. Annual Meeting British Ecological Society (poster session), Manchester, England.

Verdcourt, B. 1979. *A Manual of New Guinea Legumes,* Botany Bull., No. 11. Papua, New Guinea: Office of Forests.

Villiers, J.-F. 1984. Le genre *Calpocalyx* (Leguminosae, Mimosoideae) en Afrique. Adansonia 6:297–311.

Ward, P. S. 1989. Systematic studies on pseudomyrmecine ants: Revision of the *Pseudomyrmex oculatus* and *P. subtilissimus* species groups, with taxonomic comments on other species. Quaest. Ent. 25:393–468.

———. 1990. The ant subfamily Pseudomyrmecinae (Hymenoptera: Formicidae): Generic revision and relationship to other formicids. Syst. Ent. 15:449–489.

———. 1991. Phylogenetic analysis of ant-plant associations involving pseudomyrmecine ants. Pp. 335–352 *in* C. R. Huxley and D. K. Cutler (eds.), *Ant-Plant Interactions*. Oxford: Oxford University Press.

Wheeler, W. M. 1914. The ants of the Baltic amber. Schrift. der Physik.-Oekonom. Gesell. zu Königsberg 55:1–142.

———. 1942. Studies of neotropical ant-plants and their ants. Bull. Mus. Comp. Zool. 90:1–262.

White, F. 1979. The Guineo-Congolian Region and its relationships to other phytochoria. Bull. Jard. Bot. Nat. Belg. 49:11–55.

Whitmore, T. C. 1973. Euphorbiaceae. Pp. 105–113 *in* T. C. Whitmore (ed.), *Tree Flora of Malaya,* Vol. 2. Kuala Lumpur: Longman.

———. 1975. *Macaranga. In* H. K. Airy Shaw (ed.), Euphorbiaceae of Borneo. Kew Bull., Addit. Ser. 4:140–159.

Wilson, E. O. 1985. Invasion and extinction in the West Indian ant fauna: Evidence from the Dominican amber. Science 229:265–267.

———. 1987. The arboreal ant fauna of Peruvian Amazon forests: A first assessment. Biotropica 19:245–251.

Wright, J. B., D. A. Hastings, W. B. Jones, and H. R. Williams. 1985. *Geology and Mineral Resources of West Africa.* London: George Allen & Unwin.

Note added in proof:

Two additional genera including probable myrmecophytes should be added to the appendix (F. Breteler, personal communication, 1993):

1. *Dichapetalum* (Dichapetalaceae), a pantropical genus of about 150 species best represented in Africa, includes one species, *D. gassitae* Bret. of Gabon, with paired hairy leaf-pouches occupied by undetermined ants. This species is described in the following publication: Breteler, F. J. 1986. The African Dichapetalaceae: A taxonomic revision. IX. Agricultural University of Wageningen Papers 86-3. 74 pp.

2. An undescribed species of *Dactyladenia* (Chrysobalanaceae) from Gabon has paired, inflated stipules apparently occupied by ants. This genus, whose species were formerly placed in *Acioa,* is restricted to tropical Africa and includes 27 named species.

This brings to 33 the number of plant genera with myrmecophytic species in Africa and reinforces two findings developed in this chapter: the concentration of myrmecophyte diversity in the Lower Guinea region of the Central African block of rainforest, and the tendency, most marked in Africa, for myrmecophytism to be restricted to single species of large genera.

Contributors

Allison V. Andors, Department of Ornithology, American Museum of Natural History, New York, N.Y.

Aaron M. Bauer, Department of Biology, Villanova University, Villanova, Penn.

Kåre Bremer, Department of Systematic Botany, Uppsala University, Uppsala, Sweden

S. Cande, Lamont-Doherty Observatory, Columbia University, Palisades, N.Y.

James M. Carpenter, American Museum of Natural History, Department of Entomology, New York, N.Y.

J. A. Coetzee, University of the Orange Free State (retired) Somerset West, South Africa

Paul Colinvaux, Smithsonian Tropical Research Institute, Panama

Diane W. Davidson, Department of Biology, University of Utah, Salt Lake City, Utah

William E. Duellman, Museum of Natural History, University of Kansas, Lawrence, Kans.

Alwyn H. Gentry, Missouri Botanical Garden, St. Louis, Mo.

Peter Goldblatt, Missouri Botanical Garden, St. Louis, Mo.

J. LaBrecque, Lamont-Doherty Observatory, Columbia University, Palisades, N.Y.

Annick Le Thomas, Ecole Pratique des Hautes Etudes, Laboratoire de Phytomorphology, Muséum National d'Histoire Naturelle, Paris, France

Daniel A. Livingstone, Departments of Zoology and Geology, Duke University, Durham, N.C.

John G. Lundberg, Department of Zoology, Duke University, Durham, N.C.

John G. Maisey, Department of Vertebrate Paleontology, American Museum of Natural History, New York, N.Y.

Larry G. Marshall, Institute of Human Origins, Berkeley, Calif.

Doyle McKey, University of Miami, Coral Gables, Fla.

Samuel J. McNaughton, Biological Resources Laboratory, Syracuse University, Syracuse, N.Y.

Martin Oesterheld, Departamento de Ecología, Facultad de Agronomía, Universidad de Buenos Aires, Buenos Aires, Argentina

J. Pindell, Department of Earth Sciences, Dartmouth College, Hanover, N.H.

W. C. Pitman III, Lamont-Doherty Observatory, Columbia University, Palisades, N.Y.

Peter H. Raven, Missouri Botanical Garden, St. Louis, Mo.

Edgardo Juan Romero, Departamento de Ciencias Biológicas, Universidad de Buenos Aires, Buenos Aires, Argentina

Osvaldo E. Sala, Departamento de Ecología, Facultad de Agronomía, Universidad de Buenos Aires, Buenos Aires, Argentina

George E. Schatz, Missouri Botanical Garden, St. Louis, Mo.

Thierry Sempere, ORSTOM, Santa Cruz de la Sierra, Bolivia

Elisabeth S. Vrba, Department of Geology, Yale University, New Haven, Conn.

François Vuilleumier, Department of Ornithology, American Museum of Natural History, New York, N.Y.

Taxonomic Index

Subject Index